Advances in Intelligent Systems and Computing

Volume 508

Series editor

Janusz Kacprzyk, Polish Academy of Sciences, Warsaw, Poland
e-mail: kacprzyk@ibspan.waw.pl

About this Series

The series "Advances in Intelligent Systems and Computing" contains publications on theory, applications, and design methods of Intelligent Systems and Intelligent Computing. Virtually all disciplines such as engineering, natural sciences, computer and information science, ICT, economics, business, e-commerce, environment, healthcare, life science are covered. The list of topics spans all the areas of modern intelligent systems and computing.

The publications within "Advances in Intelligent Systems and Computing" are primarily textbooks and proceedings of important conferences, symposia and congresses. They cover significant recent developments in the field, both of a foundational and applicable character. An important characteristic feature of the series is the short publication time and world-wide distribution. This permits a rapid and broad dissemination of research results.

More information about this series at http://www.springer.com/series/11156

Nilesh Modi · Pramode Verma
Bhushan Trivedi
Editors

Proceedings of International Conference on Communication and Networks

ComNet 2016

Springer

Editors
Nilesh Modi
Narsinhbhai Institute of Computer Studies and Management
Kadi, Mehsana, Gujarat
India

Bhushan Trivedi
Faculty of Computer Technology
GLS University
Ahmedabad, Gujarat
India

Pramode Verma
Telecommunication Engineering
The University of Oklahoma
Norman, OK
USA

ISSN 2194-5357 ISSN 2194-5365 (electronic)
Advances in Intelligent Systems and Computing
ISBN 978-981-10-2749-9 ISBN 978-981-10-2750-5 (eBook)
DOI 10.1007/978-981-10-2750-5

Library of Congress Control Number: 2016954515

Printed on acid-free paper

This Springer imprint is published by Springer Nature
The registered company is Springer Nature Singapore Pte Ltd.
The registered company address is: 152 Beach Road, #22-06/08 Gateway East, Singapore 189721, Singapore

Preface

This AISC volume contains the papers presented at the COMNET 2016: International Conference on Communication on Networks. The conference was held during February 19 and 20, 2016 at Ahmedabad Management Association (AMA), Ahmedabad, India and organized by Computer Society of India (CSI), Ahmedabad Chapter, Division IV and Association of Computing Machinery (ACM), Ahmedabad Chapter. International Conference on Communication and Networks (COMNET 2016) is an open forum for researchers, engineers, network planners, and service providers targeted on newly emerging algorithms, communication systems, network standards, services, and applications, bringing together leading international players in computer communication and networks. This conference provides a forum to researchers to propose theory and technology on the networks and services, to share their experience in IT and telecommunications industries, and to discuss future management solutions for communication systems, networks, and services.

This year the conference aims to bring researchers, industrialists, and experts on a single platform to discuss the challenges and issues with the IOT revolution. Internet of Things brings many small devices such as implantable medical devices, house sensors, parking sensors, and smart city related devices work in a coordinated manner. IOT requires these devices to be addressable remotely over Internet. IOT is expected to solve problems which are considered impossible to be solved so far, for example self-driving cars and self-managed houses. As this revolution provides services never seen or expected before, they bring unprecedented issues and challenges with them as well. Security, privacy, control, protocol design, and organization policies to deal with personal and other IOT devices, integration of IOT devices with conventional networking infrastructure, standardization of IOT services are a few common challenges. This conference aims to bring such issues to front and help participants to learn solutions proposed by researchers as well as industry.

This is a major international event organized by CSI Div IV (Communication) and is hosted by one of the chapters of CSI every year. The event was held in the past in various locations like Kolkata, Hyderabad, Udaipur, Trivandrum, Coimbatore, etc.

This year, the event is hosted by CSI Ahmedabad Chapter and held at Ahmedabad Management Association. COMNET 2016 was a two-day event offering plenary sessions, technical sessions, and manufacturers' presentations.

Research submissions in various advanced technology areas were received and after a rigorous peer-review process with the help of program committee members and external reviewer, 77 papers in a single volume were accepted with an acceptance ratio of 0.51. The conference featured many distinguished personalities like Dr. Andrzej Rucinski and Dr. Sumit Chaudhary and other well-known keynote speakers. Separate Invited talks were organized in industrial and academia tracks on both days. The conference also hosted few tutorials and workshops for the benefit of participants. We are indebted to all supporting bodies and members of Computer Society of India, Ahmedabad Chapter for their immense support to make this conference possible in such a grand scale. A total of eight sessions were organized as a part of COMNET 2016 including six technical, one plenary, and one inaugural session. A total of 62 papers were presented in six technical session with high discussion insights. The total number of accepted submissions was 75 with a focal point on networks theme.

Our sincere thanks to all Sponsors, press, print and electronic media for their excellent coverage of this convention.

Kadi, India
Norman, USA
Ahmedabad, India
February 2016

Nilesh Modi
Pramode Verma
Bhushan Trivedi

Organising Committee

Core Committee

Chief Patron
Prof. Bipin V. Mehta, President CSI India

General Chair
Shri Bharat Patel, COO, Yudiz Solutions Pvt. Ltd. and Senior Member, ACM

Industry Chairs
Mr. Nikhil Jain, CEO, Elitecore Technologies
Mr. Indrajit Mitra, CEO, Gateway Technolabs

Convener
Mr. Vijay Shah, Chairman, CSI Ahmedabad Chapter

CSI DIV IV Chair
Dr. Durgesh Kumar Mishra, Chairman, Division-IV (Communication), CSI

Program Chair
Prof. Bhushan Trivedi, Director and Dean, School of Computer Studies, GLS University, Ahmedabad; Senior Member, ACM

Organizing Chair
Dr. Nilesh Modi, Chairman, ACM—Ahmedabad

Finance Chair
Jayesh Solanki, Vice Chairman, CSI Ahmedbad Chapter

Publication Chair
Dr. Suresh Chandra Satapathy, Chairman, Division-V (Education and Research), CSI

Publicity Chair
Dr. Sandeep Vasant, Secretary, CSI Ahmedabad Chapter

Organising Committee

Organising Secretary

Dr. Viral Nagori, Faculty, GLS University, Organizing Secretary

Members

Mr. Hitesh Parmar, Faculty, KS Institute of Computer, Gujarat University
Dr. Vimal Pandya, Head, Computer Department, H.K. Arts College, Ahmedabad
Dr. Gnanesh Jani, Programmer cum Lecturer, Vivekanand College of Arts, Ahmedabad
Dr. Kuntal Patel, AES Institute of Computer Studies, Ahmedabad
Mr. Ankit Bhavsar, Faculty, GLS University
Mr. Vinay Vacchrajani, Faculty, SCS, Ahmedabad University
Mr. Nilesh Advani, Marwadi Education Campus, Rajkot
Mr. Vipul Joshi, Faculty, Kadi Sarva Vishwavidyalaya

Program Committee

Program Secretary

Mr. Amit Joshi, Faculty, Sabar Institute of Engineering and Technology, Himmatnagar

Members

Dr. Promode Verma, Professor, University of Oaklahoma, USA
Dr. Kalpdrum Passi, Chair, Department of Mathematics and Computer Science, Laurentian University, Canada
Dr. Chandana Unnithan, Victoria University, Australia
Dr. Pawan Lingras, Professor, Saint Mary's University, Canada
Mustafizur Rahman, Endeavour Research Fellow, Institute of High Performance Computing, Agency for Science Technology and Research, Australia
Hoang Pham, Professor and Chairman, Department of Industrial and Systems Engineering, Rutgers University, NJ, USA
Dr. Ernest Chulantha Kulasekere, University of Moratuwa, Sri Lanka
Dr. Shashidhar Ram, Joshi Institute of Engineering, Pulchowk Campus, Pulchowk, Nepal
Dr. Subhadip Basu, The University of Iowa, Iowa City, USA
Dr. Abrar A. Qureshi, University of Virginia's College at Wise, One College Avenue, USA
Dr. Ashish Rastogi (CSI), IT Department, Higher College of Technology, Muscat-Oman
Dr. Aynur Unal, Standford University, USA
Dr. Malaya Kumar Nayak, Director IT Buzz Ltd., UK
Dr. Dharm Singh, Government Polytechnic of Namibia, Namibia

Dr. S. Das, Indian Statistical Institute Kolkota
Dr. K. Srujan Raju, CMR Technical Campus, Hyderabad
Prof. Pritee Parweker, Anil Neerukonda Institute of Technology and Sciences, Vizag
Prof. S. Ratan Kumar, Anil Neerukonda Institute of Technology and Sciences, Vizag
Prof. Y. Sunita, Anil Neerukonda Institute of Technology and Sciences, Vizag
Dr. Suberna Kumar, MVGR College of Engineering, Vizianagaram
Prof. B. Tirumala Rao, Jawaharlal Nehru Technological University, Vizianagaram
Dr. Lalitha Bhaskari, Andhra University, Vizag
Dr. Naeem Hanoon, Malaysia
Dr. Kailash C. Patidar, South Africa
Dr. V. Suma, Bangalore
Dr. H. Behera, Sambalpur
Dr. Sachi Dehuri, Balasore
Prof. Suresh Limkar, Pune
Dr. Rustam Morena, Professor, Department of Computer Science, VNSGU
Dr. Anand Kumar, Ph.D. (Comp. Sci.), M.Phil. (Comp.Sci), MCA, Dean Academics, Professor and Head, Department of MCA, M.S. Engineering College, Bangalore
Dr. Avani Vasant, Professor and Head, Babaria Institute of Technology, Vadodara
Dr. Vaishali Kaneria, Assistant Professor, Atmiya Institute of Technology and Science, Rajkot
Dr. N.N. Jani, Dean Computer Science, Kadi Sarva Vishva Vidhyalaya

Contents

About the Editors

Dr. Nilesh Modi is Professor and Head of the Department, Narsinhbhai Institute of Computer Studies and Management, Gujarat, India. He has rich experience of around 13 years in academics and IT industry. Dr. Modi is doctorate in e-Security (Computer Science and Application). Continuing his research on Cyber Security, presently he is pursuing postdoctoral research on Wireless Communication and Security and certification in Ethical Hacking. He is working as a recognized research supervisor for Ph.D. and M.Phil. programs for different universities in India. He has reviewed number of Ph.D., M.Phil., and M.Tech. theses from different universities in India and abroad. He is also serving as a manuscript reviewer and program committee member for different research journals and conferences of national and international repute in the world. He has published more than 75 research papers in international and national journals and conference proceedings. He has delivered number of expert talk on e-security and hacking in national and international conferences.

Dr. Modi is an active life member of CSI, ACM IEEE, IACSIT, IACSI, IAEng apart from his academic and industrial career. As a consultant, he is contributing to different system development projects with IT industry and actively involved in different government projects. He is also serving as a member of Board of Studies and Selection Committee in Gujarat Technological University along with different universities in the country. He is working as district coordinator for SANDHAN Program initiated by Government of Gujarat.

Dr. Pramode Verma joined the University of Oklahoma in Tulsa as Professor of Computer Engineering and Director of the T-Com program in December 1999. Besides teaching and research, his responsibilities include creating cooperative endeavors with the information technology corporations in the Tulsa region. Dr. Verma obtained his doctorate in Electrical Engineering from Concordia University in 1970, a Bachelor's degree in Engineering in 1962 from the Indian Institute of Science, Bangalore, and a Bachelor's degree in Science (Honors) in 1959 from Patna University, India. In 1984, he also obtained an MBA from the Wharton School of the University of Pennsylvania. He is the author of over

50 publications in telecommunications, computer communications, and related fields. His other professional accomplishments include editor, Journal of Telecommunication Networks, 1982–1985; and Contributor to the McGraw Hill Encyclopedia of Science and Technology (Sixth Edition).

Dr. Verma is the author of Performance Estimation of Computer Communication Networks: A Structured Approach, published by the Computer Science Press (1989), and the contributing editor of ISDN Systems: Architecture, Technology, and Applications, published by Prentice-Hall, Inc. (1990). His first book has been translated into French and published as Modeles des Performances des Reseaux in 1992 by W.H. Freeman and Company. He is a contributor to "So This is 1984: Some Personal Views by Governors of the International Council for Computer Communication", published by Elsevier Science Publishers B.V. (1984). More recently, Dr. Verma contributed to and co-edited "The Computer Communication Revolution Multidisciplinary Retrospective and Prospective," published by the ICCC Press, 1997. He has been a keynote speaker at a number of conferences, including the third annual convention of the Computer Society of India in New Delhi. His research interests include telecommunications networks, networking technology and interoperability.

Prof. Bhushan Trivedi joined the GLS University as Professor of Computer Technology in 1999. He is now serving the GLS Institute of Computer Technology as Director. Professor Trivedi obtained his doctorate from Hemchandracharya North Gujarat University in 2008, MCA degree from MS University, Vadodara in 1988 and BSc from Gujarat University in 1984. He has over 35 papers published in national and international level journals and several contributions in conference proceedings. His interest areas are effective teaching techniques, network security, and sensor networks.

A Novice Approach for Web Application Security

Jignesh Doshi and Bhushan Trivedi

Abstract Number of websites hosted increased exponentially in the past few years. More and more organizations are doing their business on web. As a result the attacks on web applications are increased. It is found that about 60 % of web resources are vulnerable. So computer security is critical and important for Web applications. There are various types of solutions exists for mitigating security risks. Developer Skills and efforts are required in most of the solutions. In this paper, the authors have proposed a model for remote database health check. The focus of model is to provide higher level security assessment. The proof of concepts has been implemented using python. The proposed model has been tested on various test scenarios. Authors have also compared model with the topmost 3 vulnerability scanners. The results were found promising and satisfactory.

Keywords Web application attacks · SQLI · Defensive coding · Hardening · Vulnerability scanner

1 Introduction

Computer Security is the biggest challenge of the current era [1–3]. Data and computer systems are key targets of attacks. As per IBM Data Breach Report, 12 % increase in security events year-to-year [4]. Top 10 web application risks remain same in the past few years [1]. Approx. 60 % of attacks are because of vulnerable application code [5].

Most common approaches used to manage web application attacks are defensive coding, hardening (filtering), static/dynamic code analysis or black box testing.

J. Doshi (✉)
LJ Institute of Management Studies, Ahmedabad, India
e-mail: doshijig@gmail.com

B. Trivedi
GLS Institute of Computer Technology, Ahmedabad, India
e-mail: bhtrivedi@gmail.com

N. Modi et al. (eds.), *Proceedings of International Conference on Communication and Networks*, Advances in Intelligent Systems and Computing 508,
DOI 10.1007/978-981-10-2750-5_1

Solutions based application adversely affect cost and developer's efforts [6]. Testing is used for building secure applications. The major problem with testing is that it requires code and web server access [6].

The authors have proposed a security Model to mitigate security risks. Our focus is to develop Model which can be used for web application database health check and act as a utility. Model which neither require developer skills nor code.

The remainder of the paper is formed as follows: Sect. 2 explains the importance of Web application Security and SQL Injection attempts. Section 3 discuss the problem statement (issues), Sect. 4 describe the proposed model, examining results and comparison is provided in Sects. 5 and 6 respectively. The conclusion is provided in Sect. 7.

2 Literature Survey

Injection attack is one of the top three attacks since 2010 [1, 7–15]. SQL Injection and Blind SQL Injection are key attacks under Injection attacks. Most commonly SQL Injection attacks are executed from application using user inputs or URLs [5, 9, 10]. The key impacts of SQLI attacks are data loss, application downtime, brand damage, and customer turnover [7, 11, 16, 17]. Blind SQL Injection attacks are used to List database information and dump data [1]. Both attack use Structure query language for execution of attacks.

Most common approaches used to manage SQLI attacks are defensive coding, hardening (filtering), static/dynamic code analysis, Intrusion detection system and black box testing [8, 6].

Web application communities have developed various approaches for detection and prevention of SQLI [11, 16–19]. Observations of various techniques (existing and proposed) are summarized in Table 1 with reference to efforts, resource requirements (Code and web server) [6, 16, 17, 20–27].

It is observed that most of the solutions require Developer Skills, developer efforts and web server/code access (refer to Table 1).

Gap: A systematic, dynamic and effective solution is required to detect and prevent SQLI [20, 21].

Table 1 Comparison of web application attack solution categories

Approach	Developer		Source code	Web server
	Skill	Effort	Required	Required
Defensive coding	X	X	X	
Static analysis	X	X	X	
Static and dynamic analysis	X	X	X	
Black-box/penetration testing	M	X	X	X
IDS	X	X		X
Hardening	X	X		X

3 Problem Statement

The authors have found that model with following functionalities is required.

(1) Any beginner can run model i.e. no or little technical knowledge is required to execute the model [6, 17, 20–22].
(2) Model work as remote penetration testing i.e. access for source code is not required [6, 17, 20–22].
(3) Web server access is not required i.e. model can be executed from remote PC without installing it on server [6, 17, 20–22].
(4) Model can work as utility [6, 17, 20–22].

4 Proposed Model: Model for Remote Database Health Check

In this research paper, the authors have proposed a novice approach for performing remote database health check (web vulnerability checks).

4.1 Objectives of Model

The objectives of model are to develop model which can work as a utility with minimum technical skills, companies of any size can perform investigations, developers can develop highly secure web applications and organizations can mitigate with web vulnerabilities.

4.2 Overview

Prototype model is developed using python and will focus on top 2 vulnerabilities (SQL Injection and Blind SQL attacks). Model diagram is described in Fig. 1.

Following subsections describe each phase of the proposed model.

(i) *Analyse Web Application* This step will verify the existence of user entered web application host name.
(ii) *Information Gathering* This step describes the process of investigating, examining and analyzing the target website in order to gather information. System Information (like Operating system name, Version etc.), Database Information (like Database Name, Version, table/column Names etc.) and Links (like number of static links, database links mailing and other links) are gathered.

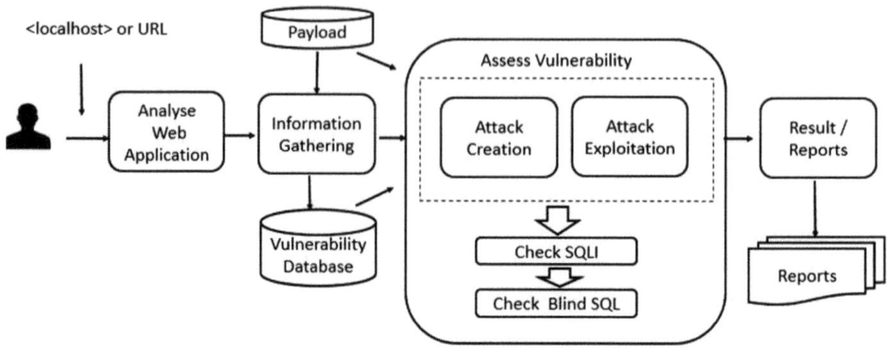

Fig. 1 Remote database health check model diagram

(iii) *Vulnerability Assessment* In this step model will check the vulnerability of web application using data gathered and rule database (payloads) for SQLI and Blind SQL Injection attacks turn by turn. This task is divided into two sub tasks. First, attacks are build using payload i.e. create injection strings using payloads. Then using identified entry points, it will execute attacks. During vulnerability check, Model will check for all types (attack vectors) of attacks. The model is using payload database. Various payloads are used for building and exploiting attacks like Login, Table and column names, attack payload, rule and words. These payloads provide scalability for any new attacks which may found in future.

The authors have prepared a prototype for implementing and testing this model.

5 Testing

5.1 Testing Environment

Figure 2 shows the test environment created for proposed model testing.

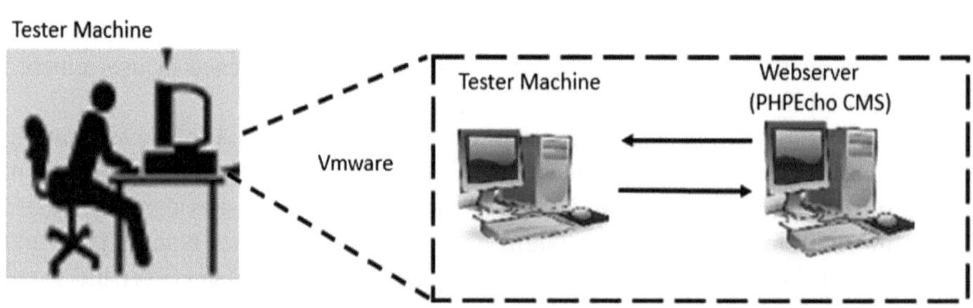

Fig. 2 Test environment

Two virtual machines named VICTIM and HACKER are created on testing machine. PHPEchoCMS web application is deployed on VICTIM machine and proposed model is installed on HACKER Machine. For testing HACKER machine is used.

5.2 *Test Scenarios*

For proof of concept verification, three test scenarios were considered.

(A) *Test Scenarios 1 PHPEchoCMS*
A Deliberately vulnerable web site is created for testing model using PHPEcho CMS. The first test scenario ran with PHPEchoCMS, a deliberately insecure J2EE web application developed. The purpose of this test campaign to verify and test the proposed model.

(B) *Test Scenario 2*
Custom web applications (developed and hosted on local host). The web site is developed using PHP and database as MySQL. The authors considered two types of websites (static and dynamic) under this scenario.

(C) *Test Scenario 3*
Purpose of this scenario is to execute unit testing of developed screen. Under this category one single login page is developed and testing is performed.

6 Results

6.1 *Results*

Testing results of above scenarios are summarized in Table 2.

Table 2 Testing results—vulnerability assessment

Test scenario	No. of components	No. of DB links	SQLI	Blind SQL
TS-1: Dynamic (PHPEchoCMS)	25+	11	TP	TP
TS-2: Custom (static)	28+	04	TN	TN
TS-2: Custom (dynamic)	127+	04	TP	TP
TS-3: Dynamic (under development)	1	4	TP	TP
TS3: (Under maintenance)	80+	5	TN	TN

TN True Negative, *TP* True positive

6.2 *Performance*

Table 3 summarize performance data of all testing scenarios. Performance data shows that proposed model is quick in assessment.

6.3 *Comparison*

Four parameters used for comparison are Vulnerability coverage (SQLI and Blind SQL), Feature (is solution GUI based), Developer Skill required and developer efforts required. The comparison between proposed model and top 10 open source tools is presented in Table 4.

It is found that

- Only eight out of top ten open source solution provide vulnerability assessment for SQLI and Blind SQL Injection, while proposed model can do for both.
- Six out of top ten open source model do not provide Graphical interface, while proposed model is menu driven

Table 3 Testing results—performance

Application	Duration (s)
TS-1: Dynamic (PHPEchoCMS)	135
TS-2: Custom (static)	119
TS-2: Custom (dynamic)	181
TS-3: Dynamic (under development)	040
TS3: (Under maintenance)	177

Table 4 Comparison—top 10 open source solutions

Name	SQLI	Blind SQL	GUI	Skill/efforts required
Grabber	X	X	No	Yes
Vega	X	X	Yes	Yes
Wapiti	X	X	No	Yes
W3af	X	X	Yes	Yes
Web scarab	X	X	Yes	Yes
Skipfish	X	X	No	Yes
Ratproxy	X	X	No	Yes
SQLMap	X	X	No	Yes
Wfuzz	X		No	Yes
Arachni	X		Yes	No
Proposed model	X	X	Yes	No

- Due to command line interface, technical knowledge is required in most of the open source solution. While proposed model does not require developer efforts for execution.

6.4 Comparison of Proposed Model with Top 3 Vulnerability Scanners

Table 5 describes comparison between proposed mode and top 3 vulnerability scanners.

Table 6 represents resource requirement comparison between Net Sparker and proposed model.

From Tables 5 and 6, we can conclude that proposed model can works with little resource i.e. works as a utility. It do not require developer efforts, skills and configuration. It is easy to use.

Table 5 Comparison—top 3 vulnerability scanners

	Wapiti	OWASP ZAP	Net sparker	Proposed model
Function	Scanner (act as fusser)	Fusser	Scanner	Scanner
Required technical skills	Yes	No	No	No
Requires source code access	No	No	Yes	No
Configuration required	Yes	Yes	Yes	No
False positive	Medium	High	Low	No
SQL injection	Yes	Yes	Yes	Yes
Blind SQL	Yes	Yes	No	Yes
Vulnerability assessment	Yes	Yes	Yes	Yes
Operations	Command line	Auto and manual	GUI	Menu driven
Report	Yes	Yes	Yes	Yes
Purpose	Audit	Detect training	Detect/exploit	Detect and exploit

Table 6 Comparison—resource requirement

	Net sparker	Proposed model
RAM requirement	1 GB RAM (min.)	<512 MB
HDD	100 MB + 100 Mb per scanning and 4.2 GB per scan	1 MB
Installation	Yes	No
Developed using	.NET	Python 2.X
Platform	Windows	UNIX/Windows

7 Conclusion and Future Work

Some investigation challenges for web vulnerabilities are exemplified in the proposed model. It provides bases for utility. The model emphasizes on the requirements of changes needed in Vulnerability risk mitigation using a light weight utility. To address challenges of multi tenancy of web application, Authors have proposed a logging mechanism, which can be useful to address known as well as unknown threats.

One of the key characteristic of Model is that it does not try to obtain sensitive data. However, it extracts weaknesses to prepare attacks and evaluate web application for vulnerability. The attack results are collected which can be used for further analysis and code fix. As mentioned Model neither need code nor server access to determine. Authors can run from any PC by giving an URL to the health check.

Authors can conclude that they have successfully tested web applications using proof of concept. The performance found was excellent. Model correctly identified vulnerability in web applications.

References

1. OWASP. Top Ten project 2013: https://www.owasp.org/index.php/Category: OWASP_Top_Ten_Project: accessed 31st May 2014.
2. Internet user statistics: http://www.internetworldstats.com/stats.htm: visited on 23rd November 2014.
3. Internet user in world: http://www.internetlivestats.com/internet-users/: visited on 23rd November 2014.
4. IBM Data Breach Statistics: http://www-935.ibm.com/services/us/en/it-services/security-services/data-breach/; visited on 23rd November 2014.
5. Eugene Lebanidze: Securing Enterprise Web Applications at the Source: An Application Security Perspective: https://www.owasp.org/images/8/83/Securing_Enterprise_Web_Applications_at_the_Source.pdf, pp. 1, 15, 32.
6. Jignesh Doshi, Bhushan Trivedi, Assessment of SQL Injection Solution Approaches, IJARCSSE, October 2014, Vol 4, Issue 10, ISSN: 2277 128X.
7. White Paper: Cutting the Cost of Application Security: An ROI White Paper: https://www.imperva.com/lg/lgw.asp?pid=349: accessed 31st August 2014.
8. Robert Richardson, "15th Annual 2010/2011 Computer Crime and Security Survey", 2011: gatton.uky.edu/FACULTY/PAYNE/ACC324/CSISurvey2010.pdf: accessed 1st December 2014.
9. Michael Howard, David LeBlanc and John Viega. "24 Deadly Sins of Software Security". "Sin 1: SQL Injection." Page 3–27. McGraw-Hill. 2010.
10. Nina Godbole and Sunit Belapure, "Cyber Security: Understanding Cyber Crimes, Computer Forensic and Legal Perspective", Wiley India Pvt. Ltd, First Edition 2011.
11. WG Hallfond, J Viegas and A Orso: "A Classification of SQL Injection attacks and Countermeasures", IEEE 2006.
12. Z. Su and G. Wassermann, "The Essence of Command Injection Attacks in web Applications", The 33rd Annual Symposium on Principles of Programming Languages (POPL 2006), 2006.

13. Open Web Application Security Project (OWASP) SQLI page: http://www.owasp.org/index.php/SQL_Injection; last visited 28th Aug 2014.
14. National Vulnerability Database (NVD) Security Checklists: http://web.nvd.nist.gov/view/ncp/repository; last visited 28th August 2014.
15. Common Weakness Enumeration: http://cwe.mitre.org/data/definitions/89.html: accessed 3rd August 2014.
16. Rahul Johri and Pankaj Sharma "A Survey on Web Application Vulnerabilities (SQLIA and XSS) Exploitation and Security Engine for SQL Injection", IEEE 2012.
17. A. Tajpour, M. Masrom and M. Z. Heydari, "Comparison of SQL Injection Detection and Prevention Techniques", 2nd International Conference on Education Technology and Computer (ICETC), 2012.
18. Chad Dougherty, Practical Identification of SQL Injection Vulnerabilities: www.uscert.gov/sites/default/files/publications/Practical-SQLi-Identification.pdf.
19. Carnegie Mellon University, Computing Services Information Security Office, Information security 101: www.cmu.edu/iso/aware/presentation/security101-v2.pdf: visited on 23rd November 2014.
20. A. Tajpour, M. Masrom and M. Z. Heydari, S Ibrahim: "SQL Injection Detection and Prevention Tools Assessment", IEEE 2010, 978-1-4244-5540-9 Aug.
21. A. Tajpour, M.J Shooshtari: "Evaluation of SQL Injection Detection and Prevention Techniques": 2010-Second International Conference on Computational Intelligence, Communication Systems and Networks: 978-0-7695-4158-7/10. doi:10.1109/CICSSyn, 2010 IEEE.
22. Diallo Abdoulaye and Al-Sakib Khan Pathan, "A Survey on SQL Injection: Vulnerabilities, attacks and Prevention Techniques", IEEE 15th International Symposium on Consumer Electronics, 2011.
23. William G.J. Halfond, Allesandro Orso, "AMNESIA: Analysis and Monitoring for NEutralizing SQL Injection Attacks", ACM, USA, 2005, pp 174–183.
24. Bisht, P., Madhusudan, P., and Venkatakrishnan, V.N., CANDID: Dynamic Candidate Evaluations for Automatic Prevention of SQL Injection Attacks. ACM Transactions on Information and System Security, Volume 13 Issue 2, (2010). doi:10.1145/1698750.1698754.
25. Sam M.S. N.G, "SQL Injection Protection by Variable Normalization of SQL Statement". www.securitydocs.com/library/3388, 06/17/2005.
26. Buehrer, G., Weide, B.W., and Sivilotti, P.A.G., Using Parse Tree Validation to Prevent SQL Injection Attacks. Proc. of 5th International Workshop on Software Engineering and Middleware, Lisbon, Portugal (2005) 106–113.22.
27. Kemalis, K. and T. Tzouramanis. SQL-IDS: A Specification-based Approach for SQLinjection Detection. SAC'08. Fortaleza, Ceará, Brazil, ACM (2008), pp. 2153–2158.
28. Stop SQL Injection Attacks Before They Stop You: http://msdn.microsoft.com/en-us/magazine/cc163917.aspx: accessed 3rd August 2014.

Correlation Between Text Book Usage and Academic Performance of Student in Higher Education Using 'R'

Shanti Verma and Jignesh Doshi

Abstract In this paper authors try to focus one of the key factors to improve student performance i.e. text book usage. In this study authors explore association between students final Semester Grades and text books usage. Scope of study is limited to only MCA program of Gujarat Technological University students and the text book of subjects that are not issued for longer period by institute library. Authors also take assumption that books are not purchased by students as they are too costly. The aim of this paper is to use correlation and regression methods to analyze the dataset containing 60 students of MCA semester III students and try to find out that does the text book usage by student affects the academic performance. The primary result of experimental analysis shows that if the text book usage increases, performance of students also increases. The results are important for the teachers to motivate students towards library as well as institutions to focus on library utilization.

Keywords Correlation · Linear regression · Academic performance · Higher education

1 Introduction

As per All India Council for Technical Education (AICTE), number of education institutes and intake have increased exponentially in past decade [1]. Poor results and quality are key challenges for universities nowadays. Each student can be classified using into various groups using learning styles [2, 3].

S. Verma (✉)
LJ Institute of Computer Applications, Ahmedabad, India
e-mail: verma.shanti@gmail.com

J. Doshi
LJ Institute of Management Studies, Ahmedabad, India
e-mail: doshijig@gmail.com

N. Modi et al. (eds.), *Proceedings of International Conference on Communication and Networks*, Advances in Intelligent Systems and Computing 508,
DOI 10.1007/978-981-10-2750-5_2

Data mining or Knowledge Discovery in Database (KDD) is a collection of advanced analytical techniques to discover knowledge form large amount of data [3, 4, 5]. Data mining techniques are applied successfully in various areas like manufacturing, production, customer relationship management, tale-communication, education for improvements [6, 7].

Many researchers have performed various types of analysis in education area [8]. In this paper, authors have applied correlation and regression methods to find relation between reading textbooks and academic performance.

This paper is organized as below: Introduction to topic is provided in Sect. 1, Objectives are defined in Sect. 2, Literature review is discussed in Sects. 3, 4 and 5 discusses experiment and results performed using correlation and regression using R programming. Conclusion is provided in Sect. 6 and Future work in Sect. 7.

2 Objectives of Study

In current scenario there is a trend of using different online available data that are not reliable. It is very important to find that how text book usage affects academic Performance of student [9]. Thus, this study aims to investigate the association between text book usage and academic performance in higher education. The objectives are:

1. To find the degree of association between text book usage and academic performance of student.
2. To determine the linear regression equation between text book usage and academic performance of student.

3 Literature Review

3.1 Related Work: Education Mining

Data Mining can be used in academic to enhance and evaluate the learning process students [10, 11, 12]. Various data mining techniques are used by various researchers in various areas for analysis like:

(a) Bhardwaj and Pal have performed research using Bays Classification to found whether new comer students will be performer or not [13].
(b) Using data analysis, Galit tried to predict the results and to warn students at risk before their final exams [14].
(c) Decision tree model was used by Al-Radiate, et al to predict the final grade of students. Three different classification methods namely ID3, C4.5, and the Naive ayes were used. The outcome of their results indicated that Decision Tree model had better prediction than other models [15].

(d) Pandey and Pal applied association rule and try to find the interestingness of student in opting class teaching language [16].
(e) As per Ayesha, Mustafa, Sitar and Khan, we may use k-means clustering algorithm to predict student's learning activities [5].
(f) Jignesh explored Association rule, chi-square and lift techniques to predict failures [17].
(g) Hijazi and Naqvi performed research using simple linear regression analysis to discover that the factors like mother's education and student's family income are highly correlated with the student academic performance or not [18].

3.2 Related Work: Correlation and Regression

Correlation technique is used by researchers to find degree association between two variables. Regression coefficients are useful to predict dependent variable value for some independent variable. Both techniques are used by many researchers to predict academic performance of student [18].

(a) Alan Chea et al. Used Data mining technique: correlation to associate various variables that influence academic performance. These association results were used to predict student's academic success [8].
(b) Norman Poh, Ian Smythe used Regression techniques to predict student performance. They predict student performance on basis of student performance based on self-efficacy, socio-economic background, learning difficulties, and related academic test results [7].

4 Experiment

4.1 Data Extraction and Transformation

In this paper authors have selected data sets from one of the Largest University "Gujarat Technological University" of Gujarat and Library log records of "L. J. institute of computer Applications", one of the. largest intake college of Master of computer Applications Program in Gujarat. The Data sample is taken from post graduate course "Master of Computer Applications" semester III of 60 students. Steps for Data Extraction, Collection and Transformation

1. Data Selection

    a. Data of 2 institutes out of 40
    b. Total 60 data collected out of 240

Table 1 Academic grade coding

Grade	AA	AB	BB	BC	CC	FF
New value	5	4	3	2	1	0

   c. Semester 3 students data taken only for two subjects Statistical Methods (SM), SOOADM

2. Data Collection

   a. University result data was collected from GTU web portal (www.gtu.ac.in)
   b. Library log collected from LJMCA Library Software

3. Data Transformation

   a. Convert student result grade as quantitative data (Table 1).
   b. From Library log records find the frequency of book issued of each student in subjects SOOADM and SM.
   c. Initially data was entered in Notepad as tab delimited file.

4. Tool Used

   a. Data Mining Tool- ‘R’

4.2 Technique: Correlation

Ho: Grades and Frequency of book used are positively correlated
Ha: Grades and Frequency of book used are not positively correlated
Steps to find the correlation value in ‘R’

1. Make a Two column tab delimited file. First Column name is Frequency of Book used as independent variable and second is Grade as dependent variable.
2. Draw scatter plot.
3. Find correlation coefficient (Fig. 1 and Table 2).

In the above scatter plot author find out that in most of the cases as frequency of book used is increased, grades also increased. In another words there is a positive correlation between SOOADM text book frequency and grades (Fig. 2 and Table 3).

In the above scatter plot author find out that in most of the cases as frequency of book used is increased, grades also increased. In another words there is a positive correlation between SM book frequency and grades.

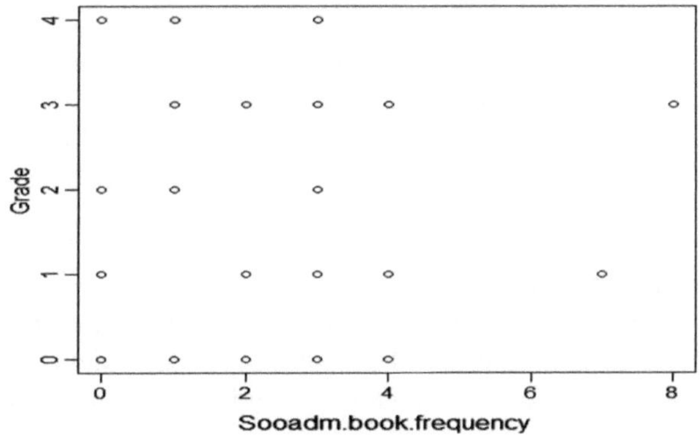

Fig. 1 Scatter plot between SOOADM text book usage and grade

Table 2 Correlation coefficient for subject SOOADM

	Sooadm.book.frequency	Grade
Sooadm.book.frequency	1.00	0.0640
Grade	0.0640	1.00

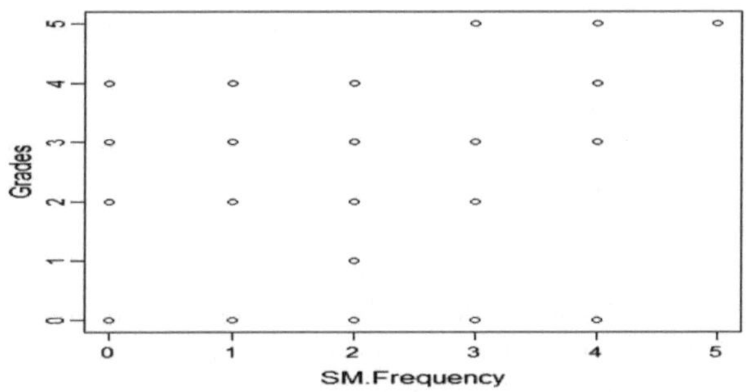

Fig. 2 Scatter plot between SM text book usage and grade

Table 3 Correlation coefficient for subject SM

	SM.Frequency	Grade
SM.Frequency	1.00	0.5270
Grade	0.5270	1.00

4.3 Technique: Linear Regression

Regression coefficients for Subject SOOADM

(Intercept)	Grade
1.55323	0.07381

Regression line for Subject SOOADM is

Grade(Y)= 1.5532 + 0.0738 * Sooadm.book.frequency (x)

Regression coefficients for Subject SM

(Intercept)	Grades
0.6372	0.4588

Regression line for Subject SM is

Grades (Y) = 0.6372 + 0.4588 * SM.Frequency (x)

Here authors find that if the value of independent variable is increased, dependent variable value also increased but not in the same rate as subject SOOADM.

5 Results and Discussion

We summarize experimental results as in Table 4.

Authors discover:

+ve Correlation

Text book usage and academic performance are positively correlated means if text book usage increase/decreases, academic performance of student also increase/ decreases.

+ve Regression Coefficients

It states that text book usage and academic performance are positively dependent to each other.

Predicted Value

Here authors take text book usage value 7 and check academic performance. For SOODAM subject value is 2.069863 means that if students use text book effectively at least 7 times in semester then he/she will get at least BC grade in final semester Exam. For subject SM predicted value is 3.8488 means if students use SM text book at lease 7 times in semester then he/she will get at least BB grade in final semester exam.

Table 4 Experimental result summary

	Correlation coefficient	Regression coefficient	Predicted value when text book usage is 7
SOOADM	0.06403358	Intercept 1.55323	2.069863
		Grade 0.07381	
SM	0.5279365	Intercept 0.6372	3.8488
		Grades 0.4588	

6 Conclusion

The objective of identifying student academic performance on data mining the text book usage is achieved. We believe that these results could be important to predict academic performance of student on the basis of text book usage.

Here authors validate the null hypothesis that text book usage and academic performance are positively correlated. It is expected that results of the research are useful for teachers to motivate students towards text book usage.

7 Future Work

We can take more variables that are dependent with academic performance and use Multivariate analysis to improve results.

References

1. All India Council for technical Education (AICTE) approval process handbook (2013–14), page 4–7.
2. Richard M. Felder, Rebecca Brent: "Understanding Student Differences": http://www4.ncsu.edu/unity/lockers/users/f/felder/public/Papers, Journal of Engineering Education, 94 (1), 57–72 (2005).
3. Richard M. Felder, G. N. Felder, E.J. Dietz: "The Effects Of Personality Type On Engineering Student Performance And Attitudes", http://www4.ncsu.edu/unity/lockers/users/f/felder/public/Papers: Journal of Engineering Education, 91(1), 3–17 (2002).
4. Galit.et.al, "Examining online learning processes based on log files analysis: a case study". Research, Reflection and Innovations in Integrating ICT in Education 2007.
5. Han, J. &Kamber, M. (2006), Data Mining: Concepts and Techniques, 2nd edition, Morgan Kauf-mann.
6. G K Gupta, Introduction to data mining with case studies, Second Edition, PHI Learning.
7. Norman Poh, Ian Smythe "To what Extent we predict Student's performance? A Case Study in colleges in South Africa", 2014 IEEE.
8. S. T. Hijazi, and R. S. M. M. Naqvi, "Factors affecting student's performance: A Case of Private Colleges", Bangladesh e-Journal of Sociology, Vol. 3, No. 1, 2006.
9. Alan Cheah Kah Hoe, Mohd Sharifuddin Ahmad, Tan Chin Hooi, Mohana Shanmugam, Saraswathy Shamini Gunasekaran, Zaihisma Che Cob, Ammuthavali Ramasamy, "Analyzing Students Records to Identify Patterns of Students' Performance", Research and Innovation in Information Systems (ICRIIS), 2013 International Conference, IEEE PP. 554–557.
10. Alaa el-Halees, "Mining students data to analyze e-Learning behavior: A Case Study", 2009.
11. Brijesh Kumar Baradwaj, Surabh Pal, Mining Educational Data to Analyze Students Performance, International Journal of Advanced Computer Science and Applications, Vol. 2, No. 6, 2011, Page 633–69.
12. Verma, Shanti. "Deciding Admission Criteria For Master of Computer Applications Program in India using Chi-Square Test." Proceedings of the Second International Conference on Information and Communication Technology for Competitive Strategies. ACM, 2016.

13. U. K. Pandey, and S. Pal, "Data Mining: A prediction of performer or underperformer using classification", (IJCSIT) International Journal of Computer Science and Information Technology, Vol. 2(2), pp. 686–690, ISSN:0975-9646, 2011.
14. U. K. Pandey, and S. Pal, "A Data mining view on class room teaching language", (IJCSI) International Journal of Computer Science Issue, Vol. 8, Issue 2, pp. 277–282, ISSN:1694-0814, 2011.
15. Shaeela Ayesha, Tasleem Mustafa, Ahsan Raza Sattar, M. Inayat Khan, "Data mining model for higher education system", European Journal of Scientific Research, Vol. 43, No. 1, pp. 24–29, 2010.
16. Q. A. AI-Radaideh, E. W. AI-Shawakfa, and M. I. AI-Najjar, "Mining student data using decision trees", International Arab Conference on Information Technology(ACIT'2006), Yarmouk University, Jordan, 2006.
17. Jignesh Doshi, Result Mining: Analysis Of Data Mining Techniques In Education, International Journal of Technical Research and Applications e- ISSN: 2320-8163, Volume 2, Issue 3 (May-June 2014), PP. 25–28.
18. Norman Poh, Ian Smythe "To what Extent we predict Student's performance?A Case Study in colleges in South Africa", 2014 IEEE.

Human Computer Interaction Through Hand Gestures for Home Automation Using Microsoft Kinect

Smit Desai and Apurva Desai

Abstract Gesture recognition has been an attractive area of research since a long time. With the introduction of Microsoft Kinect, hand gesture and body gesture recognition has become handy for the researchers. Here an innovative application has been presented which controls all electrical home appliances through hand gestures. The algorithm presented here is an assistive application useful for physically challenged and senior citizens. In this paper we have used Microsoft Kinect for image capturing along with some important computer vision (CV) and digital image processing techniques (DIP) for hand gesture recognition. Arduino Uno microcontroller and relay circuits are used for controlling electrical devices. The algorithm presented gives an accuracy of 88 %.

Keywords Microsoft Kinect · Human computer interaction (HCI) · Computer vision (CV) · Depth sensor · Feature extraction · Feature classification · Arduino uno

1 Introduction

Since a long time we have been using keyboards and mouse to interact with computers. Off late different techniques like WI-FI and Bluetooth have became more popular and handy for communicating with computers. With the development of new software applications and research in the area of Human Computer Interaction (HCI), need of a more effective communication media increased. In recent time, brain waves, voice, touch, gestures etc. are used for HCI. The use of human gesture and posture is strongly making its mark in the area of HCI. Hand

S. Desai (✉)
Sarvajanik College of Engineering and Technology, Surat, Gujarat, India
e-mail: thesmitdesai5@gmail.com

A. Desai
Veer Narmad South Gujarat University, Surat, Gujarat, India
e-mail: desai_apu@hotmail.com

N. Modi et al. (eds.), *Proceedings of International Conference on Communication and Networks*, Advances in Intelligent Systems and Computing 508,
DOI 10.1007/978-981-10-2750-5_3

gesture and facial expressions are the most popular areas of research for researchers working in this dynamic field. Hand gesture recognition is a challenging task mainly for two reasons: Localization and segmentation. Initially hand gesture recognition was done using colour marks on fingers or coloured gloves, such as [1]. More information and historical development of use of gloves and various sensors or devices used in hand gesture recognition is given in [2]. Nevertheless the use of such devices or gloves is not advocated as it restricts the convenience of the user. Constraint free hand gesture recognition has been more popular with the introduction of depth sensors. Microsoft has introduced one such depth sensor called Microsoft Kinect.

In this paper we have proposed an algorithm, which not only identifies the gesture of palm of a hand but also controls the home appliances through them. We have utilized Kinect sensor for video capturing and further Ardiuno is used as a bridge between the hand gesture recognition engine and electronic home appliances. The main motivation behind this work is to empower physically challenged and senior citizens to have complete control over their home appliances with minimal physical movements. Here in this work our proposed algorithm identifies hand gesture and it sends appropriate signals to Arduino which then controls any electrical/electronic device which is intermediately connected to a relay board. This paper is divided into five sections. In section two we have presented literature review which is followed by our proposed algorithm in section three. In section four we have presented the results of our proposed algorithm. Conclusion is proposed in the last section.

2 Literature Review

Without saying anything, human gestures can convey thousands of words and therefore human gestures should be used more often for communication. For providing such natural capabilities to a robot or a computer, researchers of Computer Vision (CV) and Human Computer Interaction (HCI) are inclined to identify and recognize human gestures. According to Pavlovic et al. [3] human gesture can be categorized as shown in Fig. 1.

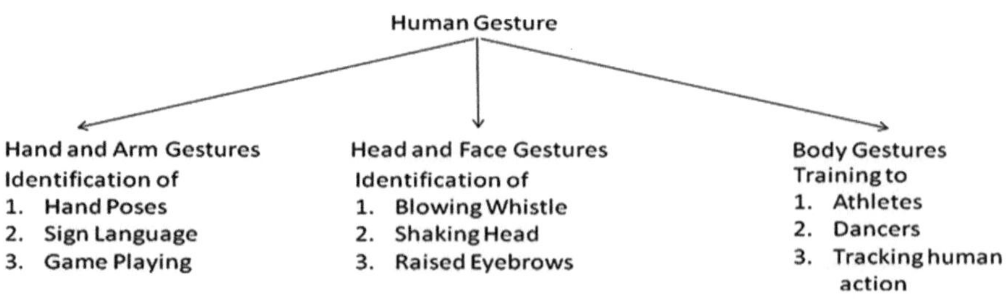

Fig. 1 Classification of human gestures

In addition to gestures, voice and touch are also used for human computer interaction. Among the human gestures, hand and arm gestures are more expressive and hence they are more popularly used in computer vision and human computer interaction. Initially, for hand gesture recognition, gloves based devices, colour bands on arm or wrist or some wired wearable devices were used. However, the use of these gears or colour bands etc. used to restrict the freedom of the user, and looses naturalness of interaction with the computer in particular and the machine in general. Obviously the techniques based on these types of gears were vision based techniques and therefore were more dependent on lighting and illumination conditions [4, 5]. In the year 2010 Microsoft introduced a revolutionary device called Kinect which was meant to be used with their XBOX 360 console. But researchers found a different and more interesting use of Kinect, which completely changed the field of CV. Kinect is the ultimate solution to combat with the problems of lighting and illumination, since it has the capability to measure depth of the object. Technically speaking Kinect is a depth sensor, which captures images in RGB mode and maps them into the depth of the scene. With introduction of Kinect, researchers came up with many interesting and innovative applications. In literature one can trace plenty of algorithms for hand gesture recognition both with and without the use of Kinect sensor. An algorithm based on gloves is presented by Agrawal et al. [6], Ren et al. [7]. In [6] researchers have made the use of red coloured gloves for localization and segmentation of hand. For hand segmentation the authors have used R colour channel threshold value of G and B channel. Whereas in [7] authors used a black coloured wristband for hand segmentation. Here it is interesting to know that in the earlier work the researchers have not used Kinect sensor where as in the later one Kinect has been used and hand has been captured from a pre-specified depth range. Since these two algorithms are based on colours, they have limitations like; if the colour is not properly captured the algorithm will not work efficiently, if the colour of glove or band is visible in the frame; in the form of background or elsewhere that part will also be captured as a part of hand.

While addressing the problem of hand gesture recognition; localization and segmentation of hand is the most crucial and important task. In almost all the research done pertaining to this problem it has been suggested to keep the gesturing hand in front of the camera. This is how the gesturing hand becomes the nearest object. There are three different approaches adopted by the researchers to locate the hand. The first approach is to find the centroid of the hand and then with some minor adjustments in depth, the hand is localized [8–10]. The second approach is to isolate the hand if it falls in a particular range of depth. Here researcher empirically decided a range of depth and the user needs to keep his hand within that range to get it captured by the camera [7, 11–13]. In the third approach researchers use mathematical model or some logical approach like finding corners of fingers, Time series curve etc. This approach has been used in [14–16]. Once, hand is localized and it is segmented the next task in the process of hand gesture is feature extraction of various hand gestures. Biswas [14] has considered various hand gestures like 'clap', 'call', 'greet', 'wave', 'yes', 'no', and 'rest' whereas Jing and Guan in [15] have considered pointing finger gesture to create an application of virtual touch screen. Against these sequential gestures one can find out applications where static hand gestures are considered which recognizes sign language, numbers and alphabets of some natural language [7, 8, 10–12, 17]. In many presented algorithms researchers have used only fingers without palm for extracting features. Such algorithms need to remove palm from the hand image. Marin [18], Ren et al. [7] used palm centroid position given by Kinect and draw a circle from there which fits the palm area of hand. By removing the area which falls into the circle, fingers can be identified. In contrast to this approach many researchers use complete hand image to extract features counting open and close fingers using some mathematical or structural information of image [8, 10, 17, 18]. Mathematical feature sets are used in [7, 8, 11, 16] where tools like Fourier descriptor, SIFT, PCA etc. are used. Whereas [6, 9, 14, 18] have suggested structural features like contours, convex hull, and depth histograms etc. Feature extraction is followed by identification of feature that is classification. In literature some popular and widely used classification techniques like; Support Vector Machine (SVM), Hidden Markov Model (HMM), K-means classifier, Naïve Bayes classifier, template matching, Finger Earth Mover Distance (FEMD) [6–10, 14, 16] are found to be used. One can find very exhaustive and useful review information on Microsoft Kinect sensor and hand gesture recognition in [19, 20] respectively.

3 Proposed Algorithm

The algorithm first captures images from the depth sensor Kinect and it localizes the hand from the frame. In the next step, hand is cropped for further processes. The image captured through Kinect requires some preprocessing before it is considered

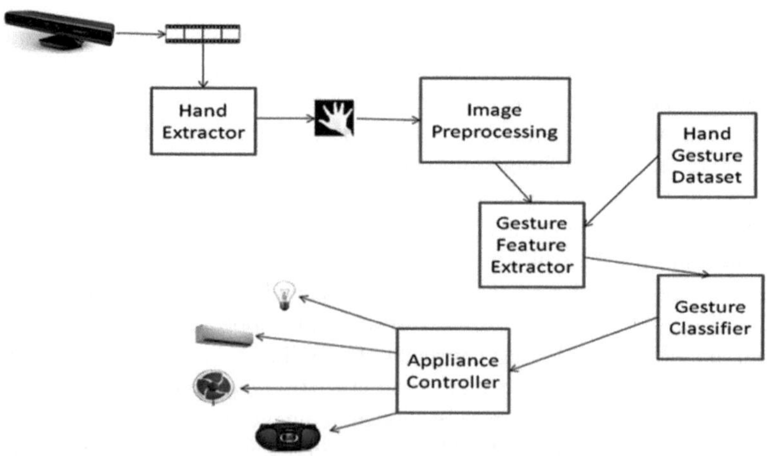

Fig. 2 Block diagram of proposed system

for feature extraction. Preprocessing task is done in the next step. In the successive step the features are extracted and it is sent to the classifier. When the classifier finds the gesture to be appropriate, signal is sent to Arduino which controls the appropriate appliance with the help of a relay board. Figure 2 is the complete block diagram of proposed system. The detailed description of each step is as follows;

4 Hand Segmentation

For segmentation of hand, we need the user to keep his hand in front of the sensor. When the user keeps his hand in front of sensor, the hand becomes the nearest object visible in the frame. In this process we have used NITE library and a predefined range of depth [11] for hand localization and segmentation. The following algorithm shows the process of hand segmentation. The output of this process is shown in Fig. 3.

Step 1: Capture depth mapped video from Kinect sensor
Step 2: Capture location (H_x, H_y) using NITE. (H_x, H_y) is the centroid of the visible hand.
Step 3: If the centroid (H_x, H_y) belongs to the predetermined range, then crop the image.

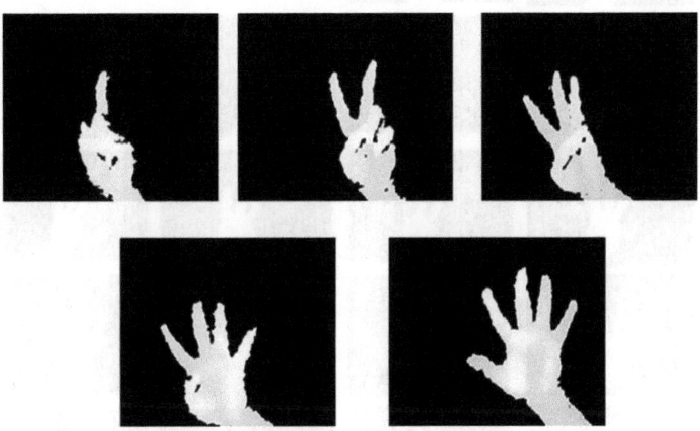

Fig. 3 Localized and segmented hand images captured by Kinect

5 Image Pre-processing

The pre-processing is a process which prepares the image for further processing. In this stage the noise, slant etc. are removed as shown in Fig. 4. Here in this problem we hardly have any noise as such but when the image is converted into binary some information is lost and discontinuity in the image is added. Also the image captured by Kinect and selected in the earlier stage is in RGB though it is appears to be in binary. Therefore it is needed to convert it into binary for finding features of the image. Here we propose removal of gaps and small holes from the image.

Step 1: RGB image obtained from pre-processing stage is converted into gray image by converting each RGB pixel value to gray scale value by performing weighted sum of red, green and blue component of respective pixel.

$$I(X_i, Y_i) = \sum ((R(X_i, Y_i) * 0.2989) + (G(X_i, Y_i) * 0.5870) + (B(X_i, Y_i) * 0.1140))$$

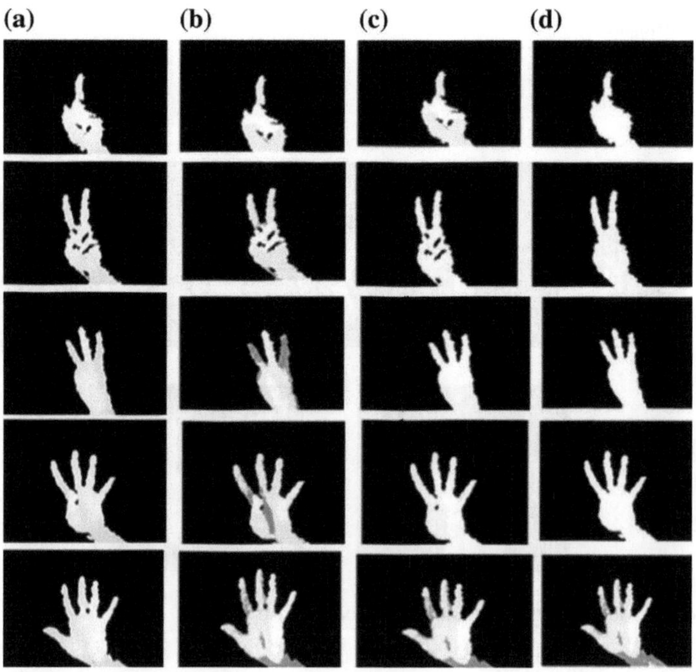

Fig. 4 Set of hand gesture 1–5. **a** Original images, **b** images with smoothen boundary, **c** dilated images, **d** images with filled in holes

Step 2: Find a global threshold value using Otsu's method [21] of gray scaled image to convert it into binary. Otsu's method minimizes intra-class variation of black and white pixels. i.e.

$$Min\,\sigma_G^2 = W_F(t)\sigma_F^2(t) + W_B(t)\sigma_B^2(t)$$

where;
σ_G^2 is an intra class (foreground and background) variation,
W_F, W_B are weights of foreground and background class respectively,
σ_F^2, σ_B^2 are variances of foreground and background class respectively,
t is a threshold value which minimizing intra class variation.

Step 3: Convert gray scaled image into binary suing threshold value calculated in step 2.

$$\mathrm{B(X,Y)} = \begin{cases} 0\,if\ G(X,Y) < t \\ 1\,if\ G(X,Y) > t \end{cases}$$

where;
B(X, Y) is a Binary image, and
G(X, Y) is gray scaled image

Step 4: Smoothen out the edge of the hand gesture image using median filter on 3×3 neighborhood.

Step 5: Perform morphological operation dilation and filling. While converting hand gesture image in binary image, small holes are created in image. To remove these holes, morphological operation dilation is performed. Filling process will fill all small holes, if any, in the image

Step 6: Crop the gesture image to the region of Interest (ROI).

6 Feature Set and Classifier

In this work the user who wishes to operate appliances through hand gestures may be of any age and gesture. Also the user may be at any distance and at any angle from the Kinect sensor. In addition to this, the user may use any hand: left or right to convey the gesture to the system. Left hand gestures and right hand gestures are mirror images of one another. In all these cases, feature set should work effectively. To consider these challenges and complexity we need to have a set of feature which is invariant of scale, place and rotation. Hence, we propose to use Hu's seven moment invariants [22] as a feature set. Hu's moment invariant are calculated as below;

$$\begin{aligned}
M_1 &= \eta_{20} + \eta_{02} \\
M_2 &= (\eta_{20} - \eta_{02})^2 + 4\eta_{11}^2 \\
M_3 &= (\eta_{30} - 3\eta_{12})^2 + (3\eta_{21} - \eta_{03})^2 \\
M_4 &= (\eta_{30} + \eta_{12})^2 + (\eta_{21} + \eta_{03})^2 \\
M_5 &= (\eta_{30} - 3\eta_{12})(\eta_{30} + \eta_{12}) + ((\eta_{30} + \eta_{12})^2 - 3(\eta_{21} - \eta_{03})^2) \\
&\quad + (3\eta_{21} - \eta_{03})(\eta_{21} + \eta_{03})(3(\eta_{30} + \eta_{12})^2 - (\eta_{21} + \eta_{03})^2) \\
M_6 &= (\eta_{20} - \eta_{02})((\eta_{30} + \eta_{12})^2 - (\eta_{21} + \eta_{03})^2) + 4\eta_{11}(\eta_{30} + \eta_{12})(\eta_{21} + \eta_{03}) \\
M_7 &= (3\eta_{21} - \eta_{03})(\eta_{30} + \eta_{12})((\eta_{30} + \eta_{12})^2 - 3(\eta_{21} + \eta_{03})^2) \\
&\quad + (3\eta_{12} - \eta_{30})(\eta_{21} + \eta_{03})((3(\eta_{12} + \eta_{30})^2 - (\eta_{21} + \eta_{03})^2)
\end{aligned}$$

where η_{pq} are normalized central moments;

$$\eta_{pq} = \frac{\mu_{pq}}{\mu_{00}^{\gamma}} \text{ where } \gamma = \frac{p+q}{2} + 1 \quad \forall p+q \geq 2$$

and,

$$\mu_{pq} = \sum_X \sum_Y (X - \overline{X})^p (Y - \overline{Y})^q f(X, Y)$$

Here it is interesting to note that M_1 and M_2 are called Hu's moments of second order where as remaining moments are called moments of order three. Thus for each of the hand gesture a feature set of seven features is calculated.

There are many classifiers being used in shape identification. These popular classifiers are Artificial Neural Network (ANN), Support Vector Machine (SVM), and kNN. Here we propose to use kNN classifier; as it is a simple classifier and does not need much prior information. We further propose Euclidian distance for classification. We have also carried out our experiment for different values of k and ultimately we decided to go with k = 1 as it gives more effective classification rate.

7 Experimental Results

For testing this proposed algorithm we have collected five gestures from fifty subjects. Thus a total of two hundred and fifty hand gestures images are considered in dataset. These subjects are of different age group and gender. Through Kinect depth sensor we collected images at 640 × 480 resolution. Kinect sensor captures images in eleven bits and it is sensitive to 2048 different levels. When hand comes within the range of 250–650 mm from the sensor it captures the hand as the nearest object. When any hand is identified in the given range it is captured and processed as described above and Hu's seven moments are calculated for it. Thus we have a

Table 1 Hu's moment of hand gestures 1–5 shown in Fig. 4

	Gesture 1	Gesture 2	Gesture 3	Gesture 4	Gesture 5
M_1	0.20365	0.2126857	0.2121163	0.2244263	0.2229706
M_2	0.00444	0.004906	0.0046875	0.0055781	0.0047563
M_3	4.32E−05	4.48E−05	4.49E−05	7.07E−05	2.04E−05
M_4	3.00E−05	2.61E−05	2.97E−05	4.27E−05	2.87E−05
M_5	4.90E−10	5.35E−10	−3.81E−10	1.32E−09	−1.23E−10
M_6	−1.92E−06	−1.76E−06	−1.40E−06	−3.18E−06	−1.40E−06
M_7	7.12E−10	3.33E−10	−1.75E−09	−8.33E−10	−1.46E−09

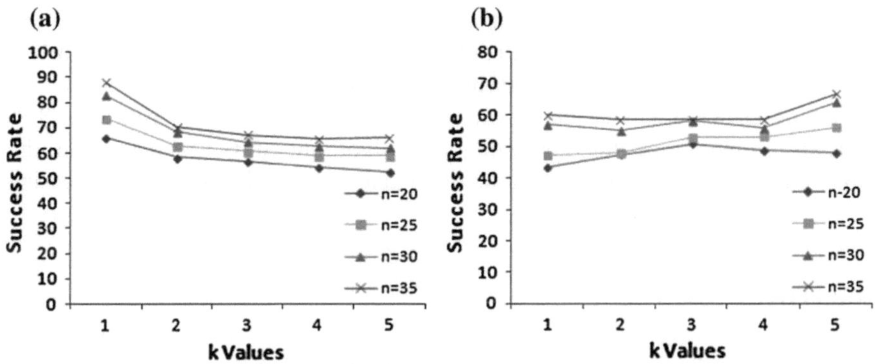

Fig. 5 Success rate of classifier for various k values and various size of train dataset. **a** Overall performance, **b** performance on unseen data

feature set of size seven. Table 1 shows a feature set of hand gesture 1–5. Further we have tested kNN classifier for various values of neighborhood values k and for different training size. Figure 5a shows a chart of different neighborhood values against total success rate for various size of training dataset and Fig. 5b is the same chart for unseen dataset. From Fig. 5b it is evident that neighborhood value $k = 1$ and training 35 size of training dataset gives the best result.

From Table 2 it is seen that gesture one is the most successful gesture whereas gesture two is the most weak gesture for identification according to this proposed algorithm. Gesture two is misidentified as three and four for six and four times

Table 2 Confusion matrix of test result when $k = 1$ and $n = 35$

	1	2	3	4	5
1	48	2	0	0	0
2	0	40	6	4	0
3	0	0	45	5	0
4	0	0	3	42	5
5	0	1	1	3	45

respectively. The identified gesture is then send to Arduino through a six channel relay board to activate different home appliances like light, fan, mobile charger or TV set. We assigned gesture one to light, two to fan, three to mobile charger and four to TV set. Gesture five is off gesture which shuts down any of the open appliance. The confusion matrix of appliance on/off is same as that of gesture recognition.

8 Conclusion

The algorithm presented here is an attempt to assist physically challenged and senior citizens who want to operate appliances with minimal efforts. This algorithm uses Kinect sensor for image capturing and various computer vision and image processing techniques. The total success rate achieved by this algorithm is 88 %. There is ample scope for improving this algorithm at two stages. The first improvement is possible in localization of hand and second is proper preprocessing.

References

1. El-Sawah A., Georganas N., Petiriu E., A prototype for 3-D hand tracking and posture estimation, IEEE transaction of Instrum Meas, vol. 57(8), pp. 1627–1636 (2008).
2. P. Premaratne, Human Computer Interaction using Hand Gesture Recognition, Springer (2014).
3. Pavlovic V., Sharma R., Huang T., Visual Interpretation of Hand Gesture for Human Computer Interaction, IEEE transaction on Pattern Analysis and Machine Intelligence, vol. 19 (7), pp. 677–695 (1997).
4. Erol A., Bebis G., Nicolescu M., Boyle R., Twombly X., Vision Based Hand Pose Estimation: A Review, Computer Vision and Image Understanding, pp. 52–73 (2007).
5. Suarez J., Murphy R., Hand Gesture Recognition with Depth Image: A Review, IEEE International Symposium on Robot and Human Interactive Communication, pp. 9–13 (2012).
6. Agrawal I., Johar S.,Santhosh J., A Tutor for the Hearing Impaired (Developed using Automatic Gesture Recognition), International Journal of Computer Science, Engineering and Application, vol. 1(4), pp. 49–61 (2011).
7. Ren Z., Youn J., Meng J., Zhang Z., Robust Part Based Hand Gesture Recognition Using Kinect Sensor, IEEE Transaction Multimedia, vol. 15(5), pp. 1110–1120 (2013).
8. Arun K., Chris Z., Joseph and J. L. Jr., Poster: Real Time Markless Kinect Based Finger Tracking and Hand Gesture Recognition, In IEEE Symposium on 3D User Interface, Orlando, USA (2013).
9. Verma H., Aggarwal E., Chandra S., Gesture Recognition using Kinect for Sign Language Transalation, In IEEE International Conference on Image Processing (ICIIP-2013) (2013).
10. Shukla J., Dwivedi A., A Method for Hand Gesture Recognition, In Fourth International Conference on Communication System and Network Technologies (2014).
11. C. Cliff and S. S. Mirfakhroei, "Hand Gesture Recognition Using Kinect," Technical Report No ECE-2013-04, Boston University (2013).
12. Du H., To T.H., Hand Gesture Recognition using Kinect, Technical Report No. ECE-2011-04, Boston University (2011).

13. Raheja J., Chaudhary A., Singal K., Tracking of Fingertips and Centre of Palm using Kinect, In IEEE International Conference on Computational Intelligence, Modelling and Simulation, Malaysia (2011).
14. Biswas K., Basu S., Gesture REcognition using Microsoft Kinect, In 5th International Conference on Automation, Robotics and Application, Wellington, Newzeland (2011).
15. Jing P., Ye-Peng G., Human Computer Interaction using Pointing Gesture based on an Adaptive Virtual Touch Screen, International Journal of Signal Processing, Image Processing and Pattern Recognition, vol. 6(4), pp. 81–91 (2013).
16. Hamissi M., Foez K., Real Time Hand Gesture Recognition Based on Depth Map for Human Robot Interaction, International Journal of Electrical and Computer Engineering, vol. 3(6), pp. 770–778 (2013).
17. Verma H.V., Eshan A., Chandra S., Gesture Recognition Using Kinect for Sign Language Translation, In IEEE Second International Conference on Image Information Processing (ICIIP-2013) (2013).
18. Marin G., Fraccaro M., Donadeo M., Dominio F., Zanuttigh P., Palm Area Detection for Reliable Hand Gesture Recognition, In IEEE International Workshop on Multimedia Signal Processing 2013 (MMSP-2013) (2013).
19. Han J., Shao L., Xu D., Shotton J., Enhance Computer Vision with Microsoft Kinect Sensor - A Review, IEEE Transaction on Cybernetics, vol. 43(5), pp. 1318–1334 (2013).
20. Mitra S., Acharya T., Gesture Recognition: A survey, IEEE Transaction on System, Man and Cybernetics - Part C: Applications and Reviews, vol. 37(3), pp. 311–324 (2007).
21. Otsu N., A Threshold Selection Method from Gray-Level Histograms, IEEE Transactions on Systems, Man, and Cybernetics, vol. 9(1), pp. 62–66 (1979).
22. Hu, Ming-Kuei, Visual pattern recognition by moment invariants, IRE Transaction on Information Theory, vol. 8(2), pp. 179–187 (1962).

Enhancement of Security in IoTSyS Framework

Hetal B. Pandya and Tushar A. Champaneria

Abstract IoTSyS is open source middleware for IoT and has the potential to provide easy access to home automation. It incorporates RESTful web services on HTTP/CoAP communication protocol. It has a gateway which communicates with IoTSyS users as well as devices. Gateway consists of an oBIX server, CoAP server and object broker for its execution. In IoTSyS, SSL security using TomCat Server is deployed between Users and IoTSyS for secure communication. However, the area of data communication between IoT Devices and Gateway is needed to be addressed. In this paper, Message-level Security instead of transport level security is used to provide end-to-end secure communication between IoT devices and Gateway. Various symmetric and asymmetric security algorithms along with different data formats such as XML, JSON and EXI are executed and compared on our experimental setup. Our results show that Blowfish encryption algorithm is more suitable for IoTSyS framework.

Keywords Internet of things · IoTSyS middleware · Message level security · Transport level security

1 Introduction

Internet of Things—In simple 3 words it can be explained as "web of things", allows interconnecting everything which we are using in our day to day life. It's like every generic thing, communicating to each other and acts accordingly. The IoT is now a day's playing leading role in various domains for fully automated systems.

H.B. Pandya (✉)
Government Engineering College, Bhavnagar, Gujarat, India
e-mail: hetalvpandya@gmail.com

T.A. Champaneria
L.D. College of Engineering, Ahmedabad, Gujarat, India
e-mail: tushar.1985@gmail.com

N. Modi et al. (eds.), *Proceedings of International Conference on Communication and Networks*, Advances in Intelligent Systems and Computing 508,
DOI 10.1007/978-981-10-2750-5_4

IoT is the combination of various empowering technologies such as sensing and communicating technology, middleware and communication protocols [1].

The communication technology includes the Two-D barcode, NFC, RFID, Wireless Sensor Network node and RFID Wireless sensor network. These devices/nodes are constrained in terms of processing power, flash memory and battery power.

Middleware provides infrastructure to leverage IoT services such as Interoperation, context awareness, device discovery and management, device abstraction, integration various processes, managing the large volume of data. There are various categories of middleware viz pervasive computing middleware, message oriented middleware, and semantic web based middleware, LBS and surveillance middleware, communication middleware etc. [2].

IoT needs different protocols for different operations. Like it requires a protocol for collecting the data from sensors, communication protocol to send the data to server infrastructure, protocols for a device to people communication, the protocol for D2D/M2M communication. MQTT, CoAP, DDS, Zigbee, Zwave, AMQP and 6LoWPAN [3] are various IoT protocols.

IoTSyS is based on REST/SOAP based web service and uses HTTP/CoAP as a communication protocol. REST on CoAP is the best approach for constrained nodes in IoT network. This combination is light weight approach for constrained nodes [4].

1.1 Motivation

Often the security problem is the major area of research in most of the domain. In a communication network, the security of data should be considered on the priority basis to avoid the malicious activity in a system by intruders. In IoT, the device to Gateway data communication should be managed in a secured way such that the confidentiality and integrity of sensitive data generated by the IoT devices should be maintained. For example in multistoried building there are many sensors like smoke detectors, motion sensors, cameras for video surveillance etc. and actuators like alarm, firefighter etc. are deployed. Let say a building is having 20 floors and 500 rooms and each room is occupied with sensors and actuators. The Gateway handling whole automated building system is residing on the topmost floor. All the data generated from IoT devices are then transferred to Gateway. Now in this case it may be possible that the data generated by smoke detectors or camera which are on the ground floor are altered and send to Gateway. Or in reverse case if as per the data received by the Gateway, it sends some information to actuators to carry out actions for say activating an alarm for fire. That data can also be altered by intruders making authority unaware of sudden fire at ground floor.

In such scenario, the data communication between IoT Devices and the Gateway must be secured in such a manner that it satisfies the CoAP and REST requirements of IoT. The term security counts a wide range of different concepts; among them are authentication, confidentiality, integrity and authorization. Employing the generic devices into the Internet makes us think about the secured data communication between the devices and gateway. Hence, a suitable security infrastructure is required which will scale to accommodate the IoT sensors and devices. In order to prevent the growth of such malicious activities, different cryptographic mechanisms can be employed.

As mentioned in [5] DTLS can be adapted for securing the channel under the CoAP protocol. However, DTLS is just a replication of TLS under User Datagram Protocol (UDP) and it is not specially meant for constrained environment.

2 IoTSyS Framework

IoTSyS is open source middleware architecture for Home and Building Automation. It consists of multiple projects based on different protocol standards such as KNX, BACnet, WMBus etc. Using 6LoWPAN and the Constrained Application Protocol, an efficient stack is provided along with oBIX to provide interoperable Web technologies in the field of sensor and actuator networks. oBIX is Open Building Information Exchange, is a standard for RESTful web services.

2.1 IoTSyS Architecture

IoTSyS Architecture is shown in Fig. 1. The Gateway is running servers such as an oBIX server, CoAP server, Object broker and couch DB as a database to store the status of IoT Devices. Using Obelix GUI, a user can be able to check the status of the sensors and actuators in the network. The user can also able to set a new status for an actuator. The copper plug-in of Mozilla Firefox can also be used to get or set the status of devices which are registered to IoTSyS Gateway. The devices are constrained devices in terms of processing power, flash memory, IP reconfigurability, battery power. It uses various web services protocol bindings such as HTTP, CoAP, SOAP and REST [4].

Various blocks of IoTSyS architecture are explained as follows:

Gateway: It is the heart of IoT system. It manages all the data communication between the IoT sensors and Actuators. It also manages the communication between the IoT devices and the Users of IoT System. It manages the state of devices, registers the device, by using various communication protocols for data communication.

Object Broker: It manages the contract registry of the entire object which is registered to the Gateway. Only registered devices can be able to communicate through Gateway.

CoAP Server: CoAP is light weight protocol specially designed for constrained nodes. It runs on 5683 port.

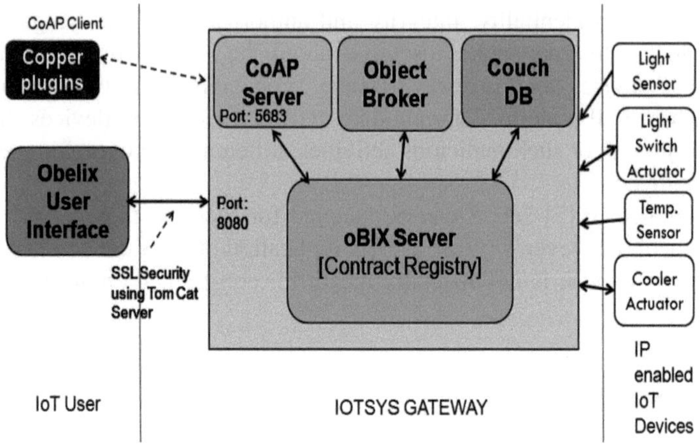

Fig. 1 IoTSyS architecture

Couch DB: It is a database system which stores the states of writable objects, such as actuators.

oBIX server: Open Building Interface Exchange (oBIX) server manages the reading and writing of data on the network. It maintains histories, alarms and watches for IoT devices.

TomCat Server: It is deployed for providing security of data communication between users and IoTSyS Gateway. It uses SSL security for proving authentication and encryption/decryption of command/data between user and Gateway.

Obelix: It is user interface designed in HTML5 using Angular JS technology. Through Obelix, interface user can be able to access Gateway and can see the status of IoT devices registered to the Gateway.

Copper Plug-in: Copper Plug-in is Firefox plug-in which is used as CoAP client and it can communicate to Gateway for CURD operation on IoT Devices through CoAP server.

2.2 IoTSyS Existing Security Mechanism

The existing system is being implemented with SSL security using Tom Cat server between client i.e. user and Gateway. Figure 2 shows the block diagram for existing security mechanism. Authenticated user can able to access the IoTSyS gateway and can able to access devices through Obelix GUI. The incoming and out coming messages between users and Gateway are transferred in security mannered under HTTP using SSL security. Users need to import the digitals certificates into their local key store, and then the tomcat server will make encryption and decryption for the requests using the keys in the digital certificates.

Fig. 2 Existing security mechanism

This security is enough when we are considering the access of Gateway by intruders. But we also need to consider the secured data communication between IoT Devices and Gateway. Few cases we have discussed earlier.

3 Limitation of Existing System

In order to prevent the growth of malicious activities in IoTSyS environment, different security concepts as authentication, confidentiality, and data integrity should be employed.

Although various techniques are available but the most common approach followed is transport layer security using DTLS in Constrained application protocol (CoAP). DTLS provides authentication, data integrity, confidentiality and automatic key management. It also supports a wide range of different cryptographic algorithms, which makes it a potential security protocol [6]. But there are few reasons why the DTLS fails to meet the requirements of IoT [7]. Such as:

1. DTLS does not support multicast communication which is one of the key functions of CoAP.
2. DTLS uses 4 rounds of handshaking signals which consume battery power of constrained devices while sending some stateless cookies.
3. If CoAP client needs Internet access that needs the CoAP/HTTP mapping process, and then DTLS handshake process remains a challenge.
4. CoAP messages cost the network only 2 transactions (1 round-trip); one message from the client (request) and the other from the server (Response). If DTLS is used, 4 round trips are required; 3 round trips for DTLS ($\sim$40–50 B) plus 1 round trip for CoAP before CoAP's actual contents are exchanged.
5. Transport layer security is hop-by-hop security. At each hop, the data packets are decrypted and further route that packet towards destination after encrypting it again.

So from the above reasons said, we can go for another approach i.e. Message level security. As and when the data is generated at IoT devices it gets encrypted and transferred to Gateway in the similar way the reverse communication between Gateway and IoT Devices is to be done. The best part is that message level security is independent of underlying communication protocol. Here no extra handshaking signals are required, so it consumes less battery power compare to DTLS. Apart from that if we need to provide access policies on the same object for multiple clients then this can be possible through message level security. It provides end-to-end security in the sense that data is encrypted at the source and decrypted at destination node only. No intermediary node in the network route can be able to decrypt the data.

We have used two methodologies for Message level, symmetric key encryption, and asymmetric key encryption. For symmetric key encryption, we have used Blowfish, JASYPT, and AES, whereas for asymmetric key encryption we have used well know RSA encryption technique.

The data communication between IoT devices and Gateway is in XML format, but this can further be serialized in binary format so that the bandwidth requirement for data transfer gets reduce.

4 Proposed Approach

For our proof of concept, we have designed the following experimental setup. Here Raspberry pi will work as data collector or we can say as resource server which manages the sensors/actuators network. Here Rpi is managing the data communication between IoTSyS Gateway and various IoT devices such as motion sensor relay switch, and moisture sensor in our case.

As these sensors/actuators do not have processing capabilities and they cannot be uniquely identified in a network, Rpi provides this facility on their behalf. Rpi will communicate with IoTyS Gateway through constrained application protocol (CoAP). This communication channel should be secured in order to prevent the data from intruders.

Secure communication using CoAP: The security in CoAP communication channel can be achieved by two approaches:

1. Datagram Transport Layer security (DTLS) at the transport layer.
2. Message layer security at the application layer.

In DTLS, it first creates a secured communication channel by exchanging handshake signals and then performs the encryption/decryption of the data. It generally used in hop-by-hop communication.

Whereas, in case of message level security, it is independent of underlying protocol (whether TCP/UDP), just use standard encryption algorithm for communication between endpoints [8].

Using our experimental setup as shown in Fig. 3, we perform various iterations using different cipher algorithms, like Blowfish algorithm, JASYPT (PBE with MD5 and TripleDES), AES algorithm, RSA and DTLS (AES-128 pre-shared keys) approaches.

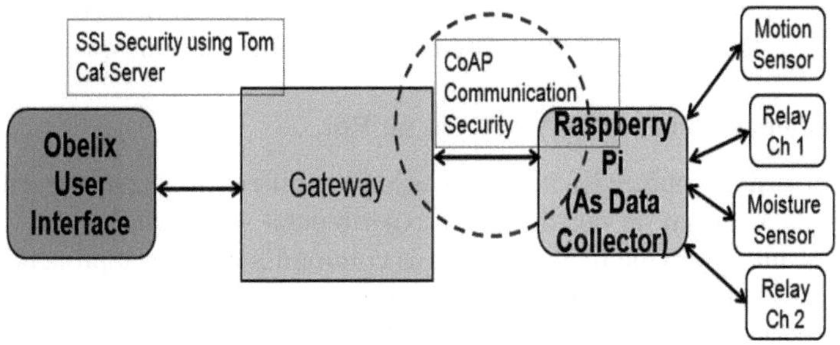

Fig. 3 Experimental setup using Rpi

Blowfish cipher is symmetric-key block cipher with 32-bit to 448-bit key length and 64-bit block size. It can be a replacement of DES. JASYPT is java library which provides high security, standards-based encryption technique. It allows password encryption, text encryption through various algorithms which are approved by JCE provider. Here we have used BasicTextEncryptor, which uses (Password-Based Encryption (PBE) With MD5 and DES. It uses 1000 iterations to generate the key. Advance Encryption Standard (AES) is a symmetric-key algorithm based on substitution-permutation method. It has 128-bit block size with 128-bit, 192-bit, and 256-bit key length. RSA is well known asymmetric key cipher. It uses a public key for encryption and private key for decryption.

Data Formats: Main purpose of oBIX is to provide the reading and writing mechanism over the network devices. All the data generated in IoT network will be converted to a common XML document. EXI is a compact form of XML which provides high performance and need low bandwidth. So it's better to use EXI instead of XML. Apart from that JSON is also lightweight data format compare to XML. So we tried to iterate the communication between IoT devices and Gateway using various data formats along with different encryption methodologies.

EXI—Efficient XML Interchange is a compact form of XML which is Schema based and uses grammar-driven approach [9]. There is EXI event and EXI stream derived from XML schema and each event has been defined with event code, for example, start document event has code 000. It cannot be read by a human.

5 Comparison of Various Security Approaches

We have considered the given performance measurements:

1. Round Trip time: It is encryption time plus communication time from device to the gateway.
2. Message Size: File size after encrypting the data with the different encrypting mechanism and with data format.

3. CPU utilization: Here we have considered the CPU utilization of Rpi while encrypting the data with the different encrypting mechanism and with data format.
4. Memory utilization: Memory utilized on Rpi.

After performing numerous iterations using the different security algorithms with 3 different data format on our designed experimental setup, we had drawn out the following results. Here the input data size is common for all the approach. The input file consists of single data generated by a sensor i.e. one file generate for single data generation from a sensor. The size of input data is 60 B. Figures 4, 5, 6 and 7 shows the analysis of various approaches we have used considering different aspects of constrained IoT devices.

All the cases with EXI data format take less time for total communication compares to JSON and XML. RSA and Blowfish prove to be better and efficient than all others. It takes on an average less time to encrypt and transfer.

But using RAS public-private key encryption the main problem is of public-private key distribution. It seems to be overhead for key distribution. We need a trusted third party to perform this job. And as the number of devices increases for a single gateway then it will be difficult to manage the public-private key pair for each device.

In parallel DTLS is also proving its capabilities near to blowfish, but it requires the handshaking signals before the actual data transfer. In case of transport layer security it necessary to provide acknowledgment from the receiver to the sender, which sometimes not necessary in some IoT cases. Like if I want to know the temperature reading after specific interval, there is no need of acknowledgment after

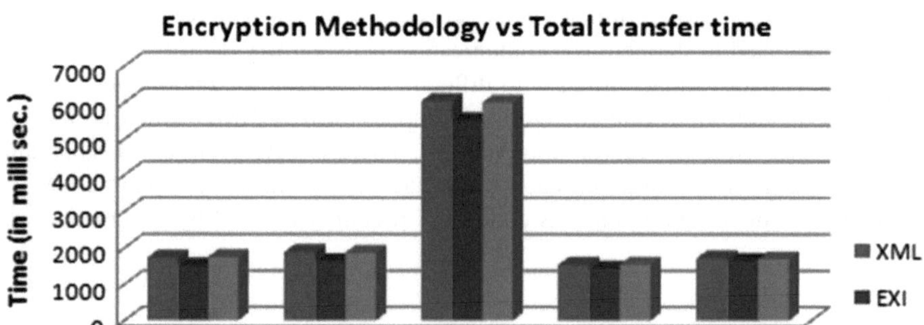

	BlowFish	Jasypt	AES	RSA	DTLS
XML	1731	1884.0625	6038.6875	1534.4375	1711.6875
EXI	1511	1635.125	5522.875	1428.75	1649.375
JSON	1749.25	1867.0625	5998.4375	1546.5	1690.125

Encryption Methodology ------------------>

Fig. 4 Encryption methodology versus round trip time

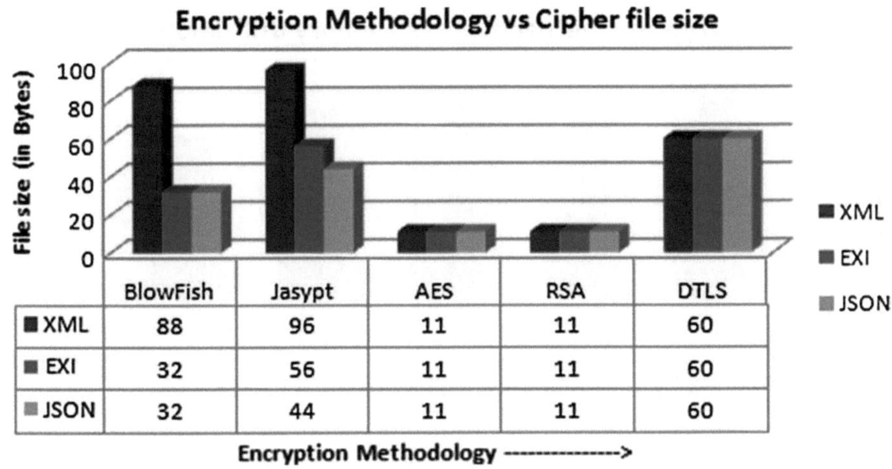

Fig. 5 Encryption methodology versus cipher file size

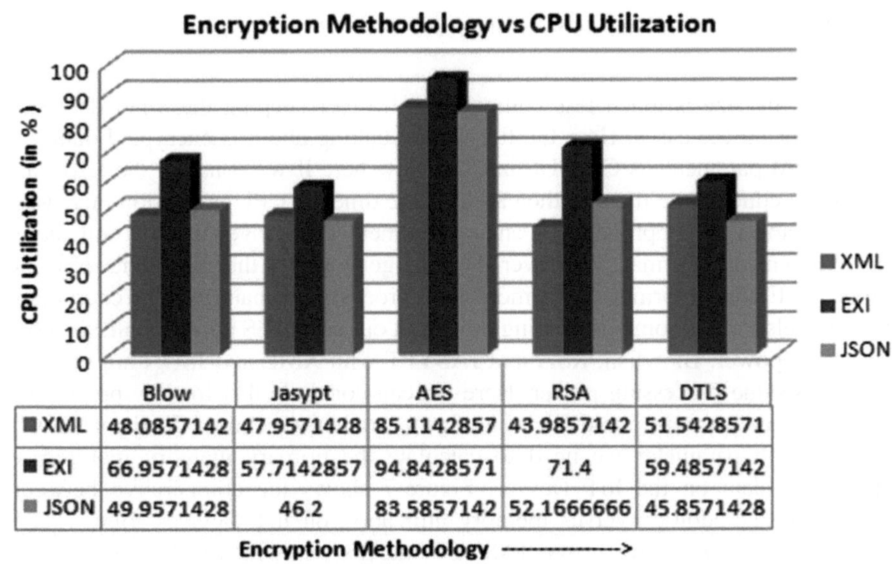

Fig. 6 Encryption methodology versus CPU utilization

receiving such readings. Due to network problem if anyone missed, then it will not affect the further processing. JASYPT is also near to Blowfish. So considering the key management in RSA and handshaking in DTLS, here we can vote for Blowfish.

The second aspect of parameter talks about the cipher file size. This is an important factor when the encrypted data is to be stored on the IoT devices for future reference. In such case Fig. 5 show that AES and RSA are better. As its

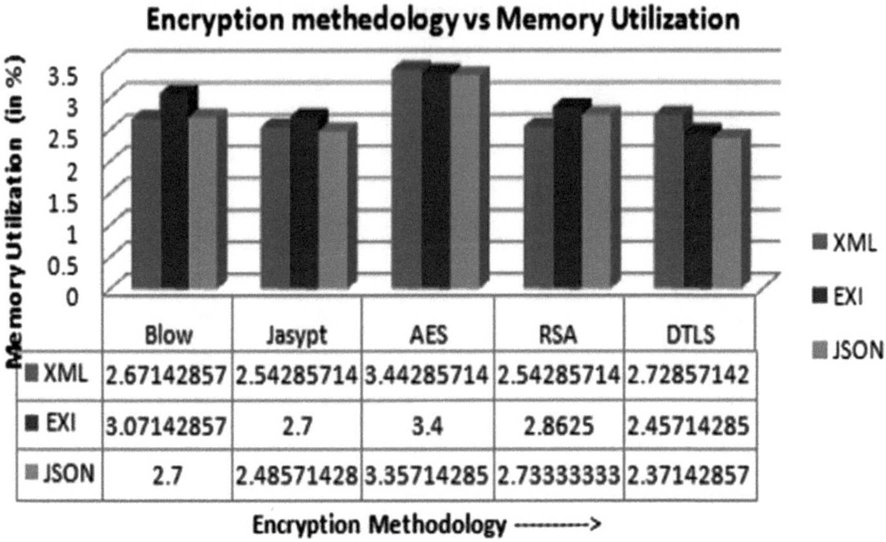

Fig. 7 Encryption methodology versus memory utilization

encrypted file size is much less compare to all other keeping the input data size constant for all the cases. Here the file we are talking about is encrypted data.

The third parameter is CPU Utilization. Now here if we consider the EXI data format for reducing the file size then at the same time the CPU utilization increases. Figure 6 shows the graph between encryption methodology versus CPU utilization with different data format. Whatever the data generate by the device needs to be mapped to EXI data format, needs more CPU processing capability. The encryption algorithm also needs some processing power, in our case AES approach needs more processing power. Blowfish, RSA and JASYPT with XML and JSON are more or less needs same processing power. Here we can conclude that for low processing power devices we must avoid EXI data format, rather we choose XML or JSON.

Our fourth parameter we need to calculate the primary memory required to process the data on the IoT Device. Figure 7 shows the comparative between encryption methodology versus memory utilization on IoT Devices. Here DTLS with EXI and JSON prove better. JASYPT and RSA stay next to DTLS.

Further, we had repeated the testing with a file having multiple data gathered from multiple sensors. In this case, the file size is bigger compare to the previous approach. Here the input file size is 1 kB.

From Figs. 8, 9, 10 and 11 shows the iteration of another scenario where we have gathered more than one data of multiple sensors and then encrypted and transfer it to the Gateway. From the graph, we can say that by increasing the input file size, CPU utilization, and cipher file size affects the most. Rest total transfer time and memory utilization change to a little extent. Here CPU utilization seems more that 100 % in some cases because our Rpi is Quad processor.

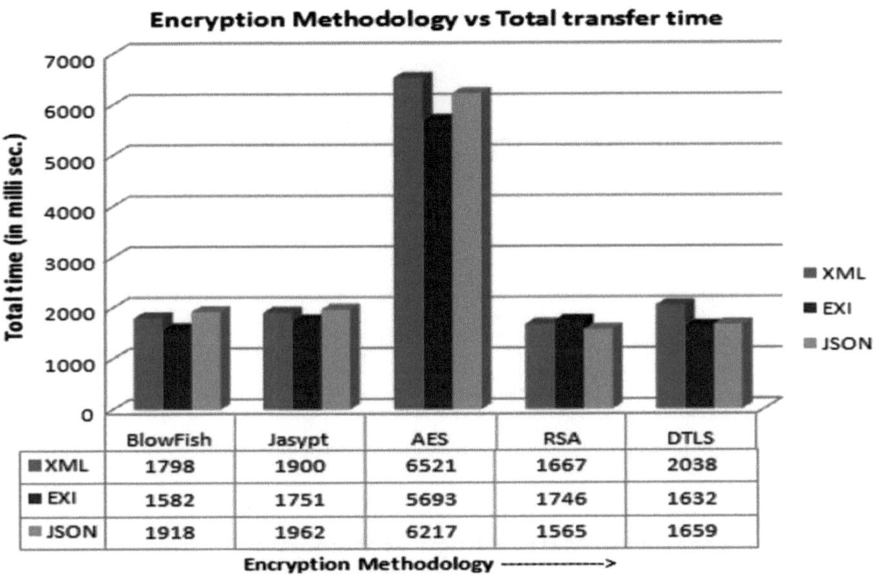

	BlowFish	Jasypt	AES	RSA	DTLS
XML	1798	1900	6521	1667	2038
EXI	1582	1751	5693	1746	1632
JSON	1918	1962	6217	1565	1659

Fig. 8 Encryption methodology versus total transfer time for multi-dataset

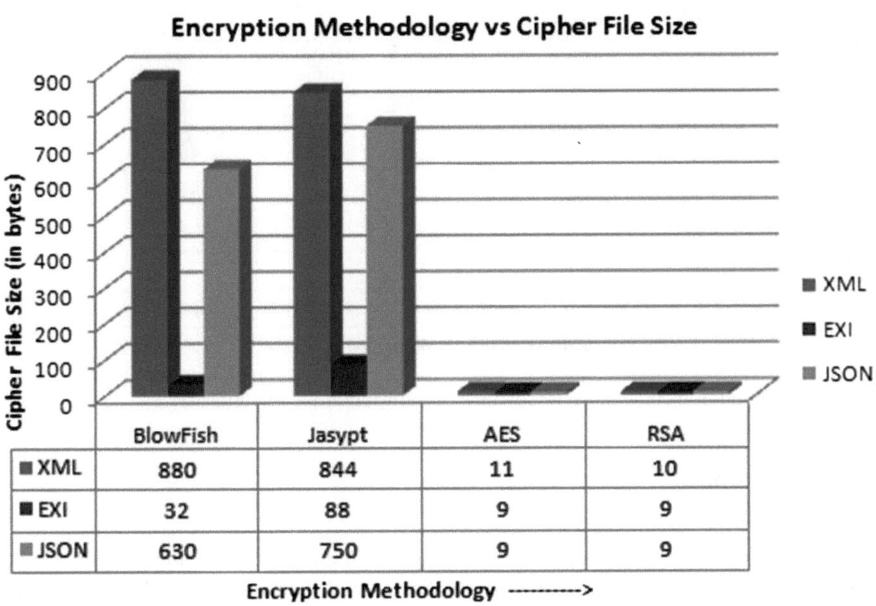

	BlowFish	Jasypt	AES	RSA
XML	880	844	11	10
EXI	32	88	9	9
JSON	630	750	9	9

Fig. 9 Encryption methodology versus cipher files size for multi-dataset

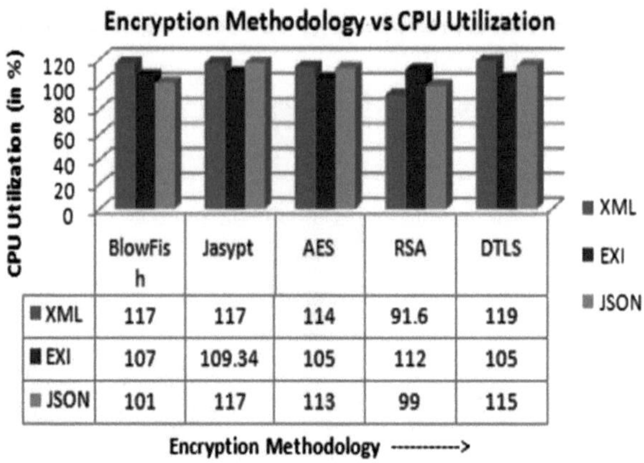

Fig. 10 Encryption methodology versus CPU utilization for multi-dataset

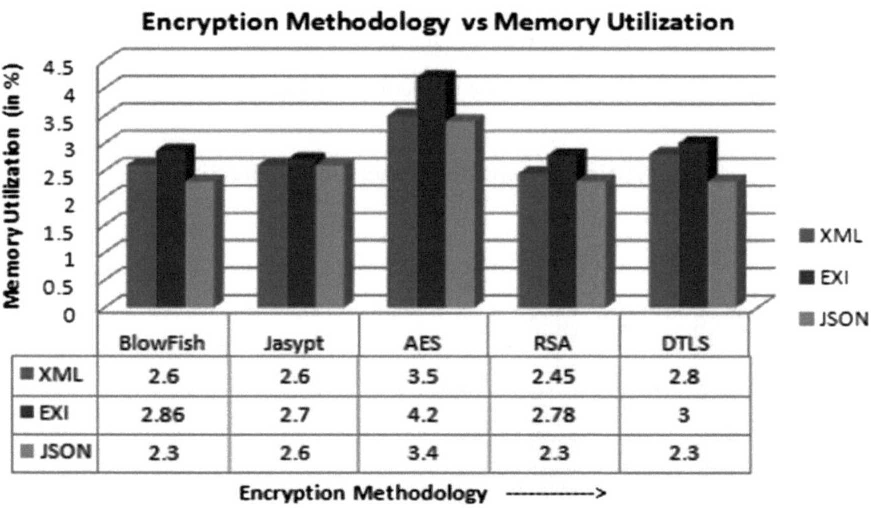

Fig. 11 Encryption methodology versus memory utilization for multi-dataset

6 Conclusion

Using our experimental setup, we have implemented various security mechanisms such as message level symmetric encryptions as well as asymmetric encryption and transport level DTLS encryption on the data communicating between Rpi network and the Gateway. From the results and the discussion, we can analyze that message level instead of transport level security will suits better. The symmetric approach using Blowfish encryption algorithm will be better for IoTSyS framework whether we use single data file or multiple dataset files. RSA is competent to Blowfish, but asymmetric approach leads to an overhead of managing the public-private key pair distribution management using third-party key manager. As per the usage of data format, either EXI or JSON can be considered lighter than XML. However, use of EXI data format may lead to more CPU utilization. So the constrained of processing power should be considered while choosing the data formats. At the same time use of EXI data format reduces the output file size and hence total transfer will be reduced. Hence we can incorporate a security layer between smart IoT devices and Gateway. While considering the constraints of data communication time between sensors and gateway, memory of intelligent devices/sensors and their constrained processing power, message level symmetric approach accomplishes the task of securing the data communication between gateway and IoT devices. Furthermore, the JSON data format is more suitable as it requires less memory storage and processing power. Further, we can add the access policies on the same object for different users. It can be done through message signature.

References

1. L. Atzori, A. Iera, and G. Morabito, "The Internet of Things: A survey," Comput. Netw., vol. 54, no. 15, pp. 2787–2805, 2010.
2. H Zhou –The internet of things in cloud: Middleware Perspective‖ CRC press.
3. IoT/M2M Protocols. [Online] Available: http://IoT.eclipse.org/protocols.html.
4. "iotsys - IoTSyS - Internet of Things integration middleware - Google Project Hosting," 15-Dec-2014. [Online]. Available: https://code.google.com/p/iotsys/. [Accessed: 15-Dec-2014].
5. Sye Loong Keoh, Sandeep S. Kumar, and Hannes Tschofenig "Securing the Internet of Things: A Standardization Perspective" IEEE INTERNET OF THINGS JOURNAL, VOL. 1, NO. 3, JUNE 2014.
6. Z. Shelby, K. Hartke, C. Bormann –RFC for CoAP [online] Available: https://tools.ietf.org/html/rfc7252).
7. T. A. Alghamdi, A. Lasebae, and M. Aiash, "Security analysis of the constrained application protocol in the Internet of Things," in Future Generation Communication Technology (FGCT), 2013 Second International Conference on, 2013, pp. 163–168.
8. "Message and Transport Security" [Online]. Available: https://msdn.microsoft.com/en-us/library/. [Accessed: 11-May-2015].
9. John Schneider, AgileDelta, Inc "Efficient XML Interchange (EXI) Format 1.0" Feb 2014 http://www.w3.org/TR/exi/.

Segmentation and Recognition of Fingers Using Microsoft Kinect

Smit Desai

Abstract Hand gesture identification is a very important part of HCI. In this paper, I have presented a very efficient algorithm for finger segmentation. Using fingers as an input medium, our interaction with the computer can become easier. Microsoft Kinect, which is a depth sensor is used to capture the image which is used for finger segmentation. Background is removed from the captured image by accepting pixels, which fall in a fixed range of depth. The image is further pre-processed and then palm area is identified and removed to obtain separate fingers. Further, to identify open fingers as gesture-kNN classifier is used. This proposed algorithm has achieved more than 90 % accuracy.

Keywords Microsoft Kinect · Gesture recognition · Finger segmentation · Centroid · Feature extraction · kNN classifier

1 Introduction

Since evolution of computers, wired input methods and devices are used for Human Computer Interaction (HCI). With development of new and innovative computer based applications and with ever evolving computing technology, researchers have developed new HCI methods. After successful use of wired and wireless HCI methods and devices, scientists are working on more complex but useful HCI methods. Voice, touch and gestures are the next three HCI mediums that the scientists are working on. Body posture, body language, facial expression, hand gestures are very powerful mediums of interaction. Normally it is seen that these medium of interaction are more powerful and convenient for human beings in general and specially for physically challenged people. However, providing understanding of such intelligence to a computer is a very challenging and complex task. Such real time gesture identification task is even more difficult. Up until now

S. Desai (✉)
Sarvajanik College of Engineering and Technology, Surat, Gujarat, India
e-mail: thesmitdesai5@gmail.com

N. Modi et al. (eds.), *Proceedings of International Conference on Communication and Networks*, Advances in Intelligent Systems and Computing 508,
DOI 10.1007/978-981-10-2750-5_5

many different hand gesture recognition techniques and methods are used for interaction with machine. Some of them have made use of colored gloves or color code on fingers such as [1] and some researchers have attempted to recognize gestures without any color code. In this paper I have presented an algorithm which recognizes open finger based gestures without any color code. I have used Microsoft Kinect for capturing hand image from 3D space. This paper is divided in seven section. This introduction section is followed by literature review. Section three is devoted to introduction to Kinect sensor which is followed by a discussion on hand capturing and its segmentation. In section six processing of segmented hand images is discussed. In the following section feature extraction and classification is described which is followed by concluding remarks.

2 Literature Review

Lately in modern computing technology hand gestures are used frequently for interaction with computers. For example, Pavlovic et al. [1] have categorized human gestures in three categories viz. one, hand and arm gestures, two, head and face gestures and three, body gestures. The first category of gestures, that is, hand and arm gestures including the palm and fingers are mainly used for applications like game playing and understanding and interpreting sign languages etc. The second category of gestures is manly utilized for applications like face recognition or emotions or actions like blowing whistle, smiling, and anger recognition. The third and last category of gestures include gestures and postures of entire human body and therefore they are more useful for teaching and training of athletes, dancers etc. Hand gestures are very useful and more and more researchers have concentrated on it. Initially hand gestures were recognized by some color coding. This technique used color gloves, bands or some indicative gears on their hand, palm, wrist or fingers for identifying the gestures. Keskin et al. [2] presented their work based on colored gloves for communicating with a healthcare system. Agrawal et al. [3] developed a gesture recognition system based on colored gloves. This system is designed as a tutor for hearing impaired people. In this paper authors have used red color gloves for localization of hands, where as in Ren et al. [4] use black colored wrist band for localization of palm. Though Ren et al. [4] use Kinect sensor, they have use used color code for hand segmentation. Further they made use of Finger—Earth Mover's Distance for measuring dissimilarities between gestures. In this work they demonstrated applications like arithmetic computation and Rock-Paper-Scissors Game playing using hand gestures. Marin [5] has presented an algorithm, which finds are of the palm using the centroid of the palm. Even though authors used a depth sensor, they used a colored band for isolation of hand like Zhou et al. [4]. There are many other work which are based on such color codes, but since this technique is based on vision, it requires proper lighting and illumination condition Erol [6] and Suwarez et al. [7]. After introduction of Microsoft Kinect researchers started using it for gesture recognition. Since it has a built in depth sensor and also gives skeletal details of human body it has become more popular. Many interesting applications of Kinect are found in literature. It consists of game playing, assistive tools and many more. Le et al. [8] presented a work where they proposed an algorithm which finds the center point of the palm and also identifies fingers. Here they segmented the palm area and then image contour is used to find moments. Raheja et al. [9] presented a work to find the finger tips and center of the palm using finger tracking algorithm. This algorithm can track both the hands. Here hand is segmented by taking a calculated threshold and NITE library [10]. Shukla and Dwivedi [11] used Kinect to capture hand image with depth information. They also used contour of the segmented hand image. For identifying hand gestures, they used convex hull and convexity defects of the hand contour as the featureset. They suggested Naiive Bayes classifier for gesture recognition. Biswas et al. [12] proposed an algorithm which recognizes eight hand gestures clap, call, greet etc. Here they capture hand gesture image through Kinect and then this image is processed for back ground removal. They used depth histogram equalizer for identify foreground. They further divided the depth image in ten bins to find features and then they identified the features using support vector machine (SVM). Jing [13] proposed an algorithm where they developed virtual touch screen tool using Kinect. This is done by identifying and tracking the fingers.

3 Suggested Algorithm

The proposed algorithm captures images from Microsoft Kinect. The captured images are in RGB color space, therefore it is converted into binary image and then smoothing of the image is done. This smoothened image may have some discontinuity which are removed by performing morphological operations. These three processes are done under preprocessing stage. Now using moments, centroid of the image is found and an ellipse which fits the hand image is considered and its minor axis is computed. This minor axis is used as the diameter of the ellipse which is then removed from the image. This circular area is calculated using the centroid as the origin. Removal of this circular area will remove the palm from the image leaving us with fingers and some noise. Here I have made one observation that most of the hand images are connected with the boundary of image and therefore when circular area is removed the forearm will remain with the boundary. I remove all the parts which are connected with the boundary of the image and we will be left with fingers. Now region of interest (ROI) is cropped for finding the featureset. After finding featureset, kNN classifier is used for identifying open fingers as a gesture. The block diagram of algorithm is shown in Fig. 1.

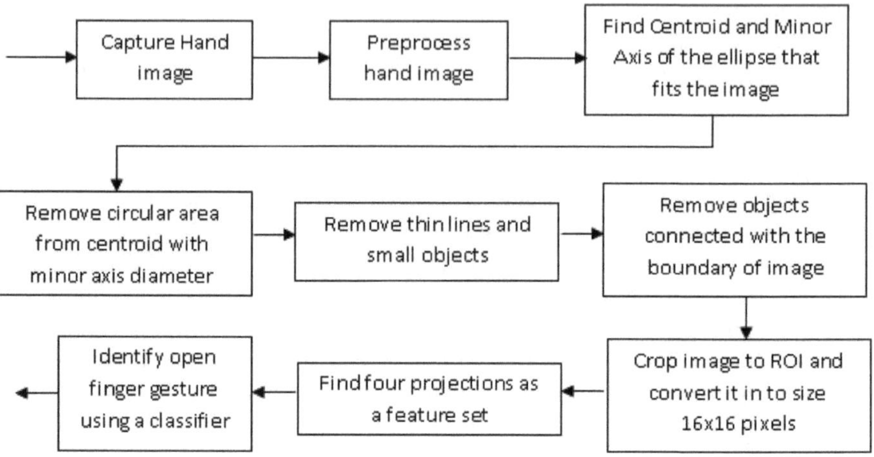

Fig. 1 Block diagram of proposed algorithm

4 Segmentation of Hand

As mentioned earlier, fingers are very reliable for HCI. When user wishes to use hand gestures as a medium for interaction, he generally keeps his hand in front of the imaging device. Here I have used Kinect as an imaging device which has an inbuilt depth sensor. This depth sensor measures the depth of an object from the device. I use this depth for segmenting hand from the body of the user. Here I suggest depth range of [250 mm, 650 mm] from the device, as region of interest (ROI) separating background from the hand as it is suggested in [4, 9, 10, 14]. That means when users' hand comes in the depth range of [250 mm, 650 mm] it will be cropped and used for further processes. This is done using NITE library as it is suggested in cliff et al. [10]. The cropped hand images are shown in Fig. 2.

As discussed, the image captured from Kinect is in RGB color space. For processing this image we need to convert this image into binary. For converting RGB image into binary, I propose to use Otsu's global threshold [15].

The next pre-process step is dilation of the image. I propose dilation because some times while converting the image from RGB to binary, some portion of the image gets separated. To connect these separated parts, dilation is used. Even after

Fig. 2 Gesture images captured through Kinect

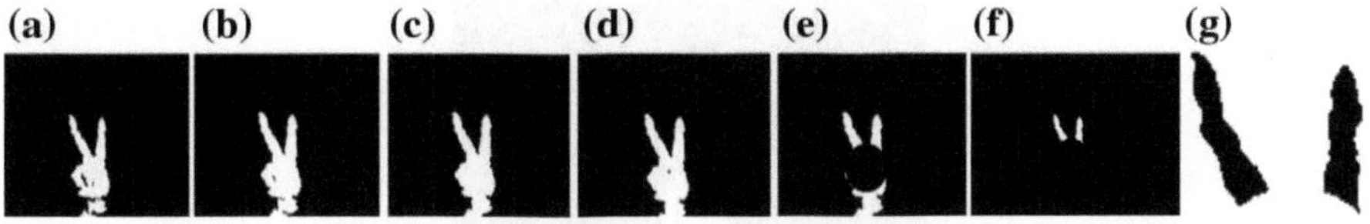

Fig. 3 **a** Original gesture image, **b** boundary of hand smoothen, **c** holes are removed from image, **d** image with centroid of the hand, **e** image with circle drawn from centroid with minor axis diameter, **f** non-finger components are removed from image, **g** fingers cropped from the image

dilation, there is a possibility to have some aberrations in the form of holes in the image and therefore in the next step we fill all such holes to make a complete image. Major stages of this process have been illustrated in Fig. 3.

5 Processing

The next task is to determine the position of the hand in the image. Ren et al. [4] and Guilo et al. [5] have used centroid for palm area identification. For estimating position of the hand, we find centroid of the hand image, which is nothing but a point where all the mass of an object is concentrated without changing its first moment. It is a statistical measure which can be derived from the raw moments of an object image. Let f(x, y) be a binary image, then its centroid is;

$$\left(\frac{\mu_{10}}{\mu_{00}}, \frac{\mu_{01}}{\mu_{00}}\right)$$

where

$$\mu_{00} = \sum_{x=0}^{width} \sum_{y=0}^{height} x^0 y^0 f(x, y)$$

$$\mu_{10} = \frac{\sum_{x=0}^{width} \sum_{y=0}^{height} x f(x, y)}{\mu_{00}}$$

$$\mu_{01} = \frac{\sum_{x=0}^{width} \sum_{y=0}^{height} y f(x, y)}{\mu_{00}}$$

Here μ_{00} is zeroth raw moment and it gives the area of object image. The other variables that we calculate for the hand image is the minor axis of an ellipse which fits the object of the hand image. I noticed that when centroid of hand image is calculated it generally falls in the area of palm. So if the palm area is separated and

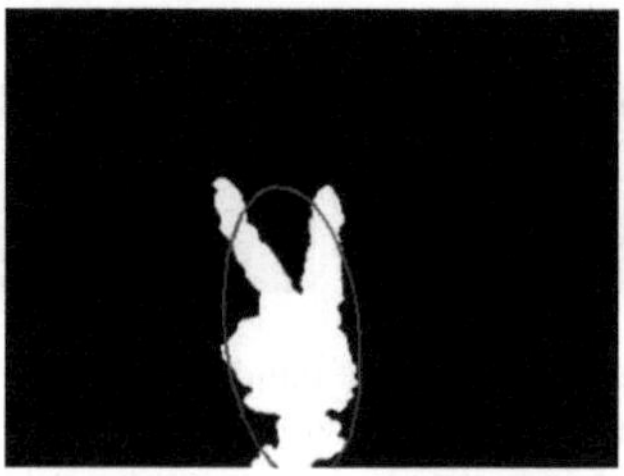

Fig. 4 Ellipse over the object to determine minor radius of the object

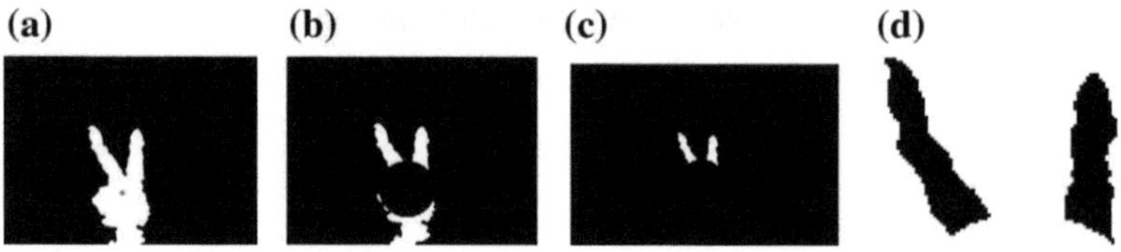

Fig. 5 **a** Image with centroid of the hand, **b** image with circle drawn from centroid with minor axis diameter, **c** non-finger components are removed from image, **d** fingers cropped from the image

removed from the hand image, we can segment fingers of the hand. Considering this observation, we remove a circle taking centroid as the origin and minor axis as the diameter. It is shown in Fig. 4.

In this process, some of the thin portions and tiny objects of the palm may still be present in the image. To remove these thin portions we perform morphological operation erosion. This process completely separates out forearm from the other parts of the hand image. I noticed one more fact about the hand images that I had captured with Kinect: Forearm of hand along with the wrist is usually also captured while capturing the image and that portion of the image is usually connected with the boundary of the image. Taking advantage of this fact I removed all the objects which are connected with the boundary of the image.

In the next stage, these images are processed for counting fingers visible in the image. This is what we are interested in. In most of the cases we get separate fingers in the images but merely counting them does not serve our purpose because in many cases we get joint fingers too. Figure 4 shows step wise process of algorithm up to getting cropped fingers and Fig. 5 is a case where cropped image has joint fingers.

Since there are possibilities of having images where joint fingers are there and under segmentation we need to identify open finger gesture applying classification technique. In the following section feature extraction and classification are described.

6 Feature Extraction and Classification

Many feature extraction methods can be traced in the literature. Feature are usually divided in three categories, structural features [5, 11, 16, 17], statistical features [4, 10, 17, 18] and hybrid feature. Since extracted fingers may be in any direction, here we have used four projection profiles as it used in [19] for Gujarati handwritten numeral identification. It is a structural feature of an object. To find four projection viz vertical, horizontal and two diagonal profiles, first extracted finger image is reduced to an image of size 16 × 16. For reduction of image size, I propose to use nearest neighbor interpolation method. Now projection profile is found for the extracted finger image. By projection profile we mean total black pixels in each of the vertical, horizontal and two diagonals of the 16 × 16 image. That comes to 94 features, 16 each for horizontal and vertical profiles and 31 each for two diagonal profiles, of an image. This projection feature set is illustrated in Fig. 6.

Many classification methods [3, 4, 11, 12, 17–19] are used for classification of hand gestures. For classification purpose I propose to use kNN algorithm with Euclidian distance. I tried out classification with different values of k, but after all experiments I found that the best result is obtained with k = 1. Here I have used a data set of five hand gestures collected from fifty people. Thus we have a total of two hundred and fifty hand gesture images. I have trained our algorithm with fifty percentage of dataset and tested it for the rest of the fifty percent of the images. Tables 1 and 2 show the accuracy of success of proposed algorithm for k = 1 and n = 25 (Fig. 7).

The results show that the proposed algorithm has achieved 91.20 % of success for unseen data and 95.60 % of accuracy for complete dataset. The algorithm gives 100 % accuracy for the training data set. We have tried out Support Vector

Fig. 6 An example of joint separated fingers

Table 1 Confusion matrix of test data set of size 25

	1	2	3	4	5
1	23	1	0	0	1
2	0	25	0	0	0
3	0	1	22	0	2
4	0	0	0	19	6
5	0	0	0	0	25

Table 2 Confusion matrix of test data set of size 50

	1	2	3	4	5
1	48	1	0	0	1
2	0	50	0	0	0
3	0	1	47	0	2
4	0	0	0	44	6
5	0	0	0	0	50

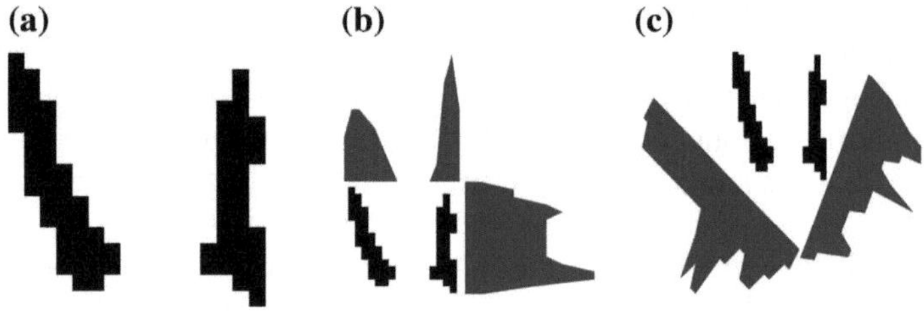

Fig. 7 **a** Finger figure converted in size of 16 × 16, **b** horizontal and vertical profile of 16 × 16 image, **c** two diagonal profiles of 16 × 16 image

Machine (SVM) as a classifier to identify open finger gestures, but it gave us a relatively lower accuracy of 65 %.

7 Conclusion and Future Work

The proposed morphological operation based algorithm gives high accuracy of identification of open fingers based hand gestures. In this proposed algorithm I have made a couple of assumptions. First, the hand must be in the range of [250 mm, 650 mm] for segmentation, and two the segmented hand must be connected with one of the boundary of the image. Here if the segmented hand image is not connected to the boundary of the image, we have cropped it to the lower boundary of the image. The main reason of non-identification of gestures is the violation of the second assumption made for the algorithm. There is ample scope of correction of the results without any assumption.

Fingers create important and expressive gestures for human communication and they are very widely used in game playing. This algorithm has open up various avenues to work in the direction of developing an interaction media for communicating with any electronic devices [20] including computers. Touch screen computers and other devices are currently market, but by extending this work, an effective interaction will be made possible just by using fingers and without the need of any touch.

References

1. Pavlovic V., Sharma R., Huang T., Visual Interpretation of Hand Gesture for Human Computer Interaction, IEEE transaction on Pattern Analysis and Machine Intelligence, vol. 19 (7), pp. 677–695 (1997).
2. Keskin C., Balci K., Aran O., Akrun L., A Multimodal 3d Healthcare Communication System, In 3DTV Conference: The True Vision - Capture, Transmission and Display of 3D Video (2007).
3. Agrawal I, Johar S., Santhosh J., A Tutor for the Hearing Impaired (Developed using Automatic Gesture Recognition), International Journal of Computer Science, Engineering and Application, vol. 1(4), pp. 49–61 (2011).
4. Ren Z., Youn J., Meng J. Zhang Z., Robust Part Based Hand Gesture Recognition Using Kinect Sensor, IEEE Transaction Multimedia, vol. 15(5), pp. 1110–1120 (2013).
5. Marin G., Fraccaro M., Donadeo M., Dominio F., Zanuttigh P., Palm Area Detection for Reliable Hand Gesture Recognition, In IEEE International Workshop on Multimedia Signal Processing 2013 (MMSP-2013) (2013).
6. Erol A., Bebis G., Nicolescu M., Boyle R., Twombly X., Vision Based Hand Pose Estimation: A Review, Computer Vision and Image Understanding, pp. 52–73 (2007).
7. Suarez J., Murphy R., Hand Gesture Recognition with Depth Image: A Review, In IEEE International Symposium on Robot and Human Interactive Communication, pp. 9–13 (2012).
8. Le V.B., Nguyen A.T., Zhu Y., Hand Detecting and Positioning Based on Depth Image of Kinect Sensor, International Journal of Information and Electronic Engineering, vol. 4(3), pp. 176–179 (2014).
9. Raheja J., Chaudhary A., Singal K., Tracking of Fingertips and Centre of Palm using Kinect, In IEEE International Conference on Computational Intelligence, Modelling and Simulation, Malaysia (2011).
10. Cliff C., Mirfakhroei S.S., Hand Gesture Recognition Using Kinect, Technical Report No ECE-2013-04, Boston University (2013).
11. Shukla J., Dwivedi A., A Method for Hand Gesture Recognition, Fourth International Conference on Communication System and Network Technologies (2014).
12. Biswas K., Basu S., Gesture Recognition using Microsoft Kinect, In 5th International Conference on Automation, Robotics and Application, Wellington, New Zealand (2011).
13. Jing P., Ye-Peng G., Human Computer Interaction using Pointing Gesture based on an Adaptive Virtual Touch Screen, International Journal of Signal Processing, Image Processing and Pattern Recognition, vol. 6(4), pp. 81–91 (2013).
14. Du H., To T.H., Hand Gesture REcognition using Kinect, Technical Report No. ECE-2011-04, Boston University (2011).
15. Otsu N., A Threshold Selection Method from Gray-Level Histograms, IEEE Transactions on Systems, Man, and Cybernetics, vol. 9(1), pp. 62–66 (1979).
16. Verma H.V., Eshan A., Chandra S., Gesture Recognition Using Kinect for Sign Language Translation, In IEEE Second International Conference on Image Information Processing (ICIIP-2013) (2013).
17. Arun K., Chris Z., Joseph, J. L. Jr., Poster: Real Time Markless Kinect Based Finger Tracking and Hand Gesture Recognition, In IEEE Symposium on 3D User Interface, Orlando, USA (2013).
18. Hamissi M., Foez K., Real Time Hand Gesture Recognition Based on Depth Map for Human Robot Interaction, International Journal of Electrical and Computer Engineering, vol. 3(6), pp. 770–778 (2013).
19. Desai A.A., Gujarati Handwritten numeral optical character recognition through neural network, Pattern Recognition, vol. 40(1), pp. 2581–2589 (2010).
20. Verma H., Aggarwal E., Chandra S., Gesture Recognition using Kinect for Sign Language Transaltion, In IEEE International Conference on Image Processing (ICIIP-2013) (2013).

References

[illegible]

[illegible] Handwritten ... Microsoft Kinect ... [illegible] pp. 2882–2886 [illegible]

[illegible] International Conference on Image Processing (ICIP), 2013 [illegible]

Randomness Evaluation of ZUC, SNOW and GRAIN Stream Ciphers

Darshana Upadhya and Shripal Gandhi

Abstract In 21st Century, Data Security is achieved through Cryptographically secure Pseudo Random Number Generators (CSPRNG). Hence, all stream ciphers uses CSPRNG. Mostly all the Hardware ciphers are developed using Linear Feedback Shift Register (LFSR). Owing to linear nature of LFSR, cipher becomes predictable and hence easily vulnerable to various LFSR based attacks. Secondly, security enhancements cannot be deployed on it because of its hardcore implementation. Therefore, our focus is to implement the various ciphers on multicore architecture without compromising throughput of the cipher. For this, we come up with rigorous literature survey of various stream ciphers like ZUC 1.6, SNOW 3G, GRAIN v1, WG-7, and DECIM v1, and also measure the randomness of above ciphers using NIST Statistical Toolkit.

Keywords Linear feedback shift register (LFSR) · Cryptography · Stream ciphers · GSM · LTE

1 Introduction

With advent of Computer Networks, "Network Security" has became a major concern. All communication protocols use cryptographic algorithms (ciphers), which are used to provide Confidentiality, Integrity and Availability for this purpose. Linear Feedback Shift Register (LFSR) has input bit, which is linear function of its previous state. Many ciphers have LFSR as their component. Study of various ciphers that use LFSR as their component was done and their mode of imple-

D. Upadhya (✉) · S. Gandhi
Department of Computer Science and Technology, Institute of Technology, Nirma University, Ahmedabad, Gujarat, India
e-mail: darshana.upadhyay@nirmauni.ac.in

S. Gandhi
e-mail: 14mcei06@nirmauni.ac.in

N. Modi et al. (eds.), *Proceedings of International Conference on Communication and Networks*, Advances in Intelligent Systems and Computing 508,
DOI 10.1007/978-981-10-2750-5_6

mentation was studied. Ciphers Hummingbird 2, Dragon-128, SOSEMANUK, ABC v2, CryptMT Version 3, Polar Bear, and Rabbit are implemented in Software while ZUC 1.6, WG-7, MICKEY-128, DECIM v2, Edon80, F-FCSR-16, GRAIN v1 and SNOW 3G have their implementation in Hardware. Our focus is to replace the Hardware implementation of various LFSR based Stream Ciphers with Software implementation along with improvement in their Speed, Memory and Randomness [1, 2]. Hence, the ciphers ZUC 1.6, SNOW 3G, GRAIN v1, WG-7, and DECIM v2, having Hardware Implementation were selected for this purpose.

Rest of this paper is designed as follows. Next Section describe details of Literature Survey. Detailed description and Working of selected ciphers is explained in Sect. 3. In Sect. 4, comparative analysis of ciphers is shown based on Speed-Up Measurement, Memory Usage and Randomness. Finally, Sect. 5 concludes this paper.

2 Literature Survey

As LFSR is linear in nature, it is possible to predict the working of LFSR after many iterations. If cipher has been implemented in hardware, then the working of cipher will be compromised. Hence, our goal is to go for software implementation of cipher so that whenever it is compromised, it can be reinitialized. We have selected five ciphers having LFSR component and with hardware implementation for this purpose. SAGE/ETSI developed ZUC 1.6 and SNOW 3G ciphers, which are used for LTE Standard and Universal Mobile Telecommunication System (UMTS) network, respectively. ZUC 1.6 cipher forms part of 128-EEA3 and 128-EIA3 protocols [3] and SNOW 3G cipher forms part of UEA-2 and UIA-2 [3], which are used for Encryption and Integrity in respective standards. WG-7 is a fast and light-weight stream cipher and is used in RFID Encryption and Authentication [4]. GRAIN v1 and DECIM v2 ciphers are selected for eStream portfolio, by eStream project, for Phase 3, Profile 2 (Hardware). Restricted hardware environments and higher security as compared to ciphers E0 and A5/1 [5], are primary design parameters for GRAIN v1. DECIM v2 is used in Hardware Applications with limited resources [6]. Brief details of ciphers, like Number of LFSR used, Length of each LFSR, their Applications, and Cryptanalytic Attacks of these ciphers are given in Table 1.

3 Working of Various Ciphers

This section briefly describes the working of selected ciphers.

Table 1 Brief details of LFSR based Ciphers

Ciphers	Number of LFSR	Length of each LFSR	Cryptanalytic attacks
ZUC 1.6	16	31 bits	Differential power analysis attack
SNOW 3G	16	32 bits	Timing attack [S-Box LookUp]
GRAIN v1	1	80 bits	Weak Key-IVs attack, related key chosen IV Attack, and differential fault attack
WG-7	23	7 bits	Distinguishing attack, key recovery and algebraic attack
DECIM v2	1	192 bits	Guess and determine attack, related key chosen IV Attack, Side channel Attack and Distinguishing Attack

3.1 ZUC 1.6 Cipher

In ZUC 1.6 Cipher [3], Key is of size 128 bits and Initial Vector (IV) is of size 128 bits. Output is a keystream of word of 32 bits at every clock. The cipher has been divided into three logical layers: LFSR Layer, Bit-reorganization (BR) Layer, and finally Non-linear function (F) Layer.

Execution of ZUC 1.6 Cipher is divided into two stages:

1. Initialization Stage: IV and Key are loaded in LFSR. Then, the cipher is clocked 32 times and its output is discarded.
2. Working Stage: Output of first clock is discarded and then for each clock, it produces a 32 bit keystream.

Figure 1 depicts structure of ZUC 1.6 Cipher.

3.2 SNOW 3G Cipher

In SNOW 3G Cipher [3], Key is of size 128 bits and IV is of size 128 bits. Output is a keystream of word of 32 bits at every clock. The cipher has been divided into two logical layers: LFSR Layer and FSM Layer.

Execution of SNOW 3G Cipher is divided into two stages:

1. Initialization Stage: Cipher is clocked for 32 times in this mode and its output is discarded.
2. Keystream Generation Stage: Output of first clock is discarded and then for each clock, it produces a 32 bit keystream.

Structure of SNOW 3G Cipher is shown in Fig. 2.

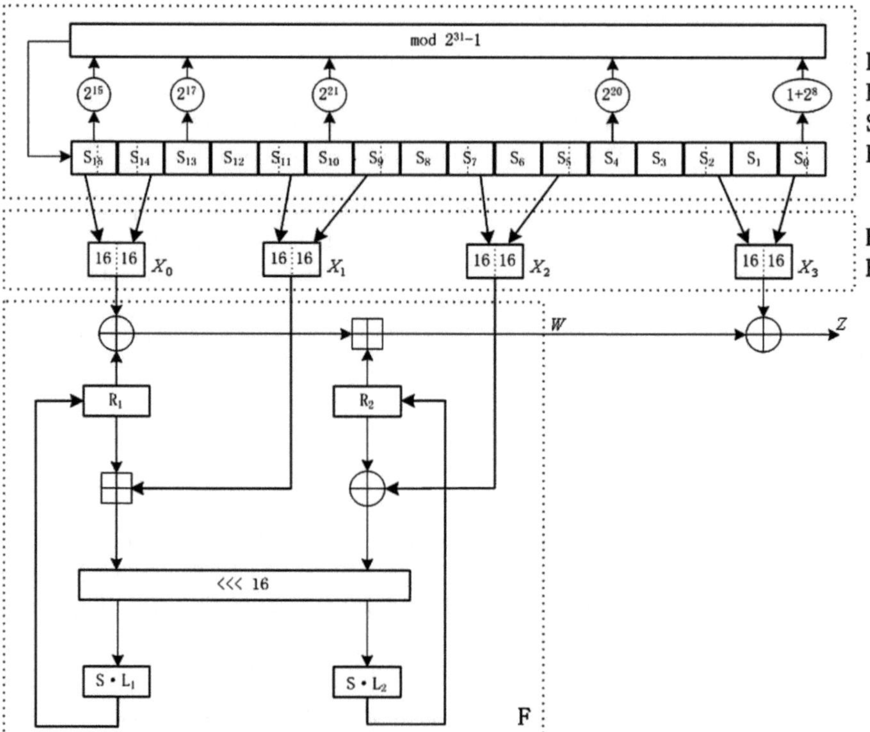

Fig. 1 Structure of ZUC 1.6 Cipher [3]

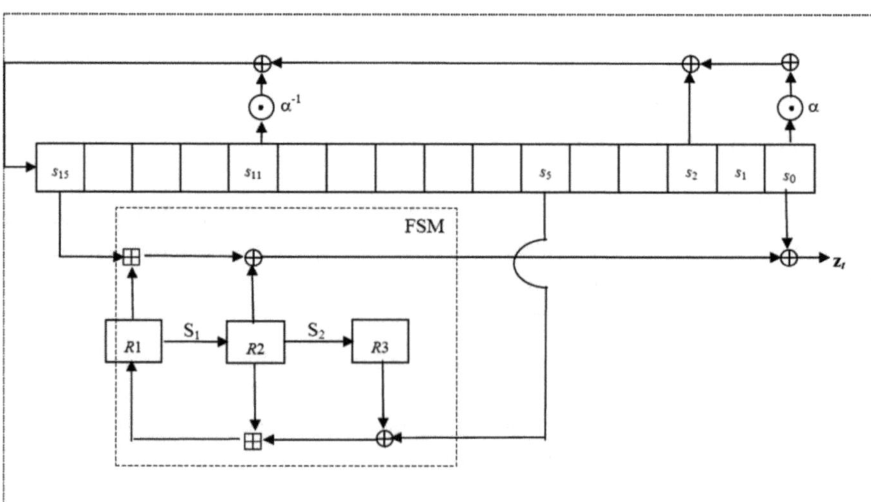

Fig. 2 Structure of SNOW 3G Cipher [3]

3.3 GRAIN v1 Cipher

In GRAIN v1 Cipher [5], Key is of size 80 bits while IV is of size 64 bits. Output is a keystream of bits at every clock. The cipher has been divided into three logical layers: LFSR Layer, Non-Linear Feedback Shift Register (NFSR) Layer, each of 80 bits and Filter Function (H). Here, $s_i, s_{i+1}, \ldots, s_{i+79}$ denotes bits of LFSR Layer and bits of NFSR Layer are expressed as $b_i, b_{i+1}, \ldots, b_{i+79}$.

$$\begin{aligned} h(x) = {} & s_{i+25} + b_{i+63} + s_{i+3}s_{i+64} + s_{i+46}s_{i+64} + s_{i+64}b_{i+63} \\ & + s_{i+3}s_{i+25}s_{i+46} + s_{i+3}s_{i+46}s_{i+64} + s_{i+3}s_{i+46}b_{i+63} \\ & + s_{i+25}s_{i+46}b_{i+63} + s_{i+46}s_{i+64}b_{i+63} \end{aligned} \tag{1}$$

$$z_i = h(x) + b_{i+1} + b_{i+2} + b_{i+4} + b_{i+10} + b_{i+31} + b_{i+43} + b_{i+56} \tag{2}$$

Execution of GRAIN v1 Cipher is divided into two stages:

1. Initialization Stage: Cipher is clocked for 160 times and its output is discarded.
2. Keystream Generation Stage: For every clock, Keystream is generated using (2).

Figure 3 shows structure of GRAIN v1 Cipher.

3.4 WG-7 Cipher

In WG-7 Cipher [4], the Key is of size 80 bits while IV is of size 81 bits. Output is a keystream of bits at every clock. It operates in following two modes:

1. Initialization Mode: Cipher is clocked for 46 times in this mode, with nonlinear permutation feedback, *WP*.

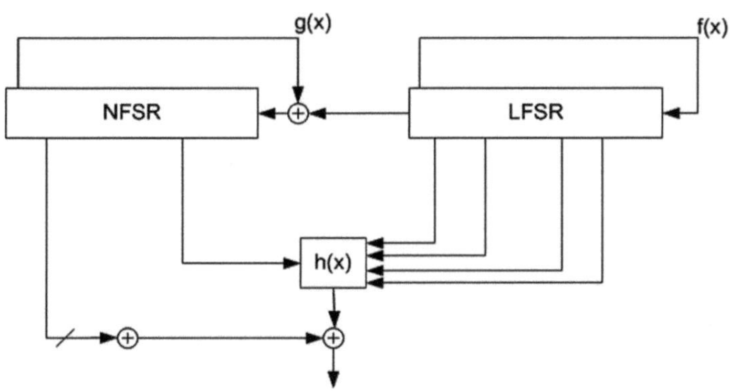

Fig. 3 Structure of GRAIN v1 Cipher [5]

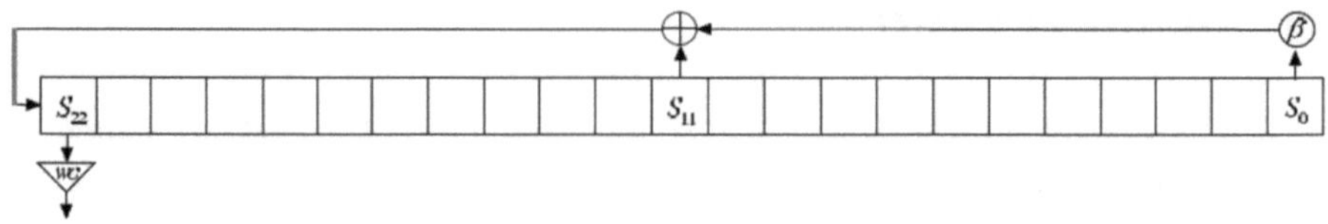

Fig. 4 Structure of WG-7 Cipher [4]

2. Keystream Generation Mode: For every clock, keystream is generated using Non-linear WG transformation, *WG*7.

The Structure of WG-7 Cipher is represented in Fig. 4.

3.5 DECIM v2 Cipher

In DECIM v2 Cipher [6], Key is of size 80 bits while IV is of size 64 bits. Output is a keystream of bits at every clock. It has been divided into three logical layers: LFSR Layer, Filter function (F) Layer, and Irregular Decimation or ABSG Mechanism Layer.

Execution of DECIM v2 cipher takes place in two modes:

1. Initialization Mode: Cipher is clocked 768 times in this mode.
2. Keystream Generation mode: Keystream is produced from bitstreams stored in buffer, for every clock.

Structure of DECIM v2 is shown in Fig. 5.

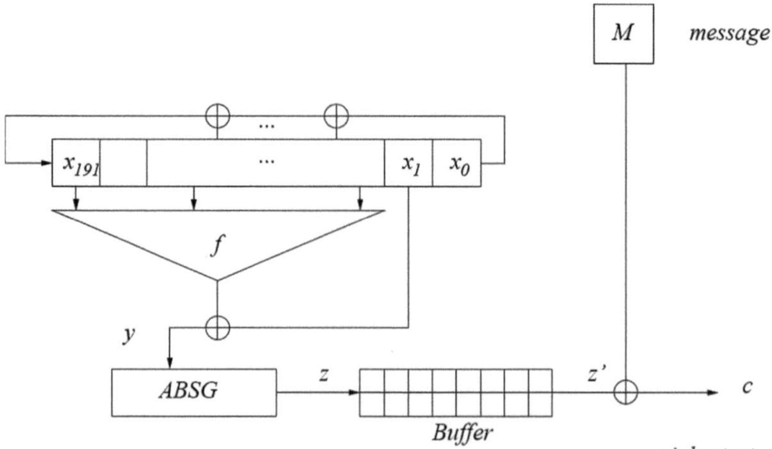

Fig. 5 Structure of DECIMv2 Cipher [6]

4 Comparative Analysis

The Comparative analysis will be done based on Speed-Up Measurement, Memory Usage and Randomness.

(a) Speed-Up Measurement and Memory Usage:
The FPGA Implementation for these ciphers has been done by various researchers on FPGA, and throughput achieved for respective frequencies of ciphers is shown in Table 2. Implementation of ZUC 1.6, SNOW 3G and GRAIN v1 Cipher has been done in C and their Execution time has been measured, which is calculated for 1,00,000 streams of 256 bits each. It is shown in Table 3.

(b) Randomness Evaluation:
Various Statistical Tests like test for Block Frequency, Cumulative Sums, Discrete Fourier Transform (Spectral), Frequency, Longest Runs, Runs, Serial, and Linear Complexity were performed using NIST Statistical Toolkit and graphical representation for ciphers ZUC 1.6, SNOW 3G and GRAIN v1 is shown in Figs. 6, 7 and 8 respectively. Here, P Value is greater than 0.01, the threshold value, and hence indicates unpredictable nature of ciphers for all the tests, except Linear Complexity test, which is shown in Fig. 9.

Table 2 FPGA implementation

Ciphers	Frequency (MHz)	Throughput (Mbps)	Area (slices)
ZUC 1.6 [3]	38	1216	1147
SNOW 3G [3]	104	3328	3559
GRAIN v1 [3]	177	177	318
DECIM v2 [7]	185	46.25	80

Table 3 Execution time of cipher in C

Ciphers	Average execution time
ZUC 1.6	1 m 04.17 s
SNOW 3G	1 m 59.95 s
GRAIN v1	1 m 20.08 s

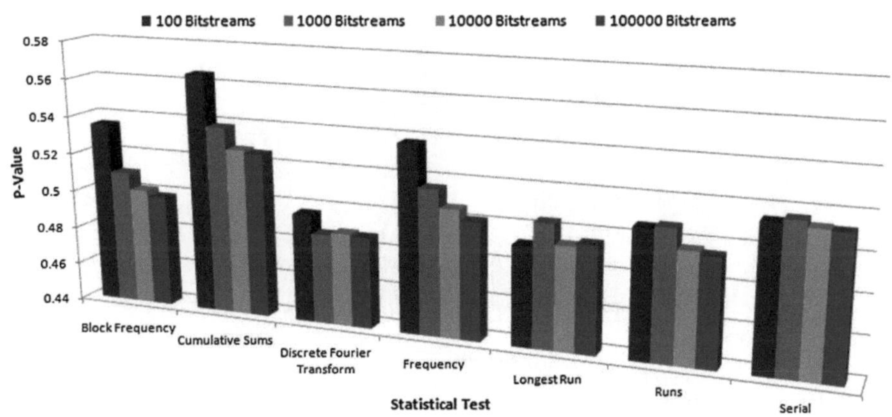

Fig. 6 NIST test results for ZUC 1.6

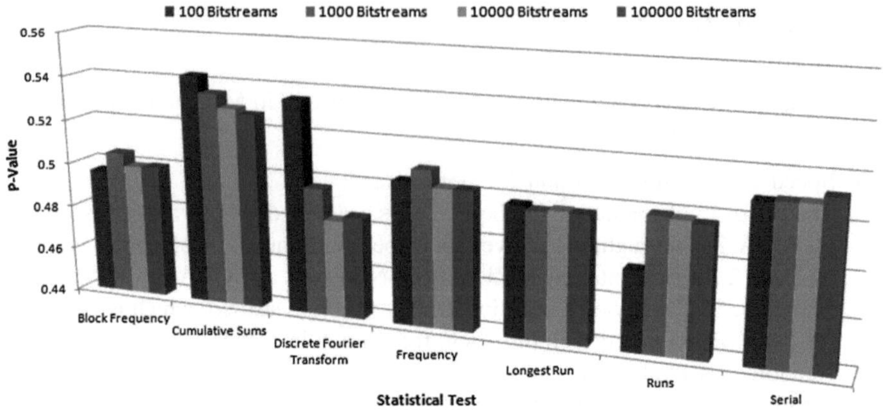

Fig. 7 NIST test results for SNOW 3G

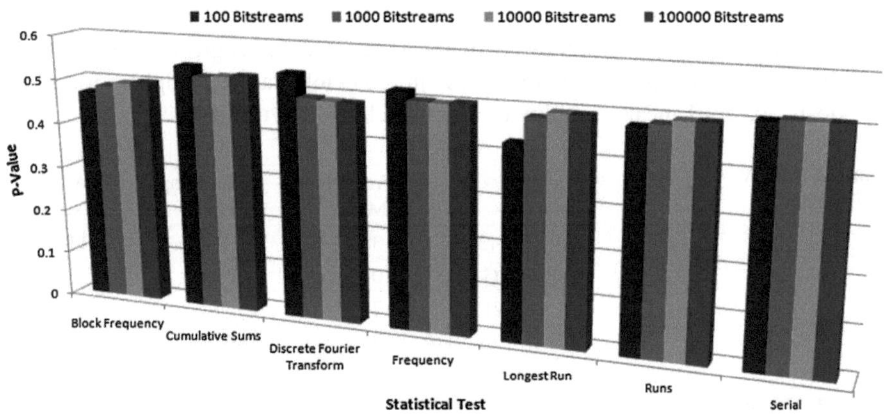

Fig. 8 NIST test results for GRAIN v1

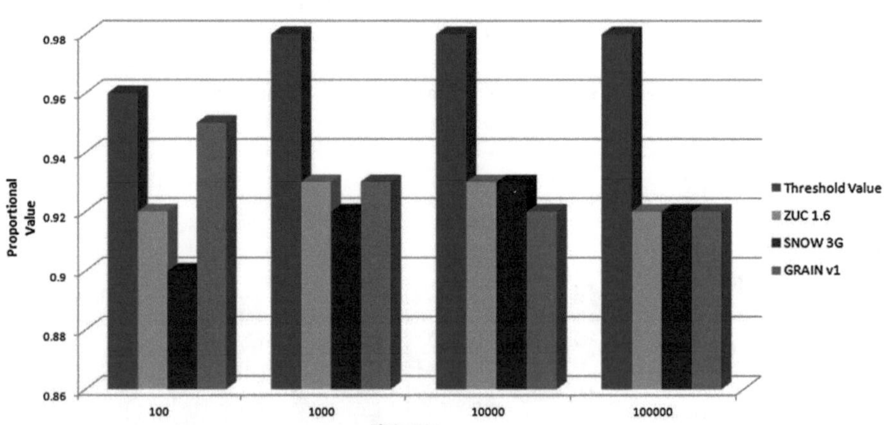

Fig. 9 Linear complexity test results

5 Conclusion

In this paper, architecture and working of five well-known stream ciphers, ZUC 1.6, SNOW 3G, GRAIN v1, WG-7, and DECIM v1 is presented. Literature Survey of these ciphers has been carried out in the terms of Hardware Utilization and Speed-Up Measurement on FPGA Toolkit. We have also implemented ZUC 1.6, SNOW 3G, and GRAIN v1 ciphers on Software platform to generate keystreams. Furthermore, various randomness parameters are measured using NIST Statistical Toolkit. Linear Complexity as key parameters has been measured in terms of proportion. Derived results indicate that the Linear Complexity of ZUC 1.6, SNOW 3G, and GRAIN v1 ciphers are lower than the threshold value suggested by NIST toolkit.

References

1. Darshana Upadhyay, Trishla Shah, and Priyanka Sharma, *Cryptanalysis of hardware based stream ciphers and implementation of GSM stream cipher to propose a novel approach for designing n-bit LFSR stream cipher*, VLSI Design and Test (VDAT) (2015), pp. 1–6.
2. Darshana Upadhyay, Priyanka Sharma, and Srinivas Sampalli, *Enhancement of GSM stream Cipher security using variable taps mechanism and nonlinear combination functions on Linear feedback shift registers.*, International conference on ICT for Intelligence Systems (ICTIS) (2015).
3. Paris Kitsos, Nicolas Sklavos, George Provelengios, and Athanassios N Skodras, *FPGA-based performance analysis of stream ciphers ZUC, Snow3g, Grain V1, Mickey V2, Trivium and E0*, Microprocessors and Microsystems 37 (2013), no. 2, 235–245.
4. Yiyuan Luo, Qi Chai, Guang Gong, and Xuejia Lai, *A lightweight stream cipher WG-7 for RFID encryption and authentication*, Global Telecommunications Conference (GLOBECOM 2010), 2010 IEEE, IEEE, 2010, pp. 1–6.
5. Martin Hell, Thomas Johansson, and Willi Meier, *Grain: a stream cipher for constrained environments*, International Journal of Wireless and Mobile Computing 2 (2007), no. 1, 86–93.
6. Côme Berbain, Olivier Billet, Anne Canteaut, Nicolas Courtois, Blandine Debraize, Henri Gilbert, Louis Goubin, Aline Gouget, Louis Granboulan, Cédric Lauradoux, et al., *Decim v2*, New Stream Cipher Designs, Springer, 2008, pp. 140–151.
7. David Hwang, Mark Chaney, Shashi Karanam, Nick Ton, and Kris Gaj, *Comparison of FPGA-targeted hardware implementations of eSTREAM stream cipher candidates*, The State of the Art of Stream Ciphers (2008), pp. 151–162.

MSECHP: More Stable Election of Cluster Head Protocol for Heterogeneous Wireless Sensor Network

Kameshkumar R. Raval and Nilesh Modi

Abstract Wireless networking of few hundreds or thousands of low-priced sensor nodes enables us to monitor a secluded area. By using clustering protocol, we get much accurate results of the sensing field with low transmission costs. In this paper we develop and analyze MSECHP, a heterogeneous aware protocol and effect of applying virtual grid on the sensor network field to evenly divide the entire WSN into the optimal number of clusters. To calculate the probability, of a sensor node to become cluster head three factors are considered. These factors are energy distribution among all the nodes of the same cluster, equal energy distribution among sensor nodes which have different residual energy and distance from the sink node. We show by the result of the simulation, MSECHP provides more stability than other clustering protocols.

Keywords WSN · Wireless sensor network · Energy efficient · Sensor nodes · Base station · BS · Sink node · Clusters

1 Introduction

Wireless Sensor Network (WSN) is a network of very small sensor nodes, which senses the data from the external environment and transmit it to other sensor node or sink node which also known as a base station using wireless communication protocols [1]. In most cases wireless sensor nodes are deployed to monitor remote areas or locations, which do not have sufficient resources for network communi-

---

K.R. Raval (✉)
Som-Lalit Institute of Computer Applications, SLIMS Campus,
University Road, Navrangpura, Ahmedabad, Gujarat 380009, India
e-mail: kameshraval@hotmail.com

N. Modi
Narsinhbhai Institute of Computer Studies and Management,
Kadi, North Gujarat, India
e-mail: drnileshmodi@yahoo.com

N. Modi et al. (eds.), *Proceedings of International Conference on Communication and Networks*, Advances in Intelligent Systems and Computing 508,
DOI 10.1007/978-981-10-2750-5_7

cation, such as wired network or consistent power supply. WSNs are mostly operated in the infrastructure less environment, therefore sensor nodes use battery as power (energy) source and it uses wireless transceiver to transmit or receive the data from sink node or other sensor nodes.

Structure of the sensor node consists of sensing unit to sense the data from the external environment, limited on board processing unit to process the data and radio transceiver to transmit or receive the data for communication. Because of the network is operated in unattended environment, replacement of the battery of the sensor nodes are not possible. So to increase the life time of the wireless sensor network, energy aware communication protocols are most needed.

In the TEEN [2] sensor networks are classified into (1) Proactive Networks, in which sensor nodes periodically sense the data and send the data towards the sink node (SN) or base station (BS) and (2) Reactive Network, in which sensor nodes sense the data and transmit it to the BS if and only if there is sudden drastic change in the data. Such protocols are well suited for real time or time critical applications.

2 Related Work

Sensor node is a low cost and tiny device having limited processing capability and limited energy. In Direct Transmission (DT) [3] method sensor nodes are directly transmitting the data sensed, to the BS. Because of the consumption of the transmission energy is depends on the distance from transmitting node to receiving node, much energy spent by those nodes which are very far from the sink node. So the node which is at longer distance from the sink node tends to dies quickly. In the Minimum Transmission Energy [3] sensor node transmit its data by multiple intermediate hops to the sink node. In the MTE method, those nodes whose physical location is just near to the BS, lose their energy faster and so they die very soon, because they have to sense their own data, not only that they also have to route the data from the other sensor nodes, situated at very longer distance from BS and forward it towards BS.

In both protocol discussed above, energy distribution among the sensor nodes of the network is not well balanced. LEACH [4] is a protocol in which, clustering scheme is introduces. Nearby sensor nodes can have same type of data, so it is wastage of energy if individual nodes are transmitting their data to the sink node. Sensor nodes which are near to each other are creating cluster. From all the nodes which belong to the same cluster, randomly any one live node is appointed as a cluster head. Periodically member nodes transmit their data to the head node of the cluster, and head node transmit only summarized data to the BS. The same process is repeated round by round. In LEACH algorithm, sensor nodes operated in autonomous mode. Much energy is wasted to manage clusters. Other limitation of the protocol is all the node of the WSN having equal energy. It is possible to increase network life time few more sensor nodes are added to the network later on. This will introduce heterogeneity of the sensor nodes (nodes having different level of residual energy).

LEACH-C [5] is another protocol in which cluster management handled centrally by the sink node. It will reduce much energy wasted for creating and managing clusters because sink node do not have resources limitation like energy and processing. Heterogeneity is also introduces in the protocol. LEACH-C provides much better results than original LEACH protocol. In the LEACH-C protocol two types of nodes are placed in the area of sensor network that are, advanced nodes and normal nodes. Advanced nodes are those which will have high amount of residual energy than normal nodes. But the energy distribution between advanced nodes and normal nodes are not balanced. In the result of the LEACH-C normal nodes dies quickly than advanced nodes. At the end of the simulation all the normal node dies quickly and all the advanced nodes remain alive for a longer period of time. LEACH multi-hop [6] is another variant of LEACH protocol in which cluster heads sends its aggregated data via multiple intermediate cluster head nodes which act as a router.

HEED [7] is another protocol in that residual energy of the sensor node is considered in the selection of cluster head node, which increases stability period of the network and increase network life time. SEP [8] protocol gives much preference to the heterogeneity of the sensor nodes and tries to distribute residual energy between advanced and normal node. The protocol increased stability period of the heterogeneous sensor node network, by increasing the probability of a sensor node to become cluster head by its residual energy. Advanced nodes get much preference to become cluster head, so the life time of the normal nodes can be increase. LPCH and UDLPCH [9] are another protocol which uses LEACH algorithm in the first round for cluster head selection, and from second round onwards those nodes which have minimum y-coordinates from the previous round's cluster head's y-coordinates are selected as a cluster heads. SECA [10] is an algorithm which propose a new way of finding better cluster head technique for energy saving.

2.1 Our Contribution

In our current research work, some assumptions considered by us are listed here. The both advanced and normal sensor nodes are energy constraint nodes. Sink node is located at the central place of the sensor network field and does not have resource limitation such as energy or processing power capabilities. The location of the sink node or base station is known well in advance. In the area of the wireless sensor network sensor nodes (Advanced and Normal) are randomly deployed. The protocol is used for such application in which the sensor nodes are static and do not change its location. In fact there are some applications of the sensor networks are there in which sensor nodes are considered to be dynamic (node can change its position). We will further extend our work for dynamic nodes which can change their position with time. The main intent of this paper is to introduce newly developed protocol called MSECHP to elect better cluster head sensor node which will provide much stability (more number of living nodes for longer time period) and increase overall lifetime of the heterogeneous WSN. We utilize the energy of the advanced node to provide more stability period to the network and the simulation result shows that it outperforms than LEACH-C and SEP protocol.

3 Radio Model

We assume simple first order radio model for energy dissipation for transmitter and receiver sensor nodes. In which transmission energy E_{TX_elec} and receiving energy E_{RX_elec} are same as 50 nJ/bit.

$$E_{elec} = E_{TX\_elec} = E_{RX\_elec} = 50\,\text{nJ/bit} \tag{1}$$

3.1 Transmission Energy

To transmit L bits of data up to d distance:

$$E_{TX}(d) = E_{elec} * L + E_{amp} * L \tag{2}$$

Now based on free space or multipath model energy used by the amplifier (E_{amp}) is calculated as follows:

$$E_{amp} = \begin{cases} \varepsilon_{fs} * d^2 & \text{if } d < d_0 \\ \varepsilon_{mp} * d^4 & \text{if } d \geq d_0 \end{cases} \tag{3}$$

where d in the formula represents distance between transmitter node and receiver node, and the value of d_0 is calculated as:

$$d_0 = \sqrt{\frac{\varepsilon_{fs}}{\varepsilon_{mp}}} \tag{4}$$

For our simulation purpose, we consider the value of ε_{fs} = 10 pJ/bit/m^2 and the value of ε_{mp} = 0.0013 pJ/bit/m^4. If we put the value of the Eq. (3) in to Eq. (2) then we can calculated the energy dissipation by the transmitter for L bits of packet up to d distance is:

$$E_{Tx}(L,d) = \begin{cases} E_{elec} * L + \varepsilon_{fs} * d^2 * L & \text{if } d < d_0 \\ E_{elec} * L + \varepsilon_{mp} * d^4 * L & \text{if } d \geq d_0 \end{cases} \tag{5}$$

3.2 Receiving Energy

Energy requires for receiving of L bits of packet is:

$$E_{RX}(L) = E_{elec} * L \tag{6}$$

4 Our MSECHP Protocol

4.1 Cluster Optimization

Rather than electing cluster head node randomly, our proposed protocol is focuses on geographic location of sensor field. We have divided the entire sensor network in to the number of clusters by considering geographic location. We have considered a virtual grid to evenly divide entire WSN into the number of clusters.

To find optimal number clusters we have used the following equation derived in [5] if the distance from the BS is more than d_0 for most of the sensor nodes of the WSN field then:

$$k_{opt} = \sqrt{\frac{n}{2\pi}}\sqrt{\frac{\varepsilon_{fs}}{\varepsilon_{mp}}}\frac{M}{d_{toBS}^2} \tag{7}$$

But if the distance is less than d_0 than following equation is used to find the optimal number of cluster, derived in [8].

$$k_{opt} = \sqrt{\frac{n}{2\pi}}\frac{M}{d_{toBS}} \tag{8}$$

Where K_{opt} is optimal number for the clusters in WSN, n is number of nodes in the wireless sensor network, ε_{fs} is energy for free space model, ε_{mp} is energy for multipath model, M is the area of the WSN field, and d_{toBS} is the mean distance from head node of the cluster to the BS.

For our simulation we have considered n = 100 nodes, ε_{fs} = 10 pJ, ε_{mp} = 0.0013 pJ. To find out average distance from cluster head node to the base station (d_{toBS}), we have used following equation derived in [8]

$$d_{toBS} = \int_A \sqrt{x^2+y^2}\frac{1}{A}dA = \frac{0.765\,M}{2} \tag{9}$$

So, if we put M = 100 in the Eq. (9), then we can find average distance from cluster head node to the base station, that is d_{toBS} = 38.25 m.

If we put the value of d_{toBS} and all other simulation parameter described above in to Eq. (8) then we can find out the optimal number of clusters, that is K_{opt} = 10.43.

So we have divided the entire sensor network in the 9 equal sizes of clusters, by applying logical grid of 3 rows * 3 columns by considering its geographic locations. Thus entire sensor network area is divided into 9 equal sizes of clusters. One more cluster is created by the sink node itself. If any of the cluster node find its distance smaller to the sink node than all other cluster head nodes, then that node is allowed to transmit the data directly to the sink node rather than transmitting the data to any other nearby cluster head.

4.2 Cluster Head Appointment

Once the entire sensor network is divided into the number of clusters by applying virtual grid, we need to appoint cluster head for each cluster. One of the node form the specific cluster is appointed as a cluster head from the set of all the living nodes (energy level is greater than 0) resided in the specific cluster. Election of the cluster head is based on the probability value of the sensor node. Probability value is an aggregation of the three different probabilities described below.

Rotational Probability

Cluster head receives the data from all cluster members, aggregates the data and sends this data to the sink node. So cluster head tends to spend much energy to receive the data, to aggregates the data and finally transmit this data to the sink node. To distribute the energy within the cluster, rotation of the cluster head is much important. A node which is elected as a cluster head for the specific round, should not become cluster head for the next $n_{counter} - 1$ round, where $n_{cluster}$ variable stores the number of nodes belongs to the specific cluster. We have used another variable $n_{counter}$ will be initialize with the same value as $n_{cluster}$. Rotational probability n_{rporb} is calculated as:

$$n_{rprob} = \frac{n_{counter} * 100}{n_{cluster}} \tag{10}$$

Initially before starting first round every node within the cluster will have probability 100 %. To elect optimal cluster head we are finding the max value of n_{rprob} and the node having highest value of n_{rprob} and residual energy is more than 0 will be promoted to be a cluster head for that specific round. After electing a node as a head of the specific cluster $n_{counter}$ will be decrement by 1, so that the probability to become a head of the cluster is reduced and that node will not become a cluster head for next $n - 1$ round. When $n_{counter}$ will become zero, then it will be again initialized with the value of $n_{cluster}$.

Energy Probability

In the heterogeneous network, the residual energy levels of all the nodes are not same. To evenly distribute the energy within the cluster, probability of those nodes which is having highest residual energy has to be increased. To calculate energy probability first of all we have to calculate average energy of the cluster.

$$E_{avgcluster} = \frac{\sum_{i=1}^{n_{cluster}} E_i}{n_{cluster}} \quad (11)$$

where, E_i is the residual energy of the node, which belongs to the specific cluster. $n_{cluster}$ is the number of nodes resided in the specific cluster and $E_{avgcluster}$ is the average energy of the cluster. Based on the residual energy and average cluster energy, energy probability is calculated as follows:

$$n_{eprob} = \frac{E_n * 100}{E_{avgcluster}} \quad (12)$$

Where, E_n is the residual energy of the sensor node. Those nodes having residual energy more than average cluster energy, its n_{eprob} is set to 100 %.

Location Probability

Importance of this probability is to find out location wise better cluster head. From all the nodes of a cluster, those nodes which are near to the sink node are better nodes to be as a cluster head, which is shown in the Fig. 1.

In Fig. 1a all the sensor nodes are submitting their data in the opposite direction of the sink node, because cluster head is appointed, is in the opposite direction of the sink node. Here cluster head tends to spend much energy to transmit the aggregated cluster data to the sink node, because cluster head is at very far from the sink node.

In Fig. 1b all the sensor nodes are using their transmission energy to transmit the data in the direction of the sink node, because cluster head is in the same direction of the sink node. Further more because of cluster head is nearer to the sink node, less transmission energy is used to transmit the data from the cluster head.

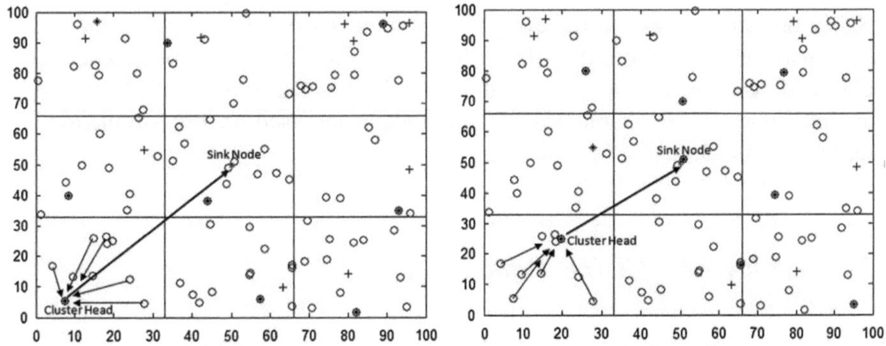

Fig. 1 **a** Cluster head nodes are far from sink. **b** Cluster head nodes are near to the sink

So, it is preferable to appoint cluster head from the set of those sensor nodes which have smaller distance to the sink node. To implement this we have introduce location probability which can be calculated as follows.

To find location probability we have find distance of all sensor nodes of the cluster, from the sink node. Then we have found the max distance from the sink node, and finally we have found the location probability as follows:

$$n_{lprob} = \frac{(n_{max-dist-sink} - n_{dis-to-sink}) * 100}{n_{max-dist-sink}} \tag{13}$$

where n_{lprob} is the location probability, $n_{max\text{-}dist\text{-}sink}$ is maximum distance from the sink node within the cluster and $n_{dist\text{-}to\text{-}sink}$ is the distance of the specific node from the BS.

To elect sensor node as a head of the cluster mean probability of rotational, energy and location probability is considered. In each round, that node which will have highest mean probability among all the nodes within the cluster is elected as a cluster head. Which will receive the data sensed by all the member sensor nodes of the cluster, aggregate the cluster data, and finally transmits this aggregated data to its destination.

$$n_{final-prob} = \frac{(n_{rprob} + n_{eprob} + n_{lprob})}{3} \tag{14}$$

5 Simulation

5.1 Simulation Parameters

We simulate a virtual grid based wireless sensor network with the field dimension 100 * 100 m in MATLAB. Out of n = 100 sensor nodes, we have considered 10 advanced nodes as well as 90 normal nodes randomly deployed across the field of WSN in our simulation. So for every sensor node X-coordinate and Y-coordinate are taken as uniformly from 0 to 100. Initial energy of the normal nodes is set to 0.5 Joules and for the advanced node it is 1.0 Joules, same as LEACH and SEP protocol. Other radio characteristics parameters are used in the simulation are discussed in the Sect. 3.

5.2 Result of MSECHP Protocol

We have run the simulation of MSECHP for more than five times and compare the results with LEACH-C and SEP protocols. The comparative result is plotted on the graph which is shown in the Fig. 2. The analytical comparisons of the first node dead in the round and number of dead nodes at the middle of the simulation, for the above protocols, for different simulations are shown in the following table.

From Table 1 we can say that in the LEACH-C protocol on an average, first node dies in the 990th round, in the SEP protocol first node dies in the 965th round and in the MSECHP protocol first node dies in the 1091st round. So we can say that MSECHP protocol provide much stability period to the network.

During the middle of the simulation, LEACH-C has 12–17 live nodes, SEP has 13–19 live nodes and MSECP has 28–46 live nodes are there. More number of live nodes in the network provides much accurate information about sensor field. So we conclude that MSECHP provides much stability than the other protocols.

Network Lifetime

There is a tradeoff between stability of the network and network lifetime. If we utilize the energy of the advanced nodes, and will give higher priority to the node having highest residual energy to be a cluster head, stability of the network increases. But because of we are utilizing much energy of the advance nodes, advanced nodes die quickly and that will reduce network lifetime. In our proposed protocol and SEP protocol to provide network stability we are utilizing the extra energy of the advanced nodes. Because of LEACH protocol do not utilize the energy of the advanced nodes, at the end of the simulation all advanced nodes will remain alive, while all normal nodes dies quickly. We have run all the protocols for 2000 round at the end of the simulation (after 2000 round) dead advanced nodes, normal nodes and total nodes are shown in the Table 2.

In Table 2, N stands for normal nodes, A stands for advanced nodes and T stand for total nodes. The statistical analysis says that in the LEACH-C protocol all the

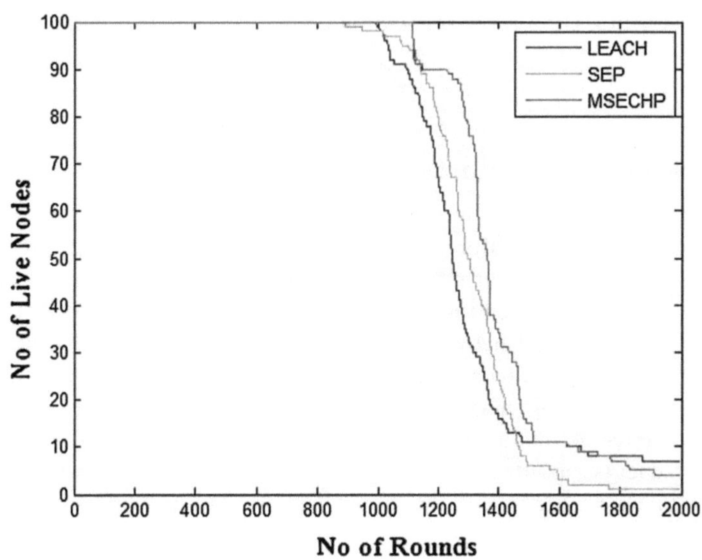

Fig. 2 Comparative analysis of LEACH-C, SEP and MSECHP protocols

Table 1 Analitical comparison of LEACH-C, SEP and MSECHP protocols

Simulation	First node dead in the round			Live nodes after 1450 round		
	LEACH-C	SEP	MSECHP	LEACH-C	SEP	MSECHP
1	996	893	1112	13	16	28
2	1075	1041	1026	16	17	46
3	1002	896	1096	15	18	28
4	943	995	1107	12	13	37
5	938	1002	1115	17	19	42

Table 2 At the end of the simulation dead nodes

Simulation	Dead nodes after 2000 rounds								
	LEACH-C			SEP			MSECHP		
	N	A	T	N	A	T	N	A	T
1	90	3	93	89	10	99	87	9	96
2	90	2	92	90	9	99	89	9	98
3	90	1	91	90	9	99	87	9	98
4	90	1	91	90	9	99	87	9	96
5	90	4	94	90	9	99	89	9	97

normal nodes die quickly which will introduce much network instability. More numbers of advanced nodes are alive at the end of the simulation. In the SEP and MSECHP protocols because of the energy of the advanced nodes are distributed we can find less number of normal nodes are dead at the end of the simulation. Compare to the SEP less number of nodes are dead at the end of the simulation, which shows MSECHP is more stable protocol which extends lifetime of the sensor network.

6 Conclusion and Future Work

We proposed MSECHP (More stable election of Cluster Head Protocol) for heterogeneous wireless sensor network, in which optimal number of cluster heads are appointed by dividing entire sensor network by virtual grid, so that in all the region of the sensor field network, we get the cluster head node which reduces transmission cost for the cluster member sensor nodes. We have also tried to evenly distribute energy among normal and advanced nodes. We have compared the simulation results of our MSECHP protocol with LEACH-C, SEP, HEED and many other energy optimization based protocol and we found better results for stability and network life time of the sensor network. In this work we have assumed that all the nodes are static. There are many applications of the sensor networks are there in which sensor node can move from one location to another location. In the future we will expand our work for the motion based sensor nodes.

References

1. Gautam, N., Sanjeev, S., Renu, V.: Data Dissemination in Wireless Sensor Networks A State-of-the Art Survey. In: International Journal of Computer Networks. Volume (4), Issue (1), 22–34 (2012).
2. Ghiasabadi, M., Sharifi, M., Osati, N., Beheshti, S., Sharifnejad, M.: TEEN a routing protocol for enhanced efficiency in wireless sensor networks. In: 2008 Second International Conference on Future Generation Communication and Networking. Volume (1), Issue (C), pp. 2009–2015 (2001).
3. Shepard, T.: A channel access scheme for large dense packet radio networks. In: Proceedings of the ACM SIGGCOMM, pp. 219–230.
4. Heinzelman, W., Chandrakasan, A., Balakrishnan, H.: Energy-efficient communication protocol for wireless microsensor networks. In: Proceedings of the 33rd Annual Hawaii International Conference on System Sciences. Volume (00), Issue (C), pp. 3005–3014 (2000).
5. Heinzelman, W., Chandrakasan, A., Balakrishnan, H.: An application-specific protocol architecture for wireless microsensor networks. In: IEEE Transactions on Wireless Communications. Volume (1), Issue (4), pp. 660–670 (2002).
6. Brand, H., Rego, S., Cardoso, R.: MH-LEACH A Distributed Algorithm for Multi-Hop Communication in Wireless Sensor Networks. In: ICN2014 The Thirteenth International Conference on Network. Issue (C), pp. 660–670 (2002).
7. Younis, O., Fahmy, S.: HEED a hybrid, energy-efficient, distributed clustering approach for ad hoc sensor networks. In: Mobile Computing, IEEE Transactions on. Volume (4), Issue (3), pp. 366–379 (2004).
8. Smaragdakis, G., Matta, I., Bestavros, A.: SEP A stable election protocol for clustered heterogeneous wireless sensor networks. In: Second International Workshop on Sensor and Actor Network Protocols and Applications (SANPA 2004), pp. 1–11.
9. Khan, Y., Javaid, N., Khan, M., Ahmad, Y., Zubair, M., Shah, S.: LPCH and UDLPCH Location-aware routing techniques in WSNs. In: Proceedings 8th International Conference on Broadband, Wireless Computing, Communication and Applications, BWCCA 2013, pp. 100–105 (2013).
10. Jau-Yang, C., Pei-Hao, J.: An efficient cluster-based power saving scheme for wireless sensor networks. In: IEEE EURASIP Journal on Wireless Communications and Networking. Volume (2012), Issue (1), pp. 1–10 (2012).

Use of ICT for Development of Smart City-Ahmedabad

Aditya Patel and Mansi Joshi

Abstract Citizens of India residing in large cities face several issues related to citizen services in their day to day lives due to rapid growth, congestion and unplanned development of cities (Mega/Metro cities). In this paper, we have studied the concept of smart city and identified various parameters/requirements for smart city (in context of Ahmedabad city) as envisioned by national level mission of Government of India for creation of 100 smart cities in India. Various ICT based information systems/software/mobile apps for providing state of the art citizen services used in top 5 smart cities of the world has been studied. Based on literature survey and data collection, major problems and issues faced by citizens of Ahmedabad city have been identified. To address the problems, Information and Communications Technology (ICT) based model has been proposed. The paper also discusses related systems and Mobile Apps which can contribute towards making Ahmedabad, a smart city and provide better citizen services.

Keywords Information and communication technology · Smart city Ahmedabad · Smart city architecture · Smart traffic · Smart energy · Smart health

1 Introduction

Development of towns and cities in human civilization has happened to improve the lives of people, provide better services fulfilling the needs of people residing at one place. In current times, with the development of civilization and technology, the needs and aspirations of people from a modern day city have increased manifold. Smart city concept is an initiative by governments of various countries which aims to improving the quality of lives, provide better citizen services, improve

A. Patel (✉) · M. Joshi
School of Computer Studies, Ahmedabad University, Ahmedabad, Gujarat, India
e-mail: aditya.patel@ahduni.edu.in

M. Joshi
e-mail: mansi.joshi@ahduni.edu.in

N. Modi et al. (eds.), *Proceedings of International Conference on Communication and Networks*, Advances in Intelligent Systems and Computing 508,
DOI 10.1007/978-981-10-2750-5_8

governance and promote economic development with planning and effective use of technology. In current times, the role of ICT is very important for development of smart city and to improve citizen services and governance using technology. ICT is a set of technologies for computing and communication to process data/information and transfer it from one system to another. In ICT, Internet of Things (IoT) allows objects to be sensed and controlled remotely enabling direct interaction between the physical world and computers for better precision. The goal of IoT is to ease the connection of heterogeneous devices and develop potential applications in various domains. ICT and IoT can help in managing some major sensitive issues such as natural disasters, urbanization and health care [1]. The accurate prediction of forest fires, earthquake, floods, etc. can guide us to a more manageable and controllable situation for recovering from these natural disasters. The automated traffic management systems can help us to determine the traffic in an area for avoiding traffic jams, making the city more urbanized and organized. The smart wearable devices for health care can help us detect and recover from the potential diseases before even they occur. These enhancements can lead to a better economic growth of the infrastructure around us making it a better place to live; making life easier than we think we can [2].

2 Problem Definition and Motivation

Ahmedabad, one of the fastest growing mega cities of Gujarat, has been recently identified by the Ministry of Urban development (MoUD), Government of India's smart city mission to be converted to smart city in near future [3]. But on the way, there are several hurdles and issues which need to be addressed to achieve this goal. One of the important issues is the efficient management of infrastructure and resources to fulfill the needs and aspirations of its citizens. Apart from availability of core infrastructure services like water, electricity, the city should provide good quality of life to its citizens and pollution free sustainable environment and resources using smart solutions. In rapidly growing city like Ahmedabad, the resources are plenty, but the allocation is improper. The water supply in one area is 24 h a day, while in another area, it is 2 h a day, and there is inefficient use, lack of monitoring and wastage of water resource.

Ahmedabad Municipal Corporation (AMC) has been focusing on improving urban infrastructure and services (i.e. water, electricity, internet connectivity, roads, Sabarmati River Front, BRTS). Still, there are many problems and issues faced by citizens of Ahmedabad, which have been identified as follows—Traffic jams and bad roads, Increase in air and water pollution, Electronic medical records and unified Health care monitoring systems, Inadequate parking zones and lack of efficient mobility and public transport system, Inadequate digitization, city automation using ICT and internet connectivity, Lack of sanitation and cleanliness in many areas, Lack of public services in a single computerized platform, No universal complaint and resolution platform for citizens, Security and safety of citizens and crime control, e-Governance and participation of citizens in decision making among others.

3 Literature Survey

3.1 Defining Smart City

The definition and meaning of smart city varies from people to people and country to country. A city can become a smart city when it comprises of a proper infrastructure in every aspect with the use high-tech heterogeneous systems based on ICT/IoT technologies for reaching a new level of urbanization. A smart city can improve the quality of life of citizens and also boost to the economical growth of a country. According to the Smart city mission [3], the core infrastructure of the smart city includes adequate resources such as water, electricity, urban mobility and public transport, affordable housing, etc. for a safe and sustainable environment. Smart solutions need to be listed for faster development of a city towards becoming a smart city. In a recent study about road spaces and population in Ahmedabad, it shows that this city has 70 cars, 250 two-wheelers per 1,000 people and 0.6 per capita road space per person, which is very less as compared to the other cities in India and other countries [4]. No designated parking space in areas makes the commuters to do random parking on streets; and BRTS project has made the roads even smaller for transport, making it more difficult for the general public to drive their vehicles which then results in traffic jams. The city authorities must take an early initiative for appropriate traffic management in the city else, this problem would be more difficult to handle in times to come.

3.2 Smart City Features and Solutions

According to the Indian government, the following are some of the smart solutions for various services and resources in cities [3]:

E-Governance and citizen services—public information, grievance redressal, Electronic service delivery, citizen engagement etc.
Waste management—waste to energy and fuel, recycling and reduction
Water management—smart meters, water quality monitoring, leakage identification
Energy management—smart meters, renewable energy, energy efficient buildings and homes
Urban mobility—intelligent traffic management, integrated multi-modal public transport, smart parking.

3.3 Study of Top Smart Cities in the World

According to Forbes, there are top 5 smart cities in the world for the following smart reasons [5]:

Barcelona, Spain [6] Famous for its environment and smart parking, this city has its own City OS for connecting and processing the information in real time of the whole city. The main objective of this OS is to do analysis of the stored data and simulate situations for solving potential problems in the city.
New York, United States Famous for its smart city lightening and traffic management, this city has an interactive platform called "City24/7" for delivering necessary information and alerts to the people across the city. The real-time information is available on easy to use smart screens, enabling citizens to retrieve it using touch, voice and audio technology. New York has plans to build nation's largest city-wide Wi-Fi networks by replacing payphones with state-of-the-art public connection points.
London, United Kingdom [7] Famous for its high technology and open data, this city has a huge data store called London Data store, one of the first platforms to make public data open and accessible to its developer community to build apps for better functioning of the city. An initiative called "The Love Clean London" has been taken to keep this city neat and clean through the complaints/reports of the people via a mobile app. The users can post pictures or texts as complaints in the app and can view the actions taken by the council.
Singapore, Republic of Singapore [8] Famous for its smart traffic management and technology, this city is building world's first Smart Nation Platform (SNP) for connecting the whole city and allowing citizens, business industries, research institutions, and government to work together and contribute ideas for co-creating innovative solutions to improve lives of citizens. The ministry of Heath, Singapore is providing a smart health app called "iHealth" on iOS platform for finding nearest health care services such as clinics, laboratories, nursing homes, hospitals, etc. along with their necessary information. One can also find health care professionals such as doctors, dentists, pharmacists, nurses, etc. on this app in an emergency situation.

3.4 Mobile Apps for Smart City

The following is the survey of mobile apps across the world which uses ICT technologies in various well-known areas to make it 'smart':

Smart App City (Smart City) [9] This app provides information about all the city services to the tourist and promotes city businesses. It is currently available in India, Spain, Chile, and will be spread across the world in near future as this app is adaptable to any city. The tourist can view the following things across the city—

Bus stops and estimated bus arrival times, City traffic to avoid traffic jams, City and tourism guide to know more about the area, Sport reservations to book events, parking's to indicate the location and occupation, Charging stations to recharge an electric vehicle, Pharmacies to find closest on-duty pharmacy shops and its timings, Gas stations to find the nearest one along with its best prices, Nearby businesses to find shops, bars, restaurants, and hotels, Platform to submit suggestions or complaints to the city council.

INRIX Traffic App (Smart Traffic) [10] This mobile app provides the best route with least delay to reach to a destination from a particular point. It updates billions of data points to provide accurate, real time traffic information. It provides insight to current traffic conditions as well as traffic forecasts for all highways, city streets, and local roads in order to help users to avoid traffic jams. It shows traffic cameras to the users for providing visuals to the upcoming route. It gives traffic alerts for nearby accidents or delays on the route. Users can save their places and routes on numerous devices for future purposes.

Kill-Your-Watts (Smart Energy) [11] A free utility app in iOS to keep track on the energy consumption of the residential electricity and provides energy reduction strategies. In order to use the app, one must sign into a free account with Green Button Connect, so that the energy consumption data will get uploaded. Users can see the hourly, daily, and monthly usage of the electricity. It is also available in form of graph and pie chart. It shows tips on how to decrease the electricity usage. One can also see the Energy usage score and carbon footprint as compared to the national average. Users can also challenge their family and friends to lower energy costs through twitter.

Google Fit (Smart Health) [12] It is a health tracking android app developed by Google. It uses user's mobile sensors to track and record various fitness activities such as walking, running or cycling. Users can set activity goals and based on the progress, the app will provide results such as calories burned in all these activities and also performance-based recommendations will be listed by the app. One can also see weekly, monthly or active time graphs along with the time spent on activities, number of goals met, and other graph details. Numerous app are connected with Google Fit such as Nike + Running, Heart Rate Plus, Aqualert, Atari Fit, etc. This app is compatible will all Android Wear Devices.

3.5 AMC Mobile Portal (AMC CCRS)

Ahmedabad Municipal Corporation (AMC) has launched a mobile app called Comprehensive Complaint Redressal System (CCRS) [13] for the people to register complaints about various facilities provided by AMC in Ahmedabad. The app provides following functionalities—Users can create and manage their profile in the app, register complaints related to drainage, water, garden, slum, cleaning and solid waste, tree cutting, etc., check status of their complaints, access nearest problem area location via Know Your Ward feature, collect information about the CCRS and call centre, view information about the professional and property tax from the AMC site.

3.6 Problems Faced by Citizens of Ahmedabad and Study of Existing System

In order to obtain information about the issues and problems faced by the citizens of Ahmedabad, primary data was collected from randomly selected citizens of Ahmedabad residing in different parts of the city using Questionnaire. Following is the summary and analysis of the responses collected from the citizens:

66 % of citizens consider the traffic management in the city as poor or not satisfactory
57 % of citizens feel that BRTS (Bus rapid transit system) project has increased traffic issues in the city
73 % of citizens are facing insufficient public parking situation, when parking zones are full or unavailable
69 % of citizens are ready to use paid parking spaces if they are available for booking
60 % of citizens face cleanliness related problems in the city
89 % of citizens face the need of public information system like kiosks or citizen service information systems/mobile apps to be available across the city
80 % of citizens think that there should be single complaint portal to address and resolve all citizen complaints and issues related to city and services provided by government authority
87 % of citizens are willing to use the citizen services mobile app, if it covers all citizen services in single app (electricity, water, telephone, internet etc.).

There are several major concerns in the city which adversely affect the lives of people. People are not aware of various web-based services by our government such as mobile apps and websites. E-waste management is also not done properly in the city as people are not fully aware of proper disposal of electronics.

4 Proposed ICT Model

To address the problems discussed in previous section, following ICT based model is proposed for the Ahmedabad city with different ICT enabled components. The model also proposes the use of IoT sensor devices for collection and monitoring of different resources of the city like water, air, temperature, energy consumption, mobility of public transport and internet usage. The model also proposed the use of crowd sourcing and crowd sensing where the citizens can push various real time information about the city from their mobile devices.

4.1 Traffic Management

As the population and number of vehicles in Ahmedabad are increasing day by day, an intelligent and computerized management of the traffic should be done by creating a system which would help monitor traffic in various areas of the city and would analyze and forecast situations where traffic jams can occur in real time. Citizens can post traffic violations and illegal parking to government using their mobiles. The concept of allotted parkings should be introduced in the areas where traffic jams occur due to reckless parking on roads.

4.2 Cleanliness and Complaint Management

There should be a single complaint portal to address and resolve all citizen complaints and issues related to city and services provided by government authority. Citizens should be able to post cleanliness and other problems with photo proofs using their mobile devices and complain should be automatically assigned to concerned municipal officer of the area. If the complaints are not solved within allotted time, there should be provision for auto escalation of complaints to higher govt. authorities.

4.3 Energy Management

An initiative should be taken for managing the energy using smart IoT enabled meters for monitoring and reducing the wastage and inefficient use of electricity. A smart meter app should display the real time electricity consumption in the house and usage of electricity by different category of devices. The rates for electricity should vary based on time of the day and demand/supply situation during peak hours. The corporation should also come up with a system which automatically manages the functioning of street and road lights according to the needs and conditions such as weather, sun light, etc.

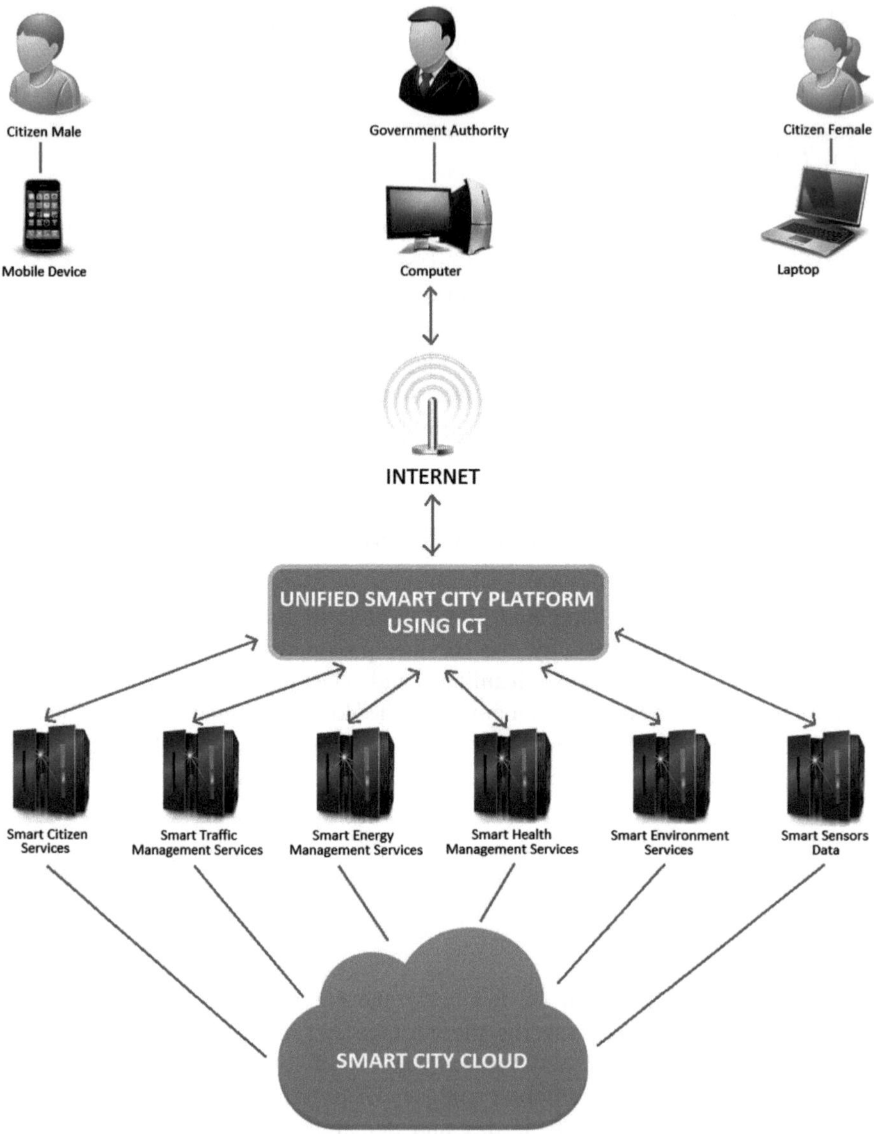

Fig. 1 Proposed smart city architecture

4.4 Unified Citizen Services Mobile App

We propose a common platform for all citizen services—Mobile App for all Citizen Services of Ahmedabad for all the stakeholders and citizens of the city. It should provide following functionality oriented to provide citizen centric services—pay the bill of various services provided such as electricity, water supply, income tax, etc. The mobile app will use the citizen unique ID for accessing all the government services.

4.5 Health Management

A mobile app for electronic and unified health records should be introduced for managing the health of the citizens. The citizens and doctors can records their health issues and diseases with the app and can later it can be analyzed by the doctor for frequency of the problem and allergies and early warning/prediction of diseases can also be given (Fig. 1).

All smart city systems and apps should be hosted on smart city cloud infrastructure providing 24 * 7 availability, reliability and efficiency of public services. The services provided by all systems and apps should be accessible by all citizens using any mobile or computing device like computer, laptop or mobile device.

5 Conclusion

In this paper, various features and requirements for a smart city has been discussed with reference to applicability of ICT. Study of various smart cities of the world and information systems and mobile apps used by them have been discussed. The results of survey regarding various issues and problems faced by citizens of Ahmedabad city have been discussed. To address the problems of citizens and provide better services, ICT based model for smart city and its various components has been proposed in the paper.

References

1. Zanella, A., Vangelista, L., Bui, N., Castellani, A., & Zorzi.: M. Internet of Things for Smart Cities. Journal of IEEE Internet of Things. Vol. 1, No. 1, 22–32 (2014).
2. O'Brien.: Why the Internet of Things is Important, http://www.ariasystems.com/blog/internet-things-important (2015).
3. Smart City Mission, Ministry of Urban development (MoUD), Government of India, http://www.smartcities.gov.in.
4. Amdavadis jostle for road space, 2015, September 3, The Times group e-Paper: http://epaperbeta.timesofindia.com/Article.aspx?eid=31819&articlexml=Amdavadis-jostle-for-road-space-03092015005006.
5. Peter High.: The Top Five Smart Cities In The World: http://www.forbes.com/sites/peterhigh/2015/03/09/the-top-five-smart-cities-in-the-world (2015).
6. BCN Smart City.: Barcelona: http://smartcity.bcn.cat/en.
7. London's City Council. Smart London Plan. London (2013).

8. Ministry of Health, Singapore. Ministry Of Health iHealth Sg. Retrieved September 8, 2015, from iTunes Preview: https://itunes.apple.com/sg/app/moh-ihealth-sg/id467311821?mt=8.
9. The App for the Smart Cities, http://smartappcity.com/en.
10. INRIX, Inc. INRIX Traffic App. from INRIX—Driving Intelligence: http://inrix.com/inrix-traffic-app.
11. KeyLogic Systems, Inc., iTunes Preview, https://itunes.apple.com/us/app/kill-ur-watts/id527704643?mt=8.
12. Google Inc., Google Fit from Google play: https://play.google.com/store/apps/details?id=com.google.android.apps.fitness.
13. Ahmedabad Municipal Corporation, AMC CRS from Google Play: https://play.google.com/store/apps/details?id=com.amc.ccrs&hl=en.

Optimization of the Neighbor Parameter of *k*-Nearest Neighbor Algorithm for Collaborative Filtering

Vimalkumar B. Vaghela and Himalay H. Pathak

Abstract Collaborative Filtering (CF) is one of the prime techniques used in the field of Recommender System. The recommender system is used for predicting the preference of the user based on his earlier preferred items. This process of predicting involves *k*-Nearest Neighbor (kNN) method, to find the users with similar type of preferences, interest or taste. In this paper, the experiments are carried out to check the influence of parameter k on the results obtained from kNN algorithm and to find the value of k for which we get the optimal accuracy for the kNN algorithm.

Keywords Collaborative filtering · *k*-Nearest neighbor · Recommendation system · Neighbor parameter

1 Introduction

Recommendation system which recommendations various items to the user, uses the CF which is the memory based reasoning variant, which is basically aimed at providing personalized recommendation. The recommendation system recommends the item to the user on the basis of his historical preferences and the interest. The involved CF technique uses the kNN for the extraction of similar users from the group of users.

In this paper, we are going to analyse the behaviour of the neighbor parameter *k* of the kNN algorithm used for finding the similar users. The kNN is used for the CF, which is been implemented for the recommendation purpose. Also, the experiments are carried out for the purpose of finding the influence of parameter *k* on the results.

V.B. Vaghela (✉) · H.H. Pathak
Computer Engineering Department, L.D. College of Engineering, Ahmedabad, Gujarat, India
e-mail: vimalvaghela@gmail.com

H.H. Pathak
e-mail: himalay.ldce@outlook.com

N. Modi et al. (eds.), *Proceedings of International Conference on Communication and Networks*, Advances in Intelligent Systems and Computing 508,
DOI 10.1007/978-981-10-2750-5_9

2 Collaborative Filtering

Collaborative Filtering is the memory-based reasoning variant [1], specifically used for providing the recommendation personalization. The CF technique recommends the user on the basis of the user's preferences and interest, in history. According to the preferences obtained earlier from the user, similar users who has the same preferences are been selected as neighbors by kNN. Then the votes of the neighbors are obtained which are then weighted by distances to make the recommendation. The weight is been calculated by distances, which means the distant neighbors' votes will be counted less and the nearer neighbors' vote will be counted more. Thus, CF is the technique to find an item according to the known preferences of the user by using the known preferences of the group for the item. Hence, CF is sometimes known as social information filtering.

For recommending the item to a new user, the CF accounts the following steps:

(a) Building a user profile by obtaining the new user's ratings for various fields, for knowing his interests and preferences.
(b) Selecting k number of nearest neighbors, based on the comparison of the new user's profile with the existing users' profile.
(c) Recommending the new user, based on the interest and preferences of the neighbors.

2.1 Detailed Study of Every Step

(a) **Building a New Profile**: The limitation of the CF is that the items without the ratings and which is to be rated are often than the items that are having known ratings and are already rated. This means that the profiles are generally suffering from sparsity and thus this result in very less match between the profiles of the users, which is not a favourable condition for making a recommendation. And the user profile is the entity with single element per item and having a huge amount of items to be rated. Every element of the entity refers the rating for an item, by the user to whom the profile belong, on the scale with positive, negative and a neutral values.

If there are large quantity of elements in the entity and if every user makes a choice for rating particular items, then there is a scope that there is some kind of matching between the two user profiles. While on the contrary, if rating for a particular group of items is made compulsory for all the users, then it will be in vain as the rating for the uncertain and the vague items are more likely to understand about the user's preference and choices.

Thus, it would be more sensible and appreciable to make the new user rate a particular optimum number of items and then should be set free whether to rate more items or not.

(b) **Comparing the profiles**: As the profile for a new user is ready after building step, next step is the measurement of its similarity with the other users' profile. For this, it would be a reasonable method of considering the entities of user profiles as the point in the space and then find the nearest neighbor based on the distance between the points which represents the user profiles in the space.

(c) **Predicting the Recommendation**: The end step of the CF is the consideration of the rating for an item, by the nearest neighbor users, for the recommendation of a rating for the user that haven't yet rated that item. The method for this is calculating the weight of the rating given by the neighbors, for an item, on the basis of the distance of the neighbor. Then, taking its average and the obtained value is the recommendation for the user who hasn't yet rated that item.

Let us take an example where the rating for a movie M by user A is to be predicted on the basis of other users P, Q, R, X, Y and Z. On the basis of the kNN algorithm, nearest neighbors of user A are user X and user Y.

As shown in Fig. 1, user X is at distance 4 from the user A and has rated 1 for the movie M. And the user Y is at distance 2 from user A and has rated 2 for the movie M. Then the recommendation for user A is obtained as:

$$(1/2(2) + 1/4(1))/(1/2 + 1/4) = 1.25/0.75 = 1.67$$

Thus, on the basis of the neighbors' ratings, the recommendation for user A to rate movie M is 1.67.

Recall and Precision are the measures for measuring the efficiency and the suitability of the recommendation obtained by the Recommender System.

Fig. 1 The neighborhood of A

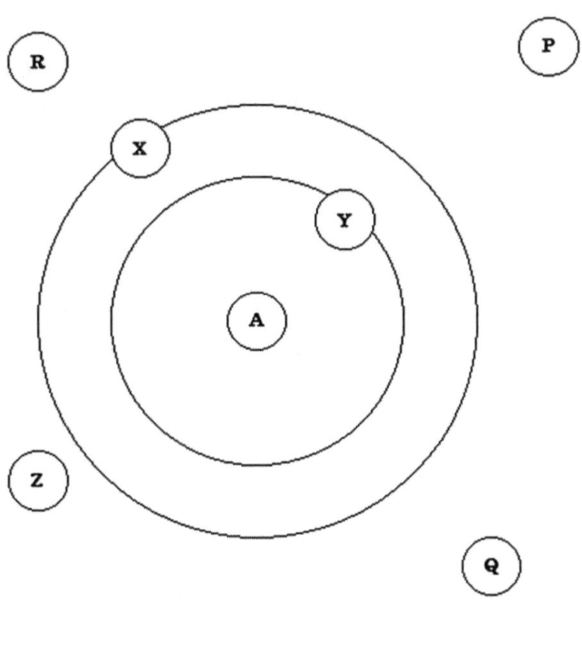

Recall refers to the percentage of the relevant recommendation that were retrieved, out of the total relevant recommendations.

$$\text{Recall} = \frac{|\{\text{relevant recommendation}\} \cap \{\text{retrieved recommendation}\}|}{|\{\text{relevant recommendation}\}|} \quad (1)$$

Precision refers to the percentage of the relevant recommendation that were retrieved, out of the total retrieved recommendations (Fig. 2).

$$\text{Precision} = \frac{|\{\text{relevant recommendation}\} \cap \{\text{retrieved recommendation}\}|}{|\{\text{retrieved recommendation}\}|} \quad (2)$$

Higher the percentage of Recall and Precision, higher the efficiency and suitability of the recommendation obtained by the Recommender System.

3 *k*-Nearest Neighbor

k-Nearest Neighbor algorithm is basically a classifier which is labor intensive in case of large data [2], which was designed in 1950s and became popular in 1960s after the availability of advanced computation powers and methods. Its wide usage is in the field of pattern recognition.

The kNN works by analogically comparing the test data with the set of training data. The data is in the form of tuples and each tuples needs n attributes for its description. Thus, a tuple can be represented by a single point in the n-dimensional data space.

The working of kNN can be described as, in case of any unknown tuple given it is describes as test tuple. Now the kNN classifier finds from the training data, the almost similar, or say, almost close tuples as the test tuple. Out of them, the *k* number of tuples that are closest to the test tuple are been selected as the nearest neighbors of the test tuple.

Fig. 2 Recall and precision

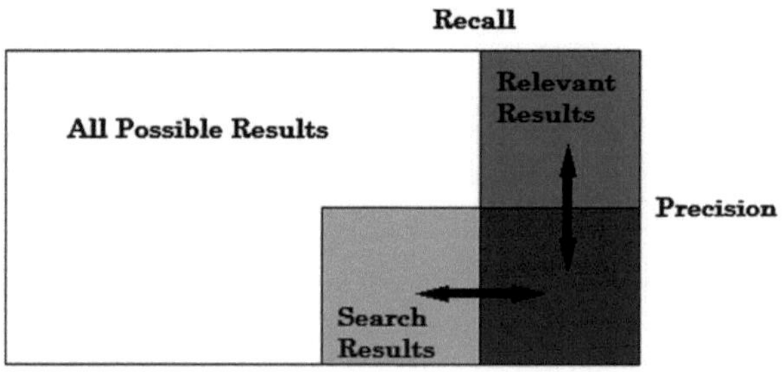

The "closeness" here refers to the distance between the points representing the tuples in the n-dimensional data space. For the measurement of the distance between two points, the metric used may be any distance metric as Euclidean distance, Cosine distance, Pearson Correlation, etc. If we consider the Euclidean distance, then the distance between the two points $A_1 = (a_{11}, a_{12}, a_{13} \ldots a_{1n})$ and $A_2 = (a_{21}, a_{22}, a_{23} \ldots a_{2n})$ can be obtained as:

$$dist(A_1, A_2) = \sqrt{\sum_{i=1}^{n} (a_{1i} - a_{2i})^2} \tag{3}$$

Generally, before the application of the distance formula, the values are generally normalized to solve the issue, in which, the larger values neglect the effect of the smaller important values. Usually, Min-Max normalization is used for the transformation of each attribute of the tuple into a definite and equal range.

$$v' = \frac{v - min_A}{max_A - min_A} \tag{4}$$

Here, v refers to the numeric value of the attribute A, v' refers to the normalized value of the numeric value of the attribute A. max_A and min_A refers to the upper limit and the lower limit respectively, of the attribute A.

On the basis of FkNN algorithm [3], we can write the pseudo code for the kNN, and it will include the following steps:

Step:1	The dataset $X = \{x_i \mid i = 1, 2, 3, \ldots, n\}$ and a value of k
Step:2	**For** $i = 1, 2, 3, \ldots, n$
Step:3	Compute the $dist()$ function for every two instances of the dataset.
Step:4	**If** i ≤ k
Step:5	Include x_i in the set of k nearest neighbor.
Step:6	**Else If**
Step:7	Delete the farthest from the set of k nearest neighbor
Step:8	Include x_i in the set of k nearest neighbor.
Step:9	**End If**
Step:10	**End For**

Table 1 Summary of the datasets

Dataset	Instances	Class	Attributes
Labor	57	2	17
Iris	150	3	4
Glass	214	7	10
Ionosphere	351	2	35
Vote	435	2	17
Soybean	683	19	35

Table 2 Accuracy for various values of *k*

Dataset	Accuracy (%) for			
	$k = 1$	$k = 3$	$k = 5$	$k = 10$
Labor	**82.456**	*91.228*	85.964	91.228
Iris	**95.333**	**95.333**	**95.333**	*96*
Glass	70.560	*71.962*	67.757	**66.355**
Ionosphere	86.324	*86.609*	**84.900**	**84.900**
Vote	**92.413**	92.643	92.643	*92.873*
Soybean	91.215	*91.361*	90.190	**87.701**

4 Experiments

4.1 Dataset and Algorithm

The kNN algorithm is been applied to all the datasets in the Weka [4, 5] tool, the 5 datasets used are from the UCI repository and 1 dataset is from standard Weka datasets. The summary of 7 such dataset is in Table 1.

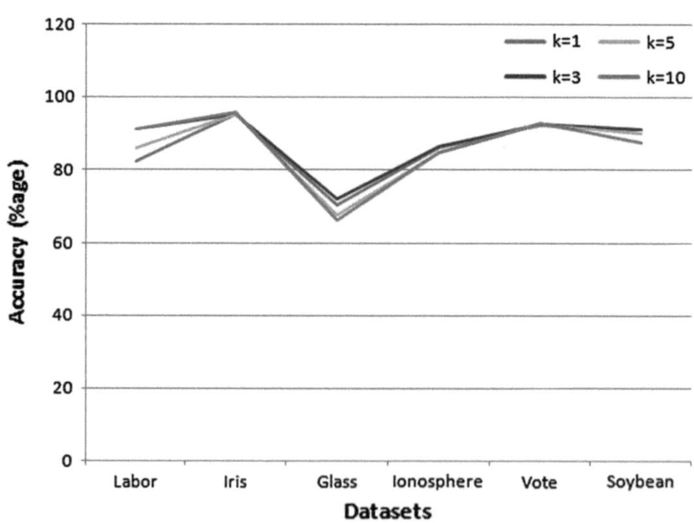

Fig. 3 Graph of accuracy (%age) → Datasets

The accuracy percentages obtained on applying kNN algorithm on different datasets with various values of *k*, are as shown in Table 2.

On the basis of the accuracy percentage obtained for the various values of *k*, the maximum accuracy for a particular dataset has been italicized, and the minimum has been bolded.

The graph obtained from the above percentages for different values of *k* for various datasets, is as follows (Fig. 3).

From the graph and accuracy percentages obtained, we can notice that for $k = 3$, the accuracy is maximum or nominally low as compared to the highest accuracy.

5 Conclusion

The Collaborative Filtering using the kNN algorithm is the widely used approach for the predictions and the recommendations by the Recommender System. Here, the kNN algorithm helps by selecting the neighbors nearest to the test case, from the training set. These neighbors are then used to recommend for the test set, or say, the target user.

On the basis, of the analysis of the behaviour of the neighborhood parameter, the experimental results and the obtained graph, we can conclude that, when we set parameter $k = 3$, i.e., when we select 3 nearest neighbors to the test case, from the data space, we gain the optimal accuracy.

References

1. Berry, Michael JA, and Gordon S. Linoff. *Data mining techniques: for marketing, sales, and customer relationship management*. John Wiley & Sons, 2004.
2. Han, Jiawei, and Micheline Kamber. "Data mining: concepts and techniques." (2001).
3. Chen, Hui-Ling, et al. "An efficient diagnosis system for detection of Parkinson's disease using fuzzy *k*-nearest neighbor approach." *Expert systems with applications* 40.1 (2013): 263–271.
4. Frank, Eibe, et al. "Weka-a machine learning workbench for data mining." *Data Mining and Knowledge Discovery Handbook*. Springer US, 2010. 1269–1277.
5. Markov, Zdravko, and Ingrid Russell. "An introduction to the WEKA data mining system." *ACM SIGCSE Bulletin* 38.3 (2006): 367–368.

The Efficient Scheme for Contention Reduction in Bufferless OBS Network

Dilip H. Patel, Kiritkumar Bhatt and Vedvyas Dwivedi

Abstract The bufferless Optical Burst Switched (OBS) Network, suffers severely from heavy contention loss. The existing reactive contention resolution schemes solve contention without any try to minimize the occurrences of contention. Therefore, we are presenting novel proactive scheme for reducing the occurrence of contention known a Dynamic Hybrid Cluster and Deflection Feedback (DHCF). In proposed DHCF scheme entire OBS network is partitioned into many small clusters and one node acts as cluster head in each cluster. The contention is minimized using clustering approach and the simulation results show improvement in Burst Loss Probability (BLP) in the range of 31–38 % in OBS network.

Keywords Contention resolution · Burst Loss Probability (BLP)

1 Introduction

The OBS is all-optical transmission network for future optical internet [1]. Burst contention occurs when two or more control packets try to reserve a same wavelength channel at the same time. Several methods have been evolved in the literature to resolve the problem of burst contention [1–3]. A comparative investigation clearly indicates [3] that they do not provide technically viable solution. Therefore,

D.H. Patel (✉)
LDRP Institute of Technology, Gandhinagar, Gujarat, India
e-mail: dilip_ec@ldrp.ac.in

K. Bhatt
Sardar Vallabhbhai Patel Institute of Technology, Vasad, India
e-mail: krbhattec1@gmail.com

V. Dwivedi
C.U. Shah University, Wadhwan, Surendranagar, Gujarat 363030, India
e-mail: provc.cushahuniv1@gmail.com

N. Modi et al. (eds.), *Proceedings of International Conference on Communication and Networks*, Advances in Intelligent Systems and Computing 508,
DOI 10.1007/978-981-10-2750-5_10

an alternative approach to control the contention loss using traffic management method in a proactive manner is very much desirable. The proactive schemes [4, 5] have been proposed to minimize the occurrence of contention and they give some improvement in occurrences of contention at the cost of generating additional delay. However, they fail to provide improvement in contention loss and delay at a high load [6].

Therefore, the Dynamic Hybrid Cluster and Deflection Feedback (DHCF) scheme is proposed in this paper to minimize the generation of number of contentions itself. In DHCF scheme, the entire OBS network is divided into group of small clusters and within each cluster one node acts as a cluster head. The contention avoidance clustering approach effectively handles the network from entering into heavy contention states. Also, the deflection routing based feedback scheme is employed in DHCF scheme along with cluster approach to further reduce the occurrence of contention. The dynamic deflection routing concept [7] is used along with clustering approach in this paper. The designated cluster head collects the information related to network resources and transfer the updates of the resources to other cluster heads within network.

The rest of the paper is organized as follows. Section 2 presents the model of cluster generation. The simulation environment is presented in Sect. 3. The numerical results are presented in Sect. 4. Section 5 conclude the paper.

2 Proposed Dynamic Hybrid Cluster and Deflection Feedback Scheme

First, the process of cluster generation is considered. The M_n is a set denoting the number of nodes in network and the P_r is a set denoting the degree of each node in the network. The N_r denotes one of node within the set $P_{r,}$ where r is the degree of node n in the given network and C_j is a set denoting the number of nodes within the jth cluster. The cluster head is decided by looking at the node that have maximum degree in the set P_r. The first cluster is formed by adding all the nodes that are one hop distance to cluster head. The entry of selected nodes in the first cluster is deleted from the set M_n with their degree from the set P_r. Once the first cluster is created then the process of selecting second cluster begins, wherein the degree in set M_n for all the remaining nodes is checked. The node that has a maximum degree is chosen as second cluster head. As long as the entire set M_n reaches zero node value, the process of new cluster formation is repeated. It may happen that the cluster has only one node. In DHCF scheme, the number of nodes in particular cluster is considered as the important design parameter. The minimum number of nodes that are taken for one cluster is four. A 12 nodes vBSN network topology is considered as shown in Fig. 1 for purpose of cluster formation. In order to make first cluster,

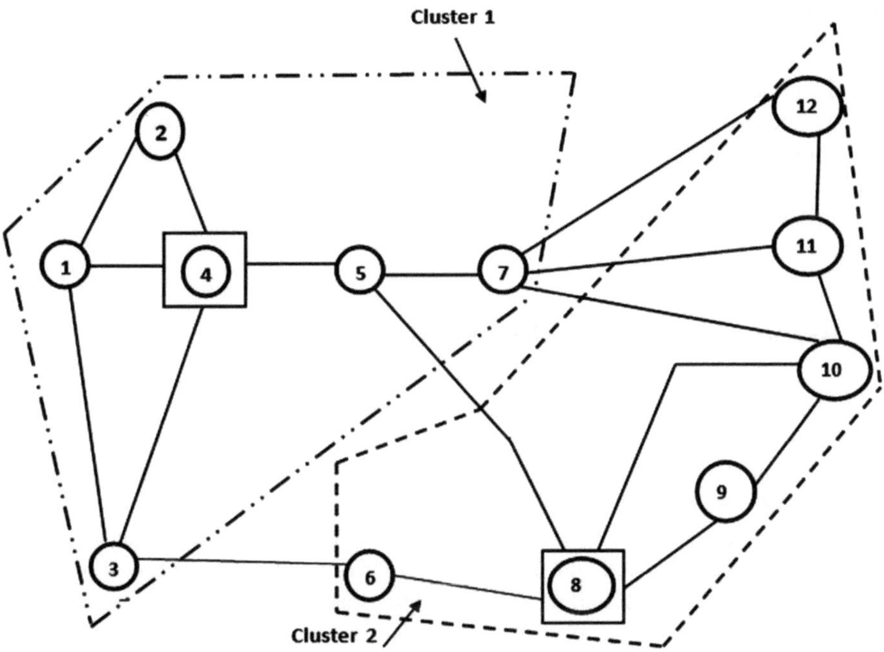

Fig. 1 The vBSN network with final two clusters

the cluster head needs to be decided. It can be observed from the set P_r that, four is the maximum degree with node 4 in a twelve nodes vBSN topology. After formation of first cluster, the remaining nodes in the set M_n, P_r and C_1 can be written as,

$$M_n = [6, 7, 8, 9, 10, 11, 12] \tag{1}$$

$$P_r = [62, 74, 84, 92, 103, 113, 122] \tag{2}$$

$$C_1 = [1, 2, 3, 4, 5] \tag{3}$$

Similarly, after formation of third cluster, the zero nodes remain in the set M_n, P_r. The cluster heads are shown in Fig. 1 by a node with a square shape. It can be observed from Fig. 1 that, the cluster 3 has only 3 nodes and it can be easily added to either the first cluster or second cluster based on their hop distance from respective cluster head. Finally, the four minimum node criterion is fulfilled in the form of two clusters.

3 Simulation Environment

The 12 node vBSN topology is considered as the OBS core network as shown in Fig. 1. The C++ code is developed for the purpose of simulation which includes all the required OBS modules [8]. The existing contention LHDR and DHRD scheme [9] is compares with DHCF. There are 28 wavelengths (10 Gb/s on each) for data transmission on each link.

4 Result Analysis and Discussion

It can be observed (Fig. 2) that the DHCF scheme outperforms the LHRD and DHRD for all ranges of traffic load. For example at very high load (load $\geq$ 0.8), the proposed DHCD scheme gives better BLP performance, about 63.5 % less BLP than DHDR and 74.56 % less BLP than LHRD.

Figure 3 shows the actual average delay against the traffic load. It can be observed that the proposed DHCF scheme outperforms other algorithms like LHDR and DHRD in terms of actual delay at all traffic loads. For example, when network load is very high (load $\geq$ 0.8), the results shows that DHCF gives better delay performance, about 31.30 % than LHDR and about 38.75 % than DHRD.

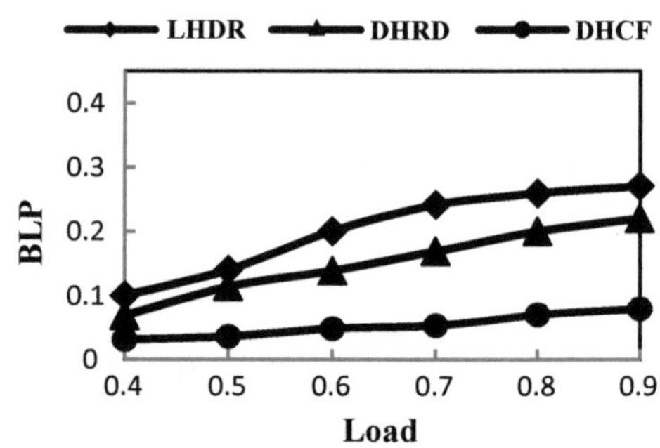

Fig. 2 Burst loss probability (BLP) versus load

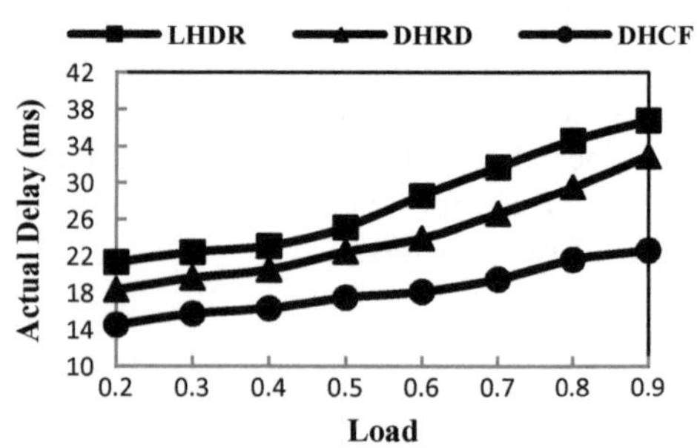

Fig. 3 Actual delay versus load on vBSN network

5 Conclusion

This paper presents a novel contention minimization DHCF scheme for OBS networks. The DHCF scheme tested on a vBSN network and based on simulation results the proposed scheme is validated. The improvement in BLP performance at very high load in our scheme is attributed to the fact that cluster based approach ensures wavelength channel which is more likely to be free or least congested on the desired path to destination. Also, the rise in the delay in the DHCF scheme is maintained in safe limit to prevent the higher BLP.

References

1. Vinay Chamola, et al.: Performance Analysis of OBS Network using Fiber Delay Line: A Simulation Approach. In International Conference on Communication, Information & Computing Technology, Mumbai, India (2012).
2. I. Chlamtac, A. Fumagall and et al.: CORD: Contention resolution by delay lines, IEEE J. Selected Areas in Communications. 14, 1014–1029 (2011).
3. Amit Kumar Garg, et al.: An adaptive reservation scheme for optical network, International Journal for Light and Electron Optics, 122(4), 281–286 (2011).
4. Lian, J., Naik, K., Agnew.: A framework for evaluating the performance of cluster algorithm for hierarchical networks. IEEE/ACM Transaction on Networking, 15(6), 1478–1489(2007).
5. Ihsan VI Haq, Henrique M. Salgado, and Jorge C.S.: Resource Aware Routing and Intelligent Wavelength Assignment for Cooperative Clustered OBS Networks. In IINESC TEC, Porto Portugal (2012).
6. Ahmed Triki, Paulette Cavignet and et al.: Efficient Control Plane for Passive OBS Network. In France Telecom - Orange Labs. Lannion FRANCE (2013).
7. Patel, D.H, Kothari, D.K.: Overview and Framework for Dynamic Deflection Contention Resolution in OBS Network. NUiCONE - Nirma University International Conference, India, 28–30 (2013).
8. T. Venkatesh, A. Sankar, A.: A Complete Framework to Support Controlled Burst Retransmission in Optical Burst Switching Networks. IEEE J. Selected Areas in Communications. 26(3), 65–73 (2008).
9. Wael Hosny, et al.: Dynamic Contention Resolution Protocol for OBS Networks. International J. Scientific & Engineering Research, 2 (2011).

Empowering Throughput Over Proactive Wireless Network Using Multistreaming

R. Dhaya, F. Abul Hasan and R. Kanthavel

Abstract The intend of this manuscript is toward enhance the throughput in a Proactive manner using SCTP, which is more reliable, message oriented and infrastructure based transport layer protocol. The conventional protocol is Transport control protocol (TCP) which thinks packet losses in erect hand over transmission time causes congestion plus as an effect thus humiliating throughput. In TCP, the 'Proactive hand off' method has been used to increase the throughput of the wireless network showing poor performance and so, it is a high need to introduce a new technique in order to increase the throughput. Here a novel method and architecture for throughput enhancement algorithm for proactive wireless network have been proposed. Simulation of projected system was accepted in Network Simulator 2. Throughput was intended for the projected method in Proactive scenario with the multihomed technique. The relative investigation of the outcome point out the throughput enrichment more than the TCP protocol for Proactive approach was 22 % and whereas throughput enrichment in SCTP, it is proved that the increase the throughput in Proactive approach is 36.8 % compared with TCP protocol.

Keywords Throughput · Network simulator · Proactive approach · Conventional protocol

---

R. Dhaya (✉) · F. Abul Hasan
Department of Computer Science and Engineering,
Velammal Engineering College, Chennai, India
e-mail: dhayavel2005@gmail.com

R. Kanthavel
Department of Electronics and Communication Engineering,
Velammal Engineering College, Chennai, India
e-mail: Kanthavel2005@gmail.com

N. Modi et al. (eds.), *Proceedings of International Conference on Communication and Networks*, Advances in Intelligent Systems and Computing 508,
DOI 10.1007/978-981-10-2750-5_11

1 Introduction

Wireless mesh networks named the peer-to-peer multi network hop form are rising to improved exposure, consistency, and easiness of arrangement. A wireless ad-hoc network can be believe as a particular wireless mesh network, anywhere a compilation of mobile nodes outlined or created a provisional network lacking the help of some recognized communications or infrastructure. Routing or direction-finding in ad hoc networks is demanding owing to the lively topology and limited properties. A perfect routing or direction finding protocol have to offer precise routing data at all required times at the same time, slaying no bring in preserving motionless ways. The majority obtainable routing protocols may pro-active or re-active [1]. Moreover elevated command for multimedia rich requests in great cellular client support has named for outlook worldwide mobile communication schemes [2]. These schemes are listening carefully to give important augmented network capacities to hold large numeral of concurrent voice and data users being wherever and all-time, with unreliable obligation of bandwidths, moderately at advanced data rates present for high-quality performance, as the nodes are mobile, need improved throughput methods. So in order to increase the throughput in the network a reliable transport protocol is very much wanted and such a protocol is SCTP [3]. SCTP is intended to deal with the boundaries and complication of TCP and UDP while conveying genuine time signaling and data or information [4]. The presentation performance and reliability of SCTP is due to its new service forces such as multi-homing, multi-streaming, improved head-of-line blocking, and improved security appearance [5].

Multihoming is one of the main significant texture in SCTP, which is utilized by the data source to propel data to a destination by unusual ways [6]. The advantage of Multihoming is potentially better survivability of the assembly in the holder of network breakdowns [7]. SCTP is stood on the TCP protocol, but included a numeral values of superior and sole texture that are not obtainable in TCP. The purpose of the proposed work is to enhance the throughput of SCTP in a Proactive manner. To adjust congestion window according in a network, while hand over process is being completed, thereby increasing the throughput of the wireless network [8], proactive approach is proposed through multihoming technique. The Proactive movement is that the congestion window is attuned according to the novel network system while the handover procedure is being finished, shunting slow start throughout this procedure to eradicate the delay in window alteration for improved throughput.

2 Existing Problem

So far there were no routing scheme developed for enhancing throughput for multipath transfer of data using TCP, owing to its limitation of strictly acknowledgement based transfer, and Head of line blocking [9].

Khurrum Aftab et al. [10], found that in TCP packet loss happens during vertical hand over transmission period as congestion and as a result thereby degrading throughput. The throughput enhancement they achieved over the TCP protocol for Proactive approach was 22 %.

Martin Hynes and Liam Kilmartin [11] offered a numeral of various protocols to hold the idea of wireless mortal mobility. They alerted on additional ornamental functionality to bear switching among the obtainable trails in order to find the overall data throughput among the device nodes.

Federico Perotto et al [12], opened an omnipresent network systems using TCP and SCTP as transport protocols and evaluated their presentation where TCP simply bears from recurrent route breakdown and disputation on the wireless channel than SCTP.

From the overall points it is identified that in TCP, the Proactive Hand off method is used to increase the throughput of the wireless network. But some packets may lose while transferring the data starting the source to the destination leads to the unordered data delivery of packets at the receiver end [13]. There is only one path between sender and the receiver in TCP, if any packet loss retransmission of packets is not possible, also TCP does not have a Partial Reliable data transfer technique [14]. As an end of the result, if the link smashes because of a path failure, data turns into out of stock until the association is re-established. The severe series preservation in TCP not only creates incomplete arranging of data unfeasible, it also roots unwanted delay on the whole data release. in addition, if a single packet is misplaced, release of following packets is steriled until the lost TCP packet is delivered, which causes head-of-line (HOL) blocking [15].

To conquer the harms or troubles in the already offered TCP and to advantage from network boundary or interface redundancy and give end-to-end network fault tolerance, SCTP that authenticates the multihoming at the transport layer in the projected wireless network is planned to boost throughput.

3 Proposed Multihomed Wireless Network

The proposed wireless network is shown in Fig. 1 with 8 nodes. Data transfers between the sender and the receiver, The Multistreaming technique offers in SCTP that the paths are independent and if one of the path failures, the other path can still deliver data.

Consider the data is transferred between the sender and receiver using the interface 1 and 2. If any packet loss occurs while reaching the interface 2, then the interface 1 takes the secondary path, between interface 3 and interface 4 to reach the receiver destination. Before the occurrence of the handoff itself, the receiver node will send the data to the sender node and the congestion window in the sender node adjusted according to that hand off in the proactive.

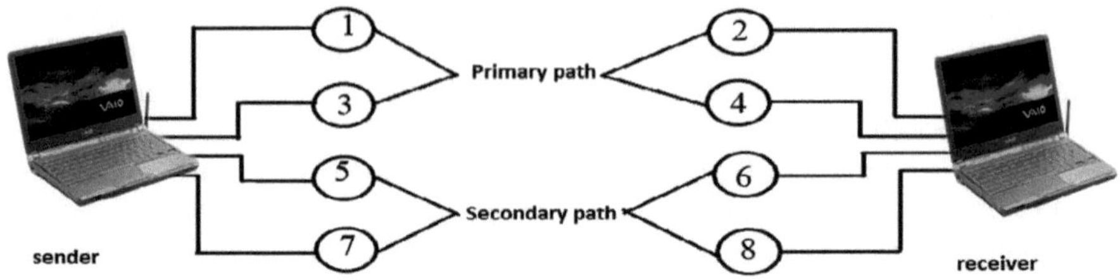

Fig. 1 Proposed system architecture for data transfer in SCTP protocol

4 Proposed Proactive Throughput Enhancement Algorithm

A brief explanation of the proposed algorithm is given below:

1. Handshaking process is taken place between the interfaces from the sender side to the receiver end.
2. Start the file transfer protocol to transfer the packets.
3. Send a data chunk from one of the interfaces among the sender node to the receiver node.

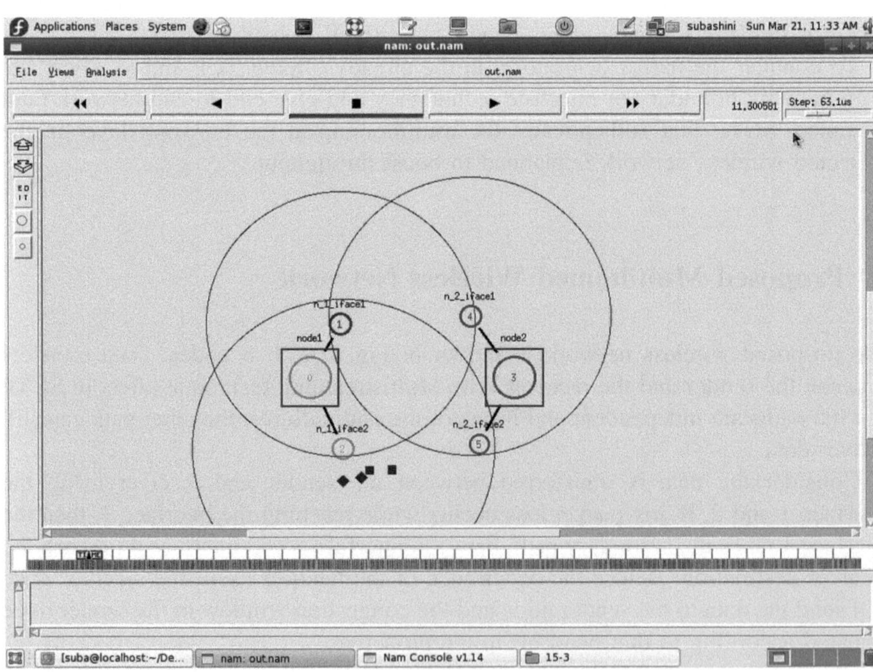

Fig. 2 Malfunction of the interface 1 and packets drop on the node 1

4. An acknowledgement is returned from the receiver node to the sender node on successful transmission.
5. If any data loss or path failure occurs,
   a. The sender node will have to find another interface to transfer data to the receiver node.
   b. The sender node will be acknowledged by the receiver node.
   c. Find round trip time and path failure time.
6. Continue the transfer protocol to transfer the packets.

5 Results and Output

If any packet loss occurs while reaching the interface 2, then the interface 1 takes the secondary path, between interface 3 and interface 4 to reach the receiver. Before the occurrence of the handoff itself, the receiver node will send the data to the sender node that the hand off is going to occur, the congestion window in the sender node adjusted according to that hand off and thus the proactive approach is

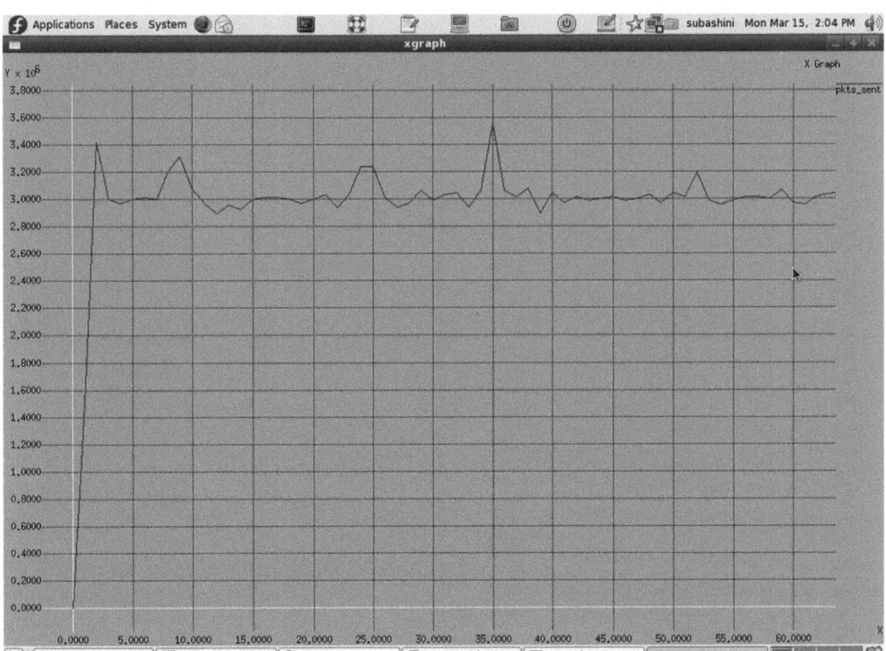

Fig. 3 Time versus packet sent for proactive in SCTP

calculated before the handoff. The Throughput estimation and the round trip time can be calculated for Proactive approach here as follows:

The Round Trip Time (RTT)

$$\text{RTT} = \text{Received Buffer Size}/\text{Maximum Throughput of SCTP}$$

Throughput is measured based on the number of packets reached in a given time.

Let the handover time be taken as 50 s. After hand over, the number of packets sent is going less times. No of packets dropped during handover time at 50 is low. No of packets received = 4.15 * 10^3 kb/s. So, proactive throughput is 22 % and the throughput 3000 kb/s. So the increasing throughput is 4150–3000/3000 * 100 % = 38.6 %. So the high throughput is achieved comparing the SCTP and TCP [10].

Figure 2 shows the output of the SCTP protocol on a Proactive approach. Node 1 and node 2 are taken as sender and receiver respectively. At first the packet is transferred from the node1 (interface 1) to node 2 (interface 4) and the acknowledgement is received by node1 from interface 5.

Figure 3 shows the malfunctioning of the interface 1 that causes the intermission in the transmission of packets. Due to this packet drop, acknowledgement from the node 2 not reaches the node 1. So the node 2 will be waiting for the packets to

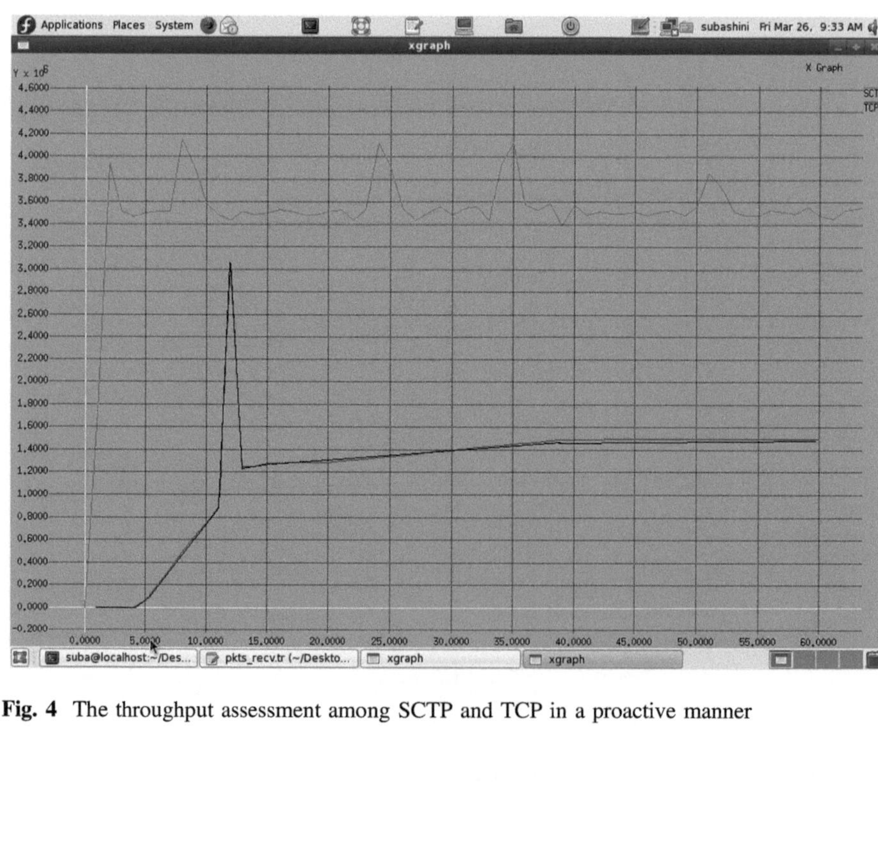

Fig. 4 The throughput assessment among SCTP and TCP in a proactive manner

Table 1 Representation of comparison of throughput enhancement over SCTP versus TCP

Protocols/methodology	Proactive	Reactive
	Throughput in %	
SCTP	38.6	52
TCP	22	17

reach. In this Proactive approach the congestion window is adjusted before the hand off. The acknowledgement sends from interface 5 to interface 2 to receive the packets.

Figure 4 shows the Time versus packet sent for the Proactive approach using SCTP. The time taken to send the packet is 10*kb/s. Here the hand over time is taken as 50. It is clearly identified that before the handoff there were some packet loss. In this proactive approach, the congestion window is adjusted before handoff.

6 Comparison of Proactive Graph Between SCTP Versus TCP

Table 1 shows the throughput. Throughput assessment among SCTP and TCP in a proactive manner. From the experiment, it is also found that in TCP the throughput is only 22 %. But in SCTP the throughput estimation is 38.6%

7 Conclusion

The throughput Enhancement over the Wireless network could be effectively improved by applying Multistreaming and Multihoming technique of SCTP. If any malfunctioning in the interface occurs, the packet takes alternate interface to reach the destination and if out of coverage happens the interface will search for the nearest node to come into coverage area and make the packets to be transferred. The outcome of projected method was too evaluated with TCP protocol. The comparative study of the outcome indicates the throughput improvement over the TCP protocol for Proactive approach was 22 % whereas throughput enhancement increase the throughput for the proposed is 40 %. Hence the throughput is greatly increased, which is evident from the graphs. Here we have analyzed the throughput enhancement for a Wireless network in a SCTP protocol by implementing, proactive approach by considering eight interfaces. In the future, the same algorithm can be applied in the Wireless Mobile Wireless Ad Hoc Networks (MANET). The number of nodes and interfaces may be increased and the proposed algorithm can be used and tested for effectiveness with different topology.

References

1. Guanhua ye, Tarek Saadawi and Myung Lee: SCTP Congestion Control performance in wireless Multihop Networks(2002). 23–57.
2. Yusuke Takemoto, Junichi Funasaka, Satoshi Teshima, Tomoyuki Ohta, and Yoshiaki Kakuda,:SCTP Performance Improvement for Reliable End-to-end Communication in Ad Hoc Networks(2002), 129–138.
3. Preethi Natarajan, Fred Baker,Paul D. Amer and Jonathan T. Leighton: SCTP: What, Why, and How, (2009), 2001–2089.
4. Dragan ZjaliC, Mario PavloviC, Marko Aralica,: stream control transport protocol (SCTP), (2004), pp. 168–198.
5. Jinyang Shi, Yuehui Jin and P. R. China: Experimental Performance Studies of SCTP in Wireless Access Networks, (2003), 139–156.
6. Yosuke Matsushita, Takahiro Matsuda and Miki Yamamoto: TCP Congestion Control with ACK-pacing for Vertical Handover", IEEE Wireless Communications and Networking Conference, vol. 3, no. (2005), 1497–1502.
7. Myung Lee, Guanhua ye and Tarek N. Saadawi: IPCC-SCTP: An Enhancement to the Standard SCTP to support Multi-homing Efficiently, (2004), 256–289.
8. Myung Lee, Guanhua ye and Tarek N. Saadawi: Performance during Handshake Process",22nd International Conference on Advanced Information Networking and Applications– Workshops, (2008), 512–627.
9. Lukasz Budzisz, Ramon Ferrús, Ferran Casadevall,:SCTP multihoming performance in dynamically changing channels with the influence of link-layer retransmissions, (2006), 199–225.
10. Khurrum Aftab, Seema Ansari and Faisal Aftab: An End-to-End Throughput Enhancement for Vertical Handovers in 3G Wireless Networks, (2009), 147–158.
11. Martin Hynes and Liam Kilmartin: Optimizing Transmission Path Selection in SCTP based Wireless Networking, (2009), 172–192.
12. Janardhan R. Iyengar, Paul D. Amer, and Randall Stewart: Concurrent Multipath Transfer Using SCTP Multihoming Over Independent End-to-End Paths" IEEE/ACM Transactions on Networking, vol.14, no.5, (2006), 512–629.
13. Shaojian Fu and Mohammed Atiquzzaman:Improving End-to-End Throughput of Mobile IP using SCTP" Telecommunications and Networks Research Lab, (2003), 219–267.
14. Mohammed Atiquzzaman and William Ivancic: Evaluation of SCTP Multistreaming over Wireless/Satellite Links, (2003), pp. 1176–1189.
15. Guanhua ye, Tarek Saadawi and Myung Lee: Independent per Path Congestion Control for Reliable Data Transmission between Multi-homed Hosts (2004), 189–215.

Control of Robot Using Neural Networks

Nikhil Nagori, Sagar Nandu and Alpa Reshamwala

Abstract The paper deals with motion control of autonomous robot. For an autonomous robot the main functionality which is to be implemented is its movement. The robot should be able to move from source to destination successfully avoiding all the obstacles in a known or unknown environment. This paper explains in detail 3 approaches for the motion control: (1) Neural Network where the problem is divided into sub problems FindSpace and FindPath (2) ANFIS (Adaptive Neuro-Fuzzy Inference System) where 6 layers are present (3) Fuzzy Logic along with neural network. Some simulation results are given which shows that Neural Network and fuzzy logic together gives better performance than Neural Network and fuzzy logic alone.

Keywords Neural networks · Autonomous robots · Robot navigation · ANFIS · Neuro-Fuzzy system

1 Introduction

A robot can be understood as a machine which can work like humans or animals. There are two types of robots Semi Autonomous and Autonomous. Autonomous robots are bots which work on their own without any human help or human intervention learning on its own. These robots can considered as having brain just like humans and animals have to make decision on its own.

N. Nagori (✉) · S. Nandu · A. Reshamwala
Computer Engineering Department, Mukesh Patel School of Technology Management & Engineering, NMIMS University, Mumbai, India
e-mail: nikhilnagori.nmims@gmail.com

S. Nandu
e-mail: sagarnandu.nmims@gmail.com

A. Reshamwala
e-mail: alpa.reshamwala@nmims.edu

N. Modi et al. (eds.), *Proceedings of International Conference on Communication and Networks*, Advances in Intelligent Systems and Computing 508,
DOI 10.1007/978-981-10-2750-5_12

To make the robot learn on its own and make decisions on its own we use various techniques or systems. Neural Network and Fuzzy logic are the techniques discussed in this paper. Neural Network in robotics can be considered as the nervous system in the humans or animals. The nervous system is the main system in the human body that sends signal to the brain and using these signals the brain makes the movement. Same way for robot to learn neural networks act as their nervous system. Fuzzy logic is logic which works on the principle of true and false or If and Else. Fuzzy logic is used for decision making. It helps neural network to make appropriate decisions.

2 Brief Description

2.1 Neural Network

The space in which robot works is called as environment and there are many obstacles. To achieve autonomy sensing and reasoning are required. Sensing is provided by sensors attached to the board used while reasoning can be achieved by devising algorithm. The motion problem is divided into two sub problem Findspace and Findpath problem. It is an iterative algorithm in which the last move of robot is stored and the next move direction is selected.

The algorithm follows:

1. Identify the object, start state, goal state and environment space.
2. Set the current object state equal to the goal state.
3. Start range finder to identify the local part of the map of the working environment space.
4. Load the first neural network, which uses the input data from range finder to do the calculations and it is iterated until all the inputs combine to obtain a single free space.
5. Start the second neural network, this network gives the direction k for the next movement step.
6. Generate the robot motion path in the direction k and go to the step 3 [1].

FindSpace problem basically can be treated or understood as to find the free space around the robot when it detects an obstacle. To solve this problem Principal Component Analysis (PCA) [2] network is used. This network takes input from ultrasonic range finder.

Ultrasonic Range scanner Sensor works on the principle of SONAR [3]. The work of PCA Network is to reduce the data to few principal components. It uses feed forward network to do the classification. Feed forward network is a network in which the data or information moves only in one direction [4] (Fig. 1).

Hidden layer in this topology is basically any layer that is not an output layer. The output of this PCA network is the few principal components differentiated from

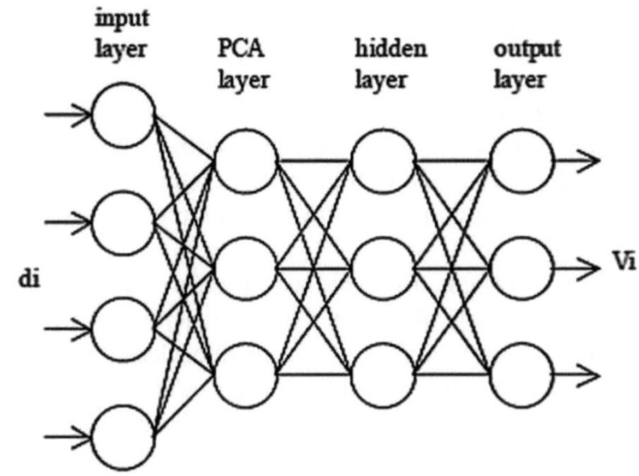

Fig. 1 PCA network topology [1]

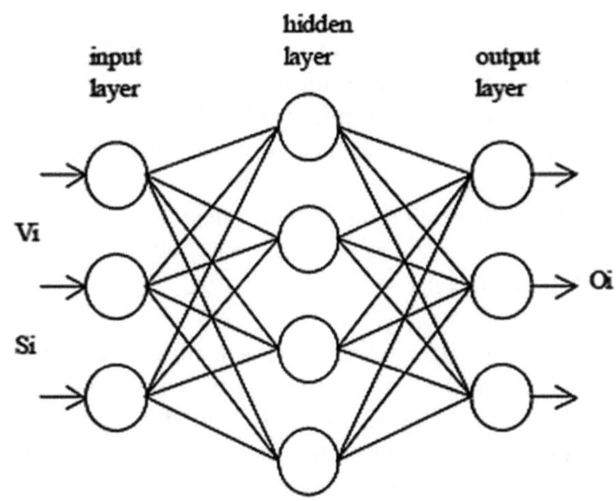

Fig. 2 MLP network topology [1]

the input data i.e. the appropriate free space. Now when we have the free space we need to find the exact direction or path for the next robot movement. This is the FindPath problem. To solve this problem MLP (Multilayer Perceptron) network is used. MLP network is layered feed forward network which is trained with static back propagation [1]. This network takes in input, the output of PCA network and gives the exact path for next Robot movement. This network has 3 layers only i.e. input layer, hidden layer and output layer (Fig. 2).

The output of this layer is then feed into the control system of Robot.

2.2 ANFIS

This approach is using ANFIS which stands for Adaptive Neuro-Fuzzy Inference System [5]. Neuro fuzzy means combination of neural network and fuzzy logic. Fuzzy logic is a type of logic that recognizes more than simple true and false values With fuzzy logic, propositions can be represented with degrees of truthfulness and falsehood.

In this system fuzzy logic is used to make explicit decision such as where to move, in which direction exactly whereas neural network is used for learning. Neural network makes the controller learn through the previous data. For an example if a robot is moving in an environment and it comes across an obstacle fuzzy logic will help the robot make decision to stop or make the next movement i.e. to move left or right or back or diagonal or in particular direction. Neural network will store this data for learning and with the help of data obtained from fuzzy logic it will find appropriate space in that direction to move.

ANFIS Approach:

The ANFIS is a hybrid system combining Artificial neural network and Fuzzy inference system. The current analysis includes four inputs. These inputs are basically 3 obstacle distances (x1, x2, x3) and a target angle (x4) where x1 = front distance, x2 = right distance, x3 = left distance, x4 = target angle. The output is a steering angle.

ANFIS has following if-then rules:

Rule: IF $x_1 = A_j; x_2 = B_k; x_3 = C_m;$ and $x_4 = D_n$

THEN $Fi = p_i x_1 + r_i x_2 + s_i x_3 + t_i x_4$

Where

$F_i = p_i x_1 + r_i x_2 + s_i x_3 + t_i x_4 + u_i$ for steering angle

$J = 1 \text{to} q_1$; $k = 1$ to q_2; $m = 1$ to q_3; $n = 1$ to q_4 and $i = 1$ to $q_1.q_2.q_3.q_4$

There are 6 layers in ANFIS model. Each node of same layer have same functions. The output of the previous layer is the input of the current layer. The layer after input is fuzzy layer and all the other layers after that are neural network layers (Fig. 3).

Layer 1:

This layer receives input from sensors in form of signals. The input from sensors (x1, x2, x3, x4) defines the obstacle in the environment space of both kind static and dynamic.

Layer 2:

All the nodes in layer 2 are adaptive nodes and have a node function

$$O_i^1 = \mu A_i(x)$$

where, x = input node i, A_i = linguistic label associated with the above node function and μA_i = member function of A_i, Typical $\mu A_i(x)$ is

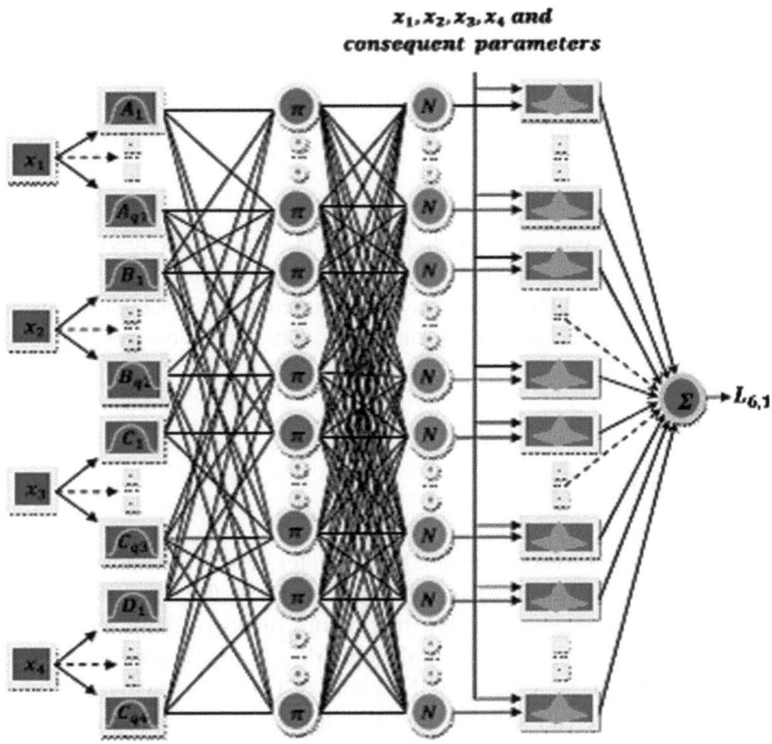

Fig. 3 Six layers of ANFIS model [5]

$$\mu A_i = \frac{1}{1 + [(\frac{x - c_i}{a_i})^2]^{b_i}}$$

Layer 3:

The nodes present in layer 3 are fixed nodes. The output of this layer is multiplication product of all the incoming signals. These nodes gives firing strength (w_i). The output (O_i) is

$$O_i = w_i = \mu A_i(x) \mu B_i(y)$$

Layer 4:

The nodes in layer 4 are also fixed node and are named as norm. The output or the value of *i*th node in this layer = firing strength (W_i)/sum of all rules firing strength

Layer 5:

The nodes present in layer 5 are adaptive nodes having node function

$$O_i = w_i f_i = w_i(p_i x + q_i y + r_i)$$

where

w_i output of Layer 3
p_i, q_i, r_i parameter set

Layer 6:
This layer contains only a single node which is a fixed node named as sum. The output of this node is the summation of all incoming signals.

The above ANFIS structure consists of six dimensional space partitions and has four regions (q1, q2, q3, q4) where each region is controlled by if-then rules of fuzzy logic [6, 5].

2.3 Neuro-Fuzzy System

In this approach an AUV is considered. Under water vehicles main work or the most important task is to avoid obstacles at any cost. Various methods are there to solve this problem but due to complex and dynamic nature of ocean these methods are not suitable so a method using Fuzzy Neural Network is used. To obtain the input sonar sensors are used. The basic working of the whole system is shown in the figure below.

From the Fig. 4 we can see that first the input is taken through sonar sensors which is then given to the local path planner and then the output of local path planner is given to Fuzzy neural network, also the state and disturbance info of the AUV is given to Fuzzy neural network which then give its output to the AUV. This final output is the path or the direction for the AUV to move ahead [7] (Fig. 5).

Local path planner basically means that to avoid obstacle we need to have certain data of the obstacle, i.e. the distance of the obstacle from the AUV which is obtained by the sonar sensors. Fuzzy neural network controller is used. This controller consist of fuzzy logic controller and artificial neural network.

From the diagram it is seen that the planner info is given to the fuzzy logic controller which handles the reference model to give the expected values, but the real values are not given. The output of the fuzzy logic controller is without

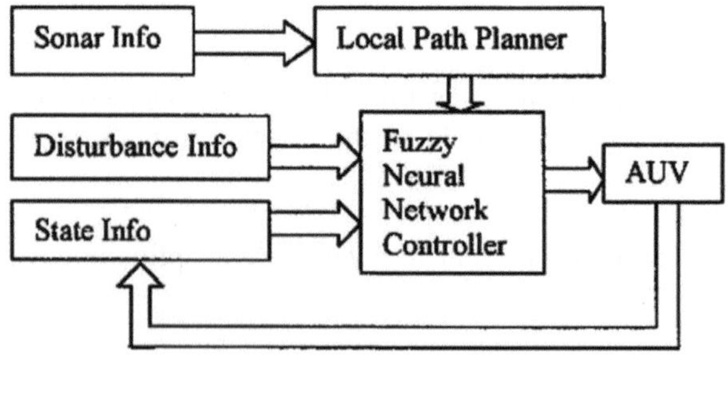

Fig. 4 Overview of whole system [7]

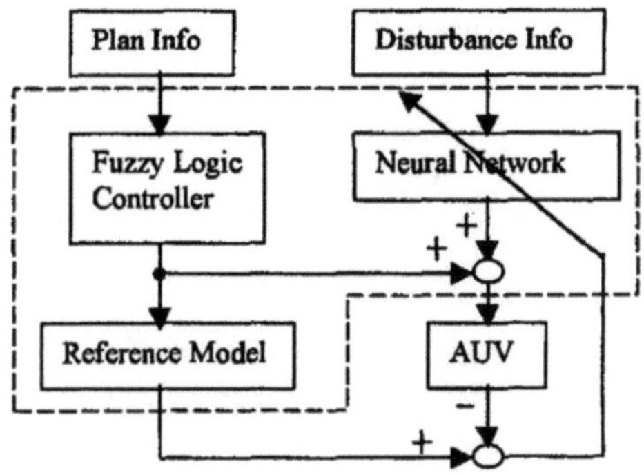

Fig. 5 Control system [7]

considering unknown or dynamic oceanic movement. For this neural network is used which has the capability of back propagation. These two together adapt well to the environment. The variation in ocean environment also known as disturbance force are considered in neural network. Thus neural network gives the real values. These real values are clubbed with expected values and an definite answer is obtained which is the desired output [7].

3 Results

The robot in the first approach is a 3 wheeled bot with dimensions (850 mm × 500 mm × 750 mm). It is built on prism platform. The steerable wheel is at the front and the on the rear side there are two passive wheels. The motors used for driving are stepper motors are used. For obtaining the input from environment ultrasonic range sensors capable of rotating from 0 to 360 are used and on the bumper 3 tactile sensors are placed. From the simulations results it can be said that the robot was able to reach the destination successfully avoiding the obstacles in a known as well as unknown environment.

The bot chassis in the second approach has dimensions (130 mm × 70 mm) and it weighs 690 gm having pay load of approx 2000 gm. The processor used is DsPIC 30F5011 at 60 MHz containing 4 kB on DsPIC and 64 MB KoreBot RAM. The motors used is 2 DC brushed servo motor. The sensors used includes 9 infrared proximity, 2 infrared ground proximity sensor, 5 ultrasonic range sensors having range of 0.2–4 m. The battery used is Lithium Polymer battery with capacity of 1400 mah. From the simulation results it is observed that The robots are able to move the surroundings using the embedded infrared sensors.

In the third approach an AUV is used to test. The main component used is Sonar sensors i.e. ultrasonic sonar sensors. The simulation results is given in. From the results it can be said that the AUV was able to avoid the obstacles using fuzzy neural network method (Table 1).

Table 1 Comparison with various scenarios [4]

S. No.	Simulation results of different methods	Length of path	Time
1.	Neural network controller	12.2	12.93
2.	ANFIS approach	6.6	7.02

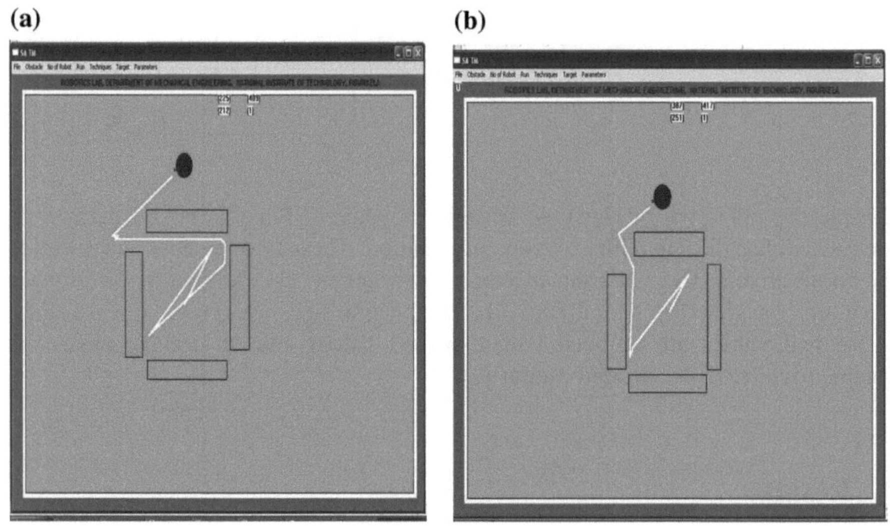

Fig. 6 **a** Navigational path using neural network. **b** Navigational path using ANFIS approach [4]

It is observed that form all the three above approaches the robot is able to avoid obstacles successfully. But in the first approach the MLP network used for second part to find free path has some disadvantages. It trains slowly and it requires more training data. So the amount of processing in the first approach will be more. From the Fig. 6a, b it can be seen that the performance of the ANFIS approach is better than neural network. The length of path in ANFIS is small than in Neural Networks and also the time is less in ANFIS. So it can be said that the first approach is less effective than the other two. The first approach is using only neural network while the other two approaches are using neural network and fuzzy logic together. So the system using neutral network with fuzzy logic are more efficient and better than system using only neural network or only fuzzy logic.

4 Conclusion

The approaches were successful in helping the robot avoid the obstacle. The system with neural network and fuzzy logic both are better than system with only neural network or fuzzy logic. The second and third approach are better than first approach. But there is scope of further optimization in the second and third approach. So there may be methods with much better performance than these which need to be found out and also further optimization techniques need to be found out to make these existing system more favorable.

References

1. D. Janglov: Neural networks in mobile robot motion. International Journal of Advanced Robotic Systems 1 (2004) 15–22.
2. Bokman Lim, S.R., Park, F.: Movement primitives, principal component analysis, and the efficient generation of natural motions. IEEE (July 2005).
3. Buragohain, M.: Adaptive Network Based Fuzzy Inference System (ANFIS) as a Tool for System Identification with Special Emphasis on Training Data Minimization. PhD thesis, Indian Institute of Technology Guwahati (July 2008).
4. Singh Mukesh Kumar, P.D.R., Kumar, P.J.: Anfis approach for navigation of mobile robots. International Conference on Advances in Recent Technologies in Communication and Computing (2009) 727–731.
5. Michael Brady, J.M.H.: Robot Motion: Planning and Control. The MIT Press (1984).
6. Xuemin Liu, Liang Peng, J.L.Y.X.: Obstacle avoidance using fuzzy neural networks. IEEE (1998).
7. S.K. Pradhan, D.P., A.K. Panda: Neuro-fuzzy technique for navigation of multiple mobile robots. Fuzzy Optimization and Decision Making (July 2006) 255–288.

Achieving Energy Aware Mechanism in Cloud Computing Environment

Komal Patel, Hiren Patel and Nimisha Patel

Abstract Cloud Computing is an emerging technology and it provides pay-per-use computing model over the Internet without giving hassle of resource management to the users. With the increasing demand and usage of Cloud applications worldwide, the issues of energy consumption, carbon emission and operational cost require special attention. Many researchers have tried to address these issues in different facets. In this paper, we first explore few research carried out in the direction of energy efficiency. We further propose (a) architecture for energy-aware mechanism in Cloud and (b) pre-processing approach by considering Virtual Machine (VM) allocation and consolidation techniques as key factors to achieving energy efficiency in Cloud Computing.

Keywords Cloud computing · Energy efficiency · Energy aware architecture · Allocation · Consolidation for VMs

1 Introduction

Availability of high-speed internet computing and large scale computational power at processor level has lead us to think a completely new model of computation viz. Cloud Computing where resources such as Processors, Memory, Storage etc. are used on rental basis and payment is made in pay-per-use form.

K. Patel (✉) · H. Patel
Computer Engineering Department, S.P. College of Engineering, Visnagar, Gujarat, India
e-mail: patelkomal8891@gmail.com

H. Patel
e-mail: hbpatel1976@gmail.com

N. Patel
Rai University, Ahmedabad, India
e-mail: nimishaa_25@yahoo.co.in

N. Modi et al. (eds.), *Proceedings of International Conference on Communication and Networks*, Advances in Intelligent Systems and Computing 508,
DOI 10.1007/978-981-10-2750-5_13

National Institute of standards and Technology [NIST] [1] categorizes the computing resources into networks, servers, storage, applications and services. These resources are provisioned to demanding users and cost to the users is calculated on the basis of the resource-usage. Further, the provisioning is elastic i.e. resources may be added if required and removed when not required. Additionally, based on the type of services offered, NIST classifies three service models viz. Infrastructure-as-a-Service (IaaS), Platform-as-a-Service (PaaS) and Software-as-a-Service (SaaS). Based on territorial deployment concerns, NIST lists four deployment models viz. Private, Public, Hybrid and Community model.

Traditionally, the concept of Cloud Computing is implemented by a group of data centers managed and operated by Cloud service providers. These data centers are equipped with large number of processing elements and huge amount of other resources such as memory, storage, and bandwidth etc. The power consumed by these computational resources is significant and has attracted the attention of many researchers. In 2012, total power consumption to the data center is around of 38 Giga Watt [GW] and that is equal to 63 % and more than the power consumption of 2011. And that is sufficient for fulfilling the energy requirement of all residential households of United Kingdom [2].

The fundamental technology which makes the idea of Cloud computing practically possible is *virtualization*. It creates virtual instances of a physical server and these instances are offered as a service to the user on a shared basis. These instances are often known as Virtual Machines (VMs). Physically, every host (sometimes known as server or node) consists of many VMs. And many such hosts comprise a data center. To address the issue of energy consumption, it is suggested to keep minimal number of host live (active or running) at any given point of time. To achieve this, one may need to migrate few VMs from one host to another based on certain criteria. Hence, VM migration can be used to address the issue of energy consumption.

In this paper, we aim to focus on the issue of reducing energy consumption by considering the communication cost between user and service provider using geographical location as one of the factors. Moreover, after selection of data center with lowest communication cost, we intend to reduce number of running hosts by applying existing VM consolidation and migration techniques.

The rest of paper is organized as follows. In Sect. 2, we explore various research carried out in the direction of energy efficiency. We also analyze strength and limitation of the approaches. We have compared few researches based on the criterion they have used to achieve energy efficiency. In Sect. 3, we present our proposed work with neat architecture, flow-diagram and algorithm. We further show the possible incorporation of VM selection and placement techniques in our proposed work. In Sect. 4, we conclude our research with mentioning our forthcoming plan of action to implement and validate the proposed mechanism.

2 Related Work

In increasing demand for internet based services, in that large amount of process as like computational data, resource management and network based communications that are significantly contribute to energy consumption. And Cloud computing is a multidirectional solution to make process and network communication easier. In [3] authors presented, the analysis a comprehensive energy consumption to consider both public and private cloud. And which includes the energy consumption in transmission and switching also data storage and data processing. And they assess three services to consume energy, namely the storage as a service; software as a service and at last service is processing as a service. In cloud switching and transmission process that represents a significant percentage to the total energy consumption in cloud computing services for storage at high and medium usage rates. These analyses tell that Cloud computing offers to save energy through mostly use virtualization and server consolidation techniques.

Energy aware cloud service provisioning approach [4] used an energy consumption model with the component part called a Trigger engine. And this engine will be used Pre-processed data [PPD] for the automatic live migration process of virtual machines to preserve the energy consumption in green cloud computing environment.

The wide recognition of cloud services and a wide spectrum and combined application can be increasingly deployed in the data center in cloud, communication in DCs is ever compact [5, 6]. As the Data center network is more complex, and switches and cables are major issues for energy consumption in data centers.

In IT companies, data centers are basic necessary for the function of communications, academic, business, government system. But data centers have concerns because they produce high energy consumption [7]. The Environment Protection Agency [EPA] report to the high amount of energy consumption is increasingly in last 5 years and their cost is $7.4 billion annually [8]. Impact of huge energy consumption for environmental is causing concern because the carbon emission of ICT is growing rapidly. In 2006, it was estimated 2 % of global carbon emission, and it equivalent to the emission of the industry [9]. In 2007, Carbon emission is 14 % of ICT for data centers and this is projected to be 18 % in 2020 [10].

The majorities of research work it provides a range of hardware and software solution the problem of energy consumption and minimize the carbon emission in the cloud. Turning on and shutting down the servers and putting them to sleep are some simple method for energy saving and that used for servers in clouds. In that use two common and popular technique are Dynamic Voltage and Frequency Scaling (DVFS) [11] and second is Dynamic Power Management (DPM) [12]. DVFS technique use for arrange the CPU power accordingly there offered load and use virtualization techniques for better resource utilization. Virtualization technique is reducing the energy consumption using live migration [13] and resource consolidation techniques. And DPM scheme use for down the power for whole servers,

Table 1 Comparison of energy Models

Author/year component	Arthi T/Shahul Hamead H 2013	Bharti wadhwa/Amandeep 2014	Ting Yang/Young Choon Lee 2014	Anton Beloglazow, Rajkumar Buyya 2012	Fransic Owusu/Colin Pattinson 2012
Compute [server]	✓	✓	✓	✓	✓
Allocation and migration		✓	✓	✓	
CPU utilization	✓	✓	✓	✓	✓
VMLivemigration	✓				
VM scheduling algorithm		✓		✓	
Energy consumption	✓	✓	✓	✓	✓

and also include their memory, buses and disks, which is make a technique more energy efficient.

The movement of virtual machines between physical nodes in data centers, and it is enables to dynamic migration of VMs according to the requirements of performance. When provided resources are not use VMs that time VMs can be resized logically and consolidate minimum number of physical nodes. While idle nodes can be turn off in sleep mode to eject them and reduce the total energy consumption for data center [14]. In heterogeneous infrastructure the formulated algorithm can be used for VM allocation and migration process. These techniques are not depend on type of VM request and not required application information and which is run on that VMs. And that can be handle strict SLA and heterogeneous VMs.

In virtual machines scheduling algorithm [15] presents a combination of the two methods. Allocation algorithm used for allocate the jobs and a migration algorithm for optimal migration of VMs. These techniques use linear integer program and that is noticeable and significant amount of energy save and it is depend on the load of the system. In [16] proposed Energy and Carbon Efficient VM Placement and Migration techniques, is use the VM allocation and migration process in federated cloud datacenters and also consider carbon footprint rate of different datacenters.

We see the various research techniques to attain the energy consumption and reduce power consumption. But some researcher tried to only reduce the energy consumption for different data centers to considering only allocation and migration techniques, but they do not reduce the communication cost. So we tried in our work to reduce the communication cost between users and service providers in cloud that will be reduce the energy consumption.

We proposed a new approach, called Energy aware mechanism in Cloud computing Environment, which is unique from the above described techniques as we have tried to optimize the problem of energy consumption and carbon emission in the cloud data centers having Pre-processing phase in that also use allocation and migration techniques to achieve maximum energy efficiency based on geographical location. Table 1 describes comparison of energy model and their approaches will be described above.

3 Proposed Approach

In this section, our contribution is Energy Aware Architecture to related cloud computing environment. And this is differ from the above mentioned techniques and we have tried to optimize the problem of Energy Consumption in the cloud data centers having Pre-Processing Phase. And this phase uses the allocation and migration technique to achieve maximum energy consumption (Fig. 1).

The cloud based services use the users via a service request processor. Users data and organization software data are store on server site based on remote location. Service request processes are manage (submission and handle) all requests for service to the server. Customized the processes to call the customer needs, and this

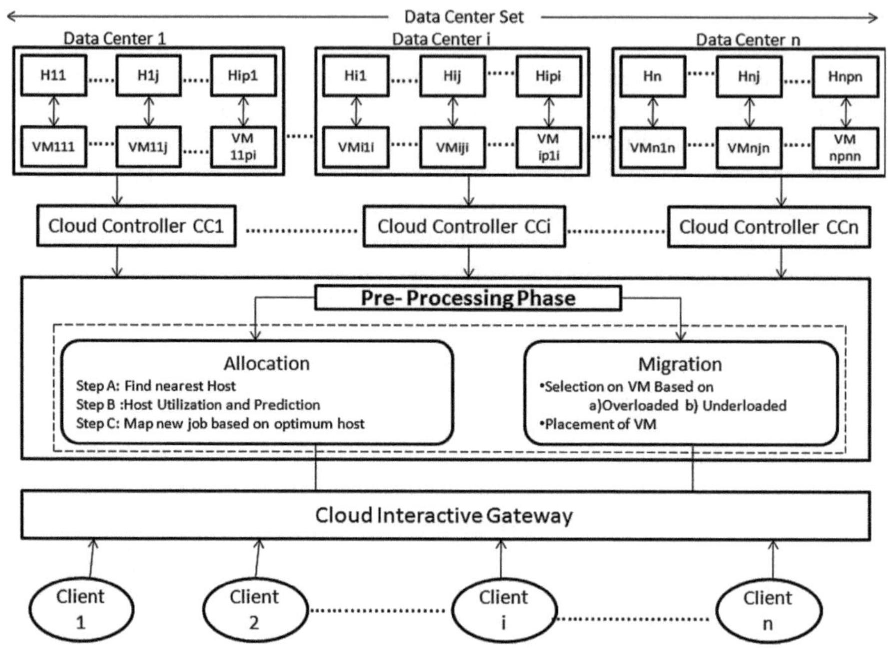

Fig. 1 Energy aware architecture [4]

process is depend on type and complexity of the customer request. An administration tool is cloud controller, and it maps the resources like services, storage to server node. And it includes centrally data and take the decision correspondence.

In server virtualization process a many users share a single server in that process, that why increase the service utilization and that time reduce the number of server and it is required for processing. And also user not need and not required any information of the task and workloads for performed by another user, utilize the server when task will be complete and they are only user on that server. During the time of less requirement server enter into sleep mode that time reduce the idle servers and reduce the energy efficiency. The Energy Consumption to reflect to profit utilization of server, consolidation of Pre-processing data is performed. Figure 2 presents our proposed mechanism.

In Pre-Processing Phase the allocation process can be designed into three phase as follow:

(1) First phase: In step (A) Fig. 2, each data centers is the distributed cloud architecture. And we find the nearest host and data centers for based on their Geo-Graphical location.
(2) Second phase: In step (B) Fig. 2, based on selected host in step A Calculate the (B.1) Current Host Utilization, for better resource utilization. (B.2) Predicted Load. Predict the host loads of every data center using the two prediction algorithm are (B.2.1) Short-Term Prediction and second is (B.2.2) Long-Term

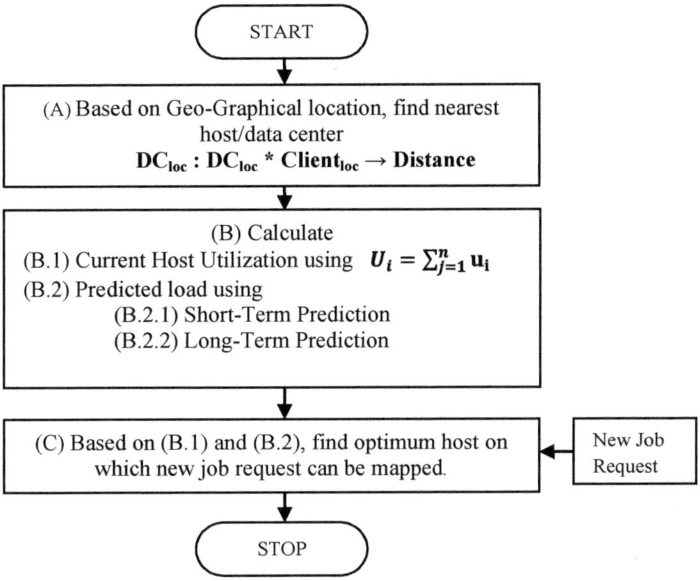

Fig. 2 Proposed mechanism

prediction. And selection of right prediction is depending on two factors (1) Time of the prediction and (2) Type of the data. And in computer system short-term prediction is the range of minute or hours and long-term prediction is the range of days, week or month.

(3) Last phase: To find optimum host in step (C) Fig. 2, we use the following equation to calculate threshold values for every host selected in step B. $T = \{T_{H1}, T_{H2}, T_{H3} \ldots T_{Hn}\}$

$$T_H = (1 - \lambda)C + \lambda P \quad (1)$$

where

T_H Threshold value
λ Constant ranging from 0 to 1
C Current host utilization
P Predicted host utilization

To find optimum host, we believe that the value of threshold should vary between two bounds viz. upper bound and lower bound. Host with threshold less than lower bound tend to go underutilized and host with threshold more than upper bound tend to go overutilized. Hence, we selected only those host whose threshold lies between these ranges:

$$\text{Optimum Host} = \text{MIN}_{T_{lower}}^{T_{upper}}\{T\} \quad (2)$$

Where set of threshold values for each selected host T = $\{T_{H1}, T_{H2}, T_{H3}\ldots T_{Hn}\}$

T_{upper} Upper bound threshold value
T_{lower} Lower bound threshold value

We further assume that detection of overloaded or underloaded hosts may be carried out using already available techniques. Based on this detection, selected VMs can be migrated using one of the existing VM migration approaches:

1. Selection of Virtual Machine
2. Placement of Virtual Machine based on Selected VM

1. Selection of Virtual Machine
   For current virtual machine the optimization process for allocation is work on two steps: (1) Select the VMs that need to be migrated, (2) Choose VMs and put on the hosts use the MBFD algorithm in [18]. When and which VMs can be migrated that is depend on the policies for VM selection. Basic thing is define the upper and lower utilization threshold values for hosts and keep the total utilization for CPU, and on host all VMs allocated between threshold values. Selection process will be use Minimization of Migration (MM) policy.
2. Placement of Virtual Machine based on Selected VM
   Virtual Machine allocation process problem can be divided in [17] two part: the (1) Insert a new requests for VM provisioning and placing the VMs on hosts, (2) Optimization of the current VM allocation. In modification, the Modified Best Fit Decreasing (MBFD) algorithms, short all VMs in decreasing order of their current CPU utilizations, and every VM can be allocate host and that supply the least of increasing order of energy consumption due to allocation. This is allowing for leveraging the heterogeneity of resources chooses based on energy efficient aware nodes first.

4 Conclusion and Future Work

In this research, we identified energy consumption as one of the key issues to be addressed and we analysed various approaches leading toward energy efficiency in cloud environment. Broadly divided into two phases, we select nearest data center to reduce communication cost in first phase for selection of hosts. In second phase, using appropriate VM selection and migration techniques, we have minimized number of active running server to reduce energy consumed. Our research is at elementary level and in future, we wise to implement the proposal and validate the results against our claim.

References

1. P. Mell and T. Grance.: The NIST definition of cloud computing (draft), NIST special publication. vol. 800, p. 145.
2. R. Brown et. al.: Report to congress on server and data center energy efficiency: Public law 109-431, Lawrence Berkeley National Laboratory, 2008.
3. Green cloud Computing. Balancing Energy in processing, Storage and transport. Jayant Baliga W.A. Ayre, Kerry Hinton, and Rodney S. Tucker, Fellow IEEE, January 2011.
4. Arthi T, Shahul Hamead H.: Energy Aware Cloud Service Provisioning Approach For Green Computing Environment, in International Conference on Energy Efficient Technologies for Sustainability(ICEETS), Nagercoil, April, 2013, pp. 139–144.
5. Y. Han, S-S. Seo, C. Hwang, J-H. Yoo, and J.W.-K Homg.: Flow-level traffic matrix generation for various data center networks. Network IEEE Operations and Management Symposium (NOMS), pp. 1–6, 2014.
6. O. Fatmi, D. Pan.: Distributed multipath routing for data center networks based on stochastic traffic modeling, in Proceedings of the 2014 IEEE 11th International Conference on Networking, Sensing and Control (ICNSC), pp. 536–541,2014.
7. J. Loper and S. Parr. Energy efficiency in Data Centers: a new policy frontier. January 2007. [online]. Available: http://www.fypower.org/pdf/ASE_DataCenter_EE.pdf. [Accessed 10 May 2011].
8. U.S. Environmental Protection Agency.: Report to Congress on server and Data Center energy efficiency. U.S. Environmental Protection Agency, Washington, 2007.
9. Kaplan, James M; Forrest, William; Kindler, Noah.: Revolutionising Data center Efficiency. McKinsey and Company, London, 2008.
10. The Climate Group.: SMART 2020: Enabling the low carbon economy in the information age. The Climate Group, New York, 2008.
11. L. Shang, L.-S. Peh, and N. K. Jha.: Dynamic voltage scaling with links for power optimization of interconnection networks, in the 9th International Symposium on High-Performance Computer Architecture (HPCA 2003), Anaheim, California, USA, 2003, pp. 91–102.
12. L. Benini, A. Bogliolo, and G. De Micheli.: a survey of design techniques for system-level dynamic power management. IEEE Transaction on Very Large Scale Integration(VLSI), San Jose, CA, April 2010.
13. C. Clark, K. Fraser, S. Hand, J. G. Hansen, E. Jul, C. Limpach, I. Pratt, and A. Warfield.: Live migration of virtual machines. In The 2nd Symposium on Networked Systems Design and Implementation (NSDI 2005), Boston, Massachusetts, USA, 2005, pp. 273–286.
14. Anton Beloglazov, Jemal Abawajy, Rajkumar Buyya.: Energy-aware resource allocation heuristics for efficient management of datacenters for Cloud computing. In Future Generation Computer Systems, 2012, pp. 755–768.
15. Chaima Ghribi, Makhlouf Hadji, Djamal Zeghlache.: Energy Efficient VM Scheduling for Cloud Data Centers: Exact allocation and migration algorithms. In 13th IEEE/ACM International Symposium on Cluster, Cloud, and Grid Computing, 2013, pp. 671–678.
16. Bharti Wadhwa, Amandeep Verma.: Energy saving approaches for Green Cloud Computing: A review. In Proc. of Recent Advances in Engineering and Computational Sciences, March 2014, pp. 1–6.
17. Anton Beloglazov, Jemal Abawajy, Rajkumar Buyya.: Energy-aware resource allocation heuristics for efficient management of data centers for Cloud computing. ELSEVIER 2012.
18. Anton Beloglazov, Jemal Abawajy, Rajkumar Buyya "Energy-aware resource allocation heuristics for efficient management of data centers for Cloud computing" ELSEVIER 2012

Information Security Emergency Plan Management System

K. Lingaraj, N. Sreekanth, Moddiudin Kaja, K.M.S. Lokesh, Keni Prashanth and V. Biradhar Nagaveni

Abstract Aiming at such problems as decentralized management, difficult querying and modification, poor information sharing, etc. of the current Information Security Emergency Plan Management (ISEPM), ASP.NET MVC design mode has been applied for designing and developing the Web-based ISEPM in realizing formulation, auditing and publicity for plans via workflow.

Keywords Emergency plan · ASP.NET MVC framework · Workflow

1 Introduction

In ISEPM, different plans are required to be formulated according to different types and levels for emergencies, and scientific management shall be adopted catering to all kinds of plans and files resources formed by plan practice [1]. Plan Management includes Plan Information Management, Plan Resources Sharing, Plan Upgrading and Updating, etc. With the upgraded information system, effective integration and

K. Lingaraj (✉) · N. Sreekanth · M. Kaja · K.M.S. Lokesh · K. Prashanth · V.B. Nagaveni
Rao Bahadhure Y Engineering College, VTU, Belagavi, Karnataka, India
e-mail: lingaraj.k10@gmail.com
URL: http://www.rymec.com

M. Kaja
e-mail: kaja_mouni@gmail.com

K.M.S. Lokesh
e-mail: lokesh.kms@gmail.com

K. Prashanth
e-mail: keniprashanth@gmail.com

V.B. Nagaveni
e-mail: nagaveni.veer10@gmail.com

N. Modi et al. (eds.), *Proceedings of International Conference on Communication and Networks*, Advances in Intelligent Systems and Computing 508,
DOI 10.1007/978-981-10-2750-5_14

management shall be carried out for plans together with the upgraded resources, which is beneficial for taking in-time call in commanding process for accident rescue; in this way, emergent, modified and formulated new plans could be coped with. At the same time, the corresponding scheduled practices shall be done, and the plans require to be modified according to the practicing process and the results so as to effectively complete emergency coping works at different phrases. Procedural and standardized Emergency Plans are the foundation for the fast and accurate Information Security Emergency Accident handling; which means, standardized documents for offering operation guidelines shall be worked out before ahead accidents happen so as to make scheduling and operation staff at the place in time. At present, these related regulations and codes are all saved in forms of texts, which are in great chaos, and inconvenient for querying and searching, difficult for modification and updating with poor information sharing [2]. In order to adapt the requirements raised by the modernized management on ISEP, a unified platform in taking management for the current operation management codes and emergency plans boasts great significance. In this Paper, ASP.NET-based ISEPMS has been designed and developed in terms of the classification for emergent accidents, combining the required features for emergency plan handling, management and application; taking emergency plans, equipments and emergency resources as the application objects and adopting such technologies as framework and workflow in purpose of standardizing the business process for handling Information Security Emergencies and improving the management level for Information Security Emergency Plans.

2 System Design

2.1 General

ISEPMS is designed to quickly and accurately start plans in case of emergent accidents occur, effectively coordinate strengths from all parties, make orderly emergency operations, control the accident developments and try all means to minimize the losses and influences rested on human beings, properties and environment. The ISEPMS designed in this Paper possesses the following features:

Procedural workflow the editing, auditing and starting works are all under-taken by workflows, which truly simulate the real workflows.

Framework-based development the most advantage for framework is to attach emphasizes, while the obtained largest multiplexing method oriented for object system is the framework.

Personalized customization and visit control role-based visit control technology has been applied so as to realize the personalized customization for users and the visit control for information access.

2.2 System Functions

Based on the requirements for Plan Management and considering related extended functions and flexibilities, Plan Management System is systematically divided into 4 sub-systems, while each sub-system is composed of several function modules, as shown in Fig. 1.

Within which, Plan Management Sub-System is available for online compilation, approval and starting for plans, displaying the current visible state for the process and the actual operation condition of the procedures; related information for publicized plans could be inquired by classified Division, Plan Type, Plan Name, Release Time or Key Words; information is allowed to be browsed and printed by such functions as Limited Access for Browsing, Printer Permissions, Reading and Printing; plan browsing and printing will be automatically kept by the System. For guidelines and instructions of accidents, special support for PDA is rendered by Plan Inquiry Module for searching accident on-site emergency plans. File Service Sub-System is available for document (for instance: codes, operation animation and demonstration) uploading, downloading and browsing, which is convenient for document delivery and sharing. Visible maintenance and management for system data are available via System Maintenance Sub-System; furthermore, updating and leveling up are also available for the system. Relying on Equipment Management, Resources Management, User Management Modules, adding, inquiring, modifying and deleting are applicable for equipment, resources and user, and assignment for user role and permission. Static Analyzing Sub-System provides comprehensive inquiring and statistic for accident information, includes: statistics according to accident type, accident level, statistics based on unit; the Sub-System is available for analyzing security accident developments and allows exporting and printing statistic analyzing reports.

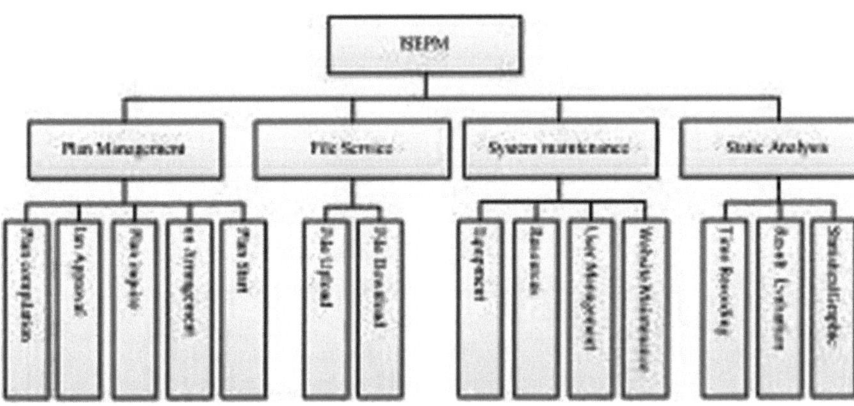

Fig. 1 Function module structure of ISEPMS

2.3 Structure Design

The wide-spread popular ASP.NET MVC framework has been applied for system development and design. In MVC mode, the model encapsulates core data, logical relations and business rules, provides the processing process for business logic. On the one hand, the model is called by controller, completing the operation process for problem handling; on the other hand, the model provides access data for view in obtaining displaying data. Since there has no relation between model and data format, a model could provide data for several views; in this way, after the model been complied for once, it could be reused by multiple views in prevention from re-compiling code. View is an interface under the MVC mode observed and interacted with users, it does not includes the handling for any business logic, it has been taken as an approach for outputting data. In MVC mode, controller works as a navigator, it calls the corresponding models and views according to the input made by users and completes the requirements of users. Controller does not output anything; it accepts users requirements and makes decisions on using models for handling problems and using views to display the feedback data after models were processed. Based on ASP.NET MVC framework, the system structure for PMS falls into 3 layers, as shown in Fig. 2.

Presentation Layer, this Layer is mainly engaged in providing logical views and completing the interaction function between the interface and users, covers information and data inputting and displaying. And this Layer is also divided into 2

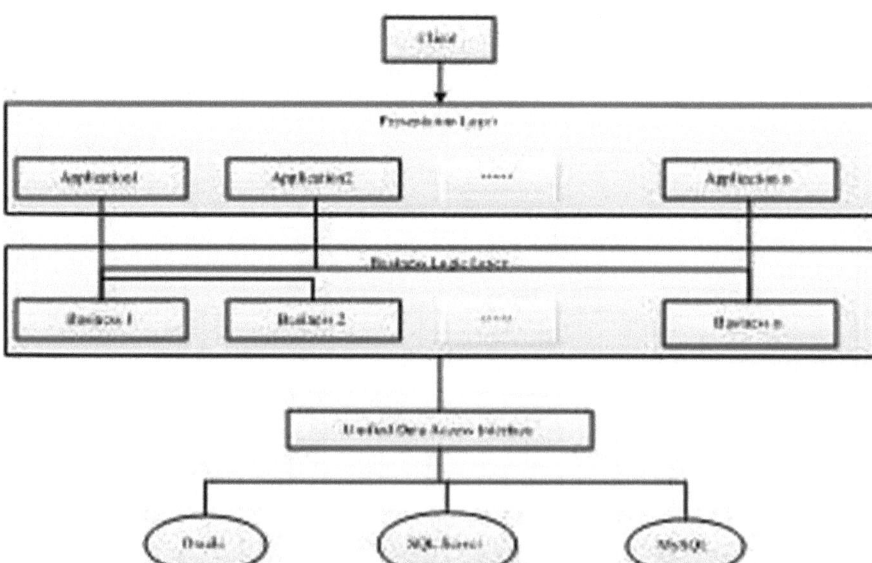

Fig. 2 The system structure for ISPMS

parts: View and Controller. Business requirements are accepted by Controller from the browser at client side, and the requirements are resolved and converted to be Model input parameters; after then, when the businesses were completed, the Controller calls the corresponded View in accordance with the handling results made by Model, generates corresponding pages and backs to the browser at client side. In application program, there are some aspx server page code and some C module codes. Business Logic Layer, this Layer is corresponded to Model part in MVC mode; the main responsibility is to encapsulate and access the lower database, encapsulate a certain business handling logic and provide the corresponded interface for user interface later. Data Access Layer, integrates all kinds of data resources distributed in different databases via data integration middleware so as to provide the unified access interface for Business Logic Layer.

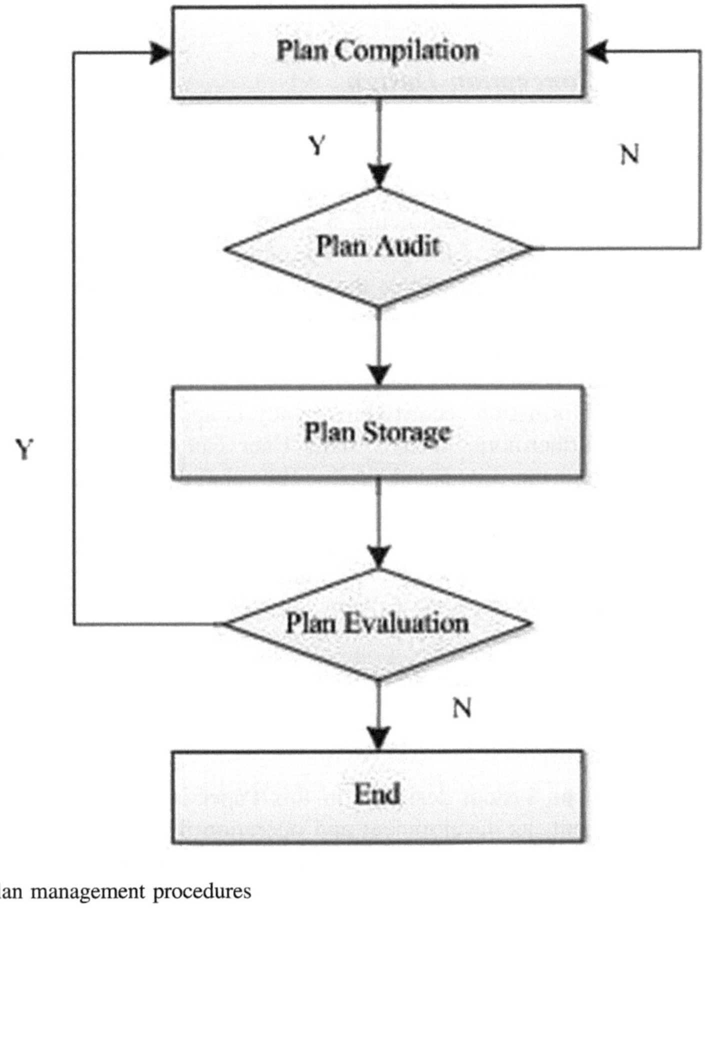

Fig. 3 Plan management procedures

2.4 Emergency Plan Management Procedures

Compilation, audit and start for plans are managed by the system in an approach of workflow. There are altogether 4 phrases for this kind of plan: compilation, audit, storage and start. Audit phrase is followed with plan compilation; the auditor attaches permission for the plans and adds the plans into emergency plan base in case of inquiring. The plans are triggered when accidents happen or there are practices, modifications will be given to these plans upon the evaluations after the accidents were settled or the practices were finished; in case that modification is required, compilation phrase will be back to, as shown in Fig. 3.

3 Database Conception Design and Run Environment of System

3.1 Database Conception Design

According to module analysis of the system, as for the settlement for emergent accidents in Emergency Aid Management System, the information saved in the database mainly includes the following 5 kinds of entities: Emergency Plan, Information Security Event, Emergency Resource, System User and System Operation Limit. The relation among these entities is: Each System User has multiple operation limits in the system in accordance with the limits distributed, and the User is available to formulate several emergency plans in terms of the limits; each emergency plan is able to be applied for several information security emergencies, and each information security emergency is applicable for several emergency resources; furthermore, each System User can call several emergency resources according to the distributed limits. The relation among these entities is shown in Fig. 4.

Therefore, targeting to these 5 entities, there are corresponding system operation limit table, system user table, emergency plan table, information security emergencies table and emergency resources table, altogether 5 data sheets.

3.2 Development and Operation Environment

The Plan Management System designed in this Paper is development based on ASP.NET environment; its development and operation firstly requires the editing environment of Microsoft Visual Studio 2005 and IIS application server; in addition, database server (such as Oracle, SQL Server, Sybase, etc.), and the System in the Paper applied SQL Server 2000 Database. Reliability is the key for ISEPMS; only reliable technologies and equipments could play the advanced functions and

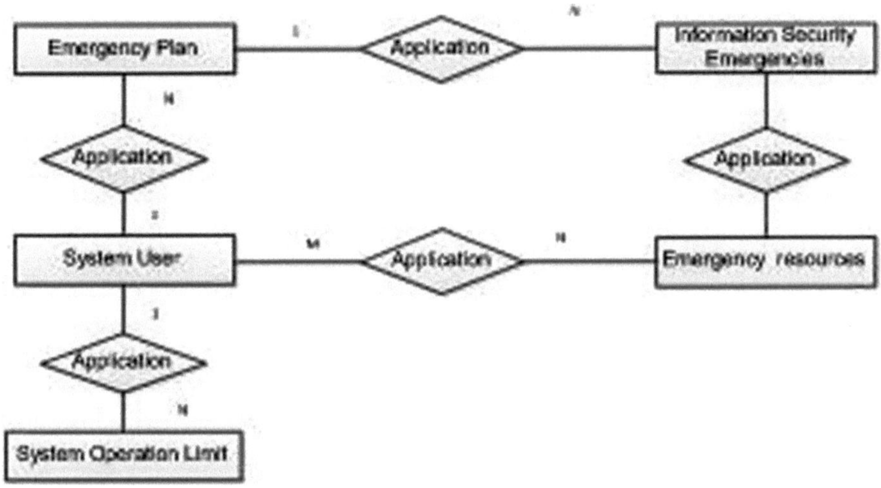

Fig. 4 Plan management procedures

advantages into full play. As a result, related international and Chinese standards and specifications were strictly abide by for system design and equipment selection, the server is IBM System X3850 considering about the maturity of the technologies and equipments; disk array was adopted for data backup, and Windows Server 2003 has been chosen as the operation system.

3.3 Login Screen

System Function Interface is for the interaction with users and finally attains for using requirements. Complete and logical system is able to bring more convenient and quick processing for users, and this is also the advantage for computer system that superior to manual artificial processing. The login screen for the system is shown in Fig. 5; before login the system, file configuration for app.config is needed and connected to the server deployed by database. In order to ensure the system safety, userIsExits has been applied so as to check users identity; the login is valid upon the true return value.

3.4 Main Interface

The main interface is available to be realized through programming, and the operation result could be yielded as shown in Fig. 6. By resorting to this system, the System User could quickly obtain the accurate Information Security Plan and make the plan to be oriented and time-based one so as to quicken the aid.

Fig. 5 Login screen for ISEPMS

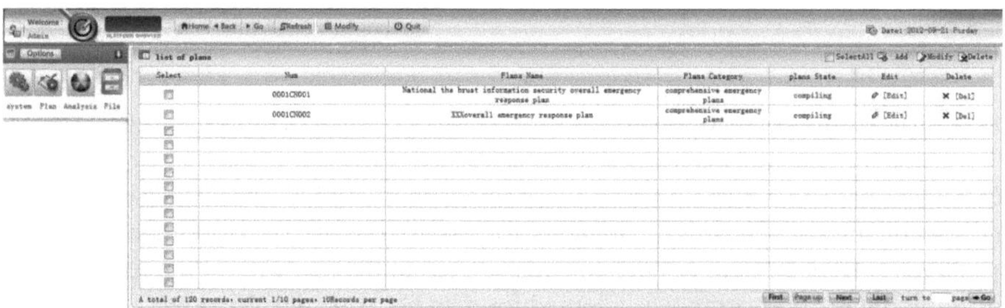

Fig. 6 The main interface for ISEPMS

4 Conclusion

Emergency plans are managed and handled for the system by means of Web; therefore, it is available for sharing resources, adding module for emergency plan, standardizing management and ordered procedures; in this way, appraisal and statistics could be available for emergency plans; classifying and related organizing are completed for information security emergency plan; the management is available for emergency plans; the emergency plan using becomes more informative and is supported with graphs. Finally, the management level for emergency plans is leveled up. Further researches and improvements are still needed for this System. Actually, ISEPMS is a complicated digital plan management system, providing functions in multiple aspects, not only covers the functions researched in this Paper. We still have gaps to be narrowed: research on data mining applied in emergency plan management system. The emergency plans in the system aim to summing up successful experiences and experiences learned from difficulties after handling schemes for emergencies. By making use of these data plus data mining, potential, effective and novel knowledge and rules could be obtained, which could provide assisted decision for plan formulation or emergency handling in the future.

References

1. Government Network- Emergency Management [EB/OL], http://www.gov.cn/yjgl/flfg_shaq.htm.
2. Hailong Zhang, Xiongfei Li, Liyan Dong.: A Study on Appraisal Method for Emergency Plan. Safety Science Journal. 147 142–149 (2009).
3. Yudong Liu, Enlai Zhao, Ya Huang, Jihui Shen.: A Study on Workflow Model of Emergency Management Based on Petrix. In: Computer Knowledge and Technology, (2010), pp. 4368–4370, (2010).
4. Zongping Lv, China's Civil Aviation University Journal: A Study on Appraisal Method for Emergency Plan. Safety Science Journal. 30–32 (2010).
5. Y. Grathwohl, F.de Bertrand de Beuvron, F. Rousselot.: A new application for description logics: disaster management. In Proc. Of the International Workshop on Description Logic 99, Linkoping, Sweden, (2009), pp. 46–49.
6. Lin, Cuimiao Zhu, Guangchang Zheng, et al.: A Study on MVC Design Mode Based on ASP. NET, Computer Engineering and Design, pp. 167–169. 30–32 (2008).
7. L Danni Zhai: A Study on Constructing Digital Plan System on Emergency Platform, China Public Security, pp. 67–169, (2008).

References

Reliability-Aware Workflow Scheduling Using Monte Carlo Failure Estimation in Cloud

Nidhi Rehani and Ritu Garg

Abstract Cloud Computing is a novel paradigm which offers large-scale resources and services through the Internet. These services are supported by huge data centers with thousands of servers. One of the core issues in Cloud is the proper utilization of computation power. Efficient task scheduling can help utilize the cloud resources up to their capacity. Moreover, in real-world scenarios, it is important to consider the reliability of computation resources at the time of scheduling since the failure of tasks can be critical to both the cloud service provider and the user. In this paper, we proposed a Cloud computing framework to model the failure characteristics of a cloud environment. We developed a Monte Carlo Failure Estimation (MCFE) algorithm that considers Weibull distributed failures in cloud, using Monte Carlo simulation method to determine the probable occurrence of failures and a Failure-Aware Resource Scheduling (FARS) algorithm that considers the reliability of task execution while assigning tasks in a workflow application to virtual machines. In order to analyze the performance of our algorithm, we compared it with the popular scheduling algorithm namely HEFT. For simulation analysis, randomly generated task graphs and task graphs for numerical real world problems like Gaussian Elimination (GE) and Fast Fourier Transformation (FFT) were considered. The simulation results show that the proposed algorithm performs better in real world scenarios where reliability is a critical issue.

Keywords Cloud computing · Reliability · Workflow scheduling · Monte Carlo simulation · Failure aware resource scheduling

---

N. Rehani (✉) · R. Garg
Computer Engineering Department, National Institute of Technology, Kurukshetra, Haryana, India
e-mail: nidhirehani@gmail.com

R. Garg
e-mail: ritu.59@gmail.com

N. Modi et al. (eds.), *Proceedings of International Conference on Communication and Networks*, Advances in Intelligent Systems and Computing 508,
DOI 10.1007/978-981-10-2750-5_15

1 Introduction

The vast development in technology today has given rise to a novel paradigm Cloud Computing, which offers resources and services to various users through the Internet on a pay-as-you-go model. These services are supported by a large infrastructure that consists of huge data centers equipped with thousands of servers and other equipments. The services offered can be classified as Infrastructure-as-a-service (IaaS), Platform-as-a-service (PaaS) and Software-as-a-service (SaaS) [1]. The cloud providers responsible for delivering these services to the users can be public, private or hybrids. The U.S. National Institute of Standards and Technology (NSIT) [2] defines Cloud Computing as a model for enabling ubiquitous, convenient, on-demand network access to a shared pool of configurable computing resources (e.g., networks, servers, storage, applications, and services) that can be rapidly provisioned and released with minimal management effort or service provider interaction.

The powerful computational resources offered by cloud are used to execute various scientific, computation-intensive precedence constrained tasks (workflow application) submitted by the users. Efficient scheduling of these tasks in the workflow application helps us to utilize the cloud resources up to their potential. Workflow Scheduling is to map heterogeneous computational tasks onto the available resources such that the precedence requirement for various tasks is satisfied along with the aim to optimize the execution time of the application. The problem is NP-Hard [3] in nature and thus requires the use of heuristics, approximations or meta-heuristics in order to achieve a near optimal solution. It becomes a challenging research area in cloud because of the added virtualization, the diversity of tasks involved and the on-demand nature of allocation of resources. Effective workflow scheduling can help to improve the performance of the application and the reliability of its execution. Researchers have proposed various workflow scheduling approaches for the allocation of user submitted tasks to distributed resources. Some popular workflow scheduling algorithms include Heterogeneous Earliest-Finish-Time (HEFT) algorithm [4], Critical-Path-on-a-Processor (CPOP) algorithm [4], Min-Min algorithm [5], Reliability-aware scheduling algorithm with Duplication (RASD) algorithm [6], Hierarchical reliability-driven scheduling (HRDS) algorithm [7] etc. HEFT algorithm is one of the most popular scheduling algorithms defined for heterogeneous computing systems. It calculates an upward rank for each dependent task in the application and then assigns these tasks to the available processors in the order of non-increasing upward rank such that the earliest finish time for each task that is allocated is minimized.

However, work on providing reliable allocation of cloud resources remains limited. The failure of resources and the corresponding tasks executing on them can lead to severe financial losses for both the cloud providers and the users. Thus, it is crucial to consider the reliability of resources while provisioning as it can help improve the performance of the application. In order to incorporate the failure of cloud resources into workflow scheduling, it is important to understand the fine grained failure and repair characteristics of cloud resources. Garraghan et al. [8], attempted to understand the failure characteristics of a large-scale cloud environment. They analyzed the Google Cloud Trace Log in order to determine the failure and repair characteristics of cloud servers and workloads, consisting of over 12,500 servers spanning over 29 days of operation. The failure of cloud servers is found to follow Weibull distribution whereas large variance is found in case of repair characteristics. Fiondella et al. [9], analyzed the cloud incidents reported in cloutage.org database and categorized the failures as outage, vulnerability, auto fail, data loss and hack.

The existing literature for estimating the reliability of cloud resources mainly considers Poisson distribution for estimating the failure of resources [6, 7, 10, 11].

According to Poisson distribution, failure rate is constant with a specific mean time between failures. Poisson distribution simplifies the failure scenario in cloud environment. Current research on cloud shows that cloud resources follow Weibull distribution for the occurrence of failures. Therefore, we consider Weibull distributed failures in our work for cloud resources. Weibull distribution is a generalization of Exponential distribution that takes into account the age of the system while calculating its failure probability.

In order to analyze the reliability of complex systems, various modeling tools are present in the literature, such as Markov analysis, Bayesian networks, Monte Carlo simulation etc. The Monte Carlo Simulation (MCS) method for analyzing a system to obtain the future patterns of its operation has gained popularity in recent years [12]. It is a computation-intensive, statistical method that has been employed in various engineering and scientific domains. We used the Monte Carlo simulation method because of its ability to correctly model a complex system and produce results highly close to the reality of complex system operation. Monte Carlo Simulations can also be parallelized easily to reduce the computing time by using a split and merge pattern for parallelization [13].

In this paper, we developed a cloud model to simulate the cloud environment using Monte Carlo Simulation (MCS) method with Weibull distributed failures. We determine the future states of the virtual machines present in the system so as to incorporate the failure of nodes into resource scheduling, leading to a more robust schedule with very less chances of occurrence of any failure, which ultimately leads to a better overall makespan (execution time) of the application and a high reliability of task execution.

2 Cloud Computing Framework and Problem Statement

2.1 Cloud Computing Architecture

Figure 1 represents our cloud computing architecture, in which we consider a compute cloud that consists of clusters of various virtual machines present in different data centers geographically distributed over a large area. A *Cloud*

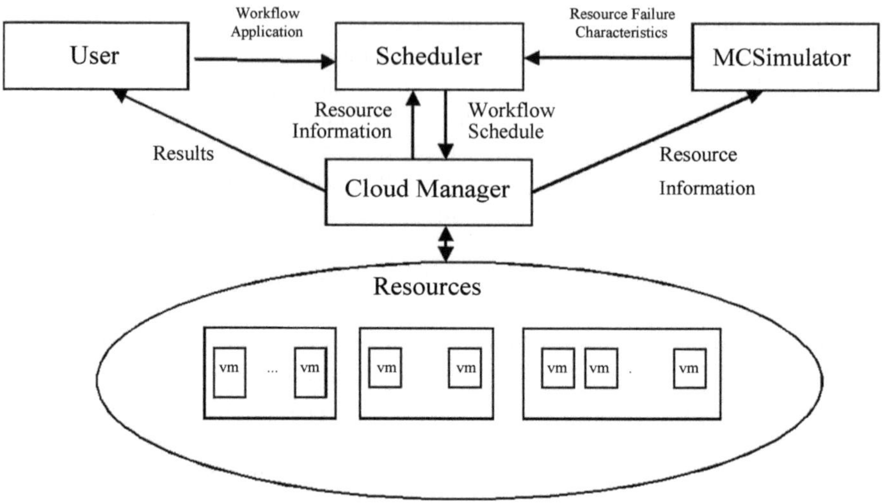

Fig. 1 Cloud computing architecture

Manager works in coordination with a *Scheduler* and a *Monte Carlo Simulator* (*MCSimulator*) in order to schedule tasks in the workflow application submitted by the user to a set of virtual machines in the data centers available with the cloud. The *MCSimulator* takes the resource information submitted by the *Cloud Manager* in order to simulate the cloud environment i.e. the failure behavior of various virtual machines present in the cloud. This failure information is passed by the *MCSimulator* to the *Scheduler*. Whenever the *user* submits a workflow application, the *Scheduler* uses the information provided by the *MCSimulator* and the *Cloud Manager* in order to generate an appropriate workflow schedule for execution. This schedule is then used to allocate various tasks in the application to the corresponding computing nodes by the *Cloud Manager*, such that the overall makespan for the execution is minimized, while taking into consideration the probable failures of various computing nodes present in the data centers.

2.2 Cloud Computing Model

In our cloud computing model, we considered a set of virtual machines present in computation servers distributed across different geographical areas. The set of virtual machines is represented as $vm \in V, (1 \leq i \leq m)$, where m is the total number of virtual machines present in the cloud. The computing speed of the virtual machine vm_i, expressed in MIPS (millions of instructions per second), is represented as $speed_i$. The bandwidth linkage between any two virtual machines is considered to be uniform. It is represented as bw, expressed in Mbps.

2.3 Workflow Model

A workflow application consists of a set of parallel, precedence-constrained tasks which can be represented by using a Directed Acyclic Graph (DAG). In this paper, we represent the workflow application submitted by the user as (DAG) $W = < T, E >$ that consists of a set of n tasks $t_i \in T, (1 \leq i \leq n)$. The weight of the task t_i, represented as $w(t_i)$, is the computation requirement of the task, expressed in millions of instructions (MI). The dependencies between the tasks is represented as a set of edges $E = e_{i,j}, (1 \leq i \leq n, 1 \leq j \leq n, i \neq j)$, where $e_{i,j}$ represents a dependency from task t_i to t_j. The dependency $e_{i,j}$ means that task t_i must be completed before the execution of task t_j can begin. The weight of an edge, $w(e_{i,j})$, is the data transfer requirement from task t_i to the dependent task t_j, expressed in MB. All the immediate parents of a task t_i are denoted as $parent_i$ and all the immediate children of task t_i are denoted as $child_i$. We assume that the workflow has only one starting task and one exit task. The parent set for the starting task and the child set for the exit task is considered to be empty.

In order to represent a workflow application, an example DAG is shown in Fig. 2 that consists of 10 tasks. The weight of the node represents the average execution time of the task. The dependencies between the tasks are shown with the help of edges, and the weight of the edge $e_{i,j}$ represents the data transfer time ($DTT(t_i, t_j)$)for the task t_i and t_j.

The execution time of a task t_i on virtual machine vm_j, represented as $ET(t_i, vm_j)$, is the time taken to execute the task t_i on virtual machine vm_j, given as:

$$ET(t_i, vm_j) = \frac{w(t_i)}{speed_j} \tag{1}$$

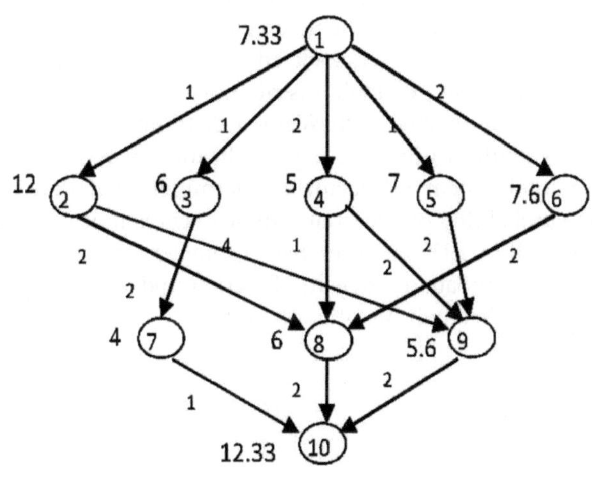

Fig. 2 An example workflow application

The Data Transfer Time (*DTT*) between two dependent tasks t_i and t_j, represented as $DTT(t_i, t_j)$, is the amount of time needed to transfer the required data between dependent tasks over the communication channel and can be computed as:

$$DTT(t_i, t_j) = \frac{w(e_{i,j})}{bw} \tag{2}$$

If two dependent tasks are executed on the same virtual machine, the Data Transfer Time (DTT) between them is assumed to be zero.

3 Monte Carlo Simulation of Cloud Environment for Failure Analysis

3.1 Monte Carlo Simulation Method

Monte Carlo Simulation (MCS) is a computation-intensive, statistical method which can be used to model the behavior of a complex system and obtain relevant information about its working. MCS has been widely employed in various engineering and scientific domains because of its ability to produce results highly close to the reality of complex system operation. The method has gained popularity in recent years because now access to vast computation resources is easily available [12]. MCS method is inherently parallel and thus can be easily parallelized using a split-and-merge pattern [13].

We used the MCS method to simulate the failure scenario in the cloud environment. Before performing the simulation, the statistical properties of the cloud environment are required. Researchers have already analyzed the cloud environment in order to determine its failure characteristics [8, 9]. Garraghan et al. [8], analyzed the failure characteristics of the cloud environment and determined the statistical properties associated with it. They concluded that failures in the cloud servers follow Weibull distribution. Conforming to this, we use Monte Carlo Simulation method with Weibull distributed failures to model the cloud.

Weibull distribution, a generalization of Exponential distribution, considers the age of a system as a factor in determining its failure probability i.e. it assumes that the probability of occurrence of an event is time dependent [12]. It was proposed by Weibull in 1950s. It uses two parameters: α (shape) and β (scale).

The probability density function (pdf) and cumulative distribution function (cdf) for Weibull distribution are:

$$f_T(t) = \frac{\alpha}{\beta}\left(\frac{t}{\beta}\right)^{\alpha-1} e^{-(t/\beta)^{\alpha}} \tag{3}$$

$$F_T(t) = 1 - e^{-(t/\beta)^{\alpha}} \tag{4}$$

The failure rate (λ(t)) for Weibull distributed failures is given as:

$$\lambda(t) = \frac{\alpha}{\beta}\left(\frac{t}{\beta}\right)^{\alpha-1} \tag{5}$$

In order to perform the simulation, a stream of random numbers between 0 and 1 is generated which represents the probability of occurrence of failure at a given time [14]. Using this random variable, we perform sampling by the inverse transform method, using Weibull distribution, in order to determine the Mean Time to Failure (MTTF) for various computing nodes in the cloud.

After sampling the random variable *r*, the Time-to-failure is computed from Weibull distribution as:

$$t = \beta(-\ln(1-r))^{1/\alpha} \tag{6}$$

Using this process, a large number of random walks through the system are performed i.e. the system is simulated a large number of times, from the beginning of system operation to a particular mission time, and then the average time to failure (*TTF*) and average time to repair (*TTR*) for each virtual machine over time is calculated. These simulations of the system can be performed in parallel.

3.2 Working of Monte Carlo Simulator

We proposed the Monte Carlo Failure Estimation (MCFE) algorithm to simulate the cloud environment using Monte Carlo Method. Algorithm 1 presents the Monte Carlo Failure Estimation (MCFE) Algorithm in detail. Prior to the submission of the workflow application by the user, the *MCSimulator* simulates the cloud environment using its statistical properties and the equations mentioned in the previous section. For this, a stream of random numbers is generated. Using the first random number from this stream, the first time to failure ($failure(1)_i$) is calculated for a particular virtual machine vm_i using Eq. (6). Subsequently, the corresponding time to repair ($repair(1)_i$) for vm_i is calculated using the repair parameters for the machine. This process is then repeated *rf* (recalculation factor) number of times using the random number stream and the average values for failure and repair of the virtual machine vm_i are obtained. The clock is then updated to reflect the failure and repair event. This process is continued for the virtual machines currently available in the system until a specific mission time is reached. A large number of such simulations are performed and the results are then averaged out in order to remove any extreme values that might arise because of the selection of extreme random numbers.

Algorithm 1: Monte Carlo Failure Estimation (MCFE) algorithm

1. Specify α and β (weibull distribution) parameters for each virtual machine currently available in the cloud.
2. for each virtual machine vm_i, do
3. for *N* simulations do
4. While *currentTime* < *missionTime*
5. for *rf* times
6. Generate a random number *r1*.
7. Compute time to failure(TTF). //according to Eq. (6)
8. Generate a random number *r2*.
9. Compute time to repair (TTR). //according to Eq. (6)
10. Calculate average TTF value.
11. Calculate average TTR value.
12. Update the value of the Clock.
13. Update system state as availability or non-availability state.

As a result, the simulator provides the probable TTFs and TTRs for the future operation of the virtual machines present in the cloud. Based on these values, the virtual machine can be in the availability state or non-availability state at a given time. When the virtual machine is in the availability state, we consider it to be in working condition and thus, schedule appropriate task on it for execution. In the non-availability state, we consider that the virtual machine has failed and thus, do not schedule any task on it. The value of number of simulations (*N*) and recalculation factor (*rf*) can be chosen appropriately. Larger values indicate more accuracy

4 The Proposed FARS Algorithm

This section presents Failure-Aware Resource Scheduling algorithm named FARS, derived from the HEFT algorithm [4]. It aims to achieve high reliability in cloud which improves the application execution performance by reducing the overall makespan and the number of task failures during the execution of the application. When the user submits a workflow application, the tasks are first prioritized and then scheduled to the currently available virtual machines while taking into account the reliability of the machine during the execution of the task.

4.1 Task Prioritization Phase

For a workflow application submitted by the user, all the tasks are prioritized for execution based on their precedence, execution time on different virtual machines, and the data transfer cost (DTT). The priority to each task is assigned as:

$$priority(t_i) = \overline{ET(t_i)} + \max_{t_j \in child_i} \left(DTT(t_i, t_j) + priority(t_j) \right) \tag{7}$$

where $\overline{ET(t_i)}$ is the average execution time for a task that can be calculated as:

$$\overline{ET(t_i)} = \frac{\sum_{j=1}^{m} ET(t_i, vm_j)}{m} \tag{8}$$

For the exit task of the application, since the child set is considered to be empty, the priority is defined as:

$$priority(t_{exit}) = \overline{ET(t_{exit})} \tag{9}$$

Based on this prioritization, the tasks are sorted in the non-increasing order of their priority and the schedule is then created by taking each task from this task-list in order and assigning it to a virtual machine.

4.2 Assignment Phase

During the assignment phase, the scheduler selects the appropriate virtual machine for the task based on the lowest Earliest Finish Time (*EFT*) of the task on each machine. For this, we first need to calculate the Earliest Start Time (*EST*) of the task as:

$$EST(t_i, vm_j) = \begin{Bmatrix} \max\left\{ ready(vm_j), \max_{t_p \in parent_i} \begin{pmatrix} EFT(t_p) + DTT(t_i, t_p), \text{ if } t_p \notin vm_j \\ EFT(t_p), \text{ if } t_p \in vm_j \end{pmatrix} \right\}, & \text{if } vm_j \in avail(k)_j \\ repair(k)_j, & \text{if } vm_j \in notavail(k)_j \end{Bmatrix} \tag{10}$$

where $ready(vm_j)$ represents the time at which the virtual machine is ready for execution of the concerned task, $avail(k)_j$ represents the *k*th availability state for the virtual machine at the concerned time and $repair(k)_j$ represents the repair time of the virtual machine for the *k*th non-availability state. We assume the virtual machine to be initially available at time 0. Using *EST*, the *EFT* for each task can thus be calculated as:

$$EFT(t_i, vm_j) = \begin{Bmatrix} EST(t_i, vm_j) + ET(t_i, vm_j), & \text{if } EST + ET + \sum_{t_k \in child_i} DTT_{i,k} \in avail(k)_j \\ repair(l)_j + ET(t_i, vm_j), & \text{if } l > k \,\&\, repair(l)_j + ET + \sum_{t_k \in child_i} DTT_{i,k} \in avail(l)_j \end{Bmatrix} \tag{11}$$

where l represents the index of the earliest availability state for vm_j for which the machine is assumed to be available during the complete execution period. The overall makespan for the workflow application is calculated as:

$$makespan = EFT(t_{exit}, vm_j) \tag{12}$$

Thus, the finish time for a task on a virtual machine is calculated while considering the precedence constraints of the workflow and the failure of the nodes. If a virtual machine is expected to fail during the execution of a task, the task will be scheduled on that virtual machine only after the virtual machine is expected to be repaired and in working condition again (availability state). This will reduce the overall makespan of the application as the probability of failed tasks is drastically reduced. It will also safeguard the application against critical losses due to failed tasks and improve the availability of the cloud resources. Algorithm 2 presents the complete FARS algorithm.

Algorithm 2: Failure Aware Resource Scheduling (FARS) algorithm

1. Determine availability and non-availability states for each $vm_j \in V$ using the MCFE algorithm.
2. Compute average execution time for all tasks. //according to Eq. (8) 3. Compute the data transfer time ($DTT(t_i, t_j)$) for each edge. //according to Eq. (2)
4. Compute $priority(t_i)$ of execution for each task. //according to Eq. (7) and Eq. (9)
5. Sort the tasks into a *task-list* by non-increasing order of their priority.
6. While the *task-list* is not empty
   a. Remove the first task t_i, from the list.
   b. Calculate the *EST* and *EFT* for each $vm_j \in V$ in the Cloud for the task. //according to Eq. (10) and Eq. (11)
   c. Select the virtual machine, vm_j with the lowest *EFT* for execution.
   d. Add task t_i assigned to virtual machine vm_j along with *EST* and *EFT* to schedule S.
7. Compute makespan for the schedule. //according to Eq. (12) 8. return S.

5 Experimental Results and Analysis

We used the CloudSim [15] toolkit to model the cloud environment and conducted extensive experiments based on simulation strategy. In this section, we present the comparative evaluation of our proposed FARS algorithm with the most popular scheduling algorithm HEFT [4], in order to verify its effectiveness.

5.1 Simulation Model

To generate precedence-constrained workflows, we have used randomly generated task graphs and task graphs that represent numerical real world problems such as Gaussian Elimination (GE) [16] and Fast Fourier Transformation (FFT) [17]. The performance of FARS algorithm is analyzed based on its comparison with HEFT algorithm, using makespan as the performance metric. First, the algorithms were evaluated based on their performance on workflows represented by random task graph and the results are analyzed as:

5.1.1 Effect of Varying the Size of Task Graph

The size of the input task graph was varied from 40 to 120, in steps of 20, as {40, 60, 80, 100, 120}. We executed the tasks on 16 virtual machines where the computation capacity of each virtual machine was generated randomly by a uniform distribution with an appropriate mean value. Figure 3 depicts the effect of varying the size of input graph on the makespan of the application. We conclude that the

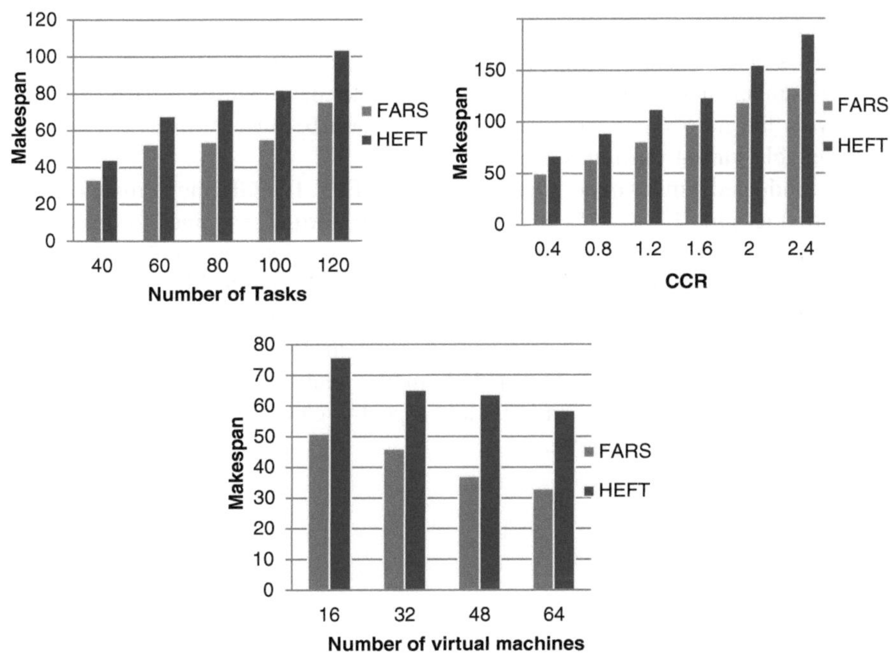

Fig. 3 Makespan for randomly generated task graphs at different number of tasks, CCR and number of virtual machines respectively

performance of our algorithm improves as the size of the task graph increases. This is because as the number of tasks increase, the number of failures occurring in the execution of the workflow also increases, which in turn increases the makespan of HEFT. FARS executes tasks based on the reliability of the virtual machine and hence, performs better when the size of the input task graph is large.

5.1.2 Effect of Varying the CCR Value

The performance of our algorithm was also analyzed by varying the CCR value from 0.4 to 2.4, in steps of 0.4. It is observed that the performance of FARS algorithm improves in comparison with HEFT as the value of CCR increases. This is because FARS algorithm ensures reliable data transfer among virtual machines along with reliable computation of tasks (Eq. 11). As the value of CCR increases, the tasks become more communication intensive.

5.1.3 Effect of Varying the Number of Virtual Machines

The number of virtual machines was varied from 16 to 64, in steps of 16, with 100 tasks. It is observed that as the number of virtual machines increase, the makespan for both FARS and HEFT decreases. With less number of virtual machines, a large variation is observed for the makespan. This is because of less reliability in case of limited number of virtual machines for HEFT. The makespan for FARS improves slightly as the number of virtual machines increase because it always selects the more reliable virtual machines for execution.

We conducted similar experiments for GE and FFT. For GE, the number of tasks was varied as {14, 27, 44, 65, 90}, considering 16 virtual machines for execution. The results found have been summarized in Fig. 4. For the simulation of FFT application, the tasks were varied from 14 to 96, as {14, 38, 65, 94}, considering 16 virtual machines for execution. The results obtained are summarized in Fig. 5.

The results for GE and FFT applications show similar performance as randomly generated task graphs. Thus, we conclude that in real world scenario, where reliability of a virtual machine is an important aspect in cloud, our algorithm performs better than the HEFT algorithm.

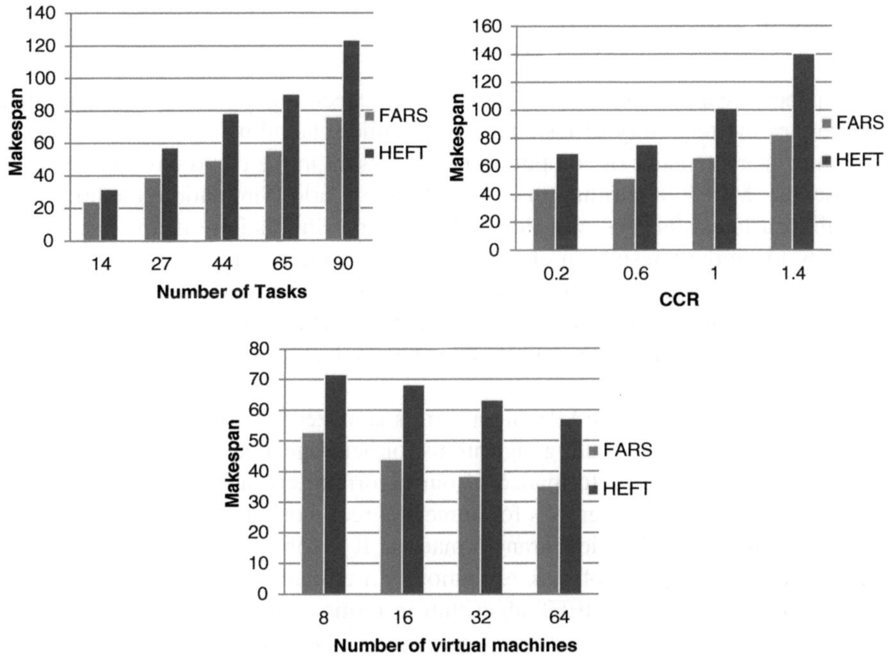

Fig. 4 Makespan for Gaussian Elimination (GE) task graphs at different number of tasks, CCR and number of virtual machines respectively

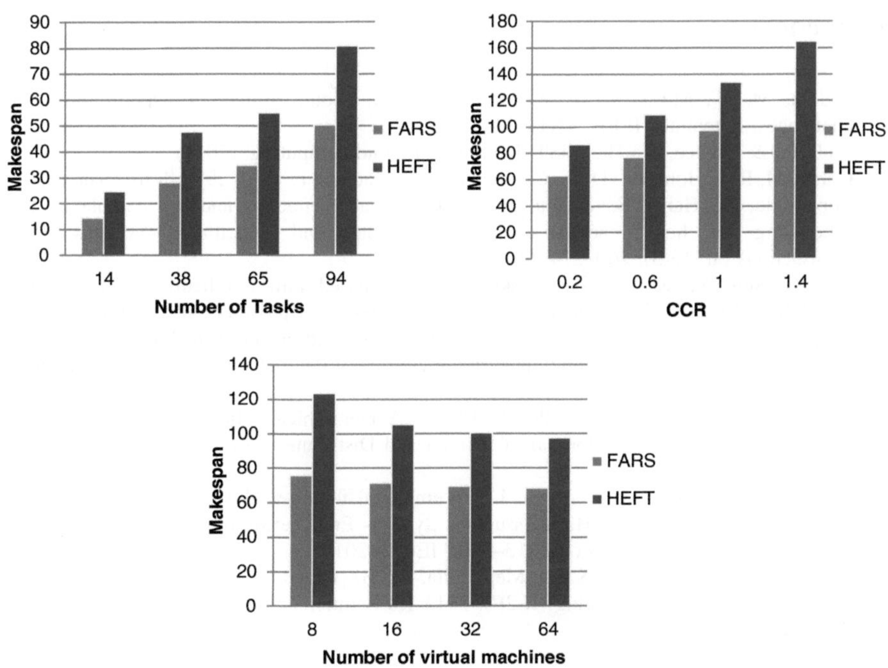

Fig. 5 Makespan for Fast Fourier Transformation (FFT) task graphs at different number of tasks, CCR and number of virtual machines respectively

6 Conclusion

In this paper, we aim at developing a workflow scheduling algorithm for cloud that incorporates the reliability of task execution while scheduling. In order to analyze the failure characteristics of computing nodes present in the cloud environment, we propose the MCFE algorithm that uses Monte Carlo Simulation method with Weibull distributed failures. We chose Monte Carlo Simulation method because of its ability to produce results in closer adherence to the reality of complex system operation. We considered Weibull distributed failures because they appropriately represent the failure scenario in cloud environment. The FARS algorithm performs reliable allocation of tasks to resources by using the information provided by MCFE algorithm.

The performance of the FARS algorithm is analyzed by comparing it to one of the most popular task scheduling algorithms for heterogeneous distributed systems, HEFT. We analyze the performance of our algorithm considering randomly generated task graphs and task graphs for numerical real-world problems like Gaussian Elimination and Fast Fourier Transformation. It is observed that in real-world scenario, where reliability of task execution is a critical issue, our algorithm significantly outperforms the HEFT algorithm in terms of makespan for task execution. In future, we plan to improve our existing algorithm by considering other important parameters in cloud environment like budget and security.

References

1. Sadiku, M. N., Musa, S. M., & Momoh, O. D.: Cloud computing: Opportunities and challenges. Potentials, IEEE, 33(1), 34–36. (2014).
2. Mell, P., & Grance, T.: The NIST definition of cloud computing. (2011).
3. Garey, M. R., & Johnson, D. S.: Computers and intractability (Vol. 29). wh freeman. (2002).
4. Topcuoglu, H., Hariri, S., & Wu, M. Y.: Performance-effective and low-complexity task scheduling for heterogeneous computing. Parallel and Distributed Systems, IEEE Transactions on, 13(3), 260–274. (2002).
5. He, X., Sun, X., & Von Laszewski, G.: QoS guided min-min heuristic for grid task scheduling. Journal of Computer Science and Technology, 18(4), 442–451. (2003).
6. Tang, X., Li, K., Li, R., & Veeravalli, B.: Reliability-aware scheduling strategy for heterogeneous distributed computing systems. Journal of Parallel and Distributed Computing, 70(9), 941–952. (2010).
7. Tang, X., Li, K., Qiu, M., & Sha, E. H. M.: A hierarchical reliability-driven scheduling algorithm in grid systems. Journal of Parallel and Distributed Computing, 72(4), 525–535. (2012).
8. Garraghan, P., Townend, P., & Xu, J.: An empirical failure-analysis of a large-scale cloud computing environment. In High-Assurance Systems Engineering (HASE), 2014 IEEE 15th International Symposium on (pp. 113–120). IEEE. (2014).
9. Fiondella, L., Gokhale, S. S., & Mendiratta, V. B.: Cloud Incident Data: An Empirical Analysis. In Cloud Engineering (IC2E), 2013 IEEE International Conference on (pp. 241–249). IEEE. (2013).

10. Mei, J., Li, K., Zhou, X., & Li, K.: Fault-Tolerant Dynamic Rescheduling for Heterogeneous Computing Systems. Journal of Grid Computing, 1–19. (2015).
11. Guo, S., Huang, H. Z., Wang, Z., & Xie, M.: Grid service reliability modeling and optimal task scheduling considering fault recovery. Reliability, IEEE Transactions on, 60(1), 263–274. (2011).
12. Zio, E.: The Monte Carlo simulation method for system reliability and risk analysis (p. 198p). London: Springer. (2013).
13. Camarasu-Pop, S., Glatard, T., Da Silva, R. F., Gueth, P., Sarrut, D., & Benoit-Cattin, H.: Monte Carlo simulation on heterogeneous distributed systems: A computing framework with parallel merging and checkpointing strategies. Future Generation Computer Systems, 29(3), 728–738. (2013).
14. Alexander, D.: Application of Monte Carlo simulations to system reliability analysis. In Proceedings of the Twentieth International Pump Users Symposium (pp. 91–94). (2003).
15. Calheiros, R. N., Ranjan, R., Beloglazov, A., De Rose, C. A., & Buyya, R.: CloudSim: a toolkit for modeling and simulation of cloud computing environments and evaluation of resource provisioning algorithms. Software: Practice and Experience, 41(1), 23–50. (2011).
16. Cosnard, M., Marrakchi, M., Robert, Y., & Trystram, D.: Parallel Gaussian elimination on an MIMD computer. Parallel Computing, 6(3), 275–296. (1988).
17. Chung, Y. C., & Ranka, S.: Applications and performance analysis of a compile-time optimization approach for list scheduling algorithms on distributed memory multiprocessors. In Super computing'92., Proceedings (pp. 512–521). IEEE. (1992).

Realization of Virtual Resource Management Framework in IaaS Cloud Federation

Anant V. Nimkar and Soumya K. Ghosh

Abstract Virtual resources can be collectively and securely managed by one or more participating stakeholder(s) in the federated cloud through a Virtual Resource Management Framework (VRMF). VRMF provides authorization, authentication and identity solutions in IaaS cloud federation. System performance must be evaluated before and after the deployment of all modules in VRMF because it may vary significantly as each module uses one or more other module(s) to perform its function. This paper presents an implementation of VRMF for managing virtual resources and then evaluate the system performance in federated clouds. The implementation uses the response time as the performance parameter for evaluating the ecosystem before and after integration of all individual modules. The integrated framework with authorization, authentication and identity services shows encouraging results.

Keywords Virtual resource management · Framework · IaaS · Cloud federation

1 Introduction

Infrastructure as a Service (IaaS) cloud delivers virtual resources (i.e. virtualized instances of switches, routers, machines and links) as a utility over the Internet. If cloud provider cannot fulfill customers' requirements, it may borrow virtual resources to form virtual networks from other cloud providers. In IaaS cloud federation, *Service Providers* (SePs) federate with *Infrastructure Providers* (InPs) and supply virtual networks out of data centers to their customers [1]. Virtual networks consist of sets of virtual nodes connected by virtual links. In IaaS cloud federation, collocation of tenants' virtual networks across data centers, is called as *Multi-tenant*

A.V. Nimkar (✉) · S.K. Ghosh
Department of Computer Science and Engineering, IIT, Kharagpur 721302, India
e-mail: anantn@sit.iitkgp.ernet.in

S.K. Ghosh
e-mail: skg@iitkgp.ac.in

N. Modi et al. (eds.), *Proceedings of International Conference on Communication and Networks*, Advances in Intelligent Systems and Computing 508,
DOI 10.1007/978-981-10-2750-5_16

Network Collocation (MNC). The management of federated virtual resources needs two special treatments. First, management of federated virtual resources must be cooperatively handled by subsets of federating stakeholders, viz. SePs, InPs and customers. For example, virtual nodes and links are collaboratively created and deleted by their respective InPs on receiving requests from SePs. Further, virtual resources need to be cooperatively configured by their SePs, users and InPs. Second, the management of federated virtual resources must be handled securely, as and when federations are established or torn down respectively.

Virtual Resource Management Framework (VRMF) [2] provides a mechanism to securely manage federated virtual resources through three local and global modules for (i) Authorization over objects (i.e. federated virtual resources), (ii) Authentication of subjects (i.e. collective stakeholders) and (iii) Identity solution for subjects and objects in IaaS cloud federation. Each module makes use of one or more other module(s). The performance of the system must be evaluated before and after the deployment of all modules in VRMF since it may vary significantly as each module uses one or more other module(s) to perform its function.

VRMF uses *Federation Access Control Model* (FACM), *Name-and-Label Space Model* (NLSM) [3] and *Caucus Authentication Protocol* (CAP) for authorization, identity solution and authentication respectively. FACM and NLSM use a new concept of subjects as subsets of federating stakeholders. The authorization of subjects over objects is carried out using NLSM, FACM and CAP as follows. CAP allows authentication of subjects composed of one or more federating stakeholder (s) using a variant of Multi-Party Computation (MPC). NLSM provides identities and security labels of subjects and federated virtual resources required in the FACM authorization process. FACM authorizes subjects to access federated objects by collective decisions of (i) comparison of security labels using Cartesian operators and (ii) federated discretionary access rights using MAC and DAC policies respectively.

In this paper, NLSM, FACM and CAP are realized as *Node-and-Path Label Distribution Protocol* (NPLDP), *Access Control Enforcement Engine* (ACEE) and *Caucus Authentication Mechanism* (CAM) in VRMF respectively. Further, the realization uses minimal part of FACM as MAC-based Enforcement Engine (MACEE) for secure multiplexing and de-multiplexing of users' traffic in virtual and physical infrastructures. NPLDP is implemented as a distributed signaling protocol to instruct physical routers for secure embedding of virtual resources. CAM is implemented as a distributed protocol running on all stakeholders' premises. Further, the simulation has been carried out on different number of users' authorization requests to evaluate the system using response times before and after integration of authorization, authentication and identity solution as suggested in VRMF.

The rest of the paper is organized as follows. Section 2 presents a synopsis on IaaS cloud federation and Virtual Resource Management Framework in Sects. 2.1 and 2.2 respectively. Section 3 presents a realization of VRMF using Name-and-Label Space Model, Federation Access Control Model and Caucus

Authentication Protocol in Sects. 3.1, 3.2 and 3.3 respectively. The simulation study and results are presented in Sect. 4. The concluding remarks are given in Sect. 5.

2 Background

This paper gives an implementation case study on collocation of tenants' virtual networks in federated clouds through *Virtual Resource Management Framework*. This section presents fundamentals of IaaS cloud federation and *Virtual Resource Management Framework*.

2.1 IaaS Cloud Federation

In IaaS cloud federation, each cloud provider can be a Service Provider (SeP), Infrastructure Provider (InP) or both. SePs supply virtual networks to their customers through a brokered collaboration among InPs. Virtual networks consist of sets of virtual nodes connected by virtual links. Figure 1 shows an ecosystem of IaaS cloud federation. It shows federations among three cloud providers, viz. InP#1/SeP#1, InP#2 and SeP#3/InP#3 and three users, viz. User#1, User#2 and User#3. The physical networks of InPs are shown in black plain-line rectangles at the bottom. The service provider SeP#1 borrows a virtual network segment $\{VN_6\}$ for User#1 from InP#3 and thus provides a virtual network $\{VN_1, VN_6\}$. Similarly, the service provider SeP#3 borrows virtual network segments $\{VN_3, VN_4\}$ and $\{VN_2\}$ from the InPs and provides virtual networks $\{VN_3, VN_4, VN_7\}$ and $\{VN_2, VN_5\}$ respectively. The blue single-dot-dash and saffron double-dot-dash ovals indicate the boundaries of controls of SeP#1 and SeP#3 respectively.

A few implementation of IaaS cloud federation are as follows. The first step towards IaaS cloud federation includes the three-phase cross federation [4], mobile-agent based cloud federation [5] and *Reservoir Model* [6]. IBM's *Virtual Data Center* (VDC) [7] is an industry implementation of IaaS cloud federation.

2.2 Virtual Resource Management Framework

Virtual Resource Management Framework (VRMF) [2] has three types of stakeholders, viz. (i) IaaS cloud providers, (ii) A broker and (iii) Users as shown in Fig. 2.

IaaS cloud can be SeP, InP or both. The data center of IaaS cloud has a controller, a physical infrastructure (PIn), a virtual infrastructures (VIn) and a set of software modules. The software modules consist of (i) local and federated *Resource Access Control* (RAC), (ii) local and federated *Identity Management* (IdM) and

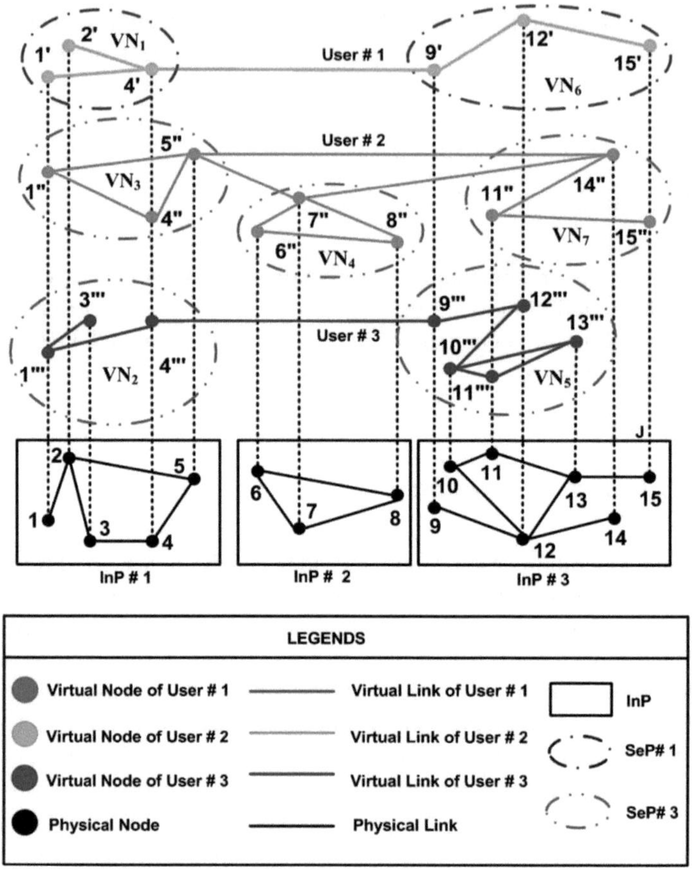

Fig. 1 An ecosystem of IaaS cloud federation among three InPs/SePs

(iii) local *Authentication Agents* (AA) and federated *Authentication System* (AS). The federations among stakeholders are established by the controllers of InPs using three standard functions—matching, discovery and advertisement—of brokered architecture. The users have only local *Authentication Agents* (AA) to participate in the authentication process. Each user also has local *Authentication Agent* (AA) for authentication of subject as collective stakeholders.

3 Implementation

The authorization, authentication and identity solutions in IaaS cloud federation are addressed by designing FACM enforcement policies, implementing CAP and deploying NLSM respectively. The details of FACM, CAP and NLSM are

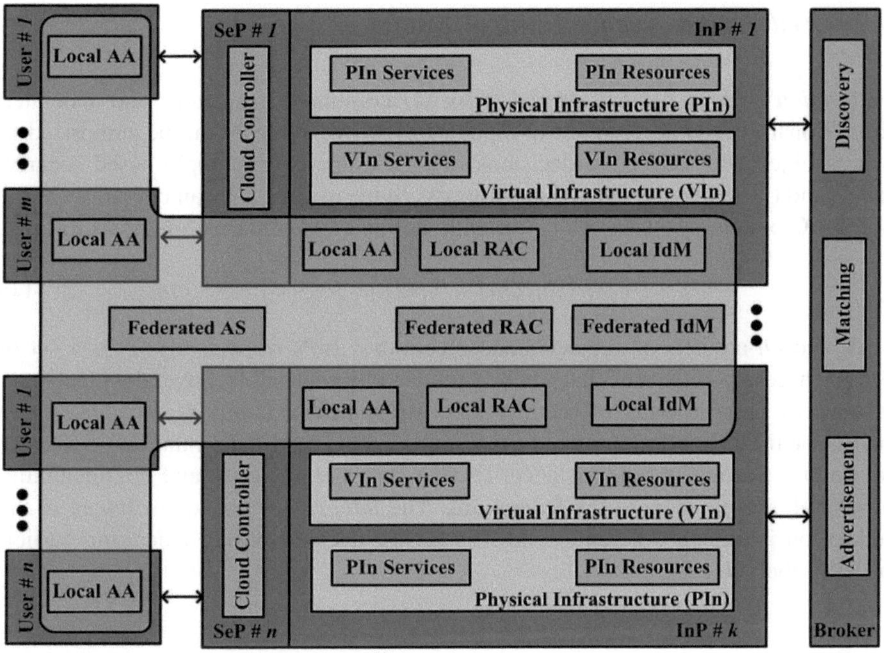

Fig. 2 Virtual resource management framework

subsequently presented in this section and their implementation details as ACEE, CAM and NPLDP are later presented in Sect. 4 respectively.

3.1 *Name-and-Label Space Model*

Name-and-Label Space Model (NLSM) uses the concepts of ordered tuple and set theories for the formation of identities and security labels respectively in IaaS cloud federation ecosystem. NLSM uses security labels of the form $\langle SL_{SeP}, SL_U, SL_{InPs}\rangle$ where SL_{SeP}, SL_U and SL_{InPs} denote security labels of SeP, its user and a set of InPs respectively. NLSM also uses three binary Cartesian operators on tuple based security labels. The binary Cartesian operator $\subsetneqq$ returns true if the first two fields of two security labels are non-empty and same, and the third field of first security label is a subset of the third field of second security label. The binary Cartesian relation $\equiv$ returns true if the first two fields of two security labels are non-empty and all the three fields are same in both security labels.

3.2 Federation Access Control Model

Federation Access Control Model (FACM) combines mandatory and federated discretionary based access control model. In FACM, subjects can be authorized to access objects by collective decisions of (i) Comparison of tuple based security labels and (ii) Federated discretionary access rights using Cartesian operators, MAC and DAC policies. FACM can be formally defined for σ InPs as 5-tuple as follows.

$$F = \{(S_i, O_i, P_i, F_i, L_i) \mid 1 \leq i \leq \sigma\} \quad (1)$$

S_i and O_i are sets of subjects and objects in σ InPs respectively. P_i is a set of 3-tuple access permissions that can be exercised by each subject over the objects. F_i is a set of ordered pairs of security label mappings of subjects and objects at a particular time instance. L_i is a *partial order set* of each individual InP's security levels at a particular time instance. FACM provides integrity and confidentiality among all stakeholders in the federations. The *integrity* of services/software in the federation is treated differently and it is higher with increasing federating participants in the subjects.

3.3 Caucus Authentication Protocol

In IaaS cloud federation, services are cooperatively executed by subjects as subsets of federating stakeholders. *Caucus*[1] *Authentication Protocol* (CAP) efficiently authenticates such subjects which cannot be handled by all identity management solutions including WS-Federation [8] and Shibboleth [9]. CAP needs three independent services, namely Caucus Server (*CS*), Role Client (*RC*) and Role Servers (*RSs*) running on an authentication server, a stakeholder as authentication initiator and all stakeholders except authentication initiator.

Caucus Authentication Protocol first authenticates all stakeholders $s \in S$ and finally the subject S using the concept of MPC [10] as follows. In the first step, an authentication initiator out of all stakeholders in the subject (i.e. $s_a \in S$) sends a request to *CS* for authentication of a subject, S. In the second step, *CS* creates *role tickets* for all stakeholders in S and then sends them to authentication initiator, RC (i.e. process running on s_a). In the third step, RC forwards *role tickets* to all RSs (i.e. processes running $S - s_a$). In the fourth step, all RCs send *role tickets* to CS to share their MPC values. Finally, CS creates an *authentication ticket* using all MPC values and then sends it to RC after its validation. Any stakeholder in the subject can use this authentication ticket for execution of federation services before the validity of the authentication ticket.

[1]The dictionary meaning of the word "Caucus" is "a small group of people who have same interests".

Designing an Enhanced Simulation Module for Multimedia Transmission Over Wireless Standards

Mayank Patel and Naveen Choudhary

Abstract Extensive use of multimedia application over the internet conduces use of video transmission. This leads the performance consideration of video transmission a popular research area for the industry and academia researches. The wireless LAN is playing important role to connect the users over the internet. As the video transmission increases it poses traffic congestion which in turn degrades the network efficiency. There are various performance parameters to be known for designing the WLAN such as throughput, end to end delay, Queue utilization, jitter, PSNR and many more. We aimed in the paper to develop simulation scenario for video transmission over IEEE 802.11n WLAN standard. For this we modify the both the fedora and NS2 settings and successfully performs the video transmission simulations.

Keywords Video transmission over WLAN · Simulation module · NS2 · H.264/SVC · IEEE 802.11n

1 Introduction

Providing real time Quality of Service over Wireless Local Area Networks is becoming a very challenging task in heavy loaded traffic scenario [1]. Currently high throughput wireless standard (IEEE 802.11n) with a high data transmission rate with 600 Mbps is used [2]. Even with high transmission data rates real time video content must be compressed before transmission. This leads the need of simulation surround for simulating and performance analysis of the video transmission over enhanced WLAN [3]. Along with the video transmission, the important performance parameters must be evaluated for industry and academia researches [2].

M. Patel (✉) · N. Choudhary
Department of Computer Science and Engineering, CTAE, MPUAT, Udaipur, India
e-mail: mayank999_udaipur@yahoo.com

N. Choudhary
e-mail: naveencl21@yahoo.com

N. Modi et al. (eds.), *Proceedings of International Conference on Communication and Networks*, Advances in Intelligent Systems and Computing 508,
DOI 10.1007/978-981-10-2750-5_17

For developing the simulation environment we choose Network Simulator–2 (NS2) as it freely available and recognized by various researches [4–7]. We can use NS2 in Microsoft Windows operating system by Cygwin [8]. But the Cygwin leads to the various limitations in the performances parameters; therefore we pick out Fedora operating system. In Fedora some parameter settings has to be modified for running NS2 properly [9]. As both are freely available therefore they can be used hazel free. For video transmission simulations and obtaining the performance parameters we have to change and modify the various libraries of NS2. For make more users friendly we also implement combination of Fedora and NS2 in virtual machine, which can be used through Microsoft windows operating system [10]. Thus we design a module for evaluating and research on multimedia transmission over WLAN.

The residual of this paper is organized as follows. We present brief procedural steps to be done for devising Fedora to support NS2 in Sect. 2. In Sect. 3 we describe the steps to support multimedia evaluation over NS2. In Sect. 4 we implement and tested our simulation environment. Finally in Sect. 5 we conclude our work.

2 Making the FEDORA Ready for NS2

Fedora-20 version has been used for developing a simulation model. Following steps are performed.

Step 1: Install fedora from the official fedora site
After fedora installation we have to install some packages to support NS2 simulations. Packages will be updated or installed in root user mode

Step 2: Enter in root user mode and the follow to update fedora operating system, command yum -y update

Step 3: To support gnuplot in fedora
yum -y install gcc tcl-devel libX11-devel libXt-devel libXmu-devel gnuplot

Step 4: For using Python script in our code

Step 4.a: yum -y install gcc gcc-c++ python

Step 4.b: yum -y install gcc gcc-c++ python python-devel

Step 4.c: yum -y install bzr

Step 5: Installing mercurial repository in fedora
yum -y install mercurial

Step 6: Installing wireless model faithfulness in GUI and GTK-based configuration system

Step 6.a: yum -y install gsl gsl-devel

Step 6.b: yum -y install gtk2 gtk2-devel

Step 7: For building dynamic analysis tools
yum -y install gdb valgrind

Step 8: Generating documentation from C++ source codes
yum -y install doxygen graphviz ImageMagick texinfo texinfo-tex
Step 9: For creating our simulation documentation
yum -y install texinfo dia texinfo-tex texi2html
Step 10: Installing flex lexical analyzer and bison parser generator
yum -y install flex bison
Step 11: To make binaries
yum -y install compat-gcc-34
Step 12: To support Datbase
yum -y install sqlite sqlite-devel
Step 13: To support database for the frame work
yum -y install sqlite sqlite-devel
Step 14: To develop XML application
yum -y install libxml2 libxml2-devel
Step 15: For creating object file and binary programs
yum -y install binutils
Step 16: Updating Fedora kernel
yum -y update kernel
Step 17: Installing kernels for gcc to run tcl script in NS2
Step 17.a: yum -y install kernel-devel kernel-headers dkms gcc gcc-c++ xorg*
Step 17.b: yum -y install gcc-c++ compat-gcc-34-c++ automake autoconf libtool libX11-devel
Step 18: For implementing X Window System
yum -y install libXext-devel libXau-devel libXmu-devel xorg-x11-proto-devel
Step 19: To support GNU and NSAM for X264/AVC Encoder
yum -y install yasm
Step 20: Installing to move Nautilus from console from Nautilus
yum -y install nautilus
Step 21: Installing FreeGult to read input/output devices on the terminal and resultant outputs
yum -y install freeglut
Step 22: Install perl module for Makefiles and build modules
perl-ExtUtils-MakeMaker
Step 23: Installing Synaptic Package Manager for cleaning operations.
yum -y install dconf* bleachbit
Step 24: Installing Wget for accessing files through HTTP and FTP.
sudo yum install wget
Step 25: To run windows based executable file in Fedora
Step 25.a: yum -y install apt
Step 25.b: yum -y install wine
Step 26: Installing GPAC multimedia frame work for to support MPEG-4
Step 26.a: Download the GPAC tar file
Step 26.b: unzip both folder and paste gpac_extra_libs-0.4.5 folder in gpac/extra_libs

Step 26.c: SU: password
Step 26.d: yum -y install freeglut
Step 26.e: yum -y install freeglut-devel
Step 26.f: cp -R/home/mayank/Downloads/gpac/usr/local/src
Step 26.g: cd/usr/local/src gpac
Step 26.h: chmod +x configure
Step 26.i: ./configure –enable-pic
Step 26.j: make lib (it will take 10 min)
Step 26.k: make apps
Step 26.l: make install lib
Step 26.m: sudo make install
Step 26.n: cp bin/gcc/libgpac.so/usr/lib
Step 26.o: ldconfig
Step 26.p: cp -r/home/mayank/Downloads/gpac/include/* /usr/include/
Step 27: Checking working of MP4Box
[root@localhost gpac]#which MP4Box
===> /usr/local/bin/MP4Box
wget "http://blog.andreas-haerter.com/_export/code/2011/07/01/install-msttcorefonts-fedora.sh?codeblock=1"-O"/tmp/install-msttcorefonts-fedora.sh"
Step 28: Downloading Microsoft Fonts in Fedora
Step 29: Installing Microsoft Fonts in Fedora
Step 29.a: chmod a+rx "/tmp/install-msttcorefonts-fedora.sh"
Step 29.b: su -c "/tmp/install-msttcorefonts-fedora.sh"
Step 30: Restart the fedora.

It will leads to successfully installation of prerequisites tools for NS2 and video simulation over the fedora.

3 Installing NS2 with Supporting Multimedia Simulations

Step 1: Step 1: Download NS2.29
Step 2: Unzip at Home directory
Step 3: $ cd ns-allinone-2.29
Step 4: For using C/C++ compiler for NS2 compilation
$ export CC=gcc34 CXX=g++34
Step 5: Installing patch for supporting NAM
$ cd/home/mayank/ns-allinone-2.29/tk-8.4.11
Step 6: $ patch -p0<tk-8.4-lastevent.patch
Step 7: Installing patch for wireless application
$ patch -p1<ns-2.29_wireless_update_patch

Step 8: yum install autoconf
Step 9: yum install compat-gcc-34-c++(this is very important as gcc3.4 is required for successful installation)
Step 10: export CC=gcc34
Step 11: export CXX=g++34
Step 12: yum install libtools
Step 13: yum install libX11-devel
Step 14: yum install xorg-x11-proto-devel
Step 15: yum install libXt-devel
Step 16: yum install libXmu-devel
Step 17: Setting all files permission to executable
Step 18: Installing NS2 in fedoara
$. /install
Step 19: set the environmental paths in.barsh file (home/.barsh)
Step 20: cd ns2.29 ->./validate
Step 21: Testing the NS2
Step 21.a: goto the terminal
Step 21.b: $ns
Step 21.c: % nam (Nam window will be opened)

Above steps will successfully install NS2 on the fedora and can be used for simulations. The output of running NS2 terminal and NAM window is shown Fig. 1.

4 Testing of Simulation Environment

After successfully implementation of NS2 over fedora, next we simulate a topology consisting of video transmission and background traffics as shown in Fig. 2.

We transmit a BUS SVC QCIF video of 352*288 with 30 frames per second over the 802.11n standard and evaluate various parameters such as throughput, end to end delay, jitter, and PSNR. Received video snapshot is shown in Fig. 3.

We transmit a BUS SVC QCIF video of 352*288 with 30 frames per second over the 802.11n standard and evaluate various parameters such as throughput, end to end delay, and PSNR. Received video snapshot is shown in Fig. 3.

Total packet end to end delay is 0.031479 s; graph is shown in Fig. 4.

Total throughput of the network is 752.52 kbps; graph is shown in Fig. 5.

PSNR of every individual frame is shown in Fig. 6. Total PSNR of received video is 32.43(dB).

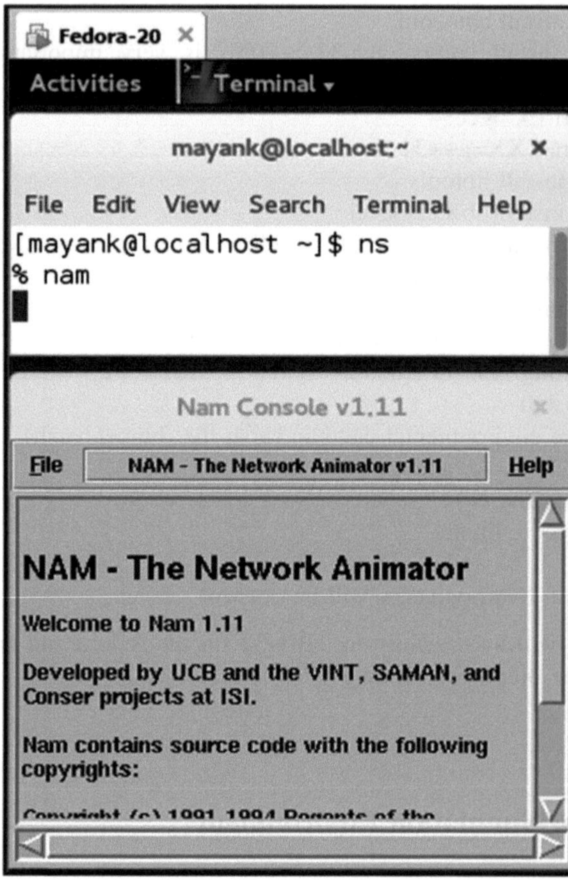

Fig. 1 Output on NS and NAM command

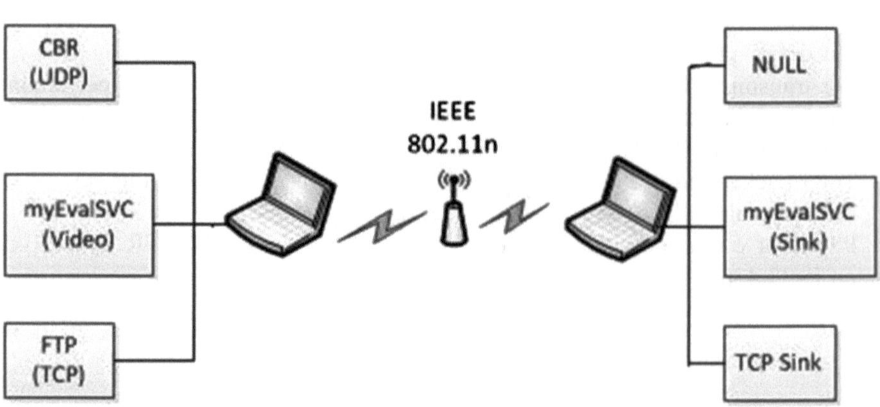

Fig. 2 Simulation topology

Fig. 3 Received video

Fig. 4 Packet end to end delay in seconds

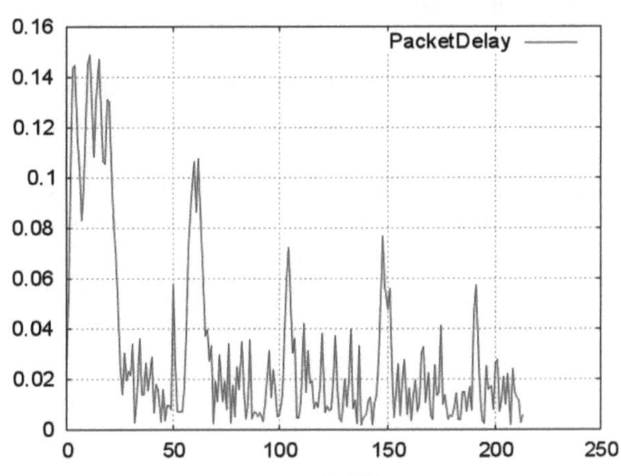

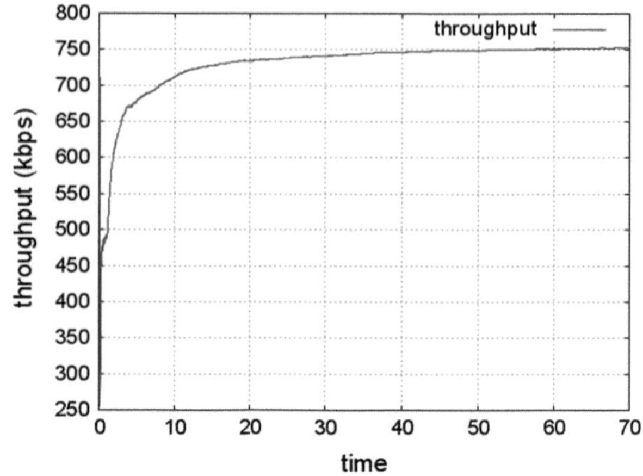

Fig. 5 Throughput of the network

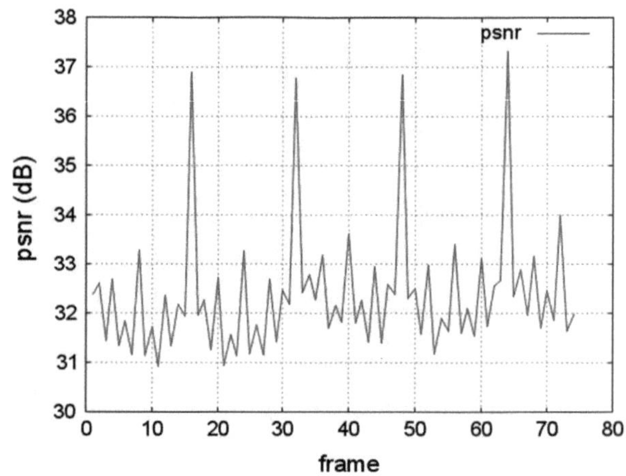

Fig. 6 PSNR of individual received video frames

5 Conclusion

As per the recent needs of performance evaluation of multimedia transmission over enhanced WLAN, we aimed to develop a module to do this. For developing such environment we install NS2 on Fedora operating system. To support simulations of NS2 we install some prerequisites tools and packages of Fedora. Some patches of NS2 have been installed to support multimedia applications. We successfully develop our simulation module and test by simulating a topology and analyzing the performances parameters. Thus our module can be used by researches to evaluate multimedia applications over wireless network.

References

1. Wilder E. Castellanos, Juan C. Guerri, Pau Arce, SVCEval-RA: an evaluation framework for adaptive scalable video streaming. Multimedia Tools and Applications, Springer, DOI: 10.1007/s11042-015-3046-y, pp. 1–25, (2015).
2. Choudhary, Naveen, and Mayank Patel. “QoS Enhancements for Video Transmission over High Throughput WLAN: A Survey.” International Journal of Research and Scientific Innovation Vol 1, Issue 7, pp. 43–50, (2014).
3. Rizk, G.G.; Zahran, A.H.; Ismail, M.H, AVIS: An Adaptive Video Simulation Framework for Scalable Video, Next Generation Mobile Apps, Services and Technologies (NGMAST), IEEE, DOI:10.1109/NGMAST.2014.12, pp. 84–89, (2014).
4. Kalvein Rantelobo, G. Hendrantoro, A. Affandi, A New Scheme for Evaluating Video Transmission over Broadband Wireless Network, Future Wireless Networks and Information Systems, Springer, DOI: 10.1007/978-3-642-27323-0_43, pp. 335–341, (2012).
5. C. H. Ke, myEvalSVC: an Integrated Simulation Framework for Evaluation of H.264/SVC Transmission, KSII Transactions on Internet and Information Systems, Vol: 6 Issue:1, pp. 378–393, (2012).
6. C.H.Ke, C.K. Shieh, W.S. Hwang, A. Ziviani, An Evaluation Framework for More Realistic Simulations of MPEG Video Transmission, Journal Of Information Science And Engineering, Taiwan, 1016–2364, pp. 425-440, (2008).
7. Hung-Chin Jang; Yu-Ti Su “A Hybrid Design Framework for Video Streaming in IEEE 802.11e Wireless Network”, Advanced Information Networking and Applications, 2008. AINA 2008. 22nd International Conference on, pp. 560–567, (2008).
8. A. Dhraief, A. Belghith, N. Montavont, J. Bonnin, M. Kassab, NS2 based simulation framework to evaluate the performance of wireless distribution systems, DOI: 10.1145/1404595.1404636 Conference: Proceedings of the 2007 Spring Simulation Multiconference, SpringSim 2007, Norfolk, Virginia, USA, March 25–29, (2007).
9. L.C. Bishnoi, D. Singh, S. Mishra, Simulation of Video Transmission over Wireless IP Network in Fedora Environment, IJCA Journal, 978-93-80864-99-3, pp. 9–14, (2011).
10. C. Yu Yu, C.H. Ke; C. Kuen Shieh; Chilamkurti, MyEvalvid-NT - A Simulation Tool-set for Video Transmission and Quality Evaluation, TENCON 2006. 2006 IEEE Region 10 Conference, DOI:10.1109/TENCON.2006.343864, pp. 1–4, (2006).

Mitigating Data Segregation and Privacy Issues in Cloud Computing

Bansidhar Joshi, Bineet Joshi and Kritika Rani

Abstract Cloud Computing refers to rendering services to the users over the internet. It includes the hardware and software in the datacenter which facilitates these services. This concept was introduced to increase the utilization of the resources available and to decrease the cost to the user because of the sharing nature of the resources. It offers a number of benefits but with those benefits come a large number of challenges. These challenges need to be tackled by the service provider in order to provide secure and reliable service. Data Storage being the main area of concern as the private data of all the users are stored at one place at all times. This data must be kept available only for that user to whom it belongs and an additional level of security must be introduces in order to prevent breach into these private sections. This paper tries to address these issues and strategies to mitigate them.

Keywords Cloud computing · Data storage · Mitigate · Private data

1 Introduction

Cloud Computing is a term for delivering services over the internet. These services maybe application services (SaaS), platform services (PaaS) or even infrastructure services (IaaS) [1]. IaaS provides basic storage and computational services to the users over the network thereby enabling the users to have more flexibility and control over their data security. PaaS provides the users the flexibility to develop, run and manage their applications without the complexity of managing and maintaining

B. Joshi (✉) · K. Rani
Jaypee Institute of Information Technology, Noida, India
e-mail: bansidhar.joshi@jiit.ac.in

K. Rani
e-mail: kritika.rani17@gmail.com

B. Joshi
Swami Rama Himalayan University, Dehradun, India
e-mail: bineetjoshi@gmail.com

N. Modi et al. (eds.), *Proceedings of International Conference on Communication and Networks*, Advances in Intelligent Systems and Computing 508,
DOI 10.1007/978-981-10-2750-5_18

the infrastructure and thus saving the infrastructure maintenance cost. This way users are in full control of their application but they cannot modify the underneath platform. SaaS is a model in which applications are hosted over the internet. The user only has the control of his application but not over how his requests or data is being handled. IaaS provides maximum flexibility and control over the data whereas SaaS provides the least flexibility and control [2, 3]. Cloud computing is a paradigm which enables resource sharing. It has following advantages:

- **Reduced Cost**: The billing model of a cloud is on pay as per usage i.e. the user has to pay only for the services that he is using. As the infrastructure is not purchased by each user, so a huge amount on the infrastructure is saved. Also the maintenance cost is reduced [4].
- **Increased Storage**: There are a number of occasions when a few more gigabytes of storage is required only for few hours. So paying for the extra gigabyte of storage at all time is wastage of money and resources. Instead, they can be acquired only at the time of requirement as the cloud can scale dynamically. Thereby saving on the storage cost and also increasing the utilization.
- **Flexibility**: With the ever growing market, enterprises are constantly moving up and adapting to these changes. But what really matters is how quickly these changes reach to the users. Cloud Computing enables the enterprises to get their products to a huge market in a very short span of time [5].

With all these advantages of the cloud, there are many issues which surround the world of Cloud Computing; data security and integrity being the most critical ones. In this paper we discussed the data privacy issues, why we need to segregate the

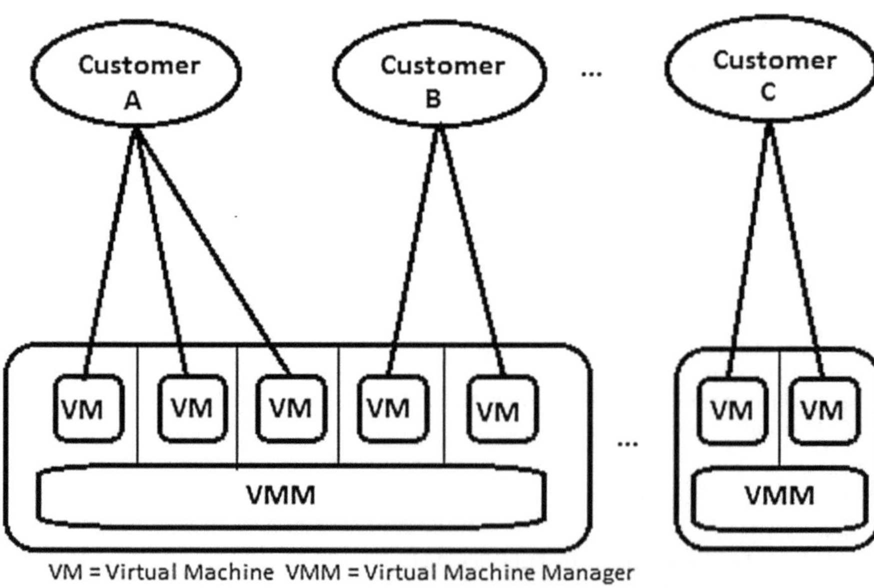

Fig. 1 Basic infrastructure of a cloud depicting resource sharing

data and how can that be accomplished. Basically what we intend to do is to classify the user's data as private or non-private and then apply additional security measures on the private data (Fig. 1).

2 Privacy Issues

With the exponential increase of cloud services, the issues of data privacy in cloud computing services are at the utmost importance. As the time is passing, numbers of service providers are increasing and due to this data stored in data centers are on a rise. This increases the risk of the data being disclosed accidentally or deliberately. Here are few of the data privacy issues which are of utmost importance:

2.1 Loss of Sensitive Data

In a cloud all the resources are shared to reduce the cost due to which the data of all the users is kept in the same container (User's data is stored in individual buckets and these buckets are stored in a common container). If this data is not managed properly and proper access rights are not set then this data can mix or may even be lost. For example in 2007, the British government misplaced personal details of 25 million child benefit records [1]. This lead to huge financial losses to the government as the records could not be tracked.

2.2 Theft

As the storage provider puts everything in a single container, there might be chances when your data can be accessed by unauthorized personnel who may misuse it. The risk of data theft increases as it goes outside our own datacenters and now it becomes the responsibility of the service provider to provide security. For example in 2014, a hacker hacked into Apple's Cloud and private files of many celebrities were flashed online [6]. Also in 2014, nearly 5 million login credentials of Google account users went public on the internet. So one must ensure that cloud service provider must take guarantee of your data in the security point of view.

2.3 Insecurity in Logical Separated Space

Earlier every organization used to store their data in a physically separated environment which used to provide them with extra security for their private data. We do not face this problem when we are using a private cloud as only one organization is in control of that cloud and the data belongs to that single organization. But once we switch to a public cloud, which is used by a number of organizations, the concept of physically separated space is replaced by a logically separated space. This means that it might be possible that data sets of two competitor companies are logical neighbors of each other and thus their data is vulnerable and insecure. Thus the risk of data being stolen is real even when the data of each and every client is logically isolated from each other.

2.4 Data Integrity and Availability

We know that the resources are shared in data centers and data may be kept in the same container. This way the data of one organization can be altered by any other organization or a disgruntled employee could alter or destroy the data. No organization can compromise its data just to reduce the cost. Every organization needs its data at all times to run their business, so we need this data separated and available at all times.

3 Methods of Separating Data

After discussing major threats to the data, we are in the need to perform some operations so as to protect our data from being misused. There are many techniques which have been already implemented. Let's first discuss these techniques along with their drawbacks and then we would propose our algorithm for the same. Here are the techniques which have already been implemented:

3.1 Logical Space Separation

Data of each user must be separated from the data of others. With all the data being stored in the same container, it becomes difficult to ensure that the data is separated. There might be times when the data overflows from one user to the other (Data Leakage Issue) [1].

3.2 Identity Access Management

Proper authentication practices must be performed. Also every user must have a defined set of permissions to access or edit or update any file. This multi-level authorization system can be proven to be very useful in organizational structure. But even this system cannot be a complete solution all by itself [2].

3.3 Encryption

This is one of the most widely used and best ways of protecting data on the internet. In this method the data is encrypted by the user using any suitable algorithm and then transferred on the network to the service provider. This encrypted cipher is stored as it is in the storage container. On requesting access to this file, only those personnel can view this file that have the key. This way the data can be protected. However in the recent past it has been observed that even the encrypted ciphers can be decrypted and thus the data can be stolen [7].

4 Proposed Solution

As we have seen above that encrypted ciphers can also be decrypted without a key and we might face data leakage issues in case of logical space separation, we propose that just separating the logical space of each and every user is not enough. What we need to do is that along with separating the logical space we also need to classify the users' data as Private or Non-Private. We used probabilistic methods for such classifications and all the private data was stored in a Private Data Bucket (PDB) which has an additional level of security i.e. additional encryption, and the rest of the data was stored in a Non-Private Data Bucket (NPDB). Now these two buckets can be encrypted using a single common key and can be stored together in a common data container. Additional level of authorization was also added for the private data bucket (i.e. proper access rights were given to appropriate users) which would grant the access to the files in this bucket (Fig. 2).

4.1 Results and Discussions

First we need to understand that the collection of datasets is a critical task because of the fact that these datasets would help us understand the performance of the system and its feasibility. This can be done in three ways: by using real traffic, by using sanitized traffic, or by using simulated traffic, with the third one being the most common way to create datasets on a test bed network. This traffic made request to upload documents and these documents were to be processed. The environment used for the deployment of the cloud was Ubuntu 10.04 and techniques of Big Data like Map Reduce were used to handle huge chunks of data because of its ability of parallel processing by dividing them into sets of

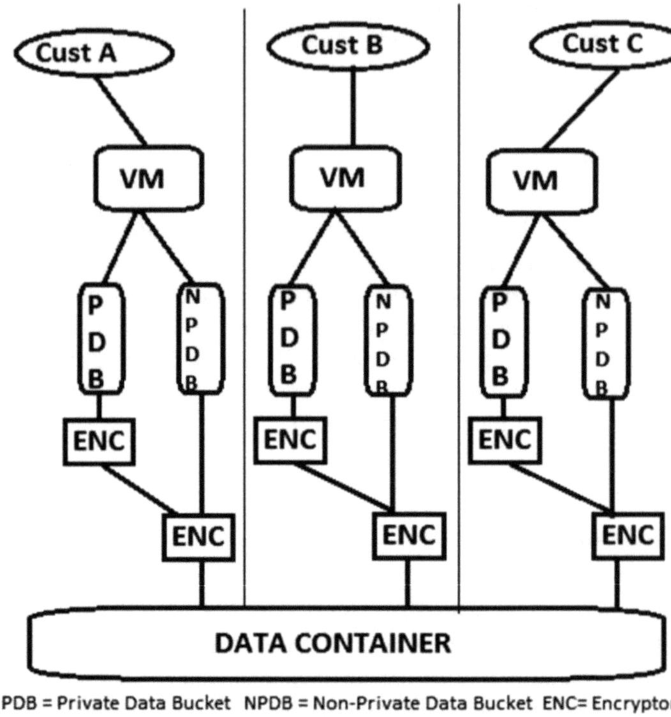

Fig. 2 Proposed architecture for the cloud by segregating private data of each user

independent tasks. This style of parallel processing is supported by some of the major frameworks like Hadoop, Google's Big Table and Spark. For this work we used Hadoop framework. The basic Hadoop framework have the following components; Hadoop Common containing all the library and utility files, Hadoop Distributed File System which stores data in a distributed manner over a network of nodes, Hadoop YARN responsible for resource management activities and Hadoop Map Reduce which is a programming paradigm for large scale data processing. With our approach we were able to classify the data as private or non-private and thus segregate it. Through the segregation of data into private and non-private buckets, it becomes easier to classify the data, not only logically, but also in terms of confidentiality. The additional level of security in terms of authentication requirements for the private data bucket ensures that the confidential data is not tampered with by the wrong hands. In simple terms, extra security ensures that one's private data remains private. The testing is done on the datasets. We used the above said techniques which enables us to handle data set of significant size. With this we are able segregate the data flawlessly in a small amount of time. Also, as the size of the dataset increased, so did the processing time, as can be seen from the graph below (Fig. 3).

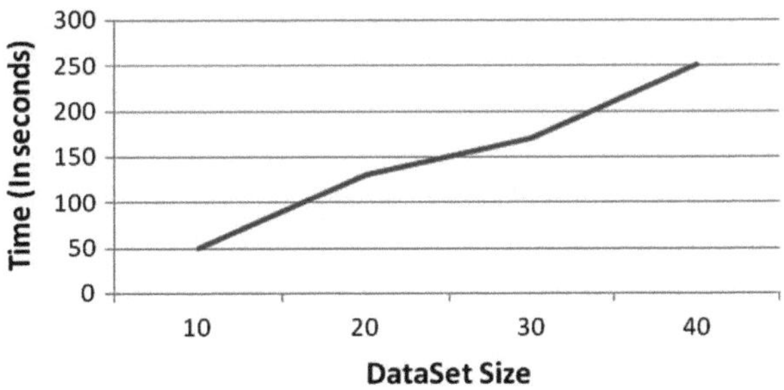

Fig. 3 Average time taken to process data

5 Future Work

We have observed that with our approach we can successfully classify the data as private or non-private. However there might be chances that few non-private files may be classified as private and vice versa. In order to attain more accurate results in the future, we may use techniques of Big Data like Machine Learning Probabilistic Methods which constantly keep on learning the parameters from the input data and thus the number of parameters and their weights constantly keep on changing thus giving us more appropriate classifications. Also better authentication practices can be adopted such as one-time passwords so as to provide better security to our private files. Even though the segregation of data had to be done manually, it leaves a wide scope of improvement in future versions of the project.

6 Conclusion

Cloud Computing is an on-demand access to the shared resources. It helps to reduce the cost, management responsibilities, maintenance, and increase the efficiency of the resources. The advantages are many but with those advantages come challenges which need to be tackled by the service provider like loss of data, theft, reliability, availability etc. This paper majorly focuses on the Data Storage issues, especially on how to secure the private and confidential data of the users. Using the above technique we can definitely prevent breach into our private data even when we face Data Leakage issues in case of logical space separation.

References

1. Armbrust M., et al.: A View of Cloud Computing, Communications, LIII (4) (2010) 50–58.
2. Kaur K., Vashisht S.: Data Separation issued in Cloud Computing, International Journal for Advanced Research in Engineering and Technology (2013).

3. Kaur P. J., Kaushal S.: Security Concerns in Cloud Computing, HPAGC 2011, Chandigarh (2011).
4. Morsy M., Grundy J.: An Analysis of The Cloud Computing Security Problem, APSEC 2010, Sydney, Australia (2010).
5. Velte A. T., Velte T. J., and Elsenpeter R.: Cloud Computing-A Practical Approach, The McGraw-Hill Companies (2010).
6. http://en.wikipedia.org/wiki/2014_celebrity_photo_hack.
7. Hashizume K., et al.: An analysis of security issues for cloud computing, Journal Of Internet Services and Application (2013).

Software Risk Measurement and Interpretation with Generated Precedence Matrix

Harshit Tripathi and Subhash Chand Gupta

Abstract Any software which is in its operational environment, the chances of it being a success is very high, however there are many chances of it being a failure too. It is established that there are several risks involved in making an awfully prime software on time and within budget. However, for it to process and be productive we need to handle such risks somehow. They need to be remunerated for a perceived reward. This paper focuses on essentially the procedures to tackle software risks, their management and mitigation.

Keywords Software risks · Risk Id · Risk impact · Risk occurrence · Rank probability · Qualitative analysis matrix

1 Introduction

After understanding the fundamentals of software risk measurement and before developing this application, we formulated/assumed a real-world situation involving a big IT software organization that has its branches across the world. When the organization wins a software project and requires its team members to work from geographically different or same locations, it becomes very difficult for the members(specifically for the senior people) to communicate with each other, especially on the significant phases of the project [1].

Keeping this objective into consideration, I have planned to develop an application through which I have assumed three important people who are associated with the project that would be able to log in securely in the application in order to

H. Tripathi (✉) · S.C. Gupta
Amity University, Noida, India
e-mail: harshittripathi18@gmail.com

S.C. Gupta
e-mail: scgupta@amity.edu

N. Modi et al. (eds.), *Proceedings of International Conference on Communication and Networks*, Advances in Intelligent Systems and Computing 508,
DOI 10.1007/978-981-10-2750-5_19

identify and evaluate risks associated with the project. The respective designations of these 3 people have been assumed as: Project Manager, Team Leader and Senior Developer.

2 Literature Review

In order to mitigate various risks of a project, first of all it was decided on which policy a particular software project is made. Policy is basically the procedure on which how a software project is made that involves gathering all the requirements for the development of the project. Then it involves deciding or forecasting demanding of the project in limited duration of time. The most important thing is to decide what actually you are going to do or what will be the outcome of the project or what will the ultimate product after the project is successfully deployed [2]. Then what you are going to decide is on which platform and what manpower and time span you have decided to allocate to each and every particular task.

Then what actually comes in a software project management is the review phase. It is in this phase our actual work starts. What basically is done in this phase is we actually find what are the various risks associated with the project on the basis of assessment done by the major people involved in the project that could be project manager, team leader, senior manager.

Then comes monitoring phase in which above personnel who have great knowledge or experience in their field and are excellent, determine what could actually be poised a major threat or risk to the project. So, what they do is give a particular Risk Id, Risk Name, Risk Impact which could be High, Medium, Low which is given to a particular risk on the basis of its effect on the successful completion of the project. Also the manager can give cause description and rate them according to their effect on occurrence of a particular risk [3] (Fig. 1).

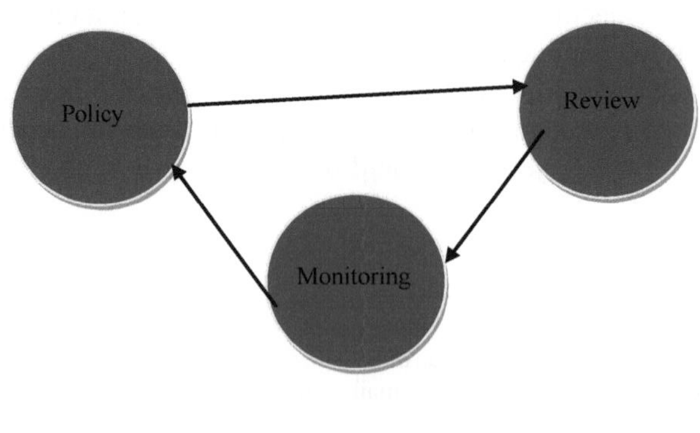

Fig. 1 Risk evaluation and mitigation mechanism in software companies

3 Methodology

The users would log the details of all possible risks associated with any project. The risk creation for any software consists of below properties.

Risk id: Every risk will be assigned a unique auto-generated identification number fora convenient reference and naming.

Risk Name: This would be the name of risk associated with the project. Any user who would be analyzing the risks will name it according to its behavior. The risk name is assumed to be entered in such a term that is understandable to all other users using this tool.

Risk Impact: The impact of a risk could be Low, Medium or high depending upon the user's perception of its impact on the project. For ex: If the risk is of High impact, it may adversely affect the successful completion of a project; may be even resulting in situations like huge overheads or compensation being paid by the company to the client.

In that scenario, the user will evaluate all the possible causes that may lead to the occurrence of that risk which will help him to devise effective measures for the risk-elimination. Similarly, the Medium and Low impact risks would be identified based upon their impact on the project in terms of time and cost. The whole significance of this parameter is to identify the rate the possibly involved risks in a project and eliminate them according to their severity by eliminating all their possible causes of occurrences [4].

Cause id: This would be an auto-generated identification of the cause of a risk in order to refer to a particular cause conveniently.

Cause Description: This would include a short description about the cause for a particular risk. For ex: if the risk is time, the possible causes could be Manpower Shortage, Technical Competency, and Project Overload etc. It is assumed that the cause description would be entered in such a manner that is understandable to all other users using the system.

Cause Probability: This would be the probability of occurrence of a particular cause on a scale of 0–10. The cause-probability of a particular risk signifies the proximity or contribution of a particular cause in the occurrence of a corresponding risk. It is assumed that this value is selected very carefully by the user based upon his/her experience and project's characteristics like domain, complexity, time etc. because this number will be later used in the calculation and determination of overall probability of occurrence of a risk based upon its inherent causes.

Rank Probability: This would be the rank or sequence no. of occurrence of a particular cause. The rank probability of a particular cause will highlight its position in the chain of occurrence of causes of a particular risk. We realistically assume that the selection of rank probability will be done by the user based upon his/her experience and project's characteristics like domain, complexity, time etc. [5] because this number will be later used in the calculation and determination of overall probability of occurrence of a risk based upon its inherent causes.

Rank Probability: This would be the rank or sequence no. of occurrence of a particular cause. The rank probability of a particular cause will highlight its position in the chain of occurrence of causes of a particular risk. We realistically assume that the selection of rank probability will be done by the user based upon his/her experience and project's characteristics like domain, complexity, deadline etc. As this figure in relation with cause probability will determine the severity or importance of a particular cause occurrence of a risk, the user needs to select it with utmost responsibility. The user should also be able to modify the inputted risk details. The editing of risks could be done in 3 ways namely [6]: *Risk Editing:* Through this feature, the user would be able to edit the *Risk Name, Risk Impact.* As we know, the evaluation, analysis and measurement of risks involved in a software system is a continuous process and involves discussion amongst the team members who are involved in the system.

Hence, there always exists a possibility to change the risk details based upon the mutual perception of team members regarding same. It may be noted that modifying the impact of a particular risk will seriously affect its overall probability of occurrence in the project and therefore, should be done with utmost caution and responsibility in order to have accurate predictions or statistics about risk evaluation Moreover, this functionality will also enable the user to delete any particular risk. Again, this is an extremely significant decision to delete the complete details of any risk and we assume that it is done after extensive discussion among the team members and a mutual decision on same. If all the users think that all the possible causes for that risk can be completely and safely eliminated, the corresponding risk should be removed from the system. This action requires utmost caution and responsibility. Moreover this functioning will also enable you to delete the risk. If all the users think that all the possible causes for that risk can be completely and safely eliminated, the corresponding risk should be removed from the system. This action requires utmost caution and responsibility.

Editing Risk-cause: Through this feature, the user will be able to modify or edit the particular cause(s) of any risk. The modification of risk-cause properties will impact in the calculation of overall risk-occurrence probability [7].

Hence, we assume that this feature will be used after mutual discussion and decision of the involved team members about alleviating or degrading any cause in the chain. Plus, this feature will also enable the users to delete particular cause(s) of the risk. Again, we realistically assume that any risk-cause will only be deleted if the team members are absolutely confident of its successful elimination.

Addition of new risk-causes: This function will enable the user to add further more causes to a particular risk. As we know that the analysis and measurement of risks involved in a software system is a continuous process, if all the team members mutually decide upon the addition of more causes to a particular risk, they can do it through adding the [8] Cause Description, Cause Probability and Rank Probability features of the cause of a particular corresponding risk.

Risk Probability: The occurrence probability will be calculated by the summation of multiplication of corresponding cause and rank probabilities of causes of a particular risk divided by the summation of rank probabilities of the corresponding causes of that risk.

This calculation will give us the probabilities of occurrence of any particular risk on a scale of 0–10 and will help us in prioritizing them for elimination. The determination of risk probability in terms of High, Medium and Low will be done based upon the below condition:

Low: 0 < Risk probability <= 3
Medium: 3 < Risk probability <= 7
High: 7 < Risk probability <= 10

Qualitative Analysis Matrix: This matrix will help me in determining the risk severity on the basis of its value it is placed in the matrix. Based on the observed risk impact and calculated risk probabilities, we will place the risks in the matrix.

Along the x-axis or horizontal axis is "Risk Probability" and y-axis or vertical axis "Risk Impact". After I have placed all the risks in the appropriate cells, it will become convenient and well classified to sort and prioritize the risks based on their severity.

4 Result

In the above sample output, it is being shown how the risks would be finally aligned in a 3 by 3 matrix on the basis of their probability and impact. The matrix quadrants would consist of Risk ID and Risk Name as the data. The above interpretation will help the organizations to prepare mitigation strategies for those risks first that fall into the quadrant of high impact and high probability by eliminating their cause(s) one-by-one in the order of priority (Fig. 2).

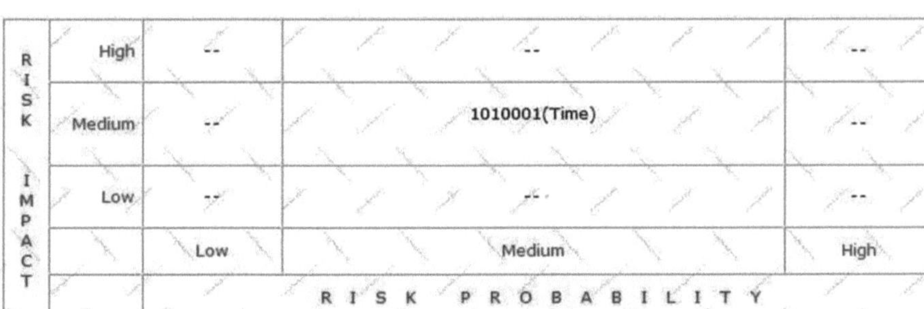

Fig. 2 Qualitative analysis matrix

5 Conclusion

There are many risks involved in creating high quality software on time and within budget. In order to successfully manage a software project one must learn to identify, assess, analyze and control the risks. Software Risk Management basically involves the process of brainstorming in identifying all the possible risks involved. This activity is usually done by the people involved in the heart of application development. But since manpower employed in this process is not found worthwhile to identify and mitigate risks especially in a company which operates in several different locations. This paper orients itself on analyzing and prioritize risk aspects and try to rectify it with proper policy and procedures, which then can easily be acknowledged and mitigated by senior officials.

References

1. Haneen Hijazi, "A Review of Risk Management in Different Software Development Methodologies", International Journal of Computer Applications (0975 – 8887) Volume 45– No. 7, May 2012.
2. Lazaros Sarigiannidis, Prodromos D. Chatzoglou, 'Software Development Project Risk Management: A New Conceptual Framework', 2011.
3. R. Dash and R. Dash, "Risk assessment techniques for software development," European Journal of Scientific Research, Vol 42, No. 4, 2010, pp. 629–636.
4. Boehm Barry W, "Software Risk Assessment", IEEE Computer Society Press, Vol. 15 (7), pp. 902–916, 1989.
5. Ayse Kucuk Yilmaz and Triant Flouris, 'Managing corporate sustainability: Risk management process based perspective', African Journal of Business Management Vol. 4 (2), pp. 162– 171, 2010.
6. Bocean C.G, Managementul protectelor in afaceri', Edituaria Universataria, Craiova, 2007.
7. Aubry M., Hobbs B, "A New Framework for Understanding Organizational Project Management through the PMO", 2011.
8. Abdullah Al, Murad Chowdhury and Shamsul Arefeen, 'Software Risk Management: Importance and Practices', 2011.

IMSS: A Novel Approach to Design of Adaptive Search System Using Second Generation Big Data Analytics

Dheeraj Malhotra and O.P. Rishi

Abstract In this present era of Big Data, different search engine users have different information requirements at different intervals of time. Thus, search results should be adapted to user's requirements [1, 2]. In this research work, we propose a novel approach to adaptive web search augmented with capabilities of carrying out Big Data Analytics using second generation HDFS. Moreover, unlike conventional personalization techniques, the proposed approach does not require additional efforts from user such as reporting feedback/ratings etc. The proposed system can be implemented in the form of Intelligent Meta Search System (IMSS Tool) to overcome the problem of irrelevant web page retrieval faced by user of generic search engines. An extensive experimental evaluation shows that the average ranking precision of adaptive IMSS tool improves with trial runs when compared with a popular search engine.

Keywords Second generation HDFS · Personalized search · Big data search system · Meta search engine · Intelligent meta search system (IMSS) tool · Adaptive web search

1 Introduction

Adaptive search when supported by HDFS-Cloud framework leads to easy and efficient analysis of Big Data available on WWW to retrieve useful personalized page ranking patterns. Search engines are known to retrieve far larger information

D. Malhotra (✉) · O.P. Rishi
Department of CSI, University of Kota, Kota, Rajasthan 324005, India
e-mail: Dheerajmalhotra@ymail.com

O.P. Rishi
e-mail: Omprakashrishi@yahoo.com

N. Modi et al. (eds.), *Proceedings of International Conference on Communication and Networks*, Advances in Intelligent Systems and Computing 508,
DOI 10.1007/978-981-10-2750-5_20

but still no search engine can index more than about 16 % of index able web [3, 4]. The issue is not just only the volume but is also the relevancy with respect to user's information needs [1, 2]. When the same query is searched by different users, even a state of art search engine returns the same result, irrespective of the user submitting the query. For example, if a user is tech savvy and usually searches for laptop/mobiles then an incomplete query search like *Blackberry* should return documents related to *Blackberry mobiles* by intermediately expanding the query rather than returning the documents of some fruit. There are various types of conventional personalized search systems as discussed in literature. However these search systems fail to satisfy the user personalized requirements without having explicit ratings/feedback from user. Moreover such systems can't handle second generation Big Data as they not just require scalability, partial failure support etc. but also need to support multiple analytic methods on varied data types, as well as the ability to respond in near real time.

2 Contribution from the Study

To the best of our knowledge, this proposed research work is the first formal attempt to design and development of adaptive search system using intelligent big data analytics and is also deployable on cloud framework. Various contributions of the proposed approach may be summarized as follows:

- The user effort for providing explicit ratings/feedback in order to use personalized search system will no longer be required.
- The proposed system will overcome the limitations of traditional mining approaches to extract useful web search and page ranking patterns from Big Databases of Search engines by providing features like Scalability, Partial Failure Support etc.
- The proposed research work discusses the design of future ready intelligent search tool i.e. *IMSS* which can well satisfy the requirements of next generation Big Data Search System such as Real time response, support of multiple analytic engines.

3 System Design

The proposed system will follow modular approach as shown in Fig. 1. Here we first accept user search query and expand the same to intermediate query based on user's preferences obtained from his search history [5–7]. Proposed system will build user profile using user's long term and short term preferences derived from browsing history of n days ago and of current day of usage respectively. Meta key word recommender is used to derive Meta keywords of search from extracted

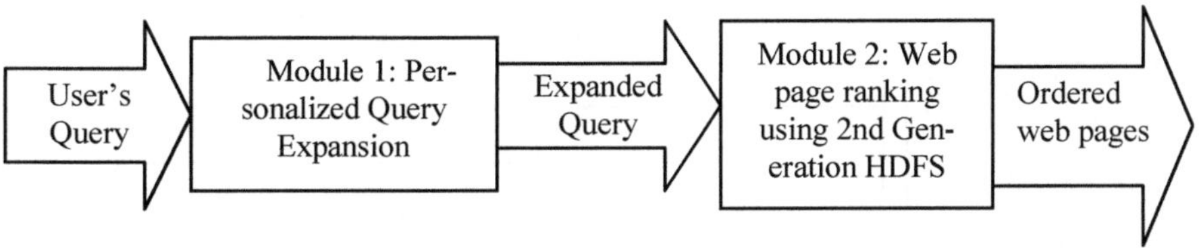

Fig. 1 Simplified design of proposed system

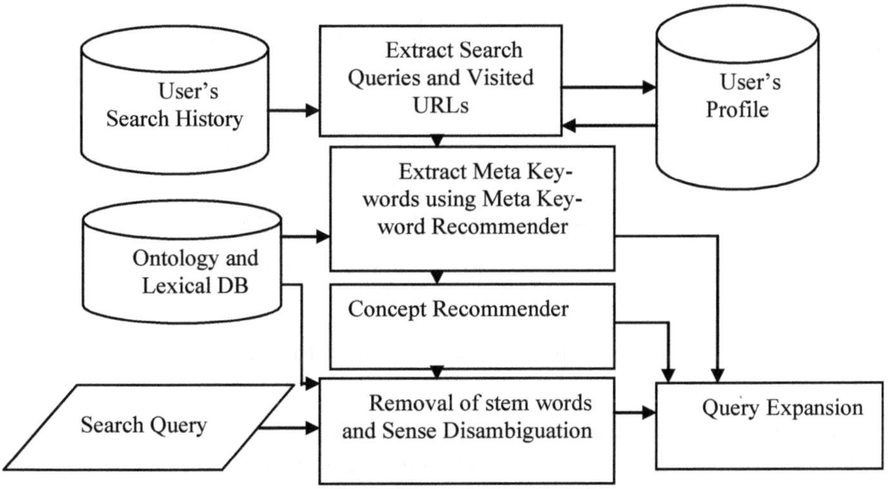

Fig. 2 Design of Module 1—personalized query expansion/modification

URLs. Similarly, Concept Recommender and Word Sense disambiguation processes are used for expanding user query into non ambiguous and more meaningful query as shown in Fig. 2. Module 2 is used for ranking of web pages obtained from backend search engines. HDFS Map() and Reduce() approach is used to calculate content relevancy vector; other relevancy vectors such as semantic relevancy vector (SRV) to determine the semantic closeness of user query with respect to web document under consideration, similarly Time Relevancy Vector is based on importance given by previous user of same web page. The detailed functionality of module 2 to determine weighted rank of candidate web page is shown in Fig. 3.

4 Second Generation HDFS and Map Reduce

There are two significant trends of Second Generation Big data Systems [2, 8] that are responsible for choosing second generation HDFS as a preferable deployment framework in proposed approach. (i) There is rapid growth in network bandwidth as

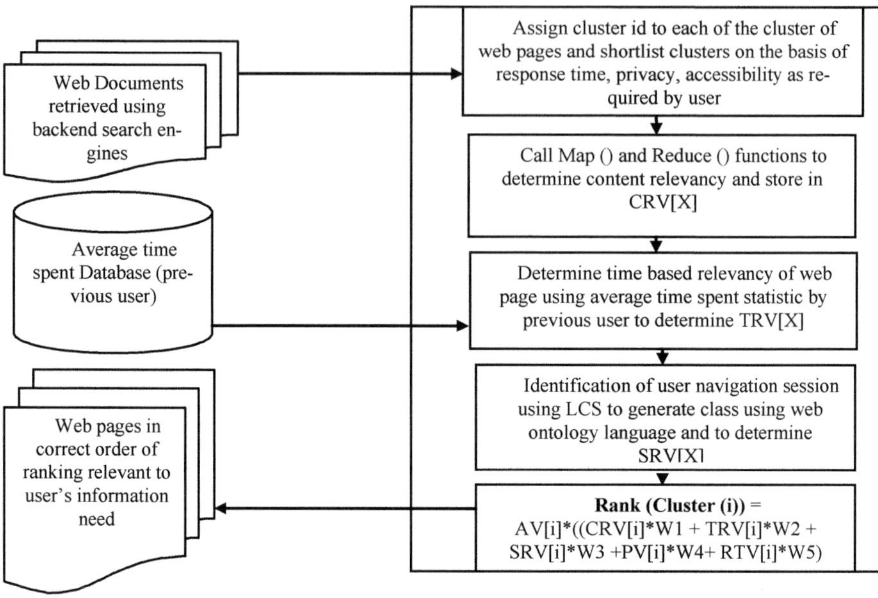

Fig. 3 Design of Module 2—web page ranking using HDFS based cloud framework

compared to hard drive bandwidth (ii) Development of In-memory computation models is urgently required to allow intermediate results to be kept in memory and hence reduces overhead of iterative analytics as suffered by conventional HDFS [9, 10].

HDFS is now adapted as long term store from which applications read their initial data and write their final results. The data layer is divided into sub layers for consistent storage and for intermediate objects separately to handle second generation of Big Data as shown in Fig. 4. In our proposed System, Map function will accept cluster ID as key and cluster log as second argument to tokenize each of web link entry in cluster log, obtained from back end search engine used by IMSS tool, to count individual occurrence of each of the keyword of search query. Extract () function is used to generate elements in list one at a time. Reduce function is coded to aggregate over all the occurrence of each keyword as provided by Map ()

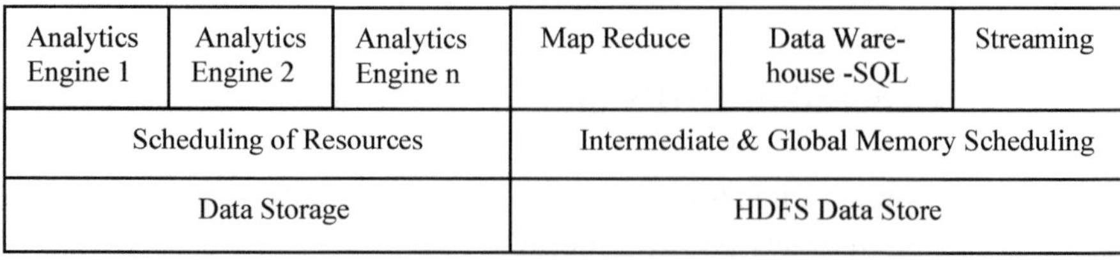

Fig. 4 HDFS deployment framework for second generation big data system

function [11] to determine keyword frequency in each of the web document and hence to determine the content relevancy vector. Map and Reduce code to be used by **Proposed System** is as follows:

```
Map (Int ID, String Log){
      List<String> X = tokenize (log)
        For each Token in X {// Token - Link extracted
                             //from back end search engine
Extract ((String) KWL, (Int) 1) // KWL - Keyword list
        }}
Reduce (String Token, List <Int> count)
              Int F = 0
             For each word in KWL {
                    F = F + 1
//F- Frequency count of each keyword
extract((string) token, (Int) F)}
```

5 Intelligent Meta Search System

In order to evaluate the proposed research design, *IMSS* tool using HDFS framework for analytics of second generation of Big Data is implemented using ASP. NET framework. The interface of IMSS tool is shown in Fig. 5. After Sign In, the inter-face of tool may allow user to select some or all of the four popular search engines like Google, Yahoo, ASK and Bing, for the purpose of intermediate web pages retrieval and search box allow user to specify search string. After clicking the Search button, tool will assign personalized rank to some of the top web links retrieved from back end search engines based on the calculation of various ranking vectors such as AV, SRV, CRV, TRV, RTV. The tool will return web links in the order of their ranking along with statistic of selected advanced search criterion. However *Take Me Fast* tab will not allow selecting any of the search criteria and will give result directly on the basis of user's history of browsing patterns stored in user's contextual database, which could be retrieved using his/her profile.

6 Comparative Precision Analyses—IMSS Tool V/S Google

In order to evaluate the effectiveness of our proposed approach, we recruited 10 human volunteers with age varied from 20 to 50 years with minimum of 5 years web search experience. 6 of them were males, 4 were females. They are asked to bring their personal laptops with installed IMSS tool followed by initial profile sign

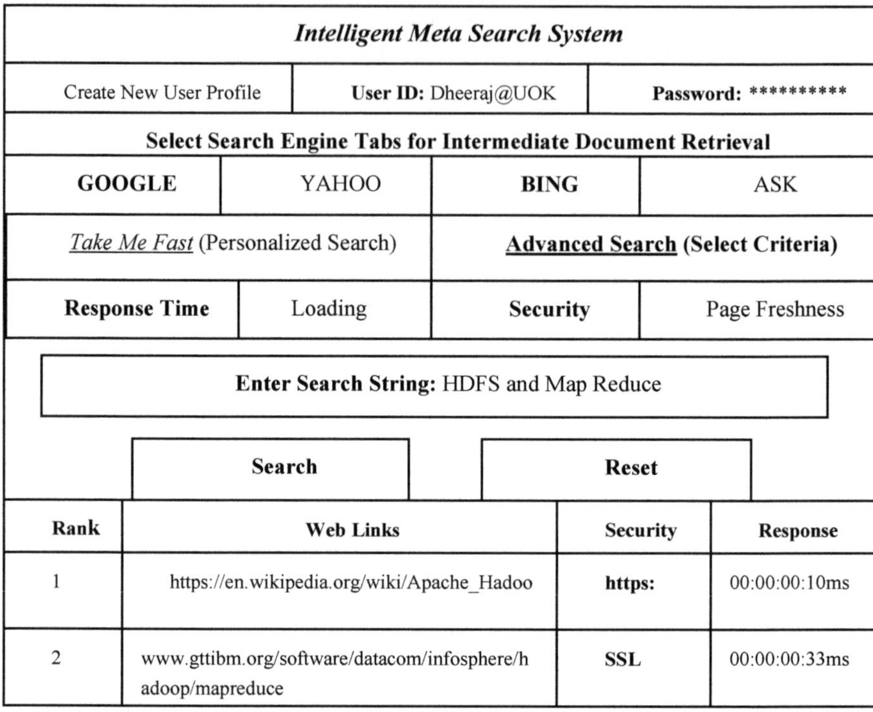

Fig. 5 Interface of IMSS tool

up process on tool, we followed following steps and asked volunteers to repeat the process for at least 4 trial runs one by one on Tool and Google:

1. In the first step, we asked volunteers to search an intentional incomplete query, for example a query like *Black Berry* rather than *Black Berry Mobiles* or *Black Berry Fruit.*
2. In the second step we asked volunteers to give points from 0(worst) to 5(best) to various precision parameters such as personalized page relevancy, page freshness, page size and response time to the top 10 links with respect to their shown rank in output of IMSS and Google.
3. After collecting data from each of the volunteer, we normalized the value of various precision parameters using expression:

$$Q_{\{ab\}} = (\mathrm{HIG}\,(P_{\{ab\}}) - P_{\{ab\}}) / (\mathrm{HIG}\,(P_{\{ab\}}) - \mathrm{LOW}\,(P_{\{ab\}}))$$

where, P_{ab} = Value of b_{th} Parameter of a_{th} web page; Q_{ab} = Normalized value of b_{th} Parameter of a_{th} web page; LOW, HIG = Lowest and Highest value of each of the parameter of precision.

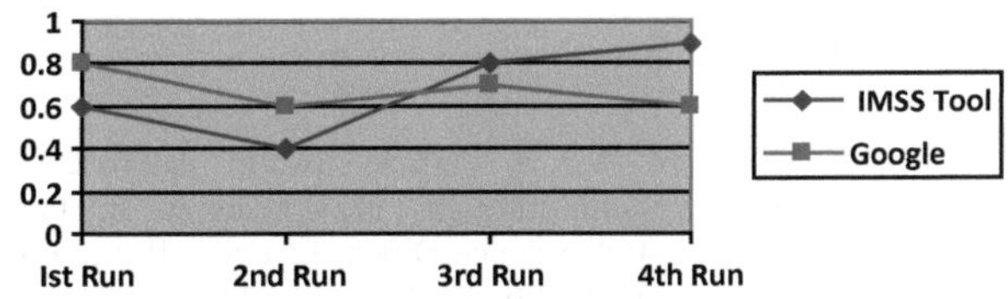

Fig. 6 Personalized precision comparison of IMSS tool with Google for Query "*BlackBerry*"

4. In the next step, we calculated the overall weighted precision of each web page retrieved by each volunteer as $N_a = \sum W_b \cdot Q_{ab}$, where, N_a = weighted precision of a_{th} web page; W_b = Weight assigned to b_{th} parameter by volunteer, usually $0 <= W_b <= 1$
5. Finally we determined overall precision by calculating average of all the weighted precisions as obtained from volunteers, Precision = AVG (N_a).

6.1 *Observation*

The graphical analysis in Fig. 6 shows that during first trial Run, precision of Google is reported as high; however with increase in number of trial runs, average precision of Tool improves slowly over Google. This is due to the fact that that Tool will build user profile and by employing personalized search can better satisfy the user for incomplete or ambiguous queries; On the other side, generic search engines try to interpret the query with all possible meanings without considering the preferences of user who searched for query and hence fails to achieve high value of personalized search precision.

7 Conclusion and Future Work

This research work present a HDFS based adaptive search framework for analytics of second generation of Big Data through implementation of IMSS Tool. The effectiveness of proposed approach is justified by experimental evaluation and comparison of personalized precision of IMSS tool over Google. The proposed approach can be applied to retail transactional or E Commerce website database as such transactional databases are also growing in the scale of Terabytes on daily basis and hence they require second generation Big data analytics system to mine useful customer buying patterns rather than conventional data mining techniques. The proposed system design can be enhanced by incorporating other advanced technologies such as Back Propagation Neural Networks, SVM etc. to further improve the precision of tool.

References

1. Wasid, M., Kant, V.: A Particle Swarm Approach to Collaborative Filtering based Recommender Systems through Fuzzy Features. In: Procedia Computer Science, IMCIP, Vol. 54, pp. 440–448, Science Direct, Elsevier, Bangalore, India, August 21–23 (2015).
2. Gebara, F., Hofstee, H., Nowka, K.: Second Generation Big Data Systems. pp. 36–41,Cover Feature Outlook, IEEE Computer Society (2015).
3. Shou, G., Bai, H., Chan, k., Chen, G.: Supporting privacy protection in personalized web search. In: IEEE transactions on knowledge and data engineering, Vol. 26, No 2, pp. 453–467. IEEE (2014).
4. Kuppusamy, K.S., Aghila, G.: CaSePer: An Efficient Model for Personalized Web Page Change Detection Based on Segmentation. Vol. 26, pp. 19–27, Journal of King Saud University, Elsevier (2013).
5. Verma, N., Malhotra, D., Malhotra, M., Singh, J.: E-commerce website ranking using semantic web mining and neural computing. In: International Conference on Advanced Computing Technologies and Applications, Elsevier Procedia Computer Science, Vol. 45, pp. 42–51. Elsevier, Mumbai, India, March 26–27 (2015).
6. Malhotra, D.: Intelligent Web Mining to Ameliorate Web Page Rank using Back Propagation Neural Network. In: 5th International Conference, Confluence: The Next generation information Technology Summit, pp. 77–81, IEEE Xplore, UP, India, September 25–26 (2014).
7. Malhotra, D., Verma, N.: An ingenious Pattern Matching Approach to Ameliorate Web Page Rank. Vol. 65, No 24, pp. 33–39, International Journal of Computer Applications, FCS, New York, USA (2013).
8. Khurana, A.: Bringing Big Data Systems to the Cloud. pp. 72–75, What's trending? Column, IEEE Computer Society (2014).
9. Tesai, C., Lai, C., Chao, H., Vasilakos, A.: Big Data Analytics: A Survey. 2:21, pp. 1–32, Journal of Big Data, SPRINGER (2015).
10. Singh, A., Velez, H.: Hierarchical Multi-Log Cloud-Based Search Engine. In: 8th IEEE International Conference on Complex, Intelligent and Software Intensive Systems, pp. 212–219. IEEE CPS, Birmingham, UK, July 2–4 (2014).
11. Son, J., Ryu, H., Yi, S., Chung, Y.: SSFile: A novel column-store for efficient data analysis in Hadoop-based distributed systems, Vol. 316, pp. 68–86. Elsevier Information Sciences, September 20 (2015).

Energy Efficient Cluster Head Selection in Energy-LEACH

Hiral Pambhar, Kausa Aghera and Naren Tada

Abstract Wireless sensor network (WSN) is all time researched area as well as new concepts are introduced frequently. Here we will consider the hierarchical clustering routing protocol called LEACH and its variant as E-LEACH (energy leach). There may be a case in E-LEACH that selected cluster heads may be accumulated in certain specific area (or not distributed) due the consideration of only residual energy. Our proposed method includes the consideration of residual energy as well as selected cluster heads to be distributed in the entire wireless network. This balance load and increase the network life time.

Keywords Load balancing · Distributed cluster heads (CHs) · Residual energy · Energy efficient protocol · Cluster heads (CHs) selection

1 Introduction

WSN is a real time environment. That includes large number of sensor nodes and base station (sink) connected with each through wireless medium [1]. The sensor nodes are configured such that it is capable to sense the environment factors like temperature, pressure, motion and etc.

This information is transmitted or forwarded (in the form of signals) to the base station (BS) where processing is carried out. In Wireless Sensor Networks (WSNs), battery, bandwidth, transmission power and processing capabilities are main constraints which cause maximum energy dissipation. So there must be some techniques which reduce energy consumption. WSNs facilitate many applications like

H. Pambhar (✉) · K. Aghera · N. Tada
VVP Engineering College, Rajkot, Gujarat, India
e-mail: hiralpambhar@gmail.com

K. Aghera
e-mail: agherakausha@gmail.com

N. Tada
e-mail: naren.tada@gmail.com

N. Modi et al. (eds.), *Proceedings of International Conference on Communication and Networks*, Advances in Intelligent Systems and Computing 508,
DOI 10.1007/978-981-10-2750-5_21

target tracking, environmental monitoring, habitat observation (Healthcare), Military monitoring, Building monitoring and so on. Most of these applications require only the aggregated value to be reported at the base station (or sink).

Routing protocols which uses clustering technique are used for having higher energy efficiency and also for increasing network life-time. This paper proposes technique for selecting cluster head in energy-efficient manner with consideration of residual energy.

2 Leach

2.1 Introduction

LEACH is a routing protocol for wireless based network. As name suggest it is cluster-based, adaptive (node can be added or removed into the cluster) as well as hierarchical (more than one hope i.e. sensor node to CH and from CH to BS). In short LEACH includes following customized features such as: randomized selection of CHs, self-configuring cluster formation, controlled aggregated data transfer using routing techniques. Concept of LEACH protocol is resides on forming clusters which includes sensor nodes as a member based on RSSI and use local CH as routers to BS i.e. all members have to forward data which is sensed, to the BS through CH. It reduces the extra wastage of energy which was there in conventional protocols that is each and every sensor node of WSNs can transmit information to BS, leading to more energy consumption due to more communication distance between sensor motes and sink as well as traffic.

Working of LEACH protocol is divided in rounds. Each round is completed when its two phases—(i) setup (ii) steady state phase are executed completely. In first step CHs and clusters are formed. Here, nodes are randomly self-elected as CHs after considering probability value P with checking whether the node was CH or not prior to current round. Those which were not CHs in prior 1/p rounds select number between 0 and 1, selected value if lower compared to threshold T(n) then nodes results into the CHs. Calculation of T(n) is taken as a formula [2]:

$$T(n) = \begin{cases} \frac{p}{1-p*(r mod \frac{1}{p})} & \text{if } n \in G \\ 0 & \text{if } n \notin G \end{cases} \quad (1)$$

According to this formula G is set, that consist of nodes that not been elected as CHs in prior 1/p rounds, P is percentage of CH which need to be considered (number of CHs to be selected depends on many factors), r is current round. Selected CHs broadcast their state using CSMA/CA (Carrier Sense Multiple Access/Collision Avoidance) protocol. Non-CH nodes select their CHs by comparing the received signals strength (RSSI) from multiple selected CHs. Once the clusters are formed, all CHs will create TDMA (Time Division Multiple Access) schedule for members within the cluster and broadcast it.

On the completion of set-up phase, a steady state phase begins, that is generally larger compared to previous step. In the second step, nodes (non-CHs) forwards data to own CHs in their assigned time slots (the allotment of the TDMA slot to the non-CHs is done by the CH) thereby keeping its antenna active (ON state) during their data communication. For rest of period they are left in sleep mode (OFF state) thereby, increasing battery lifetime. Here CH is always going to remain in active state. On receiving data information from all members within cluster from non-CHs, CHs will accumulate and aggregate this data (information) and transmit it to the BS [2].

Figures 1, 2 and 3 shows flowcharts of LEACH protocol [3], cluster creation in LEACH and various processes involve for nodes in LEACH [3] respectively.

2.2 Advantages of LEACH

The benefits of this routing protocol are such as: no more large distance interaction to BS is necessary. No need of acquiring exact area of nodes in network required, in order to establish uniform cluster creation. Moreover, overall interaction is not required to form clusters and no assumption required about present status of

Fig. 1 Flow diagram of LEACH

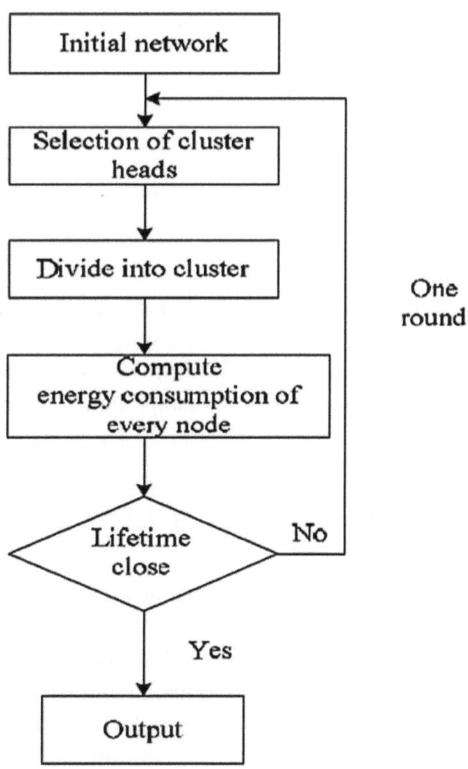

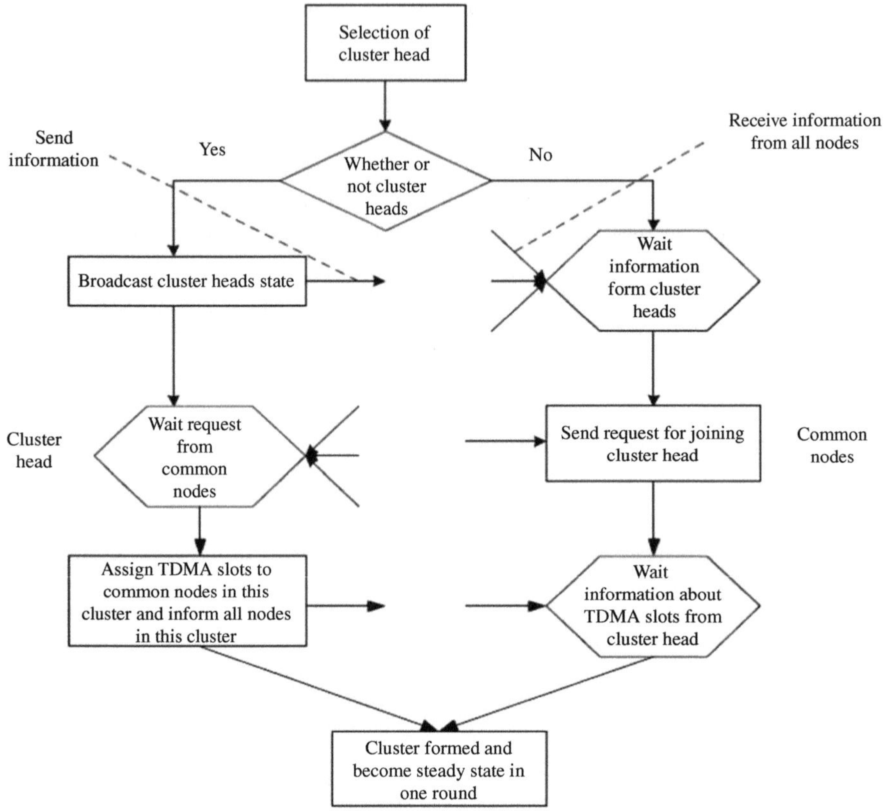

Fig. 2 Cluster creation of LEACH protocol

remaining node while cluster creation. Here aim is to have overall result of formation of clusters out of available nodes, by local decisions taken individually by every node. LEACH-protocol randomly elects CH node considering prior rounds selection of CH, So that energy of whole network is uniformly divided among all sensor nodes that can reduce energy wastage hence increases network life period [4].

2.3 Disadvantages of LEACH

Though LEACH enhance network life period with respect to conventional multi-hop and static routing, some of issues are still found. Randomly cluster heads (CHs) are chosen or elected, so optimal number and uniform distribution of CHs may not be obtained. Nodes having minimum residual energy get equal chance to become CH, along with node having maximum residual energy. Hence, there is a chance for nodes having minimum left energy may be selected as CHs sometimes

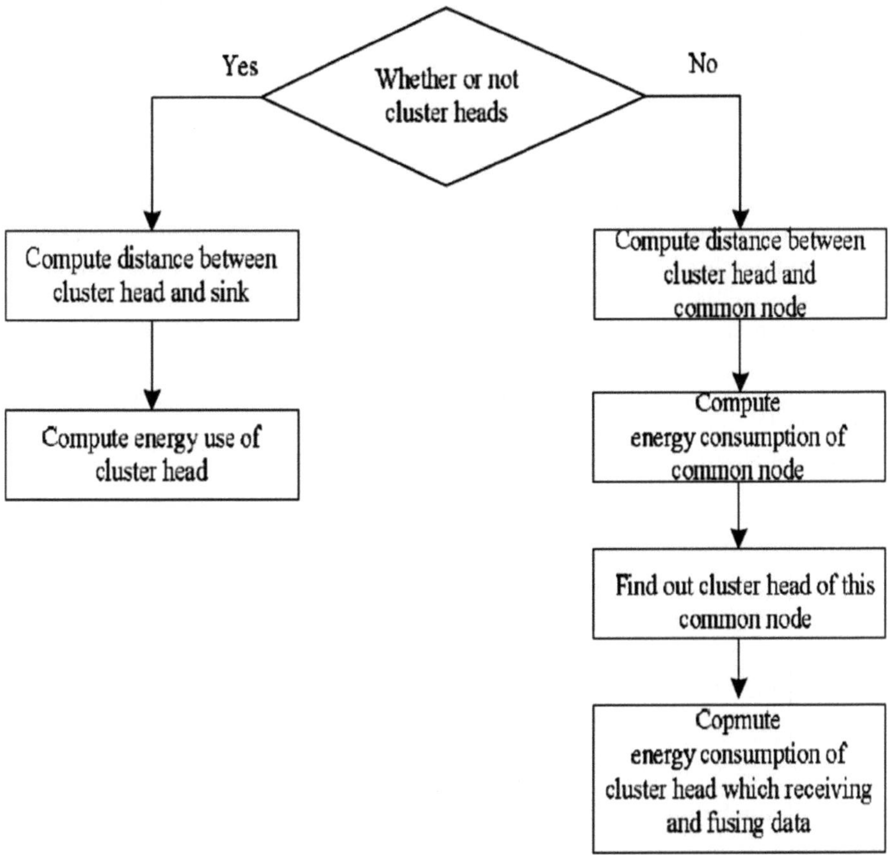

Fig. 3 Various process of nodes in LEACH

that may result into the dying of these nodes sooner with respect to other [5]. This make LEACH protocol very less energy efficient. The cluster head nodes interact with the BS in single-hop method which makes LEACH cannot be utilized for the purpose of huge-scale wireless sensor networks for limit effective communication range of the sensor nodes.

3 Energy-Leach

3.1 Introduction

As described in previous section, LEACH protocol uses normal function of probability number CH and avoids all information related to energy source of sensor nodes. This paper discusses a key variant routing protocol of LEACH called E-LEACH to optimize energy utilization of sensor nodes for giving solution to imbalance energy utilization issue [6]. The E-LEACH takes concept of rounds similar to conventional LEACH protocol. The number of CHs is a prime factor to be considered for performance of routing protocols, in hierarchical routing. Energy-LEACH [5] enhances the energy efficiency by considering remnant energy of node as basic concept in order to conform if sensor nodes can result into CH or not in upcoming round. After calculating the remnant energy of all nodes, according to required number of CH nodes equal to this number with highest residual energy is elected as a CHs. Thus nodes with low residual energy are considered as a normal nodes or non-cluster head nodes. Thus E-LEACH gives enhanced network lifetime and increased energy saving with respect to LEACH. In E-LEACH we use any shortest distance algorithm among CH nodes and select any one CH that having maximum left energy as the root node.

Figure 4 shows the flowchart of Energy-LEACH protocol [3].

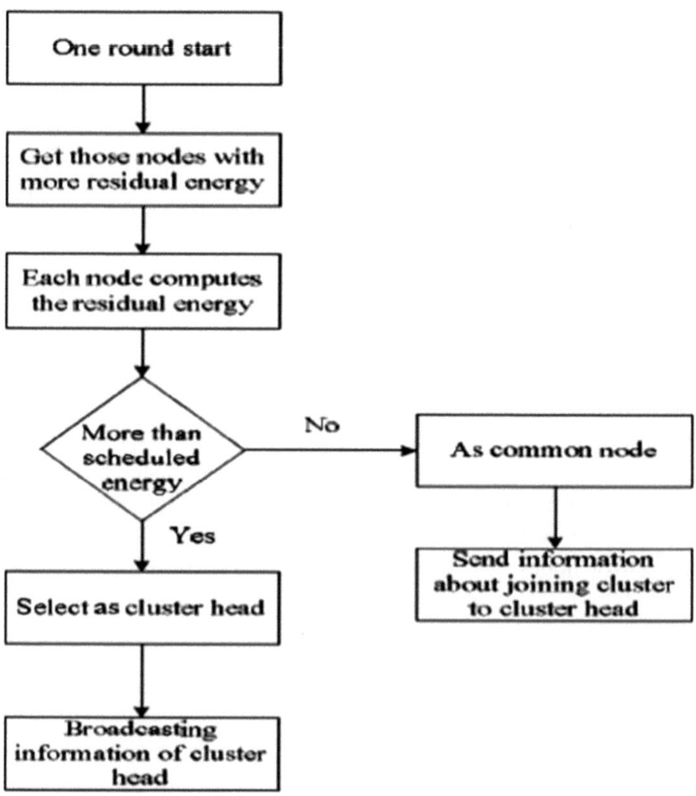

Fig. 4 Energy-LEACH protocol

3.2 *Disadvantage of Energy-LEACH*

E-LEACH protocol requires calculation of residual energy for selecting required number of that are having highest residual energy. This criterion for cluster-head selection may not always leads to balance clusters inside the network. Sometimes this also leads to imbalance if all nodes with the highest residual energy are physically located in same area, so it is difficult to cover entire network area covering all the nodes in the cluster with CH. This makes less energy-efficient because more energy is utilized to cover those nodes inside the cluster, which are located at a longer distance from this area of cluster-heads.

4 Proposed System

To overcome the above disadvantage the proposed system gives solution by considering remnant energy of the nodes within cluster. That is node having maximum residual energy within existing cluster must be chosen as a CH, for next round. With this node as cluster head with highest residual energy will again form the new clusters in network (by set up phase). This gives the uniform cluster head distribution in entire network because initially this cluster-heads are uniformly distributed to form the cluster as every nodes are having equal energy.

Here is the algorithm for the proposed system for cluster-heads selection:

1. Start round one
2. For each node in the every cluster calculating residual energy of nodes.
3. Selecting node having maximum residual energy as CH node from each existing cluster for the next round.
4. Announcing of all the new cluster-heads by the sink to every other node in network.
5. Formation of new clusters with the selected cluster-head node also known as cluster set-up phase.
6. Data aggregation by cluster-head nodes collected from non-CH nodes and sending back to BS also known as steady-state phase.

5 Conclusion

This paper, proposes a new cluster-head selection concept in E-LEACH routing protocol which is based on most used LEACH protocol which can enhance the energy efficiency and uniformly distribute load that is not balance in LEACH. Future task will be considered for how to implement the proposed algorithm for cluster-head selection process in Energy-LEACH protocol.

Acknowledgments Our thanks to Gujarat Council of Science and Technology (GUJCOST) for the grant on the work related to LEACH and VVP Engineering College for providing us an opportunity to enhance the knowledge for the same.

References

1. Pooja A. V., Naren V. T.: A new approach to routing mechanism in Wireless Sensor Network Environment. In: Nirma University International Conference (2013).
2. W. Heinzelman, A. Chandrakasan, and H. Balakrishanan.: Energy-Efficient Communication Protocol for Wireless Microsensor Networks. In: Proceedings of the 33rd Hawaii International Conference on System Sciences (2000).
3. J. Xu, N. Jin, X. Lou, T. Peng, Q. Zhou, and Y. Chen.: Improvement of LEACH protocol for WSN. In: Fuzzy Systems and Knowledge Discovery (FSKD), 9th International Conference on, (2012).
4. Jyoti Singh, Bhanu Pratap Singh and Subhadra Shaw: A New LEACH-based Routing Protocol for Energy Optimization in Wireless Sensor Network. In: 5th International Conference on Computer and Communication Technology (2014).
5. X. Fan. Y. Song.: Improvement on LEACH protocol of Wireless Sensor Networks. In: International Conference of Sensor Technologies and Applications. pp. 260–264, (2007).
6. Meenakshi Sharma, Kalpana Sharma: An Energy Efficient Extended LEACH (EEE LEACH). In: International Conference on Communication Systems and Network Technologies (2012).

MMR-LEACH: Multi-tier Multi-hop Routing in LEACH Protocol

Kausa Aghera, Hiral Pambhar and Naren Tada

Abstract Wireless Sensor Network (WSN) is an advance technology which consist a set of sensor nodes. These sensors are responsible for sensing and collecting data form environment in which they deployed. This information is further transmitted to the base station (BS) via routing protocol. Energy dissipation is major concern while data transmission is done. Various routing protocols are used to reduce energy consumption in WSN. Hierarchical routing protocols are considered to reduce energy consumption. In this paper, we proposed MMR-LEACH protocol. MMR-LEACH protocol divides WSN into various numbers of cluster layers (i.e., multi-tier) with introducing another node as a Vice Cluster Head (VCH) rather than Main Cluster Head (MCH). In MMR-LEACH, after the selection of MCH, it is responsible to select VCH based on residual energy. In this protocol, MCH is also responsible for collecting, aggregating and transmitting data from sensor nodes to BS and VCH is act as a mediator between lower layer MCH and BS for the transmission purpose. These all results into increase the lifetime of WSN.

Keywords Hierarchy · Clustering · LEACH · Multi-hop LEACH · MCH · VCH · Multi-tier · Layered clustering

1 Introduction

WSN is a certain area in which various numbers of sensor nodes are plotted. Sensor nodes are small and cheap device which sense the surrounding changes and transmit this changes to BS over a flexible network architecture. With the recent

K. Aghera (✉) · H. Pambhar · N. Tada
VVP Engineering College, Rajkot, Gujarat, India
e-mail: agherakausha@gamil.com

H. Pambhar
e-mail: hiralpambhar@gmail.com

N. Tada
e-mail: naren.tada@gmail.com

N. Modi et al. (eds.), *Proceedings of International Conference on Communication and Networks*, Advances in Intelligent Systems and Computing 508,
DOI 10.1007/978-981-10-2750-5_22

advancement in WSN, it is used in a various domains such as: environmental observation, military application, building monitoring, healthcare, home and office applications, automotive applications, etc. Energy consumption in sensing, communication and data transmission is a major challenge in WSN because the reason of battery, which cannot be recharge or replaced. Various routing protocols [1] are used for the purpose of reducing energy consumption in WSN which results into increasing the lifetime of WSN.

The routing protocols based on hierarchy [2] is well-known for reducing energy consumption. In which the whole WSN is partitioned into several clusters. Each of these clusters has one sensor node which act as a cluster-head. The responsibility of collecting and transmitting aggregated data to the base station is relay on cluster-head. LEACH and Multi-hop LEACH are an example of hierarchical routing protocols.

This paper contains detailed information of proposed MMR-LEACH protocol. In MMR-LEACH, another node selected as a Vice Cluster Head (VCH) rather than Main Cluster Head (MCH). Base station is responsible for multi-tier i.e., layered clustering. After that MCH it is responsible to select VCH based on residual energy. MCH is also responsible for collecting, aggregating and transmitting data from sensor nodes to BS and VCH is act as a mediator between lower layer MCH and BS for the transmission purpose.

The remaining portion of this paper is as follows. Section 2 contains detailed information of LEACH. Section 3 contains detailed overview of MH-LEACH protocol. Section 4 gives idea about proposed MMR-LEACH protocol. Section 5 describes overall conclusion.

2 LEACH: Low Energy Adaptive Clustering Hierarchy

LEACH [3] is a well-known example of routing protocol based on clustering hierarchy concept. LEACH introduce the concept of rounds. Working of LEACH is done in two phase. First phase is set-up phase and second phase is steady-state phase. Formation of clusters are occur in first phase and transmission of data to BS is occur in second phase. First of all cluster heads are selected from all the sensor nodes in WSN. For this purpose, random number between 0 and 1 is selected by each node. Based on the value the selection is done. Selected nodes has number which is always less than a threshold value T(n). The threshold value is calculated as [3]:

$$T(n) = \begin{cases} \frac{p}{1-p\left(r \, mod\left(\frac{1}{p}\right)\right)} & n \in G \\ 0 & others \end{cases} \tag{1}$$

where T(n) = Threshold value, n is no. of total nodes, p is the desired percentage for selecting as a cluster head, r is number of current round, G is set of all nodes which are not becoming cluster head in the past 1/p round. Each node has a chance to become cluster head once during 1/p round. At the round r = 0 all node has an equal eligibility to elect as a cluster-head. For next 1/p rounds, once selected as a cluster-head node cannot be selected again. After 1/p rounds, again each nodes are eligible for selected as cluster head.

After elected as a cluster head, advertisement message is broadcast to all nodes in WSN using same energy. After receiving message, all the non-cluster head nodes choose one cluster head and send the reply message to correspondent cluster head. Now, cluster head create TDMA schedule and broadcast this schedule to that nodes. Based on this TDMA schedule, nodes know when to transmit data. After that clusters are created and schedule for nodes is fixed in the WSN, transmission of data to the base station can begin. The cluster head transmit aggregated data to the BS. After that next round is started.

LEACH protocol's communication architecture is as shown in Fig. 1. The following Figs. 2, 3 and 4 shows LEACH protocol [4], cluster formation of LEACH protocol [4] and different processes [4] are shown respectively.

3 Multi-hop LEACH

MH-LEACH [5] is variant of LEACH routing protocol. In LEACH protocol, each cluster-head transmit data directly to BS no matter how far it is located. If the distance between cluster-head and BS is increased beyond certain level then it will

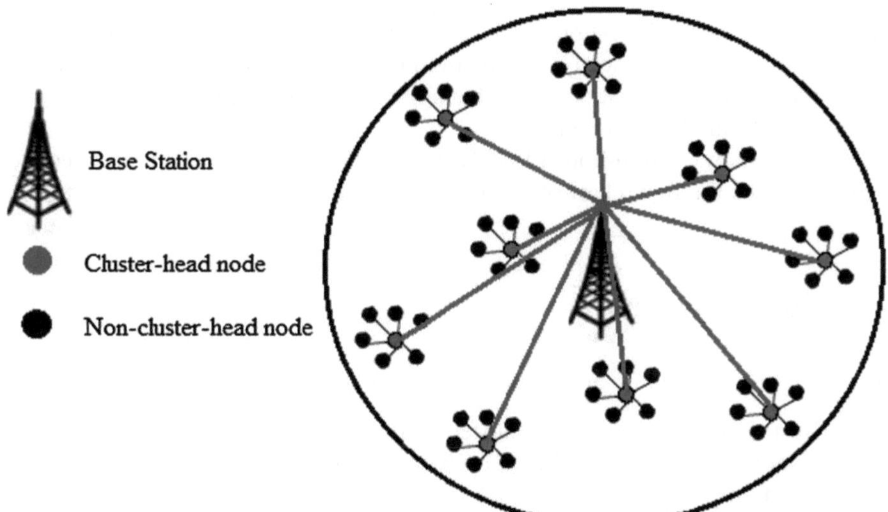

Fig. 1 LEACH protocol's communication architecture

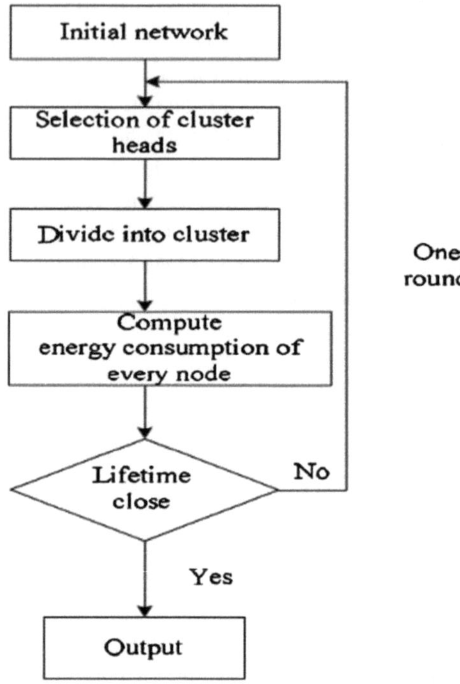

Fig. 2 LEACH protocol

consume lot of energy. To solve this problem, MH-LEACH protocol is comes into picture.

Multi-hop LEACH improve the communication mode by allowing multi-hop between CH and BS instead of single-hop. MH-LEACH protocol works in two phase. In Multi-hop LEACH protocol, set-up phase worked same as in LEACH but the main difference is in steady state phase, in which CH collects data from sensor nodes and perform aggregation. After that aggregated data is transmit directly to BS or other CH to the base station.

Figure 5 shows Multi-hop LEACH protocol's communication architecture. Figure 6 shows the flow chart of routing of Multi-hop LEACH protocol [4].

4 Proposed System

MMR-LEACH divides its operation into three phases, which are:

(i) Cluster formation with two cluster heads
(ii) Layered clustering (multi-tier) by base station
(iii) Scheduling.

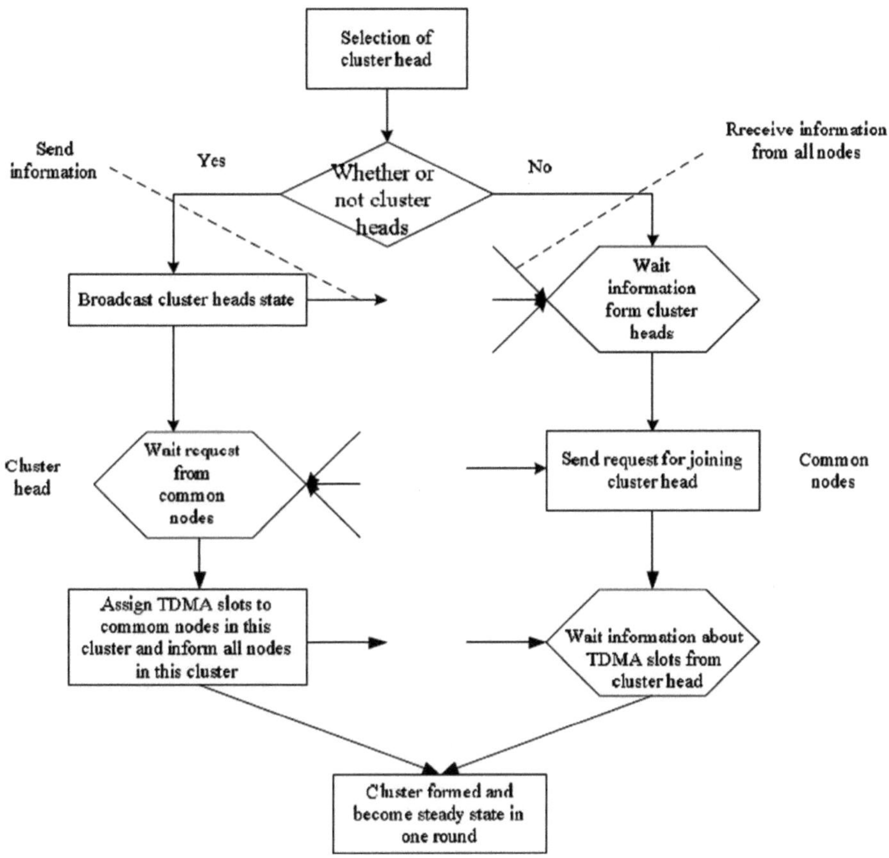

Fig. 3 Cluster formation of LEACH protocol

4.1 Cluster Formation with Two Cluster Heads

In the beginning of first phase, selection of nodes as a Main Cluster Head (MCH) is started. Selection of MCH in MMR-LEACH protocol is same as in conventional LEACH protocol. For this purpose, random number between 0 and 1 is selected by each node. Based on the value the selection is done. Selected nodes has number which is always less than a threshold value T(n). The threshold value is calculated as [3]:

$$T(n) = \begin{cases} \frac{p}{1-p\left(r\ mod\left(\frac{1}{p}\right)\right)} & n \in G \\ 0 & others \end{cases} \tag{1}$$

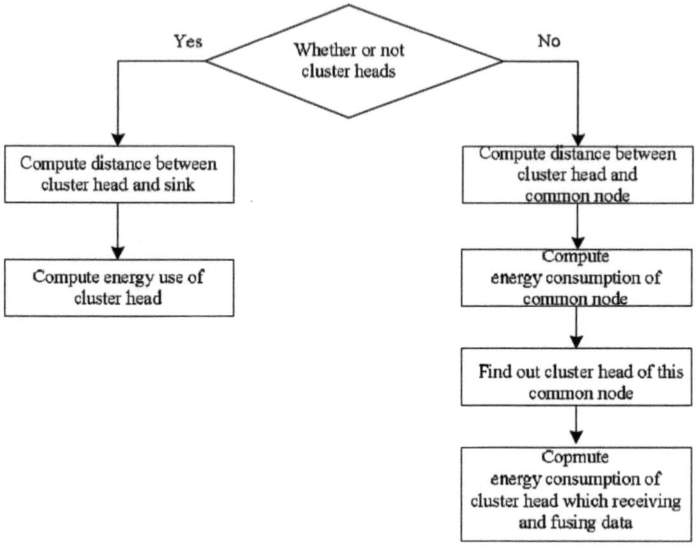

Fig. 4 Difference process of nodes in LEACH protocol

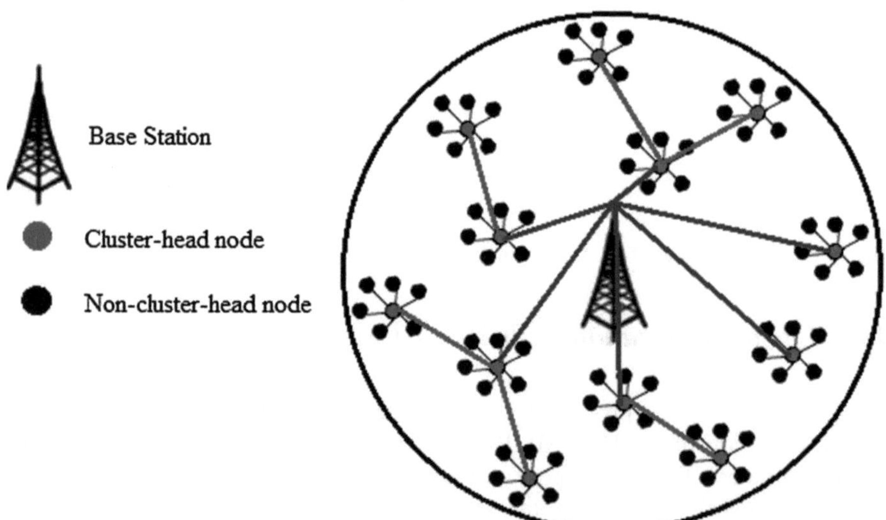

Fig. 5 Multi-hop LEACH protocol's communication architecture

where T(n) = Threshold value, n is no. of total nodes, p is the desired percentage for selecting as a cluster head, r is number of current round, G is set of all nodes which are not becoming cluster head in the past 1/p round. Each node has a chance

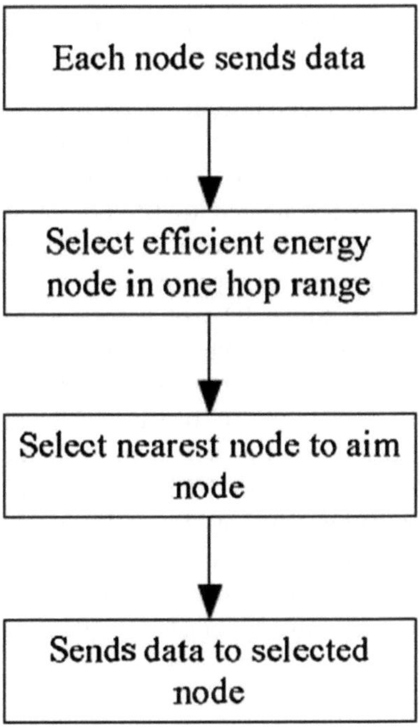

Fig. 6 Routing of multi-hop LEACH protocol

to become cluster head once during 1/p round. After elected as a MCH, advertisement message is broadcast to all nodes in WSN using same energy. After receiving message, all the non-cluster head nodes choose one cluster head and send the reply message to correspondent MCH with its T_1 value. The reason behind calculating T_1 is, the selection of VCH is done by MCH is based on this value. T_1 is calculated by each non-cluster head as [6]:

$$T_1 = \frac{E_r}{E_i} + \frac{N_a}{N_i} + \frac{S_r}{S_c} \tag{2}$$

where E_r is the remained energy, E_i is the initial energy, N_a is the total no. of neighbor, N_i is the max. no. of neighborhood sensor node at initial development, S_r is the distance to CH and S_c is the max possible strength of received signal. After receiving the value of T_1 from all its member nodes, the MCH compare those values and a sensor node with maximum value of T_1 is selected as a VCH. Now, MCH create TDMA schedule and broadcast this schedule to that nodes. Based on this TDMA schedule, nodes know when to transmit data. After that clusters are created and schedule for nodes is fixed in the WSN, transmission of data to the base station can begin. The MCH and VCH transmit aggregated data to the BS. After that next round is started.

4.2 Layered Clustering (Multi-tier) by Base Station

During this phase, ID of BS is broadcast over the flexible network channel in WSN. All VCHs which receive this message will note the base station ID. Using equal low level power, all VCHs send their ID's to BS. Layer one is formed by all VCHs which are neared to BS i.e., at single hop. Now BS broadcast message which include all layer one VCHs ID's. All VCH except VCHs of Layer one must send reply message using equal low level power with own ID's plus ID's of layer one VCHs. Since ID's of layer one VCHs are resides in the message, they will not respond to this message. Because of lower level power broadcasting, this message directly not reach to BS. For that, Layer one VCHs act as a mediator and pass the message to the BS because of the distance between them is one hop. During this all procedures, BS will record all the information such as the ID's of VCHs and MCHs, level of VCHs and MCHs and ID of mediator VCHs. BS repeats this procedure until all the VCHs in the sensor network are discovered. After all VCHs in WSN are discovered. BS use all these information like ID of VCH, Level of VCH and immediate VCH ID to form cluster of cluster head (i.e., layered clustering). Layered clustering (i.e., multi-tier concept) is shown in following Fig. 7.

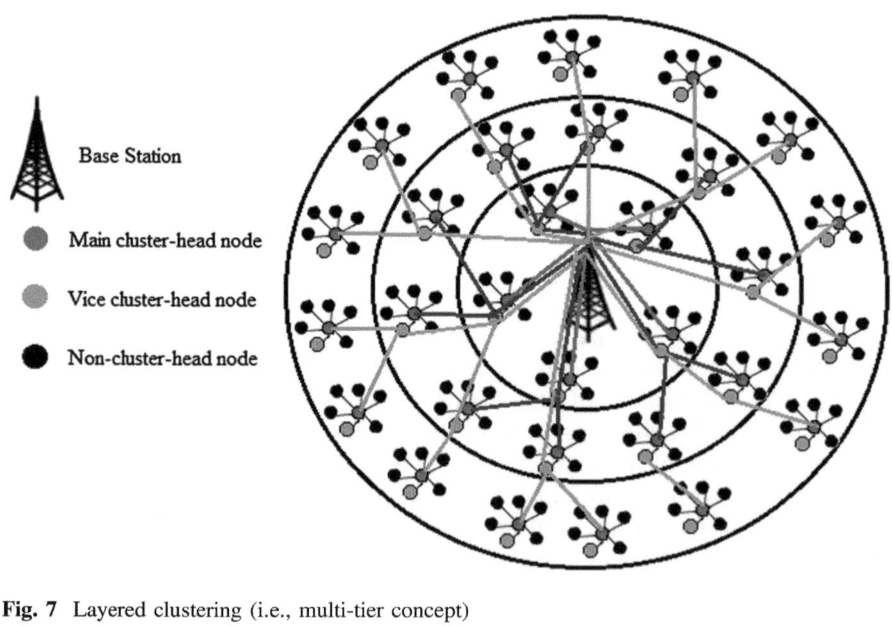

Fig. 7 Layered clustering (i.e., multi-tier concept)

4.3 Scheduling

Each MCHs and VCHs are responsible for scheduling of member nodes using TDMA (Time Division Multiple Access). All the MCHs and VCHs of lower layer send its data to the VCHs of upper layer. Upper layer VCHs allocate large time slot to their member of lower layer MCHs and VCHs because they send larger data in comparison to other nodes.

5 Conclusion

In WSN, consumption of energy is one of main challenge because of limited battery power of sensor node. As a solution of this problem MMR-LEACH protocol is proposed which uses multi-tier concept with selecting two cluster-heads. In multi-tier concept, whole sensor network is partitioned into several layers of clusters. For the data transmission, another node selected as Vice Cluster Head (VCH) rather than Main Cluster Head (MCH). MCH is responsible for collecting, aggregating and transmitting data from sensor nodes to BS and selection of VCH based on residual energy. VCH is act as a mediator between lower layer MCH and BS for the transmission purpose. The lifetime of sensor network is exceeding by MMR-LEACH protocol in comparison to conventional routing protocols. In future work, implementation of the proposed system (MMR-LEACH protocol) is considered.

Acknowledgments We would like to thank GUJCOST (Gujarat Council on Science and Technology) for sponsoring research on LEACH and CE department of VVP Engineering College to giving us an opportunity to work as a part of it and conducting our research on wireless sensor network.

References

1. J. Gnanambigai, Dr. N. Rengarajan and K. Anbukkarasi: Leach and Its Descendant Protocols: A Survey. In: International Journal of Communication and Computer Technologies, Volume 01 – No. 3, Issue: 02. (2012).
2. Pooja A. V., Naren V. T.: A new approach to routing mechanism in Wireless Sensor Network Environment. In: Nirma University International Conference (2013).
3. W. Heinzelman, A. Chandrakasan, and H. Balakrishnan: Energy-efficient Communication Protocol for Wireless Sensor Networks. In: Proceeding of the 33rd Hawaii International Conference on System Sciences, Hawaii (2000).
4. Fan X. and Song Y.: 2007. Improvement on LEACH Protocol of Wireless Sensor Network. In: International Conference on Sensor Technologies and Applications (2007).

5. Jose Henrique B. N., Antoniel da S. R., Andre R. C., Joaquim C. Jr.: 2014. MH-LEACH: A Distributed Algorithm for Multi-Hop Communication in Wireless Sensor Networks. In: The Thirteenth International Conference on Networks (2014).
6. Saeed E., Mahsa G., Ahmad H. N. and Mir Kamal M.: Energy Balancing in Wireless Sensor Networks with Selecting Two Cluster-Heads in Hierarchical Clustering. In: International Conference on Computational Intelligence and Communication Networks (2010).

Cooperative Sensors for Identifying an Impulsive Events of Asynchronous Environment

N. Prabakaran and R. Jagadeesh Kannan

Abstract Reducing the risk of person in vehicular network is essential, has premier priority to be achieved with prominent specifics such as collision detection and avoidance. Advances in image processing, falling cost on hardware and others allowed decreasing number of accidents are involved and step out from it. Today, new vehicles list safety as one of the highest priorities and use it as their main selling points. Existing series can detect obstacles in the path, and apply the brakes faster than the car user can. Introducing sensor backed scheme using light control for detecting proximity entity at emergency and decision is made in advance to avoid risk due to fatigue while driving. Falling cost of camera technology, manufacturers are started to equip their vehicles with cameras positioned at various places around the body of the vehicle. To remove any blind spots while driving, where as unpredictable traveling object found. Aim of this scheme is for detecting when a vehicle meets risk zone, drifts out of lane, or when it is within the safe stopping distance of an object ahead of it. This system is vision based, Instead of using the camera again to calculate the distance of the object to determine whether it is within the safe stopping distance, a short wave sensor used. This is done to ensure if such a system is possible, based on minimal cost and hardware usage implementation would be built.

Keywords Sensor networks · Proximity · Power optimization · Minimal cost

N. Prabakaran (✉)
SCSE, VIT University, Vellore, India
e-mail: dhoni.praba@gmail.com

R. Jagadeesh Kannan
SCSE, VIT University, Chennai, India
e-mail: dr_rjk@hotmail.com

N. Modi et al. (eds.), *Proceedings of International Conference on Communication and Networks*, Advances in Intelligent Systems and Computing 508,
DOI 10.1007/978-981-10-2750-5_23

1 Introduction

WSN environment applications such as target tracking, home automation, energy conservation monitoring etc. are using wireless sensor networks to fulfil the ubiquity. To access the data anytime anywhere it must be regulated to be available always. Wireless sensor networks act independently but supportively to route data and hop-by-hop towards base [1]. WSN assists in communication between physical and virtual through integration of sensing, communication, computing and control Based on delivery types, sensing differ in sensor networks. Figure 1 indicates the simple flow of communication among tiny nodes with intended user. The spatially distributed nodes are behaving under constraints such as event based, query based, continuous and hybrid networks [2]. WSN has data centric functional to process huge data for monitoring an environment and they have tiny sensor nodes which consist of microcontroller to control the whole system. Sensor nodes are capable of sensing and generating data according to environment without human intervention. To detect the obstacle and to assist the driver for decision making these sensor nodes are deployed in front side of the vehicle [3].

Existing traffic light systems have timers that are set at regular intervals which is taking longer time to accomplish a task. Two kinds of sensors are involved in this system, which are Simple proximity sensor and Modulated IR sensor [3, 4]. Simple Proximity sensor detects nearby objects without any physical contact and proximity sensor often transmits electromagnetic radiation (infrared) and looks for changes in the field or return signal [5]. Sensors and actuators are fixed inside to interact with each other and enable the user to monitor remotely of various devices. Camera techniques and image processing play vital role in obstacle detection with the help of internet but it is not possible in remote areas, uneven roads and too expensive

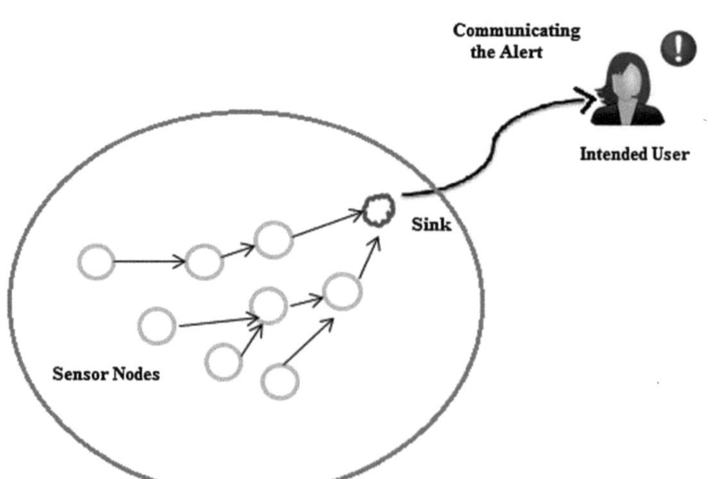

Fig. 1 Simple wireless sensor environment

when they are replaced. Low cost ultrasonic sensors can be deployed and communicated together to establish the task in which nodes work independently from external resources. Based on the behavior observed from sensor nodes the intended user gets warning through light control; direction and distance can be measured using mathematical model without the help of Internet which is irregular in updating of remote location [6].

2 Related Works

The technique about obstacle avoidance in intelligent automatic vehicles has been developed huge so far to make the vehicles travel in unknown environments securely. Bug Algorithm is the simplest obstacle avoidance algorithm [2]. When an obstacle is encountered, the vehicle fully circuits the object in order to find the point with the shortest distance to the goal and is obviously inefficient. The Bubble Band Technique requires a ring of sensor set to detect the obstacle in all directions [4]. Grids method for mobile robot to obstacle avoidance is robust and allows continuous and fast motion of the robot without stopping in front of obstacles. However increasing scalability, will take a large computational time [6]. Some of the algorithms such as the one presented in need prior knowledge to the environment, such as the obstacle's shape and size, and hence will not be suitable for unknown environments. In this paper, we address obstacle avoidance which doesn't require any pre requisite. In order to reduce computation, we define by using semi-algebraic sets. Defining the position of the vehicle as the origin, the danger zone is represented using several semi-algebraic constraints defined by the relative velocity information. The formal displacement of the object can be retrieved by using $d = v_0 t + 1/2\ at^2$ to reproduce the velocity from displacement use, $v_o = d/t - 1/2\ at^2$. Then the detection is made by comparing obstacle belongs to the region of intersection. Using single sensor to get the most of the information about the profile of obstacles and find out the passable routes. This algorithm does not require prior information in advance and results with the calculated range to determine the way to avoid the obstacles with minimum computation load [7]. We also used the projection of obstacle distance and checking danger zone constraints to determine the best direction to move around the obstacles. The combination of the danger zone concept and the distance projection information for best avoidance maneuver for multiple obstacles will be pursued in the future [8].

3 Model of Sensor Backed Monitoring

Sensing unit does the role of gathering information and processing unit has on-board memory to deal with procedures that allow the sensor node to perform sensing, execute the related algorithms and cooperate with other nodes. Hierarchical

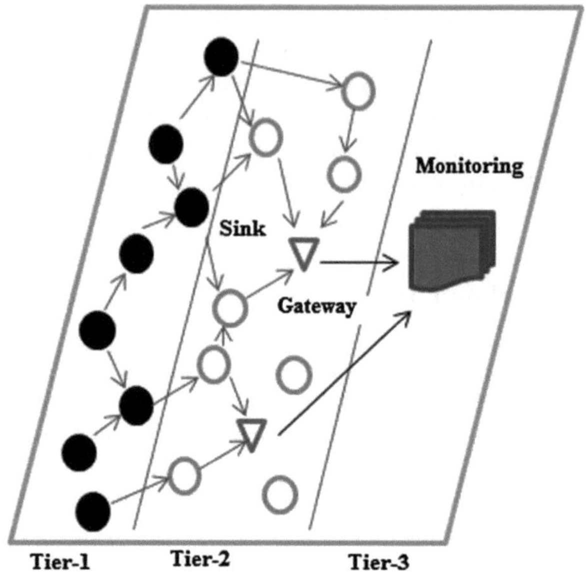

Fig. 2 Design flow of the Hierarchical architecture of wireless sensor networks

architecture of Wireless sensor networks using various categories are depicted in Fig. 2 which shows the basic communication model. Communication among nodes is possible through transceivers only and when lifespan of network is stretched longer then power generator is needed [9]. To support the well facilitated view and safety wireless sensor nodes are deployed in front side of the vehicle. Alerting a driver is vital role to avoid taking risk on persons and provide reduced risk out of extreme tiredness. This system optimizes the spreading of light from the front lights according to environments. The power usage depends upon velocity and obstacles detected on the path. To avoid frequent alert, when the vehicle passes over uneven road then the system analyses the speed of the vehicle whether it is suitable to control or not. This scheme presents a method, in which the tracking performance can be improved at minimum computational cost by utilizing the information associated with the kinematics of the light spots along with their appearance. It uses the position and size to describe a light spot and information about the light spot is used for tracking and data association. The appearance based features will allow for an appearance based comparison. The tracking is expected to perform better than a tracker which relied on kinematic information alone. As a baseline a tracking method which uses Kalman filters and data association using nearest neighbor is considered. The design flow architecture depicts the communication flow when using sensor nodes through sink and gateway [10, 11].

4 Power Optimization

When the vehicle is in safe zone then there is no need detection of obstacle, whereas on the other hand intensity is high when a curve road or any obstacle detected. Intensity of the light is initially at 65–70 % which is sufficient enough to make the road or the way visible to the driver and when the obstacle is detected, the sensor will detect that obstacle and will alert the driver through the electronic vibrator and it will allow the microcontroller to increase the glowing intensity of the light. As a result, 30–35 % of the power supplies working parallel for both, obstacle detection and light control. The decision making algorithm experiments the test-bed result based on real world instances. Based on region which falls shorter, the intensity of light beam is decided. The proximity sensors senses and objects falls in the nearby range and velocity is computed for further comparison. The positive time factor id computed using,

$$d = v_o t + 1/2\,at^2 \tag{1}$$

By rewriting into quadratic equation,

$$0 = 1/2\,at^2 + v_o t - d \tag{2}$$

which implies

$$0 = At^2 + Bt - d \tag{3}$$

After substituting a, b and c, t is equivalent to

$$T = \left[-V_0 \pm \sqrt{(V_0^2 + \;2at)}\right]/a \tag{4}$$

5 Architecture Setup and Discussion

This implemented approach has minimal cost sensors and mounted over the preferred design, inside vehicle, in such a way that they can detect the threat on the front way of vehicle. The logic behind is benefited mostly when the threat detected is in motion too. Ultrasonic proximity sensors are widely used for obstacle detection and estimation. These sensors employ sound waves to detect objects, so texture, climatic condition and atmospheric parameters do not affect the internal properties and functionalities of the sensor. This makes them ideal for a variety of applications, including the long range detection of objects and shape estimation. In the proposed scheme, five ultrasonic sensors are placed at various locations over the vehicle at a particular angle and orientation so that the entire proximity of the vehicle can be detected for the presence of obstacles.

Table 1 Algorithm for decision making and parameters configuration for experimental scenario

Algorithm for Decision Making

```
Initialize:
V← vehicle velocity; Vp← Velocity of proximity
Th← Threshold velocity value; R← Range;
Lb low intensity beam, Hb High intensity beam
While [ sensors enabled]
 do
   if [Vp -le Th ]
      R € safe , Reset from Lb to Hb
else
     echo –n " R € unsafe"
               if [Vp -gt Th ]
                  dec V;
       else  inc –o dec V;
               fi
          Enable Hb
fi done
```

Environmental Setup

Parameter	Values
Microcontroller	ATmega2560
Operating Voltage	5V
Input Voltage (recommended)	7-12V
Input Voltage (limits)	6-20V
Digital I/O Pins	54
Analog Input Pins	16
DC Current per I/O Pin	40 Ma
DC Current for 3.3V Pin	50 mA
Flash Memory	256 KB
SRAM	8 KB
EEPROM	4 KB
Clock Speed	16 MHz

Table 2 Characteristics of roads

Objects trained (percent)	Normal (in colors)	Others (in)
Road surface	Dark black	0.25
Middle of the road	White (continuous single/double line)	0.10
Side end	Yellow	0.05

The parameters mentioned in Table 1 are considered for hardware specification and logic is illustrated by means of algorithm. Role of LEDs are to glow only when the obstacle is detected through sensors. The Head light of the vehicle will be also connected with the microcontroller. The characteristics of roads are feed and trained for sensors in order to understand the environmental scenario and listed few elements below in Table 2.

Environmental scenario is to detect the obstacle and alert. To do this, instead of using cameras, we use distance measuring sensors which are placed in front face of the vehicle. The idea for not using cameras is proposed and direction and time taken to reach obstacle is done by using decimal representation after computing T. The situations where using cameras are not best choice because it will display the result in a device such as monitor to show a real time view which will mislead the focus and it's not suitable uneven roads or rough terrain. By using minimal cost sensors the environment is achieved and tested the experiment and shown below in Fig. 3.

Experimental scenario has been designed and results are illustrated based on the event detection in an uneven or asynchronous environment. Figure 4 illustrates ad hoc nodes movement are immediately down not off, when the obstacle is nearby and gradually increased only if the obstacle is crossed.

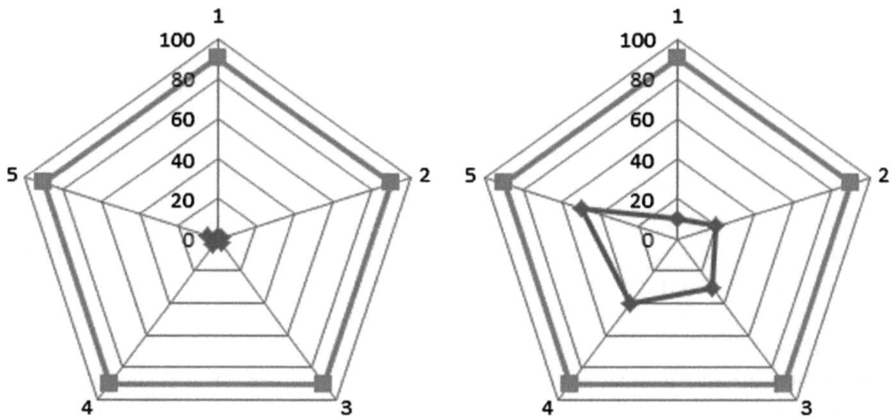

Fig. 3 90 1′ 05″, 2′, 3′, 4′, 5′ decimal representation and 90 1′ 10″, 20″, 30″, 40″, 50″ decimal representation

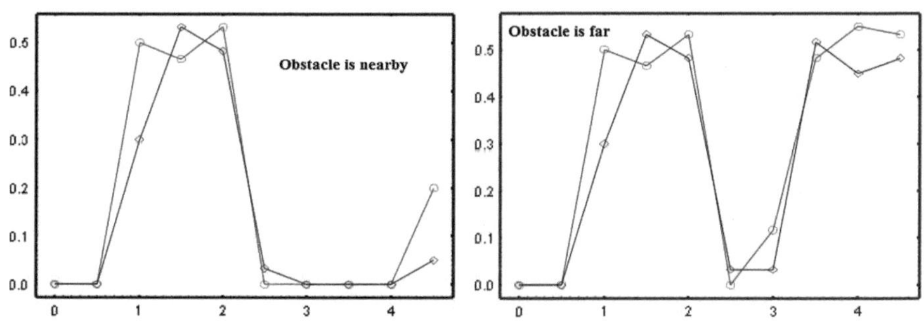

Fig. 4 Results for detection of obstacles nearby and far away

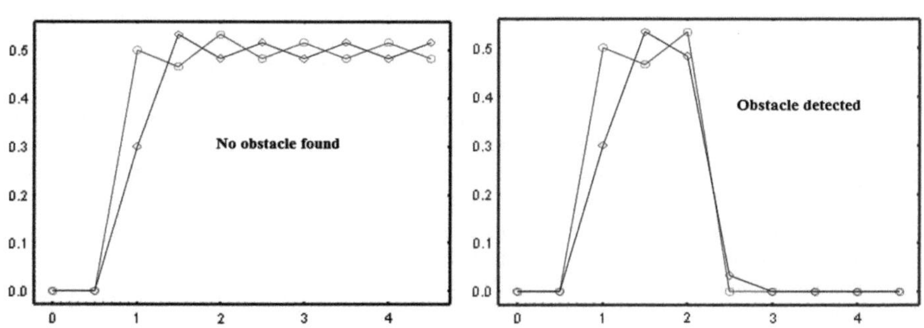

Fig. 5 Results for detection of obstacles and no obstacles found

Figure 5 illustrates there is no change in the movement of nodes if there is no obstacle and stopped if there is a raise of obstacle. All these comparisons are made with the help of trace file using network simulator 2.29.

6 Conclusion

Research into similar systems had shown that a system which detects when lane departure had occurred was possible using a forward facing camera. This scheme shows that lane detection without using a camera and alert is possible without embedded device and other external resources. Active usage of energy and performance of cooperative sensor nodes are accomplish without consuming much resources. This scheme is trustworthy for decision making over ad hoc network communication and improving consistent resource usage. Minimal hardware setup leads to power optimization in low cost and easier to handle. The collision detection calculation is done for safe stopping distance of the vehicle and being accurately able to measure the distance to the object at a ~ 20 m. However, it is showing that a mono vision based approach rather than radar based, could function to some degree at detecting objects in the path of the vehicle to detect biological objects that radar passes through. Future work concentrates on scalability of tiny nodes which must be less in number when high sensitivity of detection is needed.

References

1. Akyildiz, W. Su, Y. Sankarasubramaniam, and E. Cayirci, A survey on sensor networks," IEEE Commun. Mag., vol. 40, no. 8, pp. 102–104 (2002).
2. James Ng and Thomas Branul, Performance Comparison of Bug Navigation Algorithms journal of Intelligent and Robotic Systems, Volume 50, Issue 1, pp 73–84 (2007).
3. C.-Y. Wang, S. B. Eisenman, and A. T. Campbell, CODA: Congestion detection and avoidance in sensor networks, in Proc. ACM SenSys (2003).
4. Philippsen, R.; Siegwart, R. Smooth and Efficient Obstacle Avoidance for a Tour Guide Robot, LSA-CONF-2003-018 (2003).
5. Zhang, Q., Chen, D. and Chen T, An obstacle avoidance method of soccer robot based on evolutionary artificial potential field", Energy Procedia, Vol. 16, Part C, pp. 1792–1798 (2012).
6. Li Jigong; Lanzhou; FengYiwei; Zhu Chaoqun, Novel Path Planning Method Based on Certainty Grids Map For Mobile Robot Control Conference. Chinese (2007).
7. Ta-Chung Wang and Tz-Jian Lin Unmanned Vehicle obstacle detection and avoidance using danger zone approachtcsme Vol37/Vol37No3 (2013).
8. Chen, J.-Y.and Wang, T.-C, Semialgebraic set representation of danger zone, Master Thesis, National Cheng-Kung University, Taiwan (2010).
9. N. Prabakaran, K. Naresh, R. Jagadeesh Kannan, Fusion Centric Decision Making for Node Level Congestion in Wireless Sensor Networks advances in Intelligent Systems and Computing Volume 248, pp 321–329 (2014).
10. Wang, Q. Traffic Analysis, Modeling and Their Applications in Energy-Constrained Wireless Sensor Networks-On Network Optimization and Anomaly Detection. Mid Sweden University, Sweden ISBN 978-91-86073-64-0, (2010).
11. Susnea, I., Minzu, V. and Vasiliu, G, Simple, real-time obstacle avoidance algorithm for mobile robot, in Proceedings of the 8th WSEAS International Conference on Computational Intelligence, Man-Machine Systems and Cybernetics, pp. 24–29 (2009).

Trust Integrated Federated Architecture Ranking Service Models in Cloud Computing Environment

M. Saravanan and M. Aramudhan

Abstract Cloud Computing is an inevitable technology using by the internet users for their day to day usage, Various cloud service providers offering different services (like Software, Platform, Infrastructure, Storage etc.,..) to various customers through the Internet. Users are confused to choose best and cost reduction service provider. Our frame work designed based on user concerned, this will aid to user to select suitable and optimized cost service provider based on the necessary key performance indicators. We proposed frame work will address the key issues like User categorization, Trust analysis, Cost analysis, ranking the service providers, and Priority based selection. Hence Cost analysis based on the Service Level Agreement (SLA) cost of components, Ranking to be done based on grade values, and Priority Feedback Decision Tree applied to select best service provider from similar highest rank list.

Keywords Federated broker architecture · Scaling grade values and priority · Feedback selection algorithm

1 Introduction

Cloud computing is an interconnected computing resources to provide on demand access basis to the user, (e.g., infrastructure, platform, and software) Cloud computing identifies "five essential characteristics": On *demand self-service.* Wide *network access, Resource pooling, Rapid scalability, Measured service.*

M. Saravanan (✉)
Periyar Maniammai University, Thanjavur 613403, India
e-mail: saran2009pmu@gmail.com

M. Aramudhan
Perunthalaivar Kamarajar Institute of Engineering and Technology, Karaikal, India
e-mail: aramudhan1973@yahoo.com

N. Modi et al. (eds.), *Proceedings of International Conference on Communication and Networks*, Advances in Intelligent Systems and Computing 508,
DOI 10.1007/978-981-10-2750-5_24

1.1 Resource Provisioning

User self-provisioning: Customers buy cloud services directly from the provider, typically through internet. The customer pays on a per-transaction basis.

Advance provisioning: Customers contract in advance a predetermined amount of resources, which are prepared in advance of service. The customer pays a usage fee or a monthly fee [1].

1.2 Key Performance Indicators

Key performance Indicators used to appraise the cloud service providers [2]. Availability, Service/System availability, Cost, Performance, Capacity, Response time, Elapsed time, Meantime between failure, Meantime to repair, Throughput, Bandwidth, Processor speed, Storage capacity, Storage Types, Service/System scalability, Security, Audit, Back up etc.

2 Proposed Work

We proposed frame work will address the key issues like User categorization, Trust analysis, Cost analysis, Ranking the service providers, and Priority based selection are used to select the appropriate service provider for their requirements. Feedback Decision Tree applied to select best service provider from similar highest rank service provider list. We used scaling grades are High, Normal, and Low to categorize the resources, scaling grade values are 1.0, 0.5, 0.25, to represents key performance components.

2.1 Cloud Federated Broker Architecture

Enhanced Federated cloud model categorizing user as registered and non register, former one user always work only the particular service provider, so that our frame work helps to choose best service providers based on their requirements as usual as cost. Our previous model works only to rank the best service providers but new model added to rank the service providers based on the cost. User requirement [3] classified based on the performance indicators. Proposed frame work functions as three zones. (i) Users zone (ii) Federated cloud agent zone. (iii) Cloud provider's

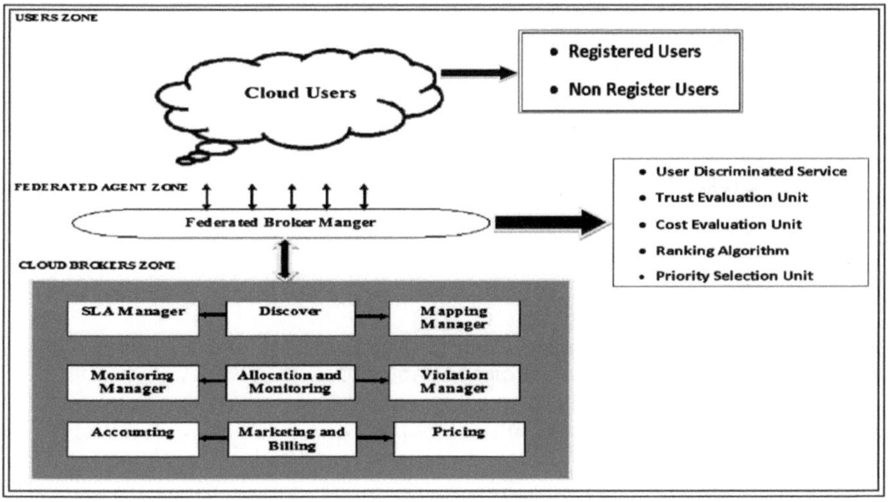

Fig. 1 Federated Broker Architecture

zone, [1] this zone have Discovery phase, Allocation and Monitoring, Marketing and Billing [4] (Fig. 1).

2.1.1 Users Zone

Users zone describe that cloud users. Users access cloud services from anywhere at any time. Some of the users try to persist with only particular service providers; they registered and regularly access same service providers without concern about cost and others in the internet world, but most of the users like non registered with any particular service providers try to identify best service providers or low cost service providers from pool of list available in the open networks.

2.1.2 Federated Agent Zone

Federated Broker Manager defined as Federated Agent Zone. Functions of proposed frame works are (i) Discriminating the users as registered and non register, (ii) Trust Evaluation, (iii) Credential evaluation, (iv) Requirement analysis, (v) Cost analysis of user, (vi) Ranking allocation to the service providers (vii) Priority selection based on user request.

F(Ł,Trust, Cost, Rank)
Ł: List of Available Service providers
Trust: Trust Evaluation
Cost: Cost Analysis
Rank: Rank the Service providers

Federated Broker Manager Algorithm

```
F( List-SP, Trust, Cost, Rank)
Step 1:
        F ———→ (List-SP)
[SP1,SP2,SP3..............  SPN ]
Step 2: F———→ (Trust)[Trust Evaluation Unit ]
Step 3: F———→ (Cost-Component)[CC1, CC2,
CC3..............  CCN ]
Step 4:F ———→ (Rank)[SPTop,SPTop-1,..............  SP1 ]
End procedure
```

3 Grading Methodology

3.1 Scaling Grade Weights

We proposed scaling grade weights are used as follows High, Normal, and Low. Each grade assigned with values (1, 0.5, and 0.25). These grade values applied in the scaling grade distribution algorithm, key performance indicators represented as user required components, tolerable range also allowed to provide dynamically by the user, here based on the tolerate value assume as maximum necessity of an user. In our algorithm assumed tolerate value as 10 % from requested component for 'High' and 5 % of requested value for Normal grade values. Thus tolerate values are not fixed, any time variable based on the user request.

3.2 Scaling Grade Values Distribution

Algorithm 2:

```
F (Rank)———> [SPTop,SPTop-1,.............. SP1]
Step 1: Get the user Requirement Components [ RC =
RC1,RC2, RC3.....RCI ]
Step 2: F———>(List-SP)[SP1, SP2, SP3.............. SPN ]
Step 3: Assign Grade Value based on user Requirement
Components   [ RC  = RC1,RC2, RC3.....RCI ]
      If ( (AR> RC  and RC <= AR1)  :  AR1 =  AR + (AR1
x10)/100 ; (Tolerate  value)
            Assign 'High' Grade value=1;
      If (AR == RC or RC <= AR1)      :  AR1 =  AR - (AR
x5)/100 ;  (Tolerate  value)
      Assign 'Normal' Grade value=0.5
             If (AR  <  RC)
            Assign ' Low' Grade value=0.25;
      Elseif
                (AR == 'Nil' )
            Assign Grade value=0;
Step 4: Compute Total Grade values for each service
provider TGV = [G1+G2+G3......GN].
Step 5:Sort the [TGV] of [SP1, SP2, SP3..............
SPN ]
Step 6: Assign the Top service provider to the user.
End procedure.
```

3.3 Total Scaling Grade Computation Methodology

Service providers are assigned rank, based on the Scaling Grade values, These values to be allocated to each components required by the user. **N** represents number of components required for user request. Tolerate value 10 and 5 % are adjustable range based on the user submission request. User also to be permitted for submits tolerable range. Available service providers listed according to the total scaling weights.

$$[SPW]_{Total} = \sum_{i=1}^{N} Wi \tag{1}$$

$[SPW_{total}]_n = \{SPW_1, SPW_2, SPW_3, SPW_4, \ldots SPW_n\}$ n—number of service providers.

$$[SPW_{Total}]_n = \sum_{i=1}^{N} Wi \tag{2}$$

3.4 Service Providers Ranking Method

Federated Broker Manager receives user requests and pulls out the required functional and non functional components and Algorithm-2 shows scaling grade values distributed to the each available components of the service providers based on the user requirements [5]. Now grade table maintaining the service providers grade list, from the grade table, grade weight total $[SG]_{total}$ and grade average $[SPW_{total}]_{avg}$ to be computed and stored in the table. Service providers are ranked based on their grade total average $[SPW_{total}]_{avg}$, if more than one service providers on similar average values of highest rank, then selection process submitted to Priority Feedback decision tree (PFDT).

$$[SPW_{Total}]_{AVG} = \left[\frac{\sum_{i=1}^{N} W_i}{N}\right] \quad (3)$$

4 Cost Analysis

```
Algorithm 3:
F (Component-Cost) ——→[CC1, CC2, CC3..............  CCN ]
Step 1: Get the user Requirement Components[ RC  = RC1,RC2,
RC3.....RCI ]
Step 2: [SLA Cost of Each Component(SCC)]
```

$$Totalcost[SP1] = \sum_{i=1}^{N}[SCC]_i$$

$$Totalcost[SP]_n = \sum_{l=1}^{N}[SCC]_l, n = 1,2,3,....R$$

```
R - Number of service providers available.
Step 3:IF (Total Cost[SP]n > User Cost)
;  n= 1,2,3......R
 Set  as 'High' ;
IF (Total Cost[SP]n   == User Cost)
Set as ' Normal' ;
IF (Total Cost[SP]n < User Cost)
Set as 'Low';
End Procedure;
```

See Fig. 2.

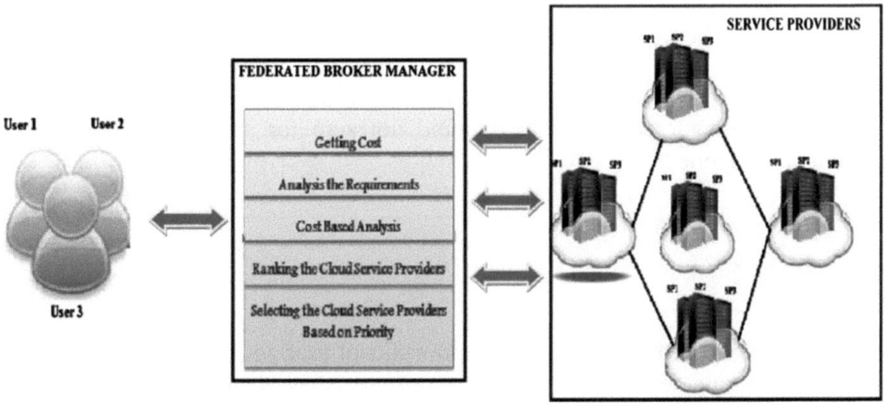

Fig. 2 Cost Analysis Architecture

5 Priority Feedback Decision Tree

Decision trees are powerful and popular tools for classification and prediction. Decision trees represent *rules*, We have introduced two set of rules into the decision tree. In our discussion, Primary level priorities (P_{α}) assigned with rule 1, Secondary level priorities (P_{β}) assigned with rule 2 (Fig. 3).

5.1 *Rule 1*

User permits to choose (N) number of priority components, if all priorities grade values are computed, then the maximum value of (γ) to be noted. When $1 = \gamma$ then respective service provider $[SP]_K$ identified and fixed the service provider suitable to the customer as they required. K value varies from 1 to R, R is the total number service provider in the similar highest rank.

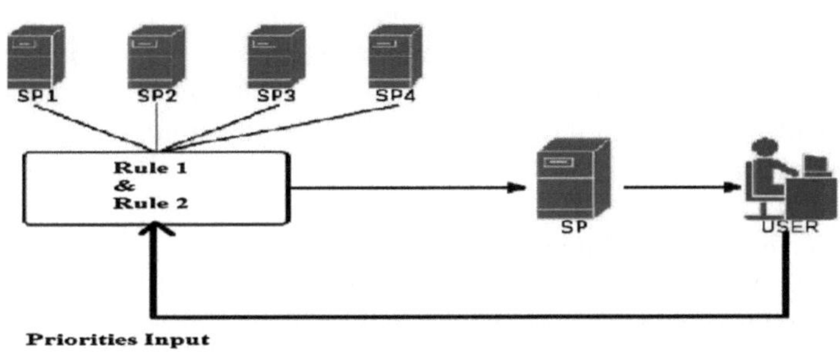

Fig. 3 Priority Feedback Decision Tree

$$[\gamma, [SP]_K] = \prod_{\alpha=1}^{N} P_\alpha, \gamma = 1, P \neq 0,\ K = 1, 2, 3, , , R \quad (4)$$

Priority scaling grade value (P) should not zero for any component of user submitted component.

5.2 Rule 2

Rule 2 to be executed when priority components of user not satisfied in rule 1. Any one of the components of primary level priority set is less than the 1 value in scaling grade value table of each service provider, then the respective service provider moves to the second level priority test. Hence number of service providers listed as R and number of user components varied from 1 to N. P_β represents priority components.

$$[\delta, [SP]_K] = \prod_{\beta=1}^{N} P_\beta, \delta \leq 0.5, P_\beta \neq 0, K = 1, 2, 3 \ldots R \quad (5)$$

K varies from 1 to maximum services providers. Hence priority values computed from rule 2 if values of $\delta \leq 0.5$, then K value identified and respective K value service provider assigned as best service provider.

6 Simulation Results and Summary

In our simulation we considered four service providers, Federated Broker Manager receives the request and requirement components from the user, check the availability of service and list out the service provider. Then each component of service provider assigned scaling grade values according to the algorithm-2, compute total scaling grade values for single service provider using $[SPW]_{Total}$ continued the process for all four service providers, Finally computed $[SPW_{Total}]_{AVG}$ for all service providers, Based on the average values arranged the service providers in the sequence, identified the largest average value from the sequence and respective service provider assigned to the user. If we get more than one service provider in similar highest rank then process redirected to Priority Feedback Decision Tree (Table 1).

From above table maximum grade mean value $[SG]_{mean}$ found for Service Provider (S3), So service provider S3 considered to be a best service provider among all service provider.

Table 1 Available Resource (AR), Scaling Grade Values (SGV)

User requirements (types)	Components	SP1		SP2		SP3		SP4	
		AR	SGV	AR	SGV	AR	SGV	AR	SGV
Availability	99.9 %	90.0 %	0.25	99.0 %	0.5	100 %	1.0	95 %	0.25
Processor speed	2.4 GHz	1.8 GHz	0.25	2.4 GHz	0.5	2.45 GHz	1.0	2.1 GHz	0.25
Memory (RAM)	2 GB	2 GB	0.5	2 GB	0.5	4 GB	1.0	2 GB	0.5
Service response time	60–100 s	60–120 s	0.5	60–100 s	1.0	60–120 s	0.5	40–120 s	0.25
Capacity	100 GB	100 GB	0.5	110 GB	1.0	100 GB	0.5	90 GB	0.25
Security	High	Medium	0.5	Medium	0.5	High	1.0	Medium	0.5
Scaling grade value total			2.5		4.0		5.0		2.0
Scaling grade value average			0.41		0.67		0.83		0.33

7 Conclusion and Future Work

Our frame work not only solves ambiguity among the cloud user also creating a healthy competition among the cloud service providers so that they able provide quality and cost reduced services to user. In future, planned to add security enhanced architecture with this frame work.

References

1. Calheiros, R., Vecchiola, C., Karunamoorthy, D., & Buyya, R. (2011). The Aneka platform and QoS-driven resource provisioning for elastic applications on hybrid Clouds. *Future Generation Computer Systems*, *28*(6), 861–870. doi:10.1016/j.future.2011.07.005.
2. http://cloudtweaks.com/2012/03/kpis-for-cloud-service-providers-customers/ "KeyPerformanceIndicators".
3. Preeti Gulia, Sumedha Sood (2013) 'Dynamic Ranking and Selection of Cloud Providers Using Service Level Agreements' International Journal of Computer Applications (0975–8887) Volume 72– No. 11, May 2013.
4. C. S. Rajarajeswari, M. Aramudhan, (September 2014) Ranking Model for SLA Resource Provisioning Management International Journal of Cloud Applications and Computing, 4(3), 68–80, July-September 2014.
5. Saurabh Kumar Garg a,∗, Steve Versteeg b, Rajkumar Buyyaa 'A framework for ranking of cloud computing services'. Future Generation Computer Systems 29 (2013) 1012–1023 journal homepage: www.elsevier.com/locate/fgcs.

Leakage Power Reduction Technique by Using Multigate FinFET in DSM Technology

Ajay Kumar Dadori, Kavita Khare, T.K. Gupta and R.P. Singh

Abstract Leakage power dissipation is the dominant contributor to total power dissipation today in CMOS integration design. Scaling is the prime thrust for development of CMOS circuits, which increases in the number of faults and leakage current in manometer scale in ultra low power circuit design. Here, in this paper we first reviewed the leakage power of various gates and highlight their merits and short come. FinFET technology completely substitute the CMOS to maintain the Mores law of scaling, next generation is of CNTFET which replaces the FinFET technology in term of scaling. We have calculated various parameters of basic logic gates like dynamic power, static power, delay, PDP and validation of results we have also implemented over C17 (ISCAS 85) benchmark circuit. Extensive HSPICE simulator on some basic gates and benchmark circuit by using SG and LP mode of FinFET and CNTFET technology at different temperature by using 32 nm Berkley Predictive Technology Module (BPTM), with supply voltage of 0.9 V at 100 MHz frequency.

Keywords Leakage controlling transistor · Low power dissipation · FinFET technology · SG & LP mode

A.K. Dadori (✉) · K. Khare · T.K. Gupta · R.P. Singh
Department Electronics & Communication, MANIT, Bhopal 462051, India
e-mail: ajaymanit0@gmail.com

K. Khare
e-mail: kavita_khare1@yahoo.co.in

T.K. Gupta
e-mail: taruniet@rediffmail.com

R.P. Singh
e-mail: prof.rpsingh@gmail.com

N. Modi et al. (eds.), *Proceedings of International Conference on Communication and Networks*, Advances in Intelligent Systems and Computing 508,
DOI 10.1007/978-981-10-2750-5_25

1 Introduction

From the history in electronics down scaling is major requirement for development of VLSI design, according to MOORE's law [1]. scaling leads to an increases in density, and also cost of the chip, power was always an important concern after speed and area of chip but with this aggressive downscaling power consumption is serious issue of concern, excessive heat dissipation shortens the battery life and the techniques used for this heat minimization leads to addition of cooling fans etc. which in turn increases the chip size. In order to overcome the problem associated with Bulk CMOS technology like Short channel effect and DIBL, Sub threshold slope degradation, threshold voltage roll off, that is as the miniaturization take place source and drain area encroaches into the channel are due to which leakage current easily from through them which make it very difficult to turn off the transistor completely [2].

FinFET is a non-planar 3D structure that contains thin vertical channel that resembles like fish's fin surrounded by gate along these sides forms Drain and Source unlike MOSFET the conducting channel gate is wrapped around the fin allowing very less leakage of current through the body during off state and providing better control over the channel, in turn threshold voltage gets lower short channel effect get decreased and performance enhances [3]. There are two types of power consumptions—Dynamic power consumption associated with active mode condition second is static power consumption associated with ideal mode condition. There are approaches to condense the power dissipation of the circuit at various design. Leakage currents with sub-threshold i.e. (source-to-drain leakage), band-to band tunneling, gate oxide tunneling, and from other current which are drawn continuously from the power supply leading to static power dissipation [1–3]. To minimize (DPS) dynamic power dissipation it is necessary to first reduce the supply voltage, but after a certain limit this reduction of supply voltage disturbs and affects the performance of the circuit [2], in order to maintain its performance it is necessary to then decrease the threshold voltage also, but it direct to leakage power dissipation. FinFET is a double gate structure device where front gate and back gate are differently connected. FinFET on the basis of these gates are classified as Tied gate and Independent gate [3]. In first type both gates are tied together giving rise to short-channel effect immunity and in other type i.e. in independent gate FinFET one gate is used for adjusting the threshold voltage and second gate is used for the switching action of transistor this lead to better control, independent gate FinFET minimizes the leakage in turn reduces power consumption to enhance the performance [4, 5].

Drain-induced barrier lowering DIBL is the major outcome effect of Short channel effect in this the high electric fields lowers the barrier from the drain which is supposedly controlled by gate only. While dealing with SCE very low V_{th} and very ultrathin t_{ox} is required to maintain speed of the device and variation to be under control because this effect can degrade the subthreshold slope of the device and can lead to change the threshold voltage (V_{th}).

In Fig. 1 L_{Gate} is the length of the FinFET wrapped over the thin silicon channel which is quite similar structure of planar FET. W is the width of the FinFET it can be defined as [6]

$$W = 2H_{fin} + T_{fin}$$

where H_{fin} is the height of the FinFET and T_{fin} is the thickness of the FinFET of thin silicon Fin respectively. Figure 2 shows the top view of the FinFET which consist of front gate and back gate which is tied over the thin silicon channel of the FinFET. In order to suppress shorter channel effect and enhance the area efficiency in FinFET, fin thickness is much smaller than fin height [7]. In SG mode both the front and back gate tied together with same supply voltage, but in LP mode front and back gate is biased independently so as to mitigate leakage power consumption.

Figure 3 represent pFinFET and nFinFET with, we consider two parallel transistors with two independent biasing front gate and back gate of FinFET

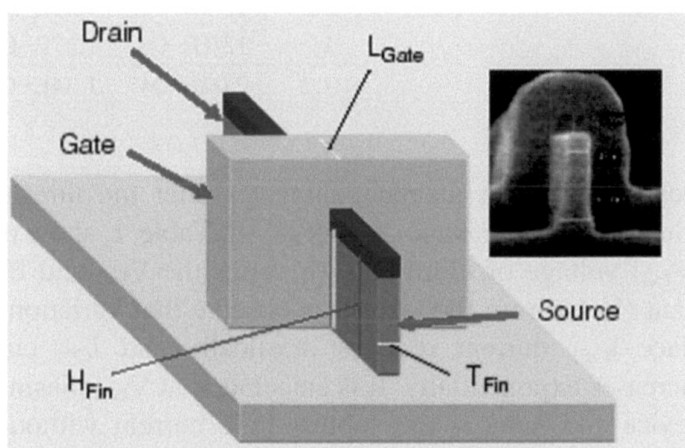

Fig. 1 3 Dimensional FinFET structure

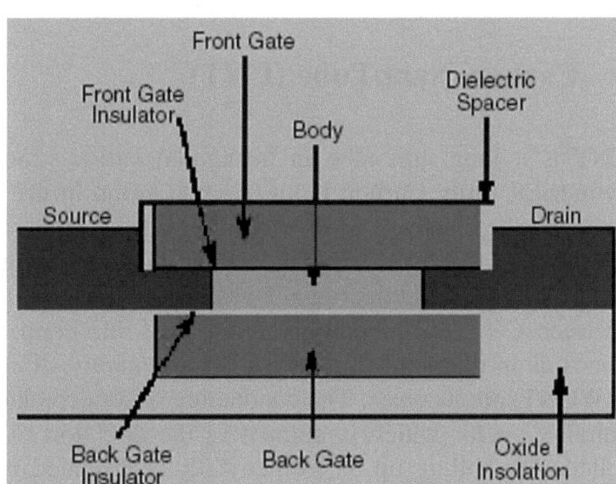

Fig. 2 Top view of FinFET

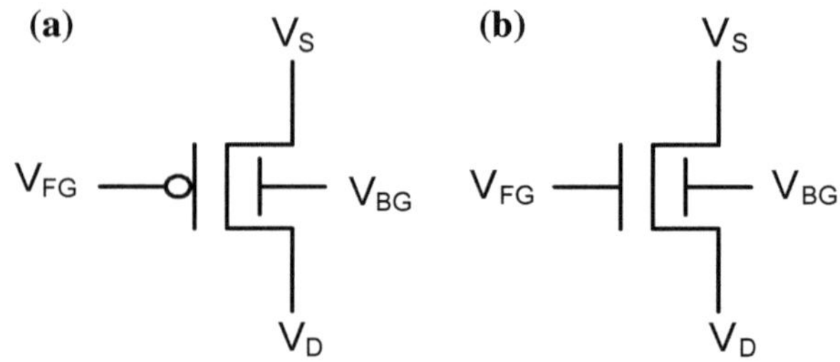

Fig. 3 **a** 4T pFinFET device. **b** 4T nFinFET device

Table 1 I_{ds} versus V_{GS} curve of nFinFET (W/L = 80/32 nm)

V_{FG}	I_{ds}			
	$V_{BG} = 0$	$V_{BG} = 0.2$	$V_{BG} = 0.4$	$V_{BG} = -0.4$
0	2.77E−08	5.31E−06	5.56E−06	5.56E−08
0.2	3.87E−05	3.87E−05	9.70E−05	2.63E−06
0.4	1.35E−04	9.70E−05	1.36E−04	5.02E−05
0.6	1.59E−04	1.11E−04	1.47E−04	7.23E−05
0.8	1.76E−04	1.20E−04	1.56E−04	8.25E−05
1	1.91E−04	1.27E−04	1.63E−04	9.06E−05
1.2	2.02E−04	1.34E−04	1.70E−04	9.77E−05

technology. Two independent gates alter the threshold voltage of the transistor which mitigate leakage current [8, 9]. Table 1, show drain current versus Front gate (V_{FG}) voltage of nFinFET, we varies the V_{FG} and Back gate voltage is kept constant (V_{BG}) from the table it is observe that variation of the threshold voltage take place I_{OFF} current reduces drastically and I_{ON} current of the FinFET device increases exponentially. It is calculate that V_{BG} biasing is more benefit for nFinFET device and reduces the subthreshold current without degradation of DIBL effect [10].

2 Carbon NanoTube (CNT)

CNT is a long thin tube in hexagonal lattice structure of carbon rolled into a cylindrical form. Carbon belongs same group in the periodic table as silicon with four valence electrons in its outermost shell. Iijima in 1991 [11] was the father of CNTs who discover large molecular that are unique for their size, shape and superior physical properties. The strength and stability is due to the SP^2-bonds between C-C (carbon-carbon) atoms and this bond in CNT is stronger than SP^3 bonds as in diamond. Single walled and multiwall carbon nanotubes (SWCNT & MWCNT) are its types. The conducting or semiconducting behavior depends on the chirality vector which is defined as the direction in which the graphene sheet is rolled. The rolling up is geometrically represented by the indices *n* and *m* which specify the diameter and the helicity angle θ of the CNT and determine its fundamental properties [12–14]. Semiconducting or metallic behavior of carbon nanotube is depends on its indexes (n, m): the tube is metallic if n = m or n − m = 3i, where i is an integer. If not follows the above condition then it is semiconducting. The possible structures are defined as armchair, zigzag and chiral. In this paper, only SWCNT and Zigzag structure (n, 0) is accounted for calculation of all parameters. An array of SWCNT is used to drive a large current. CNTs exhibit one-dimensional (1D) carrier transport which greatly reduces the scattering probability phenomena and therefore provides to a large mean free path, high current carrying capability and shows excellent thermal, mechanical and electrical properties [15].

Rolled-Up vector:

$$\mathrm{Ch} = \mathrm{na}_1 + \mathrm{ma}_2 \tag{1}$$

To determine whether a CNT is metallic or semiconducting in nature is based on indexes factor *(n, m)* (i.e. metallic if $n = m$ or $n - m = 3i$, where i is an integer. Otherwise, the behavior of tube is semiconducting.)

$$\mathrm{L} = \text{Modulus of } \mathrm{C_h} = \mathrm{a}\sqrt{\left(\mathrm{m}^2 + \mathrm{n}^2 + \mathrm{mn}\right)} \tag{2}$$

Chirality Vector & Bandgap Energy:

See Table 2.

Threshold Voltage:

It is defined as the voltage required to on the transistor and it is related with inverse of the diameter of the tube. In a CNT, the threshold voltage can be adjusted by controlling the chirality vector (i.e., the diameter) as:

$$\mathrm{V_{th}} = \mathrm{Eg}/2\mathrm{e} = \frac{\frac{\sqrt{3}}{3} a \mathrm{V}\pi}{e * \mathrm{DCNT}} \tag{3}$$

where E_g = Band-gap energy = $0.83/D_{CNT}$ *eV*, *a* = is the c-c bond distance = 2.49° A, V_π = is the carbon π–π bond energy = 3.033 eV.

Table 2 Bandgap energy of CNT with various Chiral vectors

Material	Chirality vector	Bandgap energy
CNT	10,0	0.98
CNT	11,0	0.95
CNT	13,0	0.76
CNT	14,0	0.74
CNT	16,0	0.62
CNT	17,0	0.61
CNT	19,0	0.52
CNT	20,0	0.51
CNT	22,0	0.45

Tube Diameter: The diameter of the tube is evaluated with the help of the expression [11–13].

$$D_{CNT} = \frac{C_h}{\pi} = \frac{a}{\pi}\sqrt{(n^2 + nm + m^2)} \tag{4}$$

3 Carbon Nanotube Field Effect Transistor (CNFET)

For the past several decades the scaling of CMOS offered improved performance from one technology node to the next. However, as device scaling goes beyond the 32 nm node, significant technology challenges will be faced. Currently two of the main challenges are: the considerable increase of standby power dissipation and the increasing variability in device characteristics which in turn affects circuit and system reliability. Now a day, it is nano-fabrication revolution and beyond scaling of conventional bulky CMOS, the CNT based FET (as shown in Fig. 4) technology with high thermal stability, superior controlled in process variation, excellent gate controllability and very high drive current is now achieved. Carbon Nanotube Field Effect Transistor (CNFET) devices are made by growing nanotubes on top of a thick silicon dioxide. The nanotube plays a role of channel between source and drain [14] for conduction. CNFET operates faster and it even consumes less power due to its ballistic transport and low off current properties. Hence due its extraordinary properties, a high stability, nanomemory, ultra-low power consumption devices with the replacement of Si is possible. The threshold voltage of CNFET is depends on the chirality vector and diameter of the CNT. The threshold voltage of CNFET is uniquely controlled by the variation of Dcnt [15, 16].

In this paper we analyze the behavior of different gate in FinFET and CNTFET technology we calculate the average power, delay, PDP and static power of all the gates. Two input NAND gate depend upon the input vector if either input is zero

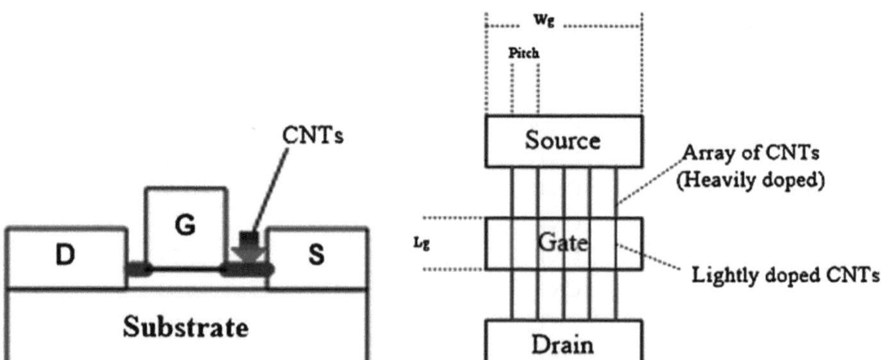

Fig. 4 Simple cross-sectional and top-view structure of CNTFET

output is one [10]. This logic gate is used to implement other gates in FinFET technology. Figures 5, 6, 7, 8, and 9 shows the FinFET based NAND, NOR, AND, XOR and XNOR gate in SG and LP mode of FinFET technology.

4 Results and Discussion

The circuit is simulated using HSPICE simulator at 32 nm at CMOS, FinFET and CNTFET technology with supply voltage of 0.9 V output capacitance is of 1 pF at temperature 25 and 110 °C with operating frequency of 100 MHz. As leakage current reduces shorter channel effect (SCE) reduces, according to the simulation

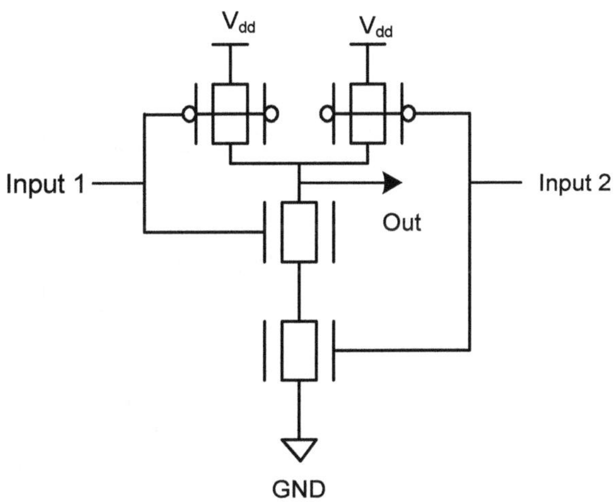

Fig. 5 Two input NAND gate

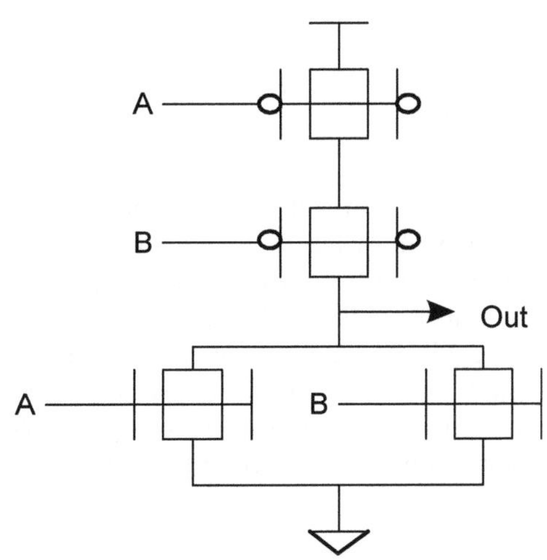

Fig. 6 Basic NOR gate

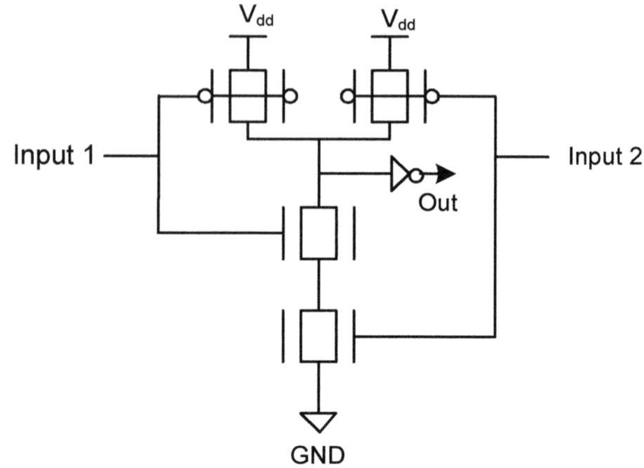

Fig. 7 Basic AND gate

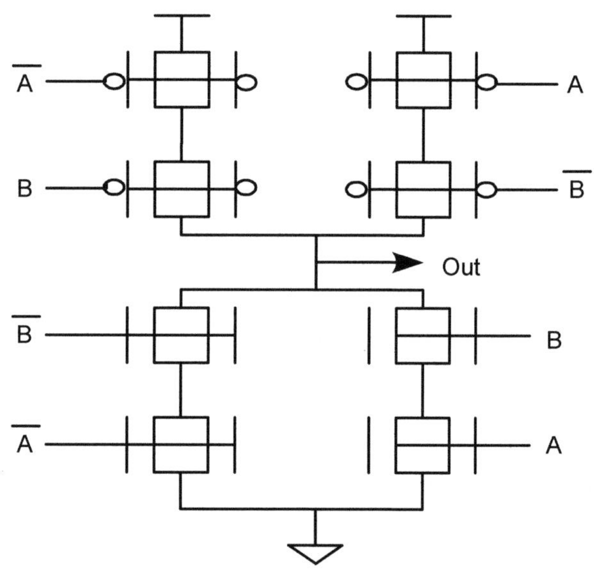

Fig. 8 Basic XOR gate

results. The power dissipation and delay of various logic gates are observed. It is clear that the CNFET based logic gates show less power dissipation than conventional FinFET devices. Therefore the CNT based device is applicable in future where there is issue of achieving less power consumption or a system with very high performance. The waveform indicates that there is no degradation in both the logic 0 and logic 1 CNTFET in terms of Average power, Delay, PDP and Static power in basic logic gate and C17 (ISCAS 85) Bench mark circuits. Simulation results shows that Fig. 10 of two input Nand gate generate proper logic level of two

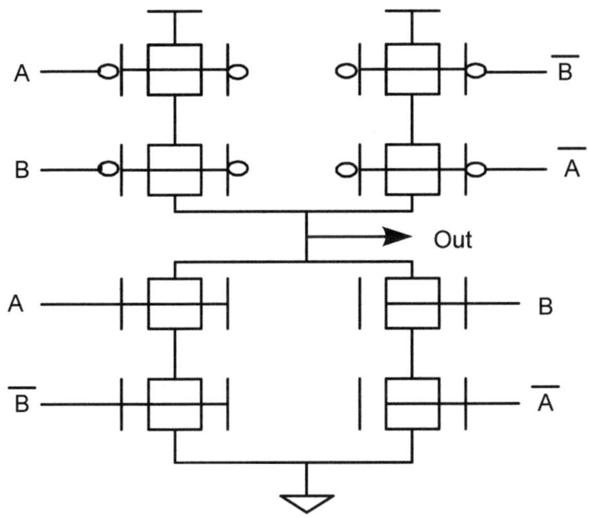

Fig. 9 Basic XNOR gate

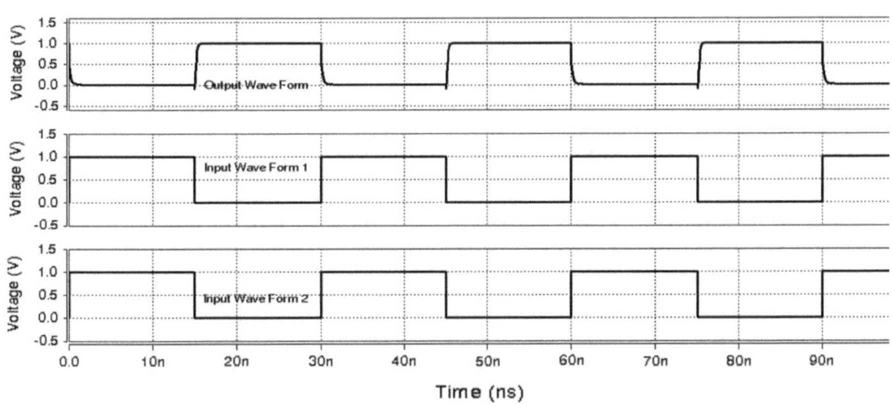

Fig. 10 Transient response of NAND gate by using CNTFET technology

input combination. In Table 3 we have observed that LP mode have lower power consumption both dynamic and static power at different temperature. In Table 4. Huge amount Dynamic power, delay and static power is saved in Nand gate dynamic power in SG mode is 0.143 μW, in LP mode 0.1017 μW and in CNTFET 3.291 nW at 25 °C. Saving of Static power is 0.024 nW in basic Nand gate in CNTFET with respect to 0.530 nW in SG mode and 0.128 nW in LP mode of FinFET technology (Fig. 11).

Table 3 Comparison of dynamic power, delay, PDP and static power in SG and LP mode of FinFET technology

Basic gates		SG mode				LP mode			
		Dynamic power (μW)	Delay (pS)	PDP (aJ)	Static power (nW)	dynamic power (μW)	Delay (pS)	PDP (aJ)	Static power (nW)
NOT	25 °C	0.126	1.009	0.127	7.868	0.0281	1.419	0.039	1.641
	110 °C	0.710	0.867	0.615	95.61	0.1882	1.332	0.250	24.75
NOR	25 °C	0.130	2.497	0.324	15.72	0.1005	10.77	1.082	3.280
	110 °C	0.627	2.361	1.480	190.4	0.2516	9.143	2.300	49.36
AND	25 °C	0.345	4.923	1.698	12.70	0.2053	14.18	2.911	0.368
	110 °C	1.785	8.678	15.47	392.1	0.5930	13.73	8.141	19.74
NAND	25 °C	0.143	5.715	0.817	0.530	0.1017	11.82	1.202	0.128
	110 °C	1.045	5.636	5.889	9.746	0.3587	11.72	4.203	2.873
EXOR	25 °C	0.270	8.315	2.245	15.55	0.1650	18.91	3.120	3.233
	110 °C	1.298	7.856	10.19	185.6	0.4490	17.58	7.893	4.834
EXNOR	25 °C	0.261	7.584	1.978	15.67	0.1614	17.24	2.782	3.276
	110 °C	1.294	7.316	9.466	184.5	0.4498	16.49	7.417	48.72
C17 (ISCAS 85)	25 °C	0.469	4.594	2.134	245.7	0.2891	19.92	5.756	94.01
	110 °C	7.431	5.321	39.54	398.5	3.9344	20.91	82.17	193.9

Table 4 Comparison of dynamic power, delay, PDP and static power in CNTFET technology

Basic gates		CNTFET mode			
		Average Power (nW)	Delay (fS)	PDP (aJ)	Static power (nW)
NOT	25 °C	2.872	3.920	11.25	0.412
	110 °C	4.242	25.06	106.3	4.735
NOR	25 °C	1.915	28.69	54.94	0.840
	110 °C	2.949	28.44	83.86	8.236
AND	25 °C	4.612	44.12	203.4	0.063
	110 °C	6.235	49.23	306.7	2.572
NAND	25 °C	3.291	24.83	81.71	0.024
	110 °C	5.267	29.67	156.2	0.273
EXOR	25 °C	7.891	34.89	275.3	0.853
	110 °C	9.268	53.29	493.8	0.993
EXNOR	25 °C	7.935	35.01	277.8	0.832
	110 °C	9.671	54.09	523.1	8.073
C17 (ISCAS 85)	25 °C	17.28	69.15	1194	7.189
	110 °C	29.29	76.08	2226	12.09

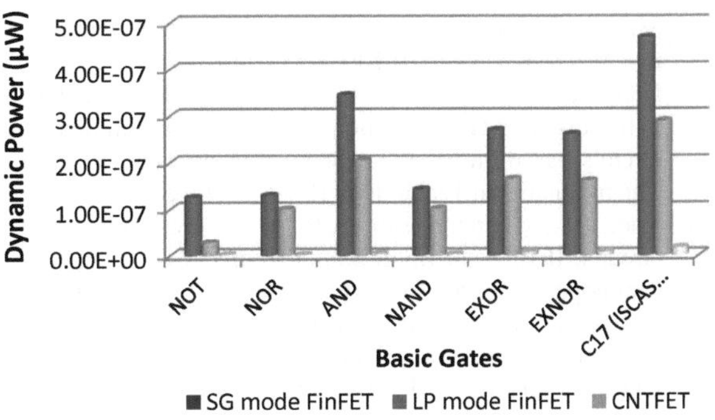

Fig. 11 Comparison of dynamic power consumption at 25 °C

5 Conclusion

In this paper we have calculate Dynamic power consumption, delay, PDP and Static power by using FinFET and CNTFET on all basic gates and C17 (ISCAS 85) benchmark circuit. For Simulation HSPICE is taken as a simulator tool. It requires a spice code (Transistor level net-list) of the desired circuit for their parameters calculation. All the circuits are mapped with 32 nm BPTM technology file. The Carbon Nanotube Field Effect Transistor (CNFET) is best idle and suitable candidate in the future implementation to achieve very low power dissipation or consumption as compared to conventional FinFET. In this paper the power dissipation of various CNFET logic gates are calculated. The conclusion resulted in this work is that the CNT based FET logic style is ideally suited to the deep submicron VLSI design technology for high performance systems. All kind of analysis with mapping of this file is shown through the flow of HSPICE design flow.

References

1. Sharifi, S., Jaffari, J., Hussein, M., Kusha, A. A., Navabi, Z.: Simultaneous Reduction of Dynamic and Static Power in Scan Structures. In: Proc. of the Design, Automation and Test, vol. 2, pp. 846–851 (2005).
2. Prakash, O.: Design and Analysis of Low Power Energy Efficient, Domino Logic Circuit for High Speed Application, International Journal of Scientific Research Engineering & Technology, vol. 1, no. 12, pp. 1–4 (2013).
3. Karimi, G., Alimoradi, A.: Multi-Purpose Technique to Decrease Leakage Power in VLSI Circuits, Canadian Journal on Electrical and Electronics Engineering, vol. 2, no. 3, pp. 71–74 (March 2011).
4. Tawfika, S.A., Kursun, V.: FinFET domino logic with independent gate keepers, Micro Electronics Journal, vol, 40, pp. 1531–1540 (2009).
5. LIAO Nan, CUI XiaoXin_, LIAO Kai, MA KaiSheng, WU Di, WEI Wei, LI Rui & YU DunShan “Low power adiabatic logic based on FinFETs” Science China, vol.57, Issn. 022402:1-022402:13 (2014).

6. Mishra, P., Muttreja, A., Jha, N.K.: FinFET Circuit Design, SPRINGER, Nanoelectronic Circuit Design, DOI 10.1007/978-1-4419-7609-3_2, pp: 23–53 (2011).
7. Nan, L., XiaoXin, C., Kai, L., Kai Sheng, MA., WU Di., Wei, WEI., Rui, LI., DunShan, YU.: Ultra-low power dissipation of improved complementary pass-transistor adiabatic logic circuits based on FinFETs, Science China, vol. 57, Issn. 042408:1–042408:13 (2014).
8. The International Technology Roadmap for Semiconductors, http://public.itrs.net/, (Nov 2003).
9. Kao, J.T., Chandrakasan, A.P.: Dual-threshold voltage techniques for low-power digital circuits, IEEE J. of Solid-State Circuits, vol. 35, pp. 1009–1018 (July 2000).
10. Mukhopadhyay, S., Neau, C., Cakici, T., Agarwal, A., Kim, C.H., Roy, K..: Gate leakage reduction for scaled devices using transistor stacking, IEEE Trans. on Very Large-Scale Integration Syst., vol. 11, no. 4, pp. 716–730 (Aug. 2003).
11. Guo,J., Lundstrom, M., and Datta, S.: Performance projections for ballistic carbon nanotube field-effect transistors," *Appl. Phys. Lett.*, vol. 80, no. 17, pp. 3192–3194, Apr. 2002.
12. Castro, L.C.,. John, D.L, Pulfrey, L., Pourfath, M., and Kosina,H.: Method for predicting *fT* for carbon nanotube FETs," *IEEE Trans. Nanotechnol.*, vol. 4, no. 6, pp. 699–704, Nov. 2005.
13. Avouris, P., Appenzeller, J., Martel, R., and. Wind, S.J.: Carbon nanotube electronics, in Proc. of IEEE, vol. 91, no. 11, pp. 1772–1784, (2003).
14. S.Iijima, "Helical microtubules of graphitic carbon", Nature, vol. 354, no. 6348, Nov. 1991, pp. 56–8.
15. Philip wong, H.S., Akinwande, D.: Carbon Nanotube and Graphene Device Physics, Cambridge University Press, New York, pp. 73–98, (2011).
16. Upasani, D.E., Shrote, S.B., Deshpande, P.S.: Analysis of Universal Logic Gates Using Carbon Nanotube Field Effect Transistor," International Journal of Computer Applications, vol. 7, pp. 29–33, (2010).

Home Automation Using Single Board Computing as an Internet of Things Application

Suneha Ashok Patil and Vishwakarma Pinki

Abstract This paper presents an implementation idea of home automation system that uses Wi-Fi. It is concerned with the automatic control of home appliances like fan, tube light, television, etc. using internet which means saving electric power and human energy. The home automation system consist of two parts central web server which communicates with single board computer (Raspberry Pi) and provides On/Off status of electronic home appliance and second part is single board computer which monitors and controls the electronic home appliance through relay switch. Information exchange format between central web server and single board computer is XML based.

Keywords Internet of Things · Home automation · Web server · Raspberry Pi

1 Introduction

Home automation explains the uses of computer and information technology to control home appliances automatically and remotely. Hence home automation is the rising field that has pulled the attention in the commercial and research field. There is continuous growth of mobile devices which leads to the demand of advanced ubiquitous mobile application development. To provide remote service access or to enable application for communication with each other web services is the most efficient way. In home automation system, it provides remote interface to the domestic appliances via Wi-Fi, to provide controlling and monitoring via a smart phone. An example of such device would be to turn On/Off particular home device when away from home.

S.A. Patil (✉) · V. Pinki
Shah and Anchor Kutchhi Engineering College, Mumbai, India
e-mail: patilsuneha278@gmail.com

V. Pinki
e-mail: vishwakarmapp@gmail.com

N. Modi et al. (eds.), *Proceedings of International Conference on Communication and Networks*, Advances in Intelligent Systems and Computing 508,
DOI 10.1007/978-981-10-2750-5_26

Internet of Things (IoTs) can be considered as interconnected network of electronic devices, controller system and the internet. The noteworthy progress of IoTs has formed an innovative aspect to the world of communication and information technologies. The IoTs concepts can be implemented across variety of home utility devices, in order to provide automated decisions, and ease handling. Adding artificial intelligence to controller devices can result in betterment of various areas such as conservation of energy, remedial activity in adverse circumstances. Hence home automation can be viewed as one of the developing application of IoTs.

2 Related Work

Today's house can be transformed into environment, where everything evolves as smarter, capable to qualify the needs of the householder [1]. Home automation will include [2] Energy and water consumption monitoring to recommend proper usage of resources, Switching on and off remotely appliances to avoid accidents and wastage of energy, Detection of intruders, observing circumstances inside storehouses. As electronic world is getting developed, the field of home automation is growing. Various means of signaling have been proposed such as Bluetooth [3], internet [4], short message service (SMS) [5], etc.

3 Methodology Used

System Architecture

IoT based home automation system will allow electronic home appliances such as fan, television, etc. to be controlled remotely using internet enabled devices such as smart phones, tablets etc. This system will support control measures like turning ON/OFF, setting up rules for alert/notification based on state of the device (Fig. 1).

- Number of handheld devices connects to the internet to access central web server.
- Central web server is connected to N Single Board Computers (SBCs) (example Raspberry Pi) through internet.
- Each Single Board Computer of N will connect to home appliances (fan, Light, etc.) through relay switch.
- Central web server will maintain the record of each home appliances.
- Example: Mobile application will send a request to switch ON devices to Central web server and SBC will pull device controller request from Central web server and delegate to relay switch.

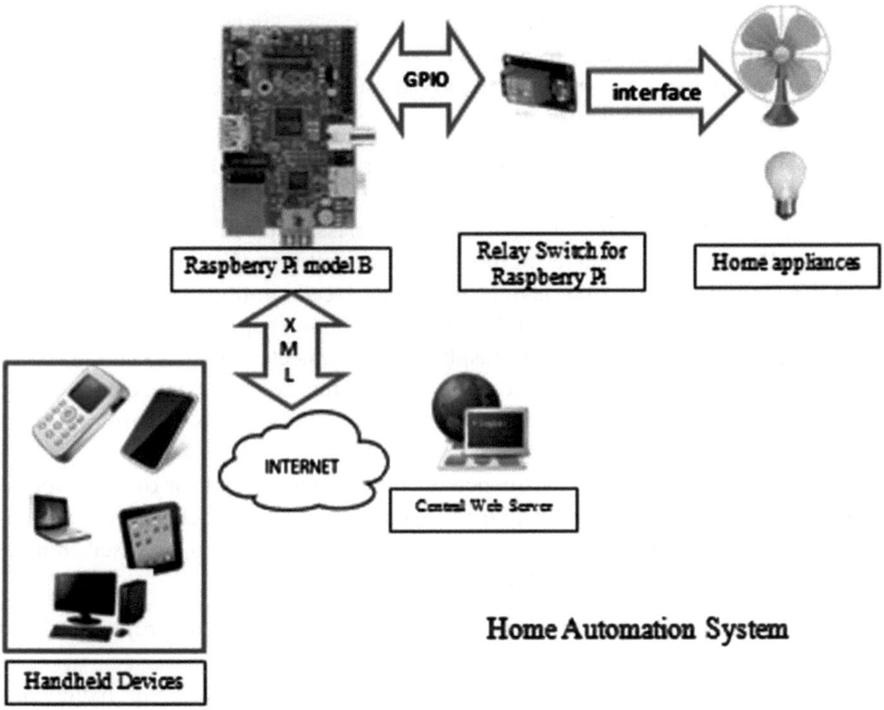

Fig. 1 System architecture

System design steps are as follow:

1. In this project 'DC Fan (12 V)' is used as 'Thing'.
2. Raspberry Pi (Model B) is used as 'Device' which controls 'DC Fan' using relay switch.
3. Java Servlets and web pages deployed on tomcat server is used as 'Service'.
4. 'DC Fan' is controlled using relay switch which is interfaced with General Purpose Input & Output (GPIO) module of 'Raspberry Pi'.
5. GPIO module is controlled using open source library (Pi4J) which provides high level Java APIs and abstracts low level Input/output operations such as writing to or reading from pin.
6. Database table is created which stores information about devices (MySQL database).
7. Java servlets are deployed on tomcat server which handles operation to:
    - Retrieve device configuration
    - Save device configuration
    - Retrieve device status
    - Save device status

8. Servlets use JDBC API to retrieve and save device status to database.
9. SMS notification services uses HTTP based APIs provided by smsc.biz.
10. Java web application listener is deployed on tomcat, which uses device database to manage SMS notifications.

Raspberry Pi has ARM processor and standard ports, such as VGA/HDMI port for display and sound, USB ports and Ethernet (LAN) port. Wi-Fi dongle which is connected to USB port of Raspberry Pi gives wireless connectivity to Raspberry Pi. Raspberry Pi provides a general purpose input-output port (GPIO) port which lets controller circuitry of external devices to be interfaced simply.

The GUI of web page is designed using Twitter bootstrap framework. This GUI act as medium between user and Raspberry Pi connected relay switch for controlling as well as viewing the present status of the devices connected to relay switch. Open JDK7 and Tomcat Web Server have been used as controller service provider.

Figures 2, 3, 4 and 5 shows sequence diagram of the system for overall process of GUI. The GUI of the system can be accessed from any hand held devices such as a PC/laptop or Smartphone using several options, using web browser, remote networking, etc. Hence it is flexible system.

- In home automation system which is developed as a part of this project 'DC Fan (12 V)' is used as 'Thing'.
- Raspberry Pi (Model B) is used as 'Device' which controls 'DC Fan' using relay switch.
- Java Servlets and web pages deployed on tomcat server is used as 'Service'.
- 'DC Fan' is controlled using relay switch which is interfaced with General Purpose Input & Output (GPIO) module of 'Raspberry Pi'.

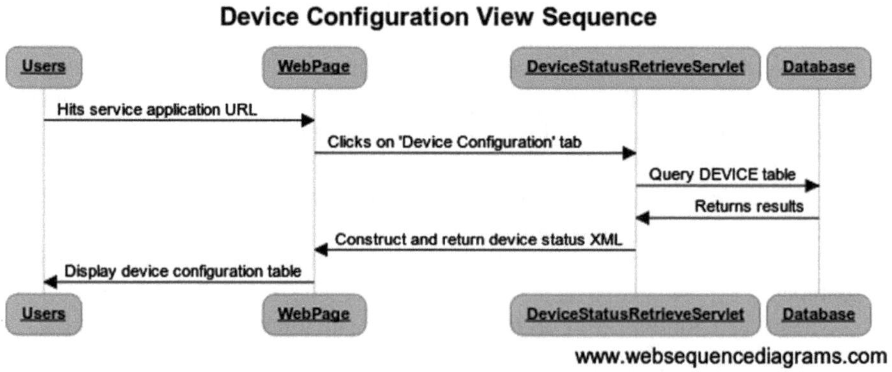

Fig. 2 Sequence diagram 1

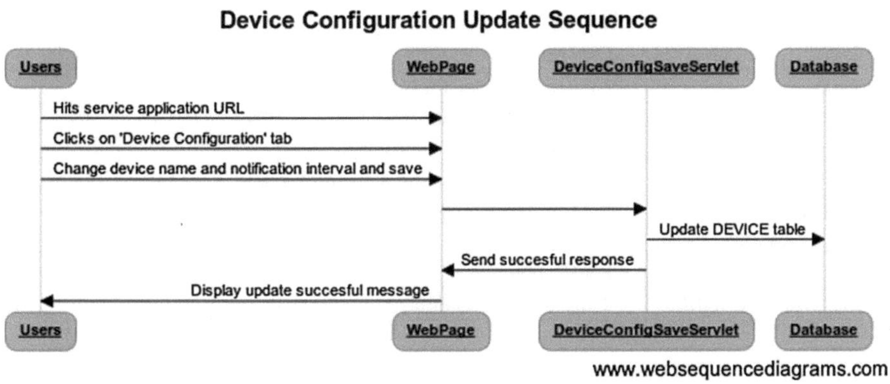

Fig. 3 Sequence diagram 2

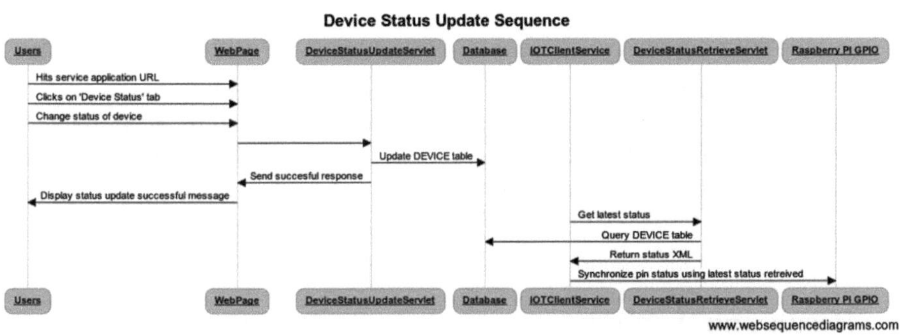

Fig. 4 Sequence diagram 3

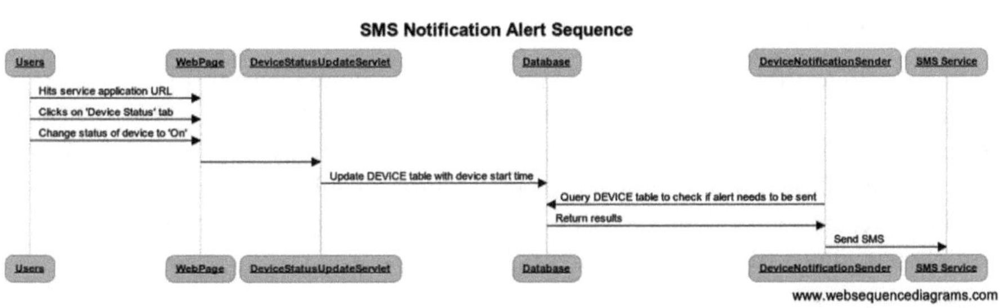

Fig. 5 Sequence diagram 4

- GPIO module is controlled using open source library (Pi4J) which provides high level Java APIs and abstracts low level Input/output operations such as writing to or reading from pin.
- Database table is created which stores information about devices (MySQL database).

Table 1 Database table design

Column	Description
PIN_ID	Unique identifier for device control pins
DEVICE_NAME	Logical name of physical device which is interfaced with Raspberry Pi (e.g. Hall Fan)
NOTIFY_INTERVAL	Interval for notification alert in minutes
STATUS	Indicate status of pin (ON/OFF).
DEVICE_START_TIME	Time when device status was changed to 'ON'. Used for handling notification alert
LAST_NOTIFY_TIME	Time when last notification alert was sent

- Java servlets are deployed on tomcat server which handles operation to:
  - Retrieve device configuration
  - Save device configuration
  - Retrieve device status
  - Save device status
- Servlets use JDBC API to retrieve and save device status to database.
- SMS notification services uses HTTP based APIs provided by smsc.biz.
- Java web application listener is deployed on tomcat, which uses device database to manage SMS notifications (Table 1).

4 Result and Analysis

The result is divided into following three parts:

1. **Hardware Result**

In order to test the functionality of the home automation system LEDs were used to indicate the change of signal of the interfaced devices. The Raspberry Pi connected to four channel 4 channel 12v ULN2003 based relay board module. Hence four pins are shown in GUI for the demonstration (Table 2).

Pin represents logical name of device interfaced with Raspberry and button to turn change status of LEDs to 'On' or 'Off'.

2. **Graphical User Interface Result**

Figure 6 shows the webpage with two tabs, first is device configuration tab and second is device status control tab.

Table 2 Mapping of GPIO output pins

Pin	Button	GPIO	GPIO output
Pin 0	ON LED 1	11	The LED1 has changed the state ON to OFF and OFF to ON
	OFF LED 1		
Pin 1	ON LED 2	12	The LED2 has changed the state ON to OFF and OFF to ON
	OFF LED 2		
Pin 3	ON LED 3	18	The LED3 has changed the state ON to OFF and OFF to ON
	OFF LED 3		
Pin 4	ON LED 4	17	The LED4 has changed the state ON to OFF and OFF to ON
	OFF LED 4		

3. Home Automation with Internet of Things

Internet of Things (IoT) is the latest buzz in the computer world. Industry experts believe that this trend will only continue to grow and develop even further in the coming few years. Home automation can viewed as application of IoT. In home automation system there multiple home appliances connect to Raspberry Pi GPIO pins. There are various types of Single Board Computers (SBC) are available in market other than Raspberry Pi such as Galileo, Arduino. For this reason data exchange which would be machine readable and interoperate across heterogeneous system is vital requirement. XML standards have evolved and accepted as de-factor standard for exchange of messages. XML provides reformatting of data for multiple devices and platforms. In this automation system the simple XML format is generated as shown in the Fig. 6 (Fig. 7).

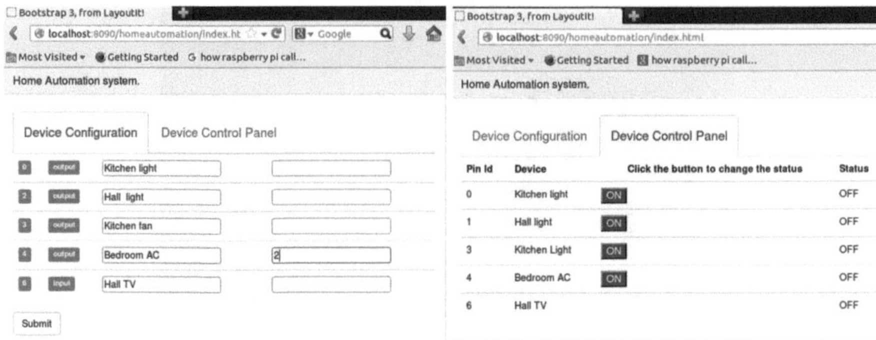

Fig. 6 The GUI for system interface for device configuration and device control panel

Fig. 7 Sample XML format generated as result and Fig. 6 shows proposed XML schema definition for exchange of messages amongst elements of IoT

5 Conclusion and Future Work

Concept of Internet of Things (IoT) was implemented for home automation system supporting control of devices over internet with Raspberry Pi model B as controller and generic XML schema for exchange of messages amongst elements of IoT has been proposed.

Standard protocol could be developed for exchange of messages between various elements of 'IoT', similar to 'Financial Information eXchange' protocol (used in stock market transactions) and ACORD (Association for Cooperative Operations Research used in insurance industry). Such protocol would further offer interoperability across 'device' implementation (e.g. Raspberry Pi, Arduino, Beagleboard etc.) and 'service' implementations (.Net, Java, php based services) and vendors of 'IoT' enabled 'things' (such as electronic appliances manufacturer like Samsung, Phillips, Onida etc.)

References

1. Various Authors, "Special issue on home automation," *IEEE Transactions on Automation Science and Engineering*. Vol. 5, Issue 1, 2008, p. 1–192.
2. Wan Nor Naema Wan Aziz, Muhammad Shukri Ahmad, Muhammad Mahadi Abdul Jamil, "Development of Novel Home Automation System via Raspberry Pi", IEEE International Conference on Control System, Computing and Engineering, 28–30 November 2014, Penang, Malaysia.
3. Sundaram, G.S.; Patibandala, B.; Santhanam, H.; Gaddam, S.; Alla, V.K.; Prakash, G.R.; Chandracha, S.C.V.; Boppana, S.; Conrad, J.M., "Bluetooth Communication using a Touchscreen Interface with the Raspberry Pi", 2013 IEEE.
4. A. ElShafee and K. A. Hamed, "Design and Implementation of a WiFi Based Home Automation System," World Academy of Science, Engineering and Technology, vol. 68, pp. 2177–2180, 2012.
5. M. S. H. Khiyal, A. Khan, and E. Shehzadi, "SMS Based Wireless Home Appliance Control System (HACS) for Automating Appliances and Security," Issues in Informing Science and Information Technology, vol. 6, pp. 887–894, 2009.
6. R. D. Caytiles and B. Park, "Mobile IP-Based Architecture for Smart Homes," International Journal of Smart Home, vol. 6, pp. 29–36, 2012.
7. U. Sharma and S. R. N. Reddy, "Design of Home/Office Automation Using Wireless Sensor Network," International Journal of Computer Applications, vol. 43, pp. 53–60, 2012.
8. K. P. Dutta, P. Rai, and V. Shekher, "Microcontroller Based Voice Activated Wireless Automation System," VSRD International Journal of Electrical, Electronics & Communication Engineering, vol. 2, pp. 642–649, 2012.
9. M. R. Kamarudin, M. A. F., and M. Yusof, "Low Cost Smart Home Automation via Microsoft Speech Recognition," International Journal of Engineering & Computer Science, vol. 13, pp. 6–11, June 2013.
10. Dr. Ovidiu Vermesan, Dr. Peter Friess, "Internet of Things: Converging Technologies for Smart Environments and Integrated Ecosystems", River Publishers, 2013.

Objective Quality Assessments of Restoration Images

Rasool Reddy Kamireddy, Shivaramakrishna Punem, Supriya Jangala, Geetha Ramakrishna Dutt Chamarthi and Kota Yedukondalu Srinivas

Abstract Picture (image) restoration is one of the significant concerns in the domain of image processing. It aims to recover the original picture (image) from its degraded observed image. After restoration, quality is another important task. A handful of various quality assessment approaches are used to evaluate the quality of a restored images, among them objective quality assessments is the best leading approach compared to others. Here our proposal mainly focuses on the analysis of restored images using several techniques of objective fidelity criteria. From simulation results we can easily examine the performance of different image restoration algorithms under different restored objects.

Keywords Image restoration · Degradation model · Image quality · Quality assessments · Objective fidelity criteria

1 Introduction

As we know that improving the quality of an image is a well-known technique for removal of unwanted artifacts in image. While performing the image enhancement operation sacrifices at least a small amount of resolution, but this cannot be

---

R.R. Kamireddy (✉) · S. Punem · S. Jangala · G.R.D. Chamarthi · K.Y. Srinivas
Malla Reddy Institute of Technology & Science, Secunderabad, India
e-mail: rasool.ellora@gmail.com

S. Punem
e-mail: shivaramakrishna.mj@gmail.com

S. Jangala
e-mail: priyalazaras@gmail.com

G.R.D. Chamarthi
e-mail: dutt.060587@gmail.com

K.Y. Srinivas
e-mail: kota_sri_2k@yahoo.com

N. Modi et al. (eds.), *Proceedings of International Conference on Communication and Networks*, Advances in Intelligent Systems and Computing 508,
DOI 10.1007/978-981-10-2750-5_27

applicable in more applications especially like in fluorescence microscope. These problems are overcome by using an advanced technique in processing of an image like image restoration. In the recent years, image restoration is a challenging problem in processing of an image and vision through computer applications. In general, the camera devices are generating the degraded images due to intrinsic (lens distortion factor, focal lens arrangement) and extrinsic (human being) parameters of a camera. The method of Image restoration operation is to estimating the cleaned image with respects to its degraded image. Image restoration is unlike differs from image enhancement because the enhancement of an image is a subjective criteria and the image restoration is an object criteria. The usage of 'restoration of an image' method in the processing of an image applications reaches to a greater level since from the last two decades [1–12].

In this proposal our discussion is about the various restoration techniques which formulates and evaluates the objective fidelity criteria of restored images. The rest of the paper organized as, in Sect. 2 discuss the various image restoration techniques, in Sect. 3 discuss the various image quality metrics, in Sect. 4 discuss the experimental results and in Sect. 5 discuss the conclusions.

2 Related Work

In this section we will discuss and review the some well known restoration techniques.

2.1 *Inverse Filtering*

Inverse filtering [13] is a familiar technique for restoration of an image and it is assumes an accurately estimated transfer function H. It generates a linear restoration filter, which satisfies the criterion of MSE (Mean Squared Error)

$$\min\left\{\left|g - h * \hat{f}\right|^2\right\} \tag{1}$$

where $\hat{f} = m * g$ and 'm' represents PSF (Point Spread Function) and then inverse reconstruction filter is given by

$$M_{inverse} = FT\{m\} = 1/H \tag{2}$$

The crucial drawback of the inverse filtering is, when H = 0 the image can't be perfectly restored even in the absence of noise because of the indeterminate form.

2.2 *Wiener Filtering*

This filtering method is developed to reduce the additive random noise in images is based on Wiener filtering [13]. The extensive drawback of Wiener filter is, the restored image get smoothed, so that an optimum linear MSE estimator is used to minimizing the mean square error and is follows

$$\min \mathrm{E}\left|(\mathrm{f}-\hat{\mathrm{f}})^2\right| \tag{3}$$

where 'f' = Original image and '$\hat{f}$' = Restored image.

The MMSE (Minimum MSE) wiener filter restoration is given [13] by

$$\mathrm{M}_{\text{wiener}} = \frac{\mathrm{H}^* \times \mathrm{SNR}}{|\mathrm{H}|^2 \times \mathrm{SNR} + 1} \tag{4}$$

where H = Degraded function and H^* = Complex conjugate of H.

The great advantage of wiener filter is, It provides a good noise performance (M = 0) when H = 0.

2.3 *Geometric Mean Filter*

Inverse filtering provides good resolution at lower frequencies, but poor at higher frequencies. Whereas wiener filter provides very good noise performance but this filter can be achieves higher smoothening restored images. Therefore, to attain the amount of resolution at higher and lower frequencies region we introduce geometric mean filter [13]. The restoration filter of geometric mean filter is

$$\mathrm{M}_{\text{Geometric Mean}} = [\mathrm{M}_{\text{inverse}}]^{\alpha}[\mathrm{M}_{\text{Parametric wiener}}]^{1-\alpha} \tag{5}$$

where the parameter $0 \le \alpha \le 1$.

$$\mathrm{M}_{\text{Parametric wiener}} = \frac{1}{H}\left[\frac{1}{1+\gamma s_n/|H|^2 s_f}\right] \tag{6}$$

where

s_n Noise power spectral density
s_f Signal power spectral density and $\gamma \ge 0$.

2.4 Constrained Deconvolution

The Constrained Deconvolution [13] is similar to inverse filtering; in this the Restoration of an image is by an unknown PSF. The constrained deconvolution of filter restoration is given by

$$M_{\text{Constrained Deconvolution}} = \frac{H^*}{|H|^2 + \gamma|C|^2} \tag{7}$$

where 'C' is Fourier transform of sampled constraint function I and 'γ' related to the Lagrange multiplier. The value of 'γ' is adjusted so that the fixed error criterion is satisfied. The constrained deconvolution restoration is more noise sensitive than wiener filtering.

2.5 Homomorphic Deconvolution

The Homomorphic Deconvolution [13] is also known as Blind Deconvolution. In Homomorphic Deconvolution, the image gets restored by unknown PSF. The blind deconvolution of a filter restoration is given by

$$M_{\text{Blind Deconvolution}} = \left[s_{ff}/s_{gg}\right]^{1/2} \tag{8}$$

where s_{ff} are an estimated PSD from an undegraded image and s_{gg} estimated PSD of blurred image.

If the noise wide image get sensed in stationary and it is a uncorrelated random process, then the estimated PSD of blurred image is given by

$$s_{gg} = |H|^2 s_{ff} + s_{nn} \tag{9}$$

The magnitude of blind deconvolution ($|M_{\text{Blind Deconvolution}}|$) is equivalent to the magnitude geometric mean ($|M_{\text{Geometric Mean}}|$) when $\alpha = 1/2$ and $\gamma = 1$.

$$|M_{\text{Blind Deconvolution}}| = \left|M_{\text{Geometric Mean}}, \ \alpha = \frac{1}{2}, \gamma = 1\right| \tag{10}$$

The utmost drawback of the blind deconvolution algorithm is computational complexity because of additional processing is required to determine the filter phase.

2.6 Block-Matching and 3D Filtering (BM3D)

The BM3D [14] filtering method is accomplished by grouping the homogeneous 2D image blocks into 3D groups and performing collaborative filtering on 3D groups. In the BM3D the corrupted image is subdivide into fragments (blocks) in sliding window manner and each fragment is processed by searching matched fragments with fixed thresholding. Finally these similar fragments are categorized together to form 3D group and then apply collaborative filtering to achieve better results. But while performing denoising using BM3D it introduces 'blocking effects' in the results.

2.7 Joint Statistical Modeling (JSM)

The characterization and formulation of local smoothness (LS) and nonlocal self-similarity (NLS) methods are mathematically challenging issues in processing of an image. In this work, the mentioned above two properties are well characterized and formulated from statistics of an image and introducing a JSM [15] for image restoration. Therefore, the JSM is including of two modules.

1. Local Statistical Modeling (LSM)
2. Nonlocal Statistical Modeling (NLSM)

$$\psi_{JSM}(u) = \tau\psi_{LSM}(u) + x\psi_{NLSM}(u) \tag{11}$$

where τ, x are represents regularization parameters.

ψ_{LSM} Represents the Local Smoothness and is given by

$$\psi_{LSM}(u) = \|D_u\|_1 = \|D_v u\|_1 + \|D_h u\|_1 \tag{12}$$

where D_h and D_v represents horizontal and vertical finite difference operators. The Eq. (12) can be acquired by putting v = 1 in Eq. (13)

$$P_{GGD}(x) = \frac{\vartheta n(\vartheta)}{2\Gamma\left(\frac{1}{\vartheta}\right)} \frac{1}{\sigma_x} e^{-[n(\vartheta)|x|\sigma_x]^{\vartheta}} \tag{13}$$

where $n(\vartheta) = \sqrt{\Gamma\left(\frac{3}{\vartheta}\right)\Gamma\left(\frac{1}{\vartheta}\right)}$ and $\int_0^\infty e^{-u}u^{t-1}du$ is gamma function, σ_x is the standard deviation and ϑ is shape parameter.

ψ_{NLSM} Represents the Nonlocal Self-similarity and is given by

$$\psi_{NLSM}(u) = \|\Theta\|_1 = \sum_{i-1}^{n} \left\|T^{3D}(Zu^i)\right\|_1 \tag{14}$$

where

Zu^i Each stack from S_u^i into 3D array i = 1, 2, …, n
T^{3D} Transform of 3D and coefficient of Zu^i
u^i Image divided by 'n' overlapped blocks.

The Eq. (14) can be acquired by processing the following steps

Step 1: The image u will be divided with N into n overlapped blocks of u^i of size b_s, where i = 1, 2, 3, …, N
Step 2: Define S_u^i the set including the 'c' best matched blocks to u^i in the searching window.
Step 3: Each block in the S_u^i stack belongs to a 3D array is represented by Z_u^i
Step 4: Denote T^{3D} as an operator of 3D orthogonal transform and T^{3D} (Z_u^i) are the coefficients of the transform for Z_u^i. Let Θ_u be the column vector of an image 'u' having all the transform coefficients with size $k = b_s * c * n$ built from all the T^{3D} (Z_u^i) arranged in the lexicographic order.
Step 5: The histogram of the transform coefficients are analyzed by

Substituting Eqs. (12) and (14) in Eq. (11) we get

$$\psi_{JSM}(u) = \tau\|D_u\|_1 + x\sum_{i=1}^{n} \left\|T^{3D}(Zu^i)\right\|_1 \tag{15}$$

Therefore JSM is able to local smoothness and non-local self similarity of natural images richly and combine the both models for improving reconstruction quality.

3 Image Quality Metrics

In this section we will discuss various image quality measurements to find out the quality of a restored image obtained from different restoration methods as we discussed in Sect. 2.

3.1 Mean Square Error (MSE)

Let f(x, y) and g (x, y) represents original image and restored image with a dimensions M * N, then Mean square error is given by

$$\text{MSE} = \frac{1}{\text{MN}}\sum_{x=1}^{M}\sum_{y=1}^{N} (f(x, y) - g(x, y))^2 \tag{16}$$

3.2 Peak Signal to Noise Ratio (PSNR)

Let f(x, y) and g (x, y) represents original image and restored image with a dimensions M * N, then Peak signal to noise ratio is given by

$$\mathrm{PSNR} = 10\log_{10}\left(255^2/\mathrm{MSE}\right) \tag{17}$$

3.3 Normalized Cross Correlation (NCC)

Let f(x, y) and g (x, y) represents original image and restored image with a dimensions M * N, then Normalized Cross Correlation is given by

$$\mathrm{NCC} = \sum_{x=1}^{M}\sum_{y=1}^{N}(f(x,y) \times g(x,y)) / \sum_{x=1}^{M}\sum_{y=1}^{N}(f(x,y))^2 \tag{18}$$

3.4 Absolute Difference (AD)

Let f(x, y) and g (x, y) represents original image and restored image with a dimensions M * N, then Absolute Difference is given by

$$\mathrm{AD} = \sum_{x=1}^{M}\sum_{y=1}^{N}(f(x,y) - g(x,y))/\mathrm{MN} \tag{19}$$

3.5 Structural Content (SC)

Let f(x, y) and g (x, y) represents original image and restored image with a dimensions M * N, then structural content is given by

$$\mathrm{SC} = \sum_{x=1}^{M}\sum_{y=1}^{N}(f(x,y))^2 / \sum_{x=1}^{M}\sum_{y=1}^{N}(g(x,y))^2 \tag{20}$$

3.6 Laplacian Mean Square Error (LMSE)

Let f(x, y) and g (x, y) represents original image and restored image with a dimensions M * N, then Laplacian mean square error is given by

$$\mathrm{LMSE} = \sum_{x=1}^{M} \sum_{y=1}^{N} [O(f(x,y)) - O(g(x,y))] / \sum_{x=1}^{M} \sum_{y=1}^{N} O(f(x,y))^2 \tag{21}$$

where

$$O(f(x,y)) = f(x+1,y) + f(x-1,y) + f(x,y+1) + f(x,y-1) - 4f(x,y) \tag{22}$$

3.7 Normalized Absolute Error (NAE)

Let f(x, y) and g (x, y) represents original image and restored image with a dimensions M * M then Universal Quality Index (UQI) is given by

$$\mathrm{NAE} = \sum_{x=1}^{M} \sum_{y=1}^{N} |f(x,y) - g(x,y)| / \sum_{x=1}^{M} \sum_{y=1}^{N} |f(x,y)| \tag{23}$$

3.8 Universal Quality Index (UQI)

Let f(x, y) and g (x, y) represents original image and restored image with a dimensions M * M then Universal Quality Index (UQI) is given by

$$\mathrm{UQI} = \frac{4\sigma_{fg}\bar{f}g}{\left(\sigma_f^2 + \sigma_g^2\right) f^2 \bar{g}^2} \tag{24}$$

where $\bar{f} = \frac{1}{M}\sum_{x=1}^{M} f_x$ and $\bar{g} = \frac{1}{M}\sum_{x=1}^{M} g_x$

$$\sigma_f^2 = \frac{1}{M-1}\sum_{x=1}^{M} (f_x - \bar{f})^2 \text{ and } \sigma_g^2 = \frac{1}{M-1}\sum_{x=1}^{M} (g_x - \bar{g})^2$$

3.9 Feature Similarity Index Measurement (FSIM)

The FSIM [16] contains 2 stages

1. The local similarity map (*PC*) is determined between f(x, y) and g(x, y).
2. Pool the similarity map into a single similarity score (GM).

Then

$$\mathrm{FSIM} = \frac{\sum_{x \subset \Omega} S_L(x) \cdot PC_m(x)}{\sum_{x \subset \Omega} PC_m(x)} \tag{25}$$

where

$$S_L(x) = S_{PC}(x) \cdot S_G(x) \tag{26}$$

$$S_{PC}(x) = \frac{2PC_1(x) .PC_2(x) + T_1}{PC_1^2((x)) + PC_2^2((x)) + T_1} \tag{27}$$

$$S_G(x) = \frac{2G_1(x).G_2(x) + T_2}{G_1^2((x)) + G_2^2((x)) + T_2} \tag{28}$$

$$PC_m(x) = \max(PC_1(x), PC_2(x)) \tag{29}$$

4 Experimental Results

In this section, we discuss the performance evaluation of restoration techniques using various objective quality metrics like AD, SC, NCC, LMSE, NAE, PSNR, UQI and FSIM. In that FSIM is achieves higher consistency than other quality metrics. The FSIM and UQI value always lies between 0 and 1. The higher FSIM and UQI means provide better quality.

The Tables 1 and 2 represents the quality assessments of various restoration techniques on SRK and Barbara256 restored images from Gaussian blur with $\sigma = 0.5$. By seeing the Tables 1 and 2, we can say that JSM is provides significant improvement in terms of PSNR, FSIM, and UIQ compared to all other state-of-art techniques like BM3D, Wiener filtering, Blind Deconvolution, Geometric Mean and all the simulations are carryout in MATLAB 8.1.0.604 in DELL VOSTRO 1014 (Figs. 1, 2, 3 and 4).

Table 1 Quality Assessments of restored SRK image from Gaussian blur

Method	AD	NCC	LMSE	NAE	PSNR	SC	UQI	FSIM
JSM	0.0028	0.999	1.094	0.005	47.107	0.998	0.887	0.997
Wiener	0.005	0.972	5.55	0.052	27.65	0.923	0.214	0.951
Blind deconvolution	0.014	0.984	3.473	0.088	25.33	0.957	0.601	0.959
Geometric mean	0.015	0.387	4.726	0.996	4.4566	0.872	0.413	0.6137
BM3D	0.0743	0.996	1.245	0.0699	37.98	0.993	0.748	0.985

Table 2 Quality Assessments of restored Barbara256 image from Gaussian blur

Method	AD	NCC	LMSE	NAE	PSNR	SC	UQI	FSIM
JSM	0.0028	0.9938	0.550	0.046	28.53	0.999	0.831	0.925
Wiener	0.026	0.9980	1.9540	0.076	26.45	0.996	0.679	0.877
Blind deconvolution	0.396	0.9889	1.265	0.145	19.97	0.987	0.387	0.739
Geometric mean	0.1159	0.9958	2.0413	0.084	25.52	0.995	0.639	0.851
BM3D	0.0032	0.9943	0.853	0.058	27.38	0.998	0.785	0.892

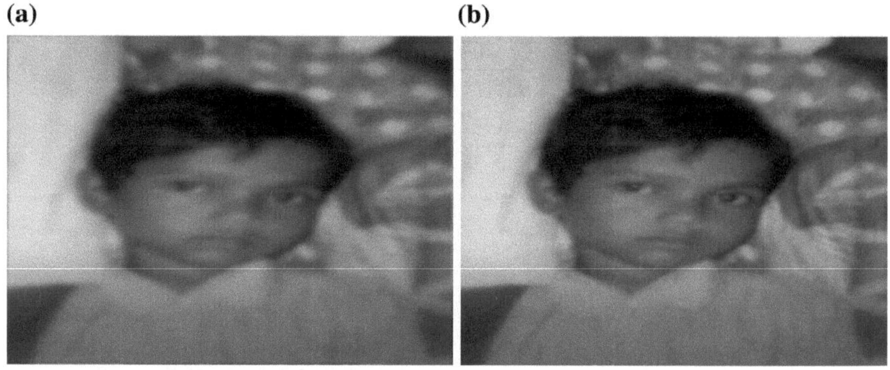

Fig. 1 **a** Blurred SRK image. **b** Restored SRK image using JSM

Fig. 2 Peak Signal to Noise Ratio (PSNR)

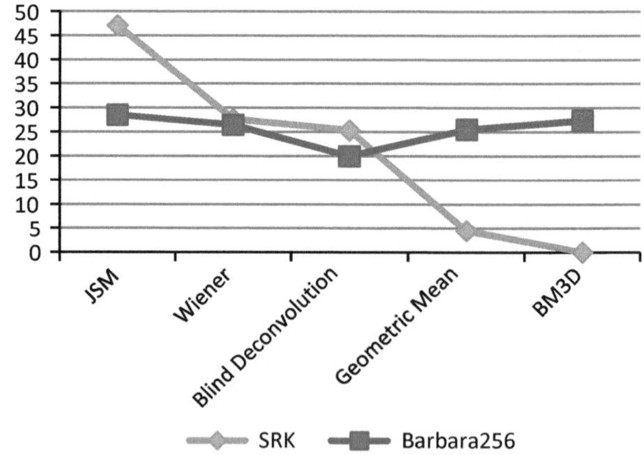

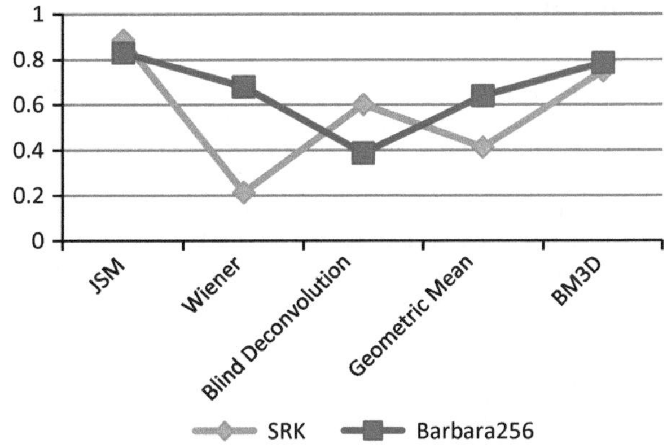

Fig. 3 Universal Quality Index (UQI)

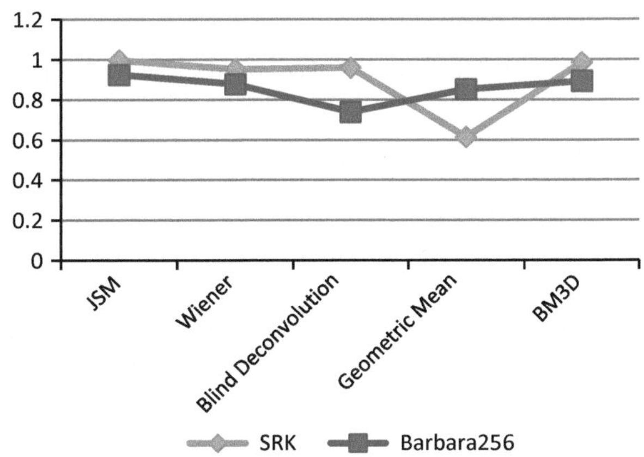

Fig. 4 Feature Similarity Index Measurement (FSIM)

5 Conclusion

In this paper, we evaluate the performance of various image restoration techniques using objective quality metrics like AD, SC, NCC, LMSE, NAE, PSNR, UIQI and FSIM. From the experimental results, we can say that the joint statistical modeling technique has higher PSNR, UIQI and FSIM compared to all other techniques. Therefore from above analysis, we can say that the joint statistical modeling is providing significant performance compared to all other state-of-art techniques. In future, this study carried out on various restoration techniques by using better image quality metrics.

References

1. M. R. Banham and A. K. Katsaggelos.: Digital image restoration. IEEE Trans. Signal Process. Mag., vol. 14, no. 2, pp. 24–41 (1997).
2. L. Rudin, S. Osher, and E. Fatemi.: Nonlinear total variation based noise removal algorithms. Phys. D., vol. 60, nos. 1–4, pp. 259–268 (1992).
3. A. Chambolle.: An algorithm for total variation minimization and applications. J. Math. Imag. Vis., vol. 20, nos. 1–2, pp. 89–97 (2004).
4. K. Dabov, A. Foi, V. Katkovnik, and K. Egiazarian.: Image denoising by sparse 3D transform-domain collaborative filtering. IEEE Trans. Image Process. vol. 16, no. 8, pp. 2080–2095 (2007).
5. Y. Chen and K. Liu.: Image denoising games. IEEE Trans. Circuits Syst. Video Technol., vol. 23, no. 10, pp. 1704–1716 (2013).
6. J. Zhang, D. Zhao, C. Zhao, R. Xiong, S. Ma, and W. Gao.: Image compressive sensing recovery via collaborative sparsity. IEEE J. Emerg. Sel. Topics Circuits Syst., vol. 2, no. 3, pp. 380–391(2012).
7. H. Xu, G. Zhai, and X. Yang.: Single image super-resolution with detail enhancement based on local fractal analysis of gradient. IEEE Trans. Circuits Syst. Video Technol., vol. 23, no. 10, pp. 1740–1754 (2013).
8. X. Zhang, R. Xiong, X. Fan, S. Ma, and W. Gao.: Compression artifact reduction by overlapped-block transform coefficient estimation with block similarity. IEEE Trans. Image Process., vol. 22, no. 12, pp. 4613–4626(2013).
9. W. Dong, L. Zhang, G. Shi, and X. Wu.: Image deblurring and superresolution by adaptive sparse domain selection and adaptive regularization. IEEE Trans. Image Process., vol. 20, no. 7, pp. 1838–1857(2011).
10. L. Wang, S. Xiang, G. Meng, H. Wu, and C. Pan.: Edge-directed single-image super-resolution via adaptive gradient magnitude self interpolation. IEEE Trans. Circuits Syst. Video Technol., vol. 23, no. 8, pp. 1289–1299 (2013).
11. J. Dai, O. Au, L. Fang, C. Pang, F. Zou, and J. Li.: Multichannel nonlocal means fusion for color image denoising. IEEE Trans. Circuits Syst. Video Technol., vol. 23, no. 11, pp. 1873–1886(2013).
12. A. Foi, V. Katkovnik, and K. Egiazarian.: Pointwise shape-adaptive DCT for high-quality denoising and deblocking of grayscale and color images. IEEE Trans. Image Process., vol. 16, no. 5, pp. 1395–1411(2007).
13. Murphy P.K.: Survey of Image Restoration Techniques. Technical Memo, Johns Hopkins Univ., Laurel, MD. Applied Physics Lab. Apr. - Aug. (1986).
14. Li Dai, Yousai Zhang and Yuanjiang Li.: BM3D Image Denoising Algorithm with Adaptive Distance Hard-threshold. International Journal of Signal Processing, Image Processing and Pattern Recognition Vol.6, No.6 pp. 41–50, (2013).
15. Jian Zhang, Debin Zhao, Ruiqin Xiong, Siwei Ma, Wen Gao, Fellow.: IEEE, "Image Restoration Using Joint Statistical Modeling in Space-Transform Domain. IEEE Transactions on Circuits and Systems for Video Technology, Volume: 24, Issue: 6, pp 915–928 (2014).
16. Lin Zhang, Lei Zhang, Xuanqin Mou and David Zhang.: FSIM: A Feature SIMilarity Index for Image Quality Assessment. IEEE Trans. Image Processing, vol. 20, no. 8, pp. 2378–2386 (2011).

Author Biographies

K. Rasool Reddy received his B.E degree in Electronics & Communication Engineering from Sir C. R. Reddy College of Engineering, Eluru, Andhra Pradesh, in 2011, and the M.Tech in Electronics & Communication Engineering from Malla Reddy Institute of Technology & Science, Secunderabad, Andhra Pradesh, in 2013. He was an assistant professor, Malla Reddy Institute of Technology & Science, JNTU Hyderabad University from 2014. His research interests include image and video processing, digital signal processing.

P. Shivaramakrishna received his B.Tech degree in Electronics & Communication Engineering from Malla Reddy Institute of Management, Secunderabad, Andhra Pradesh, in 2013 and the M.Tech in Electronics & Communication Engineering from Malla Reddy Institute of Technology & Science, Secunderabad, Andhra Pradesh, in 2015. His research interests include digital signal processing, wireless communication.

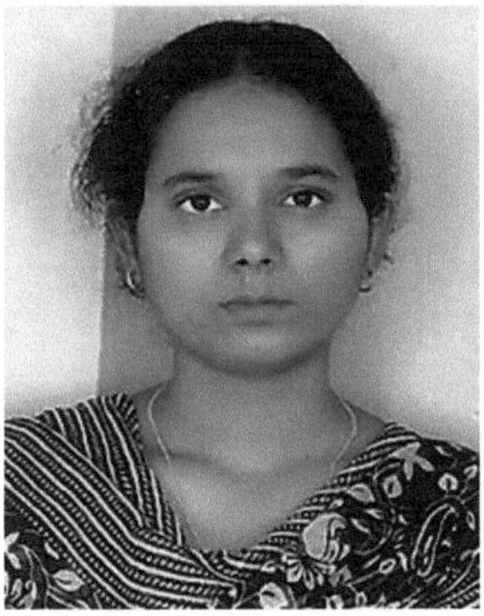

J. Supriya received his B.Tech degree in Electronics & Communication Engineering from Nalanda Institute of Technology & Science, Guntur, Andhra Pradesh, in 2009 and the M.Tech in Electronics & Communication Engineering from Malla Reddy Institute of Technology & Science, Secunderabad, Andhra Pradesh, in 2015. Her research interests include digital signal processing, image and video processing.

Ch. Geetha Ramakrishna Dutt received his B.Tech degree in Electronics & Communication Engineering from Gudlavallvru, Gudivada, Andhra Pradesh, in 2008 and the M.Tech in Digital Electronics & Communication Systems from QIS College, Ongole, Andhra Pradesh, in 2011. He was an assistant professor, Malla Reddy Institute of Technology & Science, JNTU Hyderabad University from 2011. His research interests include digital communication & signal processing, image and video processing.

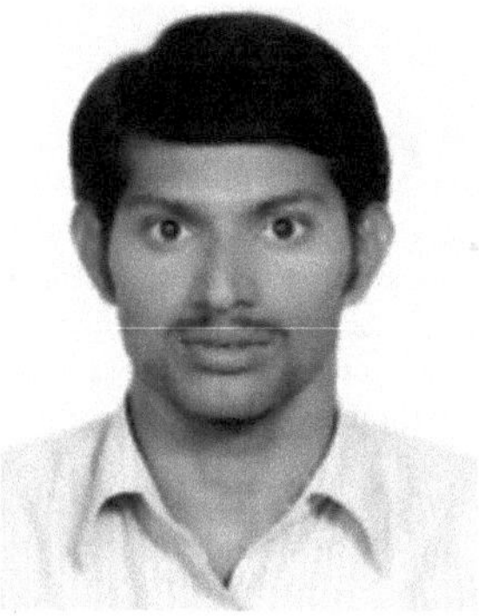

K.Y. Srinivasa Rao received his B.Tech degree in Electronics & Communication Engineering from JNTU Kakinada, Andhra Pradesh, in 1994 and the M.Tech in Signals & Systems from JNTU Kakinada, Andhra Pradesh, in 2006. He was an associate professor, Malla Reddy Institute of Technology & Science, JNTU Hyderabad University from 2006. His research interests include digital communication & signal processing, antenna and wave propagations.

miBEAT Based Continuous and Robust Biometric Identification System for On-the-Go Applications

Jayasubha Yathav, Abhijith Bailur, A.K. Goyal and Abhinav

Abstract In recent years, Biometric identification has taken a giant leap from objective security access system such as retina scan or a finger print scan to a continuous biometric identification based system and for that a single lead Electrocardiogram (ECG) signal is considered to be a good marker. However the parameters normally considered for biometrics from ECG normally requires several parameters which again depend on a good resting signal. For applications involving on-the-go Biometric identification, such systems do not provide a reliable solution. This paper describes a novel approach to a robust and continuous biometric identification system by obtaining touch based ECG as well as Photoplethesmogram (PPG) signal simultaneously from miBEAT (an open source CE certified innovative platform to develop medical grade systems) and by mapping variability features in real time common to both the signals. By validating the system on 20 healthy individuals, it was found that this system works with minimum limitations and thereby can be considered for a robust biometric identification system where higher security measures are required.

Keywords Continuous biometric identification · miBEAT · ECG · PPG · Variability

J. Yathav (✉) · A. Bailur · Abhinav
Cardea Labs, Cardea Biomedical Technologies (P) Ltd, New Delhi 110067, India
e-mail: jayasubha.j@gmail.com

A. Bailur
e-mail: abhijit_bailur@yahoo.co.in

Abhinav
e-mail: abhinav@cardeabiomedical.com

A.K. Goyal
Center of Excellence in Biomedical Instrumentation and Signal Processing,
Noida Institute of Engineering and Technology, Greater Noida 210306, India
e-mail: agoyal1008@yahoo.com

N. Modi et al. (eds.), *Proceedings of International Conference on Communication and Networks*, Advances in Intelligent Systems and Computing 508,
DOI 10.1007/978-981-10-2750-5_28

1 Introduction

According to a report by Global Industry Analysts (Global Industry Analytics, 2011) Biometrics market is estimated to cross $16 billion by 2017. Several distinctive features such as facial features, hand geometry, speech, walking manner, hand writing, prints in finger and iris have been explored to identify an individual in the conventional biometrics systems. Phua et al. [1, 2] validated exponentially uniqueness of the heart, where two heart sounds form different characteristics. Recently scientists, researchers and developers have migrated from such sporadic analytics to a continuous authentication system which involves biomedical signals such as ECG among others. Validation of ECG as the parameter is accompanied by the fact that the geometrical and physiological differences of the heart's function in different individual indicate uniqueness in the cardiac signal taken [3]. ECG features have been effectively incorporated for high precision identification and authorization of individuals to access the secured portion [4].

However, further analysis using ECG for an individual revealed obvious characteristic that may not be present in recordings from any other individuals. Agrafioti [5] proposed a fiducious feature extraction algorithm, with a 12 lead ECG system. However for a single lead ECG, the three primary aspects incorporated for biometrics are relative amplitude, angle and interval [6].

In prior studies, utility PPG signals as a biological trait has also been explored. ReşitKavsaoğlu et al. [7] proposed the application of the Photoplethysmography (PPG) signal and the time domain features acquired from its first and second derivatives for biometric identification. Moreover PPG is the optical measurement of heart which provides HR (Heat rate), rate of the flow of blood and blood oxygen saturation. But ECG and PPG systems are prone to noise and motion artifacts. Moreover these signals when captured during exercise show a different morphology. For environments with sporadic noise, the systems are bound to fail or give erroneous results at worse. This limits the system to work in places having high ambient noise and for on-the-go applications. This limits the application of ECG and PPG for basic biometric identification only.

This paper proposes a unique approach which captures ECG as well as PPG signal simultaneously from its user using a custom made hardware designed using miBEAT (an open source hardware to develop Medical Grade systems) and performs signal processing to identify unique features common to both the signals. The beauty of the system is in the fact that this system can give authentic results even in the noisiest of environment and also can be used for on-the-go applications even during exercise. ECG which is the electrical activity is the 'Cause' and the blood flow activity i.e. PPG is its 'Effect'. Because for healthy individuals, the variation in ECG will result in synonymous variation in PPG, this property can be utilized to perform biometric identification by cross verification of extracted parameters from the two signals. In the following sections, we discuss the design of customized hardware for this application and present a comparative analysis of ECG and PPG signals acquired after processing them for biometric applications.

2 Materials and Methods

The ECG and PPG signals are simultaneously recorded using customized hardware developed using miBEAT—an open source CE Certified hardware platform that relays a medical grade single-lead ECG signal adopting 2 metal based electrodes and a PPG signal employing the reflective type pulse sensor. The data is plotted and saved in real time on a smart phone having Android OS wirelessly through Bluetooth 2.0 (Fig. 1).

miBEAT hardware typically has a band pass Sallen-Key Filter with critical frequency band of 0.5–40 Hz (monitoring mode) for ECG. ATMEGA microcontroller implements a real-time Digital filtering of the order 100. The complete system is designed to work at 3.3 V/5 V obtained using low drop out voltage regulators.

The pulse sensor is attached to one of the ADC pins of ATMEGA through which PPG signal is captured from the index finger. By multiplexing the two signals at the microcontroller, the data is sent serially through Bluetooth. Both ECG and PPG signals are sampled at 250 samples per second and are of 8 bit each. The pulse

Fig. 1 Final miBEAT® board

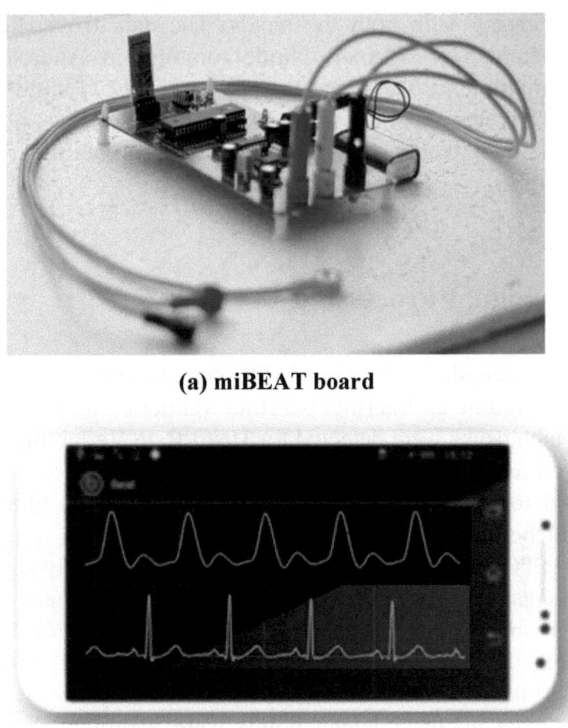

(a) miBEAT board

(b) ECG and PPG on Mobile

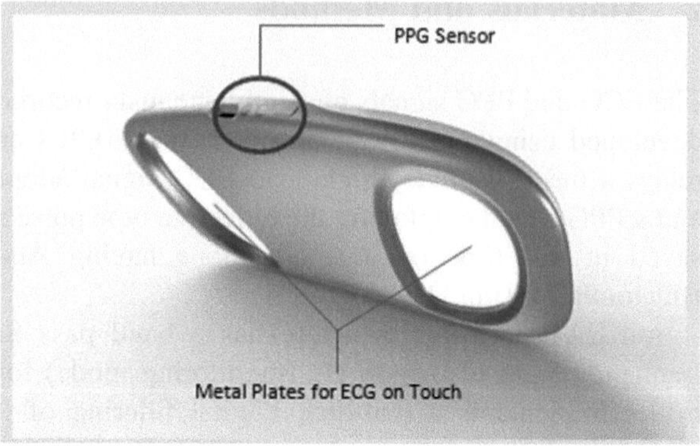

Fig. 2 Custom hardware with ECG & PPG

sensor is a reflective type 3 wire interface which can help capture blood flow rate from the index finger with ease.

The miBEAT hardware utilizing lead-I ECG system to capture ECG from touch (using thumbs) and the Pulse Sensor (from left index finger) are put in an ergonomically designed casing to capture the signals simultaneously with ease. This system is designed to allow its user to get ECG and PPG signals by merely holding the casing with both the hands. The data from Hand-Held BEAT is wirelessly transmitted to a smart phone running on Android OS where the application de-multiplexes the signal to separate the ECG and PPG signals (Fig. 2).

Data was recorded from 20 healthy individuals of the age group of 23–45 with no known history of cardiac diseases. The data was recorded during sleep, rest and exercise.

3 Signal Processing

Pre-processing of ECG and PPG signals were separately performed. Pre-processing involved notch filtering, baseline wander correction and smoothing (Fig. 3a). After which peaks were detected for Heart Rate Variability calculation from both ECG as well as PPG. A dB6 mother wavelet was used to perform wavelet decomposition and to remove baseline wander. A smoothing filter was applied to remove the glitches.

PPG signal which was acquired simultaneously with the ECG signal is processed alongside. Threshold value is set to obtain the peak of the PPG signal. Heart Rate Variability graph plotted with the peaks in the PPG allowing the analysis of HRV in time domain. Then valley HRV was derived as a reference following modal algorithm.

After all the relevant points were identified, HRV was calculated with input signal from peaks, notches and valley.

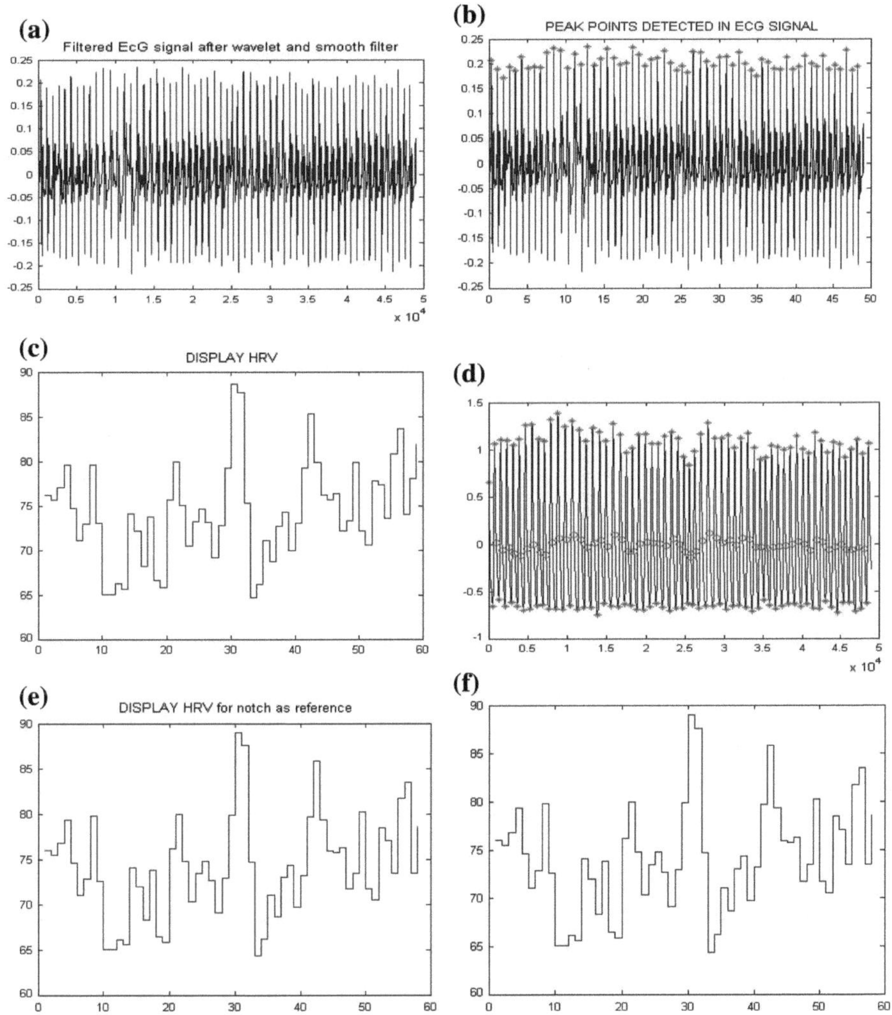

Fig. 3 **a** Represents ECG signal after the application of wavelet and smooth filter, **b** Peaks are detected and marked in the ECG signal, **c** Heart rate variability calculated employing the RR interval, **d** PPG signal with detected peaks, notches and valleys, **e** HRV from Notch Interval of PPG signal, **f** HRV from Valley of PPG signal

4 Experiment and Results

The ECG and PPG signals were recorded simultaneously using the custom hardware and signal processing was performed to calculate HRV for comparison. The table below show cases the comparison of ECG and PPG parameters to showcase the parameters that can be safely used for Biometric identification.

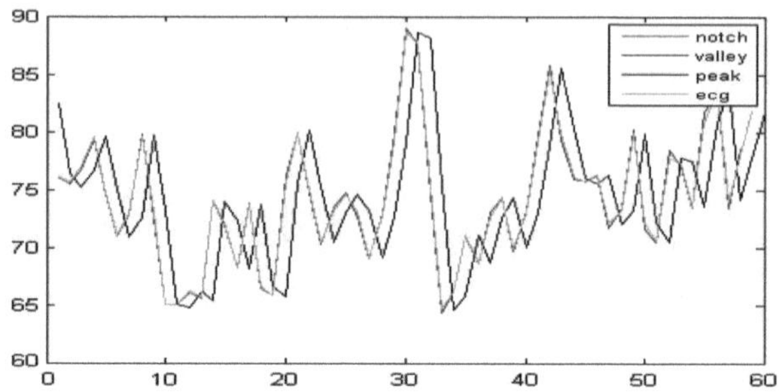

Fig. 4 HRV comparison of ECG and PPG signals for different time intervals

Table 1 Comparative analysis of the parameters that can be used for biometric identification

Parameter	Mean HRV	SDNN	RMSDD
ECG	74.32	0.0593	0.05570
PPG peaks	74.44	0.0559	0.05244
PPG valley	74.44	0.0559	0.05244
Notch	74.18	0.0555	0.053685
Average	74.34	0.0566	0.0535667
% Accuracy (ref. PPG from Notch & ECG)	99.86	99.99	99.998

As observed in Fig. 4, the HRV from ECG and PPG signals show a similar trend for various time intervals. It is relevant that if the ECG varies there will be corresponding variations in the PPG signal.

Other than Heart Rate Variability graph, absolute values for Mean Heart Rate, SDNN (Standard Deviation of NN) and RMSSD (RMS values of Standard Deviation) for ECG and PPG signals from Notch, Valley and Peaks were also calculated and tabulated as shown in Table 1.

5 Conclusion and Future Work

On-the-go biometric identification requires robust hardware and sturdy algorithms which can reliably be used in the noisiest of environment. Also the system should be easy to use and should not cause discomfort for its users.

Keeping these factors in mind, a custom made handheld system developed using miBEAT was designed to capture lead-I ECG system from touch (using thumbs) and the Pulse Sensor (from left index finger) simultaneously. The data were separately processed to calculate Heart Rate Variability for comparative analysis. It was found that signals obtained from subjects even during exercise showed similar values of HRV, SDNN and RMSSD. These factors are easy to calculate and can be deployed for real time analysis for cross verification for data captured on the go.

This paper shows promising results in utilizing ECG and PPG signals recorded simultaneously for biometric identification applications. By performing cross verification of the parameters from ECG and PPG, biometric identification systems will be robust and immune to changes in environment and hence can be safely utilized for on-the-go applications.

Acknowledgments This work was carried out using miBEAT developed by the support of Department of Scientific and Industrial Research (DSIR), Government of India.

References

1. Jain A.K, A Ross. S. Prabhakar. An Introduction to Biometric Recognition-IEEE Transactions on circuits and systems for video Technology, 14(1), 2004, 4–20.
2. Phua K, Chen J, Dat TH, Shue L (2008) Heart sound as a biometric. Pattern Recognit 41: 906–919.
3. R. Hoekema, G. J. H. Uijen, and A. van Oosterom, “Geometrical aspects of the interindividual variability of multilead ECG recordings,” IEEE Transactions on Biomedical Engineering, vol. 48, no. 5, pp. 551–559, 2001.
4. B. P. Simon and C. Eswaran, “An ECG classifier designed using modified decision based neural networks,” Computers and Biomedical Research, vol. 30, no. 4, pp. 257–272, 1997.
5. Agrafioti F (2008) Robust subject recognition using the Electrocardiogram. University of Toronto, Toronto, Canada.
6. G. Wuebbeler, et al., “Human verification by heart beat signals,” Working Group 8.42, Physikalisch-Technische Bundesanstalt (PTB), Berlin, Germany, 2004, http://www.berlin.ptb.de/8/84/842/BIOMETRIE/842biometriee.html.
7. “A novel feature ranking algorithm for biometric recognition with PPG signals”. Comput Biol Med. 2014 Jun; 49: 1–14. doi:10.1016/j.compbiomed.2014.03.005. Epub 2014 Mar 18Reşit Kavsaoğlu A1, Polat K2, Recep Bozkurt M1.

Classification of Technical and Management Metrics in Object Oriented Software Engineering

Devesh Kumar Srivastava and Ayush Singh

Abstract From the dawn of technical age, software projects have always been significantly difficult to estimate and manage well. Over costing, delays in schedule, and out front cancellations of these projects have been a common issue we are facing for over 50 years now. Poor quality of the software, when delivered, remains to be a highlighted issue. Although, several tools for management of software projects are available, when even used by experienced managers to estimate the complexity of software project raises the odds of unsuccessful completions. Project management tools have many subcategories. However, they can be classified into two major groups of tools: [1] For estimation and planning of software projects; [2] For reporting and tracking of costs and status of project while they are underway.

Keywords Object oriented programming · Software metric · Java

1 Introduction

Over nearly past 50 years, this industry has grown into one of the leading industries of this century. On global basis, software and its applications are the main tools of various corporations, govt. agencies and even allied forces. This industry employs thousands of professionals every year [3]. Because of the high costs and importance of development and maintenance of software combined with lower optimal quality,

D.K. Srivastava (✉) · A. Singh
SCIT, Manipal University Jaipur, Jaipur, India
e-mail: devesh988@yahoo.com

A. Singh
e-mail: mail.ayushsingh@gmail.com

N. Modi et al. (eds.), *Proceedings of International Conference on Communication and Networks*, Advances in Intelligent Systems and Computing 508,
DOI 10.1007/978-981-10-2750-5_29

it becomes mandatory to measure both productivity and quality of the software with a high precision. This research paper is divided broadly into 2 sections, Technical and Management Software Metrics and they are analyzed as follows.

1. For Technical Software Metrics, Object Oriented programs were studied and selected software metrics are applied to estimate their complexity to explore and compute whether each of these proposed metrics are independent of each other and effective in calculating complexity of any proposed program
2. For Software Management Metrics, Requirement Engineering is performed on several aspects of projects development. Work is broken down for providing an overview of development and 11 composite software management metrics are derived for every stage of development to support. Both sets of these metrics aim to describe a quantitative method for the prediction of difficulty it for designing, implementing, and maintaining the system. Their secondary goal is to create a mutual understanding for to initiate some important cost changes to decrease unnecessary costing over lifespan of given software

1.1 Technical Metrics

As we know, metrics are the key source of knowledge used for making decisions, a vast majority of Object Oriented metrics were proposed over a period of 10–15 years to exhibit the functioning and architecture of an Object Oriented program and also are directly related with the other extrinsic factors to measure quality [4]. As the total count of metrics which are available today is large, the sequential method to perform the calculation of the required metrics and obtain result from the resulting values becomes tedious. Also, as the count of these metrics which are proposed are bigger as compared to features like cohesion, coupling, size, polymorphism and inheritance exhibited by these metrics, our objective is to explore and compute whether each of these proposed metrics are independent of each other or is it possible to select a portion of the metrics which have equal measures and use like the preselected set.

To achieve that target, a set of 11 metrics is first selected and is defined with examples. The corresponding metric values are calculated for a standard project study and their interrelationships from the values are interpreted and recorded [5]. The faulty classes were defined based on previous investigations which were derived empirically.

2 Research Methodology

The OO metrics that were selected for the analysis of this project can be further grouped into 4 different categories which are coupling, class, reuse and inheritance metrics. Metrics taken in consideration are defined below [3, 6–10].

1. **Response For Classes (RFC)**. The class with a mathematical set of response (RFC) can be technically termed as a mathematical group of all methods which could be interpreted as a reply to any message that is received by any object made for the class.
2. **Weighted Methods for every Class (WMC)**. It is calculated by counting the total of the complexities of individual methods of that class.
3. **Data Abstraction Coupling (DAC)**. This metric formally represents the total number of all the instances of classes other than given class and within it. It is the number of all the external classes that the given classes may use.

$$\text{DAC} = \text{count of ADTs defined per class}$$

4. **Message Passing Coupling (MPC)**. This coupling metric helps us to measure the total count of all those calls by a method that are defined inside the methods of that sample class to the methods in others.
5. **Inheritance Tree's Height/Depth (DIT)**. The metric measures the degree of effect of ancestor/parent classes on the given class. The class's depth according to the tree made by estimating inheritance is directly proportional to the behavior inherited from its super class(s).
6. **Count of Subunits (NUS)**. The total count of subunits represents the count of all procedures and functions that are defined for a given class.
7. **Number Of Children (NOC)**. This metric measures the total count of immediate children in the model of hierarchy.
8. **Inheritance Dependencies**. The metric aims to exhibit characteristics of tree of inheritance. Inheritance Tree's Height/Depth = max (length of the path of the inheritance tree)
9. **Factoring Effectiveness**. Hierarchies of Inheritance can be controlled by the process named factoring. This process aims at minimizing the count of places inside the hierarchy tree of inheritance within which a selected method is executed. It can be estimated by:

$$\text{Factoring Effectiveness} = \text{Count of specific methods}/\text{Total count of every method}$$

10. **Reusability Ratio (RR)**. This metric is informally represented as

$$U = \text{Total Count of super class/Total count of all the classes}$$

11. **Specialization Ratio (S)**. This metric is mathematically represented as

$$S = \text{Count all the subclasses/Count of superclass}$$

To further understand the application of these metrics, we are going to calculate these metrics on a sample java source code which exhibits the characteristic of OOPs like polymorphism, inheritance and data abstraction to know if these metrics fulfill our demand of an accurate estimation of complexity or not.

2.1 Source Code in Java

Page 1

```
import java.io.InputStream;
import java.util.Scanner.;
class sportsman
{ static Scanner in = new
Scanner( System.in );
     public int no; private
string fullname; public void
readinput()   {
System.out.println("Please
Input Full Name:"); try {
String fullname=
in.nextLine();
System.out.println ("Please
Input Number :");
    int no=in.nextInt();}
catch(Exception ex){} }
public void showoutput()   {
```

Page 2

```
System.out.println("Your
Fullname is:" +fullname);
System.out.println ("The ID
is="
+number); }
public void display()
{
System.out.println("This
completes Sportsman Class");
} }

class golfplayer extends
sportsman
{ protected float penalties ;
protected String trophyname ;
public void readinput()
{ try{
```

<table>
<tr><th>Page 3</th><th>Page 4</th></tr>
<tr><td><pre>super. readinput();
System.out.println("Please
Input Trophy Name:"); String
trophyname= in.nextLine();
System.out.println("Please
Input Golf Club Penalties:");
float penalties =
in.nextFloat();}}
catch(Exception ex){}
} public void showoutput()
{
super. showoutput();
System.out.println ("Trophy
Name:" +trophyname);
System.out.println("Penalties
:"+penalties);
} public void display1()
{
System.out.println("This
completes golfplayer class");
} }
class badmintonplayer extends
sportsman { protected int
shuttle ; public void
readinput()
{
 try{
super. readinput();
System.out.println ("Please
Input the count of shuttles
used:") ; int shuttle=
in.nextInt();}
;} catch(Exception ex){}
} public void showoutput()
{super. showoutput();
System.out.println ("Total
Count shuttles used:"
+shuttle);
} }
class coach inherits
sportsman
{ protected int workhrs; int
sum=0;
void readinput()
{ try{
super. readinput();
System.out.println ("Please
Input no. of Work Duration:")
;</pre></td><td><pre>measure();
} catch(Exception ex){}
}
public void measure( )
{
total = workhrs*50;
}void showoutput();
{
super. showoutput();
System.out.println ("Work
Duration:" +workhrs) ;
System.out.println ("Sum:"
+sum);
} }
class hourwisecoach inherits
coach
{ protected int workhrs;
protected double inc; void
readinput()
{ try{
 super.readinput(); income();
}
catch(Exception ex){}
}public void income()
{
inc=super.workhrs*200;
}
void showoutput()
{
super. showoutput();
System.out.println ("The
Total Income is " +inc);

public static void
main(String args[])
{ golfplayer g1 = new
golfplayer();
golfplayer g2 = new
golfplayer();
badmintonplayer b1 = new
badmintonplayer();
coach c1 = new coach();
hourwisecoach s1 = new
hourwisecoach();
System.out.println ("Please
Input credentials of first
golf player"); g1.
readinput();
System.out.println ("Please</pre></td></tr>
</table>

Page 5	Page 6
`int workhrs= in.nextInt();`	`Input credentials of second`
`golf player");`	`System.out.println ("Please`
`g2. readinput();`	`Input credentials of first`
`System.out.println Please`	`hourwise coach");`
`Input credentials of first`	`h1. readinput();`
`Badminton player");`	`System.out.println`
`b1. readinput();`	`("Credentials of first golf`
`System.out.println ("Please`	`player "); g1.`
`Input credentials of`	`showoutput();`
`firstSystem.out.println`	`System.out.println`
`("Credentials of first`	`("Credentials of second golf`
`coach"); c1. showoutput();`	`player "); g2.showoutput();`
`System.out.println`	`System.out.println`
`("Credentials of first`	`("Credentials of first`
`hourwise coach "); s1.`	`badminton player"); b1.`
`showoutput();`	`showoutput();`
`coach”);`	`}`
`c1. readinput();`	`}`

2.2 *Calculation of Selected Object Oriented Metrics*

1. **WMC (Weighted Methods for every Class)**: This metric is estimated by analyzing and counting all methods present in every class.

a	b	c	d	e	f
WMC	3	3	2	3	3

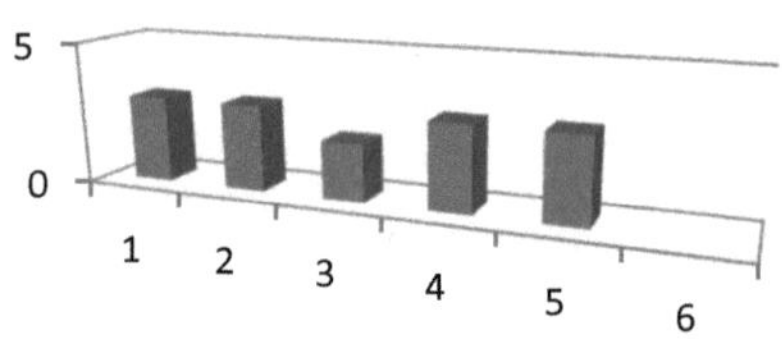

2. **RFC (Response For a Class)**: This metric is calculated by the count of every procedures that may be interpreted in unique class [11].

a	b	c	d	e	f
RFC	3	5	4	5	7

3. **NOC (Count of Immediate Children)**: The count of immediate children is calculated by counting all direct subclasses of a class.

a	b	c	d	e	f
NOC	3	0	0	1	0

4. **DIT (Inheritance tree's Depth)**: The height of the phenomena of inheritance is the basic level of a specific class within the scheme of hierarchical inheritance, and the base class being on level Zero.

a	b	c	d	e	f
DIT	0	1	1	1	2

5. **NUS (Count of all Subunits)**: The total count of subunits is basically the quantity of all the procedures and functions termed for any given class.

a	b	c	d	e	f
NUS	3	3	2	3	3

6. **DAC (Data Abstraction Coupling)**: This metric mathematically specifies the total number of instances of any specifically selected classes present in a selected class [12].

a	b	c	d	e	f
DAC	0	1	0	1	0

7. **MPC (Massage Passing Coupling)**: This metric exhibits the total number of a function/method calls or calls of procedure that were directed to any extrinsic units.

a	b	c	d	e	f
MPC	0	2	2	2	4

8. **Factoring Effectiveness (FE)**: Effectiveness of Factoring = Total count of specific methods/Count of methods = 4/14 i.e. 0.29
9. **Inheritance Dependencies (ID)**: Depth of the Inheritance tree = max (length of path of inheritance tree), according to the class specific diagram, Depth of Inheritance tree = 3
10. **Specialization Index (SI)**: The index of specialization (S) = Total count of subclasses/Total count of all the superclass. Index of Specialization = 5/2 = 2.5
11. **Reusability Ratio (RR)**: Reusability ratio = U = Total count of superclass/Actual count of classes. Reuse ratio = 2/5 = 0.4
    After studying and analyzing the above, following traits were derived with the relation of the complexity of selected programs written in java.

1. Weighted Methods for every Class (WMC). A large value of WMC leads towards larger quantities of errors. Classes having a greater count of methods are actually more software specific, and limits reusability. Study suggests that increasing the average of WMC also elevates complexity but lowers quality.
2. Response for a Class. Higher count of methods from any class which may be called through messages, the larger the complexity of that class. Programs written in java are somewhat not complex because the average value for a specific metric is less for such codes.
3. Depth of Inheritance tree. If any specific class is deep in that hierarchy, more methods will be inherited, increasing its complexity. Deep trees have greater design complexity, as more number of classes and methods are involved, but reusability also increases due to inheritance.
4. Number of Children. A high count of immediate children indicate a larger chance of malpractice of abstraction, which might be a case of misusing of sub classing ability. However, higher NOC exhibits higher reuse, as inheritance is another form of reuse but also increases complexity [4]. A class with more children requires more testing but fewer errors due to higher reuse.
5. Message passing and coupling. A higher number of passing of messages indicates higher coupling between given classes in a code. It makes them highly dependent and spikes the overall complexity of that java code and also makes the scalability and modeling difficult.
6. Data Abstraction Coupling. Complexity of the software increases as DAC increases. For Java, data is more important than methods and procedures. The data is usually not shown to the customer or user. DAC is generally not high for all programs of java.
7. Count of Subunits. As frequently as the count of methods and functions spike, classes grows more prone to error. So, complexity somehow also increases with a growth in quantity of such metric. So, the final value of NUS metric is discovered as low generally for programs in java.
8. Inheritance Dependencies. As a tree with a greater depth is somewhat more difficult in testing, a greater value of ID indicates greater complexity of programs in java. Comprehensibility might be decreased for a larger count of layers of inheritance.
9. Factoring effectiveness. A smaller count of places of implementation for an average method means that fewer mistakes were made while designing. An inheritance hierarchy with a high factor is the largest degree until which a function can be reused. A highly factored application indicates a smaller count of places of implementation for an average function with lower complexity.
10. Specialization Index. It misses the empirical and theoretical validation. If Specialization Index is increased, class maintainability becomes more difficult. The value of SI is usually high for java programs increasing usability and hence complexity.

11. Reuse ratio. If the final value of RR is coming to be zero, nothing is inherited. As this value approaches one, the tree of inheritance deepens as a chain with single root and single leaf exactly. When RR was estimated for several other programs in java, we discovered that the results were intermediate.

3 Concluding Results and Further Research

The Canonical aim of the aforementioned research study was basically to validate and verify the utility of proposed Management Metrics of Software and the applicability of carefully selected OOP Metrics to calculate the total complexity of an Object Oriented code or software. Complexities of such software and application can be measured with several types of metrics. But in this study we evaluated and classified a defined set of 11 well known OOP metrics which are measured to serialize software codes with the complexities to estimate maintainability for those programs. By this work, we may deduce that we must compromise the intrinsic attributes of software to continue maintaining a high scalability while also maintaining complexity and coupling as low as possible. Although, further in depth study may be focused on empirical validation of metrics for an environment of multi languages, we can still expect from our study and analysis that it can be further used by software project and application developers for developing a reliable, error free, maintainable Java software product.

References

1. Patrick Naughton & Herbert Schildt "Java: The Complete Reference", McGrawHill Professional, UK, 2008.
2. S. Chidamber, and C. Kemerer, "Towards a Metrics Suite for Object Oriented Design," Object Oriented Programming Systems, Languages and Applications (OOPSLA), Vol 10, 1991, pp 197–211
3. Brij Mohan Goel, Pradeep Kumar Bhatia, "Analysis of reusability of object oriented systems using object-oriented metrics ACM SIGSOFT Software Engineering Notes" Volume 38 Issue 4, Pages 1–5 July 2013.
4. Briand, W. Daly and J. Wust, Exploring the relationships between design measures and software quality. Journal of Systems and Software, 5 245–273, 2000.
5. K. Morris, "Metrics for Object-oriented Software Development Environments," Master Thesis, MIT, 1989.
6. Churcher, N.I. and M.J. Shepherd, "Towards a Conceptual Framework for Object Oriented Metrics," ACM Software Engineering Notes, vol. 20, no. 2, April 1995, pp. 69–76.
7. Roger S. Pressman: Software Engineering, A practioner's Approach, Fifth Edition, 2001.
8. S.R Schach, Object-Oriented and Classical Software Engineering. Tata McGraw-Hill, 2002 http://www.mhhe.com/engcs/compsci/schach5/student/airgourmet.java.java
9. Mahfuzul Huda, Dr. Y.D.S. Arya, and Dr. M. H. Khan. "Testability Quantification Framework of Object Oriented Software: A New Perspective." International Journal of Advanced Research in Computer and Communication Engineering, Vol. 4, Issue 1, Jan, 2015.

10. Abdullah, Dr, M. H. Khan, and Reena Srivastava. "Testability Measurement Model for Object Oriented Design (TMMOOD)." International Journal of Computer Science & Information Technology (IJCSIT) Vol. 7, No 1, February 2015.
11. Arti Chhikara1and R.S. Chhillar, "Analyzing the Complexity of Java Programs using Object - Oriented Software Metrics" IJCSI International Journal of Computer Science Issues, Vol. 9, Issue 1, No 3, January 2012.
12. Er. V.K. Jain. "The Complete Guide to Java Programming", First Edition, 2001.

Publish/Subscribe Mechanism for IoT: A Survey of Event Matching Algorithms and Open Research Challenges

Satvik Patel, Sunil Jardosh, Ashwin Makwana and Amit Thakkar

Abstract The number of sensors getting deployed around the world is increasing due to emergence of Internet of Things. It provides advanced connectivity and communication between devices which goes beyond machine-to-machine communication. Huge amount of data is expected to be generated from different locations that will be aggregated, processed and forwarded very quickly. Publish/Subscribe mechanism is powerful way to allow IoT devices to connect and communicate with each other. One of the major bottlenecks in using Publish/Subscribe systems is the efficiency of filtering incoming message. This is a very challenging problem because in a Publish/Subscribe system the number of subscriptions can be very large. There are quite a few event matching algorithms proposed in the literature to improve its efficiency. The aim of this research paper is to study and analyze how existing approaches ensure fundamental event matching requirements and discuss the open challenges and future work in the area.

Keywords Internet of things · Event matching · Event filtering · Publish/subscribe · Communication paradigm · Distributed system

S. Patel (✉)
P.I.E.T., Parul University, Vadodara, Gujarat, India
e-mail: satvik.patel@paruluniversity.ac.in

S. Jardosh
Progress Software Development Pvt. Ltd., iLab's Centre, Madhapur, Hyderabad, India
e-mail: sjardosh@progress.com

A. Makwana · A. Thakkar
C.S.P.I.T., CHARUSAT University, Changa, Gujarat, India
e-mail: ashwinmakwana.ce@charusat.ac.in

A. Thakkar
e-mail: amitthakkar.it@charusat.ac.in

N. Modi et al. (eds.), *Proceedings of International Conference on Communication and Networks*, Advances in Intelligent Systems and Computing 508,
DOI 10.1007/978-981-10-2750-5_30

1 Introduction

The IoT (Internet of Things) is network of physical objects with advanced connectivity and communication which helps achieve greater value and service [1]. Numbers of devices getting connected in such distributed systems are increasing day by day. Publish/Subscribe mechanism is powerful way to allow IoT devices to connect and communicate with each other [2]. Ability of Publish/Subscribe to supports dynamic, anonymous, many-to-many and asynchronous communication plays very important role in providing high scalability in distributed environment. The loose coupling can be classified into following three ways:

1. Space Decoupling: Publisher can share information with subscriber without knowing each other.
2. Time Decoupling: Publishers and Subscribers are not required to be actively participating in the interaction at the same time.
3. Synchronization Decoupling: Publishers are not blocked while producing the events. Occurrence of the events can be asynchronously notified to the Subscribers.

Each participant in a Publish/Subscribe based communication system can either act as a publisher or a subscriber of the information. Publisher produces information which will be consumed by subscribers. This information is denoted by the term "Event" and the act of delivering it by the term "Notification". Publisher does not directly send information to the corresponding subscribers, but it is sent indirectly according to the content of the notification. A subscriber expresses its interest for specific events by issuing subscriptions. Later on subscriber will be asynchronously notified for all events, produced by any publisher, that match their subscription [3]. Subscribers may also unsubscribe their interest for particular event (Fig. 1).

Efficient filtering of incoming messages is one of the major concerns in using Publish/Subscribe systems for large scale distributed systems [4]. This is a very challenging problem because in a Publish/Subscribe system the number of

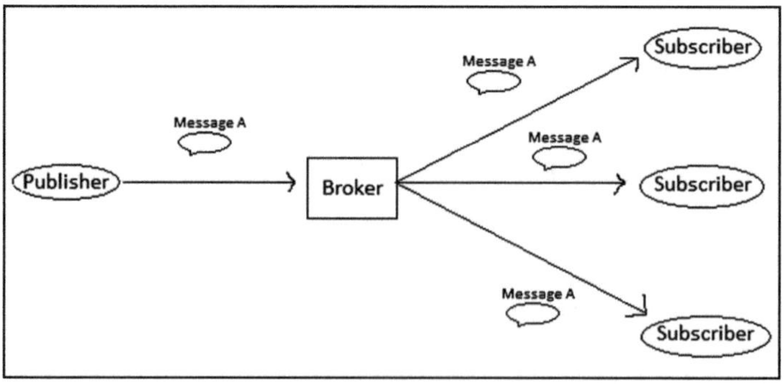

Fig. 1 Publish/subscribe model

subscriptions can grow very large. Considering high event arrival rates and large volume of subscriptions an efficient event matching algorithm is highly desirable. There are quite a few event matching algorithms proposed in the literature to improve the performance [5–10].

The outline of the contributions of this paper is as following.

1. We provide an overview of some of the key event matching techniques presented in the literature and provide a summary of related research work, open challenges and future work.
2. We present need for new and/or improved event matching mechanism for Publish/Subscribe system.

The rest of the paper is organized as follows. Section 2 depicts overview about event matching process. In Sect. 3 we discuss about existing work done in the area of event matching. Section 4 finally concludes the paper providing insight on the future work in the area.

2 Overview of Event Matching

As seen in above section Publish/Subscribe is an important communication paradigm used in IoT. Due to its ability to provide dynamic, anonymous, many-to-many and asynchronous communication between the sender and the receiver it is gaining increasing attention in large scale distributed system [11].

Event matching is one of the key aspects of Publish/Subscribe system [12]. Event matching is a process of finding set of subscriptions from large volume of subscriptions that successfully match against the occurring events. Event matching in Publish/Subscribe system is a tough challenge as these volumes of subscriptions and event arrival rate could be very high for the applications [13].

In Publish/Subscribe system the subscriptions are represented in the form of boolean expressions [14]. This offers greater flexibility to user to represent their interests in the events. Key terms involved in event matching are as follows:

1. Predicate: It is the primary unit in the boolean expression. It is a triplet consisting of an attribute A, an operator OP and an operand OD. It takes an input value X and outputs a boolean value indicative of whether or not the operator constraint is satisfied or not.

   $P^{(A,\ OP,\ OD)}(X) \rightarrow \{\text{True, False}\}$, where P is a predicate and X is a input value.

2. Boolean Expression/Subscription: A boolean expression is combination of predicates formed by using conjunction (AND) and/or disjunction (OR). A subscription S (boolean expression) can be defined over n predicates are follows:

Table 1 Subscription list

Subscription ID	Query
S1	(A = True and B = True)
S2	(A = True and B = True) or (C = True)
S3	(A = True and B = True and C = False)
S4	(A = True or D = True)
S5	(A = False or B = False) and D = False

S = P1(A, OP, OD)(X) AND P2(A, OP, OD)(X) AND P3(A, OP, OD)(X) AND…AND Pn(A, OP, OD)(X), where S is a subscription and P1…. Pn are predicates. Table 1 shows the list of five subscriptions (S1 to S5) where A, B, C and D are the attributes.

3. Event: An event is the attribute-value pair published by the publishers. E: (A = v), where E is an event, A is an attribute and v is value for an attribute A.

   Example: E1: {T = 50}, where E1 is the event occurred with attribute T with a value of 50.

4. Boolean Expression/Subscription Match/Event Match: We can say successful subscription match when predicates of a subscription have corresponding attribute-value pair occurred in the form of events.

   Example: Suppose event E1: D = True occurs then, with the help of event matching algorithm we can come to a conclusion that subscription S4 is successfully matched and its corresponding subscriber needs to be notified for this event.

With this basic overview about Publish/Subscribe system and Event Matching process, we discuss its related work and open challenges in the next section.

3 Related Work and Analysis

As discussed above one of the major bottlenecks in using Publish/Subscribe systems for large scale distributed systems is the efficiency of filtering incoming messages. Also this problem is very challenging in a Publish/Subscribe system as the number of subscriptions can be very large and event arrival rate is also very high. There are quite a few event matching algorithms proposed in the literature to improve its efficiency.

Lot of work has also been carried out in the area of indexing to efficiently identify matching subscriptions [5–10]. The subscription database is divided into subsets of predicates using any of the existing hashing technique and each predicate subset is organized using inverted list data structure. For every attribute-value pair p in the incoming event, appropriate inverted index lists are searched to identify predicates of subscription that match p, and a counting mechanism is used to determine matching subscriptions for occurring events.

It is a tough challenge to efficiently indexing Disjunctive Normal Form (DNF) and Conjunctive Normal Form (CNF) of Boolean expressions over a high-dimensional multi-valued attribute space. The objective is to quickly find the set of complex boolean expressions (including NOTs) that evaluate to true for a given assignment. Inverted list data structure can be used to efficiently match complex boolean expressions [9]. Proposed design talks about working with DNF and CNF boolean expressions with different set of algorithms. Working with combination of DNF and CNF boolean expression could be tough here and can be taken up as an extension to this work.

Many Publish/Subscribe systems are capable of handling large volume of subscriptions and high event arrival rate. In many cases these system does not prove to be effective with high dimensional and sparse database (e.g. E-Commerce Database). An efficient in-memory index (OpIndex) is scalable to the volume and updating of the subscriptions. It is also scalable to the arrival rate of events and the variety of attributes that can be subscribed. OpIndex uses a two-level partitioning scheme to organize the subscription predicates into disjoint subsets. Each of these subsets is independently and efficiently indexed to minimize the number of corresponding subscriptions accessed for event matching. In this way, OpIndex design is a highly efficient and extensible approach for subscription matching which can support complex predicate matching operators [5]. One of the drawbacks here is in its inability to handle temporal events (events of our interest occurring at different time stamp). One could extend the current work to efficiently handle the temporal events.

Sometimes users are interested in receiving up-to-date geo-textual objects according to their requested location and interest. The continuous queries issued by user could be very large, which can post challenge in efficiently matching them against geo-textual objects. A system called SOPS (Spatial-Keyword Publish/Subscribe System) is capable of efficiently processing spatial keyword continuous queries. It uses IQ-tree (Inverted File Quad Tree) data structure to efficiently match events [8]. Queries are organized in IQ-tree and geo-textual objects are matched using this index. For each new object the IQ-tree is traversed to find the queries that have object as the result. One of the drawbacks here is that the spatial information is always prioritized while index construction not considering the distribution pattern of the queries. One can extend the current work to consider distribution pattern of the queries while index construction and traversal.

The amount of geo-spatial data being generated is increasing day by day at an unprecedented scale. Users want them to be notified of interesting geo-textual objects during a period of time. Matching stream of incoming Boolean Range Continuous (BRC) queries over a stream of incoming geo-textual objects in real time can post challenge as the number of continuous queries issued by users could be very large. An index structure called IQ-tree (Inverted File Quad-tree) can be used for matching the queries with the incoming geo-textual objects. The IQ-tree integrates the Quad-tree for organizing the spatial information and the inverted file for organizing the keyword expression of the BRC queries. The IQ-tree is nothing but a Quad-tree extended with inverted files. Each node in the IQ-tree is associated with an inverted file which organizes the keyword expression of the BRC queries that are associated with the node [7]. One of the drawbacks here is that the spatial information is always prioritized while index construction not considering the distribution pattern of the queries. One can extend the current work to consider distribution pattern of the queries while index construction and traversal.

Efficiently processing continuous spatial-keyword queries over geo-textual streams is a tough challenge. Simple variant of inverted index data structure does not perform well in such cases. R-tree and IQ-tree are two indexing structure available to handle geo-textual information. They suffer from fundamental drawback like "the spatial factor is always prioritized during the index construction not considering the keyword distribution of the query set". An index structure called Adaptive spatial-textual Partition Tree (AP-Tree) can be used to effectively organizes continuous spatial-keyword queries and overcome the drawback of spatial factor always first in IQ-tree and R-tree [10]. This method will adaptively select either keyword partition or spatial partition depending on the evaluation of the cost model.

Location aware Publish/Subscribe systems have become widely available on mobile devices. Since the message and the subscription contains both the location and textual information, a high performance location aware systems are required to deliver publisher's message to relevant subscribers. We can use R-tree based index structure by integrating textual description into R-tree nodes to achieve the same [6].

Solutions proposed in the literature involve usage of the following techniques Inverted List, IQ-tree, R-tree, OpIndex and AP-tree. Below Table 2 enlists the benefits and limitation of the techniques used in the proposed solution of the literature.

Solutions discussed above suffer from one common problem that is assumption of assignment (group of events) generation. But practically assignment generation does not seem possible due to following reasons.

1. Group of events cannot be done at gateway node as the very basic feature named loose-coupling of publish/subscribe will be disturbed. Also not all data required by an application gets generated under single gateway node.
2. Group of events cannot be done at central node as well, as in real life the events occur asynchronous and independent from each other. Also events need not come together due to communication delay as well.

So in practical world generation of assignment is not possible. Hence subscribers (applications) subscribing with more than two predicates will not get successfully matched and notified. So considering events as an assignment is not a correct way to handle them.

In order to overcome the above mentioned problem one could think of designing the solution which is based on temporal (events of our interest occurring at different time stamp) and causal (Occurrence of the event influences the end result) property of the events. This area could be explored further, to successfully match subscriptions with multiple predicates against the events occurring at different time stamp.

Table 2 Benefits and limitations of event matching techniques

S. No.	Technique	Benefits	Limitation
1	Inverted Index	• Less index construction time Scalable	• Suitable for datasets with less number of attributes • Memory intensive
2	IQ-Tree (Quad Tree + Inverted File)	• Update friendly • Supports different indexing granularities for different queries • Good Performance in two-dimensional space (if tree is balanced)	• Not useful in high dimensional space • Only spatial feature is preferred during index construction • Inefficient if data resides in external storage
3	OpIndex	• Less index construction and query processing time • Less memory consumption • Scalable	• Does not support geo-spatial data
4	AP-Tree	• Supports adaptive spatial and textual partitioning • Handles large scale streaming data	• Not well-suited to the change of query workload, if AP-Tree structure built on a small proportion of the query set
5	R-Tree	• Good performance in low-dimensional spaces	• Spatial feature is preferred during index construction • Basic operations like insert, update, delete are expensive

4 Conclusion and Future Work

The Publish/Subscribe paradigm is well suited for large scale distributed systems due to its ability to provide dynamic, anonymous, many-to-many, asynchronous communication between publisher and subscriber. We can say event matching is one of the key feature of Publish/Subscribe system, where there is significant amount research scope in terms of providing better efficiency, scalability and high performance. We have discussed few of the state-of-the-art research that have been carried out in the area of event matching for Publish/Subscribe system and their related challenges. Challenges could be summarized as designing a Publish/Subscribe system based on temporal and causal property of the events, Publish/Subscribe system to support efficient event matching for geo-spatial data and Publish/Subscribe system providing high flexibility to user in registering their interest in the events.

Any enrichment to existing or new algorithm which provide better and/or flexible trade-off between scalability, expressiveness and quality of service are highly desirable and could be taken up as future work for any of the existing research. We believe this survey may prove to be an important contribution to the research community, by discussing the current state and open challenges of this important and dynamic area of the research. This would help readers interested in developing new solutions or enriching existing solutions to address challenges in event matching for Publish/Subscribe system in IoT.

References

1. Internet of Things, https://en.wikipedia.org/wiki/Internet_of_Things.
2. Banavar, Guruduth, et al. "An efficient multicast protocol for content-based publish-subscribe systems." Distributed Computing Systems, Proceedings. 19th IEEE International Conference on. IEEE, (1999).
3. Eugster, Patrick Th, et al. "The many faces of publish/subscribe." ACM Computing Surveys (CSUR) 35.2, (2003).
4. Campailla, Alexis, et al. "Efficient filtering in publish-subscribe systems using binary decision diagrams." Proceedings of the 23rd International Conference on Software Engineering. IEEE Computer Society, (2001).
5. Zhang, Dongxiang, Chee-Yong Chan, and Kian-Lee Tan. "An efficient publish/subscribe index for e-commerce databases." Proceedings of the VLDB Endowment 7.8, (2014)
6. Li, Guoliang, et al. "Location-aware publish/subscribe." Proceedings of the 19th ACM SIGKDD international conference on Knowledge discovery and data mining. ACM, (2013).
7. Chen, Lisi, Gao Cong, and Xin Cao. "An efficient query indexing mechanism for filtering geo-textual data." Proceedings of the 2013 ACM SIGMOD International Conference on Management of Data. ACM, (2013).
8. Chen, Lisi, et al. "Sops: A system for efficient processing of spatial-keyword publish/subscribe." Proceedings of the VLDB Endowment 7.13, (2014).
9. Whang, Steven Euijong, et al. "Indexing boolean expressions." Proceedings of the VLDB Endowment 2.1, (2009).
10. Wang, Xiang, et al. "Ap-tree: Efficiently support continuous spatial-keyword queries over stream." Data Engineering (ICDE), IEEE 31st International Conference on. IEEE, (2015).
11. Tarkoma, Sasu. Publish/subscribe systems: design and principles. John Wiley & Sons, (2012).
12. Kale, Satyen, et al. "Analysis and algorithms for content-based event matching." Distributed Computing Systems Workshops, 25th IEEE International Conference on. IEEE, (2005).
13. Shraer, Alexander, et al. "Top-k publish-subscribe for social annotation of news." Proceedings of the VLDB Endowment 6.6, (2013).
14. Mishra, Tania Banerjee, and ShashankSahni. "PUBSUB: An efficient publish/subscribe system." Computers and Communications (ISCC), IEEE Symposium on. IEEE, (2013).

Chronic Kidney Disease Prediction Using Back Propagation Neural Network Algorithm

Nilesh Borisagar, Dipa Barad and Priyanka Raval

Abstract In recent time Neural network system has discovered its use in disease diagnoses, which is depended upon prediction from symptoms data set. Chronic kidney disease detection system using neural network is shown here. This system of neural network accepts disease-symptoms as input and it is trained according to various training algorithms. Levenberg, Bayesian regularization, Scaled Conjugate and resilient back propagation algorithm are discussed here. After neural network is trained using back propagation algorithms, this trained neural network system is used for detection of kidney disease in the human body. The back propagation algorithms presented here have capacity for distinguishing amongst infected patients or non-infected person.

Keywords Chronic kidney disease · ANN method · Simulation

1 Introduction

Chronic kidney disease (CKD) emerges as more usual disease so there is need of early detection system so appropriate treatment cannot be delay. Chronic kidney disease, which may result in causing different levels of ill-function and loss of patient kidneys. Once a person found CKD, he/she will undergo from the disease which may reduce his/her working strength as well as live quality. Also, CKD have major role for causing other chronic diseases like high blood pressure, anemia (low blood count).

N. Borisagar (✉) · D. Barad · P. Raval
Department of Computer Engineering, B.H. Gardi College of Engineering & Technology, Rajkot, Gujarat, India
e-mail: borisagar93@gmali.com

D. Barad
e-mail: dipabarad1@gmail.com

N. Modi et al. (eds.), *Proceedings of International Conference on Communication and Networks*, Advances in Intelligent Systems and Computing 508,
DOI 10.1007/978-981-10-2750-5_31

Based on the chronic kidney disease measurement, currently most famous measuring indicator is glomerular filtration rate (GFR). GFR is most commonly used indicator in health institution for chronic kidney disease. "The physician in the health institution can calculate GFR from patient's blood creatinine, age, race, gender, and other factors depending upon the type of formal-recognized computation formulas" [1, 2] employed. Health of patient's kidney is indicated by GFR and it is also used for determining stage of severity of a patient with or without CKD.

Main goal of this paper is to provide intelligent system for kidney disease detection which can also use by normal people and provide alternative way for kidney disease detection for doctors. This intelligent system finds presence or absence of kidney disease into human body with good accuracy. The input data for system Training, validation and testing are collected from UCI machine learning repository.

2 Literature Review

Neural network's research in the field of medical is shown in Table 1 with methodology and conclusion.

Table 1 Literature Table

Title	Author	Methodology/work	Conclusion
Using neural networks in the identification of signatures for prediction of Alzheimer's Disease [3]	Lara Dantas, Meuser Valenc (2014)	Multi-layer perceptron, extreme learning machine and reservoir computing	Artificial neural network method gives better results than the support vector machine
Kidney Disease Prediction Using SVM and ANN Algorithms [4]	Dr. S. Vijayarani, Mr. S. Dhayanand (2015)	Support Vector Machine and ANN	ANN is better than SVM
Comparing performances of logistic regression, decision trees, and neural networks for classifying heart disease patients [6]	Anchana Khemphila, Veera Boonjing (2010)	Logistic regression, decision trees, and ANN	ANN is best
A Novel Method for Medical Disease Diagnosis Using Artificial Neural Networks Based on Back propagation Algorithm [7]	Jasdeep Singh Bhalla, Anmol Aggarwal	SCG and LM	Neural network aided Thyroid disease diagnosis gives promising results
Skin Diseases Diagnosis using Artificial Neural Networks [5]	Delia-Maria FILIMON, Adriana ALBU (2014)	SCG	93 % accuracy

3 ANN Method

Multilayer Neural network is used here for chronic kidney disease prediction. This system is created by three steps: training, testing, and validation. 24 attributes of CKD are considered as a neural network input. Neural network is trained using Levenberg, Bayesian regularization, scaled conjugate and resilient back propagation algorithm. Now, this trained and generalized neural network is used for prediction of new inputs.

3.1 MNN (Multilayer Neural Network)

Multilayer neural network as its name suggest, it is a system of multiple layer which are used for solving nonlinearly separable problem. So here multiple layer neural network is an advantage over single layer network because single layer neural network are only used for solving linearly separable problem. It is also called as feed forward neural network, which is collection of one or more hidden layer as shown in Fig. 1. This system of neural network is used for pattern recognition and classification problem and it is also used for prediction based on past knowledge collection.

3.2 Back Propagation Algorithm

Back propagation algorithm is a training algorithm use for adjust weight for reducing error in neural network output. After achieving network output, this predicted output is compare with expected output and based on that difference error is generated. This error propagates backward to the network. According to that error weight adjusted again for reducing error and resulting output will become closer to the expected output. This process repeats continuously up to overriding and each iteration gives high accurate result than previous iteration. This entire process inside neural network layer is called as training of neural network.

Different Back propagation algorithms are listed below:

Levenberg Marquardt
Hessian matrix [7]:

$$H = J^T J \tag{1}$$

"Performance gradient value could be calculated as" [7]:

$$g = J^T e \tag{2}$$

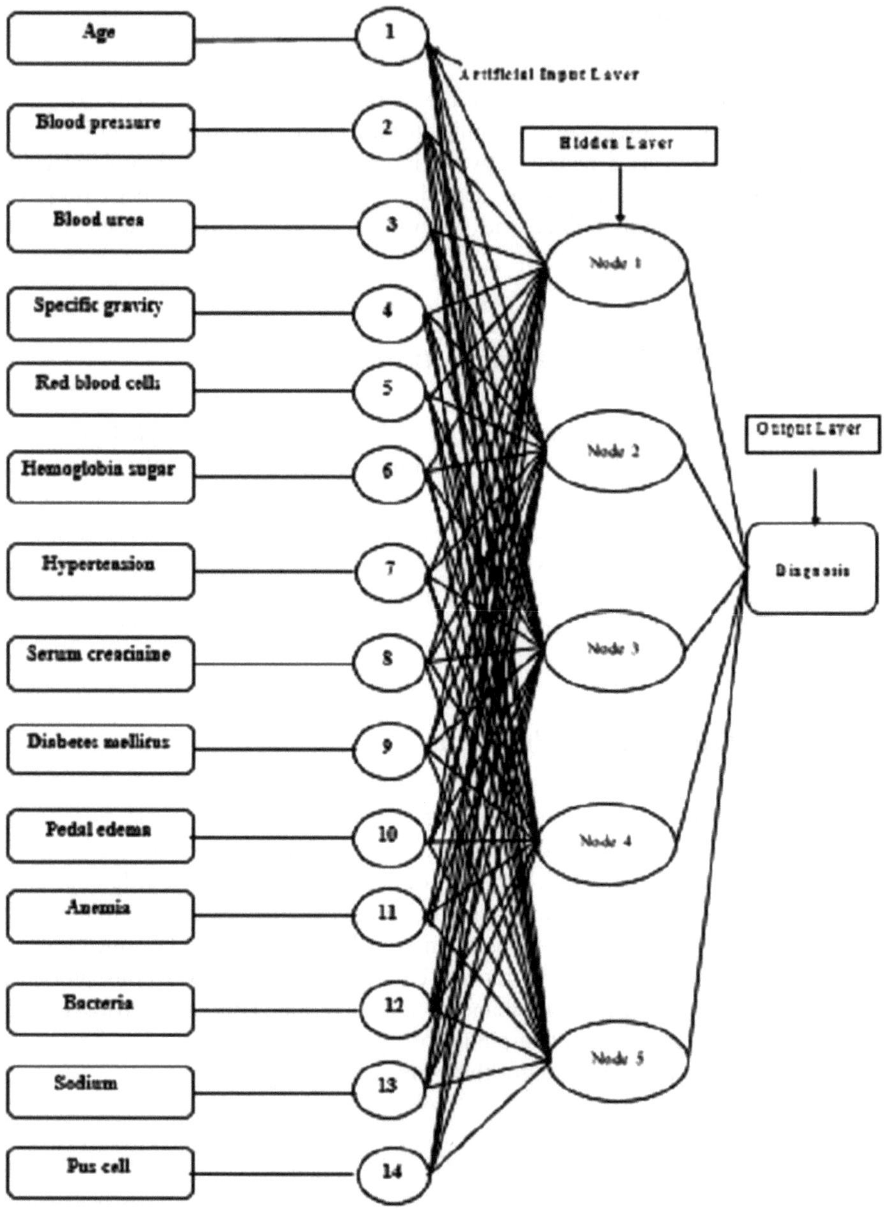

Fig. 1 Neuron model with 24 input and 5 hidden layer for CKD

"Approximation to the Hessian matrix as" [7]:

$$X_{K+1} = X_K - \left[J^T J + \mu I\right]^{-1} J^T e \tag{3}$$

Scaled Conjugate
It is depend on conjugate gradient direction but does not depend on line search [8].

Bayesian regularization
"Bayesian regularization decides the weight and bias values according to Levenberg-Marquardt algorithms" [8].

Resilient Back propagation
"Performance derivatives is calculated using back propagation with respect to the weight and bias variables X" [8].

4 Simulation

The Dataset: UCI machine learning repository is used for chronic kidney disease data source. CKD dataset have 24 attributes and two class attributes.

Implementation tool: Matlab R2013a (Trial Version) used as simulation. Before processing values collected from Apollo Hospital, Tamilnadu, missing values are replaced with most probable value and nominal values are converted into numerical values. Then after, all attributes are standardized in the range of −1 to 1 because normalization decreases training time of neural network. Input vector is given as input matrix and target is given as target matrix in Matlab. Data division used here: Training: 70 %, validation: 15 %Testing: 15 %.

Neural Network layer structure and training of network are shown in Fig. 2. Neural Network has 24 node as input, 5 node as hidden node and 1 node as output node.

4.1 Simulation Result

4.1.1 Performance with Different Number of Learning Algorithms

As shown in Table 2, all four learning algorithms are giving good accuracy for chronic kidney dataset. We are obtaining good accuracy due to preprocessed dataset. All missing values in dataset are replaced with most probable value. These all probable values are selected such way that, each value have major role for classifying all instance into the appropriate class.

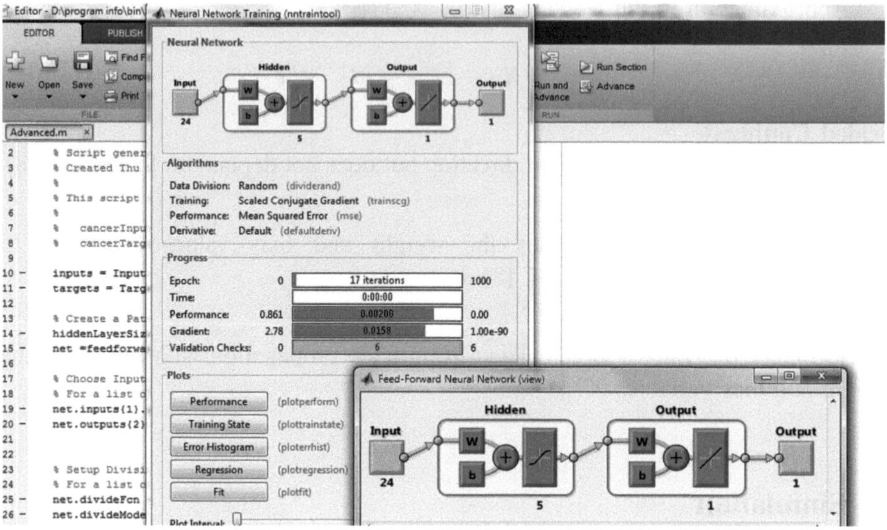

Fig. 2 Neural network training

Table 2 System performance with different learning algorithms

Chronic kidney disease	Accuracy
Levenberg marquardt	99.8
Bayesian regularization	99.5
Scaled conjugate gradient	98.7
Resilient back propagation	99.5

4.1.2 Training Time of Algorithms

As shown in Table 3, Resilient back propagation algorithm takes less time as compared to all four algorithms. Levenberg algorithm though gives results with almost same high accuracy; it takes more time compare to Scaled conjugate and resilient back propagation.

Simulation Results Evaluation are based on confusion matrix and performance plot as shown in Figs. 3 and 4 respectively.

Confusion matrix have green and red box, which shows correct and incorrect classified result respectively. Performance plot shows best validation performance at specific epochs. Here at epoch 49, network has found best performance on validation dataset.

Table 3 System Performance based on Training-time of algorithms

Chronic kidney disease	Training Time (s)
Levenberg marquardt	21
Bayesian regularization	28
Scaled conjugate gradient	4
Resilient back propagation	2

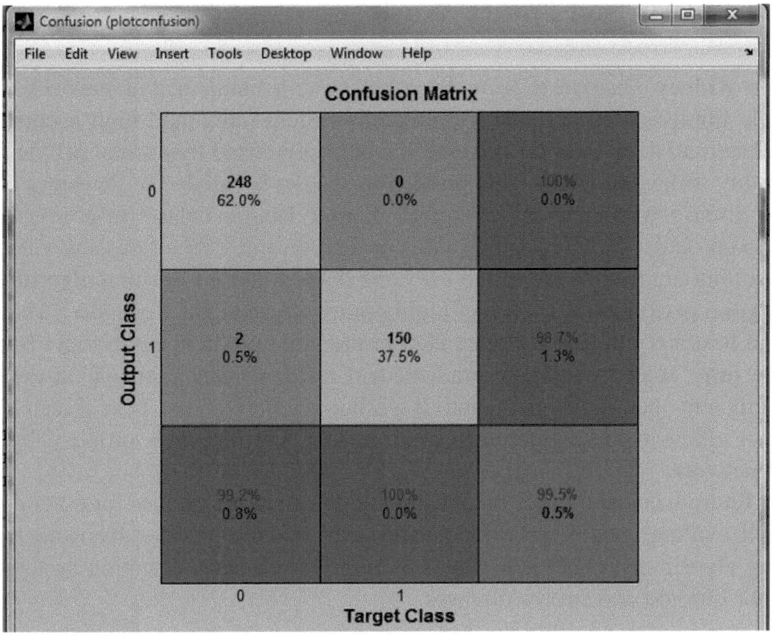

Fig. 3 Confusion matrix

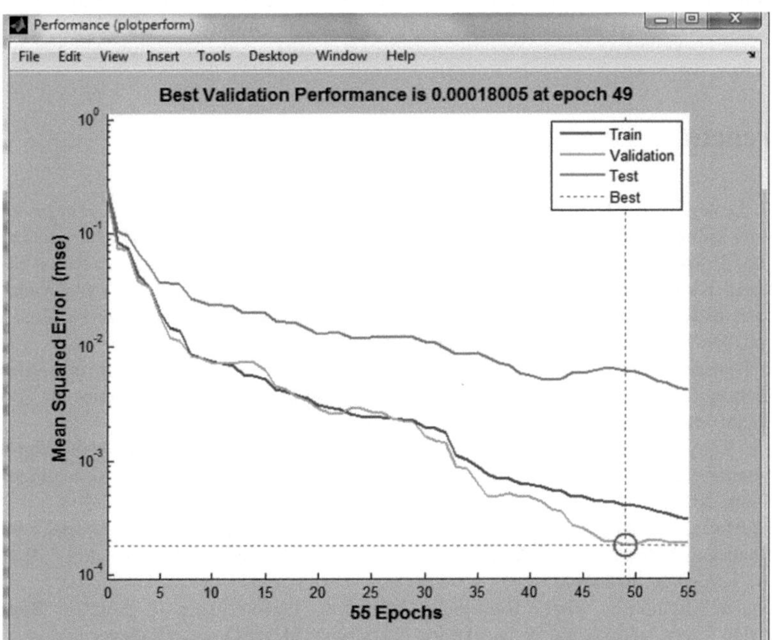

Fig. 4 Performance plot

5 Conclusion and Future Work

Chronic Kidney Disease (CKD) detection system using neural network is successfully implemented here. This prediction system has found high accuracy and can be alternative method for doctors, it can also be used by normal people to find probability of having CKD. This prediction system is capable of detecting chronic kidney disease with new set of inputs. Earlier Rough dataset is converted into highly preprocessed data by filling most probable values for all missing values. All four learning algorithms are tested on same dataset and all learning algorithms are giving good performance in case of highly preprocessed data. Levenberg Marquardt is found best algorithms for kidney dataset based on prediction accuracy. Based on training time, scaled conjugate gradient and resilient back propagation are found more efficient than Levenberg and Bayesian regularization. This research work gives an efficient and economical solution to CKD detection problem by using neural network.

Our future work will focus on developing detection techniques for other diseases and will evaluate neural network performance on that disease by using various learning algorithms. Our main aim is to improve diseases detection system especially for chronic and severe diseases.

Acknowledgments We offer our heartiest appreciation and a deep sense of commitment to our co-guide, Honorable Prof. Amit Sata (H.O.D. of the mechanical engineering Department) for his guidance, direction, encouragement and help throughout the period of research. We are also thankful to Prof. Priyanka R. Raval (Faculty of Computer Science and Engineering Department) for her valuable support and instructive comments in our research work.

References

1. A. S. Levey, P. E. De Jong, J. Coresh et al., "The definition, classification, and prognosis of chronic kidney disease: a KDIGO controversies conference report," *Kidney International*, vol. 80, no. 1, pp. 17–28, 2011. View at Publisher · View at Google Scholar · View at Scopus.
2. National Kidney Foundation (NKF), "KDOQI clinical practice guidelines for chronic kidney disease: evaluation, classification and stratification, *American Journal of Kidney Diseases*, vol. 39, pp. S1–S266, 2002. View at Google Scholar.
3. Lara Dantas, Meuser Valenca, "Using neural network in the identification of signature for prediction of Alzheimer's Disease", 20144 IEEE 26th International Conference on Tools with Artificial Intelligence.
4. Dr. S. Vijayarani, Mr. S. Dhayanand, "Kidney Disease Using SVM and ANN Algorithms", International Journal of Computing and Business Research, ISSN: 2229-6166, Volume 6 Issue 2 March 2015.
5. Delia Maria, Adriana, "Skin Diseases Diagnosis using Artificial Neural Networks", 9th IEEE International Symposium on Applied Computation Intelligence and Informatics, May 15–17, 2014, Timisoara, Romania.
6. Anchana Khemphila, Veera Boonjing, Comparing Performances of Logistic Regression, Decision Trees, and Neural Networks for Classifying Heart Disease Patients.

7. Jasdeep Singh Bhalla, Anmol Aggarwal, "Novel Method for Medical Disease Diagnosis Using Artificial Neural Networks Based on Back propagation Algorithm".
8. Mark Hudson Beale, Martin T. Hagan, Howard B. Demuth, "Neural Network Toolbox", Matlab (R2015b).
9. UCI, "Uci machine learning repository: chronic kidney-disease data set", http://archive.ics.uci.edu/ml/datasets/Chronic_kidney_disease.html

Internet of Things (IoT) Based Water Level Monitoring System for Smart Village

Timothy Malche and Priti Maheshwary

Abstract Water source is necessary and an important factor in agricultural and farm production and is a key of our quality of life as well. Monitoring water level of a water source, such as water tank or borewell etc., plays a key role in agricultural. For example if a water level drops below the threshold level for pumping in a borewell, the pump motor may get damaged due to dry running. In such case monitoring water level and controlling the water pump accordingly becomes necessary task. There are many other situations where water level monitoring is an important task. It may be used to preserve water or to study the water usage of a water source. This paper proposes a prototype system design, implementation and description of required tools and technologies to develop Internet of Things (IoT) based water level monitoring system which can be implemented in future smart villages in India.

Keywords Internet of things · IoT · Water level monitoring · Smart village · IoT application · Smart agricultural

1 Introduction

Villages in India will soon be transforming to smart villages as Government of India brings Smart Village initiative to the country. The smart village initiative will promote digital inclusion which will enable the enhanced access to services through Information Technology (IT) enabled platforms. Thus the Internet of Things (IoT) has a major role to play in Smart Village in India. In IoT enabled Smart Village every physical object, a thing, will be connected to the Internet and enable users to keep track of its status and to control it remotely. This will help users to

T. Malche (✉) · P. Maheshwary
AISECT University, Bhopal, Madhya Pradesh, India
e-mail: timothy.malche@gmail.com

P. Maheshwary
e-mail: pritimaheshwary@gmail.com

N. Modi et al. (eds.), *Proceedings of International Conference on Communication and Networks*, Advances in Intelligent Systems and Computing 508,
DOI 10.1007/978-981-10-2750-5_32

access to services provided by such objects as and when required. IoT can be used in smart village to develop Smart Agriculture, Smart Dairy, Smart Schools, Smart Healthcare and Smart Grid solutions. IoT in agriculture can be used for better management of resources used in crop production. Water is one of the important substances used in crop production. It must be saved to avoid water shortage in future. One such way to save water is to monitor and study its usage and accordingly its utilization should be managed. Monitoring water level of a water source, such as water tank or borewell etc., plays a key role in water management. Keeping track of water level in a water source can be used to preserve water and to study the water usage. Thus monitoring water level is an important task in agricultural.

2 Objectives

The main objective of this study is to develop a system to keep track of a water level of a water source from a distant location. The IoT based proposed system presented in this study will be helpful to achieve such task. The prototype system experiment of this study enables keeping track of a water source from remote location in real time. The actual implementation of the system will require changes in sensor and few other technologies and source code although the methodology and working principle remain the same.

3 Related Work

Researchers have proposed different models for agriculture sector using IoT, Wireless Sensor Network (WSN), Cloud computing and various other technologies. Few of the related work have been explained here.

Song et al. [1] explained WSN based greenhouse environment monitoring system in their study. The proposed system uses temperature, humidity, CO2 and light detection modules. The system also combines WSN technology and greenhouse control technology to provide automatic adjustment and management of greenhouse. Sakthipriya [2] in his study have proposed development of rice cropping monitoring system which is used to monitor rice crop in real time which in turn help increase rice production. The proposed system uses motes with external sensors for leaf wetness, soil moisture, soil pH, atmospheric pressure sensors attached to it. The data collected by sensors are sent to the farmer from base station via GSM modem as an SMS. Using this data farmer can decide the amount of fertilizers to be used in crop production. Dahikar et al. [3] proposed an approach based on artificial neural networks to predict crop yield by sensing soil properties and atmospheric parameters. Sonka [4] in his study explained the use of big-data technology in agriculture domain and how this technology will be helpful in cost reduction. Atzberger [5] explained challenges in agriculture sector and remote sensing applications. This study includes details about crop estimation and cropland mapping. Satyanarayana et al. [6] in their study on Wireless Sensor Based Remote Monitoring System for Agriculture have designed and implemented a WSN based system for agriculture to monitor temperature, soil and humidity condition. The system uses ZigBee and GPS technologies for the operation. Patil et al. [7] have explained the use of cloud computing and IoT in agricultural to deliver cost effective services for farmers and to manage resources efficiently. Ke [8] in his study has described system architecture of controlled smart agriculture based on IoT and Cloud Computing. Although various studies explained above have proposed few models in agriculture domain using one or more technologies; the dynamic model is still needed that provides an integrated approach to monitor various resources used in agricultural as well as to connect all the entities including farmer, resources and devices which will be helpful to develop connected agricultural system for smart village in India.

4 Methodology

4.1 System Architecture

The overall system presented in this study is subdivided into of three layers of the system architecture. The physical layer, service layer and presentation layer.

Physical Layer. Physical layer consists of sensor nodes and communication technologies. At this layer the data is collected and sent to the service layer.

Service Layer. The service layer plays an important role in the system architecture. The application/business logic is implemented here. It also stores collected data for future use. This layer also provides various tools for data analytics.

Presentation Layer. The most upper layer is the presentation layer which visualizes the information to the user and allows user to interact with the system (Fig. 1).

The working principle of the proposed system architecture for monitoring water level is shown in Fig. 2.

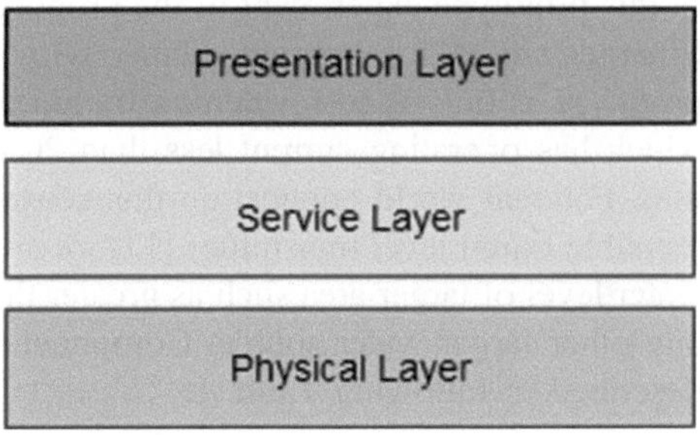

Fig. 1 The three layers of the proposed system

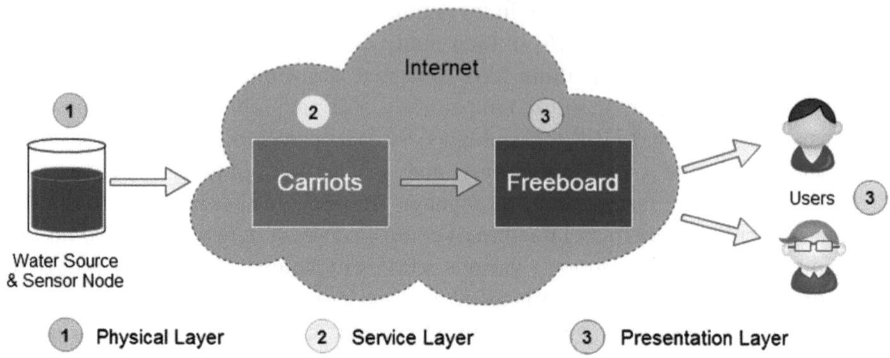

Fig. 2 Interaction between the three layers of proposed water level system

In this proposed application architecture of the system, the physical layer consists of the physical environment such as water source and WSN node to sense the water level in a water source with required network connectivity. The sensor data is uploaded to carriots [9] at service layer of the architecture which is the second layer. The carriots is a cloud application platform for IoT. Carriots has many advantages. It receive data stream from WSN in real-time and is used to store and analyze data stream. Many other operations can be performed on data streams such as generating event triggers, alerting users through SMS or social network. At the presentation layer of the system, user of the application is the main focus. This layer allows user to interact with the system. At the presentation layer of the application architecture freeboard [10] is used to visualize the data. Freeboard is a dashboard that works as a visualization tool. It provides several widgets for this purpose. Freeboard receives data stream from carriots in JSON format and then visualize it according to selected widgets. The widgets at freeboard are updated in realtime as soon as carriots is updated with data stream. The communication between carriots and freeboard is done using REST API.

4.2 Implementation of the System Prototype

In this prototype experiment of the proposed system arduino uno board along with Ethernet shield for Internet connectivity in used. A liquid level sensor in this prototype is only used for demonstration purpose. This sensor is an analog sensor which has operating current less than 20 mA and detection area of 40 mm × 16 mm. For real world application this sensor should be replaced with PT-500 submersible liquid level transmitter [11], or similar one, which can be used to measure water level of larger area such as greater than 450 ft deep water tank or borewell or any other larger water source. Components used in this prototype experiment are described in following Table 1.

Table 1 Components of the system used in prototype experiment

Hardware specification	Software and services specification
• Arduino Uno R3 • Arduino ethernet shield • Liquid level sensor • Male-female jumper wires • Water source	• Arduino Software • An account on Carriots.com • An account on Freeboard.io

The schematic diagram of the arduino liquid level sensor circuit is shown in Fig. 3 [12].

The liquid level sensor board needs about 5 V of power. The signal pin S, + (plus) and – (minus) pins on water sensor is connected to analog pin A0, pin 5 v and GND respectively on the arduino board. In this prototype the Ethernet shield is also used which is fixed above arduino board. Therefore sensor should be connected to arduino via Ethernet shield which has similar pin description hence same circuit will work. Ethernet shield only does the job of connecting arduino to Internet via wired connectivity. This prototype requires latest open source arduino software [13] found at official arduino website. Installing the arduino software will also install supported drivers. A Carriots library [14] and an account on carriots and freeboard are also required for this system prototype. Creating account on carriots will provide API key.

Carriots provide very user friendly interface which allow registration of new devices in few clicks. The newly created device will be displayed in the device list on the carriots control panel. The device registration is must as it provides Device ID. The Device ID and API Key are required by arduino firmware code. The arduino based sensor node uploads data stream to carriots is in JSON format. The data stream uploaded to carriots is used as data source to freeboard. The format of data stream is shown in following code and result of uploaded data stream is shown in Fig. 4.

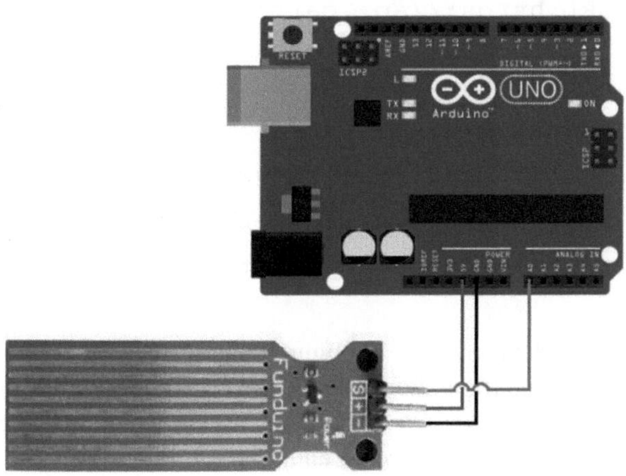

Fig. 3 Circuit diagram

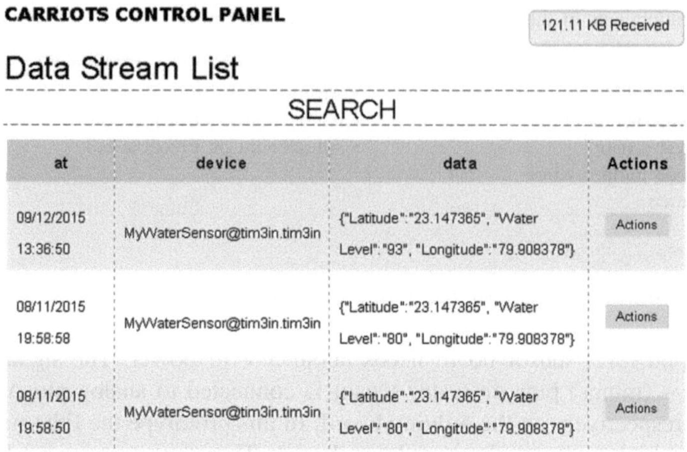

Fig. 4 Data stream received at carriots from sensor node

Example of Data stream format uploaded at carriots

```
Data {    "Latitude": "23.147365",
          "Water Level": "93",
          "Longitude": "79.908378"   }
```

Freeboard is a visualization tool which is used to visualize the sensor data in real-time. Data source can be configured at freeboard account as shown in following code.

Example of Data source configuration at Freeboard

```
TYPE: JSON
NAME: My-Water-Sensor
URL:https://api.carriots.com/devices/MyWaterSensor@tim3
in.tim3in/streams/?order=- 1&max=1
REFERESH EVERY: 5 SECONDS
HEADERS:
Accept: application/json
User-Agent: Carriots-client
Content-Type: application/json
```

As data source is configured the device will be listed on freeboard account. Data source configuration allows freeboard to receive data stream from carriots. To visualize water level from data stream which the freeboard receive from carriots a widget must be selected and configured as shown in following code.

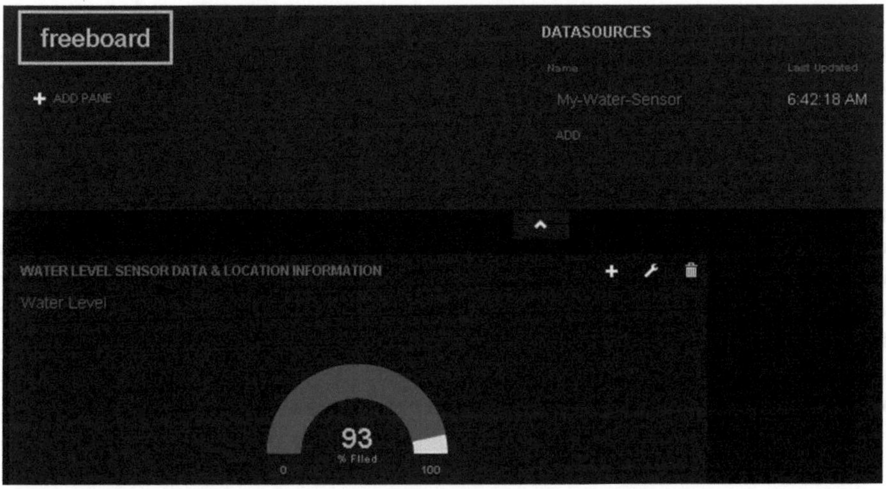

Fig. 5 Visualization of water level data showing water source is filled 93 %

Example of Freeboard widget configuration

```
TYPE: Gauge
TITLE: Water Level
VALUE: datasources["My-Water-
Sensor"]["result"][0]["data"]["Water Level"]
UNITS: %
MINIMUM: 0
MAXIMUM: 100
```

Widget configuration enables freeboard to graphically visualize water level data as shown in Fig. 5.

The widget will be updated in real time as water level in a water source changes. Therefore the exact representation of water level of water source will be displayed to the user.

5 Conclusion and Future Scope

This IoT based proposed system is used to acquire water level details of a water source in real time from any location, any device connected to Internet. This water level data can be used for various purposes for better management of water source. Monitoring water level from remote location may be very useful when it is not possible to visit location physically every time. The system can be implemented for different sources of water by replacing sensor device suitable for the condition. In future, the proposed system can be used to monitor and analyze water usage of the specific water source thus require developing such logic for the application. The system can also be used to collect and study the environmental data of water source and its surrounding area by integrating other sensor to the system. The study may include location data, water quality, temperature, humidity and various other factors. For example arduino GPS shield can be integrated in the system to obtain location data of the water source dynamically. Similarly Ethernet shield can also be replaced with WiFi shield/module for wireless connectivity if required. Concluding the proposed IoT based water level monitoring system will be helpful to collect, analyze and predict the water level detail, water usage and other information of particular water source at particular location in real-time remotely.

References

1. Song, Yongxian, Ma, Juanli, Zhang, Xianjin, Feng, Yuan: Design of Wireless Sensor Network-Based Greenhouse Environment Monitoring and Automatic Control System, Journal of Networks, vol. 7(5), 838–844 (2012).
2. Sakthipriya, N.: An Effective Method for Crop Monitoring Using Wireless Sensor Network, Journal of Scientific Research, 20(9), 1127–1132 (2014).
3. Dahikar, Snehal S., Rode, Sandeep V.: Agricultural Crop Yield Prediction Using Artificial Neural Network Approach, IJIREEICE, vol. 2(1), 683–685 (2014).
4. Sonka, Steve: Big Data and the Ag Sector: More than Lots of Numbers, International Food and Agribusiness Management Review, vol. 17(1), 1–20 (2014).
5. Atzberger, Clement: Advances in Remote Sensing of Agriculture: Context Description, Existing Operational Monitoring Systems and Major Information Needs, Remote Sensing, vol. 5(2), 949–981 (2013).
6. Satyanarayana, G. V., Mazaruddin, S. D.: Wireless Sensor Based Remote Monitoring System for Agriculture Using ZigBee and GPS, Conference on Advances in Communication and Control Systems, 110–114 (2013).
7. Patil, V.C., Gaadi, K. A. A., Biradar, D.P., Rangaswamy, M.: Internet of Things (IoT) and Cloud Computing for Agriculture: An Overview, pp. 292–296. Proceedings of AIPA, India (2012).
8. TongKe, Fan: Smart Agriculture Based on Cloud Computing and IoT, Journal of Convergence Information Technology, vol. 8(2), 210–216 (2013).
9. Carriots a Platform as a Service for Internet of Things, https://www.carriots.com/.
10. Freeboard a Dashboard for visualization of Internet of Things Devices, http://freeboard.io/.
11. Submersible Pressure Transducer for Clean Liquids, https://www.apgsensors.com/submersible-pressure-transducers/pt-500-clean.
12. Arduino liquid level indicator tutorial, http://www.learningaboutelectronics.com/Articles/Arduino-liquid-level-indicator-circuit.php.
13. Arduino official software downloads, https://www.arduino.cc/en/Main/Software.
14. Open source carriots library for arduino, https://github.com/carriots/arduino_library/tree/master/CarriotsClients.

Application of Remote Sensing for Assessing Forest Cover Conditions of Aurangabad, (MS), India

Yogesh D. Rajendra, Sandip S. Thorat, Ajay D. Nagne, Manasi R. Baheti, Rajesh K. Dhumal, Amarsinh B. Varpe, S.C. Mehrotra and K.V. Kale

Abstract The Remote Sensing has been playing an important role in mapping spatial and temporal behavior of forest cover. The mapping results are largely dependent on the user's preferences because it is location and application specific. The study deals with the use of RS techniques to know the present status of forest area undertaken in the Gautala Wildlife Sanctuary, and Bird Sanctuary, Aurangabad region. Forest cover is depleting very fast due to the conversion of forest region into agricultural or other land use. The forest cover estimation of these protected areas has been derived from forest cover map generated from LISS III satellite images of the year 1997 and 2015 using digital image classification and processing approach. The temperature of the Aurangabad district is increasing and rainfall is reducing which indicates that deforestation can be one of the associated causes for it. The classification result shows that there is a significant conversion, loss in forest cover.

Keywords Remote sensing · Forest cover · Supervised classification · Knowledge classifier · Wildlife sanctuary

1 Introduction

Satellite technology provides a synoptic view of forests and their condition on a real-time basis [1]. Satellite images are an important source for forest mapping, monitoring and understanding the function of the ecosystem, mainly through the relationships between vegetation reflectance characteristics and structural

Y.D. Rajendra (✉) · R.K. Dhumal · S.C. Mehrotra
Geospatial Technology Research Laboratory, Srinivasa Ramanujan Geospatial Chair, Dr. Babasaheb Ambedkar Marathwada University, Aurangabad, Maharashtra, India
e-mail: yogesh.rajendra@gmail.com

Y.D. Rajendra · S.S. Thorat · A.D. Nagne · M.R. Baheti · R.K. Dhumal · A.B. Varpe · S.C. Mehrotra · K.V. Kale
Geospatial Technology Research Laboratory, Department of Computer Science & IT, Dr. Babasaheb Ambedkar Marathwada University, Aurangabad, Maharashtra, India

N. Modi et al. (eds.), *Proceedings of International Conference on Communication and Networks*, Advances in Intelligent Systems and Computing 508,
DOI 10.1007/978-981-10-2750-5_33

Table 1 Satellite data used for classification

Satellite	Sensor	Path/row	Date of passing
IRS 1C	LISS III	96/58	11-Feburary-1997
IRS RS2	LISS III	96/58	01-Feburary-2015

composition. Mapping we mean as a method of depicting nature and the classification allows to estimate the true conditions [2]. Even though maps show objects with respect to their attributes, their major purpose is to represent objects in relations of their relative location [3]. A comprehensive mapping strategy is required for processing a huge amount of information gathered from various sources like satellite images, ground etc. [4]. Since last two decades, RS is being used as an operational tool for forest vegetation assessment in the country. Assessment of forest cover at the national level is now being done intermittently by FSI using different interpretation keys [5]. Satellite RS has played a pivotal role in gathering information about forest cover, vegetation type and land use changes [6–9]. In recent period, there is a daily use of RS to quantify forest cover, and its condition. In regards to the high cost of satellite data collection and processing, the role of RS has increased dramatically. In recent times, advances in RS and GIS technology has significantly increased the scientific understanding of environmental change [10]. Presently, many forest ecosystems are in high anthropogenic disturbances, which causes loss of native species. Along with deforestation, degradation is a critical issue, as well as depleting the natural resource base. Human-induced changes in ecosystems function have threatened the survival of many species and are by now scientifically well-established [11]. Deforestation can be measured in quantitative way as the decline in ecosystem state. The qualitative loss is involves a change in the ecosystem function, structure, and composition [12–14]. The review analyzed the deforestation rates and gives a summary for the government initiatives for conservation in India [15]. Many authors have been using remotely sensed data along with other parameters for the assessment and monitoring of forest in many areas [16–25]. This paper discusses a forest mapping methodology using expert knowledge classifier approach. The study deals with Linear Imaging Self Scanner LISS III images of the year 1997 and 2015. The following Table 1 shows information about the satellite data taken for study.

2 Study Area

The Aurangabad region is a mega-biodiversity hotspot of the Marathwada. The present study has been conducted in the area shown in Fig. 1. Gautala Autramghat Wildlife Sanctuary is a protected area also located in Aurangabad [26]. Aurangabad Region lying between 20°40′–19°10′ N latitude and 74°40′–75°50′ E longitude. It belongs to semi-tropical climate region and remains humid. Its average annual precipitation varies from 600 mm to about 800 mm while annual mean temperature

Fig. 1 Study area location of aurangabad region

varies from 16 to 32 °C. The altitude of the study area varies from 300 to 380 m. Summer season is extremely hot from March to June, while the winter lasts from November to January.

3 Data

The spatial data of IRS 1C, RS2 LISS III of the year 1997 and 2015 were procured from NRSC, Hyderabad for the study. The forest vector database was taken from FSI Aurangabad Division. The year 1997 and 2015 map was generated from LISS III image acquired in February 1997, 2015 respectively. The date of acquisition of images were so selected that is should be Cloud free. The IRS 1C, RS2 satellite path/row: 96/58 occurs in Aurangabad. The radiometric correction step was already taken care by the providing agency. LISS III sensor provides 4 multispectral band data. The spatial resolution for visible and NIR is 23.5 m with a ground swath of 141 km^2. The fourth SWIR band has a spatial resolution of 70.5 m with a ground swath of 148 km^2. The repetitivity cycle of LISS III sensor is 24 days. The characteristics of LISS III sensor is shown in the Table 2 [27]. The vector shape file of Aurangabad was used with the 1997 and 2015 image database for sub-setting the study area. An image interpretation key was prepared from the extensively done ground survey and visual interpretation for eight different classes. All the analysis

Table 2 LISS III sensor characteristics

Type	Linear imaging self scanner (LISS) III
Spectrum	VIS (~0.40 to ~0.75 μm) NIR (~0.75 to ~1.3 μm) SWIR (~1.3 to ~3.0 μm)
Resolution class	Medium (20–200 m)
MS resolution	23.5 m
Swath	141 km
Revisit time	24 days
On-board of	IRS 1C, IRS 1D, IRS P6 (ResourceSat-1), IRS P7 (ResourceSat-2)

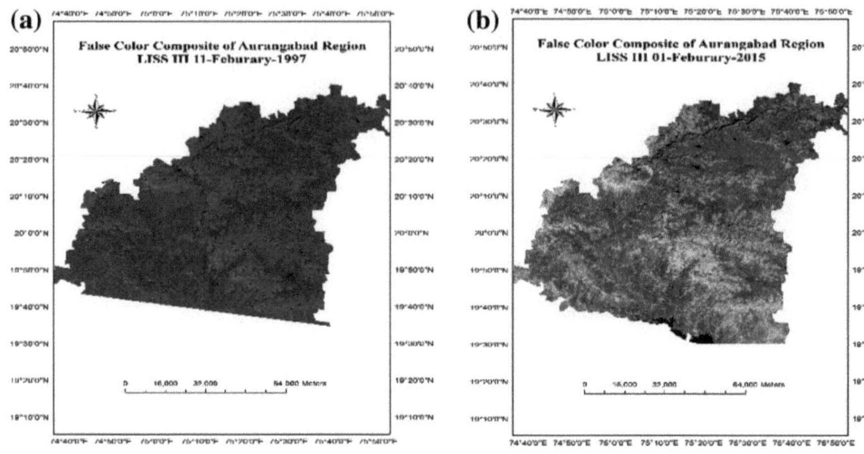

Fig. 2 **a** FCC 1997 and **b** FCC 2015 of Aurangabad Region

and data processing were performed using the ERDAS Imagine 2014 and ArcGIS 10.2v software environment. The Fig. 2a, b shows Standard FCC of the year 1997, 2015 respectively.

4 Methodology

The methodology of this study is schematically shown in Fig. 4. The entire methodology includes four phases. The first phase deals with the procurement of satellite data from NRSC. The second phase contains pre-processing for image enhancement, which includes the certain steps necessary such as radiometric and geometric correction, and image enhancement in order to improve the quality of images, which helps to the assignment of each pixel of the scene to one of the classes defined in a classification system [28, 29], and geo-referencing of satellite images, the third phase includes expert knowledge based classification, and the fourth phase includes final map generation and accuracy assessment. Standard mapping methodology using satellite images has been used in the present study. Ancillary information from Forest Department like Forest range map of Aurangabad Forest Division and Gautala Autramghat Wildlife Sanctuary was used. The reconnaissance survey of the study area was initially carried out to remain familiar with the study site. Ground truth observations were collected for preparation of the interpretation key using GPS to record the latitude, longitude and altitude information of the study sites. Remote sensing data were visually interpreted using image elements for the forest type and density classification as well as other land-use types. The whole study area was classified in 8 (Eight) classes in terms of forest density viz., Water Body, Moderate Dense Forest, Open Forest, Shrubs Forest, Fallow Land, Barren Land, Settlement, and Crop.

The knowledge-based classifier is designed to capture the intellectual processes that an expert in a particular field, in order to infer some form of information about a geographic location. The knowledge-based classification is represented as a tree diagram consisting of final and intermediate class definitions. It includes hypothesis, variables, and rule. Uncertainty is of vital importance to the knowledge-based classification [30]. A decision-tree model is developed for each of the different vegetation types under the knowledge-based classification. A rule includes the conditional statement that shows relevant variables. The variable can be raster, scalar or vector. Hypotheses in the knowledge-based classification can be specified as the classes that to be evaluated by the use of the rule. After reconnaissance work, the assessment of accuracy was done. It is the most important step to assess the reliability of generated map. No image classification is said to be complete unless and until its accuracy assessment not done. For determining the accuracy of classification result, a sample is selected on the classified image and their respective class identity is compared with the reference data. Random field samples (150 locations) collected during field survey have been used for verification of classification accuracy. Evaluation of the quality of a classification map is important in RS because, it gives proof of how well the classifier is capable of extracting the desired information from the image. In this study, the classification error matrix which is commonly used for representing classification accuracy. Another statistical technique used for the accuracy assessment is the KAPPA Statistics. KAPPA analysis calculates the producer's, user's and overall accuracy. After accuracy assessment and correction the class wise area of the image was calculated (Table 3 and Fig. 3).

Table 3 Classified area report 1997, 2015

Class name	KBC area (in Hectare) 1997	KBC area (in Hectare) 2015
Water	3275.89	11,147.9
Moderate dense forest	11,997.6	9102.41
Open forest	62,380.6	29,580.8
Shrubs forest	39,054.11	33,850.2
Fallow land	221,319	313,714
Barren land	345,445	241,088
Settlement	18,032.8	24,667.7
Crop	275,344	313,698
Total	976,849	976,849

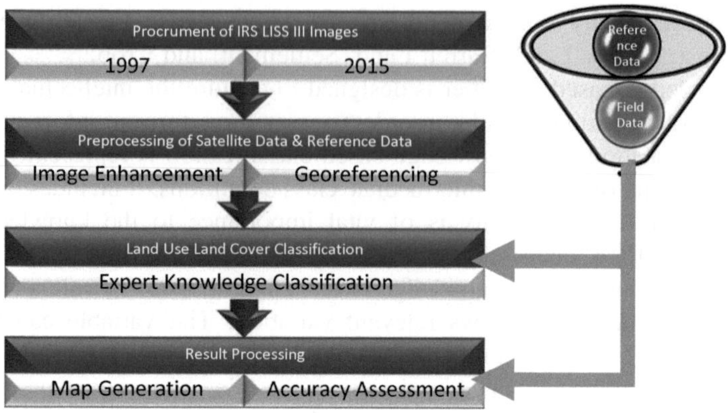

Fig. 3 Workflow of methodology

5 Result and Discussion

The results clearly demonstrate the changing scenario in forest cover of the state of Aurangabad Region, Maharashtra, India. Long-term monitoring of forests has detected forest cover change events in Gautala Biosphere. Among the eight class types mapped, Water Body, Moderate Dense Forest, Open Forest, Shrubs Forest, Fallow Land, Barren Land, Settlement, and Crop, as shown in Fig. 4a, b. The moderate dense forest is the most predominant forest type of the biosphere reserve (Tables 4, 5 and Graph 1).

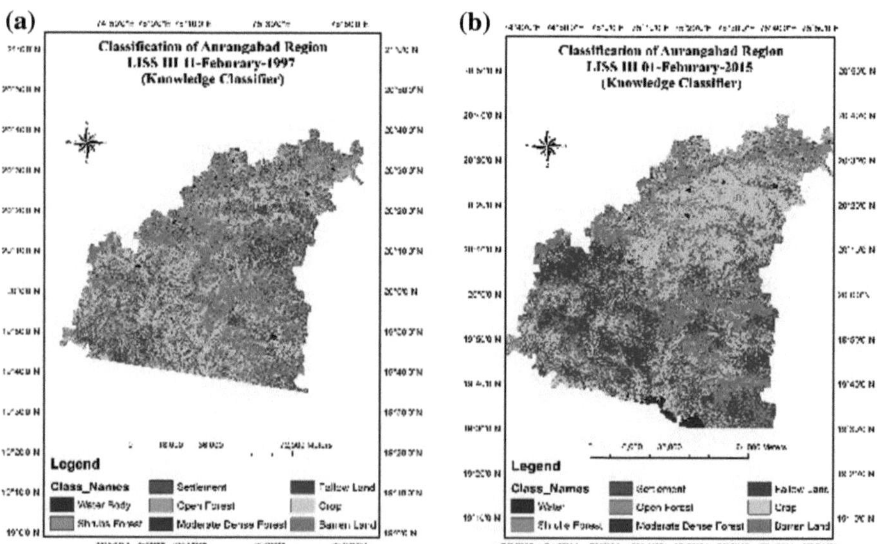

Fig. 4 **a** Classified map of year 1997. **b** Classified map of year 2015

Table 4 Knowledge classifier classification accuracy assessment report 1997

Class name	Ref. totals	Class totals	Number correct	Producers accuracy (%)	Users accuracy (%)
Crop	16	16	13	**81.25**	**81.25**
Barren land	25	23	18	**78.26**	**72.00**
Settlement	15	17	14	**82.35**	**93.33**
Shrubs forest	17	18	13	**72.22**	**76.47**
Moderate dense forest	21	20	18	**90.00**	**85.71**
Water body	18	14	14	**100.00**	**77.77**
Open forest	17	19	15	**78.94**	**88.23**
Fallow land	21	23	18	**78.26**	**85.71**
Totals	**150**	**150**	**123**		

Overall classification accuracy = **82.00 %**
Overall kappa statistics = **0.7940**

Table 5 Knowledge classifier classification accuracy assessment report 2015

Class name	Ref. totals	Class totals	Number correct	Producers accuracy (%)	Users accuracy (%)
Fallow land	37	33	30	**81.08**	**90.91**
Settlement	9	10	8	**88.89**	**80.00**
Open forest	12	12	11	**91.67**	**91.67**
Water	11	11	11	**100.00**	**100.00**
Crop	31	34	27	**87.10**	**79.41**
Moderate dense forest	9	10	8	**88.89**	**80.00**
Shrubs forest	14	12	10	**71.43**	**83.33**
Barren land	27	28	24	**88.89**	**85.71**
Totals	**150**	**150**	**129**		

Overall classification accuracy = **86.00 %**
Overall kappa statistics = **0.8329**

The analysis has addressed, the distribution of forest cover, the rate of deforestation and its change. The temporal changes in the cover provides the information about how the changes has occurred [31]. The results indicate that the loss of forest cover in the Aurangabad Region over a 19 year. The decreasing area of moderately dense forest indicates the burden on the internal forest. The increase in water bodies probably by the natural formation. Both the spatial and temporal analysis shows change in forest cover. It might be a result of fewer efforts taken by the forest department and also due to the change in climate.

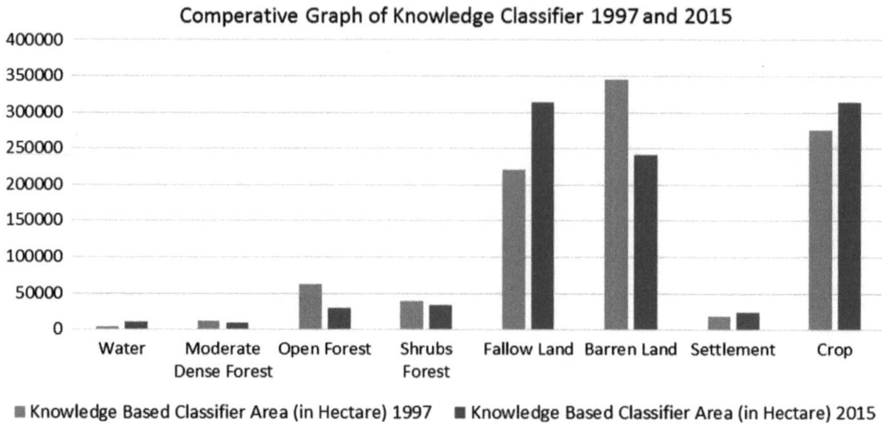

Graph 1 Classified area report 1997, 2015

6 Conclusion

The results indicates subtle loss of forest cover in the Aurangabad region over a period of 19 years. The study gives useful direction for future monitoring efforts to support organization policy and initiatives taken by the Government of India like Green India Programme. It also demonstrates the value of historical topographical maps and satellite technology to analyze the changing scenario of forests Biosphere. The study of two decades gives varying rates of deforestation in forest areas changes at different phases (1997–2015). The overall accuracies of the system using Knowledge Classifier approach of 1997 were 82 %. The corresponding accuracy by Knowledge Classifier approach of 2015 were found to be 86.00 %. We can conclude that the Knowledge Classifier (KC) is more consistent and acceptable for classifications of Forest area using the RS images. The support from local publics will be important for applying conservation actions and long-term supervision. However, Threats such as forest fires, illegal tree cutting, etc. are need to be considered for forest conservation policy.

Acknowledgments Authors thank the Dr. Babasaheb Ambedkar Marathwada University, Aurangabad, Maharashtra, INDIA for financial assistance through DST/NRDMS Srinivasa Ramanujan Geospatial Chair. Authors are thankful to the State forest department, Maharashtra, India for reference information, to UGC-DRS SAP Phase II and DST-FIST programmes for infrastructure facilities.

References

1. Lillesand, T. M., and R. W. Kiefer. (2000). RS and Image Interpretation, 736. John Wiley and Sons.
2. M. D. Behera, S. P. S. Kushwaha, P. S. Roy (2001) Forest Veg. Characterization and Mapping Using IRS-1C Satellite Images in Eastern Himalayan Region, Geo. Int, 53–62.

3. Thakker, P. S., Sastry, K. L. N., Kandya, A. K., Kimothi, M. M. and Jadav, r. N., (1999). Forest vegetation mapping using RS and GIS - a case study in Gir (Sasan) forest. *Proceedings of ISRS National Symposium on RS Applications for Natural Resources: Retrospective and Perspective.* Hyderabad, January 18–20, 1998.
4. Behera, M. D., Srivastava, S., Kushwaha, S. P. S. and Roy, P. S., (2000) Stratification and mapping of Taxus baccata L. bearing forests in Tale valley using RS and GIS. *Cur. Sci.* 78(8): 1008–1013.
5. Anonymous (2015). The State of Forest Report. Forest Survey of India. Govt. of India.
6. Houghton, R.A. and Woodwell, G.M. (1981). Proceedings of Seventh International Symposium on Machine Processing of RS Data, Indiana, West Lafayelte, 593–602.
7. Botkin D. B, Estes J. E., McDonald, R. M., and Wilson M. V., (1984), Studying the Earth's vegetation from space. Bioscience, 34, 508–514.
8. Malingreau, J. P., (1991), RS for tropical forest monitoring: an overview. In RS and GIS for Resource Management in developing countries, edited by A. S. Belward and C. R. Valenzuela (Dordrecht: Kluwer), pp. 253–278.
9. Roy, P. S., (1993). RS for forest ecosystem analysis and management. In Environmental Studies in India, edited by M. Balakrishnan (New Delhi: IBH), pp. 335–363.
10. Hadeel, A., Jabbar, M., Chen, X., (2011). Remote sensing and GIS application in the detection of environmental degradation indicators. Geo-spatial Inform. Sci. 14, 39–47.
11. Thompson, I.D., Guariguata, M.R., Okabe, K., Bahamondez, C., Nasi, R., Heymell, V., Sabogal, C., (2013). An operational framework for defining and monitoring forest degradation. Ecol. Soc. 18 (2), 20.
12. Noss, R.F., (1999). Assessing and monitoring forest biodiversity: a suggested frame-work and indicators. Forest Ecol. Manage. 115, 135–146.
13. FAO, (2011). Assessing forest degradation towards the development of globally applicable guidelines. Forest Resources Assessment Working Paper 177, Rome.
14. FSI, (2011). India State of Forest Report. Ministry of Environment and Forests, GOI.
15. Reddy, C. S., Jha, C. S., & Dadhwal, V. K. (2013). Assessment and monitoring of long-term forest cover changes in Odisha, India using RS and GIS. EM&A, 185, 4399.
16. Muley, S. V., Katpatal, Y. B., Kundal, P. P., & Khare, Y. D. (2015). Spatial Analysis of Impact of Orange Cultivation over Groundwater Regime: A Case Study of Kolar Watershed, Nagpur District, Maharashtra. Journal of the ISRS, 43(2), 395–406.
17. Reddy, C. S., Pasha, S. V., Jha, C. S., & Dadhwal, V. K. (2015). Geospatial characterization of deforestation, fragmentation and forest fires in Telangana state, India: conservation perspective. Environmental monitoring and assessment, 187(7), 1–14.
18. Satish, K. V., Saranya, K. R. L., Reddy, C. S., Krishna, P. H., Jha, C. S., & Rao, P. P. (2014). Geospatial assessment and monitoring of historical forest cover changes (1920–2012) in Nilgiri Biosphere Reserve, Western Ghats, India. Environmental monitoring and assessment, 186(12), 8125–8140.
19. Roy, P.S., S.A. Ravan, N. Rajadnya, K.K. Das, A. Jain & S. Singh. (1995). Habitat suitability analysis of Nemorhaedus goral- a RS and GIS approach. Curr Sci 69: 685–691.
20. Alam, M.S., (2011). Status, ecology and conservation of striped hyena in Gir National Park and Sanctuary, Gujarat. Ph.D. Thesis, AMU, Aligarh, India.
21. Kumar, G. P., Hemanjali, A. M., Ravikumar, P., Somashekar, R. K., & Nagaraja, B. C. (2014). Assessing the historical forest Encroachment of Kodagu region of Western Ghats, South India using RS and GIS. NRSC, Indian Space Research Organisation, Hyderabad.
22. Kumar, P., Pandey, P. C., Kumar, V., Singh, B. K., Tomar, V., & Rani, M. (2015), Efficient Recognition of Forest Species Biodiversity by Inventory-Based Geospatial Approach Using LISS IV Sensor. Sensors Journal, IEEE, 15(3), 1884–1891.
23. Das, S., & Singh, T. P. (2013), Mapping Vegetation and Forest Types using Landsat TM in the Western Ghat Region of Maharashtra, India. IJCA, 76(1), 33–37.
24. Thorat, S. S., Rajendra, Y. D., Kale, K. V., & Mehrotra, S. C. (2015), Estimation of Crop and Forest Areas using Expert System based Knowledge Classifier Approach for Aurangabad District. International Journal of Computer Applications, 121(23).

25. Y. D. Rajendra, S. C. Mehrotra, K. V. Kale, R. R. Manza, R. K. Dhumal, A. D. Nagne, A. D. Vibhute. (2014). Evaluation of partially overlapping 3d point cloud's registration by using ICP variant and CLOUDCOMPARE. ISPRS, Hyderabad.
26. Bhatt, Shankarlal C., and Bhargava, Gopal K., (2006), "Wildlife", Land and people of Indian states and union territories in 36 volumes, volume 16 MS. India: Gyan Publishing.
27. Ramanathan S, (2001), Forest Land Cover Classification Using Statistical and ANN Approaches Applied to IRS LISS III Sensor, Geo. Int, 16:2, 39–44.
28. Roy, P. S., Diwakar, P. G., Singh, I. J., & Bhan, S. K., (1994), Evaluation of microwave RS data for forest stratification and canopy characterization, IJISRS, 22(1), 31–44.
29. Rajendra, Y., Thorat, S., Nagne, A., Dhumal, R., Vibhute, A., Varpe, A., Mehrotra, S.C., Kale, K.V. (2016). Mapping forest cover of Gautala Autramghat ecosystems using geospatial technology. C2E2, 391.
30. Erdas, V.8.4. (2004). *Field Guide*, 256–259. Atlanta, GA: ERDAS.
31. Panigrahy, R. K., Kale, M. P., Dutta, U., Mishra, A., Banerjee, B., & Singh, S. (2010). Forest cover change detection of Western Ghats of Maharashtra using satellite RS based visual interpretation technique. Curr. Sci, 98(5), 657–664.

EncryScation: An Secure Approach for Data Security Using Encryption and Obfuscation Techniques for IaaS and DaaS Services in Cloud Environment

Krunal Suthar and Jayesh Patel

Abstract Now a day's a user of internet wants a freedom to access their valuable data from anywhere any time. Here, Cloud computing comes with its numerous services where user can have get anything like system power, Storage, applications and many more with just less charges. Users of Cloud mainly use it for various purpose and currently storing the data on Cloud is the very important scenario to be consider. With lots of advantage to store data on cloud and free self from burden of maintain record other side, it is very important for user that the data must be secure even it is on rest or in transition. The researcher of Cloud working on their own to deal with this issue but in most of the research proposal consider client side security related issues i.e. Integrity checking, Authentication, Versioning etc. At other side some Research model gives more importance to security data related issue of Cloud service provider i.e. Database encryption, Security based on Metadata, Data Obfuscation etc. But both of above fundamental not make strong impact for Cloud users and Service providers. To have achieve best of them here in this paper we presented a model with two techniques that's Data Obfuscation for server side to secure database details from outsiders and Encryption, authentication at client side. In this paper we have discussed implemented and testing of model towards performance and security. The proposed model ensures both users and service providers to bind trust on each other.

Keywords Cloud storage · Data protection · Integrity · Confidentiality · Encryption · Obfuscation

K. Suthar (✉)
Computer Engineering Department, Rai University, Ahmedabad, India
e-mail: krunal_bece@yahoo.co.in

J. Patel
Department of MCA, AMPICS, Kherva, India
e-mail: jayeshpatel_mca@yahoo.com

N. Modi et al. (eds.), *Proceedings of International Conference on Communication and Networks*, Advances in Intelligent Systems and Computing 508,
DOI 10.1007/978-981-10-2750-5_34

1 Introduction

The cloud computing is a virtual environment provides various services i.e. IaaS, SaaS, PaaS, DaaS to the users now days. In which users can use its virtual machine allocated to him with nominal charges and for some other reasons like to use licensed software, Store valuable information, High availability without binding of Geographical area etc. In contrast to various features provided by Cloud computing viz. elasticity, anytime access, availability, reliability, faster processing, etc.; security of user details available on cloud are still a burning issue and need to solve for wide adaption of Cloud. As a user may to reduce local burden and increase availability any where it put important data like Financial information, credential information, medical information, files etc. on cloud storage, It is very important to provide security to this data while the data is in trantion (sending from client to Cloud machine) and also when data available on cloud storage. So, using techniques like Encryption, Obfuscation we achieve a complete secure model from both the end Client and Cloud service providers.

The primary purpose of encryption is to guard the of information stored on local machine or transmitted via the network. Encryption algorithms play a especially important part in the security as well it's a key elements for data security like Authentication, Integrity and Non-repudiation. Even the data is in transmission coz of data is in unreadable format the intruder can't get anything out of that.

Changing the format or structure of data to hide actual meaning, Data obfuscation technique is used which makes reverse engineering very difficult. The good about obfuscation over encryption is that encrypted data cannot be processed until it is decrypted, but obfuscated data can be processed without de obfuscation.

If we consider encryption of data on client machine based on sensitivity of its data and then storing the information to cloud storage server, gives surety to client about secure transmission of their file on network. For Providers once it available on his premises database which contains information about lots of client in public domain using Obfuscation technique it's ensure that no any users data are misuse or tempered by unauthorized access.

This paper use encryption and obfuscation technique to provide efficient cloud storage confidentiality. Normally, Integrity or confidentiality is ensured by encryption mechanism, but for security issues in cloud encryption alone is not sufficient for information security [1]. Encryption required integrating with obfuscation technique. While Obfuscation alone is also not good for providing complete security of data in cloud storage because the unauthorized users are able to get information through attack like brute force or sometimes by reverse engineering, which break security of Cloud environment.

The paper is arranged as follows: in Sect. 2 we discuss about various security proposals, in Sect. 3 we discuss brief of proposed methodology. Section 4 we provides detailed discussion on results with security analysis followed by Sect. 5 Conclusion. Finally in last Section, we provides list of References.

2 Literature Review

Authors at [1] proposed a cryptographic technique for data security Issues in Cloud computing. In model data are Encrypted before stored on storage servers and key of file are available to data owner only; user is only approved by issuing the corresponding decryption keys by owner. Along with encryption they also used obfuscation methods to increase the confidentiality of data. Authors also proposed Algorithms are for encryption and obfuscation technique. Before storing data on Cloud premises it's encrypted or obfuscated at client side. The Proposed technique is safe to store the cloud users' data on cloud premises. Authors also argue that Encryption only or obfuscation only is not sufficient for cloud data storage.

Author at [2] presented a model for DaaS Which work to secure data which available on Cloud Machines. Proposed methodology provides two important features First Features indicate about how store data on DaaS. Second feature says that how get data from DaaS so that data confidentiality preserve. They also proposed sensitive columns mechanism for character encryption before sending it on Cloud premises, it also obfuscate Database columns at client side which contains numeric values using mathematical function before sending to Cloud storage. Main focus of proposed model to work with query over encrypted and obfuscated data. Many of the researcher are only gives idea about only obfuscation or encryption methodology for security purpose [3–7]. Some of the researcher not put focus on important criteria like efficient sharing [8] or they not shown implemented results of their proposed scheme [9, 10]. Some researcher are only provides abstract of security [11–14] or the literature about the various security issues in Cloud [15].

3 Proposed Methodology

The proposed model [16] with all the algorithms steps are presented in International conference NUiCONE 2015 organized at Nirma University and will be published in IEEE soon. So here we just provide overview of the model proposed and mainly focuses on results and analysis of proposal.

3.1 Overview

See Fig. 1.

3.2 Experimental Setup and Techniques

Here for experimental result we use ARO Encryption techniques and MONCrypt Obfuscation Technique [17]. The Client has the following Configuration Microsoft

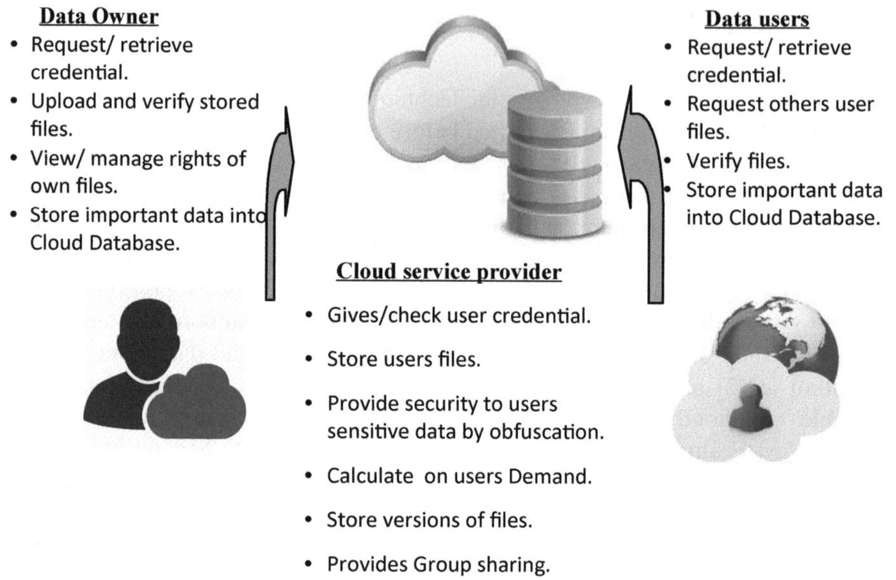

Fig. 1 Basic proposed model

windows 8 operating system 64 bit, 2.5 GHz Intel pentium processor, 4 GB RAM, 500 GB of Storage. The server having VMWare ESXi module run on 3 GHz processor with GB or Ram and 250 GB of HDD. The users upload the data via user interface form VMware Client.

4 Result Discussion

4.1 Basic Analysis

Figure 2 below shows phase wise cryptographic/obfuscation operations required. All the hash functions are performed offline and they are quite faster.

By using SHA hash function which executes in few milliseconds to compute a hash of even 1 MB file. So, overall, the overhead occurred by the cryptographic operations involved in EncryScation is very low.

In order to understand the proposed algorithms in, we consider a sample data table which stored in the cloud storage as shown in the Table 1. The data are encrypted and obfuscated by the proposed algorithms [16]. It can be noted in Table 2 that data are obfuscated. Obfuscated data in Table 2 consumes less memory in comparison with the Table 1.

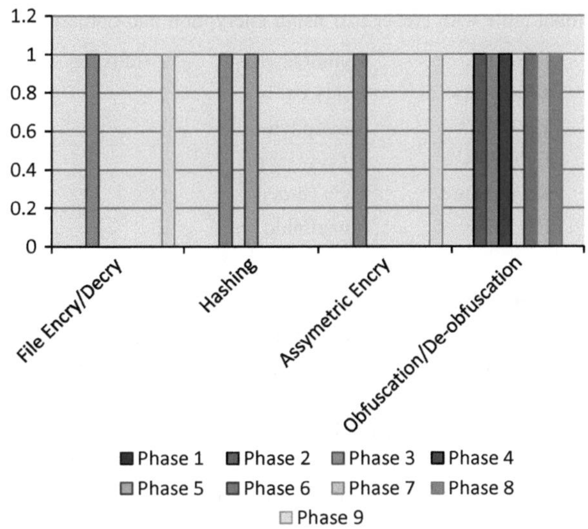

Fig. 2 Cryptographic operations/obfuscation (Phase wise)

Table 1 Transactional table with plain text

Trans_Id	Cust_Id	Item_Name	Quantity	Total_P
TId_1003	A230kum	Printer	5	10,000
TId_923	B301sus	Harddisk	13	30,000
TId_2304	C100mon	Monitor	13	22,000
TId_9087	B002lav	Mouse	9	1250
TId_0012	G123aro	Keyword	12	2400

4.2 Security Analysis

A. **Storage Correctness**: DO (or DU) can anytime request CSP for data correctness. DO issues a Dynamic random bit request to CSP, and CSP gives in form of hash of those bits having size is almost in KB. After receiving hash value from CSP, DO (or DU) compares it with new calculated hash code. If both found same, it is concluded that the integrity of data is verified.
B. **Lightweight**: Confidentiality and integrity are to be achieved through encryption and hash algorithms. Because of consume heavy computational we advise these two operations to be performed offline on the premise of DO or DU. To check integrity of data, the whole file is not transferred, but only a small sized data is exchanged between CSP and DO/DU, which is independent of the file size.
C. **Dynamism**: Granting/revoking access rights to/from DU or with the group of users is done through executing SQL query and make updation in the required Database entry.
D. **Versioning**: DO/DU is able to store multiple version of their file on CSP and can able to get it as required.

Table 2 Transactional table with cipher text using encryption and obfuscation

<table>
<tr><th>]zUdbju|p</th><th>L!pvL|xh</th><th>RymhQnwhy</th><th>Zz!dljx}&</th><th>]m\wb}ru{</th></tr>
<tr><td>]k<g3mL:?</td><td>J<y6x<5t></td><td>U,{{kx\|</td><td>2</td><td>@</td></tr>
<tr><td>]k?g5mLBU</td><td>K=3x96|?</td><td>Qm&pdvlum</td><td>)</td><td>0</td></tr>
<tr><td>]k<g6mL;@</td><td>L<z3r94v =</td><td>K{{yw rw{</td><td>)</td><td>!</td></tr>
<tr><td>]kDg3mLBC</td><td>K>#3d93u<</td><td>Uy%{drdry</td><td>Q</td><td>0</td></tr>
<tr><td>]k=g3mL9></td><td>P?{5u;4j=</td><td>Vuzglmhlq</td><td>2</td><td>0</td></tr>
</table>

E. **Data Obfuscation**: The sensitive details like Credential information, Account information etc. are obfuscated and stored on CSP Database which ensure CSP that DO/DU data are safely available on premises and no chance to tempered or misuse.

Protocol Verification Through Scyther

To verify our operational protocols, we have used Scyther [18], the tool that provides formal proofs for security protocol verification. Scyther has proven to be an effective tool for verification, falsification, and analysis of security protocols. With guaranteed termination it verify protocols with limitless number of sessions. The phases in each operation in our model are verified under various conditions in Scyther and found to be full proof against different attacks viz. man-in-the-middle attack, DoS and replay attack. Due to limitation of space, we only illustrate testing of phase Registration.

We show a claim of an attack for registration phase in Fig. 4 . In the attack, CDO#1 completes his role as CDO up to the claim. Claims are reachable and the protocols are found to be secure. We further check all our phases of Cloud storage security model in Scyther tool and displayed in Table 3 .

Table 3 Analysis of Scyther Outcomes for Proposed Protocols

Phases	Properties				
	Confidentiality	Authenticity	Integrity	Access control	Freshness
Registration	Yes (Ni,Tj,Tk)	Yes (ID_{CDO})	Yes (Ni)	Yes (K_{PU-CSP}, K_{PU-CDO})	Yes (Ni)
Pre-storage	NA	NA	NA	NA	NA
Storage	Yes (H_{CDO}, N_i, T_j, HT)	Yes (ID_{CDO})	Yes (H_{CDO}, N_i)	Yes (K_{PU-CSP}, K_{PU-CDO})	Yes (Ni)
Transmission errors	Yes(IDCDO,Ni, HT)	YES (IDCDO)	NA	NA	YES (Ni)
Manage access rights	Yes (FileID, AR, ET, HT, $K_{S-FILEID}$)	Yes (ID_{CDO}, ID_{CDU})	Yes (FileID, AR, $K_{S-FILEID}$)	Yes (K_{PU-CSP}, K_{PU-CDO})	NA
Dynamic integrity verification	Yes (FileID, Ni, Bt, Tj, C)	Yes (ID_{USER})	Yes (FileID, Ni, H_{CSP}, H_{CDO}, C)	Yes (K_{PU-CSP}, $K_{PU-USER}$)	Yes (Ni)
Data obfuscation	NA	NA	NA	NA	YES (Ni)
Versioning	Yes(ID_{CDO},Ni, FileID, Tj)	YES (IDUSER)	NA	Yes (K_{PU-CSP}, K_{PU-CDO})	YES (Ni, Tj)
Data download	Yes (FileID, Ni, Ti, Tj)	Yes (ID_{USER})	Yes (FileID, Ni)	Yes (K_{PU-CSP}, $K_{PU-USER}$)	Yes (Ni)

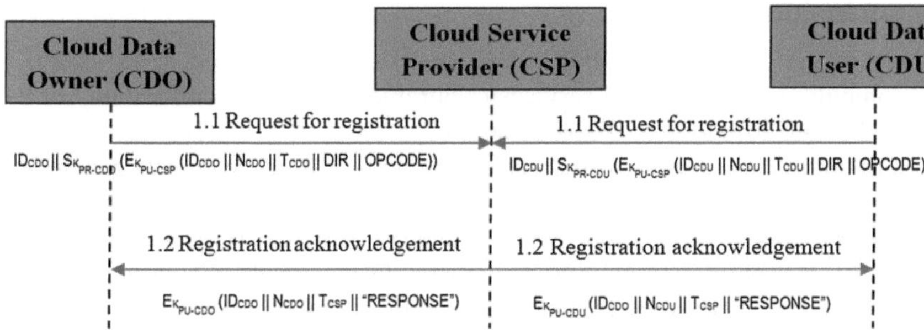

Fig. 3 Role CSP—Registration phase (Scyther outcome)

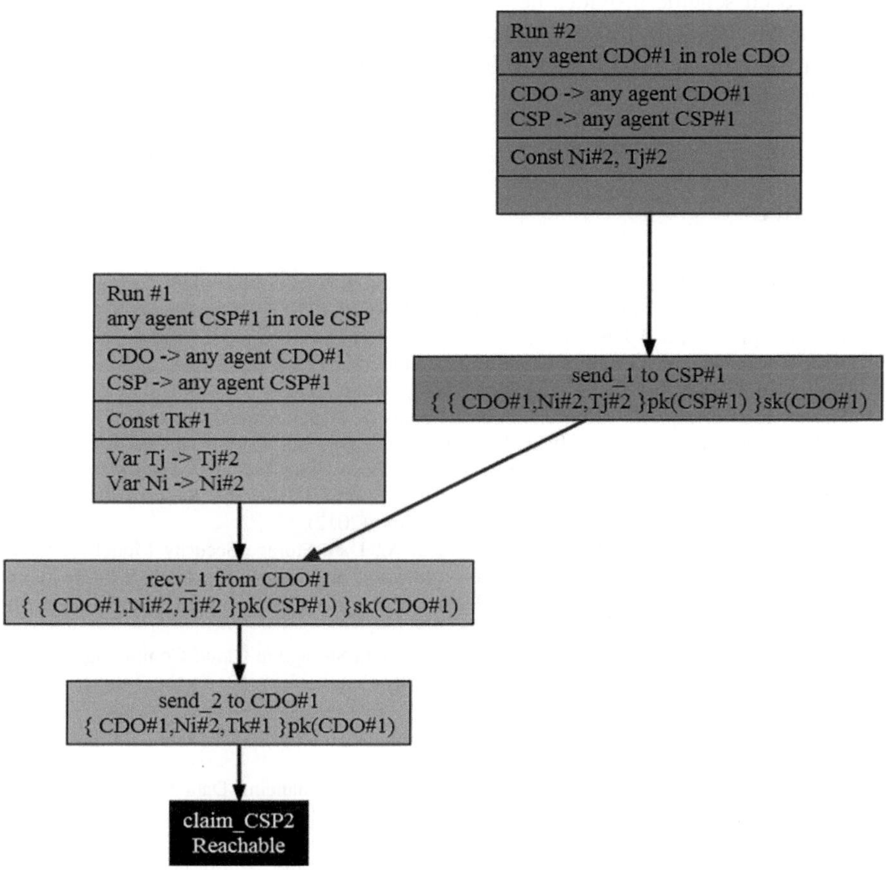

Fig. 4 Sequence diagram: Registration phase

5 Conclusion

Cloud computing provides good services to but due to lack of security many of the users are beware to adopt it. To address the problem of providing data security services to both Cloud user as well service provider, we proposed a new scheme by putting encryption and obfuscation technique works together. Data will be encrypted before sending on Cloud server based on sensitivity of data and user kept key secret gives security to data in transition coz data available in encrypted format on cloud machine makes user ensure about confidentiality.

We used obfuscation technique for security purpose at Cloud server side by which there is very less chance of tempering the data at server. We proposed an algorithm which supports all this operations and providing results with security and basic analysis. From the Model analysis using scyther security tool, it is observed that proposed scheme provides better protection to stored information on a cloud and even the data is in transition than the another available approaches which are based on encryption, obfuscation technique alone from Cloud users as well Service providers view.

References

1. Arockiam, L.; Monikandan, S., "Efficient cloud storage confidentiality to ensure data security," Computer Communication and Informatics (ICCCI), 2014 International Conference vol., no., pp. 1, 5, 3–5 Jan. (2014).
2. Atiq, R.; Hussain, M..: Efficient Cloud Data Confidentiality for DaaS. International Journal of Advanced Science and Technology Vol. 35, October (2011).
3. Halder, R.; Cortesi, A., "Obfuscation-based analysis of SQL injection attacks," Computers and Communications (ISCC), IEEE Symposium on, vol., pp. 931,938, 22–25 June (2010).
4. Hataba, M.; El-Mahdy, A.; "Cloud Protection by Obfuscation: Techniques and Metrics," P2P, Parallel, Grid, Cloud and Internet Computing (3PGCIC), 2012 Seventh International Conference on, vol., no., pp. 369,372, 12–14 Nov. (2012).
5. Patel H. B.,; Patel D. R.; Borasaniya B.; Patel A.; Data Storage Security Model for Cloud Computing, Advances in Communication, Network, and Computing Lecture Notes of the Institute for Computer Sciences, Social Informatics and Telecommunications Engineering Volume 108, 2012, pp 37–45. (2012).
6. Hyun-Suk, Yu.; Yvette, E.; Kyung K,; Securing Data Storage in Cloud Computing, Security Engineering Research Institute (Journal of Security Engineering), No. 9, No. 3, June, pp 251–260(2012).
7. Kamara S,; Lauter K,; Cryptographic Cloud Storage, IFCA/ LNCS 6054, Springer-verlag, Berlin Heidelberg, pp 136–149. (2010).
8. el-Khameesy, N.; Hossam R., A Proposed Model for Enhancing Data Storage Security in Cloud Computing Systems, Journal of Emerging Trends in Computing and Information Sciences, VOL. 3, NO. 6, June, pp 970–974(2012).
9. Wang, C.; Wang, Q.; Ren, K.; Lou, W.; Ensuring data storage security in Cloud Computing, Quality of Service, 2009. IWQoS. 17th International Workshop on, vol., no., pp. 1, 9, 13–15 July. (2009).

10. Yvette E. Gelogo, Sunguk L.,; Database Management System as a Cloud Service, International Journal of Future Generation Communication and Networking Vol. 5, No. 2, pp 71–76 June (2012).
11. Ikechukwu, U.; Omenka. U;, Building Trust and Confidentiality in Cloud computing Distributed Data Storage, West African Journal of Industrial & Academic Research, Vol. 6 No. 1 March, pp 78–83. (2013).
12. Xiaojun Yu, Qiaoyan W.; A View about Cloud Data Security from Data Life Cycle, International Conference on Computational Intelligence and Software Engineering (CiSE), pp 1–4, IEEE, Dec. (2010).
13. Mathew, A.; Survey Paper on Security & Privacy Issues in Cloud Storage Systems, EECE, Term Survey Paper, April, pp 1–13(2012).
14. Mahajan, P.; Setty, S.; Lee, S.; Depot: Cloud storage with minimal trust, 9th USENIX Symposium on Operating System Design and Implementation, pp 1–26. (2010).
15. Suthar, K., Patel, J..: Security of Cloud IAAS, DAAS Services using Encryption, Obfuscation Techniques: A Review. Technix International Journal for Engineering Research Volume 1 Issue 6, Jan (2015).
16. Suthar K., Patel J "EncryScation: A Novel Framework for Cloud IaaS, DaaS security using Encryption and Obfuscation Techniques" In 5th Nirma University International conference on Engineering(NUiCONE) Dec 2015.
17. L. ArockiamÂ and S. Monikandan " Security Framework to Ensure the Confidentiality of Outsourced Data in Public Cloud Storage" In International Journal of Current Engineering and Technology E-ISSN 2277–4106, P-ISSN 2347–5161.
18. Cremers, C., "The Scyther Tool: Verification, falsification, and analysis of security protocols". In Proc. of the 20th Int. Conf. Computer Aided Verification (CAV'08). Lecture Notes in Computer Science, vol. 5123. Springer Verlag, 414–418, 2008.

Prediction of Students Performance of an Institute Using ClassificationViaClustering and ClassificationViaRegression

Shiwani Rana and Roopali Garg

Abstract Machine Learning is the field of computer science that learns from data by studying algorithms and their constructions. In machine learning, predictions can be made by using certain algorithms for specific inputs. In this paper important classification and clustering algorithms are discussed which can be further applied to BE (Information Technology) Third Semester to evaluate student's performance. The performance of students of Digital Electronics of University Institute of Engineering and Technology (UIET), Panjab University (PU) is calculated by applying K-Means and Hierarchical Clustering Algorithms. Unsupervised Learning Algorithms like K-Means and Hierarchical clustering are discussed and for supervised learning, M5P algorithm is discussed. Further a comparison between ClassificationViaClustering and ClassificationViaRegression is done using WEKA Tool. The accuracy of grades prediction is calculated with both the approaches and a graphical explanation is presented for the BE (Information Technology) Third Semester students.

Keywords WEKA · K-Means · Hierarchical · M5P · Classificationviaclustering · Classificationviaregression

1 Introduction

Machine learning is a type of intelligent learning which provides computers with the ability to design and develop algorithms. It focuses on the advancement of computer programs that can train themselves to grow and change when exposed to new data. A machine learning program detects patterns in data and includes different combinations of logic [1] (Fig. 1).

S. Rana (✉) · R. Garg
Department of IT, UIET, Panjab University, Chandigarh, India
e-mail: shiwanirana40@gmail.com

R. Garg
e-mail: roopali.garg@gmail.com

N. Modi et al. (eds.), *Proceedings of International Conference on Communication and Networks*, Advances in Intelligent Systems and Computing 508,
DOI 10.1007/978-981-10-2750-5_35

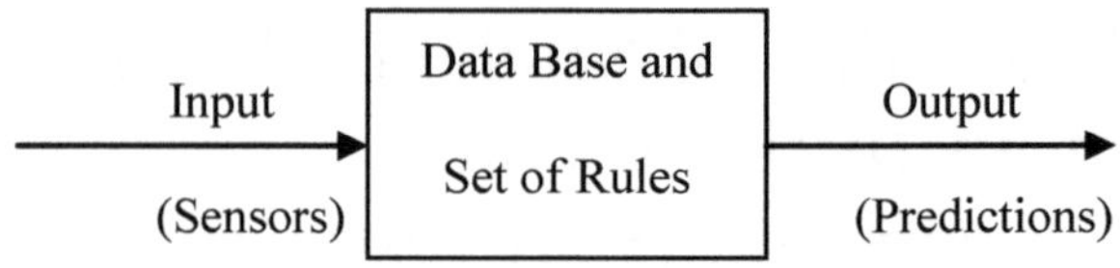

Fig. 1 Machine learning

Benefits of Machine Learning:

- With the use of machine learning, the human-computer interaction becomes easier.
- It increases customer satisfaction.
- It is simple to integrate and improves intelligent systems.

The two machine learning approaches are:

- **Supervised learning**: This type of learning algorithm is used for generating a function which is used for mapping the inputs to the desired outputs. These functions are based on the training data set. A supervised technique uses a dataset with known classification. Various algorithms like Naïve Bayes, Logistic Regression, Neural Networks, Linear Regression and Decision Trees are highly dependent on the information given by the pre-determined classifications [2].
- **Unsupervised learning**: Unsupervised Learning is also known as undirected learning. It is basically used when the output conditions are not clearly represented in the dataset [3]. The task of this algorithm is to automatically discover inbuilt patterns in the data without the prior information about which class the data could belong to. An unsupervised technique does not use a given set of unclassified data.

The paper is organized in the following section. Section 2 describes the Machine Learning Algorithms which further discusses the two Unsupervised algorithms and a Supervised Algorithm. Section 3 deals with the Implementation in which the performance of fifty eight students is analyzed with both the ClassificationViaClustering and ClassificationViaRegression approaches by using WEKA Tool, Sect. 4 deals with the Results and Comparison, comparing both the approaches, Sect. 5 describes the Output and Sect. 6 describes the Conclusion and Future Work.

2 Machine Learning Algorithms

2.1 Unsupervised Algorithms

(i) **K means Clustering Algorithm**: The process of partitioning or grouping a given set of patterns into disjoint clusters is known as clustering [4]. Big data sets can be easily clustered by using K means clustering algorithm. K Means algorithm is for clustering. It is a type of unsupervised learning where there is no idea about the class or labels for any data and need to discover the clusters without this information [5] (Fig. 2).

(ii) **Hierarchical Clustering Algorithms**: Hierarchical clustering can be done in three different ways [6]:

Algorithm of Hierarchical Clustering Algorithm (Single-linkage cluster) [7]:

1. Assign a cluster to each item, such that N clusters for N items.
2. Find and merge the pair of clusters which are closest to each other.
3. Calculate the distances between the new and each of the old clusters (using single-linkage cluster)
   (a) Start with the disjoint clustering having level $l(0) = 0$ and sequence number $n = 0$.
   (b) In the current clustering, now find the least dissimilar pair of clusters say pair (a), (b), according to d[(a), (b)] = min d[(u), (v)] where, the minimum is over all pairs of clusters in the current clustering.

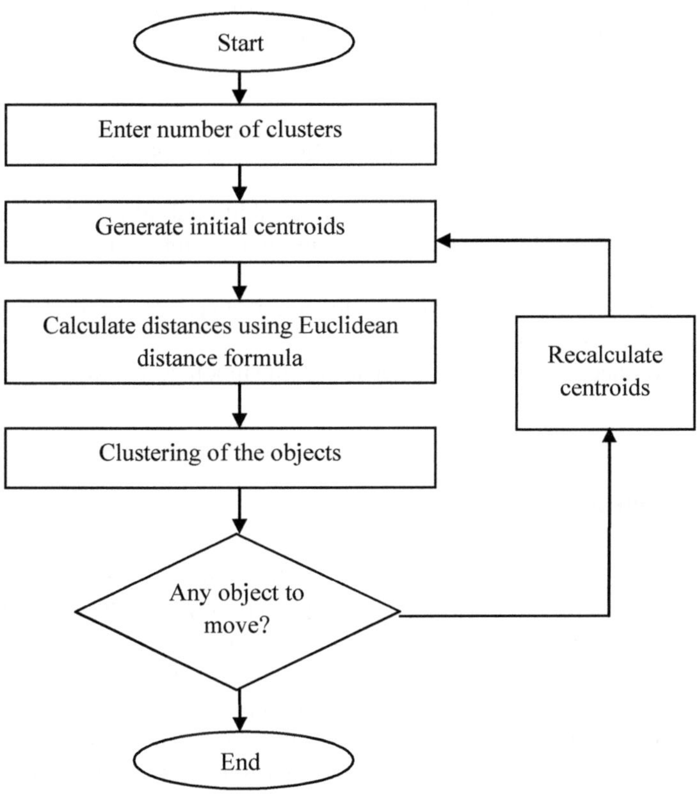

Fig. 2 Flow chart of K-Means clustering

(c) Increment the sequence number: n = n + 1 and merge clusters (a) and (b) into a single cluster to form the next clustering n. Set the level of this clustering to: l(n) = d[(a), (b)]
(d) Now the next step is to update the proximity matrix, M, by deleting the rows and columns corresponding to clusters (a) and (b) and adding a row and column correspond to the newly formed cluster. The proximity between the new cluster, denoted (a, b) and old cluster (k) is defined in this way: d[(k), (a, b)] = min d[(k), (a)], d[(k), (b)]

4. If all the objects are in one cluster then stop the process else, go to step 2.

The complexity of Hierarchical Clustering Algorithm is: O (n^2).

2.2 *Supervised Algorithms*

M5P: M5P is a special reconstruction algorithm used for classification [7]. This supervised algorithm combines a conventional decision tree with the possibility of linear regression functions at the nodes. M5P learns a "model" tree—which is a decision tree with linear regression functions at the leaves. It can be used to predict a numeric target (class) attribute.

Options of M5P algorithm are:

- BuildRegressionTree—Whether to generate a regression tree/rule instead of a model tree/rule.
- Debug—If set to true, classifier may output additional info to the console.
- MinNumInstances—It is defined as the minimum number of instances to allow at a leaf node.
- SaveInstances—This depicts whether to save instance data at each node in the tree for visualization purposes.
- UseUnsmoothed—It shows whether to use unsmoothed predictions.

3 Implementation

ClassificationViaClustering and ClassificationViaRegression are the two approaches used to analyze the performance of BE (Information Technology) Third semester students of UIET, PU, Chandigarh. The academic data of 58 students from the data set (given in the appendix) are taken which includes 8 attributes namely total marks, grade, attendance, major, minor1, minor2, institution, area and then both the approaches are applied on this data set using WEKA Tool [8]. Classification in WEKA includes some major terminologies which can be seen in Figs. 3 and 4:

```
Classifier output
Correctly Classified Instances          20               34.4828 %
Incorrectly Classified Instances        38               65.5172 %
Kappa statistic                          0.087
Mean absolute error                      0.1638
Root mean squared error                  0.4047
Relative absolute error                 83.3898 %
Root relative squared error            129.8609 %
Total Number of Instances               58

=== Detailed Accuracy By Class ===

               TP Rate   FP Rate   Precision   Recall   F-Measure   ROC Area   Class
                 0         0         0          0         0           ?        A+
                 0         0         0          0         0          0.5       A
                 0.684     0.487     0.406      0.684     0.51       0.599     B+
                 0.2       0.163     0.3        0.2       0.24       0.519     B
                 0.333     0.261     0.25       0.333     0.286      0.536     C+
                 0         0         0          0         0          0.5       C
                 0         0         0          0         0          0.5       D
                 0         0         0          0         0          0.5       F
Weighted Avg.    0.345     0.256     0.262      0.345     0.288      0.545
```

Fig. 3 Classifier output of ClassificationViaClustering

```
Classifier output
Time taken to build model: 0.04 seconds

=== Stratified cross-validation ===
=== Summary ===

Correctly Classified Instances          54               93.1034 %
Incorrectly Classified Instances         4                6.8966 %
Kappa statistic                          0.9092
Mean absolute error                      0.0923
Root mean squared error                  0.1712
Relative absolute error                 46.9798 %
Root relative squared error             54.9211 %
Total Number of Instances               58

=== Detailed Accuracy By Class ===

               TP Rate   FP Rate   Precision   Recall   F-Measure   ROC Area   Class
                 0         0         0          0         0           ?        A+
                 1         0         1          1         1          1         A
                 1         0         1          1         1          1         B+
                 1         0         1          1         1          1         B
                 1         0.065     0.8        1         0.889      1         C+
                 0         0.018     0          0         0          0.464     C
                 0         0         0          0         0          0.158     D
                 0.667     0         1          0.667     0.8        1         F
Weighted Avg.    0.931     0.014     0.907      0.931     0.915      0.967
```

Fig. 4 Classifier output of ClassificationViaRegression

Table 1 Accuracy of clustering algorithms

Type of algorithm	Accuracy using all the eight attributes	Accuracy using selected six attributes
K-Means	*34.48*	32.76
Hierarchical	29.31	*31.03*

- **TP Rate**: TP stands for true positives or the correctly classified instances.
- **FP Rate**: FP stands for false positives or the incorrectly classified instances.
- **Precision**: It is defined as the ratio of the instances of a class to the total instances classified as that class
- **Recall**: Ratio of the proportion of instances classified as a given class to the actual total in that class (equivalent to TP rate)
- **F-Measure**:

$$\mathbf{F} - \text{Measure} = \frac{2 * precision * recall}{(precision + recall)} \tag{1}$$

(i) **ClassificationViaClustering**:

For clustering 58 students, K-Means and Hierarchical clustering algorithms are used which are further used in classification for the prediction purpose. ClassificationViaClustering approach is used in WEKA Tool which combines both the classification and clustering algorithms. 2 clusters are set for both the clustering algorithms. The accuracy is described by the correctly classified instance which can vary for the same algorithm by simply varying the number of attributes [9]. For Example in Table 1, the accuracy of K-Means decreases when the algorithm runs from 8 attributes to 6 attributes but in case of Hierarchical clustering, it increases.

In Fig. 3, ClassificationViaClustering is applied using K-Means clustering algorithm which first makes cluster for characterizing the BE 3rd semester students and then predicting their grades.

(ii) **ClassificationViaRegression**:

Figure 3 shows the implementation of ClassificationViaRegression which uses M5P regression algorithm for predicting the grades. M5P actually makes a decision tree with linear regression for its implementation [10].

4 Result and Comparison

Comparison between ClassificationViaClustering and ClassificationViaRegression is shown in Table 2. After applying both the approaches over the data set of fifty-eight students of BE 3rd Semester of UIET, a striking outcome is obtained

Table 2 Comparison of the two approaches

Features/name of approaches	ClassificationViaClustering	ClassificationViaRegression
Instances	58	58
Attributes	8	8
Test mode	10-fold cross-validation	10-fold cross-validation
Time taken to build model (s)	0	0.04
Accuracy using all the attributes	34.48 %	93.10 %

which shows the accuracy of ClassificationViaRegression is much higher than that in the ClassificationViaClustering [11].

The accuracy results can be explained in more details by having a look on both the Figs. 5 and 6, which depicts the confusion matrix. Correctly and incorrectly classified instances shown in the matrix is the actual result for accuracy/prediction of grades of the BE 3rd semester students of UIET [12].

The diagonal values (aa, bb, cc, dd, ee, ff, gg and hh) are the true positive (TP) or the correctly classified instances which shows the relation between the true values and the predicted values. All the other values of the confusion matrix are incorrectly classified instances known as false positive (FP).

In case of ClassificationViaClustering, for example from Fig. 5, the value of cc shows that 13 instances are correctly predicted for B+ grade.

$$\begin{aligned}\text{Correctly classified instances (CCI)} &= \text{aa} + \text{bb} + \text{cc} + \text{dd} + \text{ee} + \text{ff} + \text{gg} + \text{hh} \\ &= 0 + 0 + 13 + 3 + 4 + 0 + 0 + 0 \\ &= 20\end{aligned} \tag{2}$$

```
=== Confusion Matrix ===

 a  b  c  d  e  f  g  h   <-- classified as
 0  0  0  0  0  0  0  0 |  a = A+
 0  0  4  1  1  0  0  0 |  b = A
 0  0 13  4  2  0  0  0 |  c = B+
 0  0  7  3  5  0  0  0 |  d = B
 0  0  7  1  4  0  0  0 |  e = C+
 0  0  1  0  1  0  0  0 |  f = C
 0  0  0  1  0  0  0  0 |  g = D
 0  0  0  0  3  0  0  0 |  h = F
```

Fig. 5 Confusion matrix of ClassificationViaClustering

```
=== Confusion Matrix ===

 a  b  c  d  e  f  g  h   <-- classified as
 0  0  0  0  0  0  0  0 |  a = A+
 0  6  0  0  0  0  0  0 |  b = A
 0  0 19  0  0  0  0  0 |  c = B+
 0  0  0 15  0  0  0  0 |  d = B
 0  0  0  0 12  0  0  0 |  e = C+
 0  0  0  0  2  0  0  0 |  f = C
 0  0  0  0  0  1  0  0 |  g = D
 0  0  0  0  1  0  0  2 |  h = F
```

Fig. 6 Confusion matrix of ClassificationViaRegression

$$\begin{aligned}\text{Correctly classified instances(accuracy in \%)} &= \frac{\text{CCI}}{\textit{Total no. of instances}} \times 100 \\ &= \frac{20}{58} \times 100 = 34.48\,\% \end{aligned} \tag{3}$$

The sum of all the remaining values in the confusion matrix gives the incorrectly classified instances.

In case of ClasificationViaRegression, for example from Fig. 6, the value of cc depicts 19 instances which are correctly predicted for B+ grade.

$$\begin{aligned}\text{Correctly classified instances (CCI)} &= \text{aa} + \text{bb} + \text{cc} + \text{dd} + \text{ee} + \text{ff} + \text{gg} + \text{hh} \\ &= 0 + 6 + 19 + 15 + 12 + 0 + 0 + 2 \\ &= 54 \end{aligned} \tag{4}$$

$$\begin{aligned}\text{Correctly classified instances(accuracy in \%)} &= \frac{\text{CCI}}{\textit{Total no.of instances}} \times 100 \\ &= \frac{54}{58} \times 100 = 93.10\,\% \end{aligned} \tag{5}$$

5 Output

X-axis represents the predicted grade and Y-axis represents the major (the marks scored in the final exam) [13]. The actual grades are shown with different colors depicting the output of the graph.

Figure 7 shows the output of ClassificationViaClustering approach in which K-Means algorithm is used for making clusters by setting the value of clusters to 2 and then applying the classification for predicting the grades of BE 3rd Semester students [14]. For Example: according to the prediction of grades, most of the students fall under B+, B and C+.

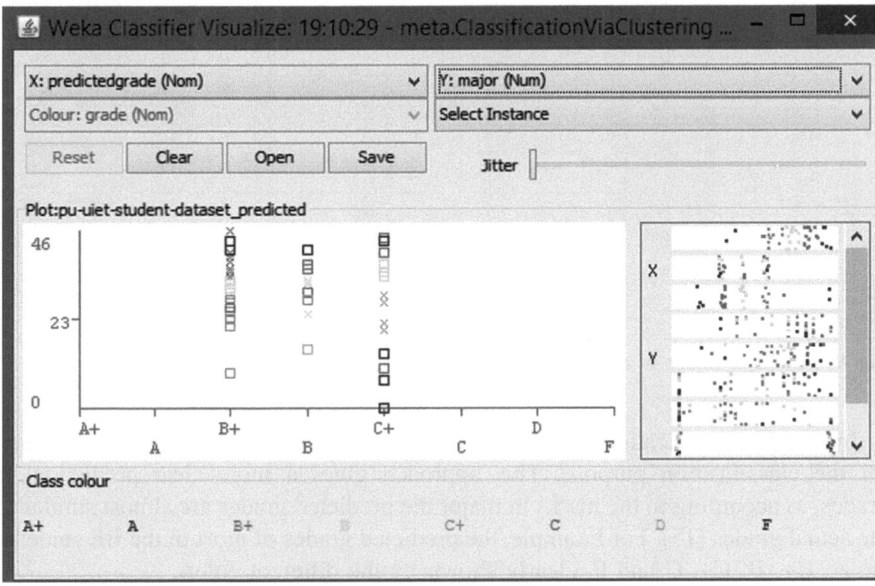

Fig. 7 Visualize classifier errors-ClassificationViaClustering

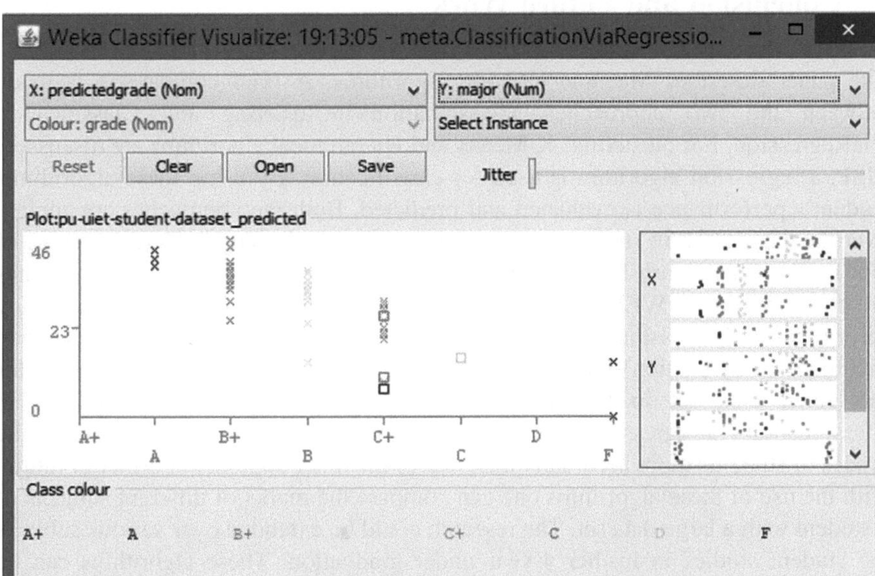

Fig. 8 Visualize classifier errors-ClassificationViaRegression

Relation: pu-uiet-student-dataset

No.	totalMarks Numeric	grade Nominal	attendance Numeric	major Numeric	minor2 Numeric	minor1 Numeric	Institution Nominal	Area Nominal
1	71.0	B+	13.0	34.0	7.0	24.0	Private	urban
2	50.0	C+	7.0	27.0	16.0	10.0	Governm...	rural
3	63.0	B	7.0	34.0	22.0	0.0	Governm...	urban
4	50.0	C+	15.0	25.0	10.0	0.0	Governm...	urban
5	72.0	B+	15.0	35.0	22.0	9.0	Private	rural
6	61.0	B	16.0	24.0	21.0	0.0	Governm...	rural
7	40.0	D	15.0	15.0	1.0	10.0	Private	rural
8	55.0	C+	14.0	9.0	22.0	9.0	Governm...	urban
9	60.0	B	14.0	14.0	10.0	12.0	Governm...	urban
10	32.0	F	5.0	7.0	20.0	0.0	Governm...	rural
11	85.0	A	15.0	41.0	29.0	16.0	Private	rural
12	27.0	F	5.0	14.0	8.0	0.0	Governm...	urban
13	61.0	B	7.0	37.0	17.0	11.0	Private	urban
14	53.0	C+	7.0	30.0	16.0	0.0	Governm...	rural
15	81.0	A	14.0	41.0	26.0	0.0	Private	urban
16	70.0	B+	16.0	33.0	21.0	7.0	Governm...	rural
17	50.0	C+	8.0	21.0	21.0	0.0	Private	rural
18	77.0	B+	16.0	39.0	22.0	10.0	Private	rural
19	78.0	B+	15.0	39.0	24.0	11.0	Governm...	urban
20	64.0	B	11.0	33.0	20.0	16.0	Private	urban
21	70.0	B+	15.0	36.0	19.0	9.0	Private	urban
22	82.0	A	15.0	43.0	24.0	17.0	Private	urban
23	68.0	B	12.0	38.0	18.0	7.0	Private	rural
24	83.0	A	15.0	39.0	0.0	29.0	Private	rural
25	73.0	B+	15.0	37.0	21.0	10.0	Private	rural
26	76.0	B+	15.0	44.0	0.0	17.0	Governm...	urban
27	73.0	B+	15.0	40.0	18.0	15.0	Governm...	urban
28	56.0	C+	13.0	27.0	16.0	0.0	Private	rural
29	72.0	B+	13.0	40.0	19.0	13.0	Private	rural
30	73.0	B+	9.0	46.0	0.0	18.0	Private	urban
31	48.0	C	13.0	26.0	9.0	1.0	Governm...	rural
32	64.0	B	13.0	31.0	20.0	2.0	Private	urban
33	61.0	B	11.0	30.0	20.0	9.0	Private	rural
34	58.0	C+	14.0	28.0	16.0	8.0	Private	urban
35	73.0	B+	18.0	35.0	20.0	9.0	Governm...	rural
36	56.0	C+	18.0	29.0	8.0	9.0	Private	urban
37	60.0	B	13.0	32.0	9.0	15.0	Private	urban
38	58.0	C+	14.0	28.0	16.0	2.0	Private	urban
39	51.0	C+	15.0	20.0	16.0	0.0	Governm...	rural
40	72.0	B+	13.0	33.0	26.0	13.0	Private	rural
41	65.0	B	7.0	33.0	0.0	25.0	Governm...	urban
42	67.0	B	15.0	35.0	17.0	8.0	Private	rural
43	50.0	C+	16.0	23.0	10.0	11.0	Governm...	rural
44	65.0	B	16.0	33.0	16.0	7.0	Private	rural
45	70.0	B+	18.0	36.0	10.0	16.0	Private	urban
46	80.0	A	15.0	43.0	0.0	22.0	Governm...	rural
47	62.0	B	13.0	34.0	15.0	0.0	Governm...	urban
48	0.0	F	0.0	0.0	0.0	0.0	Private	rural
49	70.0	B+	17.0	36.0	17.0	12.0	Private	rural
50	50.0	C+	7.0	22.0	21.0	12.0	Private	urban
51	72.0	B+	17.0	30.0	25.0	13.0	Governm...	urban
52	70.0	B+	16.0	25.0	0.0	29.0	Private	rural
53	47.0	C	14.0	10.0	23.0	0.0	Private	urban
54	78.0	B+	18.0	38.0	22.0	18.0	Private	urban
55	74.0	B+	10.0	39.0	25.0	0.0	Private	rural
56	80.0	A	14.0	39.0	27.0	0.0	Governm...	urban
57	63.0	B	8.0	35.0	0.0	20.0	Governm...	urban
58	67.0	B	13.0	32.0	0.0	22.0	Private	rural

Fig. 9 Details of marks of digital electronics students of UIET

In Fig. 8, ClassificationViaRegression approach is used. M5P algorithm is used for the classification purpose. This approach gives a more clear prediction of grades, as according to the marks in major the predicted grades are almost similar to the actual grades [15]. For Example, the predicted grades of most of the BE students are A, B+, B, C+, C and F, clearly shown by the different colors.

6 Conclusion and Future Work

The paper describes Machine learning algorithms. A brief comparison is made between the two approaches, ClassificationViaClustering and Classification ViaRegression. For clustering, K-Means and Hierarchical algorithms are discussed. MSP, a regression algorithm is used for classification. By using these algorithms, student's performance is evaluated and predicted. Both the approaches are applied over the data set of 58 students of BE (Information Technology) Third Semester of UIET, PU, Chandigarh for predicting the grades of students. Both the approaches are compared using WEKA Tool. According to the output obtained, the accuracy (or the correctly classified instances) in ClassificationViaRegression is more than that in the ClassificationViaClustering. However the time taken to build a model for the given data set in ClassificationViaClustering is less (0 s) than that in ClassificationViaRegression (0.04 s). ClassificationViaRegression predicts the grades of students more accurately then that of the other approach. Further in future, with the use of these algorithms one can compare the marks of different subjects of a student with a large data set. The research could be extended over various subjects the student studies in his/her 4 year under graduation. These algorithms can be coded in Python to analyze and discuss the data.

Appendix

See Fig. 9.

References

1. Khan, Irfan Ajmal, and Jin Tak Choi (2014). "An Application of Educational Data Mining (EDM) Technique for Scholarship Prediction." International Journal of Software Engineering and Its Applications 8(12), pp. 31–42, ISSN: 1738-9984.
2. Goyal, Monika, and Rajan Vohra (2012). "Applications of data mining in higher education." International journal of computer science 9(2): 113, ISSN: 1738-7906.
3. Ayodele, Taiwo Oladipupo (2010). Types of machine learning algorithms. INTECH Open Access Publisher.
4. Khan, Dost Muhammad, and Nawaz Mohamudally (2010). "An Agent oriented approach for implementation of initial centroids in k-means." International Journal of Information Processing and Management 1(1), pp-104–113, ISSN: 2093-4009.
5. Stephen P. Borgatti: "How to explain hierarchical clustering" http://www.analytictech.com/networks/hiclus.htm.
6. Punitha, S. C., P. Ranjith Jeba Thangaiah, and M. Punithavalli (2014). "Performance Analysis of Clustering using Partitioning and Hierarchical Clustering Techniques." International Journal of Database Theory and Application 7(6), pp. 233–240, ISSN: 2005-4270.
7. Mohammad, Thakaa Z., and Abeer M. Mahmoud (2014). "Clustering of Slow Learners Behavior for Discovery of Optimal Patterns of Learning." International Journal of Advanced Computer Science and Applications 5(11), pp. 102–109, ISSN: 2156-5570.
8. Anuradha, C., and T. Velmurugan (2014). "A data mining based survey on student performance evaluation system." IEEE International Conference on Computational Intelligence and Computing Research (ICCIC), Coimbatore, India, pp. 1–4.
9. Bydovska, Hana, and Lubomir Popelinsky (2013). "Predicting Student Performance in Higher Education." 24th IEEE International Workshop on Database and Expert Systems Applications (DEXA), Prague Czech Republic, pp. 141–145.
10. Jamesmanoharan, J., S. Hari Ganesh, M. Felciah, and A. K. Shafreenbanu (2014). "Discovering Students' Academic Performance Based on GPA Using K-Means Clustering Algorithm." IEEE World Congress on Computing and Communication Technologies (WCCCT), Tiruchirappalli, Tamilnadu, India, pp. 200–202.
11. De Morais, Alana M., Joseana MFR Araujo, and Evandro B. Costa (2014). "Monitoring student performance using data clustering and predictive modelling." IEEE Frontiers in Education Conference (FIE), Madrid, Spain, pp. 1–8.
12. Pradeep, Anjana, Smija Das, and Jubilant J. Kizhekkethottam (2015). "Students dropout factor prediction using EDM techniques." IEEE International Conference on Soft-Computing and Networks Security (ICSNS), Coimbatore, India, pp. 1–7.
13. Singh, Sushil, and Sunil Pranit Lal (2013). "Educational courseware evaluation using Machine Learning techniques." IEEE Conference on e-Learning, e-Management and e-Services (IC3e), Kuching, Malaysia, pp. 73–78.
14. Lopez, Manuel Ignacio, J. M. Luna, C. Romero, and S. Ventura (2012). "Classification via Clustering for Predicting Final Marks Based on Student Participation in Forums." International Educational Data Mining Society 42, pp. 649–656.
15. Borkar, Suchita, and K. Rajeswari (2014). "Attributes Selection for Predicting Students' Academic Performance using Education Data Mining and Artificial Neural Network." International Journal of Computer Applications 86(10), pp. 25–29, ISSN: 0975-8887.

Feature Based Object Mining and Tagging Algorithm for Digital Images

Hiteshree Lad and Mayuri A. Mehta

Abstract Object mining is the process of retrieving knowledge about meaningful objects by breaking the image into meaningful components and assigning labels to identified objects. Mining of objects from an image is nontrivial task due to representation of same object using different scales under different viewpoints and illumination changes. Moreover, occlusion and clutter reduce the probability of identification of objects from the image. In this paper, we propose a new Feature based Object Mining and Tagging Algorithm (FOMTA) that decreases the false negative rate. It also increases the probability of identification of objects under occlusion and clutter.

Keywords Object recognition · Object mining · Feature extraction · Feature vector · Object tagging

1 Introduction

Mining of meaningful components from an image is termed as the process of object mining [1]. Breaking an image into meaningful components and identification of valuable components is called the process of object recognition [2]. Object mining is to identify objects using object recognition and assigning labels to identified objects. An image with identified and labeled objects is referred as a tagged image. The tagged images are useful in several domains such as medical, agriculture, computer vision and remote sensing [1–5]. Furthermore, they are useful in image mining for mining of images having similar components from huge dataset [1].

H. Lad (✉) · M.A. Mehta
Department of Computer Engineering,
Sarvajanik College of Engineering and Technology, Surat, India
e-mail: hitilad@gmail.com

M.A. Mehta
e-mail: mayuri.mehta@scet.ac.in

N. Modi et al. (eds.), *Proceedings of International Conference on Communication and Networks*, Advances in Intelligent Systems and Computing 508,
DOI 10.1007/978-981-10-2750-5_36

It is crucial to address object mining to achieve satisfactory results due to following reasons [1–3]: (1) an object of the same scene is represented differently in images captured under varying values of scale, viewpoint and illumination conditions. Hence, an object may be misclassified or unidentified. (2) occlusion condition and clutter make object mining process more difficult. Numerous object mining techniques are available in the literature. However, some of the existing techniques do not recognize objects accurately under occlusion [1–4]. Several techniques are computationally average or slow in recognizing objects as they use Scale Invariant Feature Transform (SIFT) descriptor [4–7]. Association rule based image mining technique uses association rules that increases memory usage to store intermediate results [5]. Moreover, majority of the techniques has increased false negative rate i.e. increased rate of misclassification [1, 2, 5, 7]. In segmentation based object mining approaches, accuracy and speed of segmentation are key issues [7]. Thus, the development of an efficient object mining algorithm is required that recognizes objects accurately and decreases misclassification rate.

In this paper, we propose a new Feature based Object Mining and Tagging Algorithm (FOMTA) that increases speed and accuracy of object recognition. Moreover, it recognizes objects under occlusion condition. The overall performance of the proposed algorithm is highly dependent on the following two major steps of the algorithm: feature extraction and classification. Based on our parametric evaluation of existing feature extraction techniques, we have chosen an efficient Speeded Up Robust Feature (SURF) extraction technique [8–10]. Similarly, based on our parametric evaluation of existing classifiers, we have chosen Support Vector Machine (SVM) classifier for object classification as it classifies objects accurately [11–13]. However, SVM often misclassifies the occluded objects or partially visible objects in image. Therefore, to improve the object classification accuracy of SVM under occlusion, we use ADABOOST (Adaptive Boosting) algorithm. ADABOOST chooses strong features from feature vector generated as an output of feature extraction. Strong features are features having high values of pixel intensities. It increases the accuracy of FOMTA under occlusion condition.

Rest of the paper is organized as follows: Sect. 2 describes the existing work related to object mining. Section 3 describes the proposed object mining and tagging algorithm. Finally, conclusion and future work are discussed in Sect. 4.

2 Related Work

A large number of object mining techniques have been presented so far. Construction of matching graph is suggested for object mining by grouping images containing the same object, despite significant changes in scale, viewpoint and partial occlusion [14]. Though it extracts several features, its computational speed is not satisfactory as it uses SIFT descriptor to extract the features. A novel language model based approach uses bag-of-words (BOW) concept to retrieve image objects considering semantics related to set of visual words [6]. It is simpler in terms of computation. However, BOW suffers from mismatch problem (vocabulary problem), that is it provides irrelevant images to the user. In [15], satellite images are classified by extracting pixel intensities and organizing color pixels. In [16], common objects are recognized from set of images by assigning commonness score to each pixel. Both [15, 16] use SIFT descriptor for feature extraction. Though SIFT provides better computational accuracy, its performance under occlusion is marginal. Multiple segmentation with BOW concept is used for object recognition considering different aspects of images [7]. Quality and accuracy of segmentation is improved considering appropriate features for segmentation. In [5], common visual patterns are discovered from two sets of feature points for common object recognition. Authors observed that common visual patterns share similar local features and spatial layout. Graph is used for representation of commonness. Though the manipulation of graph is simpler, the technique is not affine invariant.

We have observed that majority of the existing object mining techniques are application specific and are based on supervised learning. Moreover, they provide optimum object recognition for specific dataset. We also observed that mining of objects from images having varying values of illumination, rotation, scale, viewpoint and clutter (noise) is simpler if global features of image are considered. However, mining of objects under varying attributes considering local features is difficult task due to representation of same object under different scale, different viewpoint, illumination changes, occlusion and clutter. Moreover, to the best of our knowledge, none of the existing approaches provide optimum recognition in the presence of occlusion. Furthermore, they use SIFT descriptor that has lesser computational speed. Hence, we propose a new FOMTA to overcome above limitations. We use SURF descriptor that is faster than SIFT [8–10].

3 The Proposed Feature Based Object Mining and Tagging Algorithm

In this section, we introduce the proposed object mining and tagging algorithm that has following features:

- Improved object recognition accuracy
- Higher computational speed
- Decreased false negative rate that is decreased number of misclassifications
- Identification of objects under occlusion and clutter

As we show in Fig. 1, the major steps involved in FOMTA are preprocessing, feature extraction and classification. The performance of FOMTA is highly dependent on techniques chosen for feature extraction and classification.

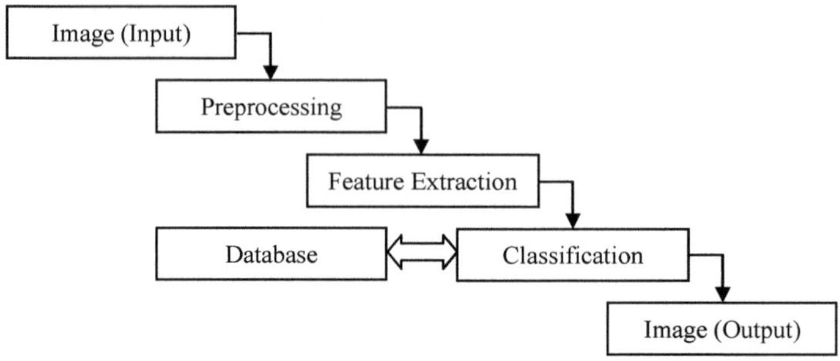

Fig. 1 Flow of FOMTA

3.1 Preprocessing

In preprocessing step, noise is removed from an input image using gaussian filter. Typically, noise is present in image due to fault in capturing device or surrounding conditions such as smoke or fog during image capturing. This type of noise is called gaussian noise that is removed using gaussian filter. After noise removal, the subsequent step is image resizing. Image is resized to one common dimension for easy evaluation of image objects. Finally, RGB to gray conversion is performed on image because gray scale image is easy to manipulate as it contains only 0–255 pixel intensity values.

3.2 Feature Extraction

An image consists of multiple features. Feature extraction technique is applied on image to generate feature vector. It extracts features based on varying values of pixel intensities and generates feature vector. Feature vector is an intermediate representation of image features that makes object identification easy.

Object recognition accuracy and speed of FOMTA are highly dependent on feature extraction technique used in it. To choose an appropriate feature extraction technique, we studied several feature extraction techniques such as Scale Invariant Feature Transform (SIFT) [8–10], Speeded Up Robust Feature (SURF) [9] and Principle Component Analysis-SIFT (PCA-SIFT) [8]. Based on our study, we have identified following parameters to evaluate the existing techniques: methodology, feature extraction speed, scale invariance performance, rotation invariance performance, illumination invariance performance and performance under viewpoint changes. The parametric evaluation of feature extraction techniques is presented in Table 1.

Methodology specifies method used to extract features from the image. Feature extraction speed refers to the computational speed of the feature extraction

Table 1 Parametric evaluation of feature extraction techniques

	SIFT	PCA-SIFT	SURF
Methodology	Construction of image pyramid	Principle component analysis	Hessian detectors
Feature extraction speed	Slow	Fast	Faster
Scale invariance performance	High	Average	High
Rotation invariance performance	High	High	High
Illumination invariance performance	Low	Low	Average
Performance under viewpoint changes	Average	Average	High

technique. Scale invariance performance describes the accuracy of feature extraction technique when images having different scales are considered for mining. Rotation invariance performance refers to the accuracy of feature extraction technique when different rotations of an image are considered for recognition. Illumination invariance performance refers to the accuracy of technique to detect objects from multiple images captured under varying lightening conditions. Viewpoint change refers to identification of objects from images captured from different viewpoints.

It is observed that amongst all feature extraction techniques, SURF has higher computational speed and gives better performance of recognition for varying values of scale, rotation, occlusion, clutter, illumination and viewpoint changes. Therefore, we have chosen SURF for feature extraction in the proposed algorithm. SURF is faster because convolution procedure is applied using box filters [8–10]. Moreover, it is applied on integral images. Major steps of SURF are interest point detection, creation of descriptor associated with each interest point and descriptor matching [9]. SURF creates stack of images without down sampling at higher levels and hence, all images in stack have same resolution [8, 9].

3.3 Classification

Classification of objects is carried out by classifier using either supervised learning or unsupervised learning [11]. Supervised learning consists of two major steps: training and testing. We use supervised learning approach for classification because it makes object identification easier via training of objects. Several classifiers based on supervised learning are available in the literature.

The number of misclassification depends on classifier used in the algorithm. Hence, to choose an appropriate classifier, we have studied various classifiers such as Naïve bias [11, 17], Iterative Dichotomiser3 (ID3) [11], Support Vector Machine (SVM) [11–13] and K Nearest Neighbor (KNN) [11, 12]. In Table 2, we present

Table 2 Parametric evaluation of classifiers

	ID3	Naïve Bias	SVM	KNN
Methodology	Decision tree	Bayes Theorem	Construction of hyper plane	Nearest neighbor
Classification accuracy	Less	More	More	More
Memory usage	More	More	Less	Less
Computation speed	Slow	Fast	Fast	Slow
Misclassification	More	Less	Less	Less
Over-fitting	More	Less	Less	More

parametric evaluation of classifiers based on the following identified parameters: methodology, classification accuracy, memory usage, computation speed, misclassification in presence of noise and over-fitting.

Methodology specifies method used for classification. Classification accuracy refers to how accurately the classifier classifies the objects. Memory usage specifies storage requirement of the classifier. Computational speed refers to the speed of classifier. Misclassification refers to classification of objects into wrong class. As number of misclassification increases, classification accuracy decreases. In the presence of noise, the data may be misclassified. Over-fitting refers to the random error or noise instead of underlying relationship [11].

Based on our study on existing classifiers, we have chosen SVM classifier because it classifies objects with higher accuracy. Moreover, amongst several classifiers, its misclassification rate is less. However, it often misclassifies occluded objects or partially visible objects of image. Therefore, to improve the accuracy of object classification under occlusion, we use ADABOOST. ADABOOST selects strong features from the feature vector generated as an output of feature extraction. The selected strong features are used to train images. Training improves accuracy of object identification.

Figure 2 shows the pseudo code of the proposed FOMTA. Set A is a set of n input images $I_1, I_2, I_3, I_4, \ldots, I_n$. Set B is a set of n output images $I'_1, I'_2, I'_3, I'_4, \ldots, I'_n$. Output images are the images with tagged objects. $FV = \{F_1, F_2, F_3, F_4, \ldots, F_k\}$ is a Feature Vector containing all features of an input image and F_j is a jth feature of an image where $1 \leq j \leq k$. FVS is a Feature Vector containing strong features of an image selected by ADABOOST.

Initially, we input set A of images to FOMTA. On each input image, four major steps are performed as follows: First, noise is removed from input image I_j using gaussian filter. After noise removal, features are extracted using SURF feature extraction technique and a feature vector FV_j containing all extracted features is generated. Subsequently, we apply ADABOOST that selects strong features from feature vector. The strong features are stored in FVS_j. Finally, SVM classifier classifies as well as tags the objects and output image I'_j is produced.

Fig. 2 Pseudo code of FOMTA

```
INPUT : Set A = {I_1, I_2, I_3, I_4, ……., I_n}
OUTPUT: Set B ={I'_1, I'_2, I'_3, I'_4, ……..., I'_n}

FOMTA (A)
  BEGIN
  B = Φ
  for j=1 to n images do
        Remove noise from I_j using gaussian filter
        FV_j = Feature_Extraction_SURF(I_j)
        FVS_j = Boosting_ADABOOST(FV_j)
        I'_j = Classifier_SVM(FVS_j)
        B=B U I'_j
  END
```

4 Conclusion

In this paper, we have proposed a new Feature based Object Mining and Tagging Algorithm. The proposed algorithm increases probability of object identification under occlusion and clutter. Moreover, it decreases misclassification rate. We have also presented parametric evaluation of existing feature extraction techniques and existing classifiers to choose an appropriate feature extraction technique and classifier respectively to use in the proposed algorithm. In future, we aim to evaluate the performance of FOMTA for images captured under varying conditions such as scale, rotation, illumination, clutter, occlusion and to compare the performance of FOMTA with existing object mining technique.

References

1. Zhao, G., Yuan, J.: Mining and Cropping Common Objects from images. In: International Conference on Multimedia, pp. 975–978. ACM Digital Library (2010).
2. Khan, I., Khan, A., Shaikh, R.A.: Object analysis in image mining. In: Computing for Sustainable Global Development. pp. 1985–1988 (2015).
3. Bane, S., Pawar, D.R.: Survey on Feature Extraction methods in Object Recognition. In: International Journal of Computer Science and Information Technologies, vol, 5, pp. 3224–3226 (2014).
4. Annasaro, H.E.: A Survey in need of Image Mining Techniques. In: International Journal of Advanced Research in Computer and Communication Engineering, vol. 2, no. 2. (2013).
5. Fatma, S.N., Nashipudimath, M.: Image mining using association rule. In: Information and Communication Technologies, pp. 587–593 (2011).
6. Hsiao, J.H., Chen, C.S., Chen, M.S.: A Novel Language-Model-Based Approach for Image Object Mining and Re-ranking. In: 8th IEEE International Conference on Data Mining, pp. 243–252 (2008).
7. Liu, H., Yan, S.: Common visual pattern discovery via spatially coherent correspondences. In: IEEE Conference on Computer Vision and Pattern Recognition, pp. 1609–1616 (2010).

8. Juan, L., Gwun, O.: A Comparison of SIFT, PCA-SIFT and SURF. In: International Journal of Image Processing, vol. 3, no. 4, pp. 143–152 (2009).
9. Oyallon, E., Rabin, J.: An analysis and implementation of the SURF method and its comparison to SIFT. In: Image Processing Online, vol. 5, pp. 176–218 (2013).
10. Rublee, E., Rabaud, V., Konolige, K., Bradski, G.: ORB: An efficient alternative to SIFT or SURF. In: IEEE International Conference on Computer Vision, pp. 2564–2571 (2011).
11. Phyu, T.N.: Survey of Classification Techniques in Data Mining. In: International Conference of Engineers and Computer Scientists, vol. 1 (2009).
12. Zhang, H., Berg, A.C., Maire, M., Malik, J.: SVM-KNN: Discriminative Nearest Neighbor Classification for Visual Category Recognition. In: IEEE Computer Society Conference on Computer Vision and Pattern Recognition, vol. 2, pp. 2126–2136 (2006).
13. Goh, K.S., Chang, E., Cheng, K.T.: SVM Binary Classifier Ensembles for Image Classification. In: 10th International Conference on Information and Knowledge Management, pp. 395–402 (2001).
14. Philbin, J., Zisserman, A.: Object Mining Using a Matching Graph on Very Large Image Collections. In: 6th Indian Conference on Computer Vision, Graphics and Image Processing, pp. 738–745 (2008).
15. Shahbaz, M., Guergachi, A., Noreen, A., Shaheen, M.: Classification by Object Recognition in Satellite Images by using Data Mining. In: Proceedings of the World Congress on Engineering, vol. 1 (2012).
16. Russell, B.C., Freeman, W.T., Efros, A., Sivic, J., Zisserman, A.: Using Multiple Segmentations to Discover Objects and their Extent in Image Collections. In: IEEE Computer Society Conference on Computer Vision and Pattern Recognition, vol. 2, pp. 1605–1614 (2006).
17. Shi, X., Manduchi, R.: A Study on Bayes Feature Fusion for Image Classification. In: Conference on Computer Vision and Pattern Recognition, vol. 8, pp. 95–95 (2003).

Exploratory Assessment Based Child Nodes Selection (EACNS): Energy Efficient Multicast Routing Topology for Mobile Ad Hoc Networks

N. Papanna, A. Rama Mohan Reddy and M. Seetha

Abstract The networks formed by the self-energized nodes that loosely coupled with no or fewer infrastructure and without a central monitoring system is said to be ad hoc networks and if mobility is the property of these nodes then that networks are referred as mobile ad hoc networks. The nodes in these networks are having limited range to couple with other nodes. Hence in order to transmit data, they establish a route between source and destination nodes through hop level nodes. The critical point of the route establishment is selection of optimal neighbor nodes in the context of Quality of Service. This is much critical if data to be transferred to multiple destinations, which can be referred as multicast routing. In order to optimize the process of neighbor node selection in multicast route establishment, here in this article we proposed an energy efficient multicast routing topology, which is based on Exploratory Assessment based Child Nodes Selection (EACNS). EACNS is a Tree-based multicast routing topology for mobile ad hoc networks distinguishing from others in its class by the defining an exploratory scale to assess the optimal child nodes towards minimal energy usage and maximal network life. The objective the model is to be loosely coupled route selection strategy to any of the benchmarking tree based multicast routing protocols explored in literature. The strategy that used to select tree based multicast route is an explorative scale that defined based on three metrics coined in this paper, which are called energy consumption ratio, reserve battery life and multicast scope. The experimental results concluding that the proposed topology is the best of in its class to minimize the energy usage and maximize the network life.

N. Papanna (✉)
Department of CSE, Sree Vidyanikethan Engineering College, Tirupati, India
e-mail: n.papanname@gmail.com

A. Rama Mohan Reddy
Department of CSE, S.V University College of Engineering, Tirupati, India
e-mail: ramamohansvu@yahoo.com

M. Seetha
Department of CSE, GNITS, Hyderabad, India
e-mail: smaddala2000@yahoo.com

N. Modi et al. (eds.), *Proceedings of International Conference on Communication and Networks*, Advances in Intelligent Systems and Computing 508,
DOI 10.1007/978-981-10-2750-5_37

Keywords MANET · Multicast routing · Energy consumption · Multicast scope · Residual energy · Energy efficient

1 Introduction

The activity of data transmission from a source node to multiple sink nodes is referred as multicast routing [1]. The source should transmit data to all of the selected sinks under single destination address [2, 3]. The traditional route discovery constraint shortest path, which is the objective of many of benchmarking algorithms [4]. In order to find shortest path between sources to multiple destinations, the route is formed by more than one tree, which magnifies the routing complexity [5]. Further that leads to extreme energy usage by the nodes in the multicast route that causes minimal life time of the network and poor quality routing in the context of quality metrics. The other significant challenge is to define multicast routing topologies for Mobile Ad hoc Network [6]. Henceforth current research involved in defining novel multicast routing topologies under multiple QoS constraints. One of that is an energy efficient multicast model.

The majority of the QoS aware multicast routing protocols [7] under multiple QoS constraints are not including the energy usage towards transmission as QoS constraint. Hence the nodes with energy consumption overhead involved in path exhaust early and destruct the route [8]. This causes multicast routing cannot be continued until the completion of data transmission. The objective of our proposal is to magnify the network route lifetime. In order to this an energy efficient routing topology is proposed.

2 Related Work

Reducing the overhead of acknowledgement process is the prime objective of the model devised by Paul et al. [9]. In order to achieve this, multilevel hierarchy and selective repeat transmission were used. Dynamic transmission range towards minimal energy usage is observed in Dynamic Ring based Multicast Routing protocol proposed by Zhou [10]. Transmission range adjusts by the process called Extending Ring Search, which major contribution of the proposal. The experiments were evidencing the scalability in multicast routing. But the model is not optimal to high mobility networks. The multicast routing model explored by Yang et al. [11] is based on stateless geographic routing and broadcasting, which also aimed to handle the virtual sinks. The routing loops and packet duplications are surpassed in the model devised by Kim et al. [12]. The experiments evincing that the packet forwarding strategy that proposed is adaptive to Tree based protocol than mesh based multicasting. The context of ad hoc networks with high mobility is the context of model devised by Vaidya et al. [13]. The disjoint path aware reliable and secure routing was concluded in this model. Aamir et al. [14] proposed an active queue management strategy towards effective buffer space sharing between neighbour nodes. The other multicast routing protocols [15, 16] are also considerable contributions. Kashihara et al. [17] proposed a multicast routing model that relies in neighbor node's information such as their location and probability of transmission to achieve the opportunistic routing and encoding in order to claim reliability. Yang et al. [18] proposed a QoS centric token bucket based MAC scheme, which is in the aim of handling contention. Sriniva et al. [19] enhanced the AODV to achieve stability and energy aware routing in mobile ad hoc networks. The assessment of signal strength is the key factor for this model. According to the receiving signal strength, the link stability is assessed and the product of the stability of the links involved in a route concludes the route stability. Gupta et al. [20] were proposed location aided routing model towards minimizing the energy utilization of nodes involved in routing. Biradar et al. [21] proposed ring mesh based multicast routing protocol that measures the reliability of the connection between any two nodes by their mobility, reserve battery and signal strength.

All of these models are aimed to maximize the network life span by identifying the nodes with sufficient energy resources or minimizing the wastage of battery life through stable routing. The context of these models is to identify route with nodes having sufficient energy resources to achieve the stable routing. But the objective of notifying a stable route with nodes using minimal energy in data transmission. In the context of this objective, the models [17, 21] are aimed to identify the neighbor nodes which can be paired under minimal energy usage. The objective of our proposal is also of the same dimension, which is to establish an optimal tree based stable multicast route with nodes that are using minimal energy to transmit data.

3 Energy Efficient Multicast Routing Topology for Mobile Ad Hoc Networks

The energy efficient multicast routing topology proposed is using the devised exploratory scale to assess optimality of the nodes towards route selection. In order to this, initial move is to identify all possible routes between source node s and set of destination nodes $D = \{d_1, d_2, d_3, \ldots d_{|D|}\}$, which is done by the conditional broadcasting of the route request [19]. The respective destination nodes responds back with route response to each route request those received. Further, upon receiving all route responses sent by destination nodes, the source node collects all possible paths between source node s and set of destination nodes D.

3.1 Energy Efficient Multicast Route Discovery

The nodes involved in the possible paths will be organized further as set of groups. The nodes those are connected directly to the nodes in D are all grouped as Egress Level to D ($EL \rightarrow \{D\}$), which will be referred further as L_1.

Then the optimal nodes from the group L_1 will be selected by the proposed exploratory scale, such that there will be a connection from L_1 to all nodes in D. The D is now successor level $sl(L_1)$ to all optimal nodes in level L_1.

Then the predecessor nodes those are directly connected to the nodes of L_1 in possible routes are all grouped as Egress Level to L_1 ($EL \rightarrow \{L_1\}$) that further referred as L_2.

Then the optimal nodes from the group L_2 will be selected by the proposed exploratory scale, such that there will be a connection from L_2 to all optimal nodes in L_1. The L_1 is now successor level $sl(L_2)$ to all optimal nodes in level L_2.

The similar process continues till successor nodes of the source node s grouped as L_n that represents $EL \rightarrow \{L_{n-1}\}$ and selects optimal nodes from L_n by exploratory scale, such that there will be a connection from L_n to all optimal nodes in L_{n-1}. The L_{n-1} is successor level $sl(L_n)$ to all optimal nodes in L_n and L_n is successor level $sl(s)$ to source node s.

4 Exploratory Assessment Based Child Node Selection

The Metrics used in exploratory scale devised to select optimal nodes in each level of set $RL = \{L_1, L_2, \ldots .L_{n-1}, L_n\}$ are:

Energy Consumption Ratio: This metric represents the average energy consumption per unit of data transmission at an egress node. This metric is the average of energy consumed per unit of transmission between nodes to all its connected nodes in successor level, which is optimal with minimal values.

Reserved Battery Life: The other key metric that defines the life time of a node involved in routing. This metric aggregates the battery life required for routing and battery life consumption in idle time and consumption due to other factors as max battery consumption (*mbc*), then subtracts *mbc* from the estimated battery life (*ebl*). The resultant value must be the positive and greater than the given threshold.

Multicast Scope: This represents the number of possible neighbour nodes in its successor level. The energy consumption ratio and max battery Life must be optimal while considering a neighbor node in successor level.

The prime objective of the EACNS is the energy efficiency, hence the metric called energy consumption ratio is prime factor that followed by the related metric Max Battery Life and then the Multicast scope will be considered. The other important constraint to select a node is that it must not have max battery life as negative value.

4.1 Assessing Energy Consumption Ratio

For each level $\{L_i \exists L_i \in RL\}$ Begin

$ec(L_i) \leftarrow \phi$ // is vector contains the energy consumption ratio of all nodes in level L_i

For each Node $\{snd \exists snd \in L_i\}$ Begin

$ec(snd) \leftarrow \varphi$ //a vector that represents energy consumption to transmit a frame to all possible neighbor nodes in $sl(L_i)$

For each neighbor node $\{nnd \exists nnd \in sl(L_i)\}$ Begin

$$ec_{snd \to nnd} = \left(\rho^{\left(\frac{d_{snd \to nnd}}{ud}\right) \otimes \lambda} \otimes \varepsilon \right) + (\varepsilon' \otimes \lambda) \text{ ......(Eq1)}$$

//Here in Eq. 1

- The outcome $ec_{snd \to nnd}$ is the energy consumed to transmit a frame snd to nnd
- $d_{snd \to nnd}$ is the Euclidian distance between snd to nnd
- ud is unit of distance can be traveled by a frame under frequency ρ
- The desired frequency is exponential of the number distance units [19]
- ε is energy required to transmit a frame under ρ frequency
- λ is the transmission loss exponent
- ε' is the energy consumption due to the overhead of other factors [20]

$ec(snd) \leftarrow ec_{snd \to nnd}$

// energy consumption between snd and nnd pushed to vector $ec(snd)$

$ec(L_i) \leftarrow \langle ec(snd) \rangle$

// average of energy consumption between snd and all nodes in $sl(L_i)$ is pushed to vector $ec(L_i)$

End

End

End

4.2 Assessing Max Battery Consumption

For each level $\{L_i \exists L_i \in RL\}$ Begin

For each Node $\{snd \exists snd \in L_i\}$ Begin

$aec = \langle ec(snd) \rangle$ //Average of energy consumption between snd and all possible neighbor nodes from $sl(L_i)$

End

$eer = aec \otimes fc$ //estimated energy required to transmit maximum number of frames fc

$mbc_{snd} = (eer + \varepsilon_{it} + \varepsilon^{'})$ ......(Eq2)

// Here in (Eq2)

- ➢ The required battery life $(eer + \varepsilon_{it} + \varepsilon^{'})$ for each connected neighbor node in $sl(L_i)$
- ➢ ε_{it} is estimated energy consumed at idle time of the node during the transmitting fc number of frames
- ➢ $\varepsilon^{'}$ is estimated energy used for other constraints during the transmitting fc number of frames

End

4.3 Assessing Multicast Scope and Reserved Battery Life

Loop 1: For each level $\{L_i \exists L_i \in RL\}$ Begin

$rbl(L_i) \leftarrow \varphi$ //A vector contains the reserve battery life of all nodes in L_i

$nnl(L_i) \leftarrow \varphi$ // A vector that contains the size of all possible neighbor nodes from $sl(L_i)$ connected to each node $\{snd \exists snd \in L_i\}$ under the given metric constraints

Loop 2: For each Node $\{snd \exists snd \in L_i\}$ Begin

$nnl_{snd} \leftarrow \varphi$

// a vector that contains all possible neighbor nodes from $sl(L_i)$ connected to snd under the given metric constraints

$mer_{snd} = 0$

//max energy used by snd initialized with 0

$tec(snd) \leftarrow ec(snd)$

// clone $ec(snd)$ as $tec(snd)$

Loop 3: For each node

$\{nnd \exists nnd \in sl(L_i) \wedge ec_{snd \to nnd} \in tec(snd) \wedge ec_{snd \to nnd} \equiv \min(tec(snd))\}$

//The condition $nnd \in sl(L_i)$ indicates that nnd must belongs to successor level to L_i

//The condition $ec_{snd \to nnd} \in tec(snd)$ represents that the minimum energy conserved by the route between snd and nnd must be available

//The condition $ec_{snd \to nnd} \equiv \min(tec(snd))$ represents that the energy conserved by the route between snd and nnd must be smallest of the set $tec(snd)$

Begin

$$tec(snd) \setminus ec_{snd \to nnd}$$

//discarding element $ec_{snd \to nnd}$ from set $tec(snd)$

$nnl_{snd} \leftarrow nnd$ // add snd to multicast neighbor nodes list nnl_{snd}

$mer = |nnl_{snd}| \otimes mbc_{snd}$

$if(ebl_{snd} - mer_{snd} \geq \tau_{mbl})$ Begin

// ebl_{snd} is estimated battery life of the node snd

// τ_{mbl} is the max battery life threshold given

$nnl_{snd} \setminus nnd$ // discard nnd from nnl

$rbl_{snd} = ebl_{snd} - (mbc_{snd} \otimes |nnl_{snd}|)$ // Reserved Battery Life rbl_{snd} is assessed

$rbl(L_i) \leftarrow rbl_{snd}$

$nnl(L_i) \leftarrow |nnl_{snd}|$

Break the loop in statement 3

End

End

End

End

4.4 Optimal Node Selection

This section explores the process of optimal nodes selection by the metrics defined in earlier sections. The process of optimal nodes selection of the each level $\{L_i \exists L_i \in RL\}$ as follows.

4.5 Normalizing the Metric Values

The initial move is normalizing [22] the metric values, since the metric energy consumption is optimal with minimal values but Reserve Battery Life and Multicast scope is optimal with maximal values. The normalization is done as follows.

The values of all these metrics should be normalized to the same order. For each metric, all available values will be normalized to the value between 0 and 1 such that 0 represents the most insignificant value and the 1 represents the most significant value, the rest of the values will be in between 0 and 1. The process is as follows:

1. The least value of the metric available will be subtracted from the each value and then the result will be division by the difference of the highest lowest values of that metric given.
2. Further, if that metric is significant with minimal value then the result of the above step (step i) will be subtracted from 1 and the result will be considered as the normal form of the respective QoS metric value.

4.5.1 Normalizing Energy Consumption Metric Value

For each level $\{L_i \exists L_i \in RL\}$ Begin

$nec(L_i) \leftarrow \phi$ // a vector that contains normalized values of the energy consumption metric for all nodes in level $\{L_i \exists L_i \in RL\}$

For each node $\{\langle ec(snd) \rangle \exists \langle ec(snd) \rangle \in ec(L_i)\}$ Begin

// average energy consumption between snd and all nodes in $sl(L_i)$

$$nec_{snd} = 1 - \left(\frac{\langle ec(snd) \rangle - \min(ec(L_i))}{\max(ec(L_i)) - \min(ec(L_i))} \right)$$

//The process of normalizing is, theminimum of the $ec(L_i)$ is subtracted from the each value of the entry $ec(L_i)$ then divides by the difference of the max and min values of the $ec(L_i)$, Every value of The resultant value is subtracted from the 1, which is since the minimal values are optimal for energy consumption metric

$nec(L_i) \leftarrow nec_{snd}$

End

End

4.5.2 Normalizing Reserve Battery Life Metric Value

For each level $\{L_i \exists L_i \in RL\}$ Begin

$nrbl(L_i) \leftarrow \phi$ // a vector that contains normalized values of the reserve battery life metric for all nodes in level $\{L_i \exists L_i \in RL\}$

For each node's reserve battery life $\{rbl_{snd} \exists rbl_{snd} \in rbl(L_i)\}$ Begin

// reserve battery life of the node snd L_i

$$nrbl_{snd} = \left(\frac{rbl_{snd} - \min(rbl(L_i))}{\max(rbl(L_i)) - \min(rbl(L_i))} \right)$$

//The process of normalizing is, the minimum $\min(rbl(L_i))$ *is subtracted from the each value of the entry in* $rbl(L_i)$ *then divides by the difference of the max and min values of the* $rbl(L_i)$.

$nrbl(L_i) \leftarrow nrbl_{snd}$

End

End

4.5.3 Normalizing Multicast Metric Value

For each level $\{L_i \exists L_i \in RL\}$ Begin

$nmc(L_i) \leftarrow \phi$ // a vector that contains normalized values of the multicast scope metric for all nodes in level $\{L_i \exists L_i \in RL\}$

For each node's multicast scope $\{| nnl_{snd} | \exists | nnl_{snd} | \in nnl(L_i)\}$ Begin

// multicasting scope of the node snd in L_i

$$nmc_{snd} = \left(\frac{nnl_{snd} - \min(nnl(L_i))}{\max(nnl(L_i)) - \min(nnl(L_i))} \right)$$

//The process of normalizing is, the minimum $\min(nnl(L_i))$ is subtracted from the each value of the entry in $nnl(L_i)$ then divides by the difference of the max and min values of the $nnl(L_i)$.

$nmcl(L_i) \leftarrow nmc_{snd}$

End

End

4.5.4 Ordering and Selecting the Optimal Nodes

Further, for each level $\{L_i \exists L_i \in RL\}$, all the nodes will be ranked differently for different metrics in descending order of values of their metric.

The node ranks given by their metric values may vary for different metrics. Hence, further we assess the covariance between the ranks given to the nodes under different metric values.

Afterward Nodes will be ordered in ascending by the rank given to prime metric called energy consumption and in second level the nodes will be ordered in ascending by their covariance.

Then select minimum nodes as a list $onl(L_1)$ from L_1 in same order formed (by step 0) such that there will be at least one node $\{o \exists o \in onl(L_1)\}$ for each destination node $\{d_i \exists d_i \in D\}$

Then select minimum nodes as a list $onl(L_2)$ from L_2 in same order formed (by step 0) such that there will be at least one node $\{o \exists o \in onl(L_2)\}$ for each optimal node $\{ol \exists ol \in onl(L_1)\}$

The step 0 is recursive for all other levels in *RL*.

4.5.5 Forming the Multicast Route

Further establish the route between source node *s* and for each node $\{ol \exists ol \in onl(L_n)\} onl(L_n)$

// between source to nodes in $onl(L_n)$

For each $i = \{n, n-1, n-2, \ldots\ldots\ldots, 3, 2\}$ Begin

Establish routes between nodes of $onl(L_i)$ to $onl(L_{i-1})$ //between nodes of any two levels in sequence

End

Further, Establish routes between nodes of $onl(L_1)$ to *D*

// between first level optimal nodes to all destination nodes *D*

5 Empirical Analysis and Results Exploration

The experimental setup was simulated using NS2, the parameters and constrictions of the simulation explored in Table 1. The quality of service metrics of each node initialized through randomly distributed values under Gaussian distribution strategy. The route discovery between source and destinations were done using MAODV [23]. Further to obtain the energy aware routes from the discovered routes was done by EACNS, which was implemented in expression language R. The performance of the proposed model was explored by simulation build by number of nodes in the range of 30–240.

The empirical study evincing that the proposed energy efficient multicast routing topology EACNS is scalable and robust. The metrics end-to-end delay, packet

Table 1 Network constraints for simulation environment

Number of nodes involved	30–240
Lower and upper bounds of the mobility	Between 0.2 m and 2.4 m/s
MAC specification	MAC 802.11 DCF
Area of network spanned	1500 × 2500 m^2
Transmission range of a node	50 m
Transmission strategy	CBR
Simulation time	In the range of 120–360 s

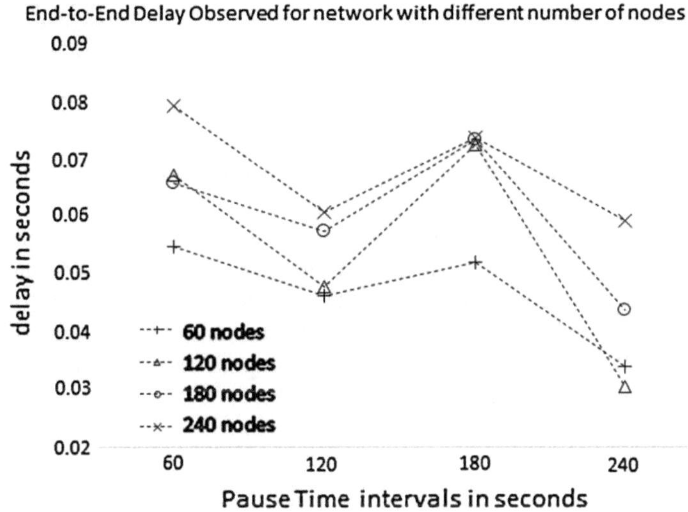

Fig. 1 End-to-End Delay observed at different intervals of the simulation

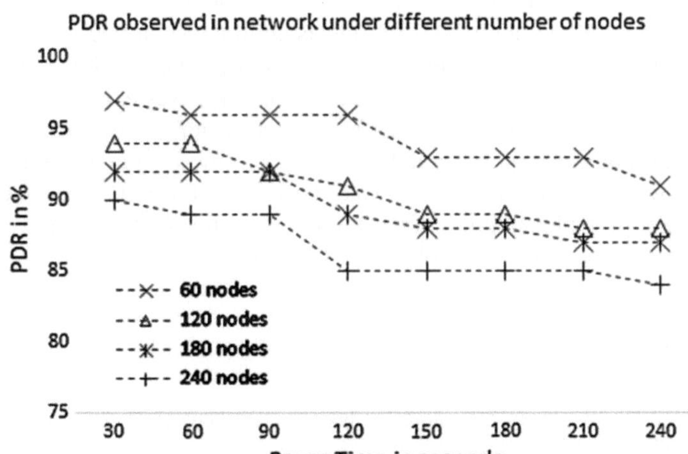

Fig. 2 PDR observed on networks build with different number of nodes

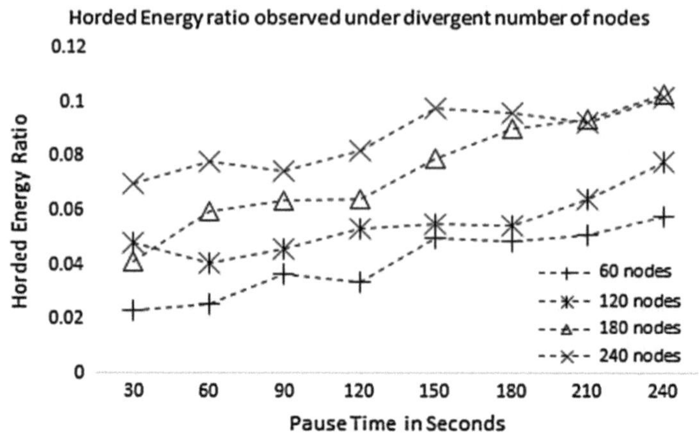

Fig. 3 Energy saving ratio (hoarded energy ratio) observed at different simulation intervals

delivery ratio and hoarded energy ratio were used in order to assess the performance of the model proposed. Metric values were observed at different simulation pause times on network build with number of nodes 60, 180, 240 respectively. The end-to end delay is considerable low (see Fig. 1), and packet delivery ratio is best in its class (see Fig. 2). The main objective of the proposal is minimizing the energy usage, which is convincingly proved (see Fig. 3).

6 Conclusion

We proposed an Exploratory Assessment based Child Node Selection (EACNS) topology for tree based energy efficient multicast routing in mobile ad hoc networks. The proposed topology EACNS selects child nodes in tree structure, which is based on the three crucial metrics called energy consumption ratio, reserve battery life and multicast scope. The impact of topology was explored by applying on the routes discovered by MAODV. The experimental results are evincing the scalability and robustness of the model proposed. The motivation given these efforts can lead us to redefine the proposed exploratory scale for high speed mobile networks. Another possible direction of the future research is to define an evolutionary strategy that uses the EACNS as objective function.

References

1. S. Sesay, Z. Y., A survey on mobile ad hoc wireless network. Information Technology Journal, 2004, 168–175.
2. D.P. Agrawal, Q. Z., Introduction to wireless and mobile systems. CA: Brooks/ Cole Publishing, 2003.

3. Luo Junhai, Y. D., Research on topology discovery for IPv6 networks. IEEE, SNPD, 2007, pp. 804–809.
4. Yang, S. C., Genetic algorithms with immigrants and memory schemes for dynamic shortest path routing problems in mobile ad hoc networks. Applications and Reviews, IEEE Transactions, 2010, pp. 52–63.
5. Huang, J., MOEAQ: a QoS-aware multicast routing algorithm for MANET. Expert Systems with Applications, 2010, pp. 1391–1399.
6. Toumpis, S., Wireless ad-hoc networks. Vienna Sarnoff Symposium,. Vienna: Telecommunications Research Center, 2004.
7. A.T. Haghighat, K. F., GA-based heuristic algorithms for QoS based multicast routing. Journal of Knowledge-Based Systems, 2003, pp. 305–312.
8. B. Wang, S. G., On maximizing lifetime of multicast trees in wireless ad hoc networks. IEEE International Conference on Parallel Processing, 2003, pp. 333–340.
9. Paul S, S. K.-H., Reliable Multicast Transport Protocol (RMTP). IEEE Journal on Selected Areas in Communications, 1997, pp. 407–421.
10. Zhou Y, L. G.-Z.,. DRMR: Dynamic Ring based Multicast Routing Protocol for Ad hoc Networks. Journal of Computer Science and Technology, 2004, pp. 909–919.
11. Yang S, Y. C., Toward reliable data delivery for highly dynamic mobile ad hoc networks. IEEE Transactions on Mobile Computing, 2012, pp. 111–124.
12. Kim Y, A. S., An efficient multicast data forwarding scheme for mobile ad hoc networks. Springer Verlang, 2005, pp. 510–518.
13. Vaidya B, Y. S., Robust and secure routing scheme for wireless multihop network. Personal and Ubiquitous Computing, 2009, pp. 457–469.
14. Aamir M, Z. M., A buffer management scheme for packet queues in MANET. Tsinghua Science and Technology, 2013, pp. 543–553.
15. Lou W, Z. W., SPREAD: Improving network security by multipath routing in mobile ad hoc networks. Wireless Networks, 2009, pp. 279–294.
16. Lee H-O, N. J.-H., Cluster and location based overlay multicast in mobile ad hoc and sensor networks. International Journal of Distributed Sensor Networks, 2014, pp. 1–11.
17. Kashihara S, H. T. (2014). Data delivery method based on neighbor nodes information in a mobile ad hoc network. The Scientific World Journal, 1-13.
18. Yang Y, W. Y. A MAC Scheme with QoS Guarantee for MANETs. International Journal of Communications, Network and System Sciences, 2009, pp. 759–763.
19. Sriniva, P.V., MECAR: Maximal Energy Conserved and Aware Routing in Ad hoc Networks, Information Systems Design and Intelligent Applications, 2015, pp., 855–864.
20. Gupta N, G. R., LAR-1: Affirmative influences on energy-conservation and network lifetime in MANET, International Journal of Computer Communication and control, 2014, pp. 284–291.
21. Biradar RC, M. S., Ring mesh based multicast routing scheme in MANET using bandwidth delay product. Wireless Personal Communication, 2012, pp. 117–146.
22. Bolstad, B. M., A comparison of normalization methods for high density oligonucleotide array data based on variance and bias. Bioinformatics, 2003, pp. 185–193.
23. Zhu, Y. &., MAODV implementation for NS-2.26. Systems and Computing Engineering, Carleton University, 2004.
24. Lu, T., Genetic algorithm for energy-efficient QoS multicast routing. Communications Letters IEEE, 2013, pp. 31–34.

Improved EAACK to Overcome Attacks in MANET and Wireless Sensor Networks

Pranjali Deepak Nikam

Abstract Remote sensor system is a situated of dispersed sensor hubs which are haphazardly conveyed in land zone to catch climatic changes like temperature, mugginess and weight. In Wireless Network MANET is a Mobile Ad Hoc Networks which is one self-configurable system. MANET is a gathering of Wireless versatile hub which is alter-ably moves starting with one area then onto the next area. Both assaults Active and in addition Passive assaults is in MANET. Security for remote system is much troublesome as contrast with wired systems. In most recent couple of years numerous security and assaults issue are face numerous scientists in MANET. Assaults like Packet dropping assault, Black-Hole assault, Denial of Service assault, wormhole assaults and Packet change assaults found in MANET. This paper proposes the overview of various types of assaults on MANET and Wireless sensor systems. This paper serves to youthful specialist to actualize new half and half calculation for secure interruption identification in MANET. It serves this purpose by using ECC with EAACK.

Keywords Mobile ad hoc networks · Wireless sensor networks · Intrusion detection system · ECC and EAACK

1 Introduction

Accumulation of Movable Mobile hub is called MANET. In Mobile Ad Hoc Network a solitary hub can be fill in as transmitter and recipient. MANET doesn't have a settled foundation. Versatile hubs can be move starting with one area then onto the next area for correspondence. Be that as it may, if there should arise an occurrence of MANET Network steering conventions assumes a vital part to figure out the most limited way and course in the middle of source and destination. Amid the information correspondence from source to destination some dynamic and

P.D. Nikam (✉)
Pune University, G. H. Raisoni Institute of Engineering & Technology, Pune, India
e-mail: pranjali.amore@gmail.com

N. Modi et al. (eds.), *Proceedings of International Conference on Communication and Networks*, Advances in Intelligent Systems and Computing 508,
DOI 10.1007/978-981-10-2750-5_38

inactive assaults ought to be get to the information. Dynamic and aloof assaults illicit access the information from the system. Dynamic assaults like spying and uninvolved assaults like disavowal of administration impact on MANET (Fig. 1).

2 Literature Survey

2.1 Watchdog

Marti et al. [9] proposed a plan named Watchdog that Watchdog plan neglects to detect malicious misconduct s with the vicinity of the accompanying: (1) vague impacts; (2) collector crashes; (3) constrained transmission control; (4) false mischief report; (5) conspiracy; and (6) partial dropping.

2.2 TWO-ACK

The working procedure of TWOACK is indicated in Fig. 2. Node A first advances Packet 1 to node B, and afterward, node B advances Packet 1 to node C. At the point when node C gets Packet 1, as it is two jumps far from node A, node C is obliged to produce a TWOACK parcel, which contains opposite course from node A to node C, and sends it back to node A. The recovery of this TWOACK bundle at node A demonstrates that the transmission of Packet 1 from node A to node C is effective. Something else, if this TWOACK parcel is not got in a predefined time period, both nodes B and C are accounted for noxious. The same procedure applies to each three sequential nodes along whatever is left of the course. The TWOACK plan effectively illuminates the collector crash and restricted transmission power issues postured by Watchdog. Be that as it may, the affirmation procedure needed in every bundle transmission procedure included a lot of undesirable system overhead.

Fig. 1 Manet organization

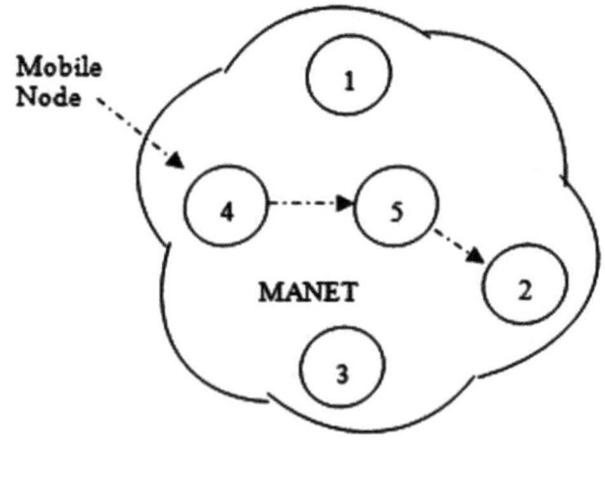

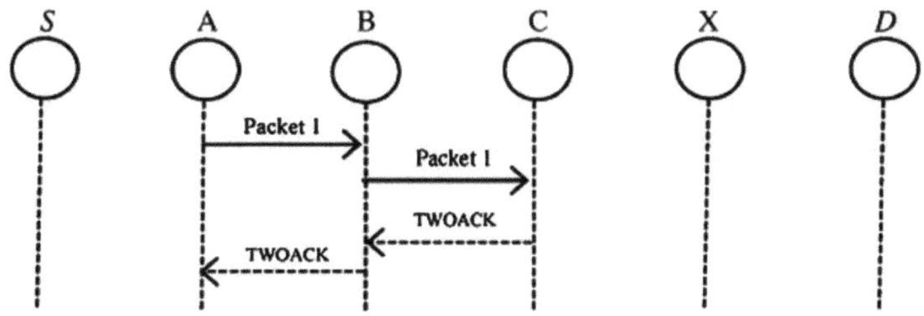

Fig. 2 TWO-ACK

2.3 *AACK*

AACK by Sheltami et al. [10] minimizes the time for data transmission over network while still keeping the network throughput constant.

2.4 *EAACK*

Hence, there is ambiguity if the acknowledgment packets are valid and authentic. Hence digital signature is introduced in Enhanced AACK (EAACK). EAACK overcomes some drawbacks of Watchdog scheme, namely, false misbehaviour, limited transmission power, and receiver collision. TWOACK and AACK overcome drawbacks, namely, receiver collision and limited transmission power.

2.4.1 ACK System

ACK is essentially an end-to-end affirmation plan. It goes about as a piece of the mixture conspires in EAACK.

2.4.2 S-ACK System

The S-ACK plan is an enhanced rendition of the TWOACK plan proposed by Liu et al. [13]. The guideline is to let each three back to back hubs work in a gathering to identify making trouble hubs.

2.4.3 MRA System

The MRA plan is intended to determine the shortcoming of Watchdog when it neglects to recognize getting out of hand hubs with the vicinity of false rowdiness report.

3 System Implementation

The proposed methodology is intended to handle three of the six shortcomings of Watchdog plan, to be specific false mischief, constrained transmission force and collector impact. In this segment, we portray our proposed plan in points of interest. In this work, we amplify it with the presentation of Elliptic Curve Based advanced mark to keep the aggressor from fashioning affirmation bundles. This venture is comprised of four noteworthy parts, specifically: ACKnowledge, (ACK), Secure-ACKnowledge (S-ACK) and Misbehavior Report Authentication (MRA) and light weight based Elliptic bend computerized mark and confirmation to keep the assailant from fashioning affirmation bundles and validate every hub as demonstrated in Fig. 3.

3.1 Key Generation

$Q = d * P'$
d = The random number that we have selected within the range of (1 to $n - 1$).
P (curve tip).
'Q' (public key) and 'd' (private key).

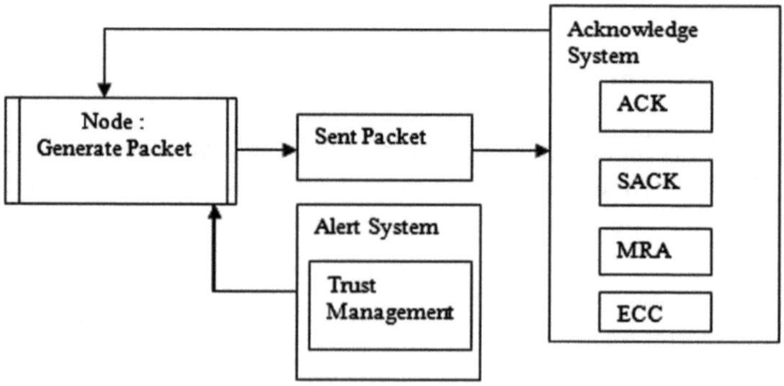

Fig. 3 System plan

3.2 Encryption

Two cipher texts will be generated let it be C1 and C2.
$C1 = k*P$
$C2 = M + k*Q$
C1 and C2 will be send.

3.3 Decryption

$M = C2 - d * C1$
M is the original message that we have send.

3.4 Proof

$M = C2 - d * C1$
'M' can be represented as '$C2 - d * C1$'
$C2 - d * C1 = (M + k * Q) - d * (k * P)$
($C2 = M + k * Q$ and $C1 = k * P$)
$= M + k * d * P - d * k *P$ (canceling out $k * d * P$)
= M (Original Message)

4 System Model

To execute the fancied recreation utilizing ns3 system test system the underneath given re-enactment parameters are needed. Our re-enactment is directed inside of the Network Simulator (NS) 3.20 environment on a stage with Fedora 19. The framework is running on a min 20 GB of HDD, 3-GB RAM and I3 processor.

5 Results

Our algorithm result is analyzed by following parameters:

1. Throughput assessment
2. End to End Delay

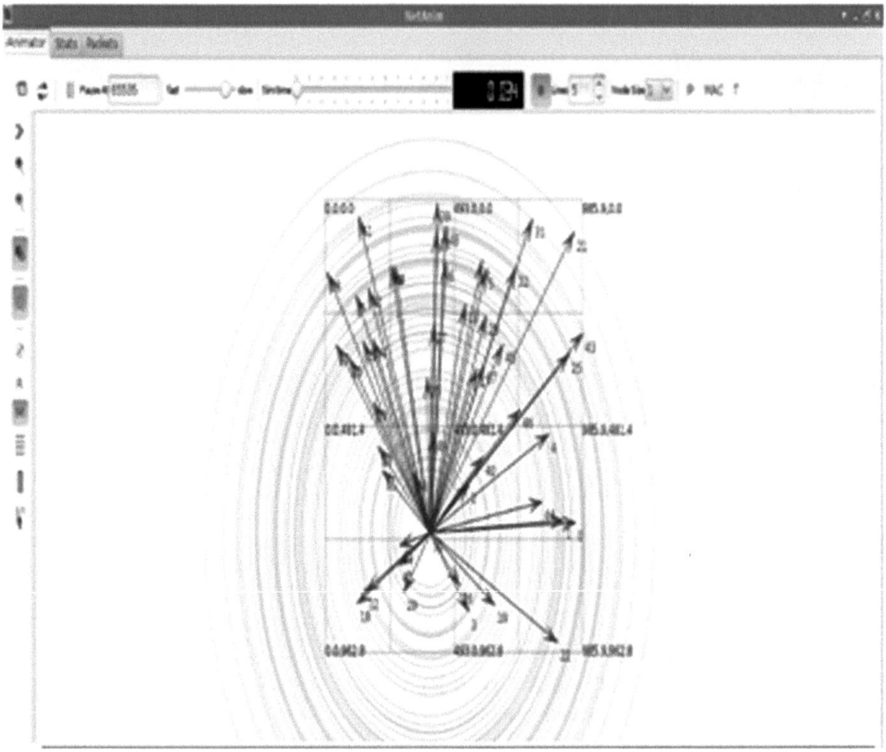

Fig. 4 Network simulation

We can observe that our proposed scheme outperforms EEACK in all test scenarios related to End-to-End delay. This is only because of introduction of strong scheme which is capable of reducing delay for every packet (Figs. 4, 5 and 6).

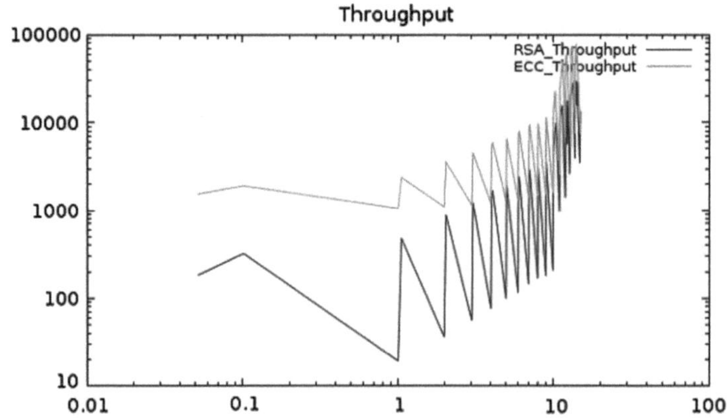

Fig. 5 Throughput analysis

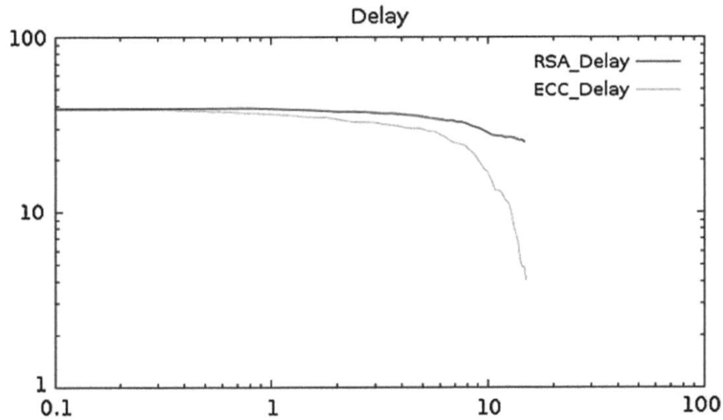

Fig. 6 Delay analysis

6 Conclusion

To change in accordance with the developing pattern of MANET in mechanical applications, it is fundamental to address its potential security issues. A noteworthy risk to security in MANET is packet dropping assault. EAACK (Enhanced Adaptive Acknowledgement) plan has been uncommonly composed that exhibits higher pernicious–behavior–location rates in specific situations. In EAACK Digital signature has additionally been utilized to keep the aggressors from starting produced affirmation assaults. In spite of the fact that EAACK defeats the issues of false rowdiness, restricted transmission force and beneficiary impact, it expands organize overhead because of utilization of digital signature. In this proposed framework Elliptic Curve Cryptography (ECC) is utilized to further lessen the system overhead brought on by digital signature.

7 Future Work

This algorithm does not work well with multipath routing. So in future work we attempt to actualize this algorithm for multipath directing.

References

1. Elhadi M. Shakshuki, Senior Member, IEEE, Nan Kang, and Tarek R. Sheltami, Member, IEEE "EAACK—A Secure Intrusion-Detection System for MANETs".
2. A. Tabesh and L. G. Frechette, "A low-power stand-alone adaptive circuit for harvesting energy from a piezoelectric micropower generator," IEEE Trans. Ind. Electron., vol. 57, no. 3, pp. 840–849, Mar. 2010.

3. M. Zapata and N. Asokan, "Securing ad hoc routing protocols," in Proc. ACM Workshop Wireless Secure. 2002, pp. 1–10.K. Elissa, "Title of paper if known," unpublished. L. Zhou and Z. Haas, "Securing ad-hoc networks," IEEE Netw., vol. 13, no. 6, pp. 24–30, Nov./Dec. 1999.
4. K. Al Agha, M.-H. Bertin, T. Dang, A. Guitton, P. Minet, T. Val, and J.-B. Viollet, "Which wireless technology for industrial wireless sensor networks? The development of OCARI technol," IEEE Trans. Ind. R. H. Akbani, S. Patel, and D. C. Jinwala, "DoS attacks in mobile ad networks: A survey," in Proc. 2nd Int. Meeting ACCT, Rohtak, Haryana, India, 2012, pp. 535–541.
5. T. Anantvalee and J. Wu, "A Survey on Intrusion Detection in Mobile Ad Hoc Networks," in Wireless/Mobile Security. New York: Springer Verlag, 2008.
6. L. Buttyan and J. P. Hubaux, Security and Cooperation in Wireless Networks. Cambridge, U. K.: Cambridge Univ. Press, Aug. 2007.
7. Nitin Goyal1, Alka Gaba2 "A review over MANET- Issues and Challenges" Dept. of computer science and engineering, JMIT, Radaur, India.
8. V. C. Gungor and G. P. Hancke, "Industrial wireless sensor networks: Challenges, design principles, and technical approach," IEEE Trans. In Electron., vol. 56, no. 10, pp. 4258–4265, Oct. 2009.
9. Y. Hu, D. Johnson, and A. Perrig, "SEAD: Secure efficient distance vector routing for mobile wireless ad hoc networks," in Proc. 4th IEEE Workshop Mobile Comput. Syst. Appl., 2002, pp. 3–13.
10. G. Jayakumar and G. Gopinath, "Ad hoc mobile wireless networks protocol—A review," J. Comput. Sci., vol. 3, no. 8, pp. 574–582, 2007.
11. D. Johnson and D. Maltz, "Dynamic Source Routing in ad hoc wireless networks," in Mobile Computing. Norwell, MA: Kluwer, 1996, ch. 5, pp. 153–181.
12. N. Kang, E. Shakshuki, and T. Sheltami, "Detecting misbehaving nodes in MANETs," in Proc. 12th Int. Conf. iiWAS, Paris, France, Nov. 8–10, 2010, pp. 216–222.
13. N. Kang, E. Shakshuki, and T. Sheltami, "Detecting forged acknowledgements in MANETs," in Proc. IEEE 25th Int. Conf. AINA, Biopolis, Singapore, Mar. 22–25, 2011, pp. 488–494.
14. V. Untz, M. Heusse, F. Rousseau, and A. Duda, "Lilith: an Interconnection Architecture Based on Label Switching for Spontaneous Edge Networks," Proc. First Ann. Int'l Conf. Mobile and Ubiquitous Systems: Networking and Services (Mobiquitous '04), Aug. 2004.
15. L.M. Feeney, B. Ahlgren, A. Westerlund, and A. Dunkels, "Spontnet: Ad Hoc Networks," Proc. Fifth Int'l Workshop Network Appliances, Oct. 2002.
16. M. Danzeisen, T. Braun, S. Winiker, D. Rodellar, "Implementation of a Cellular Framework for Spontaneous Network Establishment," Proc. IEEE Wireless Comm. and Networking Conf. (WCNC '05), Mar. 2005.
17. J. Rekimoto, "SyncTap: Synchronous User Operation for Spontaneous Network Connection," Personal and Ubiquitous Computing, vol. 8, no. 2, pp. 126–134, May 2004.
18. C. Cornelius, A. Kapadia, D. Kotz, D. Peebles, M. Shin, and N. Triandopoulos, "Anonysense: Privacy-Aware People-Centric Sensing," Proc. Sixth Int'l Conf. Mobile Systems, Applications, and Services (MobiSys '08), pp. 17–20, June 2008.
19. R. Lacuesta, J. Lloret, M. Garcia, and L. Pen˜ alver, "Two Secure and Energy-Saving Spontaneous Ad-Hoc Protocol for Wireless Mesh Client Networks," J. Network and Computer Applications, vol. 34, no. 2, pp. 492–505, Mar. 2011.
20. J. Sun, C. Zhang, Y. Zhang, and Y. (Michael) Fang, "An Identity- Based Security System for User Privacy in Vehicular Ad Hoc Networks," IEEE Trans. Parallel and Distributed Systems, vol. 21, no. 9, pp. 1227–1239, Sept. 2010.
21. FIPS 180-1 - Secure Hash Standard, SHA-1, "National Institute of Standards and Technology," http://www.itl.nist.gov/fipspubs/fip180-1.htm, Feb. 27, 2012.
22. A. Wander, N. Gura, H. Eberle, V. Gupta, and S. Chang, "Energy Analysis for Public-Key Cryptography for Wireless Sensor Networks," Proc. IEEE Third Int'l Conf. Pervasive Computing and Comm. (PerCom '05), pp. 8–12, Mar. 2005.

23. N.R. Potlapally, S. Ravi, A. Raghunathan, and N.K. Jha, "Analyzing the Energy Consumption of Security Protocols," Proc. Int'l Symp. Low Power Electronics and Design (ISLPED '03), 2003.
24. J. Goodman and A. Chandrakasan, "An Energy Efficient Reconfigurable Public-Key Cryptography Processor Architecture," Proc. Int'l Workshop Cryptographic Hardware and Embedded Systems (CHES '00), pp. 175–190, 2000.
25. S. Landau, "Communications Security for the Twenty-First Century: The Advanced Encryption Standard," Notices of the Am. Math. Soc., vol. 47, no. 4, pp 450–459, Apr. 2000.
26. J. Lo´pez and R. Dahab, "Performance of Elliptic Curve Cryptosystems," Technical Report IC-00-08, May 2000.
27. R. Mayrhofer, F. Ortner, A. Ferscha, and M. Hechinger, "Securing Passive Objects in Mobile Ad-Hoc Peer-to-Peer Networks," Electronic Notes in Theoretical Computer Science, vol. 85, no. 3, pp. 105–121, Aug. 2003.
28. S.E. Czerwinski, B.Y. Zhao, T.D. Hodes, A.D. Joseph, and R.H. Katz, "An Architecture for a Secure Service Discovery Service," Proc. ACM/IEEE MobiCom '99, Aug. 1999.
29. L. Liu, J. Xu, N. Antonopoulos, J. Li, and K. Wu, "Adaptive Service Discovery on Service-Oriented and Spontaneous Sensor Systems," Adhoc and Sensor Wireless Networks, vol. 14, nos. 1/2, pp. 107–132, 2012.
30. S. Capkun, L. Buttyan, and J.-P. Hubaux, "Self-Organized Public- Key Management for Mobile Ad Hoc Networks," IEEE Trans. Mobile Computing, vol. 2, no. 1, pp. 52–64, Jan.-Mar. 2003.
31. T. Czachorski and F. Pekergin, "Diffusion Approximation as a Modeling Tool in Congestion Control and Performance Evaluation," Proc. Second Int'l Working Conf. Performance Modelling and Evaluation of Heterogeneous Networks (HET-NETs '04), July 26–28, 2004.

An Efficient System Model for Multicasting Measured Noise Value of Polluting Industries

Naresh Kannan, Krishnamoorthy Arasu, R. Jagadeesh Kannan and R. Ganesan

Abstract In highly densely populated country like India, environmental pollution is a major problem. Specifically, the hot mix plant used for laying roadways is posing a threat to the environment by emitting heavy noise and smoke. This paper proposes an air pollution monitoring system that measures the noise generated from polluting hot mix plant and multicast the acquired noise data to the Central Pollution Control Board (CPCB) and other communication media. Initially, the pressure is acquired from sound source by microphone and then converted to electrical signals in time domain. Applying Fast Fourier Transform converts time function into the spectrum of frequency component to which A-weighting filtering technique is applied. The resulting magnitude is given as the input to the microcontroller to calculate the noise in terms of decibels. The calculated value is compared with the ambient air quality standards in respect of noise and then the appropriate outputs are displayed in LCD. The abnormalities are indicated by red, orange and green LED based on the intensity of the sound levels as heavy, medium and normal respectively. To achieve the centralized monitoring process, the results are multicast using GSM module to facilitate the authorities to take the necessary decision.

Keywords Environmental pollution · Data acquisition · Sensors · Microcontroller · Display unit

N. Kannan (✉) · K. Arasu
Vellore Institute of Technology, Vellore, India
e-mail: naresh.k@vit.ac.in

K. Arasu
e-mail: krishnamoorthy.arasu@vit.ac.in

R. Jagadeesh Kannan · R. Ganesan
Vellore Institute of Technology, Chennai, India
e-mail: jagadeeshkannan.r@vit.ac.in

R. Ganesan
e-mail: ganesan.r@vit.ac.in

N. Modi et al. (eds.), *Proceedings of International Conference on Communication and Networks*, Advances in Intelligent Systems and Computing 508,
DOI 10.1007/978-981-10-2750-5_39

1 Introduction

Pollution is the major cause of the climate change and many health effects for the human kind. In order to monitor and control the ill-effects of pollution environmental assessment board is necessary in densely populated countries like India, china and various other countries. In India, Central Pollution Control Board (CPCB) has formulated the various norms and standards to address the grievances of the public and polluting industries [1]. Hot mix asphalt is used for paving roads generate heavy noise and emits fumes with several different types of chemicals including Carbon monoxide, nitrogen oxide, sulfur, volatile organic compounds and polycyclic aromatic hydrocarbons [2]. Noise created from heating asphalt can drastically reduce the hearing capacity of human kind. To acquire the environmental parameters such as pressure, temperature, humidity, moisture level and process it for decision making, many sensors are deployed in wireless network environment and autonomously report it to central server [3, 4]. The proposed system model acquires the pressure from the sound source and report the abnormality to a centralized system.

2 Related Works

The various authors have studied about the polluting industries and their corresponding standards. [5] have broadly studied about Urban Noise Monitoring system [6]. [7] have discussed about ambient noise due to road traffic according to European regulations [7]. [6] has developed and evaluated several new noise metrics than currently used A-weighted equivalent sound pressure level L (Aeq) for more accurate assessment of exposure risks to complex and impulsive noise. [8] has proposed a new waveform profile based noise measurement system to accurately access complex noise in industrial fields instead of equal energy hypothesis (EEH) metric used in conventional SPL meter [8]. Though, the authors have been proposed a new assessment guideline for complex noise, this paper proposes a system model for measuring noise based on equal energy hypothesis since the hot mix plant does not generate the complex noise. The rest of the paper is organized as follows. In Sect. 3 the problem statement and industry standards for hot mix plant are listed, Sect. 4 deals with computation of noise and its measuring units, Sect. 5 describes the system design and its modules description for measuring noise, Sect. 6 discuss about the cost analysis and design constraints and finally concludes the benefits of noise measurement system and outline of future scope and challenges.

3 Problem Statement

The enumeration of industry standards for hot mix plant is done and a model is proposed to acquire pressure from a sound source and convert into decibel to compare with the estimated industry standards to facilitate the authorities to take the necessary decision. The CPCB has categorized the industries into three types namely Red, Orange and Green based on the highly polluting nature. In this paper, the ambient air quality standards in respect of noise of the hot mix plant that comes in red category is listed and compare it with acquired data through the proposed system model. Generally the acquired environmental parameters are analog in nature, so the suitable conversion methods are computed. Finally the cost analysis for the system model components is discussed. The flow representation for the problem statement is given below (Fig. 1).

The Table 1 shows the ambient air quality standards with respect to noise, according to CPCB norms in decibel dB (A) Leq, the time weighted average of the level of sound in decibels on scale 'A' is relatable to human hearing.

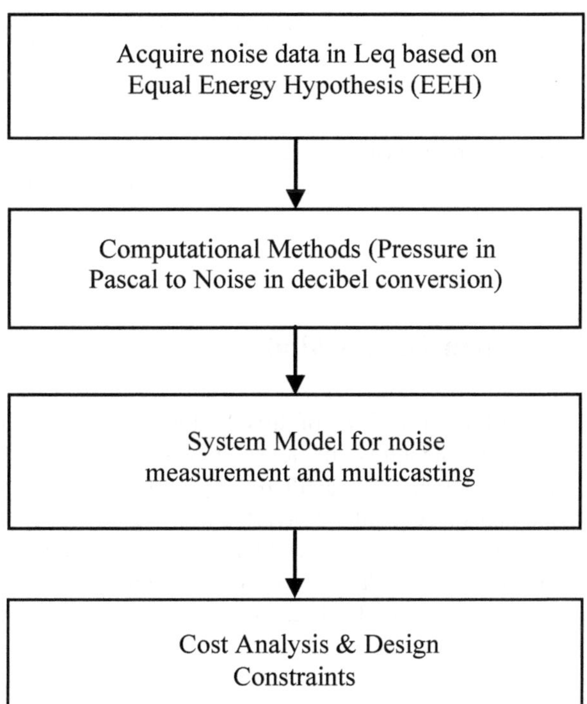

Fig. 1 Problem statement flow representation

Table 1 Ambient air quality standards in respect of noise

Area code	Category of area/zone	Limits in dB (A) Leq	
		Day time	Night time
(A)	Industrial area	75	70
(B)	Commercial area	65	55
(C)	Residential area	55	45
(D)	Silence Zone	50	40

4 Computational Methods

The main parameters to measure sound exposure to humans is Sound Pressure Level (SPL) whose standard unit is μPa or Pa(Pascal) but the range is very large based on human audibility whose range is between 20 μPa and 20 Pa, so for practical purpose logarithmic scale in dB is used commonly. The mathematical expression given below is generally meant for conversion process in transducers.

SPL is expressed as

$$SPL\_db = 20\log_{10}\left(\frac{p\_\mathrm{rms}}{p\_ref}\right) \quad (1)$$

where,

p_rms is the measured sound pressure
p_ref is the reference sound pressure being the threshold for lowest hearing of adults

5 System Design Model

The working principle of this system model is given as follows, initially the microphone acquires the noise waves which act as a input for amplifier. The amplified signal is then filtered by anti-aliasing filter, which outputs the analog signal and subsequently converted into digital signal by ADC converter available in microcontroller. Finally the FFT is applied to convert the time domain data into frequency domain data. Using the frequency data, the noise computation is done and displayed in LCD (Fig. 2).

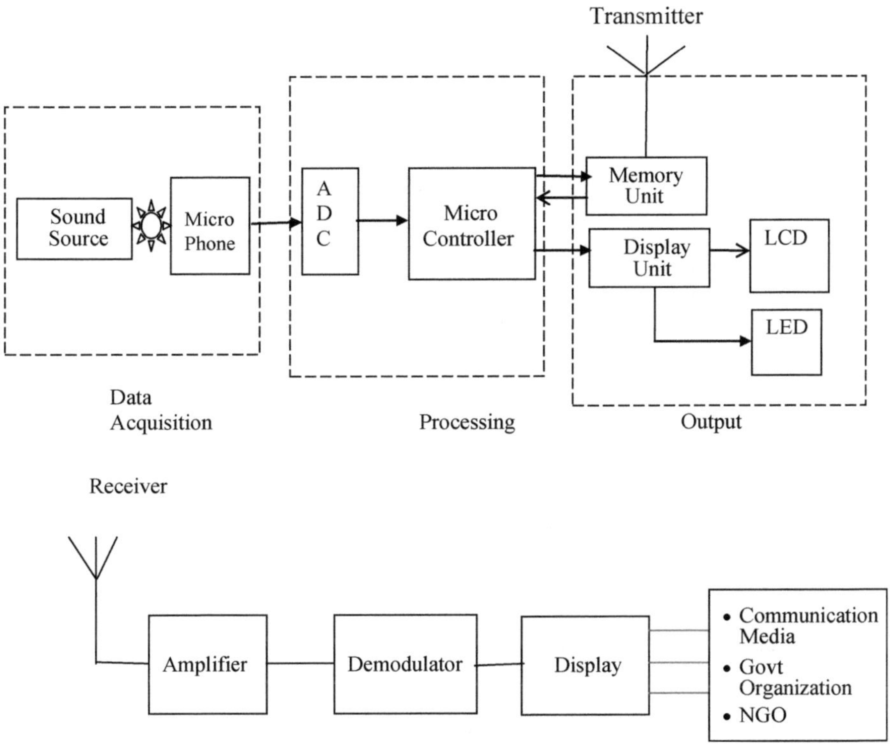

Fig. 2 System model for multicasting noise data

The pseudocode shown below provides the workflow of the system model where the noise data acquired is multicast to the various organizations based on comparison with the industry standards,

Pseudocode

```
REPEAT
   Get Pressure data sample from microphone
    Store results in memory location in microcontroller
UNTIL n samples taken
For sample value 1 to n DO
  Perform FFT
  Apply A-weighting // Get RMS and magnitude of the signal
  Apply time-weight filter to RMS value
     Compute (P_RMS)
     {
       Calculate SPL_db
  }
      If SPL_db >= Noise threshold value
      Report abnormal behaviour
      Else
        Report Normal behaviour
```

Generally, the spatial or time domain samples of noise are converted into corresponding frequency domain, by Fourier transformation equation as given below

$$\mathrm{F(u)} = \frac{1}{N}\sum_{x=0}^{N-1} f(x)e^{\frac{-2\pi ixu}{N}} \tag{2}$$

6 Cost Analysis and Design Constraints

Table 2 provides a detailed cost analysis on the required components for an efficient system model for multicasting measured noise value of polluting industries

Table 2 Cost analysis for the system

Components	Price (in ≠)
Microphone sound input module	350
Microcontroller AT89C51	150
GSM module	1200
LCD and LED	120
PCB	300
Rectifier circuits	200
Total cost per unit	2320

The purpose of doing cost analysis is to determine good investment and it involves comparing the total expected cost of each against the total expected benefits, to see whether the benefits overweigh the costs. In order to maintain the overall system cost to be minimum equal energy hypothesis (EEH) metric is preferred when compared to the wavelet transform based metric for measuring noise.

7 Conclusion

An efficient system model for multicasting measured noise value of polluting industries is proposed. The various data and technical specification of components required for the system are discussed with the aid of pseudocode and system design workflow. The cost analysis and system limitations are briefed. The proposed system facilitates the concern pollution board authorities to take necessary decision and communicate the measured noise levels of the asphalt hot mix plant to various government and non-governmental organizations in a fast and efficient manner for the benefits of public to have a pollution free life. The future scope of the paper is to implement the system in a real time environment and also compare the metric of noise based on equal energy hypothesis (EEH) and wavelet transform (WT) guidelines.

References

1. Maisonneuve, N., Stevens, M., Niessen, M. E., Hanappe, P., & Steels, L. (2009). Citizen noise pollution monitoring. In *Proceedings of the 10th Annual International Conference on Digital Government Research: Social Networks: Making Connections between Citizens, Data and Government* (pp. 96–103).
2. Kularatna, N., & Sudantha, B. H. (2008). An environmental air pollution monitoring system based on the IEEE 1451 standard for low cost requirements. Sensors Journal, IEEE, 8(4), 415–422.
3. Ma, Y., Richards, M., Ghanem, M., Guo, Y., & Hassard, J. (2008). Air pollution monitoring and mining based on sensor grid in London. *Sensors*, *8*(6), 3601–3623.
4. Prabakaran, N., Naresh, K., & Kannan, R. J. (2014). Fusion centric decision making for node level congestion in wireless sensor networks. In *ICT and Critical Infrastructure: Proceedings of the 48th Annual Convention of Computer Society of India-Vol I* (pp. 321–329).
5. Segura-Garcia, J., Felici-Castell, S., Perez-Solano, J. J., Cobos, M., & Navarro, J. M. (2015). Low-cost alternatives for urban noise nuisance monitoring using wireless sensor networks. *Sensors Journal, IEEE*, *15*(2), 836–844.
6. Zhu, X., Kim, J. H., Song, W. J., Murphy, W. J., & Song, S. (2009). Development of a noise metric for assessment of exposure risk to complex noises. *The Journal of the Acoustical Society of America*, *126*(2), 703–12. http://doi.org/10.1121/1.3159587.
7. Mircea, M., Kovacs, I., Stoian, I., Marichescu, A., & Tepes-Bobescu, A. (2008). Strategic mapping of the ambient noise produced by road traffic, accordingly to european regulations. In *Automation, Quality and Testing, Robotics, 2008. AQTR 2008. IEEE International Conference on* (Vol. 3, pp. 321–326).
8. Qin, J., Sun, P., & Walker, J. (2014). Measurement of field complex noise using a novel acoustic detection system. In *AUTOTESTCON, 2014 IEEE* (pp. 177–182).

Internet of Things Based Smart Home with Intel Edison

Shruti M. Patel and Shailaja Y. Kanawade

Abstract In order to help individuals to automate the home, Internet of Things based smart home concept has been introduced. It controls utility based power systems and other required systems inside the house with minimum requirement of an internet and a web browser and simplifies life. The main aim of Internet of Things based Smart Home is to help people to make their life easy with an automation which helps to reduce their time and energy. It operates the basic home appliances as per user requirement without human interference. This will reduce human efforts to control the things physically every time. In this paper we have proposed a system for Smart Home using Intel Edison board that provides a facility to control home appliances remotely from anywhere in the world.

Keywords Internet of things · Smart homes · Home intelligence · Sensors · Web interface

1 Introduction

A smart home is a system which is equipped with a specially structured wiring, which enable equipments to be controlled remotely or program a group of automated home appliances by providing a single command [1]. It is the system which is created to convey or share a multiple number of different applications inside or away from the home using networked devices with range. It is not crucial to have large speed of Internet access for the system components present within home; the entire system performance of a Smart Home build upon the accessibility of an available connection of internet [2]. Smart home is an Internet of Things

S.M. Patel (✉) · S.Y. Kanawade
Electronics and Telecommunication, Sandip Institute of Technology and Research Centre, Nashik, India
e-mail: sp.12th@gmail.com

S.Y. Kanawade
e-mail: kanawade.shailaja@sitrc.org

N. Modi et al. (eds.), *Proceedings of International Conference on Communication and Networks*, Advances in Intelligent Systems and Computing 508,
DOI 10.1007/978-981-10-2750-5_40

(IoT) system, which establishes an effectively managed system of residential facilities and also a family agenda with integrated wiring technology, security technology, network communication technology, automatic control technology and audio and video technology [3]. The development of newly emerging technologies in the field of electronics has brought considerable changes in the day to day life of all human beings. Apart from military and space applications, electronics today has completely revolutionized our daily life including making homes smarter and energy efficient.

This paper introduces automatic access control for home devices propelled with IoT technology. The concept is novel in terms of it provides the user to switch the option for automatic or manual control of operation depending upon the presence of person inside the home or away from the home. This paper is a demonstration of how to design and build remotely controlled multipurpose system which can switch any electronic household appliance by accessing an Intel Edison board, which is programmed to control the systems inside home, includes electrical appliances, plant water feeding, etc., when the person is away from home and enable a person to get the related information on phone. The system will provide feedback indicating the current state of the appliance stopped or functioning.

2 Related Work

The Internet of Things defines a technology that connects everyday objects like smart phones, smart watches or wearable devices, electronic tablets, actuators and sensors to Internet, in which different devices are intelligently interfaced. The components are implemented in various ways to communicate between different things or people, and between things themselves to provide inter connectivity [4]. The capability provided by the IoT makes it feasible to develop number of new applications based on it. IoT Applications are involved in many of smart "things" includes sensors, actuators, controllers or computing devices etc. [5].

The network in IoT provides a link to physical objects with the use of facilities such as network communications, cloud computing and web applications, etc. It provides a facility to devices to interconnect, communicate to receive the data on Internet, store the content of information and retrieve it as per the need, and to communicate with users, creating environment which is constantly-connected. In IoT technology, every object has unique identification and the objects are interconnected with everyone inside the infrastructure of Internet. Besides the plethora of recently emerging field of application using automation based on Internet is to grow into the emerging technology for Internet of Things, which is supposed to bring out considerable amount of inputs from different regions which are collected with large speed, by raising the demand for superior processing and storing the valuable content. Usage of web assistance can be considered as the main inter operable way of providing real time and remotely operated service, enabling applications which communicate with each other [6].

3 Proposed System for IoT Based Smart Home

This paper involves the detail design and construction of an individual control home automation system using Intel Edison board and Internet connection. The automation may be semi or fully controlled and monitors the utility grid connected home appliances. Home automation systems can be differentiated in two separate categories. One is the systems which are controlled locally and other is the systems which are controlled remotely. In the systems which are controlled locally, controller mechanism inside the home is used for accomplishing automation for home. This will permit a person to access a complete system which is automated within their home using any of one interface, a wired or a wireless. While In the systems controlled remotely, connectivity for Internet or a security system existing in the home environment is used with integration, which allows a user to control their system from devices such as smart phone, personal computer, or using telephone gadget from the provider of home security [7].

Figure 1 describes the system for Smart Home using emerging technology "Internet of Things". The smart home application consist of Intel Edison board as a heart of the system to which multiple sensors such as temperature sensor, moisture

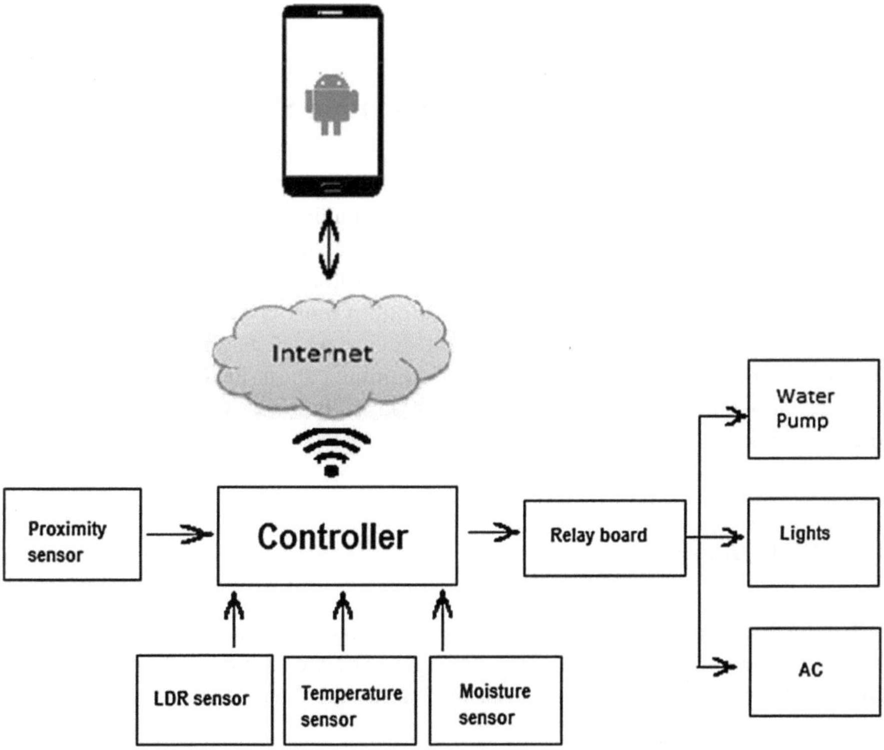

Fig. 1 Proposed system for IoT based smart home

sensor, LDR sensor and proximity sensor will be interfaced. The sensors will send the data to Intel Edison board which will further relay it to web page or Android device. The system works on two modes, one is Auto mode and another is Manual mode. In each mode user can get the required information in real time basis on Android device in terms of notification on mobile user App. The notification provided to the user contains information such as current status of appliances and the action taken place after the command sent. In manual mode user can take an appropriate action to ON or OFF the home appliance if required. In auto mode the appliances switches their state based on the input received from the sensors and user will only get the notification regarding the action taken place without any manual control. The temperature sensor will be linked with Air Conditioner to control and monitor the room temperature. A soil moisture sensor will link with watering system of garden to keep the plants moist and water them as and when required when person is away from home. Similarly, it is having occupancy sensor for room lights and energy savings with a light sensor to automate the lights. The home appliances will be controlled using a relay or any switching components like thyristor, solid state relay etc. The real time data gathered from sensors can also be used to process and provide graphical statistics for further analysis. By this interpretation user can get the record of the things operated or actions taken place on the home appliances over the time period.

4 Architecture of IoT Based Automation System

The concept of 'smart home' enhances absolute living by forecasting wellness depending on generation and detection of behavioural pattern [8]. Proposed architecture uses Wi-Fi connectivity medium between computing device and Internet. Figure 2 describes the IoT based Automation System architecture. Architecture is divided into three portions. Bottom most layer is of hardware parts, which includes all sensors and computing device which is nothing but the Intel Edison board. Middle layer provides a communication medium between user interface and hardware assembly. Topmost layer is User interface which is an Android device.

4.1 Web Interface

Web Interface consists of a server and that will consist of a database. The user's data will be stored in the centralized database. The web server only processes the request and gives response to the user and stores the data.

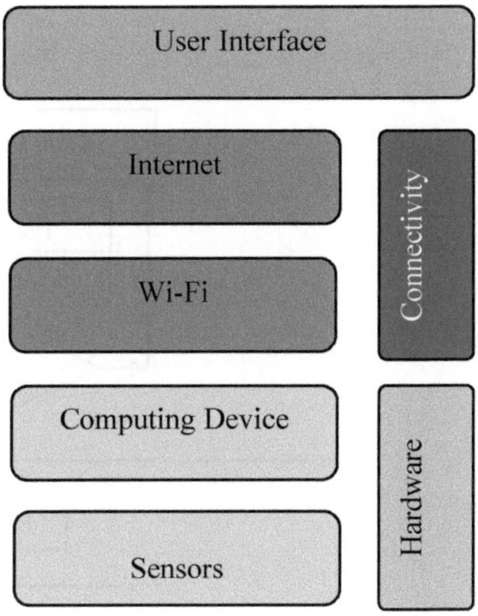

Fig. 2 Architecture of IoT based automation system

4.2 *Wi-Fi*

Wireless systems like Wi-Fi have become more and more popular in home networking systems. Also automation systems in infrastructures such as home or building, the use of wireless technologies provides several advantages which could not be achieved using a wired network [9]. Wi-Fi is a wireless a networking technology that allows electronic devices to communicate using the 2.4 GHz. There are different types of Wi-Fi speeds available, varying form 54 Mbps to 1 Gbps, categorized as IEEE 802.11 and IEEE 802.11 with sub categories extended as a/b/g/n/ac. In proposed system Wi-Fi provides communication medium between the embedded computing device and web server.

4.3 *Computing Device*

The embedded computing system works on Intel Edison board. All sensors and controlling mechanisms are connected to the Intel Edison board. Based on the provided input, it controls the home appliances. Intel Edison board is designed to provide a low power computing system. It is having features for wireless connectivity such as in built Bluetooth 4.0 as well as Wi-Fi module with on board antenna. It also provides support for many external interfaces such as SD card, UART, SPI, USB 2.0, I2C and GPIO pins. The main advantage of Intel Edison board is, it is having all the necessary facilities inbuilt inside it to provide a platform for the development of IoT based applications.

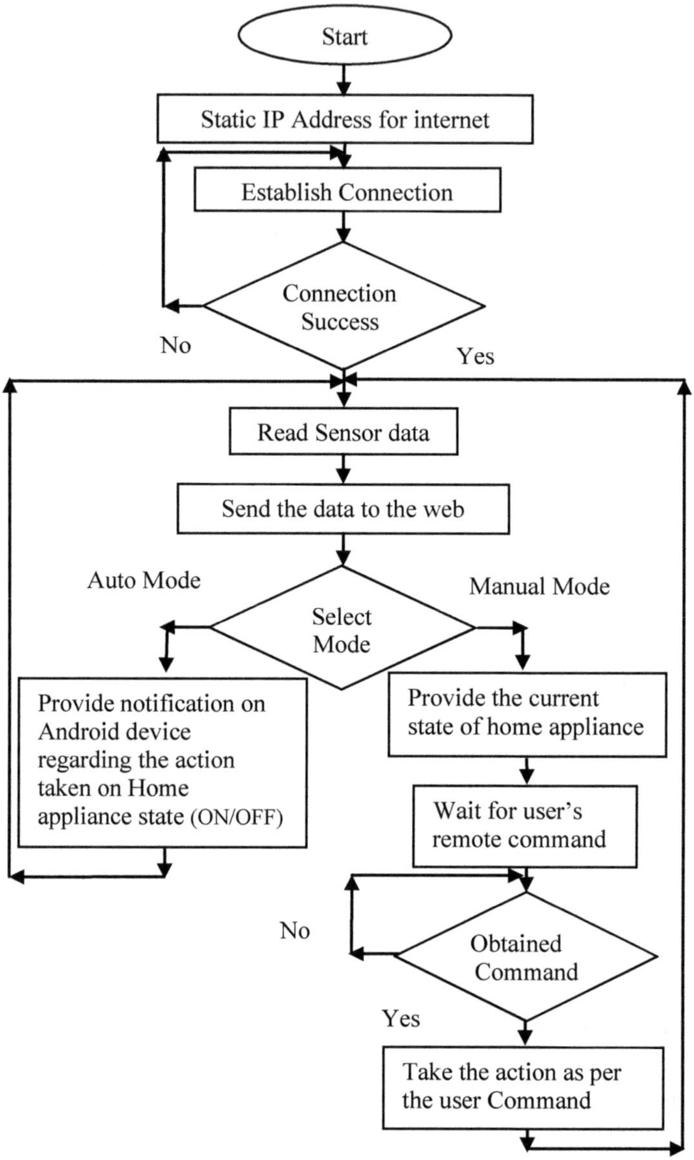

Fig. 3 Work flow of IoT based smart home system

4.4 Sensor

The Sensor component can be used widely for the implementation of electronic systems. They detect different signals which could be an electrical signals or optical signals and generates response accordingly. A Sensor detects the different physical parameters like temperature, humidity, blood pressure, light, speed, pulse signals etc. and converts them into a suitable value of signal that can be electrically measurable [10]. Sensors are the input device for a system. Different sensors are used to sense various activities inside the home related to temperature, moisture content of soil in the garden, proximity sensor to monitor the presence of person inside the home and LDR to detect amount of light inside the home. Based on the output of sensors the whole assembly will be controlled.

5 Work Flow of IoT Based Smart Home System

See Fig. 3.

6 Conclusion and Future Work

The proposed system in this paper will help to make the home more luxurious with inclusion of sensors and actuators. The smarter electronics by utilisation of low cost system for sensing is used to make smart home application. The represented system will provide a mechanism which is totally self-controlled for betterment of the operating devices in environment of monitoring stage.

This project can be further expanded to develop smart cities, by the inclusion of different sensors. Larger area can be cover up of the city and the real time data can be gathered and analysed at the location which is authorized. Results will be supportive as the sensing and information transmission is with high reliability and great accuracy with the integrated network architecture proposed. One important parameter which needs to be considered is the security mechanism, on which further work can be done for secure data transmission over long distance and authorized accessing.

References

1. Ronnie D. Caytiles, Byungjoo Park, "Mobile IP-Based Architecture for Smart Homes", International Journal of Smart Home Vol. 6, No. 1, January, 2012.
2. Juniper research, "Smart Homes ~ It's an Internet of Things Thing", Smart Home Ecosystems & the Internet of Things Strategies & Forecasts 2014–2018.
3. Liu Hongyan, "Design and Realization of Smart Home Terminal Applications Based on IOT Technology", International Journal of Smart Home Vol. 9, No. 8, 2015.
4. Rahul Godha, Sneh Prateek, Nikhita Kataria, **"**Home Automation: Access Control for IoT Devices**"**, International Journal of Scientific and Research Publications, Volume 4, Issue 10, 2014.

5. S. Pandikumar, R.S. Vetrivel, "Internet of Things Based Architecture of Web and Smart Home Interface Using GSM", International Journal of Innovative Research in Science, Engineering and Technology, Volume 3, Special Issue 3, March 2014.
6. Shiu Kumar, "Ubiquitous Smart Home System Using Android Application", International Journal of Computer Networks & Communications (IJCNC), Vol. 6, No. 1, January 2014.
7. M.B. Salunke, Darshan Sonar, Nilesh Dengle, Sachin Kangude, Dattatraya Gawade, "Home Automation Using Cloud Computing and Mobile Devices", IOSR Journal of Engineering (IOSRJEN), Vol. 3, Issue 2, Feb. 2013.
8. Hemant Ghayvat, Subhas Mukhopadhyay, Xiang Gui and Nagender Suryadevara, "WSN- and IOT-Based Smart Homes and Their Extension to Smart Buildings", *Sensors* 2015.
9. Vinay sagar K N, Kusuma S M," Home Automation Using Internet of Things", International Research Journal of Engineering and Technology (IRJET), Volume: 02, Issue: 03, June-2015.
10. Prahlada Rao, B. B, Payal Saluja, Neetu Sharma, Ankit Mittal, ShivayVeer Sharma, "Cloud Computing for Internet of Things & Sensing Based Applications", Research Gate, Conference Paper · December 2012.

Image Classification Using Discrete Block Truncation Coding

Komal Supe, Kajal Jaiswal, Almas Khan, Vijay Katkar and Premal Nirmal

Abstract Proposed paper gives a novel method for image feature extraction named "Discrete Block Truncation Coding". Discrete Block Truncation Coding is analyzed with respect to its performance for classification by applying different pre-processing methods, in different color spaces. This method is used to extract features from images in color spaces such as RGB, YUV, YCrCb, HSV. Use of different classifiers such as C4.5, Random Forest, Naive Bayes with different pre-processing methods such as Discretize, PKIDiscretize is done. Experimental results which are obtained using standard dataset proves that proposed feature extraction scheme works better for image classification.

Keywords Discrete block truncation coding · Color spaces · Pre-processing · Classifiers

1 Introduction

With the increasing use of web cameras and various other types of cameras huge volume of video data needs to be collected and processed. Video is a collection of images called as frames. Today's need is to design efficient image classification

K. Supe (✉) · K. Jaiswal · A. Khan · V. Katkar · P. Nirmal
Pimpri Chinchwad College of Engineering, Pune, India
e-mail: supe5894@gmail.com

K. Jaiswal
e-mail: kajal608jaiswal@gmail.com

A. Khan
e-mail: almas1541994@gmail.com

V. Katkar
e-mail: katkarvijayd@gmail.com

P. Nirmal
e-mail: p.nirmal@gmail.com

N. Modi et al. (eds.), *Proceedings of International Conference on Communication and Networks*, Advances in Intelligent Systems and Computing 508,
DOI 10.1007/978-981-10-2750-5_41

system. In traditional image classification system string tags were attached with images. These tags were used to classify them. But in today's systems, features are extracted from images. These extracted features are used to classify them. Features can be color, texture, edges etc. Here color features are used. While processing this data, storage and processing time are the constraints. So there is a need to use algorithms that require less processing time and less storage. From the video frames, key frames need to be identified. From these key frames features are extracted and stored. Preprocessing and classification is done on these features.

2 Related Work

Delp and Robert Mitchell [1] in the paper "Image Compression using Block Truncation Coding" introduced a new technique called Block Truncation Coding.

Kekre et al. [2] in the paper "Improved CBIR using multilevel Block Truncation Coding" used multiple thresholds to present more sophisticated techniques based on multilevel block truncation coding. This method is used by many researchers [3, 4].

Kekre et al. [5] in the paper "Multilevel Block Truncation Coding with Diverse Color Spaces For Image Classification" said that multilevel block truncation coding works better than single level block truncation coding. Multilevel block truncation coding shows more accuracy in categorizing images to their corresponding classes.

3 Proposed Mechanism

Proposed mechanism is explained with the help of the Fig. 1.

It consists of 4 units:

Input unit: The input unit consist of Dataset with many different images.
Feature Extraction unit: It receives images from input unit and Discrete Block Truncation Coding method is applied on it. It gives array of average pixel value of

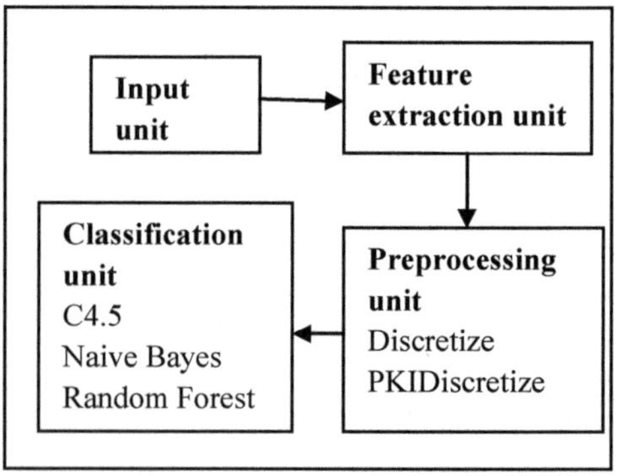

Fig. 1 Proposed mechanism

each bin. This array is the feature set. This feature set is given as input to Pre-processing unit.

Pre-processing unit: This unit is responsible for applying various (Discretize, PKI Discretize) pre-processing methods on it. Output of pre-processing unit is given as an input to Classification unit.

Classification unit: This unit is responsible for applying specific classification algorithm on received data. This unit gives the accuracy of classification of input data given to it.

2.1 Algorithm

Input : Image
b=blocking level.
p=number of bins.

Output : Feature vector of size p*b*b*3.

Algorithm discrete_block_truncation_coding()

```
for each block along the row do
        for each block along the column do
                Step 1 : Calculate average minimum and maximum of pixel
                         values
                          Say minpi.
                          Say maxpi.
                Step 2 : Divide this range into 'P' parts..
                Step 3 : Calculate Average pixel value of each part.
                          Say avg1 , avg2 , avg3 ...... avgp.
        end for
end for
end
```

Here, b is the blocking level (1, 2, 3,…). The image will be divide into b*b number of blocks. Each block needs to be divided into 'p' number of bins. Average intensity value of each bin forms the feature of feature vector.

Feature vector for all the blocks is calculated. Similarly, feature vector for all 3 planes i.e. Red, Blue, Green in different color spaces is calculated.

4 Experimental Results

Experiments are performed on I5 machine (6 GB RAM, 2.30 GHz processor) using Java, Opencv image processing library and Weka. Generic image dataset of 1000 images belonging to different categories is used to perform the experiments.

Figures 2, 3, 4, 5, 6, 7, 8, 9, 10, 11, 12, 13, 14, 15, 16, 17, 18, 19, 20, 21, 22, 23, 24 and 25 shows the experimental results obtained after applying C4.5, Naive

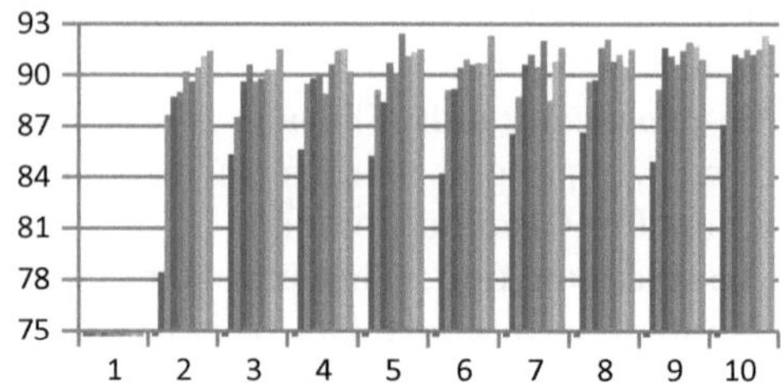

Fig. 2 Accuracy of C4.5 classifier and discretize preprocessing with RGB color space

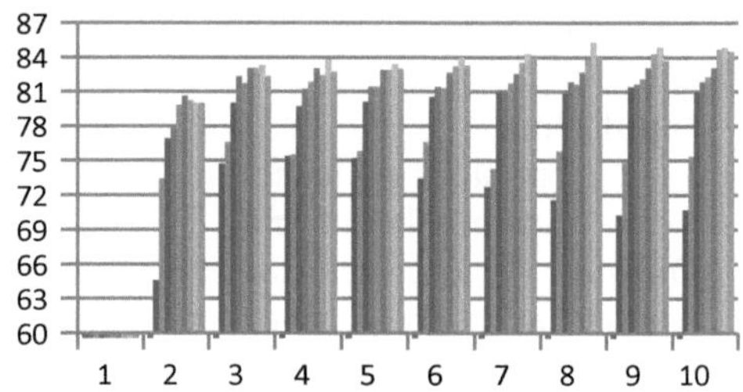

Fig. 3 Accuracy of Naive Bayes classifier and discretize preprocessing with RGB color space

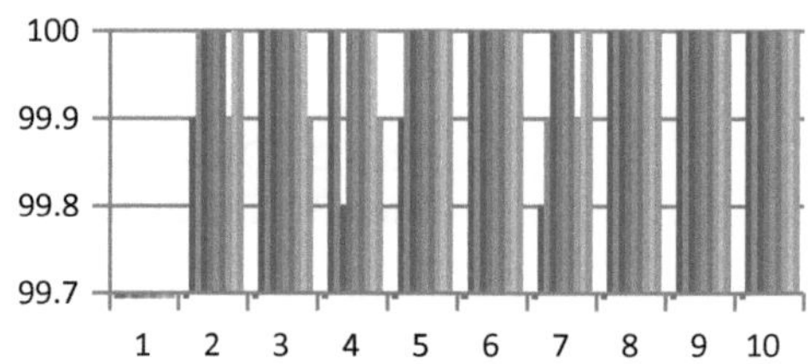

Fig. 4 Accuracy of random forest classifier and discretize preprocessing with RGB color space

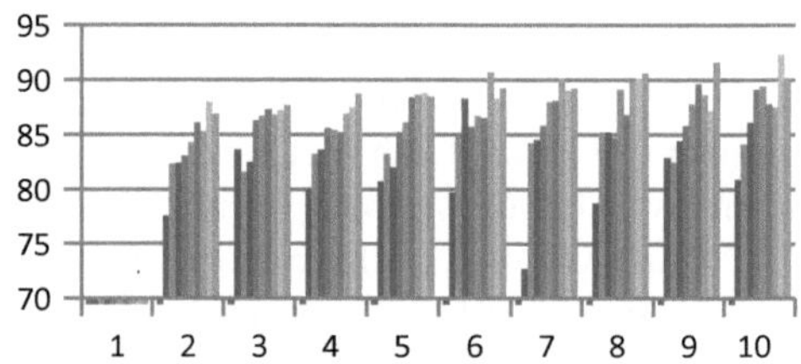

Fig. 5 Accuracy of C4.5 classifier and PKIDiscretize preprocessing with RGB color space

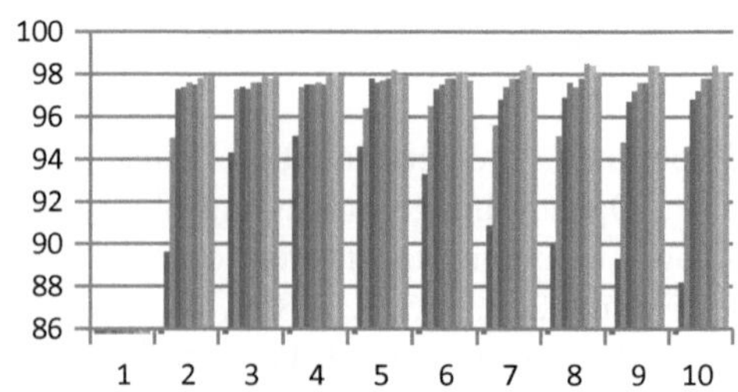

Fig. 6 Accuracy of Naive Bayes classifier and PKIDiscretize preprocessing with RGB color space

Fig. 7 Accuracy of random forest classifier and PKIDiscretize preprocessing with RGB color space

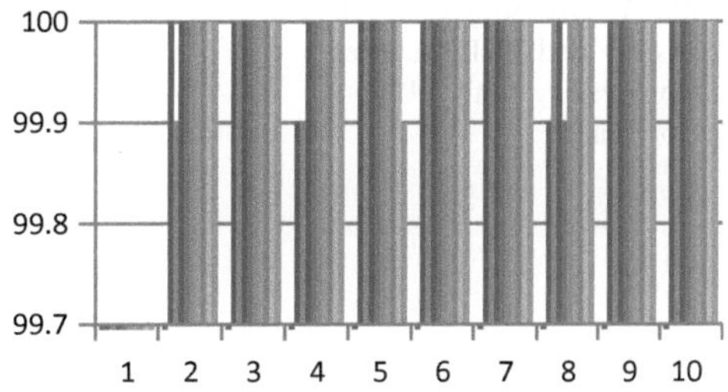

Fig. 8 Accuracy of C4.5 classifier and discretize preprocessing with YUV color space

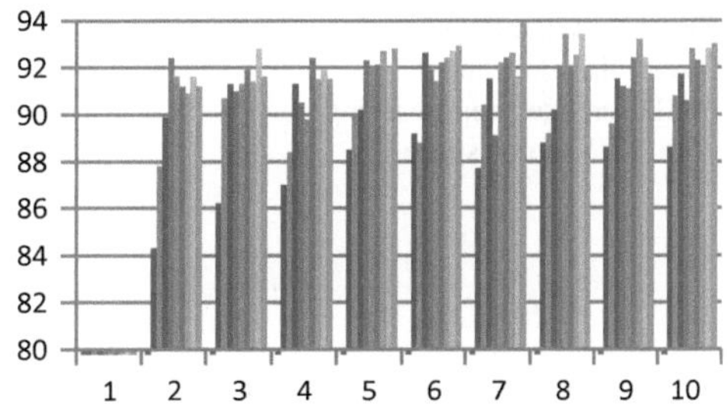

Fig. 9 Accuracy of Naive Bayes classifier and discretize preprocessing with YUV color space

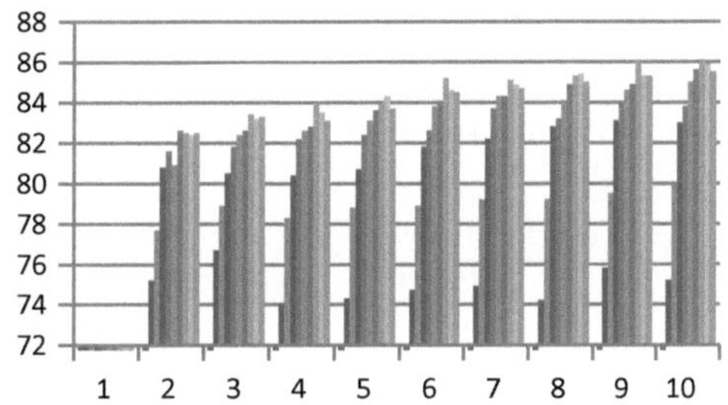

Fig. 10 Accuracy of random forest classifier and discretize preprocessing with YUV color space

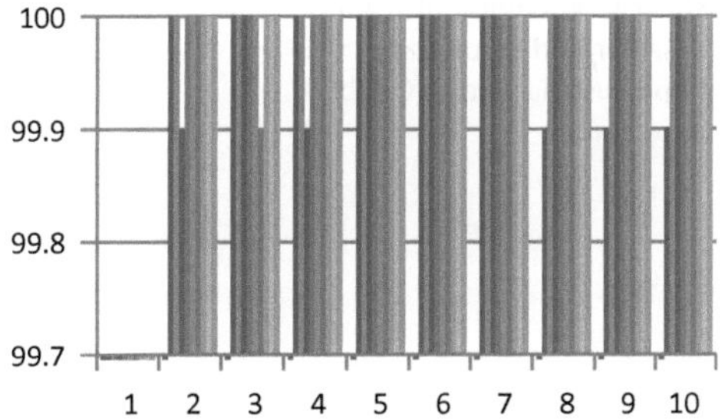

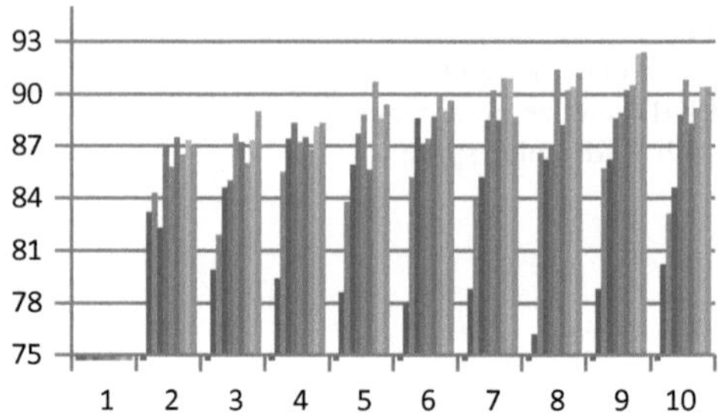

Fig. 11 Accuracy of C4.5 classifier and PKIDiscretize preprocessing with YUV color space

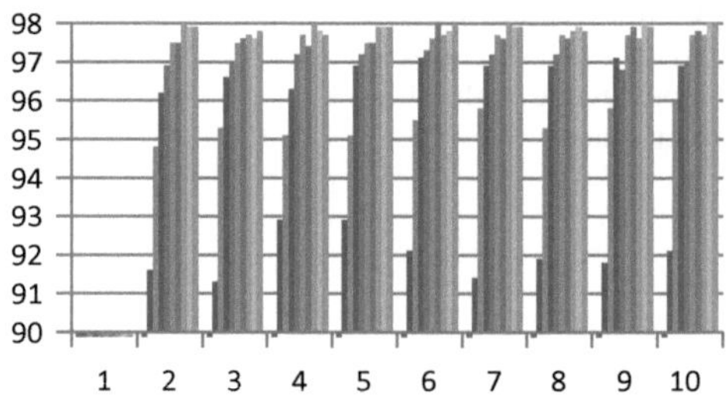

Fig. 12 Accuracy of Naive Bayes classifier and PKIDiscretize preprocessing with YUV color space

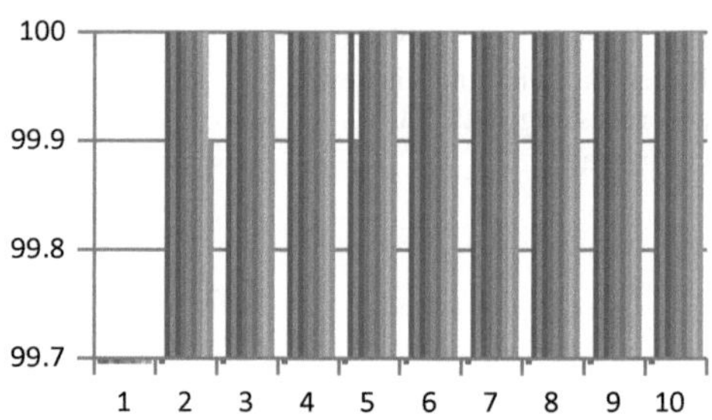

Fig. 13 Accuracy of random forest classifier and PKIDiscretize preprocessing with YUV color space

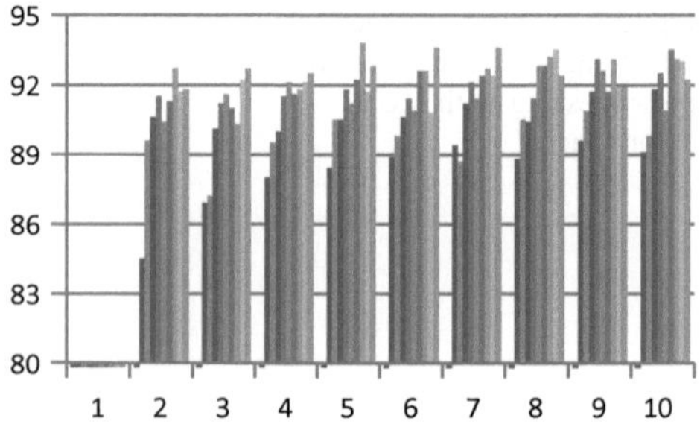

Fig. 14 Accuracy of C4.5 classifier and discretize preprocessing with YCrCb color space

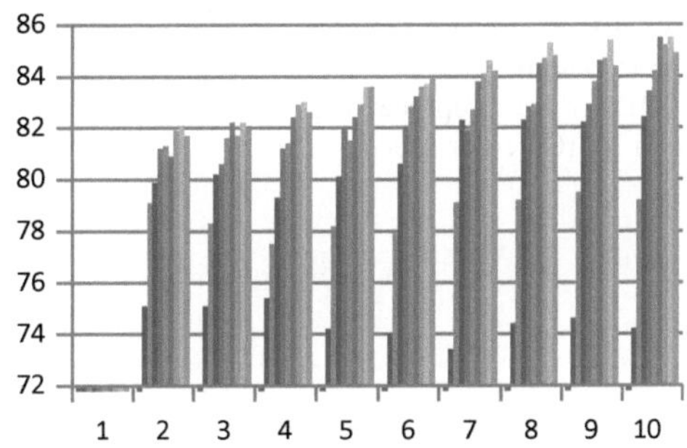

Fig. 15 Accuracy of Naive Bayes classifier and discretize preprocessing with YCrCb color space

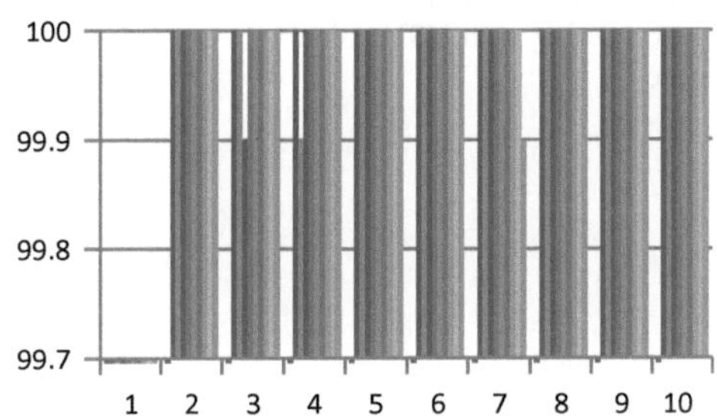

Fig. 16 Accuracy of random forest classifier and discretize preprocessing with YCrCb color space

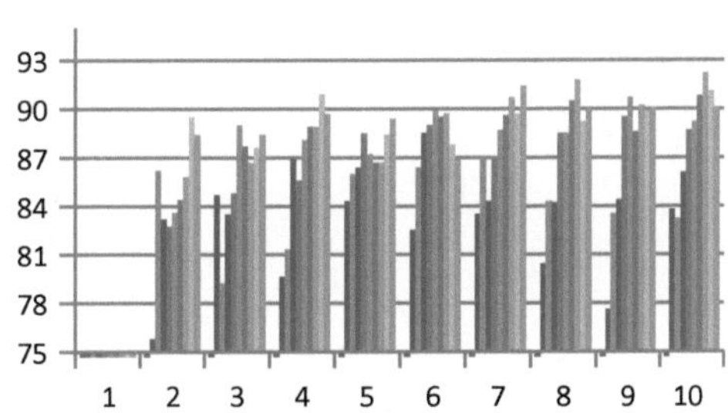

Fig. 17 Accuracy of C4.5 classifier and PKIDiscretize preprocessing with YCrCb color space

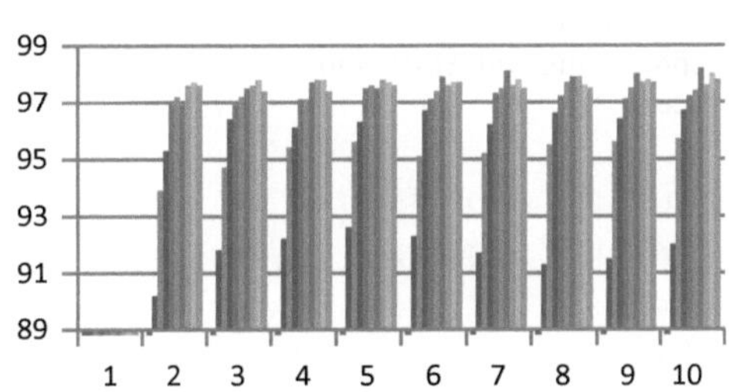

Fig. 18 Accuracy of Naive Bayes classifier and PKIDiscretize preprocessing with YCrCb color space

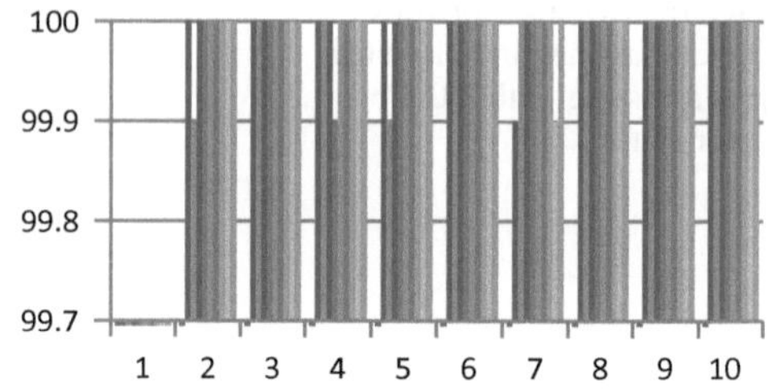

Fig. 19 Accuracy of random forest classifier and PKIDiscretize preprocessing with YCrCb color space

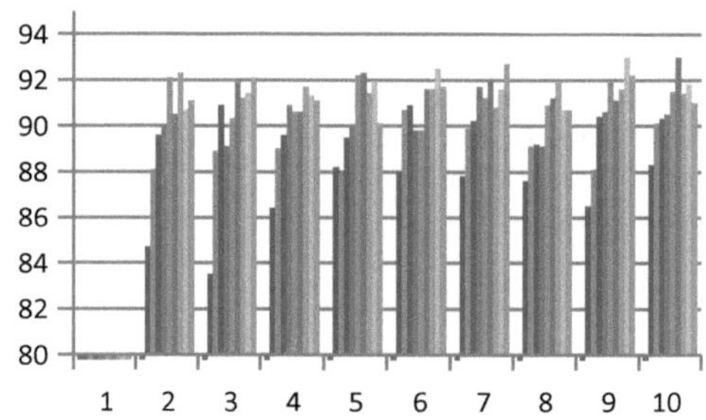

Fig. 20 Accuracy of C4.5 classifier and discretize preprocessing with HSV color space

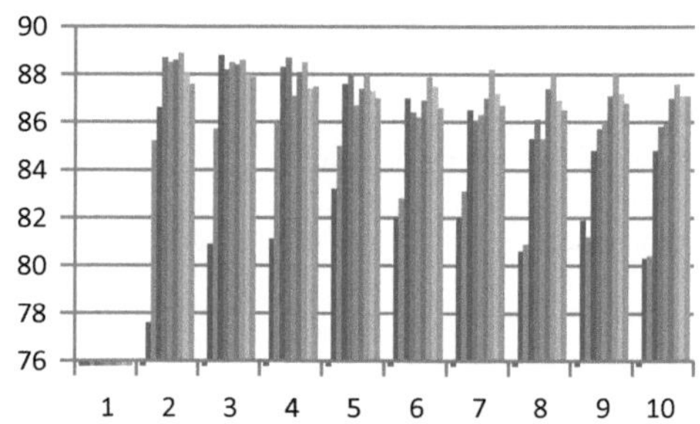

Fig. 21 Accuracy of Naive Bayes classifier and discretize preprocessing with HSV color space

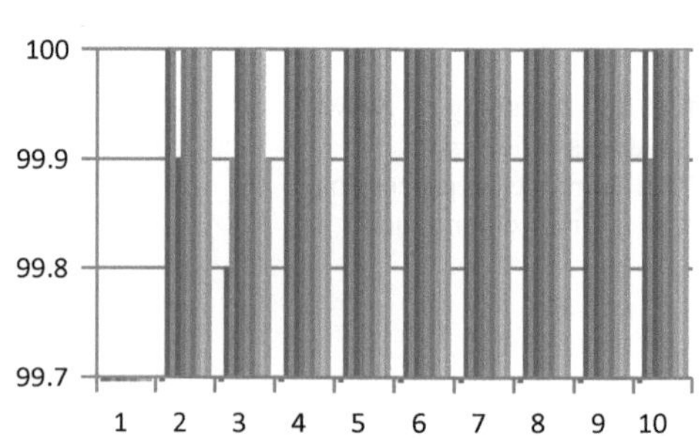

Fig. 22 Accuracy of random forest classifier and discretize preprocessing with HSV color space

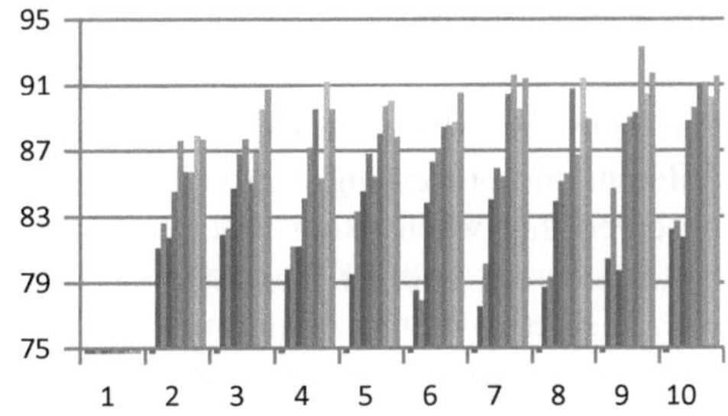

Fig. 23 Accuracy of C4.5 classifier and PKIDiscretize preprocessing with HSV color space

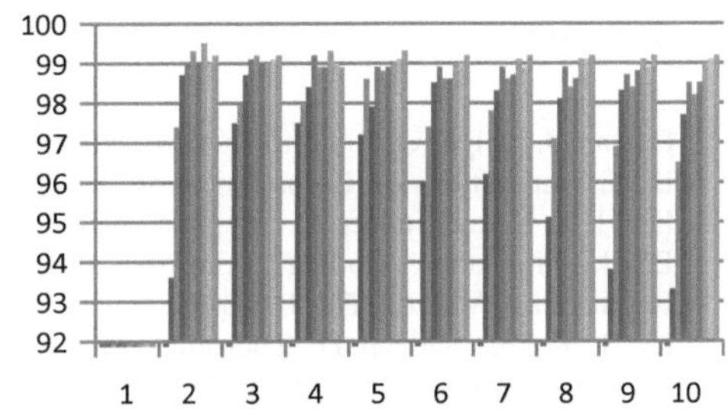

Fig. 24 Accuracy of Naive Bayes classifier and PKIDiscretize preprocessing with HSV color space

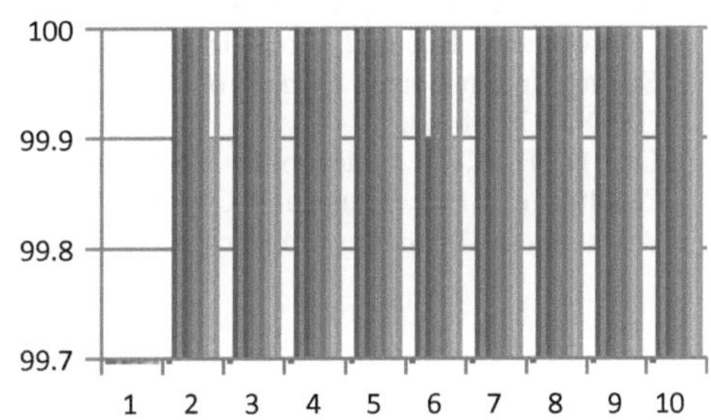

Fig. 25 Accuracy of random forest classifier and PKIDiscretize preprocessing with HSV color space

Bayes, Random Forest algorithm respectively on dataset. Same file was used as training as well as testing dataset. Here X axis represents number of bins. Y axis represents accuracy of classification. Different color bars represents blocking levels.

Table 1 gives a summarized result of performance of Discrete Block Truncation Coding in various color spaces like RGB, YUV, YCrCb and HSV.

Table 1 Performance summarization of discrete block truncation coding

Classifier	Preprocessors	
	Discretize	PKIDiscretize
Naive Bayes	Less accuracy	More accuracy
Random forest	Less accuracy	More accuracy
C4.5	More accuracy	Less accuracy

5 Conclusion

Performance of Discrete Block Truncation Coding for classification purpose with different pre-processing methods, in different color spaces is analyzed. It was observed that when using Discrete Block Truncation Coding, Naive Bayes gives better accuracy with PKIDiscretize in each color space. C4.5 gives better accuracy with Discretize preprocessing method. It was also generally observed that Random Forest gave better accuracy with PKIDiscretize in each color space.

References

1. Edward J. Delph, O. Robert Mitchell. "Image Compression Using Block Truncation Coding". IEEE Transactions on Communication, Vol. Com-27, No. 9, September 1979.
2. Dr. H. B. Kekre, Sudeep D. Thepade, Shrikant P. Sanas. "Improved CBIR Using Multileveled Block Truncation Coding". (IJCSE) International Journal on Computer Science and Engineering. Vol. 02, No. 07, 2010, 2471–2476.
3. H. B. Kekre, Sudeep D. Thepade, "Image Retrieval using Augmented Block Truncation Coding Techniques", ACM International Conference on Advances in Computing, Communication and Control (ICAC3- 2009), pp. 384–390, 23–24 Jan 2009, Fr. Conceicao Rodrigous College of Engg., Mumbai. Is uploaded on online ACM portal.
4. Kekre HB(Dr.), Thepade Sudeep(Dr.), Das Kumar Rik Kamal and Ghosh Saurav (2012). "Performance Boost of Block Truncation Coding based Image Classification using Bit Plane Slicing."International Journal of Computer Applications 47(15): pp. 45–48, June 2012 (ISSN:0975-8887).
5. Dr.. H. B. Kekre, Sudeep D. Thepade, Mr. Rik Kamal Kumar Das, Mr. Saurav Ghosh "Multilevel Block Truncation Coding with Diverse Color Spaces For Image Classification". Advances in Technology and Engineering (ICATE), 2013, International conference on, No. 07.

Preprocessing of Log Files Using Diffusion Map for Forensic Examination

T. Raja Sree and S. Mary Saira Bhanu

Abstract The increase in the number of internet users may lead to cyber crimes and attacks in network. The forensic investigator investigates the crimes by determining the series of actions taken by an attacker. Forensic examination can be performed by isolating the hard disk, physical memory, log files, etc. The information collected from the logs are huge, hence it is necessary to reduce the dimensionality of the features for the efficient investigation of attacks. The proposed method reads the web server logs and uses Diffusion Map for the extraction of relevant features. Diffusion Map helps to detect the attack more accurately than the other dimensional methods, and the computational time grows linearly.

Keywords Application layer attacks · Digital forensics · Diffusion map · Cross-site scripting

1 Introduction

Today, the internet services have become very essential for both enterprises and individuals. This growing accessibility and increasing reliance on the network have become the cause for several kinds of network threats and malicious activities which compromises the confidentiality, integrity, and availability of the network services [1]. Security is the major concern in internet based applications wherein investigation of crimes is very difficult. The attacker sends voluminous HTTP packets to the web server until the server resources get exhausted in the case of application layer attacks.

T. Raja Sree (✉) · S. Mary Saira Bhanu
Department of Computer Science and Engineering,
National Institute of Technology, Tiruchirappalli, Tamil Nadu 620015, India
e-mail: 406112001@nitt.edu

S. Mary Saira Bhanu
e-mail: msb@nitt.edu

N. Modi et al. (eds.), *Proceedings of International Conference on Communication and Networks*, Advances in Intelligent Systems and Computing 508,
DOI 10.1007/978-981-10-2750-5_42

To protect the network against the application layer threats and malicious activities, several mechanisms are in use. Intrusion Detection System (IDS) is one such mechanism that aims towards stopping the access of the network by unauthorized entities [2]. The various methods used in the existing literature for IDS are Statistical Methods [3], Machine Learning [4], Support Vector Machine (SVM) [5] etc. These methods fail to detect the attack effectively and these mechanism yields a large amount of false positives. When a security breach occurs, the forensic investigator has to prove the cyber crime before the law.

The forensic examiner collects the evidence by finding the series of action taken by an attacker. Forensic examination isolates the attacked system identification and safely protects the data viz hard disk, RAM images, Web server log files, etc. The evidence is collected and analyzed from the attacked system by several validating measures and through the log analysis [6].

The forensic investigator relies on finding the details such as where, why, when, who, what and how the attack has happened. The proposed method reads the web server log files, extract the relevant features of evidence. Diffusion Map (DM) is used to reduce the features without altering the information content. These features are processed by machine learning techniques for the identification of crimes that had occurred on the network.

The remainder of the paper is organized as follows. Section 2 discusses about the related work in digital forensics and log analysis. Section 3 discusses about the overview of the proposed system. Experimental results are presented in Sect. 4. Section 5 concludes the paper with future work.

2 Related Work

Digital forensics is the process of identification, collection and validating the digital information by preserving the evidence [7]. The forensic examiner analyzes the attack by collecting the evidence such as physical memory, disk, log files etc. either through live or dead analysis [8, 9]. Dead analysis detects the problem after ceasing all the relevant information [8]. Live forensic analysis identifies the evidence through continuous monitoring of the devices in the network since the data is evolving over time [9].

Application layer attacks play a major role in attacking the web server and their applications. Krugel et al. proposed web based attack detection by automatically retrieving the profiles such as length and structure of web server logs [10]. These profiles are compared with the incoming user requests to classify the attacks. It results in large false positives. Lee et al. proposed a method for the detection of normal or attack traffic using cluster analysis on each attack phase [11]. This method selects only few input features which result in low detection of attacks. Maggi et al. adapted a method to distinguish between the benign or malicious behavior in web based applications. The HTTP traffic response is analyzed to determine the historically modelled parameters [12]. This method needs huge volumes of well labelled data for initial training to determine the malicious behavior.

Juvonen et al. proposed a method that extracts the features of HTTP web server log files for online anomaly detection using dimensionality reduction techniques such as Principal Component Analysis (PCA), Random Projection (RP) and DM. RP is used for the analysis of large volumes of traffic, whereas DM is mainly for accurate analysis of data and for better visualization [13]. Moreover, PCA identifies the benign or malicious traffic for large amount of traffic but it does not determine the attack accurately. The consideration of anomaly detection threshold is a challenging task to identify the attacks. DM methodology identifies the attack both in online [13, 14] as well as offline [15, 16]. The online detection system uses a rule extraction algorithm for the detection of web based attacks.

The proposed method collects the evidence by extracting the relevant information for the identification of attacks from the trace log file located in the Web server. The evidence obtained is then processed by using the n-gram preprocessing and DM is applied to reduce the size of the features for efficient analysis.

3 Proposed Model

The architecture of the proposed model is depicted in Fig. 1. It consists of four stages, namely Evidence collector, Evidence Preprocessor, Evidence Repository and Forensic Intelligent Analyzer. Out of these, Evidence collection and Evidence preprocessing play a major role in the identification of attack that has happened.

- *Evidence Collector* This process is for collecting the evidence from the information sources such as network routers, switches, server and hosts which is under investigation.
- *Evidence Preprocessor* It takes the log file as the input and analyzes the log file to identify the evidence of the attack in terms of features. It preprocesses the feature set and selects a feature subset to describe the attack.
- *Evidence Repository* This is the process of storing all the preprocessed relevant information for the identification of evidences.
- *Forensic Intelligent Analyzer* The feature subset of evidence is given as input to the fuzzy expert system, which compares the newly generated log files from incoming traffic with the predetermined rules from knowledge base to generate forensic alert.

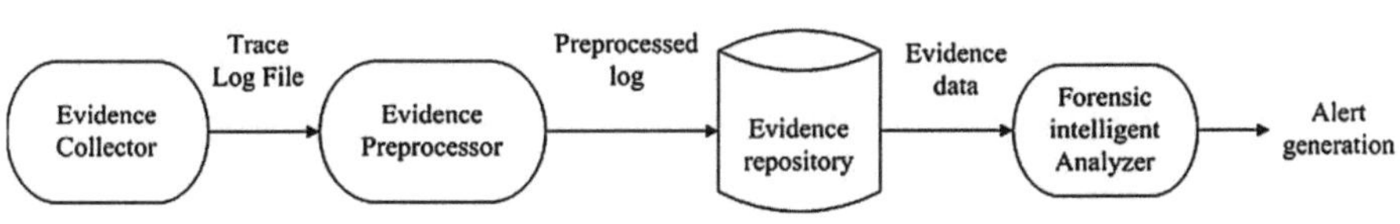

Fig. 1 Architecture of proposed model

3.1 Evidence Collector

The evidences are collected from the network sources such as router, switches, server, hosts and the internal components viz hard disk, RAM images, physical memory etc. which are under forensic investigation. The logs collected from the network play an important role in evidence collection. Application layer attacks are reflected in the various log file traces stored on Apache server. These logs are used for forensic examination to detect the application layer attacks. The various attack information stored in the trace log file is listed as follows:

- */var/log/syslog*—determines if someone is trying or has executed buffer overflow.
- */var/log/debug*—stack tracing to determine the nature of application and service based attacks.
- */var/log/ufw.log*—direct method for auditing firewall.
- */var/log/auth.log*—auditing of attacks on credentials and determines the unauthorized access.
- */var/log/dmesg*—this is not a log file, but this is used for determining anomalous activity from recent bots.
- */var/log/apache2/access.log*—useful for determining web based attacks (XSS, XSRF, SQLI, remote file inclusion, local file inclusion and flooding attacks.
- */var/log/apache2/error.log*—useful for determining web based attacks.
- */var/log/mysql.log*—useful for determining the database related attacks.

3.2 Evidence Preprocessor

The evidence of log file traces are passed as input to the preprocessor. The preprocessor is mainly used for removing the uncleaned data and for extracting the relevant features to identify the attacks. The HTTP request is extracted from the log file and then preprocessed to gather evidences.

Preprocessing of HTTP Requests
The HTTP requests are extracted and preprocessed using n-gram preprocessing technique for the reduction of noise and it filters out the static requests such as .html, . text, .pdf etc. These requests are then converted into numerical vectors to form the feature matrix. The occurrence of specific n-gram feature matrix is summed for each instance in raw log files. The HTTP requests are ASCII coded. When n = 1, n-gram analysis is performed on the character distribution with the maximum of 256 dimensions. Out of which, the ASCII code of 1-gram appears from 33 to 127 (total 95) in the HTTP requests. Similarly, the features are massively increased by varying the values of n.

Diffusion Map

The features extracted from the HTTP requests are very large, hence dimensionality reduction is applied to reduce the features. DM is a dimensionality reduction technique that maps multi-dimensional input features to the lower dimensional space with slight variation in information content. It is a non-linear geometric method that preserves the diffusion distance as Euclidean distance in the lower dimensions [13–15].

Let $y_i \epsilon L^D$, $i = 1, 2, \ldots, N$ be a input feature vector on a D-dimensional space, N is the number of input samples and D is the dimension of the HTTP requests. These input feature vectors are normalized by taking the logarithm in order to make the features comparable.

The affinity matrix is calculated by taking the pairwise distance between input points of each HTTP request [15]. This distance is measured using Gaussian kernel function given in Eq. (1).

$$\phi_{ij} = exp\left(\frac{-\|y_i - y_j\|^2}{\epsilon}\right) \tag{1}$$

The degree of each point of the affinity matrix ϕ is obtained by adding the weights connected through them to other points. The sum of the kernel matrix with its diagonal of each row is expressed as $D_{ii} = \sum_{i=1}^{n} \phi_{ij}$ and the each row ϕ is normalized by the row sums: $A = D^{-1}\phi$. Also, the matrix A is obtained by taking the transition probability between the input points. Now, the symmetric matrix $\tilde{A}D^{1/2}AD^{-1/2}$ is interpreted by substituting the original A, which is given as $\tilde{A} = D^{-1/2}\phi D^{1/2}$ [16].

The Singular Value Decomposition (SVD) of this real valued symmetric matrix is expressed as $\tilde{A} = U\lambda\, U^T$. SVD decomposes the matrix into U that corresponds to the Eigen vectors on its columns and the diagonal λ expresses the Eigen values of $\tilde{A}$. The Eigen values of the transition matrix A and the decomposition matrix are the same, and the Eigen vectors are obtained by calculating the right Eigen vectors which is given as $V = D^{-1/2}U$.

The two-dimensional coordinates are obtained by multiplying each Eigen vector column with the corresponding Eigen value. The resulting matrix contains N rows, each corresponding to the input feature (data) points and k-columns which represent the new dimensions: $Y_{DM} = V\lambda$.

Now, the low-dimensional features are obtained from the diffusion matrix, which is fed to the module that identifies the attack using Machine learning techniques.

4 Experimental Results

4.1 Datasets

The application layer attack was generated using various attacking tools, scripts, bots and the details about the simulated traffic are reflected in the access log file on an apache server. The benign HTTP traffic was generated by using the normal browsing activities carried out on different machines using valid user agents, HTTP methods and HTTP header parameters. HULK [17], HTTP DoS [18], HOIC [19] and bots are some of the attack tools and scripts that are used to launch HTTP flood attack. Most of the traffic's actual payload could not be accessed as encrypted, so it is difficult to analyze the traffic. However, access logs are taken from the web server and these logs are easier to analyze the traffic as either normal or anomalous.

4.2 Results

The raw logs obtained from the HTTP server are preprocessed by separating the relevant features. The HTTP requests are extracted and preprocessed using n-gram preprocessing, then DM is applied to obtain the weight of the reduced features. The overall weight is calculated from the reduced weight of feature obtained from DM. Figure 2 shows the low dimensional points using DM. This explains how the data points are represented in the feature space for better visualization of low dimensional points in DM and it makes the in-depth analysis easier.

The manually injected traffic falls into two aspects: (i) The attacker tries to access the important files from the server using the root password on the server. (ii) The attacker tries to inject the cross-site scripting (XSS) attacks manually when

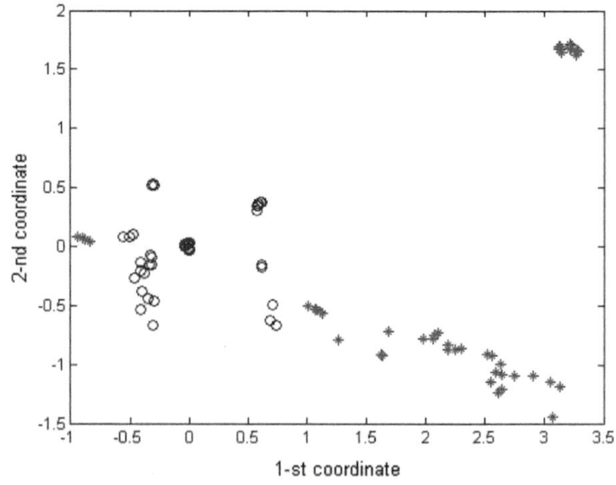

Fig. 2 Low dimensional points using DM

Table 1 Dataset considered for various tests

Test cases	Tools used
1	HOIC
2	HULK, BOT, XSS
3	HTTP DDoS
4	HULK, HTTP DDoS, XSS
5	HULK, HOIC, BOT

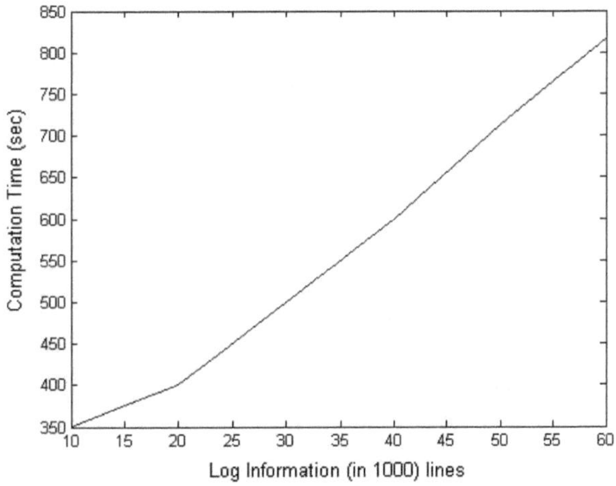

Fig. 3 Computational times of DM

the user visits the particular page. DM identifies the features accurately and the time taken for execution is high when compared to the other dimensional methods. The dataset generated using various attacking tools are considered for the different number of test cases is represented in Table 1.

The logs generated by these tools and bots are stored in apache server are used for the detection of attacks with the various tests. The subsets of the raw log data are used to test the speed and scalability of the execution for processing of large datasets. The size of the raw log data is large to test the scalability. Figure 3 shows the various computational times of DM. From the figure, it is evident that the data grows linearly in terms of scalability aspects. As the log size increases, the computational times grows linearly.

5 Conclusion

In this paper, Diffusion Map based Evidence collection and pre-processing mechanism in forensic process are proposed for the detection of application layer attacks. The attacks are generated by using various attacking tools, scripts, bots and manually injected attacks and these attacks are reflected in the access log file on an

Apache Server. The acquired log evidence is pre-processed by extracting the relevant features from the web server log files. DM reduces the size of the features which can be used for the further investigation of the attacks and crimes. DM determines the attacks accurately when compared to the other dimensional methods and the computation times grow linearly.

References

1. Scarfone, K., Mell, P.: Guide to intrusion detection and prevention systems (IDPS) NIST Special Publications 800-94,1–127 (2007).
2. Patcha, A., Park, J. M.: An overview of anomaly detection techniques: existing solutions and latest technological trends, J. Comput. Netw., vol. 51, 3448–3470, (2007).
3. Zhang, Z., Li, J., Manikopoulos, C., Jorgenson, J., Ucles, J.: HIDE: a Hierarchical Network Intrusion Detection System using statistical preprocessing and Neural Network classification, In: Proceedings of IEEE Workshop on Information Assurance and Security, pp. 85–90, (2001).
4. Govindarajan, M., Chandrasekaran, R.: Intrusion Detection using neural based hybrid classification methods, J. Comput. Netw., vol. 55, 1662–1671, (2011).
5. Hu, W., Liao, Y., Vemuri, V. R.: Robust anomaly detection using Support Vector Machines, In: Proceedings of International Conference on Machine Learning, pp. 592–597, (2003).
6. Liao, H.J., Lin, C.-H.R., Lin Y.C., Tung, K.Y.: Intrusion Detection System: a comprehensive review, J. Netw. Comput. Appl., vol. 36, 16–24, (2013).
7. Adrian T.N. Palmer, Computer Forensics, The six steps, US-CERT, (2008).
8. Liao, N., Tian, S., Wang, T.: Network forensics based on fuzzy logic and expert system, J. Computer Communications, vol. 32, 1881–1892, (2009).
9. Carrier, B.: File System Forensic Analysis, Addison-Wesley Professional, (2005).
10. Kruegel, C., Vigna, G.: Anomaly detection of web based attacks. In: Proceedings of the 10th ACM conference on communications security, pp. 251–261, ACM, (2003).
11. Lee, K., Kim, J., Kwon, K.H., Han, Y., Kim, S.: DDoS attack detection using cluster analysis. J. Expert Systems with applications, vol. 34, Issue 3, 1659–1665, (2008).
12. Maggi, F., Robertson, W., Kruegel, C., Vigna, G.: Protecting a moving target: Addressing web application concept drift. In: Kirda, E., Jha, S., Balzarotti, D., (eds.), Recent Advances in Intrusion Detection 2009. LNCS, vol. 5758, pp. 21–40. Springer, Berlin Heidelberg (2009).
13. Sipola, T., Juvonen, A., Lehtonen, J.: Anomaly detection from network logs using diffusion maps, In: L. Iliadis, C. Jayne (Eds.), Engineering Applications of Neural Networks, IFIP Advances in Information and Communication Technology, LNCS, vol. 363, pp. 172–181, Springer, Boston (2011).
14. Juvonen, A., Sipola, T., Hamalainen, T.: Online anomaly detection using dimensionality reduction techniques for HTTP log analysis, J. Comput. Netw., vol. 91, 46–56, (2015).
15. Sipola, T., Juvonen, A., Lehtonen, J.: Dimensionality reduction framework for detecting anomalies from network logs, J. Eng. Intell. Syst., vol. 20, 87–97, (2012).
16. Juvonen, A., Sipola, T.: Adaptive framework for network traffic classification using dimensionality reduction and clustering, Proceedings of the IV International Congress on Ultra Modern Telecommunications and Control Systems 2012 (ICUMT 2012), St. Petersburg, Russia, pp. 274–279, (2012).
17. HULK attack, http://github.com/grafov/hulk.
18. OWASP HTTP DDoS attack, www.exploiterz.blogspot.in/2013/07/owasp-http-getpost-ddos-attacker-tool.
19. HOIC attack tool, www.thehackersnews.com/2012/03/another-ddos-tool-from-anonymous-hoic.html.

An Efficient and Robust Image Steganographic Technique Without Stuffing Data Bits

K.S. Sadasiva rao and A. Damodaram

Abstract Steganography is the process of hiding data bits on cover or carrier file. The carrier file may be text file, image file, audio file or video file etc. If that carrier file is an image file, then that technique is called Image steganography. If color image is used as a carrier file to embed data bits, then that type of steganographic technique is called as color image steganography. Most of the steganographic algorithms are using bit replacement algorithms. In this proposed work, instead of embedding the data bits directly on carrier color image, original data bits will be mapped on the carrier file with key vectors.

Keywords Bit replacement algorithms · Steganography · Steganalysis

1 Introduction

Steganography is the process of embedding the data bits in the particular positions of the pixels in the carrier file [1]. If the carrier file used is a color image file, then that type of technique is called as color image steganographic technique. The file on which data bits will be stuffed is called as carrier file or cover file. After embedding the data bits on the cover image, the cover image is called as stego image. Most of

K.S. Sadasiva rao (✉)
Department of Computer Science & Engineering,
Sri Indu Institute of Engineering & Technology, Hyderabad, India
e-mail: karrisrini@gmail.com

A. Damodaram
School of Information Technology, Jawaharlal Nehru Technological University, Hyderabad, India
e-mail: damodarama@rediffmail.com

N. Modi et al. (eds.), *Proceedings of International Conference on Communication and Networks*, Advances in Intelligent Systems and Computing 508,
DOI 10.1007/978-981-10-2750-5_43

the image steganographic algorithms will be using the Least Significant Bit (LSB) method, because it will not much affect the quality of an image [2].

Original data bits get transmitted without embedding on the carrier or cover image, but there is a logic with which we are transferring the data bits on the carrier file, so any unauthorized user finds this cover image, it will not give any information as it is plain cover image without any data was stuffed and image quality measuring parameters are also not modified. Hence this technique is a very efficient technique. At least a single bit is not modified in the cover image, so we could not be able to identify steganographic process with any powerful steganalysis techniques, hence it is robust. So this algorithm is named as an efficient and robust image steganographic technique without stuffing data bits.

The color image is a combination of three planes Red, Green and Blue [3]. Generally in spatial domain, RGB planes are used to stuff the data bits at least significant bit positions of the pixel. Hence three bits can be stuffed in each pixel among the three planes [4]. But in the transform domain, the pixel values are converted into the transform co-efficient, and then those transformed co-efficient to be used to modify the data.

Hence either spatial domain or transform domain algorithms will replace the cover image bits with the original message bits using least significant bit algorithm or some other variations on those algorithms.

2 Related Work

Steganography is a process of hiding data in other media to transfer the secured information [1]. Actually many steganographic techniques have been implemented either in color or gray scale images. If the steganographic algorithms are implemented in gray scale images, then there is suspicion by hackers, because generally images which are under transmission are color images. Hence color images are best suitable to use for steganography process. But the color images, all the three planes RGB have been used to stuff bits. 3 bits/pixel can be added with color images, but the level of distortion is high [5].

3 Proposed System at the Sender

Hence almost all the steganographic algorithms either spatial domain or transform domain is using bit replacement algorithms. Steganalysis techniques are used to find whether the data bits are stuffed in the carrier image [6]. These Steganalysis techniques are mostly working on the statistical characteristics of an image. Whenever there is partial or slight change in the carrier image, those changes can be detected by commercial steganalysis tools, even though it is not identified by human necked eye. Most of the LSB based steganographic algorithms will be getting PSNR value is

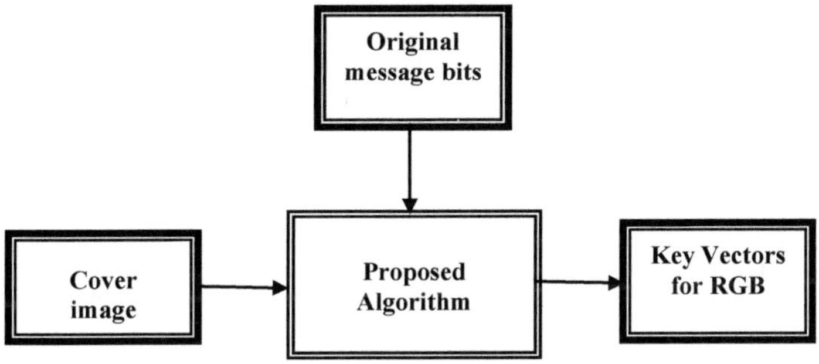

Fig. 1 Proposed system on sender side

much more than 20 (i.e., if this value is greater than 20, human eye cannot detect the change in the cover image and stego image). Here in this proposed algorithm, carrier image is transmitted as it is and there is no use of stego image. The proposed algorithm works in the following way at sender side (Fig. 1).

1. *Read a cover image which is a color image.*
2. *Display the cover image.*
3. *Extract the three planes (red, green and blue) and convert into matrices.*
4. *Read the original message bits from any input data file.*
5. *For every plane, do the following:*
    a. *Read the data bit and the MSB bit of the corresponding plane.*
    b. *Compare data bit and MSB bit*
        i. *If these bits are equal. Put 0 for 'count' filed of that pixel of that plane.*
        ii. *If these bits are not equal, then increment 'count' field by one, and compare the next MSB bit.*
        iii. *Repeat the step 'ii' until the bits are equal.*
        iv. *If the bits are not equal until all 8 bits of the pixel plane. Put count equal to '9'.*
    c. *Construct the count arrays for all three planes Red, Green and Blue.*
    d. *Stop the process.*

4 Proposed System at Receiver Side

The proposed algorithm works in the following way at receiver side (Fig. 2).

1. *Read a cover image which is a color image.*
2. *Display the cover image.*
3. *Extract the three planes (red, green and blue) from cover image and convert into matrices.*
4. *For every plane, do the following process*

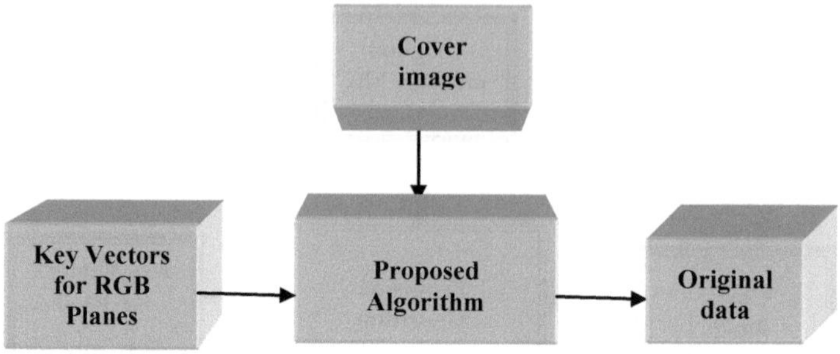

Fig. 2 Proposed system on receiver side

a. *Read the key vector.*
b. *If 'count' field for pixel is equal to '0' then directly extract the MSB bit of corresponding pixel of corresponding plane in carrier image and consider as original data bit.*
c. *If 'count' field for the pixel is equal to '1', then consider the next MSB as original data bit.*
d. *If 'count' field for the pixel is equal to '2', then consider the next MSB as original data bit.*
e. *Repeat the above process for all 8 bits of that pixel of the corresponding plane, if it is not equal to any one of 8 bits of pixel, then 'count' becomes 9, then the data bit is not available in that pixel and move to next pixel.*

5. *Extract the data bits on above mentioned way.*
6. *Display the cover image and original message bits.*
7. *Stop the process.*

Fig. 3 Cover image sent from sender side

5 Results

The following are results of an efficient and robust image steganographic technique without stuffing data bits (Figs. 3, 4, 5, 6, 7).

Fig. 4 Cover image received at receiver side

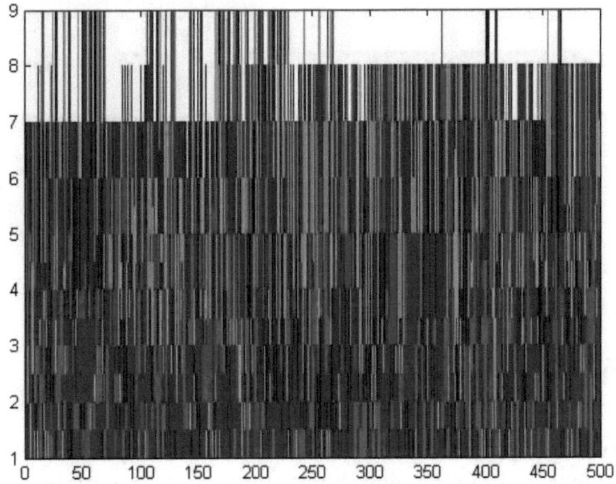

Fig. 5 *Red* key vector histogram (color online)

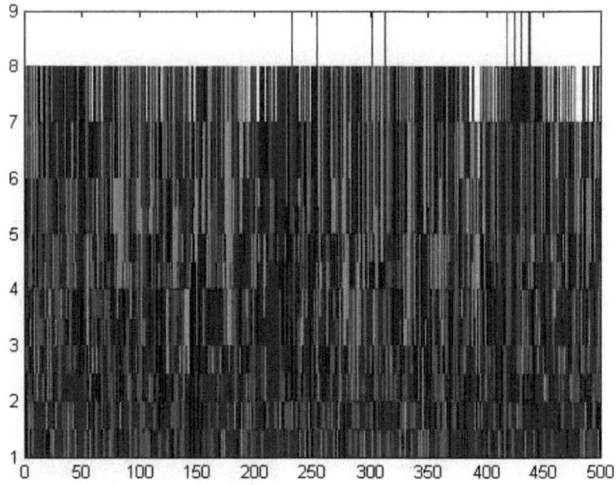

Fig. 6 *Green* key vector histogram (color online)

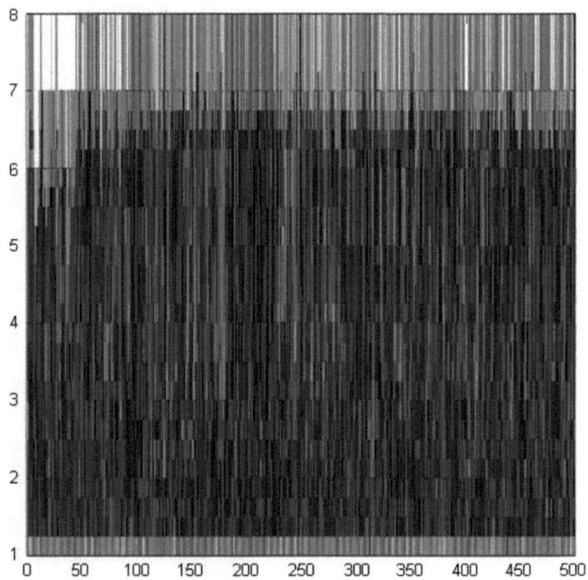

Fig. 7 *Blue* key vector histogram (color online)

6 Conclusion

In our proposed system, we have constructed and sent the key vectors for all three planes red, green and blue along with the plain cover image. Key vectors are constructed with proposed algorithm on sender side. Then after receiving on the receiver side, key vectors and input cover image are used as input to the proposed algorithm, hence it will reconstruct the original data bits. In this proposed algorithm, rather than embedding the original data bits on the carrier files, we are using a logic with which we can transfer the data bits without embedding the data bits in the cover image.

References

1. Niels Provos and Peter Honeyman, Hide and Seek: An Introduction to Steganography, IEEE Security & Privacy, 2003.
2. Saiful Islam, Mangat R Modi, Phalguni Gupta, Edge-based Image Steganography, Springer 2014.
3. Niel F Johnson, Sushil Jajodia, Exploring Steganography: Seeing the Unseen, IEEE, 1998.
4. Wien Hong And Tung-Shou Chen, A Novel Data Embedding Method Using Adaptive Pixel Pair Matching, IEEE Transactions On Information Forensics And Security, Volume 7, No. 1, February 2012.
5. Deepesh Rawat, Vijaya Bhandari, A Steganography Technique for Hiding Image in an Image using LSB Method for 24 Bit Color Image, International Journal of Computer Applications, Volume 64, 2013.
6. Abbass Cheddad, Joan Condell, Kevin Curran, Paul Mc Kevitt, Digital Image Steganography: Survey and Analysis of Current Methods, Elsevier, 2010.

Security Requirements for Internet of Things (IoT)

Shruti Jaiswal and Daya Gupta

Abstract The Internet of Things (IoT) is the tomorrow's Internet. It is being used in our everyday life where objects possessing sensing capabilities such as users, computing systems, and others are combined for convenience and economic benefits. Connecting various such devices is a challenging activity, as each device can have its architecture and security concerns. Various proposals are available in the literature to connect devices effectively, and various systems are there in the markets that are using this IoT concept. Hence dealing with all such interacting devices would be a challenging task when it comes to security. Also, various proposals are available that list the security issues present in IoT, but they are not defining the security requirements clearly for IoT. Therefore, in this paper, we are going to list the main security challenges for IoT and define the security requirements for IoT healthcare system.

Keywords Internet of things (IoT) · IoT healthcare · Security requirements · Security engineering

1 Introduction

The Internet of Things (IoT) as defined in Wikipedia [1] is the network of physical objects embedded with sensors, software, and network connectivity, which enables them to collect and exchange data. It enables sensing and controlling of objects remotely across existing network infrastructure. It is estimated that by 2020 IoT will consist of almost billions of objects. The IoT is being applied in various applications such as waste management, health care system, home automation, and many

S. Jaiswal (✉) · D. Gupta
Department Computer Science & Engineering,
Delhi Technological University, Delhi, India
e-mail: dce.shruti@gmail.com

D. Gupta
e-mail: dgupta@dce.ac.in

N. Modi et al. (eds.), *Proceedings of International Conference on Communication and Networks*, Advances in Intelligent Systems and Computing 508,
DOI 10.1007/978-981-10-2750-5_44

others. Great dependence on these areas diverted us to focus on the security issues involved in the implementation of IoT.

It is mentioned in Oxford English Dictionary that "if one thing can prevent the Internet of Things from transforming the way we live and work, it will be a breakdown in security". The common security goals are to secure the devices, network, and other capability units involved. Various security needs are found in the literature [2–4] for IoT some are Naming and Identity Management, Interoperability and Standardization, Information Privacy, Objects safety and security, Data confidentiality and encryption, Network security.

Providing security to IoT is far more complicated as compared to Internet security. Because IoT, is a mixture of various networks, that not only involves the security problems related to the mobile communication network, sensor network, and the Internet. However, problems such as privacy protection, heterogeneous network authentication, access control, management, and others are arising because of integration of different networks. Therefore, the solution to each security problems should be made.

Traditional security protocols and mechanisms may not work well for the IoT. Because of constraint imposed by the IoT devices such as limited bandwidth, low memory, limited computation, limited energy budget, and others. To address security issues, this work proposes to use the security engineering framework developed in the earlier work [5] and also proposes security requirements, suitable security algorithm specific to IoT. It first determines the security requirements during requirement phase. Then during design phase based on environment constraints, device constraints and risk measures of different security threats a suitable cryptographic algorithm is chosen to implement specified security requirements.

The rest of the paper is organized as follows: Sect. 2, presents the Security Aspect of IoT with different security issues in IoT and how it is different from network security; Sect. 3 presents the Proposed Generic Security Engineering Framework; Sect. 4 presents a case study on the system for Remote Patient Monitoring; finally, Sect. 5 concludes the paper.

2 Security Aspect of IoT

As security is the fundamental enabling factor of most IoT applications. Here in this section, various security issues are elaborated with respect to an IoT system Remote Patient Monitoring. Remote Patient Monitoring System is shown in Fig. 1 is a part of the healthcare system. In the rest of the paper this will be used as case study. Components of Remote Patient Monitoring system are:

(i) **Wireless body Area Network**. It is a body having wearable sensors capable of storing small information and sending it to a remote location. It is the patient.
(ii) **E-Health Gateway**. It would forward the packets from Wireless Body Network to another Network.

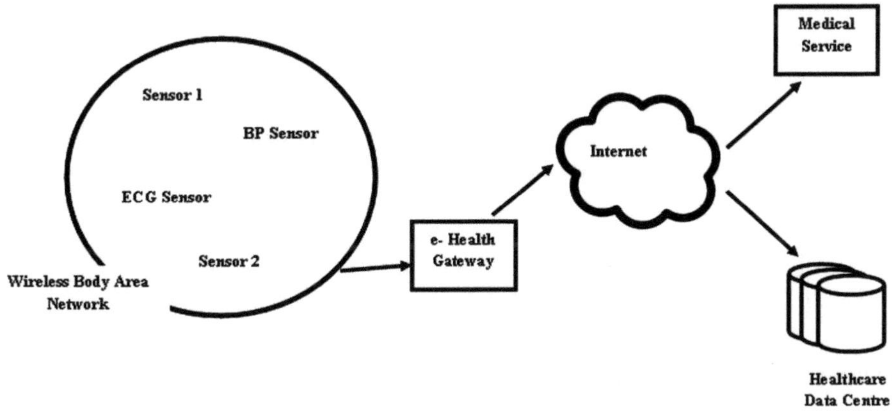

Fig. 1 Remote patient monitoring system

(iii) **Internet**. It would be our communication network that would carry information.

(iv) **Healthcare Data Centre**. It would store all the information that sensors in body generates. Data generated would be voluminous and need proper handling.

(v) **Medical Service**. Medical Facility such as monitoring devices, consultation, regular checkup, and others would be provided to the patient.

2.1 Security Issues in IOT

- **Authentication**. Authentication in IoT is very difficult as it involves heterogeneous network authentication. Things (sensors) must be identified and authenticated before joining the network. IoT requires a unique identification code or a global unique identifier (UID) for each entity in the network.
- **Confidentiality**. Need to ensure that medical information is inaccessible to unauthorized users. Also, confidential messages should not be revealed to eavesdroppers.
- **Self-Healing**. If a medical device in a Remote Patient Monitoring network fails then, the other devices must be able to provide a minimum level of security.
- **Fault Tolerance**. In case of device failure or compromise, the system should be able to provide functionality with relevant security services.
- **Resilience**. If some IoT nodes are compromised, then the system should still be able to protect the network/information from any attack.
- **Data Freshness**. For Remote Patient Monitoring network to work in efficient manner nodes must have access to recent (fresh) messages or data. For example to analyze the heart functioning of any patient Consultant needs the most recent ECG readings.

- **Anonymity**. In Remote Patient Monitoring system, some patients do not want to disclose their identity to anyone.
- **Liability**. In Remote Patient Monitoring system, accountable responsibility should be defined in case of any misuse, loss, theft or unusual event.
- **Trust**. In Remote Patient Monitoring system, users or patients need assurance that their personal and medical data will not be misused.

2.2 *IoT and Network Security*

Our research shows that IoT healthcare security is different from Internet security [6], and is far more complicated. Table 1 shows how IoT is more complex than network security.

Table 1 Comparison of IoT healthcare device security and network security

Design parameters	IoT healthcare device security	Network security
Memory constraints	The on-device memory of IoT healthcare devices is low. They mainly use embedded operating system (OS), the system software. Therefore, the system does not have enough memory to execute complicated security protocols	No such memory constraint
Speed of computation and resource constraints	Low-speed processors are available for IoT health devices. Therefore, finding a security solution that works on it is a difficult task	High-speed CPUs are available
Energy limitations or power consumption	IoT healthcare devices are available with limited battery power. They use the power-saving mode to conserve energy when sensors are idle. Therefore, the energy constraint makes finding security solution challenging	No battery problem. They are equipped with power backups
Scalability	There is a gradual increase in a number of devices. Therefore, need to select scalable security algorithm becomes a challenging task	They are connected through reliable wired links and have established wireless links also
Communication channel	IoT devices mainly connected to the network through wireless links such as Zigbee, Z-Wave, Bluetooth, WiFi, GSM, WiMax, and 3G/4G. Therefore, it is difficult to have a security protocol that works for wireless links and provides security similar to wired links	Less number of mobile devices
Security updates	To mitigate potential vulnerabilities security protocols are required to be up-to-date. Automatic updating of security protocol is difficult	They are having established system for security

3 Our Generic Security Engineering Framework

Security challenges identified from the recent research publications are mentioned in the previous section. As Internet of Thing is the extension of Internet and the existing technologies, several security mechanisms exist for known technologies. Some of them can be readily and directly applied to IoT and some may not. IoT is a network of things and these things are low in power, Memory constrained, Resource constrained, Bandwidth and low in computation power. That is why some protocols do not work in case of IoT. Identification of protocol applicable to a system having constraints related to memory and others are the key to our generic security engineering framework application to IoT.

Our proposed security framework [5] handles the concerns like low in computation power, memory, and others effectively and identifies the optimal security algorithm. Our framework first determines the security requirements during requirement phase by analyzing the existing or related systems and from the specification provided. Then, the threats/attacks possible to the system are identified and evaluated. This evaluation shows which threat is more risky and need to be handled first. In design phase based on environment constraints, device constraints and risk measures of different security threats a suitable cryptographic algorithm is chosen to implement specified security requirements. Main steps of framework is as follows:

3.1 Security Requirements Engineering Phase

- **Security Requirements Elicitation**. Security Requirements are elicited for the system, based on system requirements.
- **Security Requirements Analysis**. Various loopholes that may exist with elicited security requirements are identified related to consistency, correctness, etc.
- **Security Requirements Prioritization**. Security Requirements are prioritized based on risk measures.
- **Security Requirements Management**. Security requirements are stored for future management purposes.

3.2 Security Design Engineering Phase

- **Mapping of Security Requirements with Cryptographic Services**. Security requirements are mapped to security services like confidentiality, integrity, authentication and non-repudiation.
- **Security Design Analysis**. Various security mechanisms available are analyzed using attack analysis.

- **Security Design Constraints**. Various Design Attributes (cost, implementation platform, and others) and environment (wireless/mobile/mobile ad hoc or any other) are identified.
- **Security Design Structuring**. Various design constraints are analyzed here. An SDT (Security Design Template) is prepared.
- **Security Design Decision**. Based on previously stored attack analysis results and SDT an optimal security algorithm is chosen.

4 Case Study: Remote Patient Monitoring

Steps of our framework are now applied to the case study of Remote Patient Monitoring System. Components of the system are explained in Sect. 2.

4.1 Security Requirements Identification

Security issues are represented as security requirements. Firesmith [7] has defined security requirements as high-level requirements that give a specification of system behavior that is not acceptable. Security issues of IoT, as explained in Sect. 2.1 have been expressed here in terms of Security Requirements defined by Firesmith [7] with some newly added security requirements.

- **Identification Requirement**. The typical objective of identification security requirement is to ensure the identity of all the externals (patients, doctors, devices) before allowing them to interact with the services or resources of the system.
- **Authentication**. The typical objective of authentication security requirement in Remote Patient Monitoring system is to verify the identity of person or device with whom it is interacting.
- **Authorization Requirement**. In Remote Patient Monitoring system only authorized nodes whether the service node or a resource node in the network are accessible.

Above three requirements collectively realize the **Confidentiality**, **Anonymity** issue.

- **Immunity Requirement**. Nodes in Remote Patient Monitoring system should be able to protect themselves from infections caused by viruses, worms, and others.
- **Integrity Requirement**. In Remote Patient Monitoring system, Integrity security requirement ensures that no received medical data is modified by an intruder in transit. Also, the integrity of stored data and content should be protected from compromise. This security requirement will implement **Data Freshness**.

- **Intrusion Detection**. Compromise of any node must be detected and recorded.
- **Non-Repudiation requirements**. A node in Remote Patient Monitoring system cannot deny after sending a message. This security requirement will implement the **Liability** issue.
- **Privacy Requirements**. In Remote Patient Monitoring nodes should be able to keep its confidential data private from unauthorized access. This security requirement would implement the **Anonymity** issue.
- **Survivability Requirements**. In IoT if some intermediate nodes are compromised, then the system should still protect the network/information from any attack. This security requirement would implement **Resilience**, **Fault Tolerance, Self-Healing** security issues.
- **Security Auditing**. Administrator of Remote Patient Monitoring system should assign responsibility to some security personnel to check the state of security level in the system.
- **System Maintenance**. In Remote Patient Monitoring system, any update, fixing of bug should be followed by a security check in which violation of security mechanism is checked.
- **Physical Protection Requirements**. In Remote Patient Monitoring system devices, databases need to be protected physically using some mechanism.

Newly Added Security Requirements

- **Data Freshness**. It would take care of availability of latest data for processing.
- **Trust**. It would make people adopt the IoT-based system. It is being achieved by implementation of all other identified security requirements.

Anonymity, Liability, and *Trust* are very important factors for IoT applications to get the social acceptance. Now these defined security requirements are used to mitigate various attacks to the IoT-based system Elicited Security Requirements for Remote Patient Monitoring System is shown in Table 2.

Table 2 Security requirements for remote patient monitoring system

For Wireless Body Network	During Communication (Over the e-Health Gateway and Internet)
Trust, Anonymity Liability Physical Protection Survivability	Integrity Survivability
Healthcare Data Centre	**Medical Service**
Privacy Immunity Integrity Intrusion Detection Survivability Physical Protection Data Freshness Security Auditing	Authentication Identification Authorization Privacy Non-Repudiation

4.2 Design Decisions for Above System

Now based on various design constraints cryptographic algorithms are identified to implement the security requirements.

- First the security requirements are mapped to security services
- Then based on various design constraints as mentioned in Table 1, following algorithms are identified for implementation:

  ECC: Integrity (in things environment that is in body area network)
  Two Step Authentication: Anonymity, Identification, Authentication (other than things environment)
  AES: Privacy, Integrity (other than things environment)
  Log Maintenance: Liability, Intrusion Detection, Security Auditing, Non-Repudiation
  Data Replication: Survivability
  Antivirus: Immunity
  Timestamp: Data Freshness
  Implementing all above would achieve: Trust.

As one algorithm is not sufficient to implement all the security requirements. Therefore, prioritization of security requirements is done, so high priority security requirements would be implemented first compared to others. Due to space limitation, only important and initial steps are explained. For a clear, detailed explanation of framework refer to earlier work [5].

5 Conclusion and Future Work

As IoT is an amalgamation of various technologies, it is more prone to various security attacks applicable to involved technologies. In this paper, various security challenges in IoT Remote Patient Monitoring System are identified with various design constraints applicable to IoT healthcare system. Also, security requirements that apply to IoT are identified and are expressed for Remote Patient Monitoring system. Some cryptographic algorithms are also proposed to implement the security requirements. Further, steps of our generic security engineering framework [5] are followed in IoT with detailed explanation, for handling security issues in a structured manner. Also, device specific cryptographic algorithms to implement the identified security requirements based on various design and environmental constraints are identified.

References

1. IoT: https://en.wikipedia.org/wiki/Internet_of_Things.
2. Granjal, J., Monteiro, E., Silva, .S.: Security for the Internet of Things: A Survey of Existing Protocols and Open Research Issues. IEEE Communication Surveys & Tutorials. 17 (3), 1294–1312 (2015).
3. Catuogno, L., Turchi, S.: The dark side of the interconnection: security and privacy in the Web of Things. In: 9th International Conference on Innovative Mobile and Internet Services in Ubiquitous Computing, pp. 205–212. IEEE (2015).
4. Zhao, K., Ge, L.: A Survey on the Internet of Things Security. In: Ninth International Conference on Computational Intelligence and Security, pp. 663–667. IEEE (2013).
5. Chatterjee, K., Gupta, D., De, A.: A Framework for Development of Secure Software. CSI Transaction on ICT. 1(2), 143–157 (2013).
6. Riazul Islam, S.M., Kwak, D., Kabir, H., Hossain, M., Kwak, K.S.: The Internet of Things for Health Care: A Comprehensive Survey. IEEE Access 3: 678–708 (2015).
7. Firesmith, D.G.: Engineering Security Requirements. Journal of Object Technology. 2(1) 53–68 (2003).

Identity Based Secure RSA Encryption System

Meenal Jain and Manoj Singh

Abstract Identity Based Encryption (IBE) has emerged as a solution to the problem of trust in public keys of Public Key Cryptography (PKC) systems where most of the proposed IBE schemes are based on bilinear pairing and Elliptic Curve Cryptography. The heavy computations involved in such schemes make pairing based schemes less practical. Efficient non-pairing schemes have been designed over the famous RSA structure to bridge the practical gap between actual public key ciphers and utility of IBE in it. Yet the security threats to RSA have been retained in such proposals. This paper presents an RSA based IBE scheme which overcomes the security problems of RSA and has linear computations involved and makes key management/revocation easy.

Keywords Security · Public key cryptography · Identity based encryption · Non-pairing based IBE · Mediated RSA

1 Introduction

Internet Security has become ubiquitous today. The requirement has been well acknowledged since the invention of Internet. Internet has made the world a small place providing access to large amount of information. Crucial information can be exchanged among desiring parties through Internet, but only when there is no threat of some unwanted person getting hold of it. Such hijacking of information cannot be prevented but the harm intended can be thwarted by rendering that information meaningless for unintended receivers. This is achieved through cryptography.

Cryptography began as a symmetric version, representative techniques are DES and AES. It is a very secure style of encrypting data since the key used to "lock"

M. Jain (✉) · M. Singh
Gurukul Institute of Engineering and Technology, Kota, Rajasthan, India
e-mail: meenal.jain1985@gmail.com

M. Singh
e-mail: manojsinghq100@yahoo.com

N. Modi et al. (eds.), *Proceedings of International Conference on Communication and Networks*, Advances in Intelligent Systems and Computing 508,
DOI 10.1007/978-981-10-2750-5_45

data can only “open” the data. However, distribution of the secret key proves to be a major drawback of the Symmetric cryptography. A secure channel is required for the exchange of secret keys and the number of keys required for security is very large. Public Key Cryptography (PKC) then emerged as a solution containing separate keys for both encryption and decryption. Some popular Asymmetric Key Cryptography techniques are RSA [1] and Elgamal [2] Cryptosystems. Without compromising security, the encryption key is made public and by no means can the private key be determined from the corresponding public key due to computational impracticality. Though the number of keys involved is reduced as compared to the symmetric ciphers, key distribution and their validity are big issues.

Some approaches to reliable public key exchange are having *Trusted Centers, Controlled Trusted Centers or Certificate Authorities (CA)* for the purpose of creating *Public-Key Certificates*. These certificates prove that the published public key actually belongs to the user. The complexities associated with the whole process of issuing certificates make it a hard/heavyweight problem.

To simplify the problems in certificate management, especially in e-mail systems, Shamir [3] addressed the need of a public key that can be designed through the use of any arbitrary string related to some identity of the user further eliminating the need of public key certificates. He then proposed the concept of Identity-Based Encryption (IBE) [3] in 1984. Since then many IBE schemes have been proposed which can be broadly classified into pairing based and non-paring based. Paired IBE schemes like [4–7] are based on Weil and Tate Pairings [8, 9] which provide a mapping between the identity of a user and its public key through elliptic curve cryptography. Some non–pairing based IBE schemes are [10–13].

Practical acceptance of IBE systems is hindered by its computational complexity and cost of changing the entire software to adapt it as a substitute of PKC. In this regard, IBE systems involving a popular PKC system like RSA can be considered a practical solution. References [14–17] are such proposals. The only drawback is that instead of adding to the security of RSA, they just enhance manageability of keys in RSA while opening some vulnerabilities in the scheme.

This paper proposes an IBE system based on RSA which is able to thwart many attacks that are possible on RSA. Also, the ease of key management (a characteristic of IBE systems) is evident. There is no requirement of split keys or mediators like previous works, making it computationally more efficient.

2 Related Work

Identity Based Cryptography, as clear from the name, uses the user’s identity in place of public key and hence simplifies the management of public key no longer involving the use of public-key certificates. Shamir’s contribution [3] in this direction was an Identity-Based signature based on RSA, but a fully Identity-Based Encryption (IBE) was an open problem posed, the solution to which was the first IBE scheme by Boneh and Franklin [4]. The scheme is both reliable and provable and is based on Weil pairings over elliptical curves. Key revocation is a big issue in IBE since the key is derived from the identity of a user which practically cannot be changed. Gentry presented a solution [7] where a user creates own pair of public and private keys and obtains certificate from the Certificate Authority. The certificate serves an extra purpose of decryption key also. Though the key distribution problem is solved here through IBE yet certificates have not been eliminated.

Instead of reviewing all the IBE systems proposed till date, we focus on IBE schemes based on RSA.

2.1 IBE Combined with RSA

IBE did not gain popularity as a proper PKC due to two major reasons: first the novel constructs involved were much time and power inefficient as compared to exponential primitives of popular PKC systems like RSA [1] and ElGamal [2]; secondly these primitives of IBE were not compatible with existing PKC systems. Moreover, while IBE posed a solution to overreliance on certificates it could not much solve the revocation problem. All these issues were resolved to some extent by some researchers [14–16] through IBE systems based on RSA.

The first such proposal is due to Boneh et al. [14]. At the core of the system is a secure mediator (SEM), a semi-trusted server, which authorizes users to decrypt/sign messages. The idea is to split the RSA encryption key between the user and SEM. This not only provides security to RSA even when all the users share a common RSA modulus, but also facilitates immediate revocation of a user certificate.

Later Ding and Tsudik [15] presented a mediated IBE that could be used with both RSA and OEAP. A security proof is included for semantic security against adaptive chosen ciphertext attacks. This security depends on availability of key generation function which can generate division intractable public keys. This security is challenged by Elashry in his PhD thesis [16]. Three necessary conditions have been observed that are required for an Identity based mediated RSA (IB-mRSA) algorithm to be secure

- There should be a deterministic one to one mapping function that maps the identities of the users to their public keys
- This function should be division intractable
- The produced public keys must be co-prime with $\varphi(N)$.

Elashry [17] has proposed a letter-envelop technique to construct a secure IB-mRSA similar to a two-key encryption, one derived from user's identity and the other form SEM's identity. But only the key generation and management have been revised using the concepts of IBE while preserving the actual encryption primitive as RSA.

The review of IBE systems based on RSA and mediating approach can be summarized as: Building key generation of IBE system over a Mediated RSA assures a backward compatibility implying wide acceptance of an IBE system, but it also requires rigorous security analysis in face of attacks possible on RSA. It can be concluded that IBE has better key management but does not add to the security of RSA. Rather it may add to the vulnerability of RSA cryptosystem by making the modulus used by all users common.

3 Proposed Scheme

The structure of our proposal can be outlined as: in the setup phase a Public Key Generator (PKG) selects p and q as two safe prime numbers of bit-length equal to security parameter λ. The modulus n is computed same as in RSA, the product of p and q, and the number $\phi(n) = (p-1)(q-1)$ which is not made public by PKG. Key generation phase involves hashing of the user's identity and producing the pair of public and private key for encryption/decryption. The hash value of user's identity is computed through a simple hash design which derives a unique integer based on the identity string. If more security is desired then a secure hash algorithm can be used further. The encryption and decryption are similar to RSA with an added noise component. The encryption is the RSA ciphertext under public key e to which a random multiple of hash value I is added. Hence, decryption first needs to compute modulus of ciphertext under I and then proceed with conventional RSA decryption under private key d.

Algorithm Setup **(λ)**

Step 1: Pick safe primes $p, q \in Z_{2^\lambda}, p \neq q$
Step 2: $n = p * q$ //master key
Step 3: $\phi(n) = (p-1)(q-1)$

Algorithm KeyGen **(*ID*, *b*)**

Step 1: $I = \text{Hash}(ID)$
Step 2: Choose $e \in Z_{\phi(n)^*}$ such that $(I * x) \bmod \phi(n) = e$ for some positive x.
Step 3: If $(b = 0)$, output $d = e^{-1} mod \phi(n)$ //private key
Step 4: If $(b = 1)$, output e. //public key

Algorithm Encrypt **(*m*, *ID*, *n*)**

Step 1: $I = \text{Hash}(ID)$
Step 2: $e = KeyGen(ID, 1)$
Step 3: Pick positive random number r such that $I * r > n$.
Step 4: Output ciphertext as $c = m^e \bmod n + I * r$

Algorithm Decrypt (c, ID, n)

Step 1: $I = \text{Hash}(ID)$
Step 2: $d = KeyGen(ID, 0)$
Step 3: Output plaintext as $m = (c \bmod I)^d \bmod n$

Algorithm Hash (ID)

Step 1: Parse ID into a sequence of characters $\langle a_1, a_2, \ldots, a_L \rangle$, where L is length of ID.
Step 2: Convert each character into equivalent 256 bit integer through ASCII code as

$$\langle b_1, b_2, \ldots, b_L \rangle$$

Step 3: Compute integer equivalent of ID as $h \leftarrow (\sum i * b_i) * n$
Step 4: Apply hash and output as $H \leftarrow Secure\_hash(h)$

The *Secure_hash* used within the Hash primitive is any secure cryptographic hash like MD5 or SHA512. The time complexity v/s security trade-off exists here. The value 'h' obtained in Step 3 can also be used as the output of Hash primitive to save time since it is a unique value of known length associated to the identity string.

It can be clearly seen that above proposed scheme satisfies all the three conditions described in Sect. 2.1.

4 Performance Analysis

The major operation in encryption is the calculation of the ciphertext. In the encryption step, if fast exponentiation is used, then time complexity of "$m^e \bmod n$" sub-step is $O(\mu^2\lambda)$, where $\mu = |m|$. Decryption method is a single step which can be implemented through fast exponentiation and has bit operation complexity linear in λ. If size of result of $c \bmod I$ is also considered, an upper bound on runtime would then be estimated as $O(\lambda^2 * \lambda) = O(\lambda^3)$. The above analysis is done to estimate the bit operation complexities of the proposed cryptographic primitives. If integer operation complexities are considered then the estimates are more close to practical

Table 1 Bit operation and integer operation complexities of primitives

Primitive	Bit operation complexity	Integer operation complexity
Setup	$O(\lambda^2)$	$O(1)$
KeyGen	$O(\lambda^2)$	$O(k)$, k is the number of times x is tested
Hash	$O(\lambda L)$	$O(L)$
Encrypt	$O(\mu^2\lambda)$	$O(\|e\|) = O(\lambda)$
Decrypt	$O(\lambda^3)$	$O(\|d\|) = O(\lambda)$

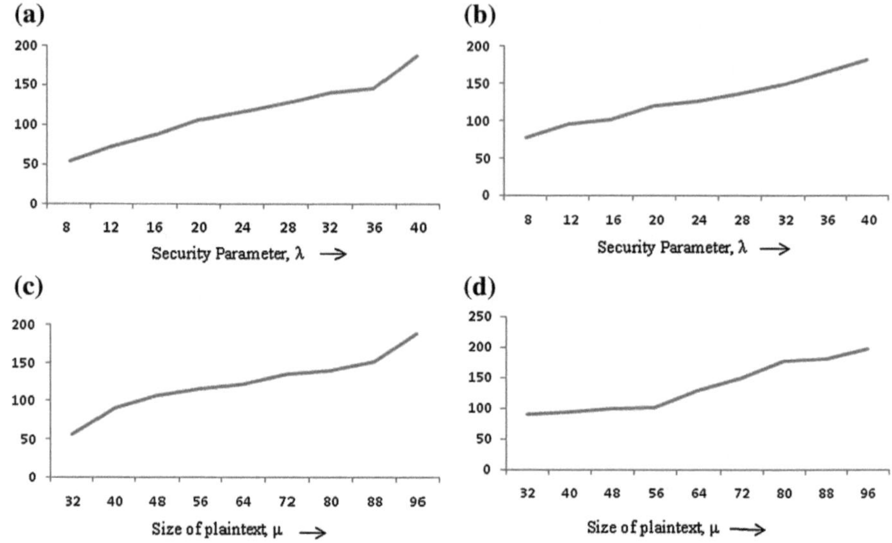

Fig. 1 **a** Growth of encryption time with increasing λ, **b** growth of decryption time with increasing λ, **c** growth of encryption time with increasing μ, **d** growth of decryption time with increasing μ

costs incurred. Table 1 lists the bit operation and integer operation complexities of all primitives.

For the empirical results, algorithms have been implemented as Java programs running on a 3.40 GHz Intel Core i3-2130 processor. The security parameter, λ is varied from 8 till 40 and the corresponding growth of setup, key generation, and encryption and decryption times is noted. Likewise, the size of plaintext, μ is also varied from 32 to 96 bits and the corresponding encryption and decryption times are noted. Figure 1 shows the plot for the growth of encryption and decryption time with increasing λ and μ. The growth is observed to be linear and conforms to the theoretical analysis.

The length of the ID, L is varied from 12 till 60 and its effect on the hash time is observed. The plot in Fig. 2 shows the growth of hash time for increasing L. The

Fig. 2 Growth of hash time with increasing length of ID

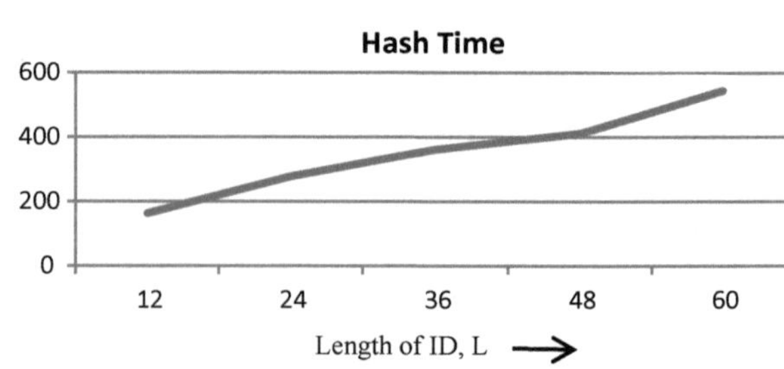

resultant plot is found linear to L, similar to the integer operation complexity of Hash primitive deduced in Table 1.

5 Security Analysis

5.1 Thwarting Attacks on RSA

Since the proposed cipher derives the basic structure from RSA cryptosystem [1], the attacks possible on RSA have to be analyzed. Though some backward compatibility has been sacrificed, this section proves that the proposal makes RSA further secure.

(i) **Factorization Attack**—The security of RSA lies in the large size of modulus. For the proposed algorithm to be secure, n should be more than 300 decimal digits, which further means that the modulus should be at least 1024 bits, which implies λ is equal to 512.

(ii) **Chosen-Ciphertext Attack**—This attack is possible on RSA due to its multiplicative property. The proposed encryption primitive uses the "addition of noise" concept which removes the multiplicative property. Hence, such attack is not possible until hash value of ID is not known.

(iii) **Coppersmith Theorem Attack**—This attack is possible only when value of e is selected low. The KeyGen primitive of the proposed algorithm does not produce low values of e, hence this attack is not possible.

(iv) **Broadcast Attack**—This attack on RSA is possible when same message with same exponent is sent to many recipients. The proposed KeyGen primitive derives value of exponent from ID of recipient; hence same exponent cannot be used.

(v) **Guessing d**—This attack requires an effort equivalent to computing inverse of public key modulo ϕ. This in turn is equivalent to factorizing n.

(vi) **Private Key Compromised**—In RSA if private key is compromised the whole cipher is cracked. The proposed scheme is safe even if d (private key) is compromised, the adversary should have access to the hash algorithm used by the authority. The hash primitive cannot be directly queried by any adversary, thus making the proposed scheme secure even if the secret key is compromised.

(vii) **Threat of Using same Modulus for All Users**—All IBE systems based on RSA have to make the modulus common for all users and protect against possible attacks by using mediation. The proposed scheme is safe against this attack because the encryption method adds a number larger than modulus to the ciphertext of plain RSA.

5.2 Thwarting Attacks on IB-MRSA

Introducing concepts of IBE in RSA leads to new vulnerabilities. Some of them have been resolved in our proposal. An IB-mRSA system is under threat if a user's public key is a factor of the product of the other users' public keys. In other words, public key of user U_1 should not be divisible by public key of other user U_2 else ciphertext under key e_1 can be decrypted by key e_2. But this is possible only due to multiplicative property of RSA, which has been undone in our proposal through noise addition.

6 Conclusion

Identity based Encryption is a good solution to the key distribution, certificate issuing and revocation problems of Public key cryptography systems. But IBE schemes have not gained acceptance as proper PKC systems due to high complexity and much change in existing software. Hence, IBE based on RSA were developed to maintain backward compatibility. But this opened more security threats in turn. This paper presents a solution how RSA can be made secure against prevalent threats and an IBE system can be constructed over it. The increased security does not cause any computation overhead. Also, a simple deterministic mapping of user ID to corresponding set of keys is proposed that can be sought as compared to costly SHA512. The proposal is more efficient than those based on mediated RSA since encryption/decryption is to be done only once.

The proposed work can be easily extended to signature schemes. A similar version using OEAP or ElGamal encryption can also be suggested. Comparison of runtime, theoretically and empirically, with other similar proposals is currently under consideration.

References

1. Rivest, R., Shamir, A. and Adleman, L.: A method for obtaining digital signatures and public-key cryptosystems. Communications of the ACM Vol. 21 (2) pp. 120–126 (1978).
2. Elgamal, T.: A Public Key Cryptosystem and a Signature Scheme Based on Discrete Logarithms. IEEE Transactions on Information Theory, Vol. IT-31, No. 4 (1985).
3. Shamir, A.: Identity-based cryptosystems and signature schemes. Advances in Cryptology, Crypto '84, LNCS, Vol. 196, Springer-Verlag, pp. 47–53, (1984).
4. Boneh, D. and Franklin, M.: Identity based encryption from the Weil pairing. Advances in Cryptology—Crypto 2001, Lecture Notes in Computer Science, Volume 2139, Springer, 213–229 (2001).
5. Boneh, D. and Boyen, X.: Efficient selective id secure identity based encryption without random oracles. Proceedings of the International Conference on Advances in Cryptology (EUROCRYPT '04), LNCS. Springer Verlag (2004).

6. Waters, B.: Efficient Identity-Based Encryption without Random Oracles. Advances in Cryptology—Eurocrypt 2005, Volume 3494 of LNCS, pages 114–127. Springer-Verlag, (2005).
7. Gentry, C.: Practical Identity-Based encryption without random oracles. Advances in Cryptography-EUROCRYPT 2006, Proceedings of the 25th Annual International Conference on the Theory and Applications of Cryptographic Techniques, St. Petersburg, Russia, pp. 445–464 (2006).
8. Menezes, A., Okamoto, T. and Vanstone, S.: Reducing elliptic curve logarithms to logarithms in a finite field. IEEE Transactions on Information Theory, Vol. 39, pp. 1639–1646 (1993).
9. Frey, G., Muller, M. and Ruck, H.: The Tate pairing and the discrete logarithm applied to elliptic curve cryptosystems. IEEE Transactions on Information Theory, Vol. 45, pp. 1717–1718 (1999).
10. Boneh, D., Gentry, C. and Hamburg, M.: Space-Efficient Identity Based Encryption without Pairings. Proceedings of the 48th Annual IEEE Symposium on Foundations of Computer Science, FOCS '07, pp. 647–657, Washington, DC, USA (2007). IEEE Computer Society.
11. Jhanwar, M. and Barua, R.: A Variant of Boneh-Gentry-Hamburg Pairing-Free Identity Based Encryption Scheme. Information Security and Cryptology, volume 5487 of LNCS, pp. 314–331. Springer Berlin Heidelberg (2009).
12. Elashry, I.F., Mu, Y. and Susilo, W.: Efficient variant of Boneh-Gentry-Hamburg's Identity-based Encryption without Pairing. 15th International Workshop on Information Security Applications (WISA 2014), LNCS, Springer-Verlag, pp. 257–268 (2014).
13. Zheng, M. Zhou, H. and Cui, G.: An Improved Identity-Based Encryption Scheme without Bilinear Map. Proceedings of the International Conference on Multimedia Information Networking and Security, pp. 374–377 (2009).
14. Boneh, D., Ding, X. and Tsudik, G.: Identity based encryption using mediated RSA. 3rd Workshop on Information Security Application, Jeju Island, Korea,. KIISC, (2002).
15. Ding, X. and Tsudik, G.: Simple identity-based cryptography with mediated RSA. Proceedings of CT-RSA 2003, LNCS 2612, Springer-Verlag (2003).
16. Elashry, I.: Pairing free identity based cryptography. Doctor of Philosophy Thesis, School of Computing and Information Technology, University of Wollongong (2015).
17. Elashry, I., Mu, Y. & Susilo, W. Identity-based mediated RSA revisited. Proceedings—12th IEEE International Conference on Trust, Security and Privacy in Computing and Communications, pp. 728–735 (2013).

Using Genetic Algorithm for Process Migration in Multicore Kernels

K.S. Shravya, Ankit Deepak and K. Chandrasekaran

Abstract Process migration is used in multicore operating systems to improve their performance. The implementation of the migration event contributes largely to the performance of the scheduling algorithm and hence decides how effective a multicore kernel is. There have been several effective algorithms which decide how a process can be migrated from one core to another in a multicore operating system. This paper looks further into the mechanism of process migration in multicore operating systems. The main aim of this paper is not to answer how the process migration should take place but it aims to answer when process migration should take place and to decide the site of process migration. For this, an artificial intelligence concept called genetic algorithm is used. Genetic algorithm works on the theory of survival of the fittest to find an optimally good solution during decision making phase.

Keywords Genetic algorithm · Process migration · Multicore operating systems · Process scheduling · Artificial intelligence · Decision making

1 Introduction

In the recent past, with the improving hardware such as ones with multiple cpu or core, there has been a need for efficient operating systems which can manage these hardwares and improve the performance. This class of operating systems have been

K.S. Shravya (✉) · A. Deepak · K. Chandrasekaran
Department of Computer Science and Engineering,
National Institute of Technology Surathkal, Mangalore, Karnataka, India
e-mail: shravya.ks0@gmail.com

A. Deepak
e-mail: adadeepak8@gmail.com

K. Chandrasekaran
e-mail: kchnitk@ieee.org

N. Modi et al. (eds.), *Proceedings of International Conference on Communication and Networks*, Advances in Intelligent Systems and Computing 508,
DOI 10.1007/978-981-10-2750-5_46

named multicore operating systems [1]. These operating systems, although currently in research stages, are gaining popularity. For process scheduling, multicore architecture has to balance load across all the cores dynamically and hence, they use a mechanism called migration [2].

Process migration is used to transfer an executing process from a source core to a destination core. Process migration can also be seen as a load balancing mechanism in multicore architecture. The steps involved in process migration are suspension of an executing process, transmission of the process from source to destination, establishment of the context of the process in the destination machine and restarting the execution in the destination machine. Process migration [3] works as a part of multicore scheduling to get an efficient execution of all the processes.

Genetic algorithm [4] is a method which aims to give a solution for optimisation problems and search problems. It is an evolutionary algorithm which tends to employ natural evolution processes of inheriting, selecting, mutating and crossing over of species. It follows three basic rules namely selection rule, which selects the parents for next generation, crossover rule which aims at combining two parents to form children for next generation and mutation rule which alters the existing parent to form the children [5].

In this paper, we try to use genetic algorithm for the described process migration. In the Sect. 2 of this paper, we look into the existing multicore operating systems and their scheduler. In the Sect. 3 of this paper we look into the mechanism of process migration in general. In the Sect. 4 of this paper we describe the basics of genetic algorithm which includes definitions associated with the algorithm and steps involved in the implementation. Section 5 of the paper aims to describe the proposed method of using genetic algorithm to optimise the mechanism of process migration. Finally in the Sect. 6, we give the outcome of the experimentation which comprised of the simulation of the proposed method.

2 Related Works

This paper continues the research in the development of multicore operating system. The paper tries to improve the scheduling of multicore kernel and the algorithm it especially targets is multicore round robin algorithm. Many research has been done for single core round robin algorithms before but the research in multicore round robin is fairly new. We try to use genetic algorithm [4] to implement process migration [2] in the round robin strategy. Genetic algorithm has been used for single core scheduling before [6].

Existing Multicore Operating Systems
Currently there are a few multicore operating systems which has been developed for research like Barrelfish [1], Akaros [7] and Baremetal [8]. Barrelfish aims at controlling the performance drop seen on increasing the number of cores in a multicore

operating system. Akaros aims to provide efficient support to parallel computing and HPC applications. It treats many-core processes as single entity and maintains asymmetry among its core for better management. Baremetal OS is an open source operating system which implements single address spacing throughout a light-weight kernel, providing high performance.

2.1 Multicore Process Scheduling

Gang Scheduling aims at running all the dependent threads and processes simultaneously on different processors [9]. Gang scheduling tries to avoid blocked waiting among the communicating processes by making sure that they are in congruent states when communication starts. Hence, this algorithm avoids the overhead of sleep and awake calls during scheduling.

Multicore FCFS uses only a single work queue. The incoming tasks are added to the queue and the front of the queue is executed by any of the free cores. This scheduling algorithm has the biggest strength in the form of zero context switch, but is vulnerable to a convoy effect.

Multicore Round Robin The whole execution timeline is divided into time slices. The algorithm is preemptive with very high context switches [10]. A very basic implementation works with individual ready queue for each core. The processes are entered into these queues on the sequence of their arrival.

Asymmetric Core Scheduling In such scheduler, each core is treated differently with a specific task performed by a specific core.

2.2 Process Migration

Process migration [2] is the mechanism of transferring process between two cores of a multicore system. Process migration helps in balancing of dynamic load and fault tolerance. Process migration starts with a migration request which is issued to a remote core. In response to the request, one of the processes executing on the core is detached from it and all the communication associated with it is redirected. The state of this process is then extracted and is transferred to the different core. Communication of the process is redirected to the core and the process restarts the execution.

3 Genetic Algorithm

Genetic algorithm, as explained earlier, is used for decision making and optimisation. It has also been used for scheduling [6].

3.1 Definitions

Individuals An individual is the entity to which the fitness function is applied.
Population Group of individuals is called population.
Trait Characteristic/Features of an individual.
Fitness value It is the individual's fitness function value.
Genome All traits of an individual collectively constitute is genome.
Diversity Variation of traits among individuals.
Fitness Functions This is the objective function which is to be optimised.
Selection Selection is the process of choosing or selecting individual genomes of a population, useful for producing new offspring.
Mutation Mutation is the process of maintaining diversity in population by altering the values of genes. User sets the value of the mutation probability function and care must be taken to set it low in order to avoid random search when the probability value is set high.
Crossover It is the process of producing a child by intermixing various parents' traits and is a crucial process in determining best trait.
Generation Every new population which is generated after each iteration by genetic algorithm is known as generation.
Inheritance Genetic algorithm uses the concept of inheritance to represent the process of transferring selected genes from parents to their children.

3.2 Algorithm

The following steps are involved in genetic algorithm execution:

- *Step 1*: The algorithm is started by defining a initial population of individuals.
- *Step 2*: The process of mutation and cross over are applied to generate child solutions from the selected parents solution.
- *Step 3*: After every generation the population which can give the highest fitness function value is used as the parent population for the next generation.
- *Step 4*: Steps 2 and 3 are iterated until a termination criteria is reached. The termination conditions may vary based on the needs of user for the best solution.

4 Genetic Algorithm for Process Migration

The paper tries to use genetic algorithm to make decisions like when a process migration should take place, which core should be the source and which core should be the destination, etc.

4.1 Initialisation of Genetic Algorithm

Chromosome (C) Each process (P) will act as the building block of this algorithm

$$C_i = P_i \tag{1}$$

Individual (I) During the implementation the process queue of each core

$$I_i = \{P_a, P_b, P_c \ldots P_n\} \tag{2}$$

Population (Ppl) The set of all the queues for all the core forms the population

$$Ppl = \{I_a, I_b, I_c \ldots I_n\} \tag{3}$$

Crossover (x) Crossing over will take through process migration and will an important part in the algorithm

$$I_a * I_b = \left(I'_a, I'_b\right) \tag{4}$$

Mutation (M) Mutation takes place in the form of the execution of the process. Also, to better the average turnaround and waiting time, smaller processes can be moved ahead in the queue

$$M(I_a) = I'_a \tag{5}$$

Fitness Value (f) The factor which defines the fitness of an individual is the time it is going to complete the execution

$$f_i = Execution\ time\ remaining\ (I_i) \tag{6}$$

Fitness Function (F) The fitness of the population will be the reverse of maximum of the time needed to complete the execution of all the processes in the queue of each core. The aim of the algorithm is to minimise this function (Fig. 1)

$$F = 1/MAX(f_i) \tag{7}$$

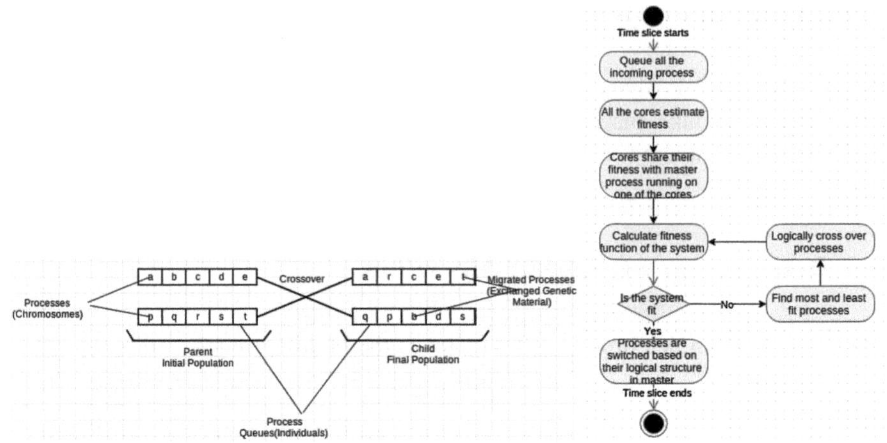

Fig. 1 Proposed algorithm

4.2 *Initial Population*

Since this algorithm is designed for multicore round robin scheduler, the initial population will be same as the initial processes in the queue, which can be added in a FCFS manner.

This algorithm will not have a constant population in the practical scenario because processes can join the queues at random times and will leave the queue after their execution. During each iteration, we need to update the population accordingly.

4.3 *Crossover*

Whenever there is a change in the work load of the system, each core will calculate their fitness value. For simplicity, an efficient fitness value can be the time remaining to complete all the processes currently on the queue of the core. This fitness function can be calculated like the remaining time of execution is found for the shortest remaining job first algorithms.

Theoretically, the crossover, here can be done by two different methods. The first method will be to actually switch process from its current queue to the next queue and the fitness function is calculated after the switch and the next iteration is initiated. The disadvantage of this approach is that every process migration will be accompanied with an overhead and hence will be infeasible.

The other and more feasible approach is to initiate an highest priority process, the aim of this process is to estimate the fitness of all the queue. The individual processes now work together to calculate the fitness function of the population.

During crossover, these processes now share the fitness value of the process to be migrated, and the required crossover processes are finally migrated to their destination core found during that iteration of the genetic algorithm.

During the iteration, the process should be migrated in this manner; Suppose A is the core with a very high remaining time, can be said to be unfit, and B is the core with low remaining time, then A chooses to migrate a process with remaining time closest to the |Fitness (A) – Fitness (B)|/2 from its queue to B in the first iteration, similarly all the iteration tries to make the unfit individual fitter in the next generation.

4.4 Selection

In each iteration, all the records of the process migration can be stored in a list at the end of the iteration. The implementation of this data structure can be crucial to reduce time take by the process of the migration. The data structure used for this implementation can ideally be a priority queue with the priority given based on the fitness of the population. Implementation using a priority queue will allow easy access to the fittest population.

4.5 Termination

The generation can be stopped after a chosen time, T. In the implementation T can be a linear function of time slice with a constant k where k will be below 0.1 for ideal implementation. Another method to stop the generation can be the number of iterations.

The termination will be marked by the beginning of the scheduling time slice. At the termination of the algorithm, following actions will take place:

- Processes migrate to the core as per the arrangement of the fittest population.
- The priority queue will be reset, to prepare it for the next iteration.
- The system scheduler will be resumed to the round robin.

5 Experimental Results

During our experimentation, we scheduled 20 randomly generated processes using three multicore scheduling algorithms, namely, multicore FCFS, multicore round robin and the proposed algorithm. For the purpose of our of experiments, we implemented all the three algorithms. On of the best way to test a multicore scheduling algorithm is to modify the existing kernel of Baremetal OS. The

Fig. 2 Total time taken for all the process to get executed across all the cores

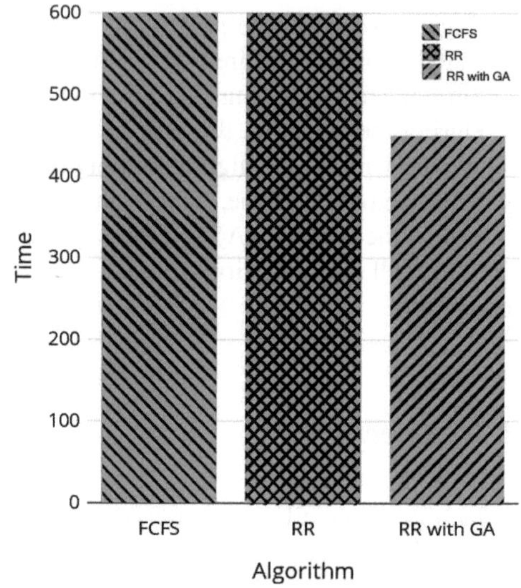

operating system has built in system calls for context switch. The processes can be made to change queues using the os_smp_enqueue and os_smp_dequeue system calls (Fig. 2).

The first observation was that the maximum of the execution time across all the cores showed similar values in FCFS and round robin algorithms but as the goal of the proposed algorithm was to reduce the average latency across cores and bring fairness in execution across all the cores, the results were seen to be better than that of both.

The proposed algorithm improved the difference in the time of the fastest and slowest core, nullifying the latency induced by slower cores, and hence, ended the

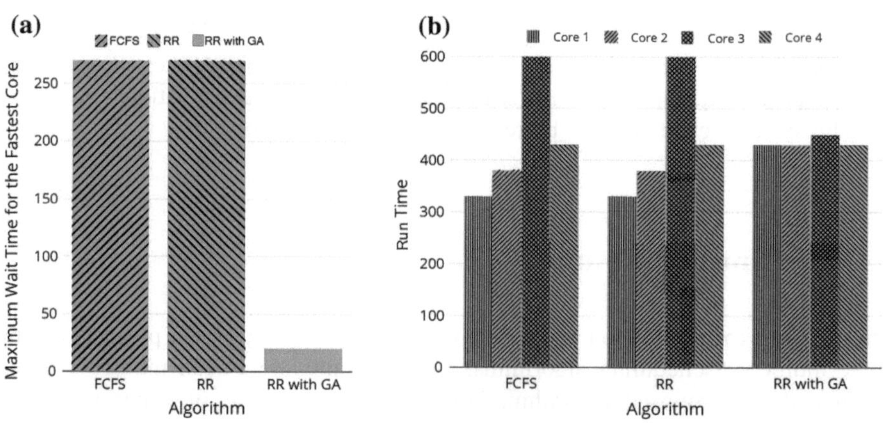

Fig. 3 **a** Time difference of fastest and slowest core, **b** execution time across all the cores

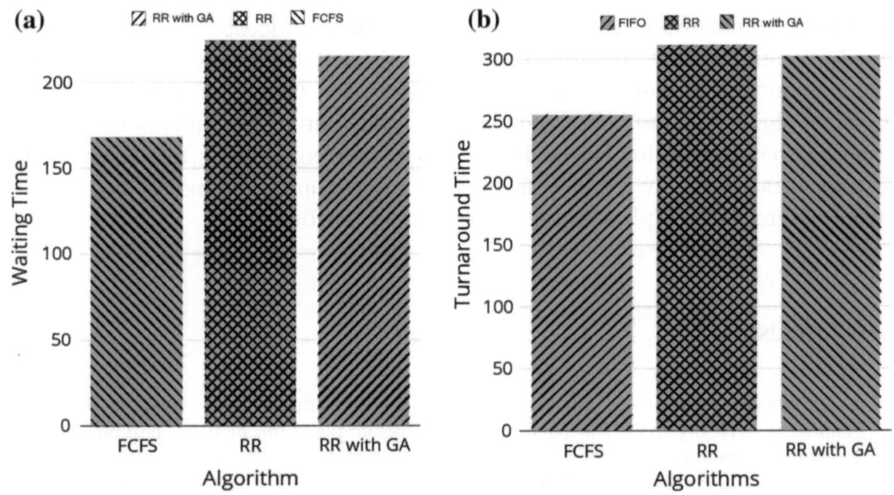

Fig. 4 **a** Average waiting time, **b** average turnaround time

execution faster than the other two. The algorithm didn't give any uniform speed up or slowdown in all the four but the behaviour across the four cores can be summarised as the process migration made the slower cores faster at the cost of the faster cores so that the overall execution is faster than the previous algorithms (Figs. 3 and 4)

Apart from the fairness across the cores the algorithm also showed a slight improvement in the execution time of individual processes from the round robin. We observed the value of the average waiting time and average turnaround time for the set of processes. FCFS algorithm shows the best performance here because the initial processes have very low waiting time, in spite of these the problems of convoy effect and unfairness still remains in all FCFS algorithms but the proposed algorithm bettered the performance of simple round robin algorithm.

The experimental analysis showed satisfactory results for the algorithm and showed it to be an improvement to multicore round robin algorithm in terms of fairness among cores.

6 Conclusion

We proposed an algorithm which uses genetic algorithm to event process migration in multicore round robin algorithm. Our simulations gave agreeing results. One of the property of genetic algorithm is that it does not try to give the best solution but it tries to give an optimally good solution. So, it can be said that the result of this research can be improved even more depending on the implementation of genetic algorithm. Further, the algorithm can be implemented along side the different variants of round robin algorithms which has been proposed before. We have already mentioned that the aim of the paper is not to propose another process migration algorithm but to look for the best time when the existing process migration algorithms can be implemented, also, the aim of the paper is not even to propose a new scheduling algorithm but to implement process migration in the existing ones because we feel with the increasing number of multicore systems, process migration will play important role in the schedulers.

References

1. A. Schpbach, S. Peter, A. Baumann, T. Roscoe, P. Barham, T. Harris, and R. Isaacs. Embracing diversity in the Barrelfish manycore operating system. In Workshop on Managed Many-Core Systems, (2008).
2. Zarrabi, A.; Samsudin, K.; Ziaei, A., "Dynamic process migration framework," In Information and Communication Technology (ICoICT), International Conference of, pp. 410–415, (2013).
3. F. Douglis and J. Ousterhout. Transparent Process Migration: Design Alternatives and the Sprite Implementation. Software—Practice and Experience: 757785, (1991).
4. Pengfei Guo; Xuezhi Wang; Yingshi Han, "The enhanced genetic algorithms for the optimization design," in Biomedical Engineering and Informatics (BMEI), 3rd International Conference on, pp. 2990–2994, (2010).
5. Chaiyaratana, N.; Zalzala, A.M.S., "Recent developments in evolutionary and genetic algorithms: theory and applications," In Genetic Algorithms in Engineering Systems: Innovations and Applications. Second International Conference On, pp. 270–277, (1997).
6. Maktum, T.A.; Dhumal, R.A.; Ragha, L., "A genetic approach for processor scheduling," In: Recent Advances and Innovations in Engineering (ICRAIE), 2014, pp. 1–4, 9–11 (2014).
7. B. Rhoden (2015, November 23), Akaros, http://akaros.cs.berkeley.edu/akaros-web/news.php.
8. Ian Seyler (2015, November 23), Baremetal, http://www.returninfinity.com/baremetal.html.
9. Ishikawa, Y., "Highly Efficient Gang Scheduling Implementation," In Supercomputing. IEEE/ACM Conference on, pp. 43–43, (1998).
10. Grunewald, W.; Ungerer, T., "Towards extremely fast context switching in a block-multithreaded processor," In EUROMICRO 96. Beyond 2000: Hardware and Software Design Strategies., pp. 592–599, (1996).

An Extensive Conception of Reusability in Software Component Engineering

Devesh Kumar Srivastava and Priyanka Nair

Abstract In early 1960s, intricacy of software systems led to a call for the emergence of the concept of Software Reuse. Rather than building software applications from genesis, software reuse consents creating software systems from existing software. Efficient software reuse programs implemented by the firms may increase their productivity and value, thereby giving the organizations headway. Several reuse metric and models reign the software industry. Reuse assessment commit to high quality and economic system development. Despite its commencement as a potent vision, software reuse has botched to become a part of the typical software engineering practice. The paper is an attempt to articulate the notion of software reuse and the concerning issues. Reusability facet has been conferred analogous to OO paradigm and agile development. Here the concept of reuse has been addressed as a combination of artifacts as well as individual components.

Keywords Software reuse · Reusability · Reuse approaches · Software reuse metrics · Agile software development · Object oriented paradigm

1 Introduction

Reconstruction of new systems pertaining to changing requirements is not viable. Software components can be used time and again for creating new systems and applications. Components can be integrated into software systems. Everything associated with a software that can be reused is termed as software reuse. Software Reuse leverages the project structure and cost effective issues of software engineering. However, Reusability is difficult to maintain and its inclusion in new systems is even more severe [1]. The NATO Software Engineering Conference,

D.K. Srivastava (✉) · P. Nair
Department of CSE/IT, Manipal University, Jaipur, India
e-mail: devesh988@yahoo.com

P. Nair
e-mail: priyankanair@live.com

N. Modi et al. (eds.), *Proceedings of International Conference on Communication and Networks*, Advances in Intelligent Systems and Computing 508,
DOI 10.1007/978-981-10-2750-5_47

1968, gave the prime valuable coverage to the bottlenecks of software engineering. From amongst various experts who attended the conference, McClory in his working paper proposed the notion of necessity and adequacy of reusable component factory. He contended the effectiveness of using component libraries for various system processing and computations [2, 3]. The code level reusability is coherent as compared to the conception of specification and design reusability which is challenging [1]. The problem of dealing with software reuse is the radical fixate with additional proposition of measurement of reusable components.

2 Approaches of Software Reuse

To realize software reuse work, conventional approaches are employed. On the frontier, the classification is primarily based on component level and process level. Reuse based on object of reuse or component is the *Compositional Reuse approach* whereas process reuse fall under *Generative Reuse approach.* In sync, these approaches serve as reuse aid to the system [3, 4].

Compositional Reuse appropriates the notion of reusable objects that are unaltered during reuse. It is a bottom up system development. Combinative accessions of simpler components frame obscure and complex objects [1]. Components that are compatible with reuse support features are archived in repositories. Retrieval is a key feature here. Components are dispersed segments which benefits the developers to achieve high productivity. *Generative Reuse* is reuse of process rather than product. Parsers and Lexical analyzers are based on generative approach. Reusable pattern generation is taken into account before assimilating objects of reuse into the program [1].

3 Types of Reuse

Reusability scrutinized over domain scope can be categorized in two forms: *Vertical Reuse* and *Horizontal Reuse. Vertical Reuse* is generative in nature. However it has not yet been widely accustomed in software business industry. In software development it has an impending and extensive connotation [5]. However, *White Box Reuse* is strenuous to maintain. It is an elemental form of *Vertical Reuse*. Code is modeled as the reused entity for white box reuse. The access to the source code and implementation is required herewith. Reuse is met with alteration and adaptation as the core [6, 7] *Horizontal Reuse* is widely accustomed across applications. It follows compositional reuse approach. *Black Box Reuse* forms the domain component of horizontal reuse. Component reuse is carried out without modifications. It employs Commercial Off the Shelf (COTS) which is a third party application. They are economical and reliable. COTS components are incorporated in already built software in order to provide additional services. However, they are employed for general applications [8]. It is well appropriated as Black box reuse as they are perceived only in terms of input and output without taking into account the functionality.

4 Reuse Assessment

In order to identify the effectiveness of various reuse methods, it is imperative to quantify and assess them. Various software pertinent metric can be employed as quantitative index to measure the reusability in terms of software assets: *product* and *process*. Some of the reuse metric models have been taken herewith.

4.1 Cost/Productivity Metric Model

There is an additional cost associated with software reuse. Reuse cost is viewed as an investment. Reuse incurs added cost to the traditional software development process. The cost model was based on cost benefit analysis [6]. The two models for cost and productivity commit to the cost of *reusing* software components and the cost of *developing* objects of reuse. The software reused is decisive and reliable thereby conforming to the black box properties. Apropos the properties, enough documentation related to the objects of reuse is available but the size remains non-existent. Negligible cost is associated with the reuse of components [9]. For estimating the relative size of reusable components, it is required to measure the size of object of reuse with the hypothesis that they are built from scratch. The relative size, R of reusable components is hereby articulated as:

$$R = \frac{S_R}{S_R + S_E} \tag{1}$$

where,

S_R estimated size of reusable components

S_E effective size of reusable components which is a regulated consolidation of altered and new source code.

The higher order cost model estimates the cost of developing objects of reuse. Let C_D be the relative cost of developing the software product corresponding to all current code and b is the cost relative to all new code, of using the reused code in the new product. C_D and b for all new code is assumed to be 1. The relative cost of software development is presented as follows:

$$C_D = 1\,(1 - R) + bR \; Or \; C_D = R\,(b - 1) + 1 \tag{2}$$

where,

R: proportion of reused code in the product $(1 - R)$: proportion of all new code.

According to Gaffney and Durek, when only source code is reused b = 0.85 whereas when requirements, design and code are reused b = 0.08. It is because in the former case all other phases are required to be endured [8].

The productivity, P is the inverse of *cost metric*

$$P = \frac{1}{C_D} \tag{3}$$

or

$$P = \frac{1}{R(b-1)+1} \tag{4}$$

Developing Software with reusable component incurs more cost as compared to developing software without reusable objects [6, 9].

Consider a small module of an ongoing Omnicare Healthcare Management Project. We will use the Gaffney and Durek Model to calculate the relative cost and productivity of the project module with reuse components.

Table 1 indicates the development cost corresponding to different stages of development of the module. Prototyping or design phase incurs the maximal cost with respect to other development stages (Fig. 1). Relative reuse cost can be calculated by taking into account the objects of reuse with additional activities. If we take source code as our object of reuse then requirement analysis and testing are to be performed as accompanying tasks. Here b = 0.33 (0.07 + 0.26) similarly, when

Table 1 Relative development cost corresponding to different phases of developing the module

Development phase	Relative development cost (phase wise)
Requirement analysis	0.07
Prototyping	0.42
Implementation	0.25
Validation and testing	0.26

Fig. 1 Cost of developing the software corresponding to different phases of module development

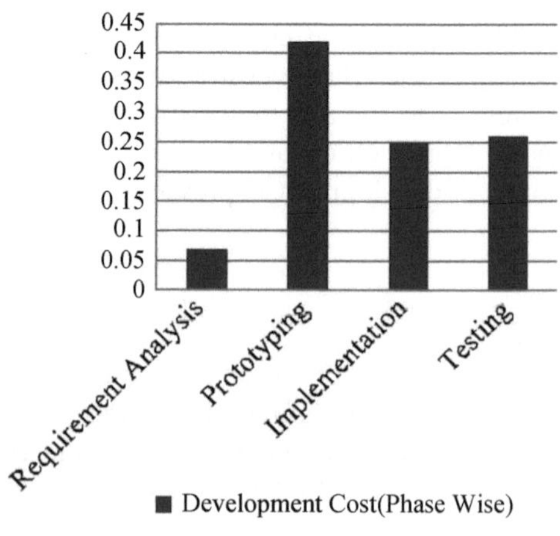

Table 2 Integration cost of reusable component

Object of reuse	Accompanying tasks to be completed	Relative cost of component with reuse (*b*)
Requirement	Prototyping, Implementation, Testing	0.93
Prototype	Requirement Analysis, Implementation, Testing	0.58
Code	Requirement Analysis, Testing	0.33
Requirement, Prototype, Code	Testing	0.26

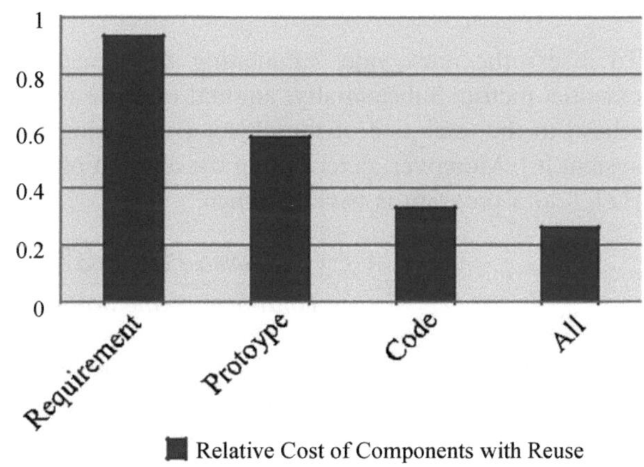

Fig. 2 Relative cost of components with reuse (cost of integrating reusable components)

requirement is reused then the relative reuse cost will be 0.42 + 0.25 + 0.26; i.e. b = 0.93. Table 2 shows the relative reuse cost of development. With requirement taken as the object of reuse holds the highest relative reuse cost (Fig. 2).

Suppose there are 10,000 lines of code in the original software application. Assuming 4700 lines of code is reused. Here $C_D = 1$ (*relative cost of development of all new code is assumed to be* 1). The proportion of reused code R is 0.47 ($R \leq 1$). From Fig. 2, b = 0.33 for source code taken as the reused component. *b* is the integration cost of reusable component. Using the Gaffney and Durek model, we can calculate the cost and productivity of development of system with source code as the object of reuse.

$$C_D = 1\,(1 - 0.47) + 0.33\,(0.47) = 0.69 \tag{5}$$

Taking source code as the object of reuse, the relative cost and productivity of development of the module is estimated as 0.69 and 1.4 respectively. Cost of developing the module relative to all new code is assumed to be 1.

4.2 Maturity Metric Model

Maturity metric model is used to assess the implementation and effectiveness of systematic reuse activities. The advancement of reuse programs is measured on an ordinal scale [10]. Kolton and Hudsen Reuse Maturity Model is a five level model that directs an organization for effective reuse of activities in order to achieve maximal performance.. The levels are: *initial/chaotic*, *monitored*, *planned*, *coordinated* and *ingrained.* At the onset of any program, organizations are traceably between *initial/chaotic* and *monitored* level. Post *ingrained level*; reuse is viewed as a unified part of the system [1, 10].

4.3 Percent Reuse

To assess the reuse rate, calculating the percentage of reuse, is viewed as an essential metric. Substantially, amount of reuse is the ratio of the amount of object reused to the total size of the object considering the life cycle of the program or system [6]. Moreover, determining the amount of reuse on account of *lines of code* (*LOC*) in a program is trivial. Hence,

$$\%Reuse = \frac{Reused\ LOC\ in\ a\ Software}{Total\ Size\ of\ the\ Software\,(LOC)} * 100 \tag{6}$$

Consider a software application with 10,000 lines of code. Assuming 4,700 lines of code of the same application is reused to develop a new software product. The percentage of reuse is calculated as follows:

$$\%Reuse = \frac{4700}{10000} * 100 = 47\% \tag{7}$$

Higher percentage of reuse is indicative of better reuse rate.

5 Agile Development and Reuse

The extension of reusability concept in agile development is complex. The agile development is the software development methodology that focuses on continuous improvisation with effective communication between people. However, there is minimal documentation which makes reusability critical. The major limitation with reuse in agile environment is the difficulty in continuous redesign, due to paucity of application-specific artefacts. [11] Reusability can be employed with agile software development in three ways. The methodologies used for incorporating reusability in agile development *Component Based Development (CBD)*, *Refactoring* and *Reusable Architectures. CBD* validates the component in conformance with the suitability for reuse. *Refactoring* reorganizes and remodels an existing code. It can be viewed as a template or design that can be employed in varied scenarios pertaining to requisite applications. Architectural patterns may be used to develop *reusable architectures* [11, 12].

6 OO and Reuse

In the 3rd International conference on software reuse it was substantiated that object oriented paradigm does not validate to be the necessary and sufficient condition to support reusability. Some of the credos of OO obstruct the reuse and hence there is need to be very cautious while incorporating its features [13]. Despite the restraints, OO approach complements features that support software reuse. The Object Oriented paradigm considerably enhances the productivity with reuse in an elemental role. The assortment and contrast of programming languages becomes a key consideration when incorporating reusability to improve the productivity [3, 9]. OO braces both types of reuse. *Inheritance* backs White Box reuse whereas *Client-Supplier relationship* supports black box reuse [10]. Much of the assistance of Object Oriented paradigm to reusability is on probation. However, some of the precepts of OO approach need to be scrutinized with respect to violations to the reuse support.

7 Issues with Reusability

Software reuse work can only be accomplished with the employment of any or both of the *reusable assets*: product and process. However, building software with the reusable assets raises certain methodological and technical concerns. Issues are often related to spotting of the reusable assets and identifying their conformance to the current requirements. For ensuring the adaptation of these reusable assets to the current needs automation of the reusable components is met employing OO features [3]. There is very little tool support for locating the reusable components and maintaining a component prospectus. However certain tools like CASE tools are viewed as a way to improve and promote reuse in software projects in organizations. Computer Aided Software Engineering (CASE) tool is used for retrieval of reusable components from a software catalogue [14]. Reusable Components are fragments that are stored in repositories. Another major issue that crops up is the reuse barrier. Efficient retrieval is necessary for development of reusable software. The repository keeps changing constantly which makes it difficult for the developer to foresee the occupancy of object of reuse [15]. Also, there is an additional cost associated with development of software with reusable components. It is more challenging to reuse specification and design as compared to the code reuse. To reuse specification/design a reserve of solutions is required to be searched in a problem-oriented demeanor [16].

8 Conclusion

The above sections elucidate the imperative aspects of software reuse. Software reuse reinforces software productivity and quality. Software repository is essential for maintaining the catalogue of reusable software components. Reusable assets are conferred in terms of product and process. To carry out software reuse, component level and process level reuse approaches are prevalent in industry. From amongst the various software metrics employed for the measurement of different reuse techniques, cost/productivity metric model, maturity assessment and percent reuse estimation have been taken up. The cost is negligible when reusing the components. However an additional cost is incurred when developing software with reusable components. Despite the efficacy of reusability, there are issues raised while developing software with reusable components. With Agile development, reusability facet becomes complex with the limitation of documentation. However reusability can be incorporated with the agile environment employing various methodologies. Also, even though OO paradigm has been widely used for supporting conception of reuse, it has not been empirically proven to be the necessary and sufficient condition for reuse support. Reusability in Agile method is an open area for researchers for further improvement.

References

1. Sametinger, J *Software Engineering with Reusable Components.* Springer Science and Business Media. (2013).
2. Naur, P., Randell, B., & Committee, N. S. *Software Engineering*: Report of a conference sponsored by the NATO Science Committee, Garmisch, Germany, 7–11 Oct. 1968. NATO Software Engineering Conference, (October 1968), 231. http://doi.org/10.1093/bib/bbp050 (1968).
3. Jalender, B., Govardhan, D. a., & Premchand, D. P. (2010). A Pragmatic Approach To Software Reuse. Journal of Theoretical and Applied Information Technology (JATIT) Vol, 14, 87–96. Retrieved from http://www.jatit.org/volumes/research-apers/Vol14No2/3Vol14No2.pdf.
4. Czarnecki, K. Overview of Generative Software Development. Unconventional Programming Paradigms, 3566, 326–341. http://doi.org/10.1007/11527800_25. 2005.
5. Jamwal, D. *Software Reuse : A Systematic review.* Proceedings of 4th National Conference; IndiaCom, 1–7, (2010).
6. Marshall, J. J., & Downs, R. Reuse readiness levels as a measure of software reusability. International Geoscience and Remote Sensing Symposium (IGARSS), 3(1), 1414–1417. http://doi.org/10.1109/IGARSS.2008.4779626, (2008).

7. Dosch, W., Lee, R.Y., Wu C. *Software Engineering Research & Applications*. Springer, 172 (2006).
8. Galorath, D. *Software Reuse and Commercial Off-the-Shelf Software*, Galorath Incorporation, El Segundo, CA. 1–22. Retrieved from http://www.compaid.com/caiinternet/ezine/galorath (2007).
9. Tripathy, P., Naik, K. (2014). *Software Evolution and Maintenance*. John Wiley & Sons.
10. Soora, S. K. *A Framework for Software Reuse and Research Challenges*, IJARCSSE, 4(10), 441–448, (2014).
11. Spoelstra, W. J. T. *Reusing software assets in agile development organizations—a management tool*. University of Twente, Hengelo, (2010).
12. Singh, S., & Chana, I. *Enabling Reusability in Agile Software Development*. International Journal of Computer Applications, 50(13), 33–40. http://doi.org/10.5120/7834-1132, (2012).
13. Patidar, R., & Singh, P. V. OPEN ACCESS A Survey of Software Reusability, 4(8), 96–101, (2014).
14. Sharma, A., Grover, P. S., & Kumar, R. Reusability assessment for software components. ACM SIGSOFT Software Engineering Notes, 34(2), 1. http://doi.org/10.1145/1507195.1507215. (2009).
15. Shiva, S. G., & Shala, L. A. S*oftware Reuse: Research and Practice.* Information Technology,. ITNG '07. Fourth International Conference on, 603–609. http://doi.org/10.1109/ITNG.2007.182, 2007.
16. Sharma, K., Agnihotri, N., & Hooda, M. Software Reusability: Possibilities From The Existing Software, 97–99. (2013).

Opportunistic Location Update—A Novel Cost Efficient Reactive Approach to Remove Pauses in Cellular Networks

Kalpesh A. Popat and Priyanka Sharma

Abstract In the era of mobile communication, every user is free to move from one place to another without worrying about the connectivity to the network. Whenever a use with a cellular phone changes its location, his location must be updated to the database so that the same can be reached later on. Some devices remain almost stationary as they don't move out of a specific area like an office or a home. At the same time, some devices move frequently as the users are carrying them while travelling. In every case, if location updates are not properly performed, users will not get connectivity all the time. This paper explains some of the most widely used location update strategies in mobile communication. As mobile communication is a kind of wireless communication, we have tried to introduce a novel location update strategy by modifying the concept of opportunistic networking. The paper concludes with some of the innovative ideas of doing location updates.

Keywords Cellular networks · Location update · GSM · Location update cost reduction · Opportunistic location update

1 Introduction

True location information is very difficult to manage in cellular networks for all the users at all the time. Mobile computing allows user to access the network at all the time from every place. In GSM based networks, the geographical area is divided into a hexagonal areas called cells. Every cell has a BTS and BTSs are connected

K.A. Popat (✉)
Marwadi Education Foundation's Group of Institutions, Rajkot, Gujarat, India
e-mail: kppopat@gmail.com

K.A. Popat · P. Sharma
Gujarat Technological University, Ahmedabad, Gujarat, India
e-mail: pspriyanka@yahoo.com

P. Sharma
Raksha Shakti University, Ahmedabad, Gujarat, India

N. Modi et al. (eds.), *Proceedings of International Conference on Communication and Networks*, Advances in Intelligent Systems and Computing 508,
DOI 10.1007/978-981-10-2750-5_48

with each other through BSC. A set of BSCs are connected with MSC. As this paper is exclusively about the location updates, we are not discussing GSM architecture in detail. The important factor is the mobility. Whenever there is mobility, there is a change in location. In a cellular network, whenever a user changes its location, his device's new location must be updates with the cellular network database. So in future, if a call or a SMS or a data packet needs to be forwarded to that device, cellular network can find the new location easily. There was a standard way of doing so in data networks with the concept of home agent and a set of foreign agents. Such scenarios are applicable with IP based networks. Here we are discussing location updates with exclusive cellular networks where mobility is frequent and the only way we can find the location is through the BTS to which the device is currently connected. The next topics discuss few of the most widely used location update strategies. A novel location update strategy called opportunistic location updates is proposed [1].

2 Location Update and Its Cost Calculation

Location update has two major factors to be considered, Location Inquiry Cost and Location Update Cost. Location Inquiry Cost − Paging Cost is mainly network initiated while Location Update Cost is network as well as device initiated. The location update cost can be calculated as below [2].

$$\text{Total Cost} = \text{C} * \text{N}_{\text{LU}} + \text{N}_{\text{P}} \tag{1}$$

2.1 Distance Based Location Update Strategy

This is the most fundamental strategies among all others for Location Update LU-procedure. The distance is the primary factor for decision making. BTS records the distance travelled by each device since its last update. The distance is recorded in terms of number of cells travelled through. LU process is initiated when a device travels through more number of cells than predefined threshold for distance D. To keep the track, with each cell boundary crossing, the counter for a device is incremented by 1 and this process continues till counter becomes equal to D. The last location is always available so paging cost is less but with each cell change, there is a little processing so processing overhead is there. Figure 1 shows an example of this strategy with D = 2. Here a device starts mobility with cell A and it moves from this sequence of cells A → B → C → D → E → F cells. The LU occurs when device reaches E [3].

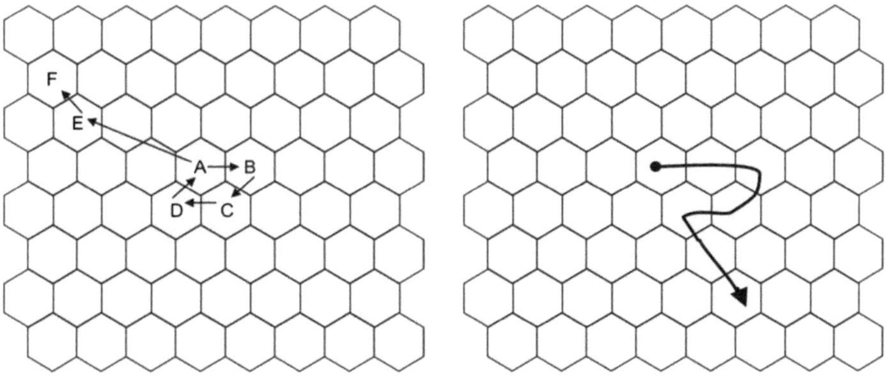

Fig. 1 Distance based location update

2.2 Time Based Location Update Strategy

In this method, LU occurs at a regular fixed interval say T seconds. Doesn't matter whether LU is needed or not but it occurs for every device after every T seconds. BTS maintains a timer for every device since its uptime. The advantage is it is easier to implement than distance based but the disadvantage is that when there is no mobility, then even for that device LU is performed every T seconds [4].

2.3 Movement Based Location Update Strategy

In this method, BTS records the number of turns the device takes during its mobility. When this information becomes equal to movement threshold M, LU is performed. Figure 2 is an example of LU with M = 4. Here a device starts mobility with cell A and it moves from this sequence of cells A → B → C → D → E → F cells. The LU occurs when device reaches E. The main disadvantage is if a device moves in straight direction without any turns, this method fails [5].

2.4 Profile Based Location Update Strategy

Here every device's mobility pattern related profile is maintained either at BSC or MSC level. The main idea behind mobility pattern related profile is to analyse the user's habits to be mobile his devices. For example, during office hours, a user's device will be at a cell where his working place is situated. At the same time, during nights, a user's device will be at a cell where his home is situated. Such information is not permanent and fixed for every day, but in majority of the cases it is as it reflects the daily schedule. Based on such information, LU process can be scheduled at a particular time, especially when user is predicted to move [6].

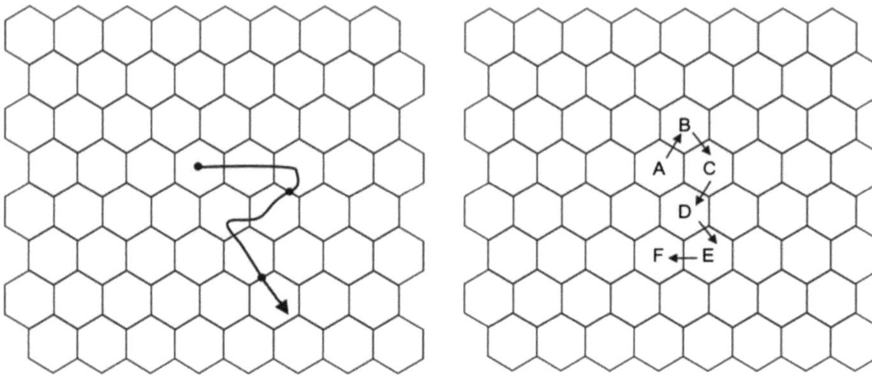

Fig. 2 Movement based location update

3 Opportunistic Location Update

Location update is one of the most crucial issues to be considered. LU is the necessity to provide smooth connectivity to the users while they are mobile. The entire process must be as light as possible and as efficient as possible. Opportunistic Location Update—OLU is a special category of location update strategies where the resources required to perform LU are dynamically decided and allocated as per the current scenario of the cellular network [7].

OLU is a process which mainly focuses on two improvements: (1) It follows the concept of opportunistic communication. (2) It tries to reduce pauses in cellular communication. Opportunistic networks are different than of cellular networks. Here we are introducing the base of opportunistic networks to the cellular networks. In a cellular network, every BTS has a large number of mobile devices under its roof. BTS is responsible to manage such devices. At present, in GSM, under a single BTS, a set of devices are there, but they don't form any kind of network. In our proposed work, we want them to form an opportunistic network when needed and if possible. As the owner of every device is different, and every device is meant for some specific purpose, we can't expect every device to play a predefined dedicated role which is the main reason why opportunistic network is chosen. Those devices who are presently inside the roof of a BTS as well as which are willing to be a part of such network can take part, otherwise rest of the devices will work as independent devices of cellular networks.

3.1 Opportunistic Communication

Figure 3 is a simple example of how an opportunistic network works.

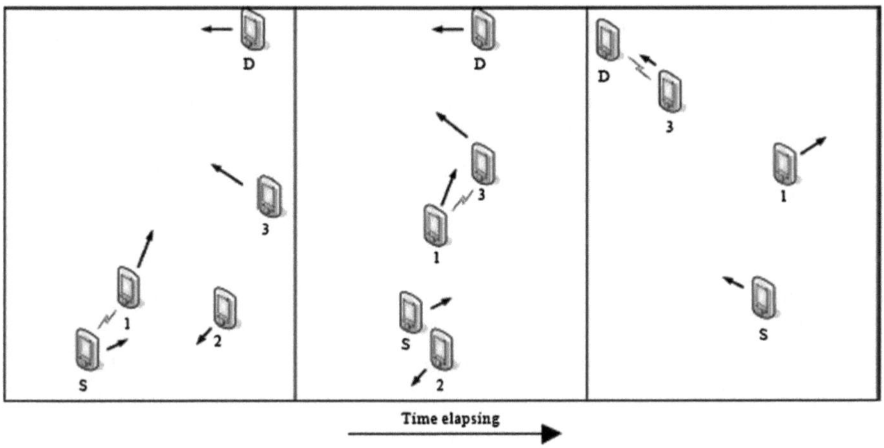

Fig. 3 Opportunistic network

In this example, node S wants to communicate with node D. Unfortunately, there is no route between two and so communication can not be performed in a traditional network. If this network is an opportunistic, then even if direct route is not available, source may try to communicate and there is a possibility that it works. Here in our example, S sends data to nearest neighbour node 1. Initially node 1 cannot communicate with node 3. But it keeps data with it, until it reaches near to node 3. Then it sends data to node 3. Same thing will be done by node 3 to deliver data to node D. In this way, even if S and D are not connected through a common route all the time, they communicated. Such a network is called an opportunistic because, whenever a node gets an opportunity, It tries to send packets towards the destination. It is all about the mobility pattern, no guaranteed but it is based on best effort to ensure delivery [8].

3.2 Pauses in Cellular Communication

Cellular networks are becoming more and more popular not only for the purpose of voice calls but also for the Internet connectivity. Voice call related communication is continues while data communication (Internet) is more kind of discrete. In earlier days of cellular networking, handovers were used to disconnect calls, with handovers management strategies; nowadays it is possible to perform handovers without disconnections with an extremely negligible pause. The main purpose here is to avoid pauses in cellular communication while doing data transfer mostly to and from the Internet. There are several reasons which may introduce pauses in communication. We mainly focus on pause due to lack of knowledge of the latest location of a specific device [9].

3.3 Opportunistic Location Update

As explained earlier that in a cellular network, opportunistic network is formed across the devices under the roof of a same BTS, here we are showing the procedure for a single BTS. BTS with high density should follow such opportunities while BTS with less density should avoid.

Parameters:

Description	Value
Total nodes under a BTS	n
Total nodes with active mobile data connection	m out of n
Total nodes which are willing to be a part of opportunistic network	p out of m

Location update ways:

Description
Traditional location update without opportunistic networking
Enhanced location update with opportunistic networking

Those devices which don't participate in forming an opportunistic network have to get their locations updated as per the traditional location update strategies. The main reason behind keeping both the kind of LUs is that we can't force every device to adapt new way of cellular networking.

3.4 Opportunistic Location Update Implementation

As explained in 3.3, at any particular moment, there will be p devices that are using mobile data as well as willing to participate in opportunistic networking. The main advantage of this method is it is very light weight. Every device decides whether to send request to BTS to update its location or not. In another side, if it is highly needed to retrieve location, BTS can follow conventional location update strategy.

Situation 1:

Suppose a device is currently communicating through BTS1. Being a mobile, it is moving towards the boundary of BTS1. It will perform SSLM—Signal Strength based Link Management related prediction that as the strength of signal is getting continuously weaker, soon it will be disconnected from BTS1. Now it will flood this information as a packet to the opportunistic network. If this packet reaches to the BTS1, it initiates LU process immediately. This packet is called Location_Update_Need packet.

Situation 2:

Suppose a device is currently communicating through BTS1. Being a mobile, it is moving towards the boundary of BTS1. Now if all of sudden it feels disconnected from BTS1 and not connected to any other BTS yet, it floods the network with a packet called Location_Update_Change. If this packet reaches to the BTS1, it will try to reach back to the device to detect temporarily disconnection. At the same time, if this packet reaches to the BTS2 (New BTS), BTS2 will start location initialization and location set procedures.

Signal is measured in dBm which is the power ratio of radio power per one milli watt. The unit is called decibel. A signal with −60 dBm is perfect, and −112 dBm is worst which may drop a call completely. If we get signal about −87 dBm, Android shows us perfect signal with all the four bards high. To implement our algorithm, we can use such measurements to find whether we are moving opposite to BTS or we are in a dead spot. Dead spot is a area within BTS where signals can't reach. Some dead spots are intentionally made like jammers. In such situation, opportunistic networking works as they are of local frequencies. Device can inform BTS that even if it is switch on, it is not being connected using our algorithm.

4 Simulation

Simulation is performed in Qualnet and with different scenarios as shown below.

No. of devices	No. of calls	No. of data connections	No. of cell change	No. of LU	No. of OLU	Success rate (%)
50	10	15	280	114	89	98.34
100	20	30	318	135	96	97.75
150	30	45	467	187	113	95.33
200	40	60	492	215	128	92.11
250	50	75	511	288	137	98.45
300	60	110	538	366	144	93.64
350	70	225	598	396	189	94.33
400	80	250	613	411	203	96.92

Figure 4 shows a graph of time versus OLU for scenario 1.

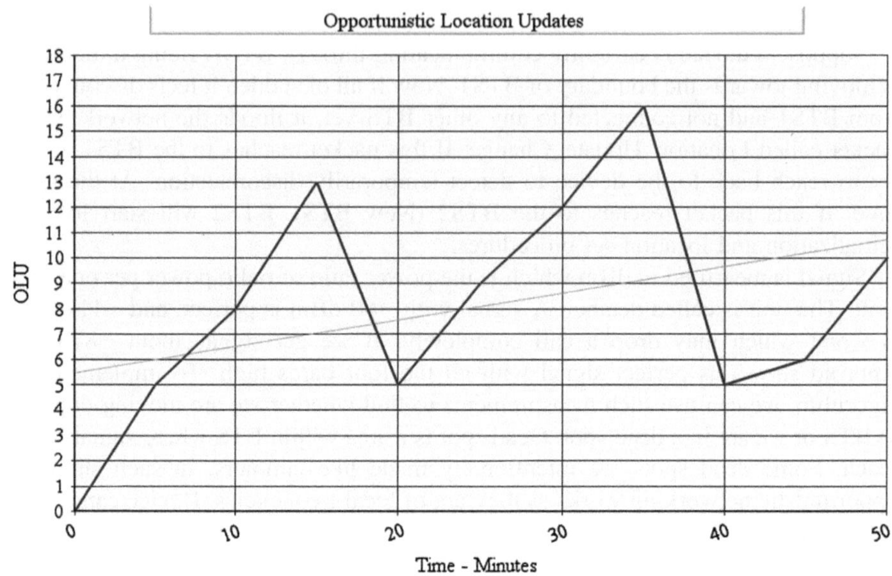

Fig. 4 Opportunistic location updates

5 Conclusion

Opportunistic networks are next generation of wireless networks with mobility. Due to the fact that they belong to a category of computer networks where even if a route is not available, communication performed, they are becoming popular. Cellular networks are nowadays essential to use by everyone. But at the same time, there is no drastic change in it other than increasing infrastructures to increase coverage. This scheme would be the significant change as a new technique is introduced. If this scheme is implemented in real life scenario, there will be significant amount of performance improvement can be found.

References

1. J. Zhang and J. Mark, "A local VLR cluster approach to location management for PCS networks", *WCNC. 1999 IEEE Wireless Communications and Networking Conference (Cat. No.99TH8466)*, 1999.
2. E. Clayirci and I. Akyildiz, "User mobility pattern scheme for location update and paging in wireless systems", *IEEE Transactions on Mobile Computing*, vol. 1, no. 3, pp. 236–247, 2002.
3. I. Akyildiz, J. McNair, J. Ho, H. Uzunalioglu and Wenye Wang, "Mobility management in next-generation wireless systems", *Proceedings of the IEEE*, vol. 87, no. 8, pp. 1347–1384, 1999.
4. M. Ilyas and I. Mahgoub, *Mobile computing handbook*. Boca Raton: Auerbach Publications, 2005.

5. H. Abdul-Kader, "Location Updating Strategies in Moving Object Databases", *IJCTE*, pp. 65–70, 2009.
6. G. Pollini and Chih-Lin I, "A profile-based location strategy and its performance", *Proceedings of PIMRC '96 – 7th International Symposium on Personal, Indoor, and Mobile Communications*.
7. Y. Zhu and V. Leung, "Optimization of Sequential Paging in Movement-Based Location Management Based on Movement Statistics", *IEEE Trans. Veh. Technol.*, vol. 56, no. 2, pp. 955–964, 2007.
8. I. Martinez-Arrue, P. Garcia-Escalle and V. Casares-Giner, "Location Management Based on the Mobility Patterns of Mobile Users", *Wireless Systems and Mobility in Next Generation Internet*, pp. 185–200.
9. M. Ali, M. Khan and M. Alam, "A Profile-Based Two-Level Pointer Forwarding Cache Scheme for Reducing Location Management Cost in Wireless Mobile Networks", *International Journal of Computer Applications*, vol. 3, no. 7, pp. 12–18, 2010.

Fuzzy Analytic Hierarchy Process for Software Durability: Security Risks Perspective

Rajeev Kumar, Suhel Ahmad Khan and Raees Ahmad Khan

Abstract Software development is a field which is filled with different types of risks. In nowdays, secure software development is very difficult task. Security risk mitigation is the activity which aims to identify and clear most of the security threats before it could harm the system software. This paper is focusing on identifying and mitigating security risks which are affect the duration of secure software after development. A hierarchical structure of durability risk factors with respect to security in software development is established. This paper aims to apply Fuzzy Analytic Hierarchy Process (FAHP) during the pre-negotiation stage to identify security risks factor. This paper aims to apply Fuzzy Analytic Hierarchy Process (FAHP) during the prenegotiation stage to identify security risks for better assessment. With the help of this prioritization, it may be helpful to developers for better management performance at late stage of development life cycle. After applying this prioritization, organizations might improve longevity of secure software.

Keywords Software security · Software durability · Security risks · Secure serviceability

1 Introduction

The emergence of software durability introduces new options for small medium initiatives, as well as the industry begins to generate decent sales and exploration of latest marketing techniques [1–3]. The long lasting program has evolved the best

R. Kumar (✉) · R.A. Khan
Department of Information Technology,
Babasaheb Bhimrao Ambedkar Central University, Lucknow, Uttar Pradesh 226025, India
e-mail: rs0414@gmail.com

R.A. Khan
e-mail: khanraees@yahoo.com

S.A. Khan
Department of Computer Application, Integral University,
Lucknow, Uttar Pradesh 226026, India
e-mail: ahmadsuhel28@gmail.com

N. Modi et al. (eds.), *Proceedings of International Conference on Communication and Networks*, Advances in Intelligent Systems and Computing 508,
DOI 10.1007/978-981-10-2750-5_49

way how the application is bought and furnished in an awfully cost-effective way. Apart from pushing corporate business to the subsequent level of progress, durability also has strong talents in the schooling sector especially in terms of constructing a flexible environment [4–8].

There are so many proposed methods for using multi-standards choice-making methodology by using AHP, so one can enable the prioritization of safety dangers. Also, there has been a sufficient of study that combines fuzzy logic and safety, which is a general approach of combining ambiguous range into the decision-making approach, with analytic hierarchy procedure for safety threat comparison [9–11]. Those threat assessment ways are largely applied to multiple fields corresponding to prioritization of protection causes. Lots of the FAHP approaches advocate each and every safety chance aspect in a framework is joined as a fuzzy measure, which is a combination of the probability and related results of data [12, 13].

For increasing software services this paper is using an indirect contribution for security risks factors in software development process to enhance the durability of software. In order to efficaciously transition to this new flexible environment, software based on the trustworthy environment must be developed with a certain level of software security besides fulfilling their functional requirements for secure design [13–15]. Software security implies how the system should behave besides having all the essential functionalities with secure services. Fulfilling both functional and non-functional security requirements is a demanding task especially when stepping into a new field with respect to durability.

2 Background and Definitions

There are typically many challenges in security engineering. Software security is usually a qualitative measure. Customers often specify their security requirements verbally making the requirements capturing process less effective [16–18]. Besides that, security requirements often clash with each other. Designing software that is extremely secure will inevitably cause a negative impact on the software usability. This kind of development of software design can be enhanced by Fuzzy Analytic Hierarchy Process. Priority analysis is one of the clash resolution techniques to resolve the trade-off between conflicting security requirements [19].

The relative priority and importance of security requirements play a significant role in development of software. However, it is often difficult for investors to directly provide the priorities for all security requirements due to the multifaceted relationship between each other with respect to risks for durability of software. The nature of prioritizing security risk factors for the requirement of durable software is a multi-criteria decision-making (MCDM) problem [20, 21]. MCDM is a research of methods and procedures which define about evaluating multiple clashing criteria and derives a way to come to a negotiation. This set of criteria often differs in the degree of importance. AHP has been a tool that is widely used and adopted by decision makers and researchers to aid in priority analysis [22, 23].

Factors of software security risks should be specified in the early stage of development. These security requirements captured beforehand can then modify the software developed by security engineers. Nevertheless, during the implementation and deployment phase, security developers have to make sure that the stated security risks factor must be present in the software [24, 25]. This paper provides weights of software security risks factors for durability with its ranking which is evaluated from FAHP technology. This paper also evaluates security risks assessment methodology that measures the performance. Thus new researchers want to use fuzzy logic with AHP for evaluation process of priorities as it is an effective process and is considered by fuzziness.

3 Priority Assessment of Security Risk Factors for Software Durability

Fuzzy AHP is chosen for security risk factors because it is capable of controlling vague judgmental inputs given by the participants. Fuzzy AHP is also capable of converting qualitative inputs into quantitative results, in the form of weightage and ranking which is a better assessment of security risks [26]. The weightage and ranking of security risk factors can easily help participants to analyze trade-off and choose the development guidelines that are opposite to the associated security risk factors for durable software services.

The top of the hierarchy consists of the ambition for administering the test, followed by an accumulation of accessible choices to accomplish the accurate goal.

The choices can be further divided into sub-criteria if required. The problem domain for this paper is on durable security design of software development environment for the corporate sector. Functionality is not involved because it is expressed as a totality of essential functions that the software product provides [27]. As for the remaining security risk factors, they can only be measured when the functionality of a given system is present. Thus, achievement of functionality

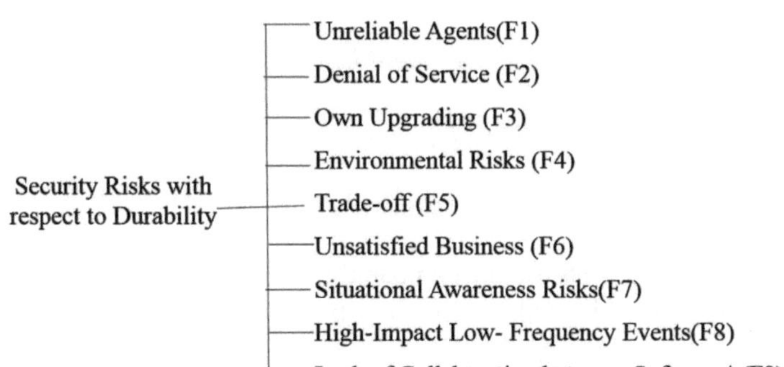

Fig. 1 Hierarchy modelling of security risks for durability

Table 1 Sampling fuzzy pair-wise comparison matrix based on collected expert's judgment

		Attribute 1	Attribute 2	Attribute 3	Attribute 4	…	Attribute n
$\eta_{ij} =$	Attribute 1	(1, 1, 1)	F_{12}	F_{13}	F_{14}	…	F_{1n}
	Attribute 2	F_{21}	(1, 1, 1)	F_{23}	F_{24}	…	F_{2n}
	Attribute 3	F_{31}		(1, 1, 1)	F_{34}	…	F_{3n}
	Attribute 4	F_{41}			(1, 1, 1)	…	F_{4n}
	⋮	⋮	⋮	⋮	⋮	⋮	⋮
	Attribute n	F_{1n}	F_{2n}	F_{3n}	F_{4n}	…	(1, 1, 1)

becomes the pre-requisite for the fulfilment of the remaining risk factors of security shown in Fig. 1 as a hierarchy model.

There is a total of nine evaluation criteria. As indicated earlier, the researcher has proven that AHP is one of the best arrangement techniques in a small-scale MCDM problem, such as the one presented in this paper. Figure 1 shows an illustration of AHP hierarchy model of security risks durable software.

This paper is used TFN membership function and also proposes pair-wise comparisons to generate the fuzzy judgment matrix. Table 1 presents a example of the fuzzy judgment matrix which is further used in AHP technique.

The persons who participated in this evaluation include university students and developers who have experiences in security risks [28–30]. Thirty-one participants were chosen to ensure the consistency of AHP testing. As previously discussed, under the condition when comparison matrix size (number of criteria to be evaluated) is 9 × 9, the group size threshold to achieve an acceptable level of consistency is thirty-one participants [31–34]. This means that as long as the number of participants are more, one can ensure that the aggregated results will be consistent and revision of judgment is not needed (Table 2).

The Eqs. (1–3) shows the least possible, most likely and the extreme possible assessment of a fuzzy number (l, m, h). According to Saaty Scale Triangular fuzzy numbers $[\eta_{ij}]$ are established as the following:

$$\eta_{ij} = \left[l_{ij}, m_{ij}, h_{ij}\right] \quad \text{where } l_{ij}, \leq m_{ij}, \leq h_{ij} \; [1/9, 9]$$

$$l_{ij} = min(J_{ijk}) \tag{1}$$

$$m_{ij} = \left(J_{ij1} \cdot J_{ij2} \ldots J_{ijk}\right)^{1/k} \tag{2}$$

$$h_{ij} = max(J_{ijk}) \tag{3}$$

In above equations J_{ijk} shows the comparative importance of measures A_i and A_j given by expert k. In this step, the method proposed is used for defuzzification process.

Table 2 Fuzzy pair-wise comparison matrix based on collected expert's judgment

Risk factors (D)	D1	D2	D3	D4	D5	D6	D7	D8	D9
D1	(1, 1, 1)	(0.25, 1.64, 5)	(0.17, 1.37, 5)	(0.14, 1.38, 9)	(0.20, 1.58, 9)	(0.14, 1.32, 7)	(0.12, 0.87, 4)	(0.25, 1.20, 5)	(0.14, 1.13, 9)
D2		(1, 1, 1)	(0.33, 1.49, 5)	(0.33, 1.67, 7)	(0.14, 1.30, 5)	(0.25, 1.16, 7)	(0.20, 0.86, 9)	(0.12, 0.57, 9)	(0.12, 0.57, 9)
D3			(1, 1, 1)	(0.20, 1.06, 5)	(0.14, 1.50, 9)	(0.33, 1.13, 7)	(0.20, 0.86, 9)	(0.17, 1.37, 5)	(0.14, 1.38, 9)
D4				(1, 1, 1)	(0.33, 2.43, 7)	(0.25, 1.20, 5)	(0.14, 1.13, 9)	(0.12, 0.87, 4)	(0.14, 1.22, 5)
D5					(1, 1, 1)	(0.14, 1.22, 5)	(0.14, 0.66, 4)	(0.12, 0.57, 9)	(0.20, 1.06, 5)
D6						(1, 1, 1)	(0.12, 0.57, 9)	(0.17, 1.37, 5)	(0.14, 1.38, 9)
D7							(1, 1, 1)	(0.12, 1.20, 5)	(0.14, 1.13, 9)
D8								(1, 1, 1)	(0.12, 0.57, 9)
D9									(1, 1, 1)

Following is the construction of comparison matrix, defuzzification takes places to produce a quantifiable value based on the calculated TFN values. The defuzzification method applied in this paper inserted values shown in Tables 3 and 4. Alpha cut method formulated in (4) which is commonly referred. Alpha cut enables one to describe a fuzzy set as a composition of crisp sets. Crisp sets simply describe whether an element is either a member of the set or not. Formula (4) shows the algorithm of alpha cut.

$$\rho_{\alpha,\beta}(\eta_{ij}) = \left[\beta \cdot \eta_{\alpha}(l_{ij}) + (1-\beta) \cdot \eta_{\alpha}(h_{ij})\right] \tag{4}$$

where $0 \leq \alpha \leq 1$ *and* $0 \leq \beta \leq 1$

Such that,

Table 3 Sampling total fuzzy pair-wise comparison matrix by alpha cut

		Attribute 1	Attribute 2	Attribute 3	Attribute 4	…	Attribute n
$\eta_{ij} =$	Attribute 1	1	D_{12}	D_{13}	D_{14}	…	D_{1n}
	Attribute 2	D_{21}	1	D_{23}	D_{24}	…	D_{2n}
	Attribute 3	D_{31}		1	D_{34}	…	D_{3n}
	Attribute 4	D_{41}			1	…	D_{4n}
	⋮	⋮	⋮	⋮	⋮	⋮	⋮
	Attribute n	D_{n1}	D_{n2}	D_{n3}	D_{n4}	…	1

Table 4 Final fuzzy pair-wise comparison matrix by alpha cut after defuzzification

Security risk factors (D)	Dl	D2	D3	D4	D5	D6	D7	D8	D9
Dl	1	2.08	1.97	2.92	2.48	2.53	1.45	2.00	1.51
D2		1	1.86	2.69	1.94	2.40	1.19	0.99	0.993
D3			1	1.87	1.71	2.44	1.19	1.97	2.92
D4				1	2.20	2.00	1.51	1.45	2.08
D5					1	2.08	1.51	0.99	1.87
D6						1	0.99	1.97	2.92
D7							1	2.00	1.51
D8								1	0.99
D9									1

$$\eta_{\alpha}(\mathrm{l_{ij}}) = \left(m_{ij} - l_{ij}\right) \cdot \alpha + l_{ij} \tag{5}$$

$$\eta_{\alpha}(\mathrm{h_{ij}}) = h_{ij} - \left(h_{ij} - m_{ij}\right) \cdot \alpha \tag{6}$$

α and β in these equations are used for preferences of stakeholders. These two values varies between 0 and 1.

The next step is to determine the eigenvalue and eigenvector of the fuzzy pair-wise comparison matrix. The purpose of calculating the eigenvector is to determine the aggregated weight of particular criteria. Assume that ρ denotes the eigenvector while λ denotes the eigenvalue of fuzzy pair-wise comparison matrix η_{ij}.

$$\left[\rho_{\alpha,\beta}\left(\eta_{ij}\right) - \lambda I\right] \cdot \rho = 0 \tag{7}$$

Table 5 Weightage and priority of selected criteria with overall percentage

S. No.	Security risk factors	Weightage	Percentage (%)	Ranks
1	Unreliable agents (Fl)	0.200	20.00	1
2	Denial of service (F2)	0.157	15.70	2
3	Own upgrading (F3)	0.114	11.40	4
4	Environmental risks (F4)	0.115	11.50	3
5	Trade-off (F5)	0.105	10.50	5
6	Unsatisfied business (F6)	0.100	10.00	6
7	Situational awareness risks (F7)	0.086	8.60	7
8	High-impact low-frequency events (F8)	0.061	6.10	9
9	Lack of collaboration between software/s (F9)	0.062	6.20	8

Formula (7) is based on the linear transformation of vectors, where I represent the unitary matrix. By applying Formulas (1–7), the weightage of particular criteria with respect to all other possible criteria can be acquired.

The aggregated result in terms of weightage is tabulated in Table 5. The results obtained are ordered as follows: Unreliable Agents (0.200), Denial of Service (0.157), Environmental Risks (0.115), Own Upgrading (0.114), Trade-off (0.105), Unsatisfied Business (0.100), Situational Awareness Risks (0.086), Lack of Collaboration between Software/s (0.062), and High-Impact Low-Frequency Events (0.061).

4 Discussion

By facing the actual situations, the security risks are classified in many categories, which causes in the software development. These hidden risks are identified in this work. The hierarchy for security risk factor related to durability is established and their weightage is calculated. Assessment method to enhance security of software for a specific period, is also established in this work.

These security risk factors are ordered according to their impact level on the overall risks of software durability. This method also provides a new methodology for the transforming qualitative measures in forecasting the development risks. This method will be suitable for the security risk assessment for the software development process.

5 Conclusion

Prioritization of security risks plays an important role in helping software developers to focus on fulfilling higher priority security within constant maintenance. However, stakeholders often specify security risks qualitatively using some accepted language. These qualitative measures are often not used for prediction. Even if security risks can be prioritized, most of the security engineers only look into one risk attribute at one time, neglecting the contradicting effects among each security factor. The fact is that most of the software security are required to have multiple security risk factors at the same time.

This paper addresses the problem of capturing stakeholders' security requirements and achievement of multiple security risk factors in a flexible environment. This paper has made several contributions toward the establishment of a hierarchy which is useful in software development industry. Security developers able to pinpoint the essential security risks factors ensures the successful implementation of software. This will enable security developers to concentrate on the most important security risks and achieve high satisfaction among stakeholders with a constant maintenance. However, it is to be noted that the implementation of proposed approach is based on security risks case study.

Acknowledgments This work is sponsored by UGC-MRP, New Delhi, India under F. No. 43-391/2014 (SR).

References

1. Kazman R., Klein M., Barbacci M., Longstaff T., Lipson H, and Carriere J., 1998. The Architecture Trade-off Analysis Method. In Engineering of Complex Computer Systems, 1998. ICECCS '98. Proceedings. Fourth IEEE International Conference on, 1998, pp. 68–78.
2. Bengtsson P., Lansing N., Bosch J., and Vliet H. V., 2004. Architecture-Level Modifiability Analysis (ALMA). Journal of Systems and Software, vol. 69, pp. 129–147, 2004.
3. Zhu L., Aurum A., Gorton I., and Jeffery R., 2005. Tradeoff and Sensitivity Analysis in Software Architecture Evaluation Using Analytic Hierarchy Process. Software Quality Journal, vol. 13, pp. 357–375, 2005/12/01 2005.
4. Rosado D. G., Gutiérrez C., Fernández-Medina F., and Piattini M., 2006. Security Patterns and Requirements for Internet-based Applications. Internet Research, Vol. 16, pp. 519–536, 2006.
5. Takabi H., Joshi J. B. D., and Ahn G., 2010. Security and Privacy Challenges in Cloud Computing Environments. Security & Privacy, IEEE, vol. 8, pp. 24–31, 2010.
6. Liou T. S. and Wang M. -J. J., 1992. Ranking Fuzzy Numbers with Integral Value. Fuzzy Sets Syst., vol. 50, pp. 247–255, 1992.
7. Global Information Assurance Certification Paper, Available at http://www.giac.org/paper/gsec/835/clark-wilson-security-model/101747, last visit Jan 09 2016.
8. Martin A. P. and Khazanchi D., 2006. Information Availability and Security Policy. Pro. Twelfth Americans Conference on Information Systems, Acapulco, Mexico.

9. Banerjee S., Mattmann C. A., Medvidovic N., and Golubchik L., 2015. Leveraging Architectural Models to Inject Trust into Software Systems. Available at http://sunset.usc.edu/~mattmann/pubs/sess05.pdf last visit Jan 08 2016.
10. Harvey E. and Prasad A., 2004. The Effect of Piracy on the Market Penetration of Subscription Software. Journal of Business, vol. 77, no. 2, 2004.
11. Parker D. B., 1992. Restating the Foundation of Information Security. Proceedings of the Eighth International Conference on Information Security, Netherlands, 1992, pp. 139–151.
12. Chou C.C., Liu L.J., Huang S.F., Yih J.M., and Han T.C., 2011. An Evaluation of Airline Service Quality using the Fuzzy Weighted SERVQUAL Method. Applied Soft Computing, vol. 11, pp. 2117–2128, 2011.
13. Karlsson J., Wohlin C., and Regnell B., 1998. An Evaluation of Methods for Prioritizing Software Requirements. Information and Software Technology, vol. 39, pp. 939–947, 1998.
14. Karlsson J., and Ryan K., 1997. A Cost-value Approach for Prioritizing Requirements. Software, IEEE, vol. 14, pp. 67–74, 1997.
15. Kumar R., Khan S. A., and Khan R. A., 2015. Revisiting Software Security: Durability Perspective. International Journal of Hybrid Information Technology (SERSC) Vol. 8, No. 2 pp. 311–322, 2015.
16. Vaidya O. S., and Kumar S., 2006. Analytic Hierarchy Process: An Overview of Applications. European Journal of Operational Research, vol. 169, pp. 1–29, 2006.
17. Andreopoulos B., 2004. Satisficing the Conflicting Software Qualities of Maintainability and Performance at the Source Code Level. In WER-Workshop em Engenharia de Requisitos, 2004, pp. 176–188.
18. Liu X. F., 1998. A Quantitative Approach for Assessing the Priorities of Software Quality Requirements. Journal of Systems and Software, vol. 42, pp. 105–113, 1998.
19. Bachmann F., Bass L., Klein M., and Shelton C., 2005. Designing Software Architectures to Achieve Quality Attribute Requirements. Software, IEE Proceedings, vol. 152, pp. 153–165, 2005.
20. Koziolek A., 2012. Research Preview: Prioritizing Quality Requirements Based on Software Architecture Evaluation Feedback. In Requirements Engineering: Foundation for Software Quality. vol. 7195, B. Regnell and D. Damian, Eds., ed: Springer Berlin Heidelberg, 2012, pp. 52–58.
21. Firesmith D. G., 2003. Common Concepts Underlying Safety. Security and Survivability Engineering, Technical Note CMU/SEI- 2003-TN033, Software Engineering Institute, Pittsburg, Pennsylvania, 2003, pp. 1–75.
22. Rogers R. L., 2004. Principles of Survivability and Information Assurance. © Carnegie Mellon University Pittsburgh, PA 15213, 2004.
23. Siddiqi J., 1996. Requirement Engineering: The Emerging Wisdom. Software, IEEE, vol. 13, p. 15, 1996.
24. Trustworthiness and Integrity: What It Takes and Why It's So Hard, Available at: http://josephsoninstitute.org/business/blog/2011/01/trustworthiness-and-integrity-what-it-takes-and-why-it%E2%80%99s-so-hard/ last visit Jan 02 2016.
25. Khan S. A. and Khan R. A., 2010. Securing Object Oriented Design: A Complexity Perspective. International Journal of Computer Applications, 2010 Oct, 8(13), pp. 8–12.
26. Kumar R., Khan S. A., and Khan R. A., 2015. Durable Security in Software Development: Needs and Importance. CSI Communication, 2015, pp. 34–36.
27. Kumar R., Khan S. A., and Khan R. A., 2014. Software Security Durability. International Journal of Computer Science and Technology, vol. 5, no. 2, April–June, pp. 23–26, 2014.
28. Khan S. A., and Khan R. A., 2012. A Framework to Quantify Security: Complexity Perspective. International Journal of Information and Education Technology, Vol. 2, No. 5, October 2012.
29. Saaty T. L., 1986. Axiomatic Foundation of the Analytic Hierarchy Process. Manage. Sci.32 (7), 841–855, 1986.
30. Saaty T. L., 1990. How to Make a Decision: the Analytic Hierarchy Process. Eur. J. Operational Res. 48(1), 9–26, 1990.

31. Khan S. A., and Khan R. A., 2014. Security Assessment Framework: A Complexity Perspective. Computer Fraud & Security, Elsevier, July 2014.
32. Khan S. A., and Khan R. A, 2015. Security Improvement of Object Oriented Design using Refactoring Rules. I.J. Modern Education and Computer Science, vol. 2 pp. 24–31, 2015.
33. Littlewood B., Strigini L., 2000. Software Reliability and Dependability: a Roadmap. Proc. ICSE, the 22nd International Conference on Software Engineering, pp. 1–12, 2000.
34. Meyer B., Zurich E., 2006. Dependable Software. Software Computing eds, Juirgkohlas, Lecture Notes in Computer Science, Springer-verlag, 2006.

Author Biographies

Rajeev Kumar is pursuing Ph.D. in Information Technology from Babasaheb Bhimrao Ambedkar University (A Central University), Vidya Vihar, Raibareli Road, Lucknow and he has been completed his master's degree in Information Technology from same University. Mr. Kumar is the member of many National and International bodies such as ACM-CSTA, IBM-TechTarget, IAENG, and BVICAM etc. Hisresearch interests are in the areas of Software Security, Security Testing, and Software Risk and currently working in the area of Software Security Durability.

Suhel Ahmad Khan has earned his Doctoral Degrees from Babasaheb Bhimrao Ambedkar University, (A Central University), Vidya Vihar, Raibareli Road, Lucknow and he is currently working as an Assistant Professor in the Department of Computer Application, Integral University, Lucknow, UP, India. Dr. S. A. Khan is young, energetic researcher and has completed a Full Time Major Project funded by University Grants Commission, New Delhi, India. He has more than 5 years of teaching & research experience. He is currently working in the area of Software Security and Security Testing. He has also published & presented papers in refereed journals and conferences. He is a member of IACIT, UACEE, and Internet Society.

Raees Ahmad Khan has earned his Doctoral Degrees from Jamia Millia Islamia, New Delhi, India and he is currently working as a Professor and Head in the Department of Information Technology, Babasaheb Bhimrao Ambedkar University, (A Central University), Vidya Vihar, Raibareli Road, Lucknow, India. Prof. R. A. Khan has more than 13 years of teaching & research experience. His area of interest is Software Security, Software Quality and Software Testing. He has published a number of National and International Books, Research Papers Reviews and Chapters on Software Security, Software Quality and Software Testing.

Sorted K-Means Towards the Enhancement of K-Means to Form Stable Clusters

Preeti Arora, Deepali Virmani, Himanshu Jindal and Mritunjaya Sharma

Abstract Clustering is a mechanism of arranging data points of a data set in groups which are similar. The similarity is found on the basis of some metric like Euclidian distance, density, etc. One of the major data clustering methods is the K-Means in which initial centroids are selected randomly. Here, we have presented an effective and modified algorithm, the Sorted K-Means which determines initial centroids after sorting the data points and provides stable clusters using lesser number of iterations. The tool used for implementation is Matlab to find initial centroids so as to reduce number of iterations for K-Means cluster formation. As a final result, stable clusters are obtained with minimized complexity.

Keywords Clusters · K-means · Centroid · Sorting · Iteration

1 Introduction

Clustering is the procedure of forming groups of similar objects in such a way that objects that belongs to single group are more similar to each other than to objects that belongs to other groups. Clustering has a number of applications particularly in the field of market research, pattern recognition, data analysis, etc. Cluster analysis [1, 2] is an activity that involves analysis of the formed clusters to determine the

P. Arora (✉) · D. Virmani · H. Jindal · M. Sharma
Bhagwan Parshuram Institute of Technology, New Delhi, India
e-mail: erpreetiarora07@gmail.com

D. Virmani
e-mail: deepalivirmani@gmail.com

H. Jindal
e-mail: himanshujindal1994@gmail.com

M. Sharma
e-mail: mritunjay.sharma2@gmail.com

N. Modi et al. (eds.), *Proceedings of International Conference on Communication and Networks*, Advances in Intelligent Systems and Computing 508,
DOI 10.1007/978-981-10-2750-5_50

togetherness or link of data points analyzed based on certain metric such as Euclidian distance, density, etc. Various clustering methods or algorithms have been developed to help firms, organizations and businesses to easily do analysis of the big data and see the trends and patterns in various fields. There are many different procedures for clustering, like, density-based clustering (like DBSCAN), hierarchical clustering, centroid-based clustering (like K-Medians), distribution-based clustering etc. One of the prominent clustering based on centroid is the K-Means [3, 4] algorithm which arranges clusters based on the arithmetic mean of the Euclidian distances between the data points.

1.1 The K-Means Algorithm

K-Means clustering segregates 'n' observations into 'k' clusters or groups, where each object or data point belongs to the cluster with the closest mean. For this, it chooses 'k' initial centroids arbitrarily and the clusters are formed based on the proximity of the data objects with these centroids. It performs the following steps to form the clusters:

Input:k: the number of clusters or groups **D**: data set of 'n' objects
Output: Formed k clusters.
Algorithm:

1. Input k value and data set.
2. If k = 1, then Exit.
3. Else
4. Select k objects from D randomly as initial cluster centroids.
5. Assign each point d_i in D to the closest centroid.
6. Calculate and update new cluster centroids.
7. Repeat from step 5 until centroids no longer move.

1.2 K-Means++ Algorithm

K-Means++ algorithm is used for providing initial centroids to the K-Means algorithm. Centroid object is determined by using weighted probability distribution, where every point is considered with probability proportional to square of the distance of that point from most adjacent centroid. The steps for algorithm is as follows:

Input:k: the number of clusters to be formed **D**: a dataset containing n objects
Output: A set of k clusters.
Algorithm:

1. From data set D, select one data point as the center randomly.
2. For each data point d_i in D, compute its distance from the chosen nearest center.
3. Using a weighted probability distribution, choose new center.
4. Find the k centers by repeating steps 2 and 3.
5. Proceed using standard K-Means from these k centers.

Though we conclude that the K-Means algorithm is easy to implement it has certain drawbacks. Firstly, it chooses initial centroids randomly, which leads to formation of unstable clusters, that is, K-Means will give different clusters than the ones formed previously for same data set and the same number of clusters. Though K-Means++ is successful in providing starting centroid objects, since it uses probability density function, the issue of randomness and instability is still not solved.

The remaining parts of this paper are organized as Sect. 2 comprises of the proposed algorithm (Sorted K-Means). Section 3 gives the implementation and results. It also gives the comparison of the K-Means with the Sorted K-Means. Section 4 contains the conclusion. Section 5 defines the future scope of the work done. References used are provided finally.

2 Proposed Work

So, we have designed a new algorithm, the Sorted K-Means, which gives stable clusters in lesser number of iterations.

Sorted K-Means Algorithm

The idea for Sorted K-Means is to choose initial centroids with the following procedure instead of taking them at random. After we get initial centroid data points from this algorithm we apply Simple K-Means algorithm for the rest of the operation to reduce the complexity.

Input:k: the number of desired clusters **D**: a data set containing n object points
Output: A set of k clusters.
Algorithm:

1. Read the input data set and value of k.
2. If k == 1, then Exit.
3. Else
4. Calculate the Euclidian distance, x_i of each data point d_i from origin.
5. Sort the data points d_i based on x_i.
6. Calculate average number of data points the cluster can hold, nod = n/k .
7. Divide the sorted data points d_i into 'nod' groups and name them as g_i .
8. Calculate arithmetic mean a_i of every group g_i.
9. Proceed with K-Means using a_i as initial centroids.

In this algorithm, by employing sorting [5] only once we can reduce much number of iterations which also yields stable output. We can use any cost-effective sorting procedure based on the type of data like merge sort [5, 6] or quick sort [5, 7]. Saving much iteration on a large dataset like big-data reduces the cost for clustering. Thus we name it as Sorted K-Means algorithm.

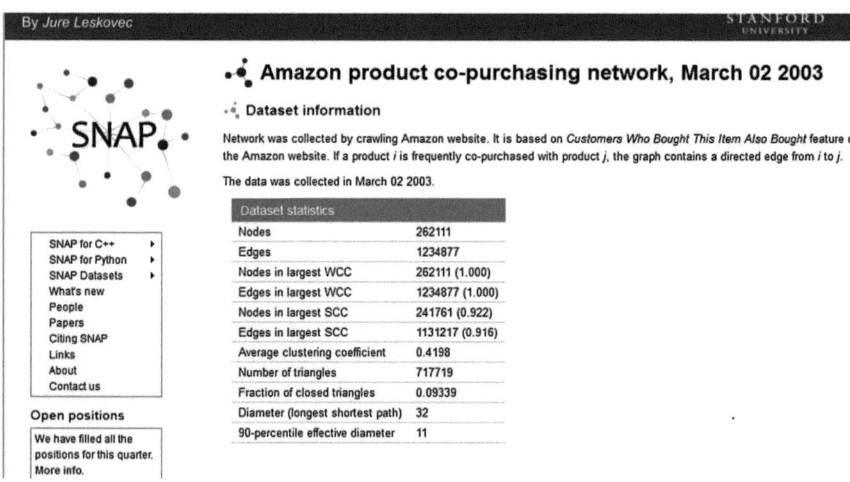

Fig. 1 SNAPData set repository

Table 1 Description of data set

Column No.	Name	Range	Type of value
1	FromNodeId	0-220686	Integer
2	ToNodeId	0-261944	Integer

3 Implementation and Results

For implementing this proposed algorithm we have used a specimen dataset. Amazon product co-purchasing network from March 02 2003 [8] Total Observations taken is 1,048,572 with the following description (Fig. 1 and Table 1).

With the implementation of the Sorted K-Means algorithm on MATLAB Fig. 2a, b shows Reduced and Stable number of iterations as compared to the Simple K-Means algorithm.

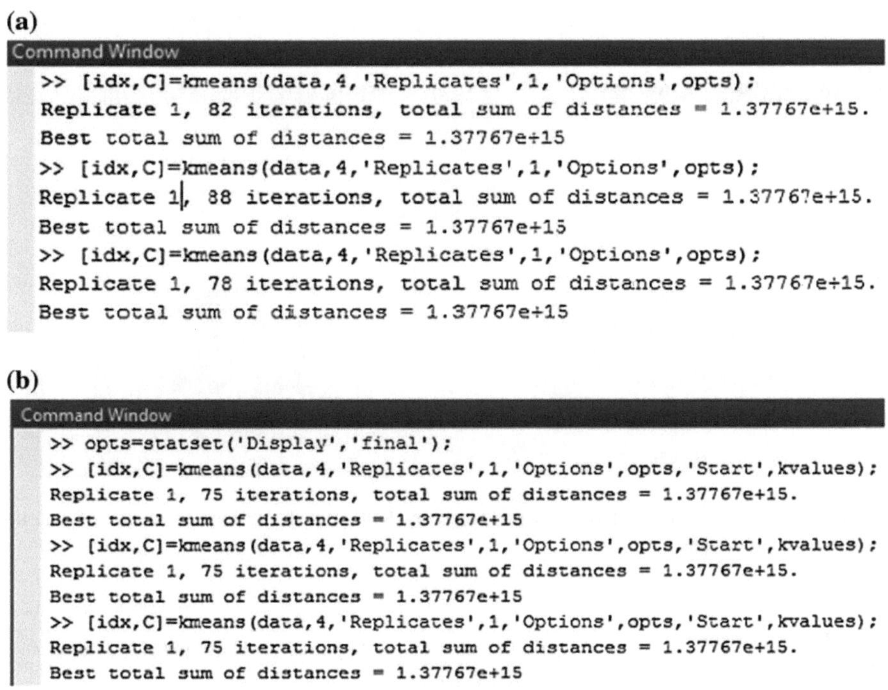

Fig. 2 **a** K-Means. **b** SK-Means

Hence, through Fig. 3a, b and Table 2 we can show the continuous execution of Simple K-Means and Sorted K-Means. Sorted K-Means is consistently showing 75 iterations. While for the same dataset and k value, Simple K-Means is using

(a)

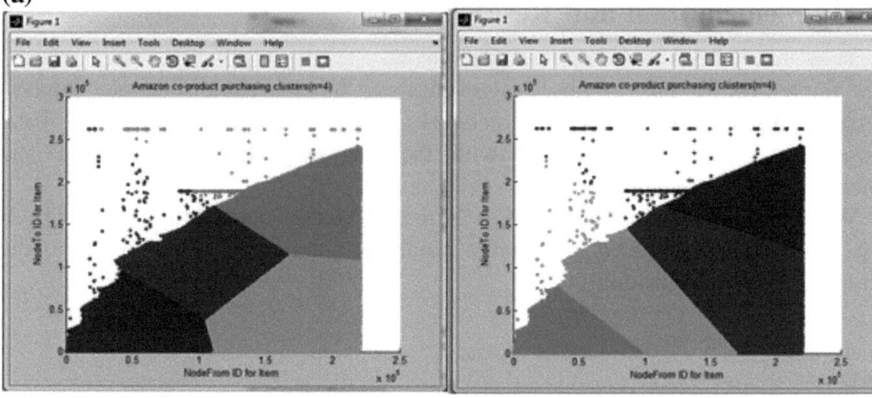

(b)

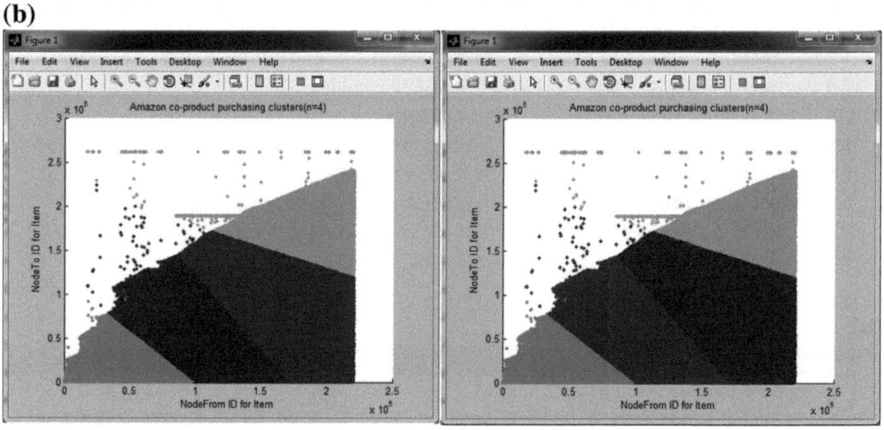

Fig. 3 **a** Various instances of outputs produced by K-Means. **b** Various instances of outputs produced by SK-Means

Table 2 Comparison of simple K-Means and sorted K-Means

Algorithm results	Data points	Clusters (k)	Iterations
Simple K-Means	10	2	3
Sorted K-Means	10	2	1
Simple K-Means	50	4	5
Sorted K-Means	50	4	2
Simple K-Means	100	6	17
Sorted K-Means	100	6	9
Simple K-Means	1,048,572	4	88
Sorted K-Means	1,048,572	4	75

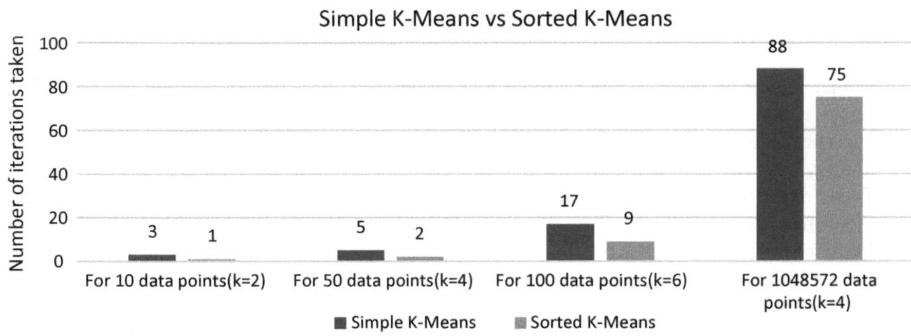

Fig. 4 Comparison between K-Means and SK-Means

different number of iterations for each execution, that too is higher than 75. A comparison of results for both algorithms has been shown in Fig. 4 for various length datasets which shows that Sorted-K-Means is better algorithm as compared to Simple K-Means in terms of stable output each time for a given set of inputs.

4 Conclusion

As seen it is perceptible from the results, the Sorted K-Means algorithm is much more efficient and stable than Simple K-Means. Even when we are using K-Means++ algorithm to determine the initial centroid objects, the Sorted K-Means algorithm is still resulting stable and fixed clusters with lesser number of iterations. Although picking the initial centroid points is performed by the K-Means++ algorithm, still it fails to give stable clusters and the algorithm presented in this paper, the Sorted K-Means algorithm, gives stable clusters with reduced iterations. Hence, the Sorted K-Means clustering algorithm can be effectively and efficiently used to form stable clusters with less number of iterations. Thus, it provides an easier, faster and stable way of doing predictive analysis.

5 Future Scope

In future, this work can be continued and extended to minimize the space and time complexity used during the process and also refining accuracy and precision of the resulting clusters. This improved algorithm is of great use since it may lessen iterations used for clustering on big-data which may comprise of millions of data points. A computation of each of this record of a big data takes a lot of time. These computations are greatly minimized. We have implemented our algorithm successfully on both types of data i.e. real and non-real. Also, the stable and non-changeable output clusters produced may help in predictive analysis of data.

References

1. Daniel T. Larose, Data Mining Methods and Models. pp. 294–296 John Wiley and Sons, New Jersey (2006).
2. Veronica S. Moertini. Introduction to Five Data Clustering Algorithms. Vol 7, No. 2. Integral (2002).
3. Tapas Kanungo, David M. Mount, Nathan S. Netanyahu, Christine D. Piatko, Ruth Silverman, and Angela Y. Wu. An Efficient k-Means Clustering Algorithm: Analysis and Implementation. Vol 24, No 7. IEEE Transactions on Pattern Analysis and Machine Intelligence. (2002).
4. K.A. Abdul Nazeer, M. P. Sebastian. Improving the Accuracy and Efficiency of the k-means Clustering Algorithm. Vol. I, WCE, Proceedings of the World Congress on Engineering. (2009).
5. Pankaj Sareen. Comparison of Sorting Algorithms (On the basis of average case). Vol 3, Issue 3 International Journal of Advanced Research in Computer Science and Software Engineering. (2013).
6. Rohit Yadav, Kratika Varshney and Nitin Kr. Verma. Analysis of Recursive and Non-Recursive Merge Sort Algorithm. Vol 3, Issue 11. International Journal of Advance Research in Computer Science and Software Engineering. (2013).
7. D. Abhyankar and M. Ingle, Engineering of Quick Sort Partitioning Algorithm. Vol 2, No.2. Journal of Global Research in Computer Science. (2011).
8. SNAP Data Repository, https://snap.stanford.edu/data/amazon0302.html.

Target Tracking Accuracy in Context of Energy Consumption in Wireless Sensor Network

Niteen Patel and Mehul S. Raval

Abstract Target tracking using wireless sensor network (WSN) is an important constituent of application like surveillance. The energy is scarce and a very important resource in WSN and therefore, it forms the focal point of this proposal. The goals of the paper is; (1) Analyze the tracking accuracy for a target moving in WSN with multiple clusters; (2) Understanding type of relation between the energy consumption and tracking accuracy in WSN. In the posed experimental scenario, nodes are uniformly distributed in a cluster and they collaborate to track the target moving with a variable velocity. The distance between target and a nearest sensor node is measured and then sent to cluster head for estimating target location. Accuracies of target locations and tracking is calculated from these distance measurements. In this paper we examine effects of varying noise power on the tracking accuracy, effect of cluster size and node density on the target location accuracy. We also focus on energy consumption with respect to change in noise variance and node sensing range. We establish that increase in cluster size improves target tracking accuracy at the cost of increase in energy consumption. Target detection accuracy is dependent on cluster size, node distribution within cluster and also on accuracy of distance measurements.

Keywords Cluster · Distance measurement accuracy · Energy efficiency · Target tracking accuracy · Wireless sensor network

N. Patel (✉)
Electronics and Communication Engineering Department, Sarvajanik College of Engineering and Technology, Surat, India
e-mail: niteen.patel@scet.ac.in

M.S. Raval
Institute of Engineering and Technology, Ahmedabad University, Ahmedabad, India
e-mail: mehul.raval@ahduni.edu.in

N. Modi et al. (eds.), *Proceedings of International Conference on Communication and Networks*, Advances in Intelligent Systems and Computing 508,
DOI 10.1007/978-981-10-2750-5_51

1 Introduction

Target tracking problem in WSN is alike to establishing manual surveillance in a desert (surveillance area). The deputed soldier (sensor node) has limited resources, but has to survive for a long time. Soldier has to interact with other deputed personnel (other nodes in cluster) located in neighborhood and transfer information. Soldier reports (to cluster head) and track intruder based on the measurements carried out. Wireless sensor network provides better alternative for target location tracking because, nodes are relatively small, inexpensive and consumes low power. Efficient use of node energy is one of the most important factors of target tracking as sensor nodes are battery powered and. may be deployed in battle fields, forest or hostile terrain.

Design of a large scale tracking system for wireless sensor network has number of challenges like; (1) sensor nodes have limited energy due to small form factor; (2) Deploying large number of sensor node requires mechanisms for placement and coordination; (3) inadequate signal to noise ratio (SNR) at node due to poor signal strength; (4) node has to handle time synchronization and stamping of sensor data which reveals target positions. Two important attributes of target tracking systems are (a) tracking accuracy; (b) computation and communication cost.

The survey carried out by Demigha et al. [1] provides state-of-the-art in energy-efficient target tracking schemes. Energy efficient protocols for target tracking are classified in two main categories; one exploits sensing related capabilities, and other controls communication cost among sensor nodes. Energy consumption and data accuracy are inversely related to each other. In sensing related techniques, management of a sensor and length of sampling period determines a data accuracy. On other hand communication related techniques select data reporting node, and use adaptive topologies to achieve accuracy. This shows that data accuracy depends on network coverage, connectivity, and topology configuration.

Pantazis et. al. [2] presented survey on power control issues specifically for wireless sensor network. They classified methods in active and passive sets. Passive set covers physical layer, fine grain, back bone and distribution types while network, MAC and transport layer techniques are classified as active sets. Zhu et al. [3] proposed novel way to reduce dimensionality of a transmitted data. Lin et al. [4] proposed an adaptive energy-efficient multisensory scheduling scheme for collaborative target tracking in WSNs. The scheme calculates the optimal sampling interval to satisfy tracking accuracy and selects the cluster of tasking sensors according to their joint detection probability. Authors assumed that all sensing nodes have identical properties, like sensing distance, measurement noise covariance and detection probability. Similar idea is implemented by Onel et al. [5] where, information-controlled transmission power (ICTP) adjustment are carried out.

Clustering plays important role in energy savings because collectively sensor nodes can track the target more effectively. Many researchers have proposed schemes for cluster organization. Wang et al. [6] proposed concept of maximum entropy clustering in which the sensing field is divided for parallel sensor deployment optimization. Zhou et al. [1] proposed practical system based on Zigbee, which utilizes received signal strength to estimate location of target by triangulation. Different type of nodes like reference node, end devices, router and coordinators are constructed.

Shih et al. [7] described excellent method to derive energy model for wireless sensor network, specifically by conducting experimental work on DSP hardware and measuring actual power dissipation in a radio chip. They concluded that inefficient hardware architecture is responsible for large processing power and it dominates power required for communication among nodes. Authors derive energy dissipation model and specify energy in terms of electronic circuit dissipation and receive transmit energy dissipation per bit of transmit received data.

Wang and others [8], dealt with problem of target tracking by constructing framework using maximum likelihood (ML) estimation and Kalman filter (KF). Input measurements (Z_k, k time instants) when fed to a system, results in estimate of target location (x, y) and velocity. ML estimates calculated based on a priori knowledge and given measurements, $P(X|Z_k)$ are related using Baye's rule. This estimate is used to upgrade KF innovations to predict future state. In recent past, researchers improved energy efficiency through improved clustering techniques, like [9] discusses technique of self configurable clustering to improve energy efficiency. It does so through well-timed failure detection of a cluster head. Authors in [10] proposes joint optimization model involving protection for cluster head. They propose two appropriate sub-utility functions, to achieve energy efficiency.

Specifically for target tracking in WSNs, [11] proposes a novel coordinated and adaptive information collecting strategy (CAICS). Information utility based on sensing capability of nodes is used for their selection. Results are shown using parameters like execution/simulation time, number of active nodes contributing in aggregating process. However, this research does not cover nodes range measurement variability, which is addressed in our proposal.

The proposed work in this paper estimates target locations at successive time instances through collaborative sensor nodes by measuring distance from the target. Clusters are formed for collaborative processing and the effect of a noise variance on target localization accuracy is derived and analyzed. We investigate impact of cluster size, node density and sensing range on target tracking accuracy. Moreover we evaluate the correlation between the energy consumption, tracking accuracy and a node sensing range.

The flow of the paper is set as follows; Sect. 2 covers problem formulation, with Sect. 3 discussing energy model and estimated target error. The experimental results are discussed in Sect. 4 and conclusions are drawn in Sect. 5.

2 Problem Formulation

In this work, we consider tracking a vehicle moving through two dimensional sensor field. The trajectory of a target is estimated from time series of measurement carried out by the sensor nodes in a wireless sensor network. In the collaborative target tracking, usually cluster head processes the measurements and estimate target location. Following are important experimental assumptions in this work.

The framework has a single acoustic non-cooperative target moving randomly in a sensor field. Sensor nodes do not have any prior knowledge about vehicle trajectory. Coordinates of all sensor nodes in a field is known. Target is a point source with emanated acoustic signal following an isotropic pattern. Sampling instances are synchronized among all the collaborative sensor nodes and nodes take a measurement only when the target is within effective sensing range. Sensor nodes are densely deployed so that during each sampling period, there are at least three sensors in the target's vicinity. A sensor node has computing capabilities to execute collaborative work like data aggregation. All nodes are capable of communicating wirelessly with each other in the predefined range. The necessary routing and MAC layer protocols are defined and utilized for communication.

In the target tracking problem, state describes target's motion and the task is to estimate the state at time instant $k+1$ given the state of the target at time instant k is known. The model accommodates a noise term to account for randomness in the target motion. The target state is an either two dimensional or four-dimensional vector that consists of either the two-dimensional position of the target, $[x, y]$ or in addition to position, the velocity of target. The target state vector is defined by

$$X = [x, y, dx/dt, dy/dt]^{\mathrm{T}}. \tag{1}$$

and it evolves in time according to

$$X(k+1) = FX(k) + v(k), \tag{2}$$

where $X(k)$ the target state vector at time k, F is the transition matrix, and $v(k)$ is the independent Gaussian-distributed noise process with zero mean and covariance matrix Q. All sensors measure amplitude of the acoustic signal. $\lambda_i^{(t)}$ is the sensor characteristics, which includes sensor node position $\gamma_i = [\gamma_{xi}, \gamma_{yi}]$. σ_i^2 is known additive noise variance of the sensing process; where, i = 1 to N is number of sensor nodes in vicinity of target γ_i.

$$\gamma_i = [\gamma_{xi}, \gamma_{yi}]. \tag{3}$$

$$\lambda_i^{(t)} = [\gamma_i, \sigma_i^2] \tag{4}$$

γ_i and $\lambda_i^{(t)}$ are independent of time. Measurement by sensor node i, is given by

$$z_i = \frac{a_t}{\|X - \gamma_i\|^{\alpha}} + w_i, \tag{5}$$

where a_t is random variable representing amplitude of signal at target, α is known attenuation co-efficient, $\|\cdot\|$ depicts second norm, w_i is zero mean Gaussian random variable having measurement variance σ_i^2.

The problem is to localize stationary signal source using set of the sensor measurements. At least three independent distance measurements are required to uniquely localize target on a two dimensional plane. Assuming $\alpha = 2$ in signal propagation model, received signal model can be given as

$$z_i = \frac{a_t}{\|X - \gamma_i\|^2} + w_i \tag{6}$$

Considering absence of noise term, above equation leads to simplified version

$$\|X\|^2 + \|\gamma_i\|^2 - 2X\gamma_i^T = \frac{a_t}{z_i} \tag{7}$$

For node i = 1, (7) becomes,

$$\|X\|^2 + \|\gamma_1\|^2 - 2X\gamma_1^T = \frac{a_t}{z_1} \tag{8}$$

There will be set of equations for i = 2 to N.
Subtracting (8) from Eq. (7)

$$\|\gamma_i\|^2 - \|\gamma_1\|^2 - 2X(\gamma_i - \gamma_1)^T = a_t\left(\frac{1}{z_i} - \frac{1}{z_1}\right) \tag{9}$$

$$-2X(\gamma_i - \gamma_1)^T = a_t\left(\frac{1}{z_i} - \frac{1}{z_1}\right) - (\|\gamma_i\|^2 - \|\gamma_1\|^2) \text{ reduces to}$$

$$c_i X = f_i, \tag{10}$$

where

$$c_i = -2X(\gamma_i - \gamma_1)^T \text{ and} f_i = a_t\left(\frac{1}{z_i} - \frac{1}{z_1}\right) - (\|\gamma_i\|^2 - \|\gamma_1\|^2) \tag{11}$$

For a given k sensors, k − 1 linear constraints expressed in the matrix form can be obtained. To uniquely determine the location of target, at least three equations are required which are solved using least squares estimation (LSE). In this case solution is obtained with the help of pseudo inverse of c_k.

$$X = \left[(c_k^T c_k)^{-1} c_k^T\right] f_k \tag{12}$$

All sensors may not collect equal information and geometry of sensor placement and distance of target to sensor node plays an important role. However, the problem formulation is generic and it is independent of the distance measurement technique, e.g. in (7), $\frac{a_t}{z_i}; i = 1, 2, 3, \ldots, N$, represents square of distance from target to ith sensor node, if amplitude of received signal is measured as zi. Even if any other technique for distance measurement is used, then also it can be easily replaced with measured distance di, and same equation can be written as

$$\|X\|^2 + \|\gamma_i\|^2 - 2X\gamma_i^T = d_i^2,\ i = 1, 2, 3, \ldots, N, \tag{13}$$

where d_i is the distance between target and ith sensor node. Accordingly (8) can be written as

$$\|X\|^2 + \|\gamma_1\|^2 - 2X\gamma_1^T = d_1^2 \tag{14}$$

Subtracting (14) from (13)

$$\begin{aligned} \|\gamma_i\|^2 - \|\gamma_1\|^2 - 2X(\gamma_i - \gamma_1)^T &= d_i^2 - d_1^2 \\ -2X(\gamma_i - \gamma_1)^T &= d_i^2 - d_1^2 - (\|\gamma_i\|^2 - \|\gamma_1\|^2) \end{aligned} \tag{15}$$

Reduces to

$$c_i X = f_i \tag{16}$$

where, $c_i = -2X(\gamma_i - \gamma_1)^T$ and $f_i = d_i^2 - d_1^2 - (\|\gamma_i\|^2 - \|\gamma_1\|^2)$ giving X *as* $X = piv(\boldsymbol{c_i}) f_i$, pseudo inverse of c_i is given as $piv(\boldsymbol{c_i}) = \left(c_i^H c_i\right)^{-1} c_i^H$, where, H is Hermitian of a matrix. Therefore,

$$X_{estimated} = \left[(c_k^T c_k)^{-1} c_k^T\right] f_k \tag{17}$$

In robust tracking problem, the task is to accurately estimate future target coordinates in spite of noisy distance measurements. Thus in the given problem formulation, distance is derived from node measurement and collaborative processing. A least square solution is derived from over determined system comprising of multivariate equations.

3 Energy Model and Estimated Target Error

This paper also focus to establish and analyze correlation between network energy consumption and tracking accuracy. It is found from independent energy model [7] in WSN, communication consumes significant proportion of energy compared to energy spent by node CPU in computations. A long distance transmission needs more energy compared to short distances. Average energy consumption for radio communication E_{radio} can be modeled as addition of energy consumption during transmission E_{tx} and reception E_{rx}.

$$E_{radio} = E_{tx} + E_{rx} \tag{18}$$

Energy consumption during reception depends only on bits received per unit time, i.e. on data rate. Total energy consumed for receiving K bits, is given by E_{rx}

$$E_{rx}(K) = K * E_{rx-elec} \tag{19}$$

where, $E_{rx-elec}$ is energy consumed per bit during reception.

The total energy consumed during transmission is due to energy consumed in processing a bit and energy consumption by an amplifier. Later is a function of the distance and energy consumption is directly proportional to square of the distance between target and the source. So energy consumed while transmitting K bits information is given by $E_{tx}(K, d) = K\, E_{tx-elec} + E_{amp}(K, d)$

$$E_{tx}(K, d) = K\, E_{tx-elec} + K\, d^2 E_{amp} \tag{20}$$

In above equation, $E_{tx-elec}$ is energy consumed per bit by transmitting electronics and E_{amp} is energy consumption by amplifier used in transmission per bit per m^2. Typical values as used by the researchers are shown in Table 1.

One of most important attribute of target tracking is the accuracy with which the algorithm estimates the target location with respect to its actual value. In this paper the cumulative mean square error (MSE) between estimated and actual target position is taken as a measure of a target accuracy. MSE is computed as per (21).

$$MSE = \frac{1}{M} \sum_{i=1}^{M} \|X - X_{estimated}\|_2^2 \tag{21}$$

Table 1 Typical values of energy consumption in sensor communication

Reference	$E_{tx-elec}$ nJ/bit	$E_{rx-elec}$ nJ/bit	E_{amp} $nJ/(bit * m^2)$
[4]	45	135	10
[6]	50	50	0.1

4 Experimental Results

We have developed a full simulation framework to analyze the target tracking in the WSN. The user can define the number of sensor nodes and size of the sensor field. Using this frame work, we simulate scenarios with 300 nodes deployed in 100 m × 100 m sensor field, which is further divided into clusters. We use clusters to cover area of 20 m × 20 m, 25 m × 25 m and 50 m × 50 m for varying degree of analysis. Each cluster has cluster head which receives distance measurements from sensor nodes located in vicinity of the target. We use 15 m as sensor node measurement range in the simulation. Cluster head processes data received from sensor nodes, and apply LSE target detection algorithm as per (17) to estimate the target coordinates in the sensor field.

For uniformity of analysis on energy consumption analysis, we assume that 128 bytes of data is exchanged between the sensor nodes and cluster head. Target track is generated by a state model specified in Sect. 2. Tracking error is measured in terms of MSE in meters and it is plotted with reference to time. The MSE plot denotes error magnitude (in meters) at each time instance.

We intend to study following through simulation: (a) Impact of additive noise variance on tracking accuracy (b) Effect of cluster size on tracking accuracy (c) Impact of additive noise variance on energy consumption in multi cluster WSN (d) Impact of node sensing range on energy consumption and target tracking accuracy.

4.1 *Impact of Additive Noise Variance on Tracking Accuracy*

Figure 1a showcase results for sensor field with cluster size 25 m × 25 m and low additive noise variance of 0.05. The estimated track is shown with RED color which overlaps the actual target trajectory. It is evident that target can be accurately tracked. Also as seen in Fig. 1b MSE tracking error is of order 0.5 or less. However, with increase in additive noise variance from 0.1 to 0.9, MSE of target localization gradually increases and results are tabulated in Table 2. It should be noted that MSE degradation at time k + 1 is calculated with respect to time instant k.

4.2 *Effect of Cluster Size on Tracking Accuracy*

We now analyze effect of variations in cluster size on tracking accuracy. We simulate scenarios with 300 nodes deployed in 100 m × 100 m sensor field. Sensor field is further divided in to clusters of size of 20 m × 20 m, 25 m × 25 m

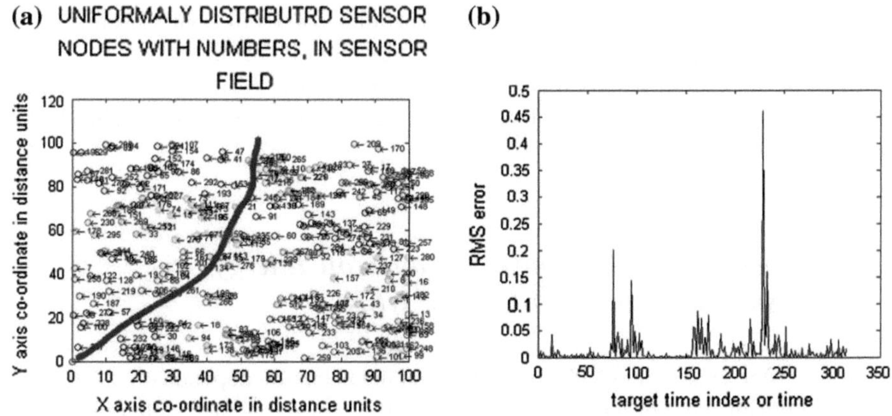

Fig. 1 **a** Low error variance for 100 m × 100 m sensor field, 300 sensor nodes, 15 m of distance measurement range. **b** Plot of MSE corresponding to track in Fig. 1a

Table 2 Incremental MSE of target localization

Additive noise variance (A)	MSE (B)	% of degradation of MSE $\frac{B_{K+1}-B_K}{A_{K+1}-A_K} \times 100$
0.05	0.013	Reference
0.075	0.034	0.816
0.1	0.067	1.332
0.2	0.224	1.568
0.3	0.489	2.657
0.4	1.091	6.015
0.5	1.284	1.929
0.6	2.153	8.692
0.7	2.485	3.322
0.9	4.585	10.492
1	7.706	31.213

Table 3 Cluster size and MSE

Sr. No.	Cluster size (A)	MSE m^2 (B)	% improvement $\frac{B_{K+1}-B_K}{A_{K+1}-A_K} \times 100$
1	20	10.48	Reference
2	25	1.28	87.75
3	50	0.74	92.91

and 50 m × 50 m with distance measurement range of 15 m. During simulation, noise variance is kept as 0.5. The results are tabulated in Table 3.

It can be seen from results that for a large cluster size, MSE reduces considerably. This is a result of collaborative processing as many nodes participate in the state estimation.

4.3 *Impact of Additive Noise Variance on Energy Consumption in Multi Cluster WSN*

In multi-cluster sensor field MSE of target tracking accuracy depends on cluster size and range of sensors. For a benchmark, we use 128 bytes as a basic data unit among nodes. We assume that target does not leave boundary of sensor field. Total 315 points target trajectory is generated by the nodes arranged in various clusters as shown in Fig. 2. The clusters are shown with different colors in Fig. 2. The parameters values of the energy dissipation model are taken as per Table 1. Energy model described in Sect. 3 by Eqs. (19) and (20) are used during the analysis. Accordingly, cost of 128 bytes of data transmission and reception is 6.63 exp −05 J for 20 m × 20 m cluster. During simulation, numbers of such transactions are calculated by varying additive noise variance. Typically measurement error varies from 0.2 m to several meters. Simulation experiments are designed to observe effects of variation in measurement error and cluster size on energy consumption. Using parameters defined in Table 4 several iterations are performed to analyze impact of additive noise variance on energy consumption in multi-cluster WSN.

One must note that total energy consumption in the network do not change with distance measurement error variance as it does not affect the number of

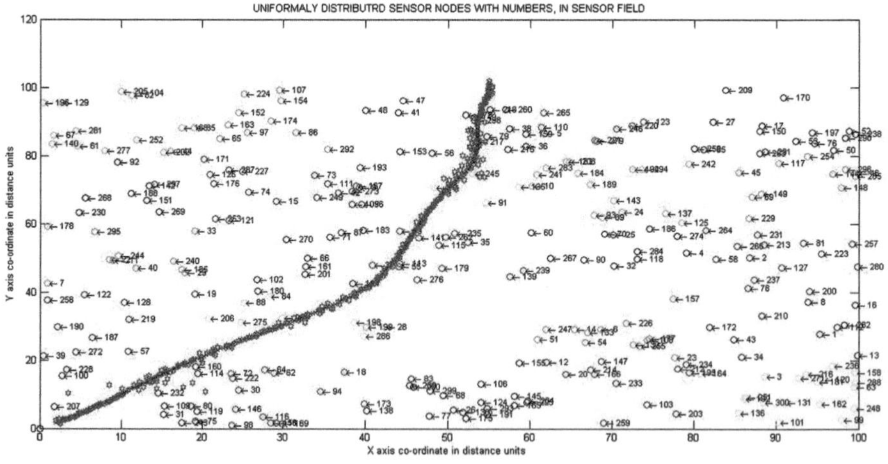

Fig. 2 Node locations and target movement, for parameter specified in Table 4

Table 4 Parameters for computing energy consumption

Overall squared error	Error variance	MSE	Energy consumed per 128 byte in J
424.35	0.2	1.34	6.102
401.00	0.2	1.27	6.102
513.64	0.2	1.63	6.102

Table 5 Distance measurement error and MSE

Total energy consumed in J	Distance measurement error (m)	MSE (B)	% change $\frac{B_{K+1}-B_K}{B_K} \times 100$
6.10	0.2	1.63	49.62
6.10	0.3	3.2	39.43
6.10	0.4	5.3	51.49
6.10	0.5	11.02	84.16
6.10	1.5	69.61	Ref. value

communications instances in WSN. In Table 4, the overall squared error is defined as sum of squared error due to all target positions. It remains almost constant as error variance does not change.

Table 5 depicts experimental results when target moves for specific time interval within sensor field. During its movement, target is observed by many sensor nodes at discrete time interval. For example in this case, 315 states of target positions are observed. Based on a distance, cluster head estimates the location and calculates square of errors, between actual and estimated positions. This squared error when averaged over all the points, gives MSE. The energy utilized during the estimation of target position is computed and shown in Table 5.

One must note that MSE varies as target moves through the WSN. Actually, MSE is a function of the number of nodes contributing to target state estimation. Possibly, more the number of target sensing nodes, better is the state estimation. The Fig. 3 shows squared error at different time and it can be observed that error is large between time points 35–53. This happens as target neighbourhood has a lesser number of sensing nodes during that time points.

The percentage change in last column of Table 5 is computed as follows: $\frac{(69.61-11.02)}{69.61} = 84.16$.

4.4 *Impact of Node Sensing Range on Target Tracking Accuracy and Energy Consumption*

We now analyze the impact of node sensing range on tracking accuracy and energy consumption. We use additive noise variance of 0.5 and energy cost per 128 bytes

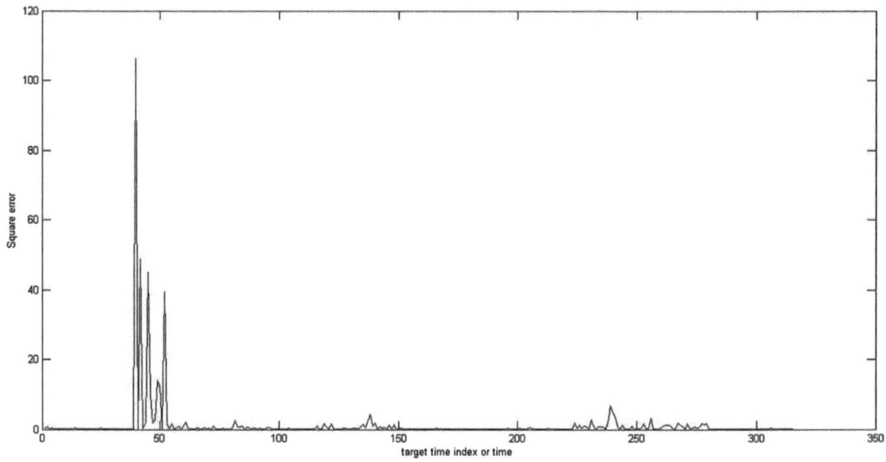

Fig. 3 Squared error versus target time index

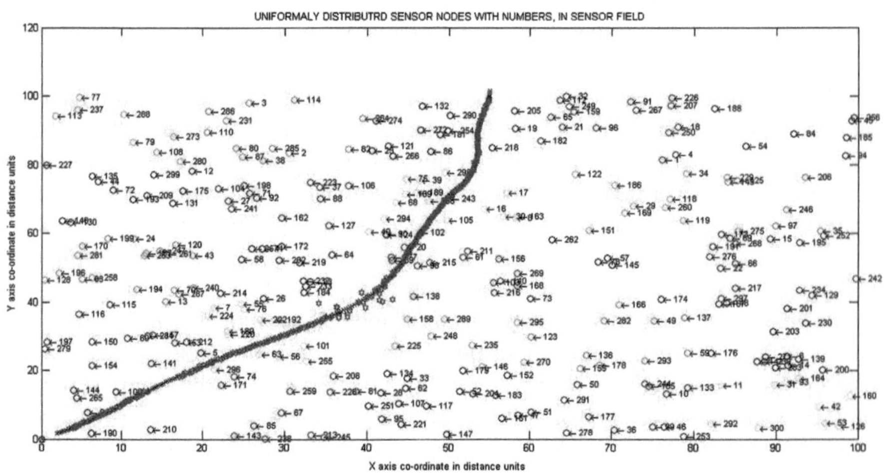

Fig. 4 Sensor node distribution

of transmission and reception is 6.63 exp −05 J. The node distribution is as shown in Fig. 4.

Table 6 shows results in terms of MSE in target tracking for different values of sensing range of node. The variation in MSE and over all squared error is very large therefore, we used logarithmic scale for better representation

Figure 5 shows log of MSE against sensor range as per Table 6. It can be seen that log of MSE is inversely proportional to a range of sensing node. It can also be observed from Table 6 that consumption of energy in WSN increases with increase

Table 6 Sensor range measurement and energy consumption

Sensing range (m)	MSE m^2	log (MSE)	Total consumed energy (J)
6	2620	3.42	1.02
7	1841	3.27	1.39
8	1244	3.09	1.96
9	795	2.9	2.41
10	249	2.4	2.93
11	17	1.23	3.56
12	6	0.75	4.21
13	2	0.27	4.91
14	2	0.34	5.64
15	2	0.33	6.35

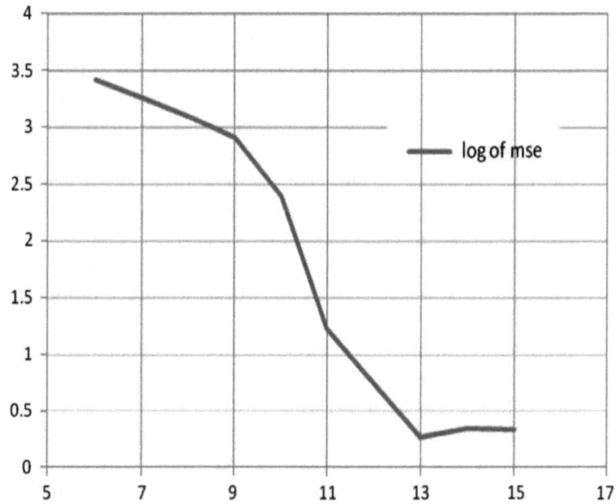

Fig. 5 y axis: Log of MSE (sq. m) versus × axis: sensor range (m)

in sensing range of a node. This is due to the fact that higher sensing range gives opportunity for more number of nodes to participate in the collaborative processing resulting into higher data transmission to cluster head.

5 Conclusion

Collaborative target localization is important for many applications. We have investigated impact of noise variance on target tracking accuracy. Large noise variance causes large error in target state estimation. The accuracy of the target tracking increases as; (1) size of clusters grows; (2) with increase in measurement range of a sensor node. Moreover, high node density improves the target localization accuracy. Paper also examines impact of additive noise variance and node sensing range on energy consumption pattern for a WSN. It was observed that since number of communication do not vary with noise variance, energy consumption remains unaltered in spite of changes in noise levels. However, increase in node sensing range increases the energy consumption across the network. This is due to increase in number of communication transactions.

References

1. Zhou Yan; Zhu Jiaxing: Implementation of particle filter for personal dynamic positioning based on ZigBee network. Computer Science and Information Technology (ICCSIT), 2010 3rd IEEE International Conference on, vol. 2, pp. 521–524, 9–11 (2010).
2. Pantazis, N.A.; Vergados, D.D.: A survey on power control issues in wireless sensor networks. Communications Surveys & Tutorials, IEEE, vol. 9, no. 4, pp. 86–107, 2007.
3. Hao Zhu; Schizas, I.D.; Giannakis, G.B.: Power-Efficient Dimensionality Reduction for Distributed Channel-Aware Kalman Tracking Using WSNs. Signal Processing, IEEE Transactions on, vol. 57, no. 8, pp. 3193–3207 (2009).
4. Jianyong Lin; Wendong Xiao; Lewis, F.L.; Lihua Xie: Energy-Efficient Distributed Adaptive Multisensor Scheduling for Target Tracking in Wireless Sensor Networks. Instrumentation and Measurement, IEEE Transactions on, vol. 58, no. 6, pp. 1886–1896, (2009).
5. Onel, T.; Ersoy, C.; Delic, H.: Information Content-Based Sensor Selection and Transmission Power Adjustment for Collaborative Target Tracking. Mobile Computing, IEEE Transactions on, vol. 8, no. 8, pp. 1103–1116, (2009).
6. Xue Wang; Junjie Ma; Sheng Wang; Daowei Bi: Distributed Energy Optimization for Target Tracking in Wireless Sensor Networks. Mobile Computing, IEEE Transactions on, vol. 9, no. 1, pp. 73–86, (2010).
7. Shih, E.; Calhoun, B.H.; SeongHwan Cho; Chandrakasan, A.P.: Energy-efficient link layer for wireless microsensor networks. VLSI, 2001. Proceedings. IEEE Computer Society Workshop on, vol., pp. 16–21, (2001).
8. Xingbo Wang; Minyue Fu; Huanshui Zhang: Target Tracking in Wireless Sensor Networks Based on the Combination of KF and MLE Using Distance Measurements. Mobile Computing, IEEE Transactions on, vol. 11, no. 4, pp. 567–576, (2012).
9. Demigha, O., Hidouci, W. K. Ahmed, T.: On Energy Efficiency in Collaborative Target Tracking in Wireless Sensor Network: A Review. Communications Surveys & Tutorials, IEEE, vol. 15, no. 3, pp. 1210–1222 (2013).
10. Zhixin Liu; Yazhou Yuan; Xinping Guan; Xinbin Li: An approach of distributed joint optimization for cluster-based wireless sensor networks. Automatica Sinica, IEEE/CAA Journal of, vol. 2, no. 3, pp. 267–273, (2015).
11. Izadi, D.; Abawajy, J.; Ghanavati, S.: An Alternative Clustering Scheme in WSN. in Sensors Journal, IEEE, vol. 15, no. 7, pp. 4148–4155, (2015).

Security in Mobile Ad Hoc Networks

Pimal Khanpara and Bhushan Trivedi

Abstract Due to the proliferation of mobile devices, Mobile Ad hoc Networks (MANETs) are increasing in popularity. However, security of such networks is an important issue as MANETs are vulnerable to various attacks occurring at different layers of TCP/IP protocol suite. This paper focuses on the Network layer vulnerabilities as this layer is responsible for one of the basic MANET functions, routing. This paper describes the existing detection approaches for Network layer attacks. The comparison of the existing security solutions for Network layer attacks is also presented in this paper. Finally, the paper describes the scope of further research.

Keywords Mobile ad hoc networks · Security · Network layer · Point detection mechanisms · Intrusion detection schemes

1 Introduction

Mobile Ad hoc Networks are increasing in popularity but due to the basic characteristics of MANETs, they are vulnerable to various types of attacks. The operation of the network can be compromised by attacking at different layers of the network model. MANETs have many peculiar characteristics like limited battery and computational power, a lack of the centralized control entity, participation of

P. Khanpara (✉)
Institute of Technology, Nirma University, Ahmedabad, India
e-mail: pimal.khanpara@gmail.com

B. Trivedi
GLS Institute of Computer Technology, Ahmedabad, India
e-mail: bhtrivedi@gmail.com

N. Modi et al. (eds.), *Proceedings of International Conference on Communication and Networks*, Advances in Intelligent Systems and Computing 508,
DOI 10.1007/978-981-10-2750-5_52

network nodes in the routing process, dynamic topology, mobility and short-term network services. Due to this, they are vulnerable to route level attacks. Unlike conventional networks, the job of routing is shouldered by the nodes themselves in MANETs and that introduces lot of additional issues. In this environment, conventional security measures like encryption and authentication fail to provide complete protection as conventional solutions are computation heavy and are not designed for battery limited and memory limited devices.

In the last few decades, many researchers have proposed large number of intrusion detection system prototypes for MANETs which are mainly classified into two categories: Point Detection Algorithms and Intrusion Detection Systems (IDSs). The following section presents a survey of Point Detection Algorithms and IDSs proposed in the literature to provide protection against network layer attacks in MANETs.

2 Network Layer Attacks

There are two main categories of Network Layer attacks in MANETs: Passive attacks and Active attacks. In passive attacks, the attacker does not try to affect the normal operation of the routing protocol but tries to get some valuable information about the network. Passive attacks in MANETs are categorized as: Eavesdropping, Traffic Analysis and Location Disclosure. In active attacks in MANETs, attackers try to disrupt the functioning of the network by altering, forging, dropping, fabricating or injecting data or control packets in the network. Active attacks are more severe compared to passive attacks as they can degrade the performance of the network significantly or bring down the network. Active attacks are mainly categorized as routing attacks, packet dropping attacks, Sleep Deprivation attacks, Black Hole attacks, Grey Hole attacks, Sybil attacks and Rushing attacks.

3 Point Detection Algorithms

This section presents a survey of different approaches proposed in the literature to defend from major network layer attacks. As shown in Fig. 1, protection mechanisms for Network Layer are classified based on the number of attacks they can detect. Point detection algorithms can detect only a single type of attack at Network Layer. The other category, intrusion detection systems can identify a range of attacks. Point detection algorithms are further divided according to the type of attack they detect. Table 1 shows the analysis of existing point detection mechanisms for network layer attacks.

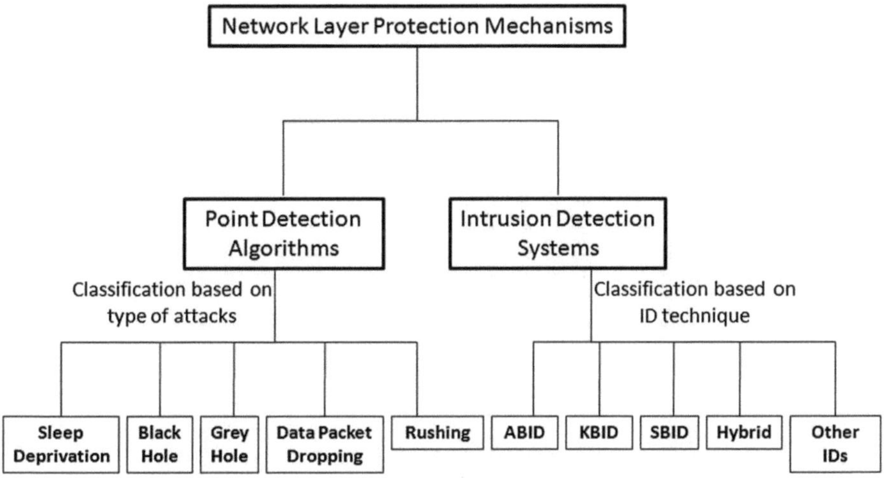

Fig. 1 Network layer protection mechanisms

4 Intrusion Detection Algorithms

Intrusion Detection Systems (IDSs) can detect a range of attacks. This section reviews the existing IDSs and challenges faced by them in MANETs. There are three main categories of IDSs: Knowledge Based Intrusion Detection Systems (KBIDs), Anomaly Based Intrusion Detection Systems (ABIDs), Specification Based Intrusion Detection Systems (SBIDs). In MANETs, some IDSs are combinations of two or more types of intrusion detection techniques and are known as Hybrid Intrusion Detection Systems.

KBIDSs are also known as misuse detection systems. They use and maintain a knowledge base consisting of patterns or signatures of well-known attacks. They add new rules in the knowledge base for unknown attacks. Compared to other IDSs, KBIDSs have very low false positive rates of detection. The limitation of KBIDSs is that they are able to detect only those attacks whose signatures or patterns are available in the knowledge base. Moreover, it is tedious to keep the knowledge base up-to-date for maintaining information about attacks.

ABIDs are also known as behavior based IDSs. They observe the anomalous activities to detect the intrusion and work in two phases: Training Phase and Testing Phase. Training phase is used to model the normal expected behavior of the network. Testing phase compares the current behavior model of the network with the expected behavior model. The main advantage of these systems is that they try to exploit unknown attacks. The drawback of ABIDs is that they are prone to generate false alarms.

SBIDs use explicitly defined specifications to monitor the operations performed at the network layer. Initially, they extract the specifications that specify the correct functionality of the network. In the next step, the system monitors the execution of

Table 1 Point detection algorithms

Protocol name	Architecture	Attacks detected	Detection technique	Corrective measures	Routing protocol	Data gathering mechanism	Remarks
FAP (Yi et al. [1])	Distributed	Sleep deprivation caused by malicious route requests	Priority queue of route requests	Exclude attackers	AODV	Single node monitoring	May suppress legitimate nodes
None (Martin et al. [2])	Not specified	Sleep deprivation	Energy signature, multilevel authentication	Not specified	Not specified	Requests to SSH server	Analyzes the effect of sleep deprivation attack on real systems
LIP (Hsu et al. [3])	Not specified	Sleep deprivation	Local broadcast authentication	Not specified	Not specified	Observation by nodes	Lightweight; helps to prevent packet injection and impersonation
None (Sarkar and Roy [4])	Hierarchical	Sleep deprivation	Based on cluster head's decision	Not specified	Not specified	Observation of packet forwarding	It is not specified how to determine threshold value for packet forwarding
TOGBAD (Pedillia et al. [5])	Centralized, hierarchical	Black hole	Topology graph	Not specified	OLSR	Topology graph	Not feasible for reactive routing
None (Medadian et al. [6])	Distributed	Black hole	Finding safe path	Not specified	AODV	Neighbors' observation	May generate false alarm in highly dynamic MANETs
None (Zhang et al. [7])	Distributed	Black hole	Verifying sequence number of route reply	Not specified	AODV, SAODV	Intermediate nodes' observation	Increased overhead, lack of security checks for sequence request and reply packets

(continued)

Table 1 (continued)

Protocol name	Architecture	Attacks detected	Detection technique	Corrective measures	Routing protocol	Data gathering mechanism	Remarks
None (Xiaopeng et al. [8])	Distributed	Grey hole	Checkup, proof and diagnosis algorithms	Not specified	DSR	Proof from forwarded packets	Specific to DSR
None (Wei et al. [9])	Distributed	Grey hole	Aggregate signature algorithm	Not specified	Not specified	Aggregate signature algorithm	A certificate authority is assumed to be present
None (Yang et al. [10])	Not specified	Grey hole	Historical evidence	Not specified	Not specified	Neighbors' observation	Historical trust values are used to make detection decision
None (Sharma and Garg [11])	Not specified	Sybil	Considered RSS, node speed	Not specified	Not specified	Node speed observation	Threshold value of speed is 10 m/s
None (Abbas et al. [12])	Not specified	Sybil	Localization process	Not specified	Not specified	Localization process	Once a node is registered, no further localization is performed
None (Tangpong et al. [13])	Not specified	Sybil	Exchanging observed information	Exclude attackers	Not specified	Cooperative monitoring	No central authority is needed
None (Hashmi and Brooke [14])	Not specified	Sybil	Authentication agent	Not specified	Not specified	Verification by authentication agent	Uses hardware id for authentication
RAP (Hu et al. [15])	Distributed	Rushing attack	Mutual authentication protocol	Not specified	DSR	Neighbors' observation	Specific to DSR
SRP (Papadimitratos and Haas [16])	Not specified	Rushing attack	SMT protocol	Not specified	Not specified	SMT protocol	Effectiveness of SRP is not checked against routing attacks in MANETs

(continued)

Table 1 (continued)

Protocol name	Architecture	Attacks detected	Detection technique	Corrective measures	Routing protocol	Data gathering mechanism	Remarks
None (Gonzalez et al. [17])	Distributed	Packet dropping	Adaptive policies	Not specified	Not specified	Distributed management overlay	Adaptable protection of routing protocols
SCAN (Yang et al. [18])	Distributed	Packet dropping	Information cross validation	Exclude attackers	AODV	Collaborative monitoring	Specific to reactive routing process
None (Shu and Krunz [19])	Not specified	Packet dropping	Correlation between lost packets	Not specified	Not specified	Public auditing architecture	Increased overhead

Table 2 IDS Algorithms

Algorithm name	Architecture	Attacks detected	Intrusion detection technique	Corrective measures	Routing protocol	Data gathering mechanism	Remarks
None (Liu et al. [20])	Distributed	DoS	Bayesian game theory based anomaly detection	Not specified	Not specified	Lightweight and heavyweight monitoring systems	Use of two IDS
None (Sun et al. [21])	Distributed	Routing disruption attacks	Anomaly detection using Markov chain classifier	Not specified	Not specified	Audit data sources	Can detect local intrusion
None (Jabbehdari et al. [22]	Not specified	DoS	Anomaly detection using neural networks	Not specified	Not specified	Trace output	Specific to DoS attacks
AIDP (Nadeem and Howarth, [23])	Clustered, hierarchical	DoS	Anomaly detection	Exclude intruders	General; tested on AODV	Routing information	Specific to DoS attacks
AFIDS (Chaudhary et al. [24])	Not specified	Black hole	Fuzzy based anomaly detection	Exclude intruders	AODV	Network monitoring	Performance depends on the accuracy of fuzzy inference engine
None (Komninos et al. [25])	Not specified	Not specified	Knowledge based detection	Not considered	Not specified	Audit data trails	Not tested against attacks
IDAR (Alattar et al. [26])	Distributed	Pattern matching	Signature based detection	Not specified	OLSR	Logs generated by OLSR	High bandwidth and memory requirement

(continued)

Table 2 (continued)

Algorithm name	Architecture	Attacks detected	Intrusion detection technique	Corrective measures	Routing protocol	Data gathering mechanism	Remarks
AODVSTAT (Vigna et al. [27])	Distributed	Resource depletion, packet dropping	Knowledge based detection	Not specified	AODV	AODV routing packets, data packets	Detects the attacks against AODV only
None (Tseng et al. [28])	Distributed	DoS	FSM based SBID	Not specified	OLSR	OLSR information	Specific to OLSR
EFSM (Orset et al. [29])	Distributed	Sybil, modification, fabrication	Extended FSM based SBID	Not specified	OLSR	OLSR information	Specific to OLSR protocol
None (Stakhanova et al. [30])	Not specified	Behavioral specification	Specification based detection	Not specified	AODV, DSR	Network traffic flow	Specific to AODV and DSR
CRADS (Joseph et al. [31])	Not specified	Rushing, medication, spoofing, packet dropping	Hybrid intrusion detection	Not specified	OLSR	Data collected from physical, MAC, network layer	Cross layer approach
GIDP (Nadeem and Howarth [32])	Clustered, hierarchical	Various network layer attacks	Hybrid intrusion detection	Exclude attackers	General	Network characteristics; performance matrix	Tested using AODV
None (Yi et al. [33])	Clustered, hierarchical	Routing loops, DoS	Other IDS	Generate alarm	DSR	DSR specifications	Specific to DSR

the operations with respect to the given specification. If it finds any deviation from the specification then it detects it as intrusion.

Table 2 shows the analysis of various intrusion detection systems proposed in the literature.

5 Conclusion

The existing intrusion detection techniques for Mobile Ad hoc Networks are categorized as either Point detection or Intrusion detection schemes. They focus on specific attacks, capable of detecting one or multiple attacks. Many existing reactive security mechanism for MANETs are studied in this paper. From their comparison, we can say that none of these schemes are able to defend against all possible attacks. Moreover, most of the schemes also add additional overhead and complexity in the normal functioning of the network. Hence, there is scope to propose a new security solution for MANETs that is robust and lightweight.

References

1. Yi, P., Dai, Z., Zhong, Y., Zhang, S.: Resisting Flooding Attack in Ad Hoc Networks. In: IEEE International Conference on Information Technology Coding & Computing, pp. 657–662 (2005).
2. Martin, T., Hsiao, M, Dong, H., Krishnaswami, J.: Denial-of-Service Attacks on Battery Powered Mobile Computers. In: IEEE International Conference on Pervasive Computing and Communications (PerCom) (2004).
3. Hsu, H., Zhu, S., Hurson, A. R.: LIP—a Lightweight Interlayer Protocol for Preventing Packet Injection Attacks in Mobile Ad Hoc Networks. In: International Journal of Security and Networks, Vol. 2, Nos. 3/4, pp. 202–215 (2007).
4. Sarkar, M., Roy D. B.: Prevention of Sleep Deprivation Attacks using Clustering. In: IEEE ICECT, Vol. 5, pp. 391–394 (2011).
5. Padilla, E., Aschenbruck, N., Martini, P., Jahnke, M., Tolle J.: Detecting Black Hole Attack in Tactical MANETs using Topology Graph. In: IEEE Conference on Local Computer Networks (2007).
6. Medadian M., Yektaie M. H., Rehmani, A. M.: Combat with Black Hole Attack in AODV Routing Protocol in MANETs. In: IEEE Asian Himalayas International Conference on Internet (2009).
7. Zhang, X. Y., Sekiya Y., Wakahara, Y.: Proposal of a Method to Detect Black Hole Attack in MANETs. In: IEEE International Symposium on Autonomous Decentralized System ISADS (2009).
8. Xiaopeng, G., Wei, C.: A Novel Grey Hole Attack Detection Scheme for Mobile Ad-Hoc Networks. In: IFIP International Conference on Network and Parallel Computing (2007).
9. Wei, C., Xiang, L., Yuebinand, B., Xiopeng, G.: A New Solution for Resisting Grey Hole Attack in Mobile Ad Hoc Networks. In: IEEE Conference on Communication and Networking, China (2007).
10. Yang, B., Yamamoto, R., Tanaka, Y.: Historical Evidence Based Trust Management Strategy against Black Hole Attacks in MANET. In: IEEE ICACT, pp. 394–399 (2012).

11. Sharma, H., Garg, R.: Enhanced Lightweight Sybil Attack Detection Technique. In: IEEE Confluence, pp. 476–481 (2014).
12. Abbas, S., Merabti, M., Jones, D.: Signal Strength Based Sybil Attack Detection in Wireless Ad hoc Networks. In: IEEE DESE, pp. 190–195 (2009).
13. Tangpong, A., Kesidis, G., Hsu, H., Hurson, A.: Robust Sybil Detection for MANETs, In: IEEE ICCCN, pp. 1–6 (2009).
14. Hashmi, S., Brooke, J.: Towards Sybil Resistant Authentication in Mobile Ad hoc Networks. In: IEEE Secureware, pp. 17–24 (2010).
15. Hu, Y., Perrig, A., Johnson, B.: Rushing Attack and Defence in Wireless Ad Hoc Networks Routing Protocol. In: ACM Workshop on Wireless Security, pp. 30–40 (2003).
16. Papadimitratos, P., Haas, Z. J.: Secure Message Transmission in Mobile Ad Hoc Networks. In: Elsevier Journal of Ad Hoc Networks, Vol. 1, No. 1, pp. 193–209 (2003).
17. Gonzalez-Duque, O. F., Hadjiantonis, A. M., Pavlou, G., Howarth, M.: Adaptive Misbehaviour Detection and Isolation in Wireless Ad Hoc networks Using Policies. In: IFIP/IEEE International Symposium on Integrated Network Management, pp. 242–250, NY, USA (2009).
18. Yang, H., Shu, J., Meng, X., Lu, S.: SCAN: Self-Organized Network-Layer Security in Mobile Ad Hoc Networks. In: IEEE Journal on Selected Areas in Communications, Vol. 24, No. 2, pp. 261–273 (2006).
19. Shu, T., Krunz, M.: Privacy-Preserving and Truthful Detection of Packet Dropping Attacks in Wireless Ad Hoc Networks. In: IEEE Transactions on Mobile Computing, Vol. 14, issue 4, pp. 813–828 (2014).
20. Liu, Y., Comaniciu, C., Man, H.: Modelling Misbehaviour in Ad Hoc Networks: a Game Theoretic Approach for Intrusion Detection. In: International Journal of Security and Networks, Vol. 1, Nos. 3/4, pp. 243–254 (2006).
21. Sun, B., Wu, K., Xiao, Y., Wang, R.: Integration of Mobility and Intrusion Detection Wireless Ad Hoc Networks. In: Journal of Communication Systems, Wiley International, Vol. 20, No. 6, pp. 695–721 (2007).
22. Jabbehdari, S., Talari, S. H., Modiri, N.: A Neural Network Scheme for Anomaly Based Intrusion Detection Systems in Mobile Ad Hoc Networks. In: Journal of Computing, Vol. 4, No. 2, pp. 61–66 (2012).
23. Nadeem, A., Howarth, M.: Adaptive Intrusion Detection & Prevention of Denial of Service Attacks in MANETs. In: ACM International Wireless Communication and Mobile Computing Conference, Leipzig Germany (2009).
24. Chaudhary, A., Tiwari, V., Kumar, A.: Design an Anomaly Based Fuzzy Intrusion Detection System for Packet Dropping Attack in Mobile Ad Hoc Networks. In: Proceedings of IEEE IACC, pp. 256–261 (2014).
25. Komninos, N., Vergados, D., Douligeris, C.: Detecting Unauthorized and Compromised Nodes in Mobile Ad Hoc Networks. In: Journal of Ad Hoc Networks, Elsevier, Vol. 5, No. 3, pp. 289–298 (2007).
26. Alattar, M., Sailhan, F., Bourgeois, J.: Log-based Intrusion Detection for MANET. In: Proceedings of IEEE IWCMC, pp. 697–702 (2012).
27. Vgina, G., Gawalani, S., Srinivasan, K., Belding-Royer, M., Kemmerer, A: An Intrusion Detection Tool for AODV Based Ad Hoc Wireless Networks. In: IEEE Annual Computer Security Application Conference ACSAC (2004).
28. Tseng, H., Song, T., Balasubramanyam, P., Ko, C., Levitt, K.: A Specification-Based Intrusion Detection Model for OLSR. In: International Symposium on Recent Advances in Intrusion Detection (RAID), pp. 330–350 (2005).
29. Orset, J. M., Alcalde, B., Cavalli, A. R.: An eEFSM-Based Intrusion Detection System for Ad Hoc Networks. In: International Conference on Automated Technology for Verification and Analysis, pp. 400–413 (2005).
30. Stakhanova, N., Basu, S., Zhang, W., Wang, X., Wong, J.: Specification Synthesis for Monitoring and Analysis of MANET Protocols. In: International Symposium on Frontiers in Networking with Applications (2007).

31. Joseph, J., Das, A., Seet, B., Lee, B.: CRADS: Integrated Cross Layer approach for Detecting Routing Attacks in MANETs. In: IEEE Wireless Communication and Networking Conference (WCNC) (2008).
32. Nadeem, A., Howarth, M.: A Generalized Intrusion Detection and Prevention Mechanism for Securing MANETs. In: IEEE International Conference on Ultra Modern Telecommunications and Workshops, St Petersburg Russia (2009).
33. Yi, P., Jiang, Y., Zhong, Y., Zhang, S.: Distributed Intrusion Detection for Mobile Ad Hoc Networks. In: IEEE Application and Internet Workshop (2005).

Design of Ultra Low Power Voltage Controlled Ring Oscillator

Bhavana Goyal, Shruti Suman and P.K. Ghosh

Abstract This paper presents a voltage controlled ring oscillator using sub-threshold source coupled logic (STSCL) which runs at very low voltages. STSCL variable high resistance load devices are implemented by shorting drain and bulk terminal of PMOS transistors. It consumes extremely low power which is due to reduction in dynamic and static power dissipation. Proposed ring VCO is implemented at 180 nm technology and 1 V power supply. It gives low power consumption of 6.75 μW at 124 MHz oscillation frequency and 0.65 V control voltage. This type of ultra low power ring VCO can be used in biomedical applications like pacemakers, etc.

Keywords VCO · Ring VCO · Inverter · SCL · STSCL

1 Introduction

Voltage Controlled Oscillators (VCOs) are the most important element in all wired or wireless communication systems. In addition to communication systems, VCOs are integral part of biomedical applications like pacemakers etc. [1]. The VCO gives an oscillatory output voltage where output signal is a linear function of input control voltage [2]. VCOs can be realized by two main topologies: the ring structure and inductance (L)—capacitance(C) structure. Ring structures provide inferior phase noise performance (because of their low quality factor), lower dissipated power,

B. Goyal (✉) · S. Suman · P.K. Ghosh
ECE Department (CET), Mody University of Science and Technology, Lakshmangarh, Sikar, Rajasthan, India
e-mail: goyal.bhavana1991@gmail.com

S. Suman
e-mail: shrutisuman23@gmail.com

P.K. Ghosh
e-mail: pkghosh.ece@gmail.com

N. Modi et al. (eds.), *Proceedings of International Conference on Communication and Networks*, Advances in Intelligent Systems and Computing 508,
DOI 10.1007/978-981-10-2750-5_53

less design complexity, smaller layout area and wide tuning range compared to LC structures.

A ring oscillator consists of three or more odd number of inverter stages in a negative feedback loop configuration [3]. For delay (t_d) of a single inverting stage the frequency of oscillation (f_{osc}) for N number of identical stages [4] can be calculated as:

$$f_{osc} = \frac{1}{2Nt_d} \tag{1}$$

Main performance parameters of VCOs are power dissipation, center frequency, phase noise, tuning range and linearity. At present, low power consumption is one of the most important design parameters of VCO because in all applications it consumes more power as in a frequency synthesizer; the key power-hungry circuits are the voltage-controlled oscillators. For enhanced battery life of handheld and wireless devices low-power circuit techniques are becoming increasingly important. In CMOS logic, delay is highly increased due to power supply scaling and output swing is also reduced. For SCL circuits operating in subthreshold region [5], the power dissipation and operation time is controllable. This allows better logic swing and performance even at low supply voltage.

2 Power Consumption Sources

To achieve low power dissipation, overall leakage current must be minimised. Static dissipation (gate leakage and sub-threshold leakage) and dynamic dissipation are two main leakage components. Dynamic power consumption $(P_{dynamic})$ occurs during switching and transition of logic levels. It varies quadratic proportional to the supply voltage (V_{DD}) as:

$$P_{dynamic} = fC(V_{DD})^2 \tag{2}$$

there f is the frequency of operation of logic circuit and C denotes the output capacitance of the circuit. When no input provided on a logic gate the static power consumption occurs due to the flow of leakage from supply to ground, only the sub-threshold leakage current (I_D) which flows through the channel of the MOS devices is given by Eq. (3) where I_{D0} is saturation current and n denotes the subthreshold slope factor of transistor.

$$I_D = I_{D0} e^{\left(\frac{(V_{gs} - V_{th})}{n} V_T\right)} \left[1 - e^{\left(\frac{-V_{DS}}{V_T}\right)}\right] \tag{3}$$

where V_{gs}, V_{th} and V_{DS} are respectively the gate to source voltage, threshold voltage and drain to source voltage of transistor and $V_T = kT/q$ is the thermal voltage at temperature T, k is the Boltzmann constant and q is electronic charge.

Prime method of reducing power dissipation is power supply scaling. Due to power supply reduction dynamic power dissipation is reduced but gate delay increases. To maintain delay performance threshold voltage is also scaled down. But as Eq. (3) suggests that subthreshold leakage current is increased due to low threshold voltage, to neutralize the effect of threshold voltage scaling gate to source voltage is also reduced due to which the device enters in weak inversion regime.

3 Source Coupled Logic

A basic SCL inverter is shown in Fig. 1 where NMOS transistors $M1$ and $M2$ form input differential pair. Transistors $M1$ and $M2$ are loaded with PMOS transistors $M3$ and $M4$, respectively. PMOS transistors work in linear region as variable

Fig. 1 Source coupled logic inverter

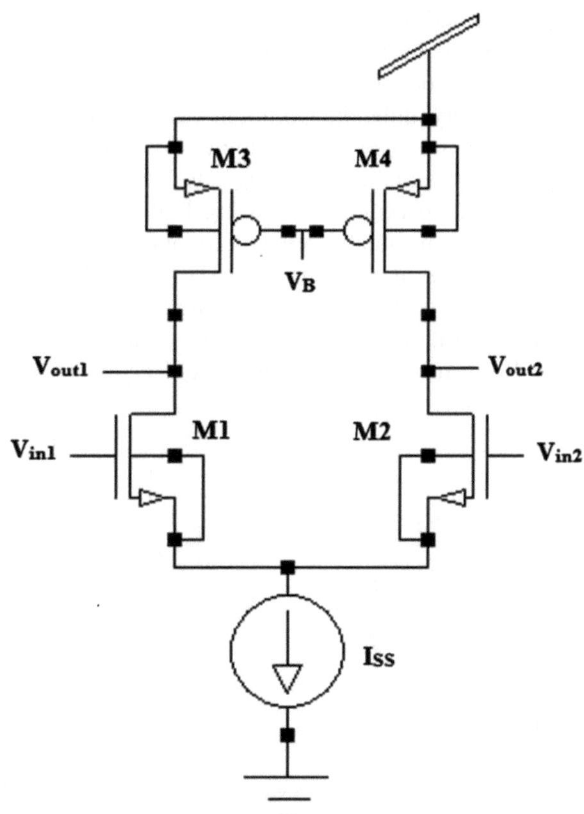

resistors controlled by bias voltage (V_B).The output swing (V_{Swing}) can be defined by the Eq. (4) where R_d is the resistance of PMOS load device and I_{SS} is a bias current source.

$$V_{Swing} = V_{out1} - V_{out2} = 2R_d I_{SS} \tag{4}$$

Output swing of SCL inverter must be high to completely switch the input differential pair of the next stage. In strong inversion region, output swing should be greater than $\sqrt{2n}V_{Dsat}$ (here n is the subthreshold slope factor and V_{Dsat} is the overdrive voltage when device operated in saturation region) [6]. For weak inversion region this must be larger than $4nV_T$. Moreover, the SCL circuit gives high noise immunity due to differential logic implementation.

4 Subthreshold Source Coupled Logic

For ultra power applications, leakage must be at minimum level. In subthreshold region, current density for MOS devices is very low due to which subthreshold source coupled logic enables the gates to run at low supply voltages. Low supply voltage also reduces the dynamic power dissipation. Supply voltage scaling results output swing degradation due to which CMOS circuit operation in subthreshold region is very difficult. Optimum performance can be achieved using the SCL topology in place of CMOS logic. For sub-threshold operation of SCL circuit, output swing with thermal voltage (V_T) equals to $4nV_T$ which approximate 150 mV for n = 1.5. To acquire this swing high load resistance is required according to Eq. (4) so some modification necessitates in PMOS load device. High resistive PMOS Load is implemented by shorting bulk terminal to drain terminal which causes a reverse bias formation of diode as shown in Fig. 2. High resistance value ensures operation of SCL circuits in weak inversion region at very low bias currents. Resistance value in subthreshold region (R_{dSub}) is obtained as:

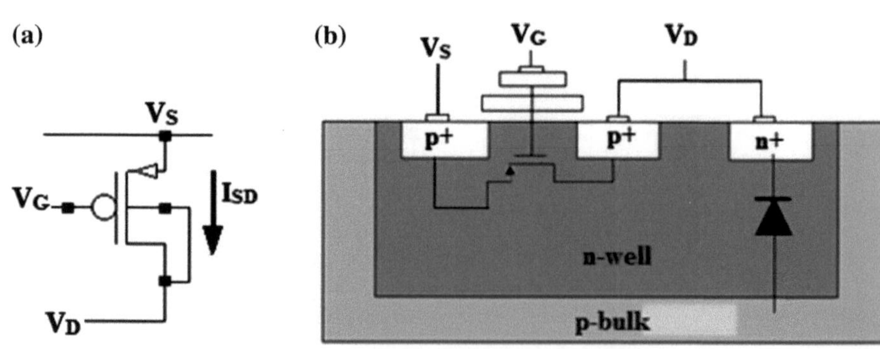

Fig. 2 **a** STSCL PMOS load device. **b** Cross section of PMOS load device showing the parasitic components

$$R_{dSub} = \frac{\frac{(n_P V_T)}{I_{SD}}\left(e^{\frac{V_{SD}}{V_T}} - 1\right)}{(n_P - 1)e^{\frac{V_{SD}}{V_T}} + 1} \tag{5}$$

Here n_P denotes the subthreshold slope factor of PMOS transistor. I_{SD} and V_{SD} are the source to drain current and voltage of the transistor, respectively. From the above equation, R_{dsub} can be controlled by the bias current.

Exponentially dependence of resistance on the source to gate voltage allows tunability of the resistance values over a wide range without modifying internal parameters like size of the devices. For STSCL circuit sub-threshold leakage and gate leakage are negligible compared to CMOS logic, due to differential logic style. The speed of operation in an SCL gate is mainly limited by the time constant (τ_{SCL}) at the output node which is given by:

$$\tau_{SCL} = R_d C_L = V_{Swing} C_L / 2I_{SS} \tag{6}$$

Here C_L is the load capacitance.

The delay for the sub-threshold SCL is inversely proportional to the bias current. In the choice of bias current trade-off exists between delay and power consumption. The most important parameter for an STSCL gate to run properly and consume less power is to control the amount of bias current (I_{SS}) flowing through the logic. For the operation of circuit in subthreshold region, bias current (I_{SS}) must be reduced to a very low value but for a controlled operation in that region biasing circuit is required. The main advantage of using STSCL is the possibility of running the gates in sub-threshold region. This allows the impact of having any sub-threshold leakage current, to be negligible.

5 Proposed Delay Cell

Proposed delay cell uses a single ended differential pair configuration which as shown in Fig. 3. Here current mirror technique is used for load implementation in which gate of the transistors $M3$ and $M4$ are interconnected hence same current will flow through both the transistors where current is controlled by the control voltage (V_{ctrl}). Transistor $M6$ behaves as a bias current source in which current is also controlled by control voltage with the help of current mirror formed by transistors $M2$ and $M6$. Transistor $M4$ is a STSCL PMOS load devices for the delay stage where substrate terminal is shorted to drain terminal making the PMOS transistor a high variable resistance device. The NMOS transistor $M5$ performs inverting operation. Here each stage has to drive only a single gate which results in a low parasitic capacitance at the output. So delay caused by charging and discharging of capacitor is reduced. As load device works in subthreshold region, a small voltage

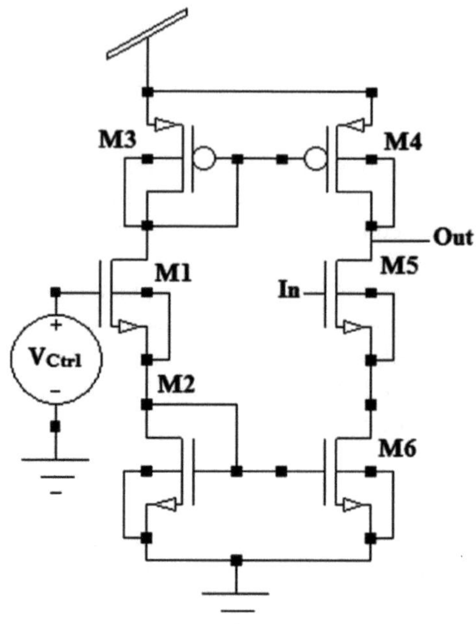

Fig. 3 Proposed delay cell

supply is required for proper operation due to which power consumption is highly reduced maintaining overall good performance.

6 Simulation Results

The proposed three stage ring VCO shown in Fig. 4 is simulated at 180 nm CMOS technology and output is analyzed for different control voltages. The oscillation frequency range is also observed as well as the corresponding power dissipation to find the circuit limitations and the maximum frequency that could be achieved in an STSCL-based ring VCO. Output waveform of proposed circuit is shown in Fig. 5. Simulation results are summarized in Table 1 which shows that power consumption is low, output range and frequency of operation.

Comparative analysis between current starved [7] and proposed ring VCO is displayed in Table 2. From table it is clear that proposed circuit consumes less power as compared to conventional current starved ring VCO.

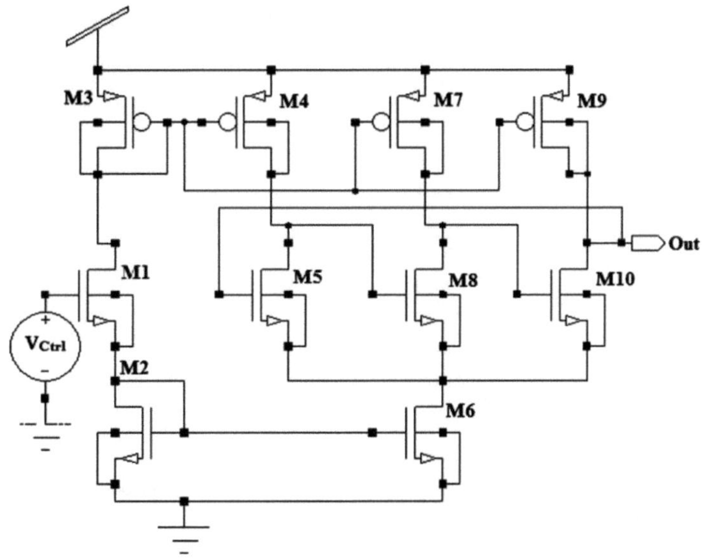

Fig. 4 Proposed three stage ring VCO

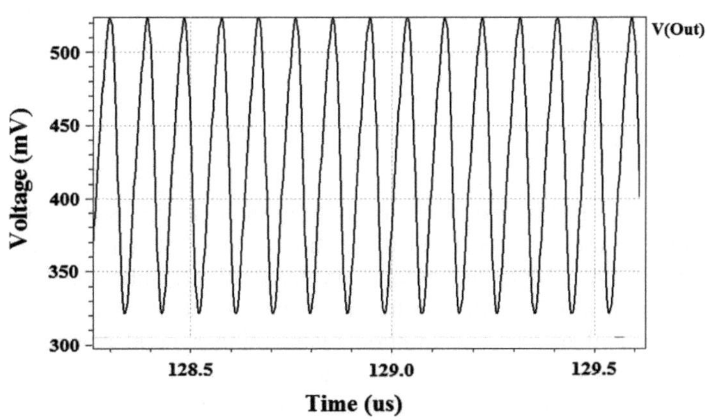

Fig. 5 Output waveform of proposed three stage ring VCO

Table 1 Simulation results of proposed three stage ring VCO at 180 nm technology

Control voltage (V)	Frequency (MHz)	Power consumption (μW)
0.20	7.442	0.175
0.25	10.70	0.304
0.30	17.01	0.521
0.35	25.39	0.872
0.40	36.32	1.432
0.45	52.49	2.302
0.50	73.39	3.569
0.55	0106	5.060
0.60	0119	6.190
0.65	0124	6.750

Table 2 Comparative analysis of conventional current starved and proposed ring VCO at 180 nm technology and power supply 1 V

	Conventional current starved ring VCO	Proposed ring VCO
Frequency (MHz)	1220.50	123.9
Power consumption (μW)	000206	6.750
No. of transistors used in three stage ring VCO	14.0000	10.00

7 Conclusion

The proposed ring VCO is designed using subthreshold source coupled logic where high resistance compact STSCL PMOS load devices are used. This circuit is simulated in 180 nm CMOS technology and simulation results satisfies the basic concept of VCO in which oscillation frequency varies linearly with supply voltage. It consumes a small power consumption maintaining a good performance level. This VCO can be used for very low voltage and low frequency applications like day distributed sensor networks, bio-medical applications etc.

References

1. L. S. Wong et al., "A very low-power CMOS mixed-signal IC for implantable pacemaker applications," *IEEE J. Solid-State Circuits,* vol. 39, no. 12, pp. 2446–2456, Dec. (2004).
2. Jubayer Jalil, Mamun Bin Ibne Reaz, Mohammad Arif Sobhan Bhuiyan Labonnah Farzana Rahman, and Tae Gyu Chang, "Designing a Ring-VCO for RFID Transponders in 0.18 μm CMOS Process," *Hindawi Publishing Corporation,* vol. 2014, Article ID 580385, Jan. (2014).
3. Bhawika Kinger, Shruti Suman, K. G. Sharma and P. K. Ghosh "Design of Improved Performance Voltage Controlled Ring Oscillator" 5th *IEEE International Conference on*

Advanced Computing & Communication Technologies, Rohtak, (India), *IEEE Digital library*, ISBN No. 978-1-4799-8487-9, pp. 441–445, Feb. (2015).
4. Shruti Suman, Monika Bhardawaj Prof. B. P. Singh, "An Improved performance Ring Oscillator Design", in International Conference on Advance Computing and Communication Technology, Rohtak, India, *IEEE digital library,* pp. 236–239, Jan. (2012).
5. E. Vittoz, "Weak inversion for ultimate low-power logic," in *Low-Power Electronics Design,* C. Piguet, Ed. Boca Raton, FL: CRC Press (2005).
6. P. R. Gray, P. J. Hurst, S. H. Lewis, and R. G. Meyer, *Analysis and Design of Analog Integrated Circuits, 4th edition,* New York: Wiley (2000).
7. Gudlavalleti Rajahari, Yashu Anand Varshney, and Subash Chandra Bose, "A Novel Design Methodology for High Tuning Linearity and Wide Tuning Range Ring Voltage Controlled Oscillator", *Springer-Verlag Berlin Heidelberg* 2013, CCIS 382, pp. 10–18, (2013).

A Dynamic Session Oriented Clustering Approach for Detecting Intrusions in Databases

Indu Singh, Poornima and Nitish Kumar

Abstract Modern era applications involve interaction with sophisticated database systems to store information. This stored information is under serious safety threat, not only from an outsider, but also from a malicious insider, who has full information about the structure and functioning of the system. This has led to the incorporation of an intrusion detection mechanism, which provides access control for databases. However, mapping individual access patterns becomes infeasible due to the abundance of users. A solution to this problem is to classify the users into groups and identify their legitimate access patterns. This paper provides one such approach of intermingling intrusion detection system with data mining to form clusters of role attributes in order to identify roles dynamically in a session.

Keywords Intrusion detection · Session management · Role based access control · Clustering · Anomaly detection

1 Introduction

Since the commencement of the 21st century, automation has been on an all-time high. Several data oriented fields such as those based on the management of the database systems (DBMS) have reaped its obvious benefits [1]. The sophistication of these databases allows the storage of every form of information including

I. Singh (✉) · Poornima · N. Kumar
Department of Computer Science and Engineering, Delhi Technological University, New Delhi, India
e-mail: indu.singh.dtu14@gmail.com

Poornima
e-mail: vpoornima.1603@gmail.com

N. Kumar
e-mail: nitishkumar8663@gmail.com

N. Modi et al. (eds.), *Proceedings of International Conference on Communication and Networks*, Advances in Intelligent Systems and Computing 508,
DOI 10.1007/978-981-10-2750-5_54

important data. However, this data is under a constant threat of attack by not only an outsider, wishing to gain access to crucial information for his unethical intentions but also by an insider, who may be any user internal to the system and having the rightful access to information, but an intention of gaining access to the restricted content [2].

An anomaly detection system, which is a part of Host based Intrusion Detection System (HIDS), monitors normal data access patterns and any deviations from these are identified as anomalous behavior. Since the retrieval of log files used for anomaly detection can become cumbersome [3], sessions are now being used to reduce the retrieval process by organizing the access patterns on a sessional basis. The need for sessions, however arose when in order to achieve more user interactivity, services were developed which not only seeked a dynamic interaction with the user, but also needed to keep a track of the users activities in order to increase the quality of service [4].

A simpler way of restricting access is grouping users together on the basis of roles which they can take in the session and identifying the access patterns for each role. Such access scheme is known as Role Based Access Control (RBAC), which is a powerful way to implement the administrators' view of the organizational structure [5].

The identification of roles dynamically in a session is a non-trivial task that can be simplified by the incorporation of data mining for role classification. Moreover, algorithms such as k-means clustering help in increasing the successful role identification rate. However, when k-means is used alone for the purpose of identification of anomalous behavior, it suffers from the drawback of being too sensitive to border conditions since any data object with a large value may distort the formation of the cluster. In order to overcome this problem, classification techniques are used with probability distribution schemes such as Naïve Bayes Classifier to enhance the detection rates [6].

In this paper we have proposed an innovative approach for the incorporation of RBAC in an anomaly detection based IDS, by making use of clustering to map the user roles with the related attributes which the user can access under that role. Our approach is motivated by two factors. The first factor is the need for a system which allows users to switch his role dynamically without compromising the security. The second factor is the incorporation of the clustering technique for the identification of the user's role, based on his query, an approach which has not yet been fully explored.

The rest of the paper is organized as per the following sections: in the next section, we provide an overview of the existing approaches to solve the problem of intrusion. In Sect. 3, we provide a detailed description of our proposed approach. We have given the algorithm and its related description in Sect. 4. Following which we evaluate the performance of our algorithm and analyze its impact in Sect. 5. We end with our conclusions as well as the scope for future work in Sect. 6.

2 Literature Review

The overviews of some of the approaches to deal with the advent growth of intrusive attempts are given in this section.

Chung et al. [7] proposed the Misuse Detection System for Database System (DEMIDS). Audit logs were used to derive user profiles, which were then used to determine the misuse of authority by the insiders. Distance measure was used to quantify the knowledge about data structures and semantics of the system, and was also used to search for frequently occurring itemsets in the user profiles.

Qin [8] proposed Q-clustering algorithm for clustering query attributes generated by the user based on a similarity function to compute the similarity between the query and each of the existing clusters. Cluster with the largest similarity was selected. The algorithm was determined to be more accurate than the conventional clustering approaches. However, its complexity posed a problem for real time applications.

Panda et al. [9] demonstrated an approach for the efficient mitigation of the insider threats. Different sequences of transactions were created to reduce the risk associated with the data being accessed, and the sequence associated with the least risk was termed as the safe sequence. In the process of detection, a huge emphasis was laid on the sensitivity of data. Hence the model faced a significant limitation in the cases where the information sensitivity was difficult to quantify.

A strong incentive for the minimization of the false rate detection in intrusion detection systems was provided by Rezk et al. [10]. The approach was based on the assumption that the number of intrusion attempts was significantly lower than the general access attempts in any system and focused on increasing the efficiency of the models used for data dependency representation through log file mining. Experimental results showed that the approach was able to reduce the false positive rate.

For detecting the anomalous patterns, Kamra et al. [11] extracted useful information from the structure of the parse-trees. The role with the maximum probability was chosen in the detection phase, which then deployed the Bayesian classification methodology, while the training phase allowed the user to train the system on the basis of sampling the probability of access. Though the system excelled in intrusion detection, it laid no emphasis on its adequate response.

Rao et al. [12] presented a model for increasing the efficiency of intrusion detection by using the f-triplet format profile generation, while attribute level information was incorporated in the profile to reduce the false negative rate. This was achieved by altering the sequence of consecutive select commands and mapping the attribute subset access pattern. It was demonstrated that reducing both false positives and false negatives could improve the overall performance of the database IDS.

3 Proposed Approach

In this section we present the design and implementation of our proposed system including the system's architecture.

3.1 Architecture

The proposed system architecture is shown in Fig. 1. In order to ensure user authentication, we propose the incorporation of one-time password (OTP) security mechanism. When user requests a login access to the system for the first time, an OTP would be sent to user's trusted mobile device. Upon the verification of the OTP, the authentication module would then provide the session certification in the form of cookies to the user and would store the details of the user as well as the session in the certification and access log table. Once the user receives his session certification, the access to the Query Generation Interface (QGI) is granted.

The QGI combines the query of the user with session credentials and sends it to the Query Context Parser (QCP). The QCP extracts keywords from the query using string matching, keeping the session credential part intact. Now, the extracted keywords would be forwarded to the Similarity Mapping Module (SMM).

Our proposed system applies clustering in both the learning as well as the detection phase. During the learning phase, access would be granted only to the trusted users by the administrator. As they use the system, queries are fed into the Clustering Engine where clustering of the keywords is done by associating full attribute names of the form 'table_name.attribute_name' with user roles. In the detection phase when the control flow reaches the SMM, a similarity function is

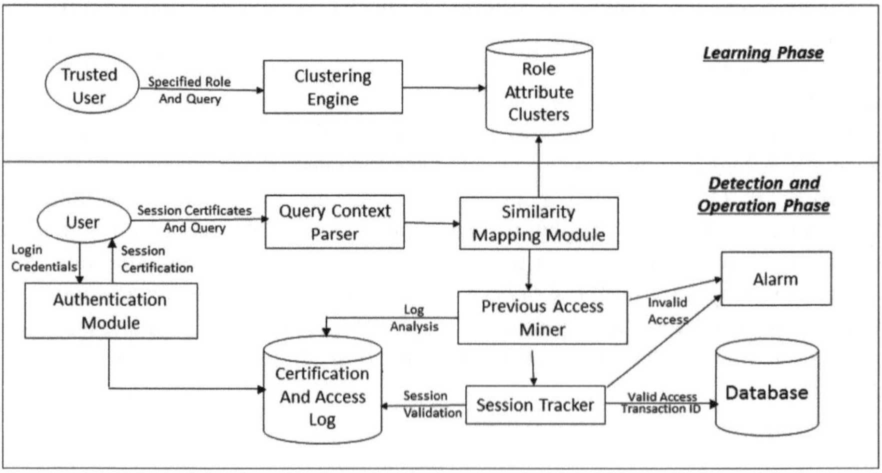

Fig. 1 System architecture of the proposed system

applied to associate the similarity of each role which the user can access with the type of query filed by the user.

The similarity function is given by:

$$\text{Similarity}, \delta(\text{Rol}_\text{i}) = \frac{\text{Rol}_\text{i}\text{\_Att} \cap \text{Que\_Att}}{Que\_Att} + \text{P}(\text{Rol}_\text{i}) * \text{CAI}$$

$$\text{P}(\text{Rol}_\text{i}) = \frac{Hr\_Lvl(Rol_i) * f(Rol_i)}{\sum_i Hr\_Lvl(Rol_i) * f(Rol_i)}$$

$$\begin{aligned}\text{CAI} &= \text{Cluster Associativity Index}\\ &= \frac{\text{Rol}_\text{i}\text{\_Att} \cap \text{Que\_Att} - (\text{Que\_Att} - \text{Rol}_\text{i}\text{\_Att})}{\text{Rol}_\text{i}\text{\_Att}}\end{aligned}$$

where Rol_i is set of roles belonging to the user, f is the frequency of access and Hr_Lvl is the hierarchical level of each user role specified by the administrator to help in the division of authority within the roles.

After applying the similarity function on each role that the user can access, the role with the highest similarity is found. The identified role along would then be forwarded to the Previous Access Miner (PAM). The PAM would then search the logs to extract the information of the records which cannot be accessed by the user due to his access of a similar query at a time before the threshold time period.

Upon the analysis by PAM, if no inconsistencies are found, one last check for intrusion would be performed by engaging the Session Tracker (ST) module to access the certification log and point out any sessional inconsistencies. Upon successful verification, it would then issue a transaction id and grant the database access. In case of any inconsistencies found during any of the above mentioned analysis operation, the transaction id would not be generated and an alarm would be raised.

3.2 *An Illustrative Approach*

For the purpose of illustrating this example, consider a clinic with various roles defined as: Doctor, Patient, Receptionist, Pathologist and Nurse. Suppose the tables in the database are Patient_Info, Patient_Test, Test_Details, Ward_Info and Staff_Info. The attributes which can be accessed by Doctor are all the attributes of Staff_Info, Patient_Info.DischargeDate, Patient_Info.Allergies, Patient_Info.Illness, by Patient are all the attributes of Patient_Info, by Receptionist are all the attributes of Ward_Info, Staff_Info, and Patient_Info.NextAppointment and by Pathologist are all the attributes of Patient_Test, Test_Details, Staff_Info and Patient_Info.Illness.

There may be a scenario where several roles may be played by a single user. For example, suppose there is a shortage of staff and the assignment of work is such that the receptionist has to work as a part time pathologist. Let us consider two queries and their results: (Fig. 2).

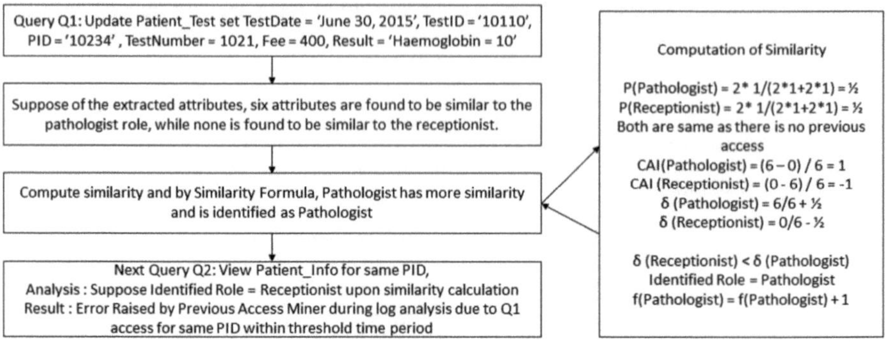

Fig. 2 Flow of two queries Q1 and Q2 depicting the raising of alarms and computation of similarity

4 Proposed Algorithms

In this section we narrow down our approach by providing two algorithms, one for the detection phase and another one for the learning phase. The algorithms are as follows:

4.1 Learning Phase

Let Query be represented by Que, Attribute by Att, Table by Tab, Role repository by RolRep, Attribute List by Att_List

```
1. Initialise Rol = R and Att_List = NULL
2. for each Que for Rol
3.   Extract Att, Tab
4.   Link Att and Tab in the form Tab.Att for every extracted Att and add to Att_List
5.   if (command == select) type = read
6.   else type = write
7.   Group Att_List by type to form Rd_Att_List and Wr_Att_List
8.   if (row({Rol}) is not in RolRep)
9.       insert ({Rol, {Rd_Att_List}, {Wr_Att_List}} in RolRep)
10.  else
11.      update row({Rol, {Rd_Att_List}, {Wr_Att_List}})
12.  Perform K-means clustering with Rd_Att_List and Wr_Att_List for Rol
13. end
```

The above mentioned algorithm is applied during the learning phase in which a trusted user from each role filed queries into the system. For each query filed by the trusted user, attribute and table accessed for that query are extracted and appended to the list of attributes which can be accessed. If the user query contains the keyword 'select', then it is a read query and the type is set to read else it is set to write. The output is the clustering of the read attributes and write attributes with the user roles.

4.2 Detection Phase

Let UserID be represented by UID, SessionID by SID, Access Timestamp by T, Session Timeout time by t, Similarity function for role Rol by δ(Rol), Expected Role which can access the Query by ExpRole and Role for the current user which is most suitable context for the Query by IdRole

```
Authenticate UID
Initialize SID = new SessionID
SID.t = getLargestRole(UID).t
Save SID in SID_Log
while user feeds Que
    extract Att,Tab from Que
    for each Rol in UID.Rol
        get CAI
        evaluate δ(Rol)
     ExpRole = Rol with max(δ(Rol))
    update P(ExpRole) set f(ExpRole) +=1
    extract target_tuple from Que
    T = extract max(log(UID,{UID.Rol-ExpRole},target_tuple))
    if(Curr_Timestamp–T >TTH and δ(Rol)> δTH)
        if(SID_Log(SID,UID))
            assign TID
            TID.t = ExpRole.t
            allow DB access
            store TID in TID_Log
        else reauthenticate
    else raise_alarm
    if(SID.expire) then reauthenticate
end
```

After authentication of OTP, a session timeout time is initialized on the basis of the highest role that the user is allowed to take. Following which the user is allowed to query. The expected role is assigned as the role with the highest similarity. The access frequency of that particular role by the user is increased which in turn helps to reduce the chances of a user switching to another role in the middle of the session when a query is found to be equally similar to several roles which the user may take. In order to maintain the consistency of the transaction, the transaction details are stored in the log to rollback in case of failure.

5 Performance Evaluation and Analysis

In this section, we provide the analysis of our proposed approach and compare it with the approach in which the roles are identified only on the basis of k-means clustering of the roles and attributes. For the purpose of evaluating the performance, we generated the hospital datasets and tested the approach by firing legitimate as well as non-legitimate queries as a part of the transactions.

5.1 True Positive

The graphical comparison in Fig. 3 depicts that, as we increase number of queries in the detection phase, the positive identification rates increase. This is because when the number of queries is less, there may be a number of roles which may be associated with the extracted attributes which could lead to the wrong identification of the roles.

5.2 Number of Queries in Training Phase Versus True Detection Rate

Figure 4 shows the graphical representation of the effect of increasing the number of queries in training set. When the number of queries in the training set is low, the detection rate is also low. This is because the roles get associated with their corresponding access attributes in the learning phase. A poor formation of the cluster in the learning phase could in turn lead to poor detection in the operation phase.

5.3 False Negative

In Fig. 5, comparison of false identification of non-anomalous transactions between the 2 approaches is given. Initially, k-means clustering performs better than the

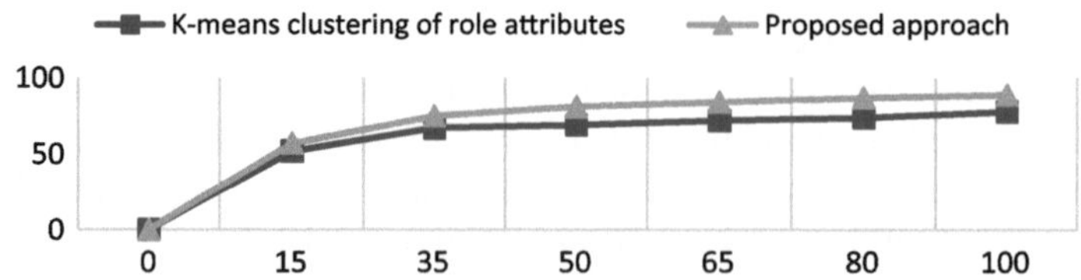

Fig. 3 Detection rate comparison of K-means clustering and proposed approach with respect to number of queries

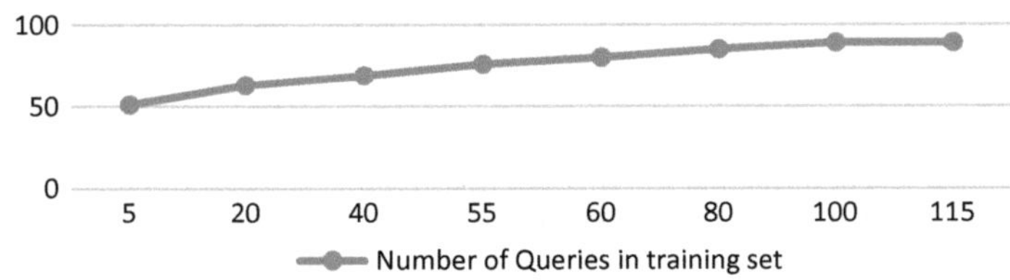

Fig. 4 Graphical representation of the effect of increasing the number of queries in training set

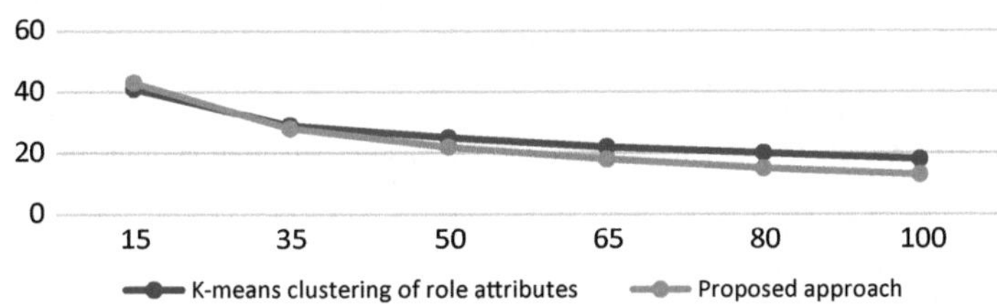

Fig. 5 False negative comparison of K-means clustering and proposed approach with respect to number of queries

proposed approach because it is not dependent on the frequency factor for identification of user roles. When the number of queries is less, the frequency factor tends to be biased towards the previous role access which affects user role mapping.

6 Conclusion and Future Work

In this paper we have proposed an approach for the application of intrusion detection in systems involving high user interactivity. Clustering was used to map the roles with their associated attributes and for dynamic identification of roles, the factor of cluster associativity along with the history of its reference was used. We have mitigated the problem related to session management by ensuring maintenance of proper log files and the safe generation of the OTP. We have compared our proposed approach with the conventional techniques of role attribute clustering alone and arrived at a conclusion that our system's performance is a mass improvement over the existing approaches.

As a part of the future work, we plan to extend this work by making the system adaptable to complex queries and for multivalued attributes. We would also like to reduce dependency on the frequency factor when the number of queries is less.

References

1. Bertino, E., Byun, J.W., Kamra, A.: Database security. In: Security, Privacy, and Trust in Modern Data Management, pp. 87–101. Springer, Berlin Heidelberg (2007).
2. Denning, D.E.: An intrusion-detection model. In: Transactions on Software Engineering Vol. 2, pp. 222–232. IEEE (1987).
3. Scarfone, K., Mell, P.: Guide to Intrusion Detection and Prevention Systems. In: National Institute of Standards and Technology (NIST), pp. 800–94 (2007).
4. Fernandez, E.B., Pernul, G.: Patterns for session-based access control. In: Proceedings of the 2006 conference on Pattern languages of programs. ACM (2006).
5. Ferraiolo, D., Cugini, J., Kuhn, D.R.: Role-based access control (RBAC): Features and motivations. In: Proceedings of 11th annual computer security application conference (1995).
6. Bradley, P.S., Fayyad, U.M., Reina, C.: Scaling Clustering Algorithms to Large Databases. In: KDD, pp. 9–15 (1998).
7. Chung, C.Y., Gertz, M., Levitt, K.: Demids: A misuse detection system. In: Integrity and Internal Control in Information Systems, pp. 159–178. Springer US (2000).
8. Zhong, Y., Zhu, Z., Qin, X.L.: A clustering method based on data queries and its application in database intrusion detection. In: Proceedings of 2005 International Conference on Machine Learning and Cybernetics. Vol. 4. IEEE (2005).
9. Yaseen, Q., Panda, B.: Organizing Access Privileges: Maximizing the Availability and Mitigating the Threat of Insiders' Knowledgebase. In: The 4th International Conference on Network and System Security (NSS). IEEE (2010).
10. Rezk, A., Ali, H., El-Mikkawy, M., Barakat, S.: Minimize the false positive rate in a database intrusion detection system. In: International Journal of Computer Science & Information Technology Vol. 3, No. 5 (2011).
11. Shebaro, B., Sallam, A., Kamra, A., Bertino, E.: PostgreSQL anomalous query detector. In: Proceedings of the 16th International Conference on Extending Database Technology, pp. 741–744. ACM (2013).
12. Rao, U.P., Singh, N.K., Amin, A.R., Sahu, K.: Enhancing detection rate in database intrusion detection system. In: Science and Information Conference (SAI), pp. 556–563. IEEE (2014).

Cognitive Decision Making for Object Recognition by Humanoid System

Ashish Chandiok and D.K. Chaturvedi

Abstract Object recognition signifies an important module in cognitive computer vision structures, such as in a robotic visual system, smart video surveillance arrangements, or mechanical and home service robot interfaces. This paper contributes to the research by a thorough analysis of computer vision based on cognitive decision capability by proposing an architectural framework and its process for consistent real-time application. Secondly, the paper explains past research of computer vision for intelligent decision making. At last, the authors presents the hardware and software setup for object recognition done by the humanoid.

Keywords Computer vision · Cognitive · Decision making · Object recognition

1 Introduction

Computer vision represents a leading technology of Cognitive Computing, which intentions at the analysis and understanding of observable facts [1]. It deliberates as a procedure initializing from an image or set of image frames and causing in a rational decision of the world sight. The difficulties of image recognition are at the central of contemporary struggles to support a machine to make 'logical' connections with the external world [2]. Camera Sensors are used to obtain facts from its environment that can occur in the form of image sequences. This data is then

---

Ashish Chandiok—Cognitive decision making and its computer vision technology is emerging as an important choice of future.

A. Chandiok (✉) · D.K. Chaturvedi
Faculty of Engineering, Dayalbagh Educational Institute, Dayalbagh, Agra, Uttar Pradesh 282005, India
e-mail: achandiok@gmail.com
URL: http://www.dei.ac.in

D.K. Chaturvedi
e-mail: dkc.foe@gmail.com

N. Modi et al. (eds.), *Proceedings of International Conference on Communication and Networks*, Advances in Intelligent Systems and Computing 508,
DOI 10.1007/978-981-10-2750-5_55

handled in a direction to reach at features vector of internal representation, and empowering the intelligent machine to interact with the environment to compute the interpretation for a robot [3]. The essential characteristics depictions create the knowledge representation or the learning models of knowledge-based computer vision technology [4]. According to the precise method, the complication of the handling steps is grasped by framing and reviewing each cognitive problem on three jointly self-governing levels—the levels of perception, learning and action [5].

1.1 Past Research Development of Computer Vision

The discipline of computer vision has progressed from various innovative prototype moves over the last 60 years. It had its beginning in the 1950s when the leading efforts were commenced to use the *computing methods and tools* to process images [6]. During the period 1970–1980 computer vision was primarily deliberated as *pattern recognition* [7, 8]. In this approach, an object is designated by a feature vector. The resemblance of objects is well-defined by the computable amount of the matching of the feature data that identify the objects. The pattern recognition method rapidly met numerous essential problems. In specific, the difficulty of segmenting an image into substantial chunks that might be categorized to be unsolvable. Ultimately it became usually conventional that computer vision necessitates an understanding of the domain objects that signifies in the image. This result to an adjustment of the perspective to the point that vision was an application field for Artificial Intelligence and Cognitive techniques.

This modification procured in the 1980s, when novel methods establish in Cognition for encoding expert systems, in particular, means of *knowledge representation and implication* [9, 10]. The anticipation was that it would be promising using these techniques to deliver the domain knowledge needed for the investigation and understanding of images.

The Knowledge representation approach similarly faced obstacles that restricted its attainment. Above all, the task to attain and validate the essential realm knowledge demonstrated to be feasible only for restricted areas. The segmentation difficulty is unable to resolve with this approach. The primary cause is that maximum Artificial Intelligence techniques are relatively complex to imperfections of the image segmenting. Preliminary segmenting denotes still currently a fundamental problem because of which many favourable procedures flops.

Another method discusses that inferencing an image needs shifting from the 2D pattern of grey scales or colour features to the three-dimensional patterns of the objects. Numerous methods were indicated using the objective to restructure the form of imaged objects by image characteristics such as texture, contour, motion, etc. This technique also fails because it is impossible to reconstruct images from static scenes. The camera is needed to take picture frames from all sides to solve unknown ambiguity in the shape of the object.

A significant breakthrough in the research on computer vision occurs by the introduction of *active vision* [11]. An active vision system is unique that can handle the perspective of the camera focus in a direction to study the situation and recovering information hidden in it. Models of dynamic vision structures typically involve a robot mounted camera that can interface among perception and added cognitive skills, such as learning, thinking, planning, and acting.

The *object recognition using knowledge representation* approach also progress simultaneously [12, 13]. The preliminary fact is the postulation that object recognition embraces the contrast of the objects with interior illustrations of objects and scenes in the image Knowledge-based system. This recognition depends on features of two-dimensional images instead of three-dimensional approach. The progress of object recognition methodology and algorithms has shown the capability of linguistic investigations and answering from image scene, tracking of collision free navigation of robots, identification of objects from the visual scenes.

2 Proposed Architectural Framework and Layout of Computer Vision for Cognitive Decision Making Agents

Authors propose a Cognitive Computer Vision architectural framework that represents in Fig. 1, typically contains following primary components: illumination source, a camera, active vision image capture board like frame grabbers or video cards, personal computer, cognitive software system and motors for action. The

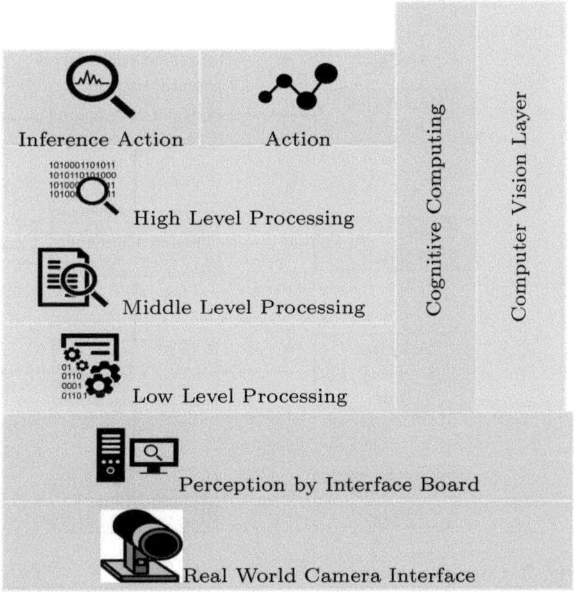

Fig. 1 Computer vision framework

hardware and software tools must have the capability of vision processing like colour pixel detection and classification, edge detection, the centre of mass and blob discovery, shape detection and recognition, face finder, and stereo vision supports. The illumination source must be intended sensibly to deliver consistent image removing the presence of distinctions. Visual capturing is a software program that assists handlers to upload images from digital and computer vision web cameras or scanners that are either attached directly to the computer machine or the net. Choice of the frame grabber focus on the camera yield, three-dimensional and grey level resolutions requisite and the handling capability of the computer vision board itself. The procedure of transforming graphic pictures into a quantized binary mathematical form known as digitization. In this process, an image is allocated into a two-dimensional matrix of unit cells comprising image elements well-defined as pixels with the help of a vision board called a frame grabber. A computer or dedicated image processing microprocessor system is used to offer storage of pictures and computational capability with cognitive programs. Also, the cognitive computing system is provided with a high-level reasoning and knowledge representation ability, which supports in human based visualizing images and the properties of various active analysis routines.

3 Process of Proposed Architecture for Cognitive Computer Vision

The architecture contained by computer vision observed in cognitive decision making is represented in the Fig. 2. Cognitive Image understanding is defined as a process of four stage sub-processes which in each circumstance necessitate precise representations.

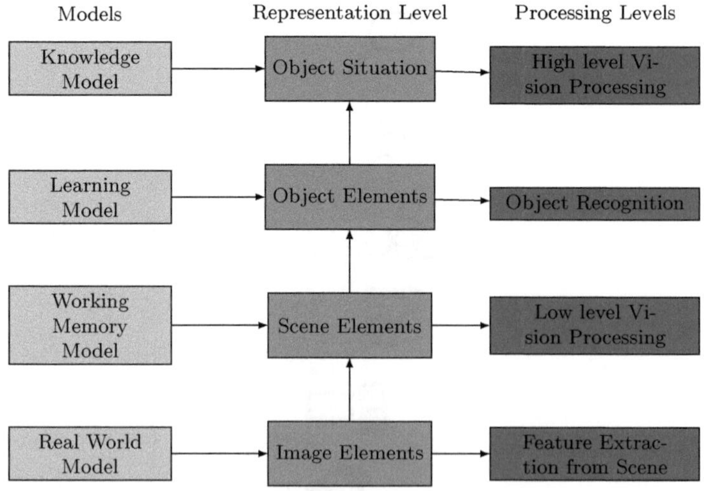

Fig. 2 Cognitive computing process system

(1) Perception: In the Perception stage, the image is grab by the cognitive agent and performs feature extraction to form the quantitative representation of the environment, that the system can understand. In this stage, colour, as well as intensity value of each pixel, are stored, for the determination of image elements like edge boundaries, matching areas, texture, etc.
(2) Low-level Processing: In the Low-level processing, interpretation targets at inferring image elements using scene elements. Processes of the particular level are to solve an essential job of image understanding: the mining of real world features from image descriptions. It comprises in actual the retrieval of three-dimensional object shapes using blob detection techniques.
(3) Object Recognition: In this stage, objects are recognized in the image data mined so far, and due to the scene elements. A critical part at this point is using the a priori knowledge of which presentations are shaped by the camera if objects perceive from diverse views. This a priori knowledge characterized by the object representation and learning models of the knowledge base.
(4) High-Level Processing: The higher level vision understanding encapsulates advance handling stages that target at sensing object and time surpassing relations, like multiple object shapes recognition, weird situations and locations understanding, logical motion systems, etc. Similar to object recognition, a priori knowledge of what require detecting exhibits a significant part. The content of the consequential explanation depends not only on the scene or the conforming image but also on the query or the situation in which the consequences is to be cast off.

3.1 Cognitive Interface for Humanoid

3.1.1 Hardware Interface

The Humanoid Mimzy uses intelligent decision making as a stage for testing the object recognition methods as in Fig. 3. The robot has a stereo vision RGB camera mounted on its platform. Both are placed at a height of approximately 0.7 m, directing down at an angle of 60° respect to the perspective. In the described experiments, objects are placed on a table with a background to learn. The two Camera has a similar high definition resolution of pixels and determines the object put on the table.

3.1.2 Software Setup

The computer vision software setup is developed in Visual Studio 2013 using C-Sharp language for windows forms. The detection of hardware interface is done using Aforge.net library. The learning stores the feature vector in the XML files for

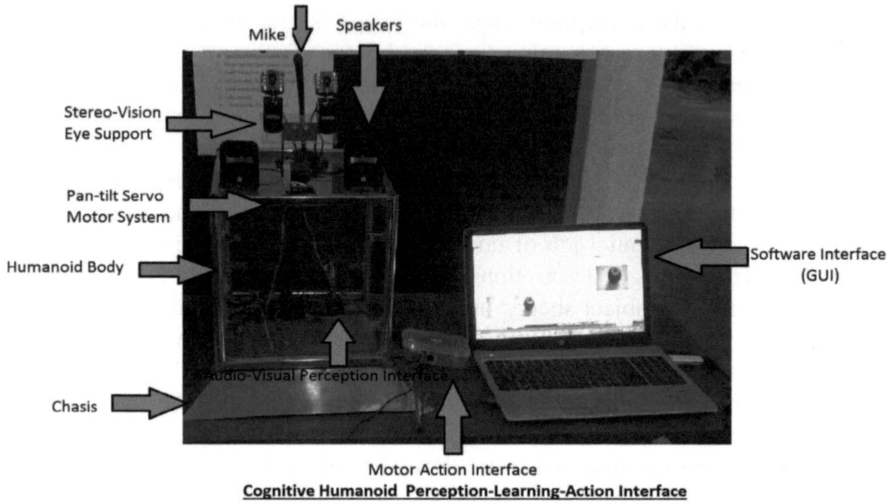

Fig. 3 Hardware developed by authors for cognitive agent showing object recognition having cognitive decision making capability of (*perception, Learning, Action*)

each object. The logic of object recognition implements on colour palette histogram matching between the query image and learnt images.

3.2 Methodology

On visualizing the world and identifying object as a human being, we choose to do three cognitive capability: Perception, Learning and Action.

3.2.1 Perception

The computer vision world for recognizing objects focuses on three relevant templates:

(1) Colour template: The colours linked with an object are an important local feature that helps to focus quickly our attention and identify it. For example, a tomato, a face, football will have a restricted vector of colours that are linked with each of them and support to signify that object visually to an individuality.
(2) Visual texture template: The colours connected with an object have a respective surface for its uniqueness for recognition. For example, an apple is red with small lighter colored spots, glass is mainly translucent and has reflections, the furniture of my home is brown colored wood with dark knots like texture.

(3) Shape template: The colour and texture linked with an object must appear in a particular form that gives the object its resemblance and distinct look. So if a surface that has the colour and visual texture of an apple but has the shape of a bottle, undoubtedly needs a closer gaze since it would likely be a very different bottle.

Please confirm if the section headings identified are correct.

This work focus, real-time object recognition done by the agent, using the color template as a perception feature since it is mostly used by human to recognize distinct objects in the real world in fast and easy manner.

3.2.2 Learning

The learning phase is a significant step to implement, it captures an image of a background and utilize the similar situation with the object image to add it to learnt object. The intelligent logic extracts the colour features of the back environment and learning object, and the variance is the vector of the object to learn. It then sends that vector out to a knowledge base, along with the object identification name to build the corpus.

3.2.3 Action

In the action phase, the agent captures the image of the object to be analyzed, and then compares the feature vector of each object signature learned during the learning phase. Any signature that has a feature vector that is most likely match the query object represents the alike as shown in the Fig. 4. When the object is not in the knowledge base, then the intelligent agent show negative response as

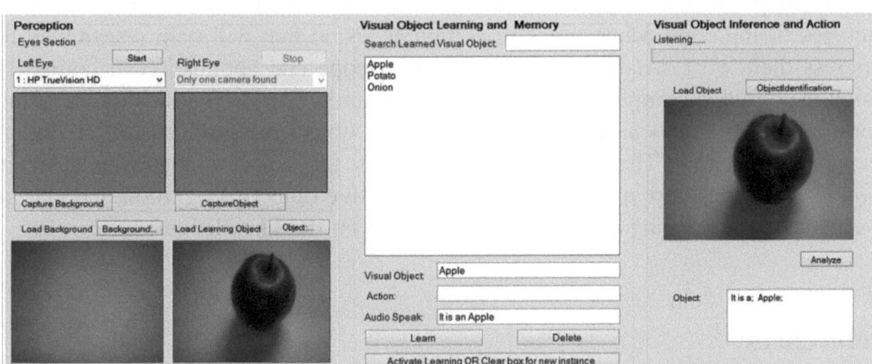

Fig. 4 Application developed by authors for cognitive agent showing object recognition having cognitive decision making capability of (*perception, Learning, Action*). The object recognition utilize computer vision tools of active vision and using colour template matching. An apple is learnt and recognized by the cognitive agent correctly

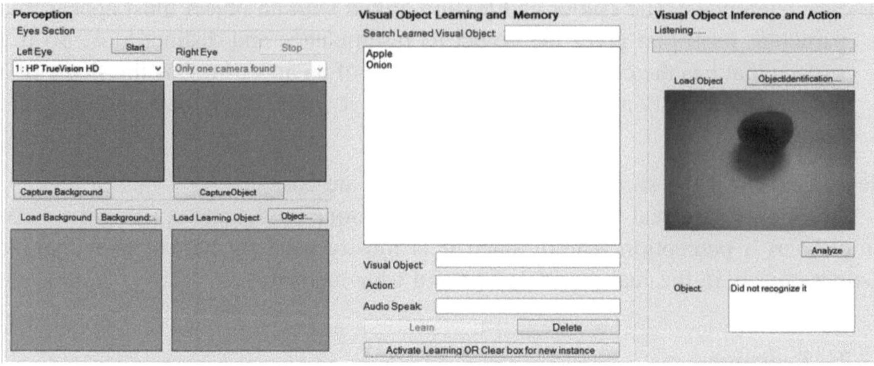

Fig. 5 A potato is perceive by the cognitive agent which it has not learn. It gives the response that it did not recognized

visualizing in Fig. 5. It works well with objects that have unique color attributes, but unwell with objects that have color features that match other everyday objects. Potato, Onion, Apple, for example, gives useful output.

4 Conclusions

Over the previous years the creation of autonomous robots using interaction with computer vision computing devices for cognitive decision making unrelenting to be a flourishing part of the research. This paper is an effort to deliver the future researchers in the field of human-computer interaction by proposing an architectural framework for utilizing computer vision tools and a cognitive model for solving real world problems. The author discusses the identified advantages and disadvantages of the core technologies in computer vision that will help researchers to progress their work. Also, the paper lists, the scope and application of computer vision in cognitive computing domain for the business world. The illustration of the work, the paper shows an experimental work for object recognition and intelligent decision for action. The prototype applications for object recognition systems demonstrate a desktop application capable of active object recognition and cognitive perception, learning and actionability based on the knowledge base.

References

1. Hartley, R., Zisserman, A.: Multiple view geometry in computer vision. Cambridge university press. (2003).
2. Fischler, M. A., Firschein, O. (Eds.).: Readings in Computer Vision: Issues, Problem, Principles, and Paradigms. Morgan Kaufmann. (2014).

3. Rautaray, S. S., Agrawal, A.: Vision based hand gesture recognition for human computer interaction: a survey. Artificial Intelligence Review. 43(1), 1–54 (2015).
4. Han, J., Shao, L., Xu, D., Shotton, J.: Enhanced computer vision with microsoft kinect sensor: A review. Cybernetics, IEEE Transactions. 43(5), 1318–1334 (2013).
5. Katz, D., Venkatraman, A., Kazemi, M., Bagnell, J. A., Stentz, A.: Perceiving, learning, and exploiting object affordances for autonomous pile manipulation. Autonomous Robots. 37(4), 369–382 (2014).
6. Meer, P., Mintz, D., Rosenfeld, A., Kim, D. Y.: Robust regression methods for computer vision: A review. International journal of computer vision. 6(1), 59–70 (1991).
7. Wold, S.: Pattern recognition by means of disjoint principal components models. Pattern recognition. 8(3), 127–139 (1976).
8. Ballard, D. H.: Generalizing the Hough transform to detect arbitrary shapes. Pattern recognition. 13(2), 111–122 (1981).
9. Marr, D., Poggio, T.: A computational theory of human stereo vision. Proceedings of the Royal Society of London B: Biological Sciences. 204(1156), 301–328 (1979).
10. Marr, D.: Vision. WH Freeman, San Francisco. (1982).
11. Aloimonos, J., Weiss, I., Bandyopadhyay, A.: Active vision. International journal of computer vision, 1(4), 333–356 (1988).
12. Jain, A. K., Vailaya, A.: Image retrieval using color and shape. Pattern recognition. 29(8), 1233–1244 (1996).
13. Alexe, B., Deselaers, T., Ferrari, V.: What is an object?. In Computer Vision and Pattern Recognition (CVPR), 2010 IEEE Conference. 73–80 (2010).

Comprehensive Trust Based Scheme to Combat Malicious Nodes in MANET Based Cyber Physical Systems

N. Bhalaji and Chithra Selvaraj

Abstract A large scale Wireless Sensor Network (WSN) or Mobile Ad hoc Network is to be definitely integrated into Internet as a backbone of Cyber Physical System (CPS), it is indispensable to believe that Cyber physical systems are free from security challenges, such as the detection of malicious attacks. A trust based model is attributed as an important door to defend a large distributed sensor networks in CPS. Trust is perceived as a critical tool to detect malicious node attacks in distributed computing and communication entities, detection of unreliable entities, and uphold decision-making process of various protocols. In this paper, Trust is invoked between participating nodes to improve the performance of Cyber physical systems by improving the degree of cooperation among them. The proposed schemes are also used to establish reliable path in packet forwarding and route finding. The realism, robustness and effectiveness of the proposed model is validated through a broad set of simulations.

Keywords Trust · Cyber physical system · Security · Malicious attacks · Multicast routing protocol · ODMRP

1 Introduction

Cyber-physical systems merge digital and analog devices, interfaces, networks, computer systems, and they link the natural and unreal physical world. The intrinsic unified and diverse amalgamations of behaviours in these systems make their analysis and design a demanding task [1].

N. Bhalaji (✉) · C. Selvaraj
Department of Information Technology, SSN College of Engineering, Kalavakkam, Tamilnadu 603110, India
e-mail: bhalajin@ssn.edu.in

C. Selvaraj
e-mail: chithras@ssn.edu.in

N. Modi et al. (eds.), *Proceedings of International Conference on Communication and Networks*, Advances in Intelligent Systems and Computing 508,
DOI 10.1007/978-981-10-2750-5_56

Cyber-physical systems (CPS) are concrete and engineered systems whose functions are observed, harmonized, controlled and incorporated around the nucleus of computing and communication. Just as the internet changed how humans communicate with one another, cyber-physical systems will transform how we interact with the physical world around us. Many stately challenges await in the economically vital domains of transportation [2], health-care [3], manufacturing, agriculture [4], energy, defense, aerospace and buildings. The design, construction and verification of cyber-physical systems put forward a massive amount of scientific challenges that must be addressed by an inter-disciplinary community of researchers and academicians. Advances in CPS will enable capability, adaptability, scalability, resiliency, safety, security, and usability that will surpass the simple embedded systems of today. New smart CPS will drive novelty and contest in sectors such as those mentioned above.

Massive progress has been accomplished in evolving Cyber physical systems during the course of preceding few years. Preliminary technologies are investigated that have spanned an ever-rising set of application fields, endowing breakthrough triumphs in many contrast fields. At the same time, the demand for modernisation in these areas [5] continues to flourish, and is urging the need to step up primary research to keep stride.

Conventional probing outfits are incompetent to manage the full complexity of CPS or amply envisage system performances. For example, trivial outcomes that trip the current electric power grid—an ad hoc system can intensify with startling rate into prevalent power breakdowns. This circumstance epitomises the want for appropriate science and technology to put forward design for the deep interdependencies amid engineered systems and the natural world. The trials and projections for CPS are thus extensive and massive. Novel communication between the cyber and physical constituents demand new architectural models that rewrite form and function. They assimilate the continuous and discrete, compounded by the ambiguity of open environments. Conventional present-day accomplishment assurances are inadequate for CPS when systems are large and spatially, temporally, or hierarchically disseminated in patterns that may quickly alter. With the better autonomy and association possible with CPS, greater possibilities of safety, security, scalability, and reliability are demanded, placing a high premium on open interfaces, modularity, interoperability, and verification [6–8].

Sensors and RFID tags are used as a smart nodes CPS which constitute nerve end of the cyber physical systems and acts as an interface for the data transmission between cyber and physical environments. Smart nodes entrenched with sensors deploy dynamic wireless multihoc environments equipped communication medium like Bluetooth, WiFi, Zigbee protocol etc. in near future huge number of nodes possessing sensing capability and vast number of mobile devices may amalgamate for form a strong CPS. The CPS formed out of the above combination yields intelligent services to the living community and change the way of their perception towards cyber and physical entities around them. The core communication in CPS is handled by flexible networks like MANET.

The MANET used for the establishment of CPS poses new challenges which are of different from challenges arises in conventional networks. In the later one nodes handles packet forwarding, route selection and data handling whereas networks deployed for the establishment of CPS consists of different class of nodes which are often smart entities. Further architecture constraints, energy utilisation and most importantly low power processing are the highlighting factors which needs more attention in these emerging services which forms the CPS. Wireless nature of MANET make them more vulnerable to non-cooperating behaviour of participating nodes and it is very much possible that nodes may be captured by an adversary which may lead to capture and alteration of data packets. Malicious nodes which became part of the network may damage them by causing falsified routing belief and network partitioning.

2 Generic Trust Functions

Collecting Information: This process involves accumulation of information about nodes participating in the networking functionalities. Primarily the behaviour of the node is monitored to analyse the trustworthiness of it. The decision is arrived based on the information collected either through direct or indirect experience i.e. first-hand experience or through recommendations from it peer members.

Ranking and Routing: Once the above information are collected from the nodes they are categorised into trustworthy nodes and non trustworthy ones. This is achieved based on the trust value obtained with each and every node derived from their present and past functioning. Trust values obtained also acts as an indicator for route selection. Reliable route is chosen where there is less number of malicious nodes prevails.

Node Selection: The nodes with greater trust values are preferred for data transfer. When there is a choice for a node to select its successor among pool of nearby nodes it performs the selection depending upon their trust values.

Transaction: The nodes with higher trust values are chosen for packet transfer and as well given priority in route selection.

Observation: In the above process the node weighs the transaction based on their own experience and collects information about transactions from its nearby nodes too.

3 Trust in Dynamic Networks

Trust is used to identify malicious nodes and employed to guide decision making activity of many protocols in a MANET which is inevitable for performing particular ser of tasks which ends up in increasing collaborations among nodes present. it becomes is necessary to introduce trust among nodes to eliminate malfunctioning nodes from routing path and data transmissions. One way of enhancing security in CPS is to utilise trust to evaluate the trust worthiness of each node participating in the network functionalities. Such trust incorporation into networks not only eradicates the participation of malicious nodes but also increases the overall performance of the networks [9]. The delicate part of computation and judgment of trust is a very challenging task in general and dynamic nature of MANET adds bit more complexity into it. The trust calculations may be initiated between any nodes in the MANET scenario based on direct experience, indirect experience (recommendation) and combination of both [10].

Trust is a security mechanism in conditions where many entities communicate and interact. This trust based security mechanism is derived from the human relationship. Here trust between any nodes cannot be evaluated as it is done in normal societal scenario and needs special attention as stated above in human based society trust is measured up on their activities over the time. The human tend to believe in other human under uncertain conditions depending up on his direct or indirect experience.

Trust is one of the most complex phenomenon in social, business and in digital world. Lot of issues are encountered while imposing and measuring trust in unpredictable networks such as MANET used for establishing CPS. These includes difficulty in evaluating trust in rapidly changing environments and to categorise nodes based on the calculated trust. Wireless networks possess various challenging feature such as energy constraint, dynamic routing and restricted security. CPS based on MANET is prone open to different types of attacks which are introduced due to malicious nodes such as packet dropping attack, Blackhole attack, grey hole attack, duplication and replica attacks.

3.1 Trust Computation

Trust computation leads to various degrees of trustable and non-trustable nodes. In this article trust computation is quantified between 1 and -1. The negative number (-1) represents the degree of distrust, where as positive number represents the complete trust. The number 0 is assigned for the new entrants or unknown node. In this trust model two types of trust are calculated between trustor and trustee nodes.

Trust is a notion of human behaviour. In this article the definition of "Trustor" node refers to the node that implements the trust evaluation and "Trustee" node refers to the node that is evaluated. Another term mentioned in this test is "Recommender". Such recommender node is the one who provide honest recommendation on a specific trustee to the trustor when demanded.

Direct Trust Direct trust value is calculated basing on the direct experience that trustor nodes possess over trustee node. This direct experience could result in either positive or negative way. The quantity of experience may be unlimited but the computable trust value falls in the range between −1 and +1. To obtain values in the above range hyperbolic sin function a = sinh (b) is used where 'a' represents trust value and 'b' stands for node's direct observation.

In real time a trustor may have several experiences over a trustee node and each experience may impact the trust calculation in either way. The direct trust is calculated as below.

$$\mathrm{DT} = \sinh \sum_{i=1}^{n} P_i\, E_i A_i C_i \tag{1}$$

where,

$$\mathrm{P_i} = \frac{No.\,of\,Packets\,received\,in\,application\,layer\,of\,Trustee\,node}{No.\,of\,Packets\,send\,by\,application\,layer\,of\,Trustor\,node}$$

$$\mathrm{E_i} = \sum\nolimits_{i=1}^{n} \frac{Consumed_i}{Send_i + Received_i + \tau}$$

Energy plays an important role in the successful deployment of MANET in CPS. Energy consumption model is defined as above which dissipates the real scenario where the nodes are being used for the data communication and path finding. In the above sending and receive i indicated the energy consumed by ith sensor while sending and receiving a packet. Consume i denotes the total energy cost of consumption the trust values of the sensor nodes. τ Denotes energy consumption required for the survival of the node.

$$c_i = \frac{No.\,of\,Control\,packets\,received\,successfully}{No.\,of\,Control\,packets\,forwarded\,successfully}$$

When a trustor node doesn't possess enough direct experience on a trustee node, the truster node enquires a third node for recommendation. Let's assume that third node has some trust value IT (indirect trust) on the trustee node basing on its own observation.

Recommended Trust is calculated as

$$\mathrm{RT} = \frac{1}{n} \sum\nolimits_{i=1}^{n} DT * \mathrm{IT}_i \tag{2}$$

where

RT Recommended Trust
DT Direct Trust
IT Indirect Trust

To ensure convincible recommendations a trustor node may enquire more than one third node for recommendations.

Comprehensive Trust calculation

$$CT = DT + (1 + |DT|) * RT \tag{3}$$

where

$-1 \leq \text{DT} \leq 1$
$-1 \leq \text{RT} \leq 1$

When a trustor node obtains direct and recommended trust in this way, a formulae to combine both the values is required to balance the relationship between direct trust and recommendation trust. Impact of recommendation trust depends upon how much direct experience value does the trust verifier holds. If the trust verifier node has no direct experience over a trust prover then the recommendation trust is solely believed.

4 Simulation Results

In this article, an event driven Network simulator [11] is constructed to simulate trust assisted based routing in multicast MANET based CPS. Each experiment is considered as an average of ten different runs and each run is implemented with randomly selected source and destination. The 502.11 DCF is hired as a MAC protocol, and on-demand multicast routing protocol (ODMRP) [12] is as routing protocol. Hundred nodes are randomly located randomly in 1000 * 1000 m simulation area. Sixty traffic pairs with Poisson packet interval are generated. The routing protocol discovers up to 5 routes between source and destination. Maximal route length is 10 hops. The mobility model is the random waypoint model. The velocity chosen falls between 0 and 10 m/s. The average pause time is 250 s. Recommendation trust is handled not more than three hops. The entire schedule of simulation is 2500 s.

Figure 1 depicts the throughput performance results for the traditional ODMRP protocol and Trust enhanced ODMRP protocol in the presence of 5 malicious nodes. The malicious nodes were designed to drop the data as well control packets and the results indicate that the throughput of the traditional protocol rapidly drops with the increase in time when compared to the proposed nodes. When there are no malicious nodes present in the scenario they almost share same throughput thus standing as evidence that trust calculations have very lesser impact over the

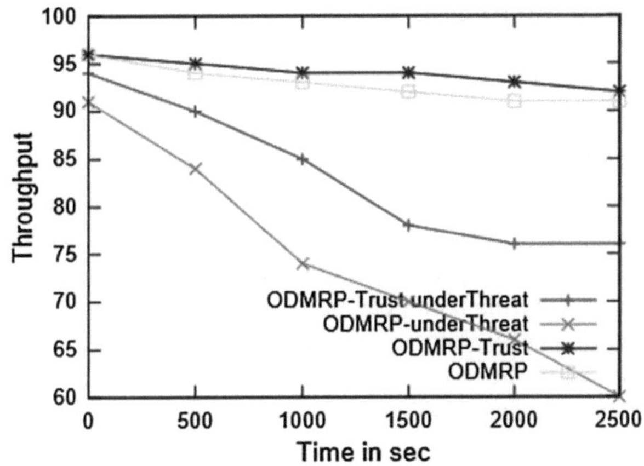

Fig. 1 Network throughput comparison

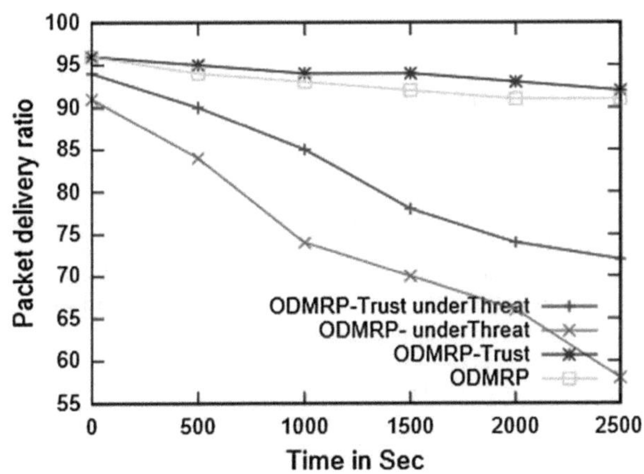

Fig. 2 Network packet delivery ratio

performance of the protocol. In an average 20 % improvement is obtained during the overall simulation duration and it is a very significant improvement considering the dynamic nature of the MANETs.

Figure 2 exhibit the performance analysis of ODMRP protocol's packet delivery ratio with and without trust extensions. The results indicate that the packet delivery ratio of the proposed and existing protocols seems to be same when there are no attackers in the simulation scenario but the situation drastically change as soon as the malicious nodes are brought in. Packet delivery ratio gains an 21 % improvement by employing the trust based route selection when exposed to 5 malicious nodes which drop the packets unintentionally.

5 Conclusion

Trust factors play important role in securing the cyber-physical systems which need to be incorporated from the design phase itself. CPS is opening up exceptional challenges for research and development in several domains. Since MANETs need to be integrated in cyber world to initiate data communication, this article illustrates the necessity of trust assessment in MANET based cyber physical systems. In particular the merits of trust embedded cyber physical systems functioning are simulated through hiring a on demand based multicast routing protocol and the results yielded are immensely encouraging towards further exploration. In future trust based systems will be further compared with the other security providing mechanisms and dedicated protocols may also be deployed for better understanding and evaluation.

References

1. Sanislav, Teodora, and Liviu Miclea. "Cyber-Physical Systems-Concept, Challenges and Research Areas." Journal of Control Engineering and Applied Informatics 14.2 (2012): 28–33.
2. BretHull, Vladimir Bychkovsky, Yang zhang, Kevin Chen, Michel Goraczko, "CarTel: A Distributed Mobile Sensor Computing Systems," in the 4th ACM Conference on Embedded Networked Sensor Systems, Boulder, 2006, pp. 125–138.
3. Insup Lee, Sokolsky. O, "Health Cyber Physical Systems," in 47th ACM/IEEE Design Automation Conference, Anaheim, 2010, pp. 13–18.
4. Meng Zhijun, Zhao Chunjiang, Wang Xiu, Chen Liping, Xue Xuzhang, "Field multi-source information collection system based on GPS for precision agriculture", Transaction of the CSAE, vol. 19, no. 4, pp. 13–18, Jul. 2003.
5. Chaudhary, D. D., S. P. Nayse, and L. M. Waghmare. "Application of wireless sensor networks for greenhouse parameter control in precision agriculture." International Journal of Wireless & Mobile Networks (IJWMN) Vol 3.1 (2011): 140–149.
6. Quanyan Zhu, Craig Rieger and Tamer Basar, "A Hierarchical Security Architecture for Cyber-Physical Systems", IEEE 4th International Symposium on Resilient Control Systems (ISRCS), 2011.
7. Nayot Poolsappasit, Rinku Dewri and Indrajit Ray, "Dynamic Security Risk Management Using Bayesian Attack Graphs", IEEE Transactions on Dependable and Secure Computing, 2012.
8. Teodor Sommestad, Mathias Ekstedt and Pontus Johnson, "Cyber Security Risks Assessment with Bayesian Defense Graphs and Architectural Models", Proceedings of the 42nd Hawaii International Conference on System Sciences, (HICSS), 2009.
9. C Selvaraj., Anand, S.: "Peer profile based trust model for p2p systems using genetic algorithm", Peer-to-Peer Networking and Applications, Vol 5(1), (2012), pp. 92–103.
10. N. Bhalaji, Dr. A. Shanmugam "Defense Strategy Using Trust Based Model to Mitigate Active Attacks in DSR Based MANET" Journal of Advances in Information Technology, Vol 2, No 2 (2011), 92–98, May 2011.
11. NS-3. [Online]. Available: http://www.nsnam.org/index.html.
12. S. Deering, D.L. Estrin, D. Farinacci, V. Jacobson, C.-G. Liu, and L. Wei,. The PIM Architecture for Wide-Area Multicast Routing. IEEE/ACM Transactions on Networking, vol. 4, no. 2, Apr. 1996, pp. 153–162.

A Review on Wireless Mobile Communication Systems Generations and Integration

G.C. Manna and Bhavana Jharia

Abstract Computers, computational devices and data acquisition systems are confined at a location. Captive data has limited utility. Data communication among devices, for shared usage and further processing, revolutionized the world through Information and Communication Technologies. Data communication systems piggy backed on the telecommunication network and went on with it from wireless to wireline domain. GPRS/EDGE, EVDO/WCDMA, mobile WiMAX/LTE are the three generations of mobile wireless access technology domains in the local area. The backhaul transport system consists of optical fibre links and nodes with route switching systems called Routers. Routers also acts as add/drop nodes where data processing devices, called servers, are connected which serves application services data to user community. Pure wireless data communication networks has also been evolved as wi-fi network which covers very small area but provide very large bandwidth and functions in ISM band. Wi-Fi networks are considered equivalent to fourth Generation access network. The generation of technologies has been designed based on commitments of data throughputs and environmental conditions. The present paper analyses the throughputs available in practical conditions for GPRS, EDGE, EVDO, WCDMA and WiMAX. The paper also discusses integration of all the three generations in a heterogeneous environment. The 3GPP recommendations for machine type communications with cellular network as backhaul has been discussed as part of 5G network.

Keywords GPRS · EDGE · EVDO · WCDMA · OFDM · Wimax · LTE · Wi-Fi · 4G · 5G

G.C. Manna (✉)
BRBRA Institute of Telecom Training, BSNL, Jabalpur, Madhya Pradesh, India
e-mail: gcmanna@yahoo.com

B. Jharia
Department of Electronics & Communication Engineering, Ujjain Engineering College, Ujjain, Madhya Pradesh, India
e-mail: bhavanajharia@yahoo.co.in

N. Modi et al. (eds.), *Proceedings of International Conference on Communication and Networks*, Advances in Intelligent Systems and Computing 508,
DOI 10.1007/978-981-10-2750-5_57

1 Introduction

Wireline communication between sender and receiver are guided through defined media. Radiated Energy from transmitter is radiated in wireless communication systems which attenuates following inverse square law. Also, several copies of the same transmitted signal reaches the receiver through direct, reflected, refracted and diffracted paths which constitutes a complex waveform profile. The profile moves when there is relative motion between transmitter and receiver. Signal to noise ratio at the receiver limits the channel capacity according to Shannon limit. Depending upon environmental conditions, distance of receiver from transmitter, receiver speed etc., dynamic coding at transmitter is required to be done to optimize data throughput. All these impairments has been addressed in different access technologies like GSM, CDMA and OFDM. Generic network architecture has been added at the end of this topic.

2 Wireless Impairments

Path loss is the difference between transmitted and received power.

$$\text{Path loss} = \text{Transmitted power} + \text{Antenna gain} - \text{Received power.}$$

In practice, we use the transmitting power of antenna as 40 dbm per RF and transmit antenna gain of 17 dbi. Path loss is the reduction in signal strength when it propagates from Transmitter to Receiver. Various factors reduces the receiver signal strength viz. antenna height, distance between Transmitter and Receiver, obstacles such as trees, buildings etc. Many researchers like Hata, Okumara, Walfisch, Ikegami etc. has laid down specific formula which are suitable for different environmental conditions and frequency ranges.

A general pathloss model is given as

$$\mathrm{PL(dB)} = \mathrm{PL(d_0)} + 10\,\gamma\,\log(\mathrm{d}/\mathrm{d_0}) + \sigma$$

where, $PL(d_0)$ is the path loss at reference distance d_0 usually taken as 100 m for outdoor conditions, γ is the path loss exponent and $\gamma > 2$, σ is the standard deviation of received signals for small and large scale fading and d is the distance between Transmitter and Receiver. Losses are measured in dBm and d in meters. In free space $\gamma = 2$ and in near open field area, γ has a value nearly 3.

Multiple paths exist between a pair of transmitter and receiver; one may be direct path and others reflected. This constitutes an RF channel. Hence, same signal transmitted from a transmitter reaches the receiver at different instant of time and hence necessarily at different phases. Change in phase depends on frequency and hence wireless channel is called frequency selective. For example, if the time difference between direct path and a reflected path is 1 μs, the path difference is

$3 * 10^8 * 1 * 10^{-6} = 300$ m and if frequency used for this channel 1 MHz, the two signals will constructively combine and if the frequency if 500 kHz, they will combine destructively. So, for every channel, there is a range of frequency for which channel response is flat between 3 db points on both sides. This bandwidth is called channel coherence bandwidth. If the bandwidth used is less than channel coherence bandwidth, there will be no loss of signal due to frequency selective fading. For outdoor propagation, normally delay spread is assumed between 1 and 3 μs and for this; the coherence bandwidth is about 450 kHz at 900 MHz band operation. When there is a relative motion between transmitter and receiver, the channel profile moves. An average speed of 30 km/h vehicle movement is taken into consideration in existing technologies. When relative velocity between transmitter and receiver exceeds about 450 km/h, effect of Doppler frequency spread become appreciable. This leads to the effect of channel coherence time. For most practical purposes, when transmission is fixed base station oriented, we may ignore this effect.

Signal to noise ratio (S/N) at the receiver is the most dominant factor. This is used to calculate effective bandwidth which will be available at S/N threshold using Shannon Channel limit. S/N is estimated by the transmitter at regular intervals during active communication for dynamic coding. Channels may be multiplexed by time division as in GSM, code division as in CDMA or frequency division as in OFDM. We explain below the data throughputs practically available in these technologies through several experimentation carried out by the authors, under different carrier to interference ratio conditions.

3 GSM (GPRS/EDGE)

Mobile communication was introduced for voice in sub GHz band. For GSM-900, downlink is from 935 to 960 MHz and uplink is from 890 to 915 MHz. The 25 MHz bandwidth is used for transmitter and receiver with 20 MHz band gap. Each RF channel has 200 kHz bandwidth which works in TDM mode and provide 8 physical channels through 8 Timeslots. Because, it has low band width, frequency selective fading is not applicable. For data services, GSM introduced GPRS service with GMSK modulation and 4 coding schemes viz. CS-1 to CS-4 depending upon S/N value and theoretically achieves 9–21.4 kbps for one timeslot i.e. channel. If a handset or dongle can use all 8 channels, a top speed of 171 kbps shall be achievable. In Evolved GPRS or EDGE service, with MCS coding scheme and 8PSK modulation, helped to generate a top speed of 59.2 Kbps per timeslot and maximum of 473.6 kbps in 8 time slots or channels. In Evolved EDGE, 16 QAM and 32 QAM coding scheme were used to enhance data speed up to 1 mbps but not much commercially successful.

Measurements taken in a very dense city is shown in Fig. 1. In dense cities, the Transmit stations are placed as close as at 500 m separation. Transmit signal strengths are adjusted in such a way that signal strength is good everywhere in the

Fig. 1 Throughput versus C/I at RLC plane

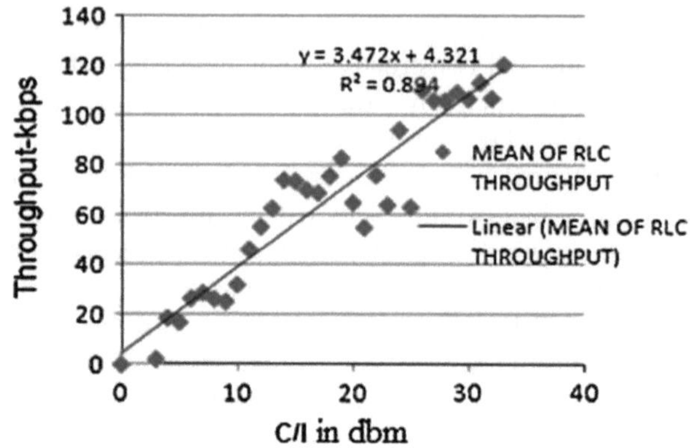

sector i.e. between −55 and −75 dbm whereas GSM defined threshold is −95 dbm. C/I in dense city situation is usually not good due to presence of numerous scattering objects. Engineering handset with 4 timeslots were used for the measurement. Throughput at radio link control (RLC) plane with corresponding C/I has been plotted in the Fig. 1. Irregularity in throughput is observed at values between 7–10 and 15–20 of C/I. They are due to coding scheme change by system at different C/I in EDGE system. It is observed that a peak value of 120 kbps is available at 32 dbm C/I for 4 TS structure which can be extrapolated to conclude that 240 kbps speed can be practically achievable in EDGE system with 8 TS [1].

4 CDMA (EVDO/WCDMA)

For CDMA, downlink is from 869 to 889 MHz and uplink is from 824 to 844 MHz. Thus 20 MHz bandwidth is available in CDMA with 25 MHz band gap. Each RF channel has 1.25 MHz width which works in Code Division Multiplexing mode and provides 64/128 code channels. CDMA suffers from frequency selective fading and hence it uses RAKE receiver for receiving six dominant signals through different paths. CDMA introduced 2000 1x which provided 144 kbps top speed per channel in normal mode. A CDMA variant, called Evolution Voice and Data Optimised (EVDO) was developed with 16 codes to provide 3.2 Mbps throughput when one complete RF is dedicated for data services.

Figure 2 shows C/I in terms of Ec/Io plot for a cluster of base stations. Mostly the tests were carried out close to the base stations. C/I varied from −8 to 0 dbm, 0 to 5 dbm and 5 to 10 dbm with respective counts as 1715, 5741 and 2867 respectively.

For data processing, user count was set to 1 to ascertain maximum throughput of the system. The available data for throughput was averaged in steps of 1 dbm and corresponding throughput was also averaged. As shown, at minimum average Ec/Io of −6.195, throughput is 134.21 kbps and at maximum Ec/Io of 11.7325,

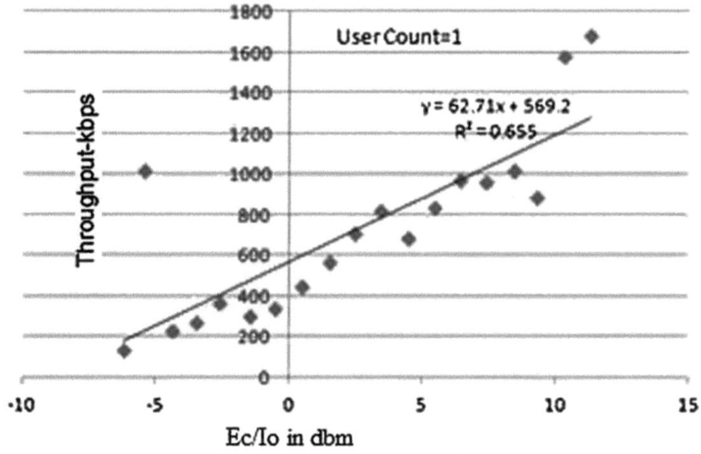

Fig. 2 Throughput versus Ec/Io

throughput is 1676.545 kbps. Trend line shows an increase of 62.71 kbps per dbm and 596.2 kbps at zero SINR [2].

Wideband CDMA (WCDMA) has been designed based on IMT 2000 recommendations. It is more popularly known as Third Generation (3G) mobile technology. This is the most dominant technology from 2000s and till date. It is based on 5 MHz duplex bandwidth which works in 2000 MHz band. It's uplink frequency range is 1939–1979 MHz and downlink frequency is 2129–2169 MHz. This employs up to 512 Walsh codes, however, for data part, it uses a section of code tree with less number codes. In general, it uses only 16 codes for data and effectively 15 codes are available for users. Roughly 900 kbps is maximum throughput specified per code with 64 QAM modulation and up to 5/6 RS codes in very good environmental conditions.

It uses 6 finger RAKE receiver to accommodate multipath channels. WCDMA receiver sensitivity is specified as −105 dbm. However, for data, a better throughput is expected when signal strength is more than −90 dbm. Ec/Io is the chip power to total interference power from all other codes working in same band and measured in dbm in absolute scale. Figure 3 above shows the result of Ec/Io versus throughput in dense city environment. In the experiment, the engineer's data set used which was capable of handling 5 codes at a time and hence theoretical maximum speed it can achieve is 4.5 Mbps under ideal conditions. In the experiment, total about 73,000 data was collected with over 10,000 effective data during which measurements were taken. Throughout measurements, the Received Signal code power level was better than −90 dbm and mostly in range of −80 to −70 dbm throughout. The Fig. 3 shows steady increase of datarate from −15 to −10 dbm, but it drops from −10 to −5 dbm. The drop may be due to the reason when the test set is nearer the tower, received power is high, Ec/Io is also high and hence system wants to move to higher coding and modulation zone but fails resulting in lower

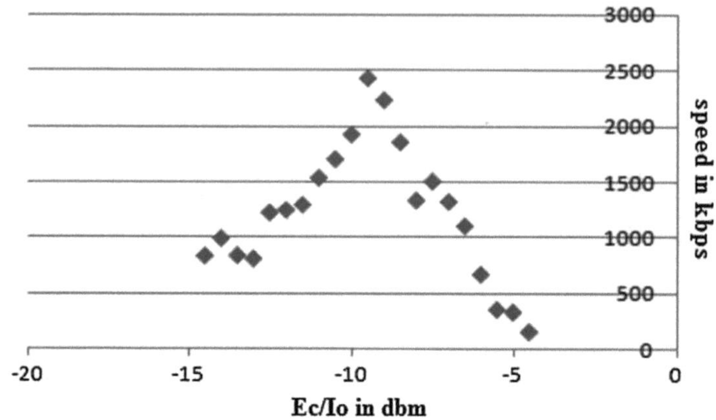

Fig. 3 Throughput variation in WCDMA

throughput. Peak data rate achieved is 2.5 Mbps whereas theoretical maximum is 4.5 Mbps. In other experiments also, similar behavior of the technology is noticed, however, the maximum achieved speed is dependent on RSCP where it is found to be 1239 Mbps at −9 Ec/Io when RSCP was −91 dbm.

5 OFDM (WiMAX/LTE)

Orthogonal Frequency Division Multiplexing (OFDM) technology has 5 MHz bandwidth which is divided into number of subcarriers e.g. 256/512 etc. This works similar to multi carrier modulation, each carry one symbol at a time. The gain obtained is that flat fading is obtained in respect of each sub-carrier. It operates at different frequencies, but the predominant practical implementations like Worldwide interoperability for Microwave Access (WiMAX) and Long Term Evolution (LTE) work respectably in 2600 MHz and 2300 MHz bands respectively. However, multipath introduces Inter Symbol Interference (ISI) which is mitigated through use of Guard Period at the end of each symbol. Guard period depends on symbol duration and usually it is 1/8th part of symbol duration. It also introduces Inter-Carrier Interference (ICI) for which trail part of the signal is fed back to the leading part of the signal in the guard period of the previous signal. All these factors make OFDM channel virtually immune from fading and multi-path effects. UMTS Advanced promises a maximum data speed of 50 Mbps in uplink and 100 Mbps in downlink direction through use of 8 × 8 MIMO, 64 QAM ½ rate coding, advanced error correction etc. These have been implemented by IEEE 16 m recommendation as WiMAX and 3 GPP forum recommendation as LTE for mobile communication.

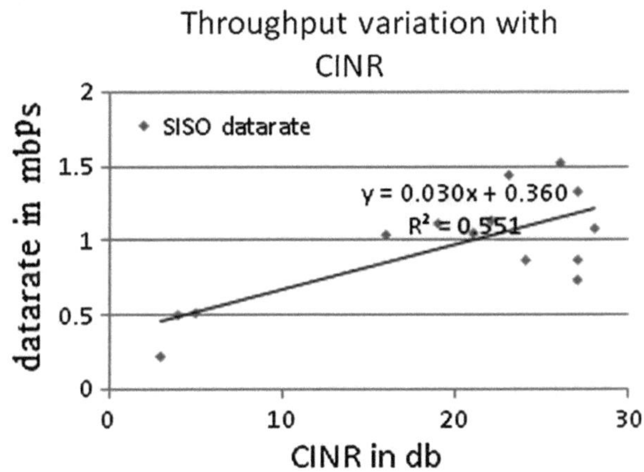

Fig. 4 Throughput variation in OFDM

For WiMAX, download data throughput obtained is from 2 transmitters at base station and single receiver at dongle receiver. Download throughput for 1 × 1 SISO configuration has been considered through a conversion formula using Shannon Channel capacity for 2 × 1 MIMO and MIMO gain factor for each CINR value. It is observed from Fig. 4 that in most of the observations, CINR is 20 where throughput is near 1 Mbps for 90 carriers (1/4 th part of traffic carriers of 5 MHz allocated bandwidth) due to dongle capacity limitation. A distance of nearly up to 3 kms was considered in the measurement drive [3].

5.1 4G System in Legacy Network

A complex heterogeneous network consisting of 2G, 3G and 4G network are shown in the Fig. 5. User Equipment (UE) connects to Base Transceiver Station (BTS), Node B (NB) and evolved Node B (eNB) with respective generations. Base Station Controller (BSC) used in 2G connects to many BTSs and it divides data and speech where speech part is send to Mobile switching Centre(MSC) and data part to Serving Gateway Support Node (SGSN) which works as Router for packet network. Traffic from 3G UE is sent through Radio Network Controller (RNC) and Media Gateway (MGW) where data and speech are sent through appropriate signaling conversion. Short Message Service Controller (SMSC), Wireless Access Protocol Gateway (WAP GW) for low speed internet, Service Delivery Platform (SDP) for miscellaneous audio-video and other services are the user service equipments of the network. Home Location Register (HLR) keeps all fixed and dynamic information of each customer in the network in conjunction with Visitor Location Register (VLR) which normally remains integrated with MSC. SSTP is versatile Signaling Transfer Point which acts as gateway for speech traffic to Ethernet network. In Long Term Evolution (LTE)-4G implementation of IMT

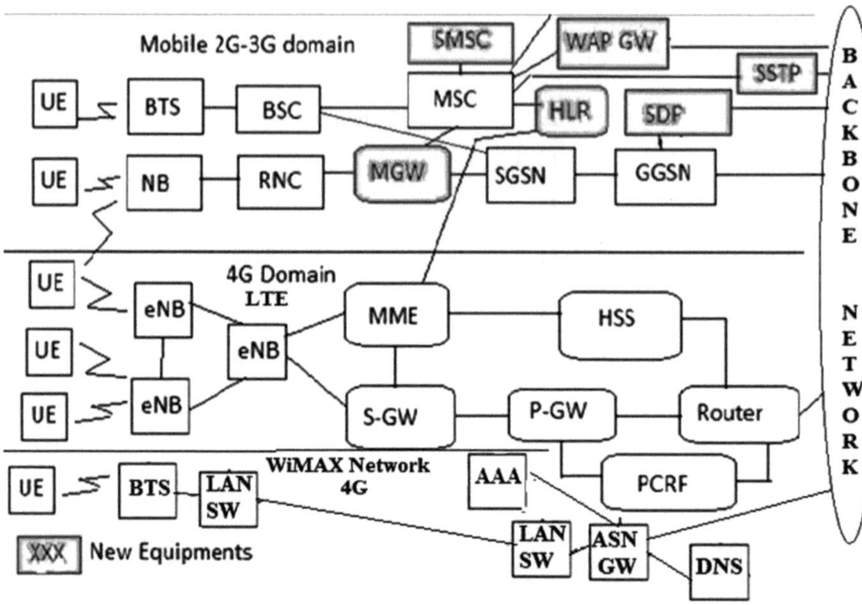

Fig. 5 Mobile network subsystem architecture

Advanced Recommendation, eNB combines the role of NB and RNC. Mobile Management Entity (MME) plays most important role of keeping track of the mobile, particularly for managing speech traffic and data traffic. The basis of such monitoring depends on capability of the UE handset i.e. it can function simultaneously with 2G/3G network for speech path and 4G network for datapath OR the UE can solely depend on 4G network where Voice over LTE (VoLTE) type of service will be available for speech purpose.

Serving Gateway (S GW) routes traffic to Packet gateway (P-GW) where traffic is controlled for each user based on directions from Policy Charging and Rules Function (PCRF) unit. Home Subscription Server (HSS) is the master database for subscribers which supports IP Multimedia Subsystem (IMS). WiMAX UE or Dongle provides mainly data services through WiMAX BTS and Access Service Network Gateway (ASN GW). The user services are verified through Authentication, Authorization and Accounting (AAA) server and traffic is routed based on route IP address available from Domain Name Server.

5.2 5G Network and M2M Communications

The need for speed at the wireless backhaul will be an important criterion during 2020 and beyond for which bands above 6 GHz has already been identified by different countries. This will provide high speed data communication between devices with low latency requirements. On the other hand, there will be machines with very low throughput, occasionally transmit and receive with very low but long-life power supply requirements. These equipments will be of RFID type, connected through wireless personal area network (WPAN) and connect to a Gateway (GW). 3GPP uses Machine Type Communication (MTC) Gateway which connects to a 2G/3G/4G type cellular networks. However, the GWs can also connect to a WiMAX type of network deploying Wi-Fi type of communication deploying low power wireless communication type of platform. Legacy networks may deploy MTC AAA and other support services for effective control of the 5G service requirements. The servers for 5G shall be connected to back bone network like other web based services.

6 Conclusion

Through generations of mobile technology, the coding and modulation techniques were instrumental for providing more bits per Hertz but under no circumstances, it is possible to enter Shannon Channel limit region. It is now the turn of Multiple Input Multiple Output (MIMO), fifth Generation (5G) access and cloud eNB technologies to rule the road of mobile broadband network. Performance of newly developed technologies and devices for PAN and MTC will decide the strength of future mobile network.

References

1. S.B. Mule, G.C. Manna, Neeta Nathani: Assessment of Spectral Efficiency about 900 MHz using GSM and CDMA Technologies for Mobile Cognitive Radio: IEEE International Conference on Pervasive Computing (ICPC), (2015).
2. S.B. Mule, G.C. Manna, Neeta Nathani; Comparison of Spectral Efficiency of Mobile OFDM-WiMAX Technology with GSM and CDMA for Cognitive Radio Usage: ICAC3'15, ELSEVIER.
3. S.B. Mule, G.C. Manna, Neeta Nathani: Measurement of Spectral Efficiency of mobile OFDM-WiMAX Technology at 2600 MHz band in Green Field Area: IEEE International Conference on Pervasive Computing (ICPC), (2015).

Reducing the Cold-User and Cold-Item Problem in Recommender System by Reducing the Sparsity of the Sparse Matrix and Addressing the Diversity-Accuracy Problem

K.R. Bindu, Rhama Lalgudi Visweswaran, P.C. Sachin, Kundavai Devi Solai and Soundarya Gunasekaran

Abstract Recommender Systems are a subclass of information filtering systems that seek to predict the preferences of a user or the preference that a user would give to an item. The most common problem faced by these systems is the lack of data. Such a situation leads to a matrix that is extremely sparse thus reducing the accuracy of prediction. Cold-start problem is one such problem that is faced by the recommender systems when a new user or a new item enters the system. We are hoping to reduce the cold-user and the cold-item problem by reducing the sparsity of the sparse matrix with the help of Iterative Local Least Squares algorithm and a hybrid of Heat Spreading algorithm and Probability Spreading algorithm.

Keywords Sparsity · Heat spread · Iterative local least squares · Cold-start · Cold-user · Cold-item

K.R. Bindu (✉) · R.L. Visweswaran · P.C. Sachin · K.D. Solai · S. Gunasekaran
Department of Computer Science and Engineering, Amrita School of Engineering, Amrita Vishwa Vidyapeetham, Amrita University, Coimbatore, India
e-mail: j_bindu@cb.amrita.edu

R.L. Visweswaran
e-mail: rhamavis@gmail.com

P.C. Sachin
e-mail: sachinpc.1993@live.com

K.D. Solai
e-mail: kundu450@gmail.com

S. Gunasekaran
e-mail: soundarya251293@gmail.com

N. Modi et al. (eds.), *Proceedings of International Conference on Communication and Networks*, Advances in Intelligent Systems and Computing 508,
DOI 10.1007/978-981-10-2750-5_58

1 Introduction

The amount of information that is being retrieved is enormous. We have abundant information, but we starve for knowledge, the knowledge required to use the information commercially. The information needs to be processed for it to be useful. Recommender Systems process the information such that appropriate items can be recommended for different users. Several approaches have been proposed by researchers to make useful recommendations to users. Every approach has its own strengths and weaknesses. Hence hybrid recommenders are used commonly.

Two commonly used recommendation approaches are Collaborative filtering [1] and Content-based filtering. Collaborative filtering method assumes that users with similar taste will rate items similarly. The similarity between the target user and other users are computed and items are recommended.

In case of Content-based filtering, the recommendations made are based on a comparison between the content of the items and the user profile [2]. In this paper, when we say Recommender system we are referring to Collaborative filtering based Recommender systems.

The Recommender system involves a matrix that represents users and their preferences for items. To provide recommendation for a target user, similarity between that user and the other users are computed. One of the problems that Collaborative Filtering based Recommendation systems face is the cold-start problem, which is having a lack of initial rating for users or items. In this paper we try to reduce the cold start problem by the sparsity of the sparse matrix [3]. We use techniques such as heat spread [4], probability spread [4] and iterative local least squares [5] to dense the sparse matrix that we derive from dataset. We also diversify the items without compromising the accuracy of the results and hence, bring a balance between the diversity and accuracy of the recommended items.

2 Proposed Method

Since the recommendation approach we are using in this paper is Collaborative filtering, we make use of a very simple data. The users, the objects recommended to the users, relationship of the user and the object and the ratings of each object, o, given by each user, u, will be used in this method. The relationship of the user and the object can be mapped into a bipartite graph in which each relationship can be denoted by the graph's edge. This graph is represented by an adjacency matrix, a, object $\times$ user where a_{ij} is the relationship between ith user and the jth object.

$$a_{ij} = \begin{cases} 1, & \textit{if jth user has rated ith object} \\ 0, & \textit{else} \end{cases} \tag{1}$$

The ratings matrix, R, also describes the relationship of the user and the object, however the edge value in the bipartite graph is equal to the ratings given by the user. i.e. R_{ij} is the rating given to the jth object by the ith user.

$$R_{ij} = \begin{cases} r, & \textit{if jth user has rated ith object} \\ 0, & \textit{else} \end{cases} \tag{2}$$

Due to the limited data, the adjacency matrix is sparse. For recommendations to be more accurate, we need it to be dense. We make use of various methods to do the same.

2.1 Addressing Diversity-Accuracy Issue

As accuracy increases, diversity decreases and vice versa. A balance needs to be struck between diversity and accuracy. As accuracy increases, the recommendations become more and more relevant. Then the question arises, why do we need to address diversity? As the system becomes more accurate, it recommends items that are highly relevant to the user and does not diversify the recommendations, this can result in the user looking into the same items over and over again. The main purpose of a recommendation is to introduce the user to the huge market available as well as to please the user. To do this, recommendations should be diverse. But as the accuracy decreases, it is difficult to please the user. Hence a perfect balance is required.

First we build an object × object matrix to spread the resource allocated to each user. To increase the accuracy, this object × object matrix needs to be column normalized. To increase the diversity of recommendations, the same should be row normalized. A hybrid of the two can strike a balance in whichever proportion we need. Each element of this object × object matrix, W_{ij}, is given by

$$W_{ij} = \frac{1}{\sqrt{k_i k_j}} \sum\nolimits_{l=1}^{u} \frac{a_{il} a_{jl}}{k_l} \tag{3}$$

where k_i and k_j are the number of users who possess object i and j respectively and k_l is the no of object possessed by the lth user. Here we give equal importance to diversity as well as accuracy. This spread of objects results in resource redistribution. The multiplication of distribution matrix W and ratings matrix R will give a matrix R' that is denser than the original matrix R.

2.2 Filling Up the Missing Values

Even though we have processed the ratings matrix R, the pre-processed rating matrix R' is still sparse. To fill up the missing values, we find K similar users and the average of their ratings is used. Generally, for a total of u users, $\sqrt{u}$ similar users are considered.

$$K = \sqrt{u} \tag{4}$$

To find the similarity, we are using the distance measure and compute a user user similarity matrix. To construct this matrix we take a user × object matrix as input. I.e. Transpose of R'.

$$Sim(u_i, u_j) = \frac{\sum_{l=1}^{o} a_{il} a_{jl}}{\sqrt{k_i^2 + k_j^2}} \tag{5}$$

Sim is used to compute K similar users and the missing values of R'^T can be filled up by the row average of these K users. The resultant ratings matrix, R_f is a user × object matrix which can be used to suggest recommendations.

2.3 Recommendation

We use user-based collaborative filtering to make recommendations in this paper. User-based collaborative filtering makes use of the similarity between users. This might sound similar to the previous method, iterative local least squares; however, this method is different. The user-based collaborative filtering makes use of cosine similarity between users. The cosine similarity between two users U_i and U_j is given by

$$Sim(u_i, u_j) = \cos\theta = \frac{u_i \cdot u_j}{|u_i| \times |u_j|} \tag{6}$$

To find the recommendation for the given user U_i we find top k similar users, from the top k users we find the frequency of occurrences. To find the frequency we consider only those recommendations whose value exceeds a certain threshold t. Based on the highest frequencies we make the recommendations. Here we take k as 3 and t also as 3.

3 Evaluation Metrics

In many instances, recommendation systems recommend the items to the user that they may use instead of predicting the rating of those items. Some of the evaluation methods that we have adopted from the literature, for recommendation quality measures are as follows.

3.1 Recall

$$recall = \frac{|\{relevantitems\} \cap \{retrieveditems\}|}{|\{relevantitems\}|} \tag{7}$$

The recall rates vary differently for different datasets. It should be noted that the *Restaurant* dataset [6] and the *Music* dataset [7] follow a similar trend of decreasing initially and then increasing. However, the final value remains lower than the initial value (Fig. 1).

3.2 Precision

$$precision = \frac{|\{relevantitems\} \cap \{retrieveditems\}|}{|\{retrieveditems\}|} \tag{8}$$

The overall precision has increased for *Restaurant* dataset [6] and *Music* dataset [7]. And has reduced drastically for the *Movielens* dataset [8]. It should also be noted that a continuous increase in precision is only in the *Music* dataset [7] (Fig. 2).

3.3 F-Measure

$$Fmeasure = 2 \cdot \frac{precision \cdot recall}{precision + recall} \tag{9}$$

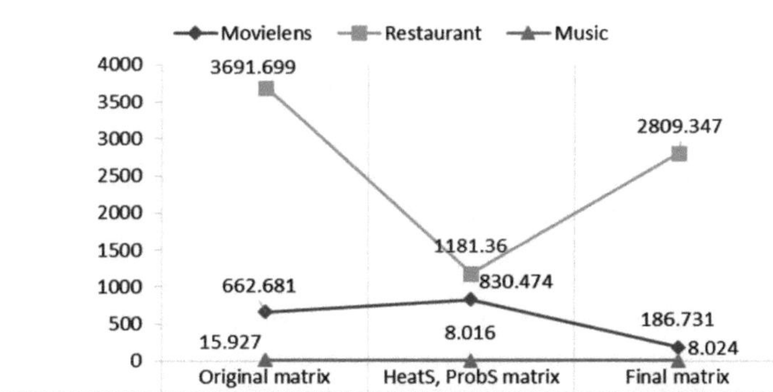

	Original matrix	HeatS, ProbS matrix	Final matrix
Movielens	662.681	830.474	186.731
Restaurant	3691.699	1181.36	2809.347
Music	15.927	8.016	8.024

Fig. 1 Recall rates at different steps, for the datasets taken

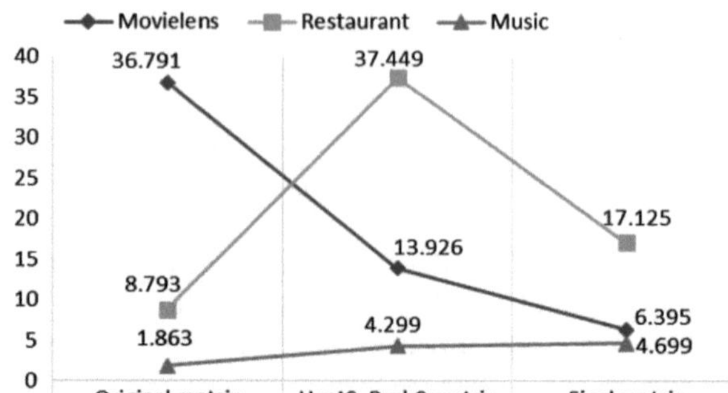

	Original matrix	HeatS, ProbS matrix	Final matrix
Movielens	36.791	13.926	6.395
Restaurant	8.793	37.449	17.125
Music	1.863	4.299	4.699

Fig. 2 Precision rates at different steps, for the datasets taken

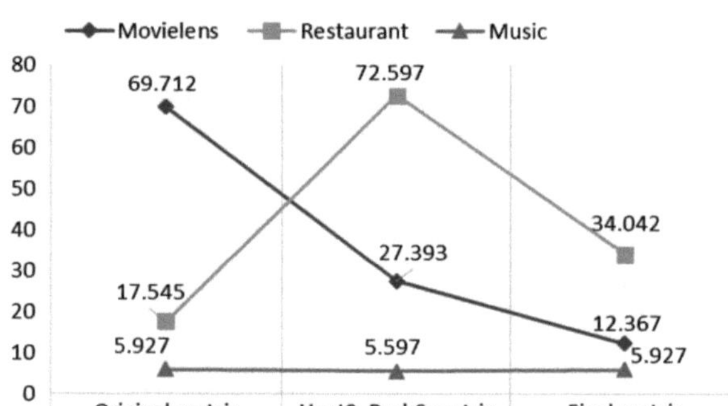

	Original matrix	HeatS, ProbS matrix	Final matrix
Movielens	69.712	27.393	12.367
Restaurant	17.545	72.597	34.042
Music	5.927	5.597	5.927

Fig. 3 F-measure at different steps, for the datasets taken

High recall implies that our system returned most of the relevant items whereas high precision means more relevant items than irrelevant. In our system we have computed the F-measure [9] by giving equal weightage to both precision and recall. If we want measure the system's exactness, we have to give higher weightage for precision. However, if we want to measure the system's completeness, we have to give higher weightage for the recall. Here, we are trying to bring a balance between both precision and recall through our method, hence have given an equal weightage (Fig. 3).

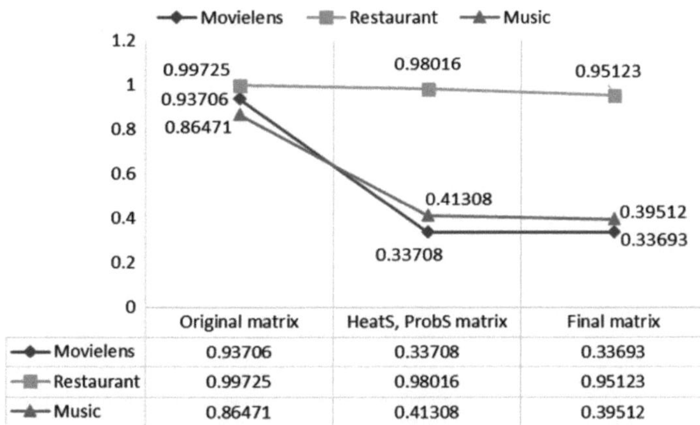

Fig. 4 Sparsity at different steps, for the datasets taken

3.4 Sparseness

There is a decrease in the sparseness for all the three datasets. Even though the *Restaurant* dataset [6] shows a slow decrease in sparseness, there is a huge improvement for the other two datasets. It should also be noted that the *Restaurant* dataset [6] has the highest sparsity (Fig. 4).

4 Results

The proposed method has been demonstrated on 3 different datasets and has been evaluated. The *Movielens* dataset [8] consists of 100,000 ratings (1–5) from 943 users on 1682 movies. A part of *Kaggle Song* dataset [7] where 5019 users have rated 23 songs. Finally, *the Entrée Chicago Recommendation data* dataset [6] 138 users rating on 2550 restaurants.

The performance of the proposed method depends on the sparseness of the initial data which includes the number of cold-users and the number of cold-items. The dataset with larger number of cold-items (*Movielens* [8]) has resulted in lower F-measure and the dataset with the lesser number of cold-items (*Restaurant* [6]) has resulted in a higher F-measure.

5 Illustration of the Proposed Method

See Table 1.

Table 1 An example using a sample matrix that has a cold-user and a cold-item

Object and User relationship	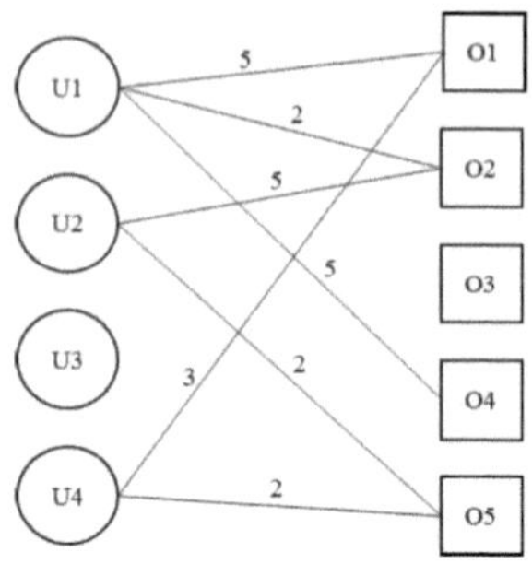	
Initial Sparse Matrix	$a=\begin{bmatrix}1&0&0&1\\1&1&0&0\\0&0&0&0\\1&0&0&0\\0&1&0&1\end{bmatrix}$ $R=\begin{bmatrix}5&0&0&3\\2&5&0&0\\0&0&0&0\\5&0&0&0\\0&2&0&2\end{bmatrix}$	Precision: 0.250 Recall: 0.750 F-Measure: 0.375 Sparseness: 0.650
Diversity-Accuracy Algorithm	$W=\begin{bmatrix}0.417&0.167&0&0.236&0.250\\0.167&0.417&0&0.236&0.250\\0&0&0&0&0\\0.236&0.236&0&0.333&0\\0.250&0.250&0&0&0.500\end{bmatrix}$ $R^{'}=\begin{bmatrix}3.599&1.335&0&1.751\\2.849&2.585&0&1.001\\0&0&0&0\\3.317&1.180&0&0.708\\1.750&2.250&0&1.750\end{bmatrix}$	Precision: 0.533 Recall: 0.667 F-Measure: 0.592 Sparseness: 0.400
Iterative Local Least Square	$\text{Sim}=\begin{bmatrix}2.828&1.789&0&1.265\\1.789&1.414&0&1.109\\0&0&0&0\\1.265&1.109&0&0.943\end{bmatrix}$	
Final Matrix	$R_f=\begin{bmatrix}3&2&0&3&1\\1&2&0&1&2\\2&3&0&1&3\\1&1&0&0&1\end{bmatrix}$	Precision: 1.200 Recall: 1.500 F-Measure: 1.333 Sparseness: 0.250
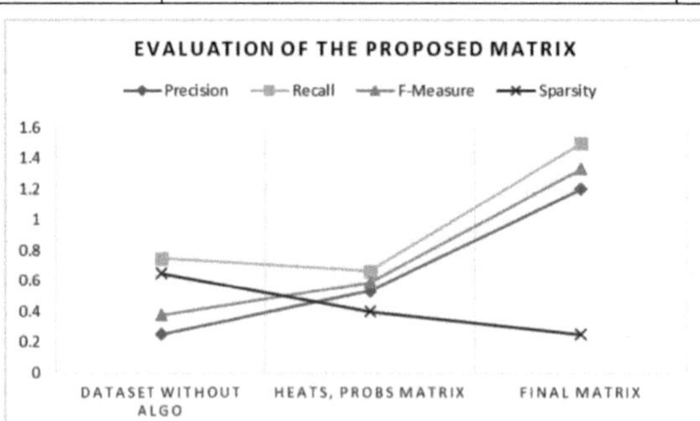		

6 Related Works

Recommender systems have been an important research area since the mid-1990s and have been attracting e-commerce companies like Amazon, Netflix to enter the scenario. The recommender systems, with time, have improved in terms of accuracy and efficiency. However one of the main problems still remains: the cold-start problem. To solve this issue some of the authors ask a series of questions to the new user so that a study of the user taste can be done based on the answers given by the user. However this is a tedious activity for the users and causes them displeasure. Some other authors have proposed using social data from social networking sites [10]. SNA (Social Network Analysis) is one of the methods that can make use of such social data and hence to an extent reduce cold-start problem.

One such method uses Foursquare API [11]: a location based social networking website to enable the generation of recommendations [12]. Few other methods are, using latent factor models that maps users and items into a dense and reduced latent space. It captures their most salient features. Some of models include probabilistic latent semantic analysis (PLSA), [13] principal component analysis (PCA) [14] and singular vector decomposition [15]. However, the main disadvantages are that the learned latent space is not easy to interpret and many latent factor models depend on other users, which may be lacking in case of sparse dataset. In recent works variant of Latent-Dirichlet allocation was used for making recommendations [16].

7 Conclusion and Future Work

This paper intends to reduce the cold-start problem of recommender systems by addressing the sparse data. However, this method does not completely eliminate the same. The cold-item problem can be further reduced by making use of the detailed information relating to each item and thus can easily classify the item.

References

1. Bobadilla, J., Ortega, F., Hernando, A., Gutiérrez, A.: Recommender systems survey. Knowledge-Based Systems 46, 109–132 (2013).
2. Huang, Z., Zeng, D., Chen, H.: A comparative study of recommendation algorithms in e-commerce applications. IEEE Intelligent Systems 22(5), 68–78 (2007).
3. Sparse matrix—Wikipedia, the free encyclopaedia, https://en.wikipedia.org/wiki/Sparse_matrix.
4. Zhou, T., Kuscsik, Z., Liu, J.-G., Medo, M., Wakeling, J., Zhang, Y.-C.: Solving the apparent diversity-accuracy dilemma of recommender systems. Proceedings of the National Academy of Sciences 107(10), 4511–4515 (2010).
5. Liu, J., Chengcheng, Y., Zhang, Z.-K.: A two-step Recommendation Algorithm via Iterative Local Least Squares. arXiv preprint arXiv 1206(3320) (2012).

6. Index of /ml/machine-learning-databases/entrée-mld, https://archive.ics.uci.edu/ml/machine-learning-databases/entree-mld/.
7. EMI Music Data Science Hackathon—July 21st—24 hours, https://www.kaggle.com/c/MusicHAckathon/data.
8. The GroupLens Research Project at the University of Minnesota, http://www.grouplens.org/datasets/movielens.
9. Manning, C., Raghavan, P., Schütze, H.: Evaluation in information retrieval. In: Introduction to information retrieval 1. Cambridge university press, Cambridge (2008) 154–158.
10. Ramage, D., Dumais, S., Liebling, D.: Characterizing Microblogs with Topic Models. ICWSM, Washington, DC, vol. 5, pp. 130–137 (2010).
11. Castillejo, E., Almeida, A., Lopez-de-Ipina, D.: Alleviating cold-user start problem with user's social network data in recommendation systems. Sensors 15, 315–316 (2011).
12. Aarthi, S.: A Heat Diffusion Method for Mining Web Graphs for Recommendations Using Recommendation Algorithm. International Journal of Engineering Research and Technology 2(8) (August 2013).
13. Hofmann, T.: Probabilistic latent semantic indexing. In: Proceedings of the 22nd annual international ACM SIGIR conference on Research and development in information retrieval, pp. 50–57 (1999).
14. Jolliffe, I.: Principal component analysis. John Wiley & Sons, Ltd, (2002).
15. Klema, V., Laub, A. J.: The singular value decomposition: Its computation and some applications. Automatic Control, IEEE Transactions on 25(2), 164–176 (1980).
16. Ricci, F., Rokach, L., Shapira, B., Kantor, P.: Recommender systems handbook 1. Springer, NewYork (2011).

Differential Voltage Controlled Ring Oscillators—A Review

Tripti Kackar, Shruti Suman and P.K. Ghosh

Abstract The differential ring voltage controlled oscillator (DRVCO) is a key component in various fields of communication and transmission systems with increased modularity and excellent controllability. This paper analyzes the different DRVCOs cell topologies and their performance have been interpreted in terms of frequency range, power consumption, phase noise and technology used, which allows the designers to opt the most suitable differential ring VCO for specific applications.

Keywords Voltage controlled oscillator (VCO) · Differential voltage controlled ring oscillator (DVCRO) · Delay stages

1 Introduction

Voltage controlled oscillator (VCO) is an essential block of electronic systems in which output frequency is linearly varied by input control voltage. The major applications of VCOs are optical transmission, clock generation, radio frequency integrated devices (RFID) transponders [1] and data recovery circuits and also in medical domains [2]. CMOS based ring VCOs are widely used as they have high tuning range. The design of circuits with enhanced speed, power and larger bandwidth is one of the greatest challenges for designers today. To successfully deal with it, the best choice is a ring VCO, because it generates multiple phases.

T. Kackar (✉) · S. Suman · P.K. Ghosh
ECE Department (CET), Mody University of Science and Technology,
Lakshmangarh, Sikar, Rajasthan, India
e-mail: tripti.kackar@gmail.com

S. Suman
e-mail: shrutisuman23@gmail.com

P.K. Ghosh
e-mail: pkghosh.ece@gmail.com

N. Modi et al. (eds.), *Proceedings of International Conference on Communication and Networks*, Advances in Intelligent Systems and Computing 508,
DOI 10.1007/978-981-10-2750-5_59

The odd number of stages is connected such that the output of last stage is fed back to the input of first stage. In order to have sustained oscillations, it must satisfy necessary conditions i.e. the total phase shift around the loop must be 360° and the loop gain must be equal to unity as stated by Barkhausen. The advanced demands of VCO circuits comprise miniaturization, less power dissipation and lower phase noise. The frequency of oscillations for N stages can be given by Eq. (1) [3]:

$$f_{osc} = \frac{1}{2Nt_d} \tag{1}$$

where N is the total number of stages in a ring oscillator and t_d is the delay at each stage given by $t_d = C_{eq} * R_{eq}$. In general, VCOs has output frequency as an function of applied control voltage which can be given by Eq. (2) [4]:

$$\mathrm{f_{out}} = \mathrm{F_o} + \mathrm{K_{VCO}V_{Ctrl}} \tag{2}$$

where $\mathrm{F_o}$ is the frequency of oscillations of VCO, K_{VCO} is the gain of the VCO which checks variation over control voltage ($\mathrm{V_{Ctrl}}$) which is input to the VCO determining it's operating frequency. The ring VCOs can be categorised as single ended and differential configurations [5].

Figures 1 and 2 shows the simplified cascaded structures in ring form of basic types of VCOs for N number of stages. The differential ring oscillators are advantageous as they reject common mode noise and avoid usage of bypass and coupling capacitors together with high gain stability at high frequencies. Although single ended configurations consumes less area but are less efficient in terms of noise. The odd numbers of stages are well suited for single ended oscillators and will always oscillate whereas differential configurations can have both odd as well as even number stages. The differential configurations with even number of stages are useful for generating quadrature or multiphase outputs [6].

The objective of this paper is to analyze the records on differential VCOs using delay stage of each oscillator circuit. The remaining paper is categorized as:

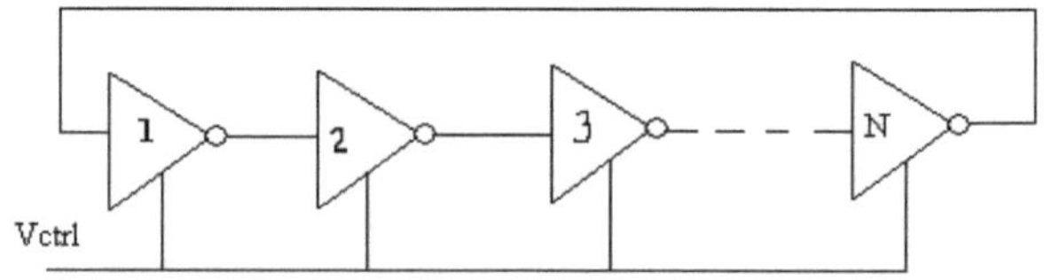

Fig. 1 Single ended ring VCO

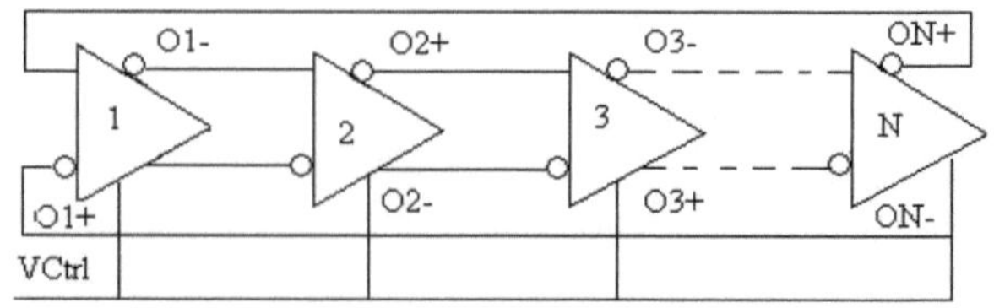

Fig. 2 Differential ring VCO

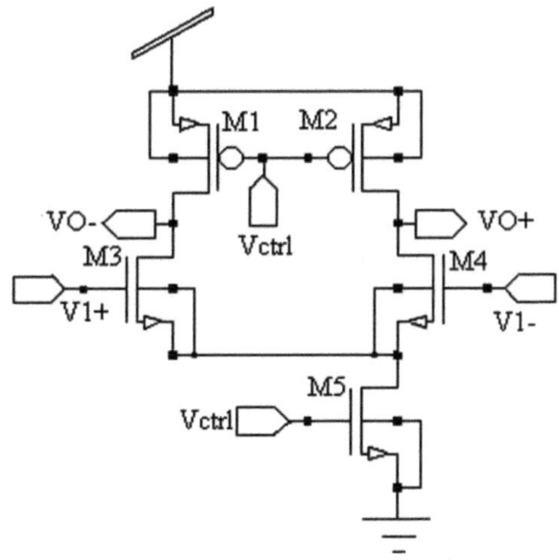

Fig. 3 Conventional delay cell for DRVCO

Section 2 focuses on fundamental basics of DVCRO. The improved versions of differential delay cells and results are summarized in Sect. 3. Section 4 finally gives future scope and conclusions.

2 Differential Voltage Controlled Ring Oscillator

The differential ring VCO in semiconductor industries is extensively used for all integrated circuit applications in differential mode of operation. The circuit designing is done generally based on ring oscillator topology with inverting and non-inverting stages which is helpful for low power consumption and closed loop structure with positive feedback enhances the speed of the design.

The schematic of conventional delay cell is shown in Fig. 3. Each delay stage comprises of two transistors as a load M1 and M2, two input transistors M3 and M4 and one current source transistor M5 is added at the tail. This delay cell network fails to satisfy the oscillations criterion at two stages due to hindered stability conditions and insufficient gain observed which is thereby, overcome with the help of positive feedback.

3 Different Delay Stages for Differential Voltage Controlled Ring Oscillator

There are several methods by which differential VCOs can be designed. The design perspectives are implemented in relevance with fundamental parameters. Few of the differential voltage controlled ring oscillators delay stages are briefly discussed as follows:

3.1 High Speed and Low Power Delay Cell

With the help of CMOS technology, the delay cell technique [7] gives wider band of frequencies for operation but also allows larger gain, linearity, spectral purity of signal in compensation with power and noise due to glitches and switching at the circuit nodes. The VCO is designed using fast slewing saturated differential delay cell [8] gives fast rise and fall times and leads to full switching operation. The average noise power P_n for any random process is expressed in Eq. (3):

$$P_n = \lim_{T\to\infty} \frac{1}{T} \int_{-T/2}^{T/2} v_n^2(t).dt \tag{3}$$

where V_n is the noise voltage given by $v_n^2(t) = 4\,KTR$. Here, K is the Boltzmann's constant, R is the resistance value and T is switching on time for each cycle which is defined in Eq. (4) as,

$$T = \frac{C_L V_{DD}}{I} \tag{4}$$

Here, C_L is the load capacitance at the output of each node, V_{DD} is the power supply voltage and I is the current in the device during operation. Hence, fundamental noise analysis done for ring oscillator yields average noise power as the function of time constant, delay at each stage and oscillation frequency.

The four-input differential delay cell is shown in Fig. 4. Firstly M9 and M10 forms differential input block, then pair of load transistors M2 and M3 forms latch block. Now the transistors M6 and M7 which is a controlled circuitry for CMOS latch stage to strengthen it resulting in decreases delay time. The acceleration block with transistors M1 and M4 are added to the load so as to generate negative skewed

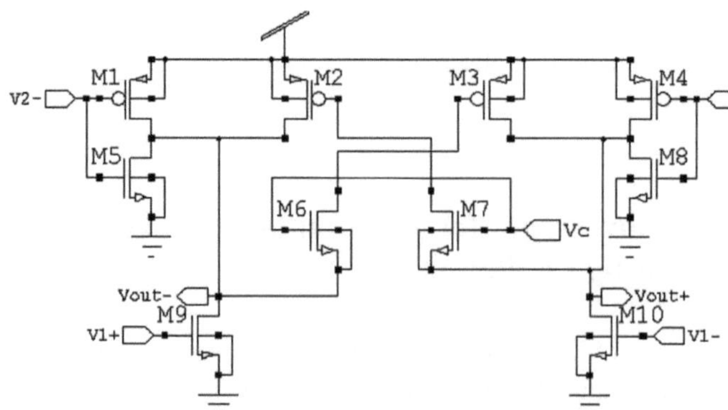

Fig. 4 Delay cell for high speed DRVCO

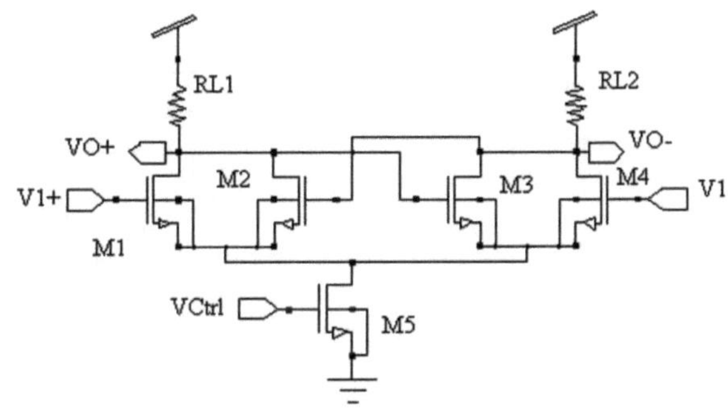

Fig. 5 Delay cell for low power DRVCO

signal which precharges the output nodes which ensures fast operation. This arrangement reduces rise time of the output as well as phase and supply noise. Similarly, transistors M5 and M8 NMOS are added to pre-discharge the output nodes decreasing fall time and achieves higher oscillation frequency, also hinders the frequency limitation problem.

The NMOS transistors provide positive feedback and sufficient delay to accomplish oscillatory criteria as shown in Fig. 5 [9]. The circuit can be best suited for optical communications applications as it exhibit good transient stability as shown in Eq. (5) and superior characteristics.

$$H(s = j\omega) \approx -\frac{G_o}{\left(1 + \frac{S}{\omega_o}\right)} \approx \frac{g_{m1R}}{1 - g_{m2R}} \tag{5}$$

where ω_o is the frequency at the pole, G_o is the open loop gain of the cell, g_{m1} and g_{m2} are the transconductance of transistor M1–M4 and M2–M3 respectively and R is the resistance of the symmetric load. The measured phase noise at 1 MHz offset is −141 decibels per hertz.

3.2 Low Noise Multiloop Delay Cell

In order to achieve the maximum desired frequency, the multiloop architecture using coarse and fine controls is reported in [10] as shown in Fig. 6. This can be done by adjusting positive feedback cross coupled pair of NMOS transistors M9 and M10. The phase noise and power spectral functions are determined using impulse sensitive function (ISF) and noise modulation functions (NMF). By varying the load PMOS using gate voltage of transistors M3 and M4 in ratio of M3 and M4 coarse tuning by control terminal V_c can be done and by changing the current in transistor M11 of latch network fine tuning using control V_F can be obtained. According to the linear time invariant theory, the phase noise can be closely stated as in Eq. (6) [11]:

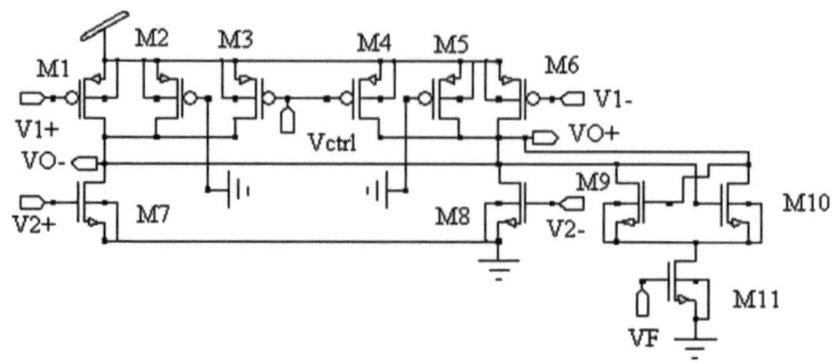

Fig. 6 Delay cell for low noise multiloop DRVCO

$$L(\Delta\omega) = 10\log\left[\frac{2KTF}{p}\left(1+\left(\frac{\omega_o}{2Q\Delta\omega}\right)^2\right)\left(1+\frac{\Delta\omega 1/f3}{|\Delta\omega|}\right)\right] \tag{6}$$

where K is the Boltzmann's constant having value 1.38×10^{-23} J/K, P is the average signal power dissipated, T is the absolute temperature, F is fitting parameter for figure of noise, $\Delta\omega 1/f3$ is corner angular frequency between $1/f_3$ and $1/f_2$ components of flicker noise and Q is quality function of the oscillator which is expressed as given by Eq. (7):

$$Q = \frac{f_o}{2\Delta f_{-3db}} \tag{7}$$

When Q increases due to decrease in −3 db bandwidth represented by Δf_{-3db} results in sharpening of the peak in magnitude response. The measured phase noise at 1 MHz offset is −103.4 decibels per hertz.

3.3 Wide Bandwidth and Pseudo Differential Delay Cell

The day by day rise in demands for higher spectrum of frequencies allows devisers to explore new methods for adjusting the range requirements of the DVCRO systems. This delay cell as referred in [12] proposes a method for increasing the frequency of the CMOS ring VCOs. The approach of delay cell was implemented in which diode connected load M1 and M4 is used in differential pair in order to have large output swings as shown in Fig. 7. A control voltage (V_{Ctrl}) is applied such that it adjust the frequency of the oscillator by regulating the current in the transistor M7.

In pseudo differential delay cell [13] as shown in Fig. 8, input to the cell is given through CMOS differential push pull inverter. It also avoids the need of common

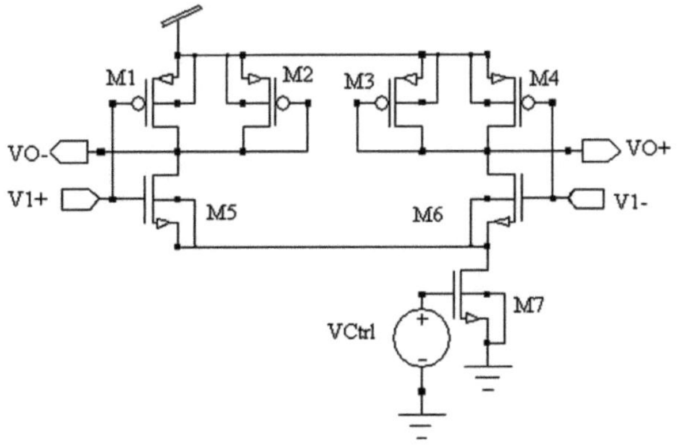

Fig. 7 Delay cell for wide bandwidth DRVCO

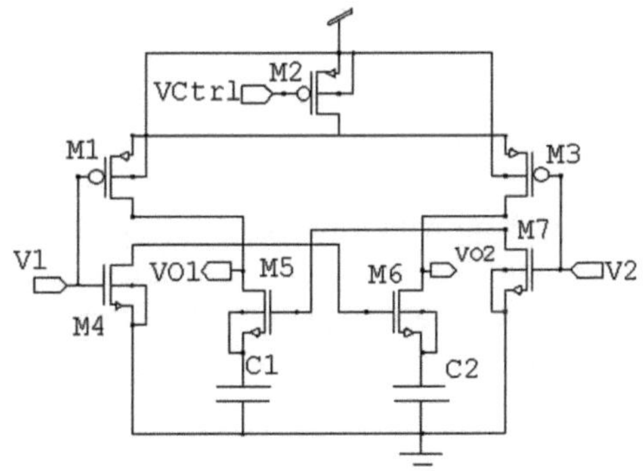

Fig. 8 Pseudo differential delay cell for DRVCO

tail transistor as a current source and gives fine frequency range. The operating frequency is obtained obtained as given by Eq. (8):

$$F_{osc} = \frac{\sqrt{3}}{2\Pi}\frac{(R_{\lambda} - g_{mp2})}{C_T} = \frac{\sqrt{3}}{4\pi}\frac{(g_{mn1} + g_{mp1})}{C_T} \tag{8}$$

where g_m the transconductance parameter of transistors for NMOS and PMOS, C_T is the total capacitance at the output node and R_{λ} is the resistance due to channel length modulation CLM effect. The delay cell design determining the best possible configuration for the ring oscillators having the least power consumption and precise delay with lesser sensitivity to the variations in the temperature and supply voltage for frequencies of GHz along with higher figure of merit. These Systems also realized with RFID transponders, WSN and 802.11 protocols thereby signify maximum data rate transfers at high speed process. The measured phase noise at 1 MHz offset is −162.4 decibels per hertz.

The results of different differential ring VCOs cells at different technology are shown in Table 1.

Table 1 Summary of simulated results

References	[7]	[9]	[10]	[12]	[13]
No. of transistors used in single delay stage	10	05	11	07	07
Power (mW)	12	1.09	60	1.09	2.47
Supply voltage (V)	2.5	1.8	1.5	1.2	1.5
Technology (μm)	0.25	0.18	0.13	0.5	0.18
Bandwidth (GHz)	0.85–2.1	1.6–2.6	7.3–7.86	1.6–2.6	0.5–2.54

4 Conclusions

The review of various differential ring VCOs delay stages has been presented in this paper, and finally concluded that the with the advancement in technology, there would always be the explosive requirement of efficient system devices in respect to the desired results, with tradeoffs between transistor sizing, frequency, speed, delay and power consumption of the circuits. The multiple feedback architectures are useful for low noise characteristics and maximum signal swings. The differential stages along with saturated load are providing the useful characteristics in terms of maximum frequency and low output phase noise. These characteristics of differential VCOs are explored when used at high frequencies at low power supply applications.

References

1. Rezvan Dastanian, Ebrahim Abiri, Mohammad Reza Salehi, Saeed Ghorbani, "A temperature compensated CMOS differential ring oscillator with low power consumption for RFID applications", International Journal of Engineering & Technology Sciences, Vol.03, No. 1, February 2015, pp. 45–54.
2. Ahmet Tekin, Mehmet R. Yuce, and Wentai Liu, "Integrated VCOs for Medical Implant Transceivers", *Hindawi Publishing Corporation,* 2008.
3. Stephen Docking and Manoj sachdev, "A method to derive an equation for the oscillation frequency of a ring oscillator", IEEE Transactions on circuits and systems, Vol. 50, No. 2, February 2003, pp. 259–264.
4. B. Razavi, Design of analog CMOS integrated circuits, Tata McGraw - Hill, Third edition, 2001.
5. Ashraf Mohamed Ali Hassen, "Analysis and design of high performance ring voltage controlled oscillator", International Journal of Computer Applications, Vol. 70, No. 20, May 2013, pp. 5–10.
6. Zuow-Zun Chen and Tai-Cheng Lee, "The design and analysis of dual-delay-path ring oscillators", IEEE Transactions on Circuits and Systems, Vol. 58, No. 3, March 2011, pp. 470–478.
7. Kuo-Hsing Cheng, Shu-Chang Kuo and Chia-Ming Tu, "A low noise, 2.0 GHz CMOS VCO design", IEEE Transactions on Circuits and Systems, Vol. 1, No. 1, December 2004, pp. 205–208.

8. Vikalp Thakur and Virendra Verma, "Low power consumption differential ring oscillator", on International Journal of Electronics and Communication Engineering, Vol. 6, No. 1, September 2013, pp. 81–92.
9. Harikrishnan Ramiah, Chong Wei Keata and Jeevan Kanesan, "Design of low-phase noise, low-power ring oscillator for OC-48 application" IETE Journal of Research, September 2014, pp. 425–428.
10. Hai Qi Liu, Wang Ling Goh, Liter Siek, Wei Meng Lim, and Yue Ping Zhang, "A low-noise multi-GHZ CMOS multiloop ring oscillator with coarse and fine frequency tuning" on IEEE Transactions on very large scale integration(VLSI) Systems, Vol. 17, No. 4, April 2009, pp. 571–577.
11. Asad A. Abidi, "Phase noise and jitter in CMOS ring oscillators" IEEE Journal of Solid-State Circuits, Vol. 41, No. 8, August 2006, pp. 1803–1816.
12. Haipeng Zhang Hao Li Yang Wang, "A tunable CMOS ring VCO in a wide frequency range," Key Laboratory of RF Circuit & System, Ministry of Education, School of Electronics & Information, Hangzhou Dianzi University Hangzhou, China, July 2011, pp. 6475–6478.
13. Jubayer Jalil, Mamun Bin Ibne Reaz, Mohammad Arif Sobhan Bhuiyan, Labonnah Farzana Rahman, and Tae Gyu Chang, "Designing a ring-VCO for RFID transponders in 0.18 μm CMOS process", Hindawi Publishing Corporation, The Scientific World Journal January 2014, pp. 1–6.

An Advanced Web-Based Bilingual Domain Independent Interface to Database Using Machine Learning Approach

Zorawar Virk and Mohit Dua

Abstract Interface to Database is basically a system which is designed on the basis of Natural Languages. These type of Interfaces to Database form a stepping stone in the fields of Intelligence, Medical Science, Database Mining, Search Engines etc. The paper proposes a system namely an Advanced Web-based bilingual Domain Independent Interface to Database using Machine Learning approach that takes punjabi and hindi language as the input. Some of the main features that have been incorporated into the developed system are domain independency, bilingual approach and a user friendly web interface. Also important features such as acceptance of queries with spelling mistakes, multiword keywords along with other functions like the auto complete function have been introduced. Hence the main objective is to make the system more user friendly, efficient along with increasing its scope without compromising on complexity.

Keywords Natural language interface · Information retrieval · Mining of database · Domain independent · Machine learning

1 Introduction

With the ever rising growth in the volume of data that is being generated every hour, the area of digitizing the information or storing of the information in databases are gaining prime importance. As a result, organization or institutions that provide services like storing of the information and other services related with it need a mechanism for providing an easy access to the database. A normal person without the knowledge of query language is unable to access this information. Due

Z. Virk (✉) · M. Dua
Department of Computer Engineering, National Institute of Technology, Kurukshetra, India
e-mail: zorawarvirk@yahoo.ca

M. Dua
e-mail: er.mohitdua@gmail.com

N. Modi et al. (eds.), *Proceedings of International Conference on Communication and Networks*, Advances in Intelligent Systems and Computing 508,
DOI 10.1007/978-981-10-2750-5_60

to this an interface to these databases needs to be designed in the natural languages itself so as to facilitate the access to such kind of stored information among a large number of people. The given system consists of an interface that has been developed for Punjabi as well as Hindi Language. The input natural language query will be transformed into an SQL query and then executed on the database. The results will be given in the natural language that was used for input itself. This system does data matching based on the concept of similarity functions which leads to an increase in the hit rate thereby increasing the overall performance as well as accuracy of the entire system.

The major problem that haunts these type of interfaces is the data mining problem. This problem deals with the accurate and efficient retrieval of only the useful entities out of a large raw data present in the database. Therefore the prime motive behind this paper is to implement a Interface for databases making use of Punjabi as well as Hindi as the Natural Language.

2 Related Work

Natural Language Interfaces to Databases (NLIDBs) prototype had appeared in late sixties and early seventies. Since then a number of systems have been developed. Here, we discuss some of them with their feature and techniques which we have incorporated in this system. LUNAR system was introduced in 1971 and answers the questions about samples of rocks brought back from the moon [1]. LUNAR system comprised of the two databases, one database was for the chemical analysis and other one was for the literature references. LUNAR system had an impressive performance. It neatly managed to handle about 90 % of the requests by the users without any error [1]. JANUS was a kind of a system that had the ability to interface to multiple systems (databases, expert system and graphics device) [2].

Dua et al. [3] presented a detailed overview about the various state of the art NLIDB systems developed in the last four decades. The article reviewed all the merits and demerits of various architectures used to implement the interface to database. Khalid et al. [4] described QA system where help of the information extraction module was taken to make the training data of classifier and this system was designed keeping in view only the English language. English Language Front (ELF) system [5] is like many other commercial systems, claims a rather good performance. The system reads the schema of the database and then creates a set of rules that are used during semantic parsing, when the natural language input is converted into SQL query for the relational database system.

Llopis et al. [6] discussed and classified various issues involved in making NL interfaces accessible to everyone. One of the latest addition to this area is the DHIRD [7] system. This system is designed for accepting queries in both English as well as the Hindi language. Stanford parser [8] for English language and Hindi Shallow parser [9] for Hindi were used respectively.

3 Proposed Work

The natural language query is processed by this system and finally converted to an SQL query whose execution returns the answer to the given query in the input language itself. The database where the answers are searched is in tabular form and has its data values in Hindi, Punjabi with their corresponding field names in English. This system analyses the input query and uses Machine learning to predict the labels T, AF and QF, where T is the name of the table, AF is the Query Answer Field and QF is the Query Question Field. The final SQL query is constructed as:

Select AF from T where condition (QF)

3.1 Phases in Machine Learning

Machine Learning concept used in for the developed system has been categorized under two categories namely the Training phase and the Testing phase. In the training phase we train the classifier in the following manner which is shown below in Table 1. For a given question, SQL query is manually generated for it and mapped to one of the possible permutation of select, from and where classes. Also a feature vector space matrix for the same question is generated by using the entity detection and feature extractor. In this way, a training set for classifier is generated. Once the classifier is trained using the given set of questions, then the testing of it is done to check out how accurate are the results.

Under the testing phase, the trained classifier is put to use to check the efficiency of answer searching capability of classifier for a given query. The question is tagged using Hindi/Punjabi part of speech (POS) tagger. The tagged input is then fed to the feature extractor that extracts features out of it. These features provide meaning to the question in a language understandable by the classifier. The features used are shown in Table 1.

The feature vector is then provided as an input to the classifier that generates the one of the permutation of select, from and where classes i.e. table look-up labels.

Table 1 Example of feature extraction table

Qword	Kab/Kya/Kiska/Kahan/Kaun etc.
Qnouns	First three nouns of the question
Qverb	First verb in the question
QAdjective	First adjective in the question
QConjuncts	First preposition/subordinating in the question
QCompounds	First XC tag in the question
Quantifiers	Last quantifier in the question (*kama* (less), *jyAdA* (more), *bahuwa* (lots), etc.)
Cardinal	Last cardinal number in the question
Frequency	Number of special symbols like $, %, quotes etc.

By using these labels, the SQL query is generated. Finally, designed NLIDB retrieves the query result from database using the query and provides the output to the user.

4 Architecture of the Implemented System

The architecture of the implemented system consists of 5 main modules namely Language Identification module, Lexical and Syntactic module, Domain Identifier module, Query Decoder module, Query Executor module along with the Domain databases and Train database which is used for the training of the given system. The architecture has been given in Fig. 1 along with the detailed explanation of the different modules.

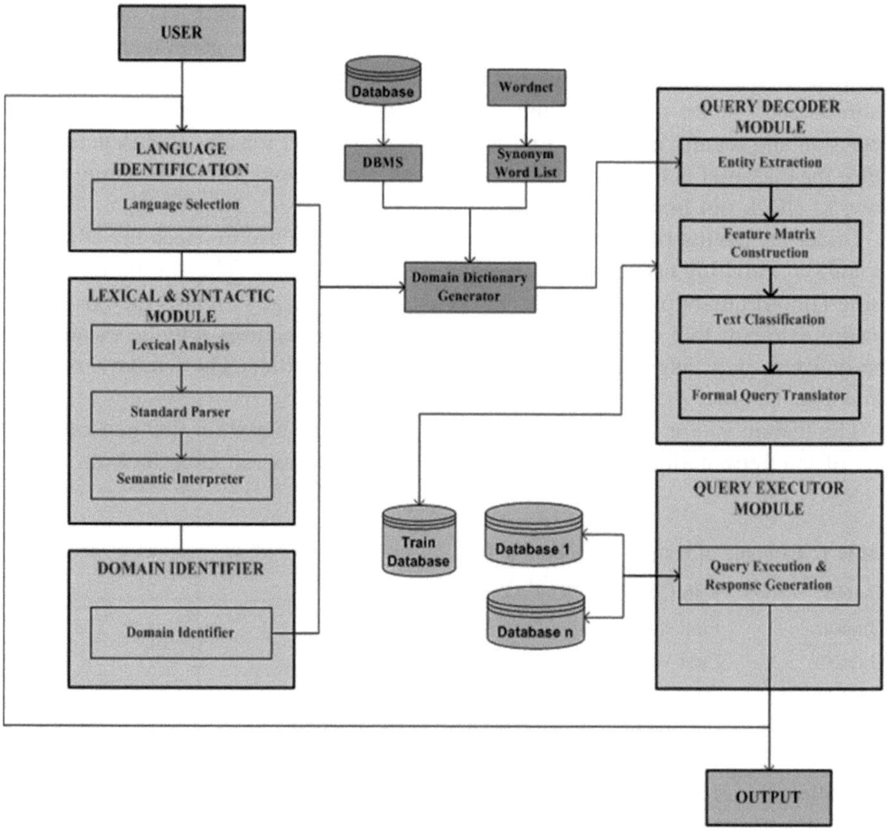

Fig. 1 Architecture of developed system

4.1 Language Identification

In the Language Identification phase, user is given the option to choose the language with which it is going to implement the system. Language here refers to the natural language in which the user wants to proceed for input of the query so that the database as well as the DBMS, WordNet are taken into consideration accordingly. The two natural languages with which the system has been implemented are Hindi and the Punjabi language.

4.2 Lexical and Syntactic Module

Lexical and Syntactic module is responsible for all the tasks related to preprocessing of the query, semantic analyzing are dealt in the module. This module further has three components namely Lexical Analysis, Standard Parser and Semantic Interpreter. The work of lexical analysis is to break the query into different useful tokens. Parsing of the input query and generation of the syntax tree is done by the standard parser. The output of the parser acts as input to the semantic interpreter where discarding of all the useless tokens takes place. Matching of the useful tokens with their corresponding English word is also performed by the semantic interpreter.

4.3 Domain Identifier Module

Domain Identifier module is responsible for the selection of the database. This selection of the database is done automatically based on the tokens, keywords, fields name that appear in the query that has been input by the user. The output of the database module is then sent to the domain dictionary generator which based on the input from the language identification phase and the domain identifier module is responsible for the fetching of the correct data or information from the given domain in the selected language. Synonym word list, DBMS depend on the crucial information that is passed by the domain dictionary generator.

4.4 Query Decoder Module

Query Decoder Module is responsible for the translation of the natural language now domain dependent query tokens received from the lexical and syntactic module through the domain dictionary generator phase into the formal SQL query. The query translator module consists of Entity Extraction phase, Feature Matrix

Construction phase, Text Classification phase, Formal Query Translator phase. Here, Entity Extraction phase deals with the detection of the Known Words that are also called the Entities. These are the entities that are known or common to the database. This phase helps in determining the important words or entities using which the query can be generated.

In feature matrix construction phase a vector space feature matrix is constructed based on arranging the entities under the respective categories using the threshold value. In the text classification phase, the matrix so designed is compared with the training data set. Classification is done using the matrix along with the concept of K-Nearest Neighbors (KNN) Algorithm [10]. This is done so as to find the possible closest match of the entities to the queries. After the text classification phase comes the formal query translator phase which is responsible for creation of a formal language query with all the information that has been gathered till now.

4.5 *Query Executor Module*

Query Executor Module is responsible for executing the query that has been passed by the formal query translator phase on the database as well as also for the correct display of the information or the results in the natural language that was selected before the query was given as the input. This module also deals with the establishing of the required connections using the connection manager.

5 Implementation and Results

Selection of the language in which the user wants to enter the query is to be done by the user itself. For selection of the language a user interface with a drop down menu has been provided. The language selection phase is shown in Fig. 2.

After the language has been selected then comes the main interface where the user enters the query in the query input field. The entered query is first converted into proper SQL format query on click of the convert button. Then the query is executed when the user clicks on the Execute button and the result is displayed. This has been shown with the help of an example.

Input Query: सभी विधार्थीयो की विभाग नाम बताओ

Translated Query: select department.name from stu

Successful queries field is present which displays the list of all the queries that have been successfully tested. A view of the interface has been illustrated in Fig. 3.

Some of the queries in punjabi and hindi language for different database domains have been given in the Table 2.

Fig. 2 Language selection menu

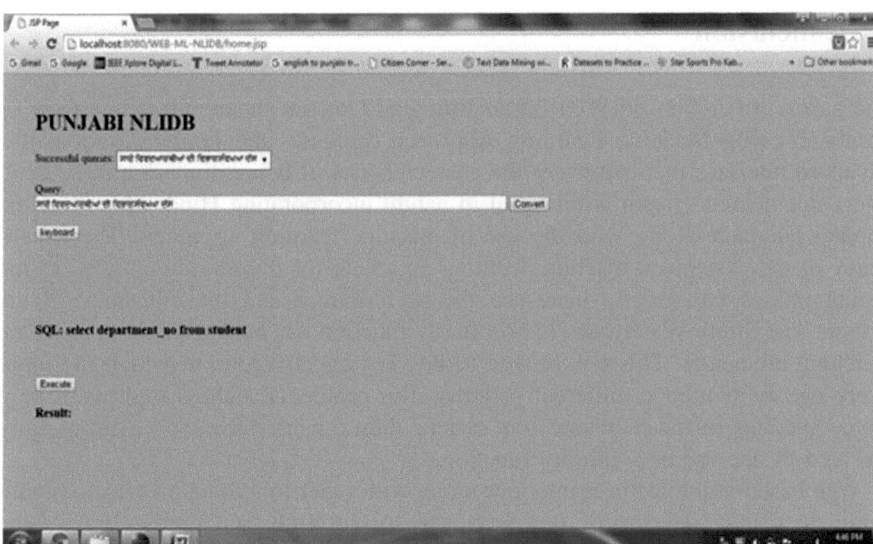

Fig. 3 Punjabi language input interface

Table 2 Examples of tested queries

User query in natural language	Translated query in SQL format	Database domain
ਸਾਰੇ ਕਰਮਚਾਰੀਆਂ ਦਾ ਕਰਮਚਾਰੀ ਸੰਖਿਆ ਦੱਸੋ	select emp.emp_no from emp	Employee database
ਉਹਨਾਂ ਕਰਮਚਾਰੀਆਂ ਦਾ ਨਾਮ ਅਤੇ ਸ਼ਹਿਰ ਦੱਸੋ ਜਿਹਨਾਂ ਦਾ ਵੇਤਨ 1000 ਤੋਂ ਜਿਆਦਾ ਅਤੇ ਵੇਤਨ 2000 ਤੋਂ ਘੁਟ ਹੈ	select emp.name,emp.city from emp where emp.salary > 1000 and emp.salary < 2000	Employee database
ਸਾਰੇ ਵਿਦਿਆਰਥੀ ਦੱਸੋ	select * from stu	Student database
सभी विधार्थीयो की जन्मतिथि, अनुकर्मान्क बताओ	select stu.dob,stu.roll_no from stu	Student database
उन सभी कर्मचारियो के कर्मचारीसंख्या ,वेतन बताओ जिनका नाम 'दीपक' है	select emp.emp_no ,emp.salary from emp where emp.name = 'deepak'	Employee database

6 Conclusion

The paper, An Advanced Web-Based Bilingual Domain Independent Interface To Database Using Machine Learning Approach discusses the implementation of a advanced interface that inculcates the characteristics of being domain independent. The implemented system is bilingual in nature incorporating Hindi as well as the Punjabi language along with the use of machine learning approach. The dependency of the system on machine learning approach for the training as well as the testing phases has greatly increased the performance and the efficiency of the system. The Smith Waterman [11] similarity function has been used to increase the matching efficiency. This has proved to be very effective due to which the input query can be written in different patterns. The problems such as multiword keyword, spelling mistakes where the system didn't work have been successfully resolved by the use of similarity function.

Web based graphical user interface along with other functionalities such as auto complete feature, storing of the previously successfully run queries have been added so to make the designed system more user friendly.

References

1. W. A. Woods, R. M. Kaplan, and B. N. Webber: The Lunar Sciences Natural Language Information System: Final Report. In: BBN Report 2378, Bolt Beranek and Newman Inc., Cambridge, Massachusetts, 1972.
2. P. Resnik: Access to Multiple Underlying Systems in JANUS. In: BBN report 7142, Bolt Beranek and Newman Inc., Cambridge, Massachusetts, 1989.
3. Mohit Dua, Shivani Jindal, Rajender Kumar: An Architectural Overview of Natural Language Interface to Knowledge Base. In: International Conference on Computation of Power, Energy, Information and Communication, pp. 437–441, IEEE (2014).
4. Khalid, M.A., Jijkoun, V., de Rijke, M.: Machine Learning for Question Answering from Tabular Data. In: 18th International Workshop on Database and Expert Systems Applications, pp. 392–396, DEXA (2007).
5. W. S. Luk, S. Kloster: ELFS: English language from SQL. In: ACM transactions on database systems, Vol. 11, No. 4, pp. 447–472, December (1986).
6. Miguel Llopis, Antonio Ferrandez: How to make a Natural Language Interface to Query Databases Accessible to Everyone: An Example. In: Computer Standards & Interfaces, vol. 35, no. 5, pp. 470–481, (2013).
7. R. Kumar, M. Dua, S. Jindal: Domain-Independent Hindi Language Interface to Relational Database. In: International Conference on Computation of Power, Energy, Information and Communication, pp. 81–86, IEEE (2014).
8. Stanford Parser, http://nlp.stanford.edu/software/index.shtml.
9. Hindi Shallow Parser, http://ltrc.iiit.ac.in/analyzer/hindi/.
10. Gongde Gou et al.: KNN Model-Based Approach in Classification. In: Lecture Notes in Computer Science, pp. 986–996, Springer (2003).
11. R. Mott: Maximum Likelihood Estimation of the Statistical Distribution of Smith-Waterman Local Sequence Similarity Scores. In: Bulletin of Mathematical Biology, pp. 59–75, Springer (1992).

Comparison of ABC Framework with AHP, Wiegers Method, Cost-Value, Priority Groups for Requirements Prioritization

Sita Devulapalli, O.R.S. Rao and Akhil Khare

Abstract ABC framework for requirements prioritization for software products development proposed by the author is compared with some of the well known methods in the literature—AHP, Cost-Value method, Wiegers method and Priority Groups. Simplicity, Scalability, Closeness to Practical development, Amenability to Change Management and Release Planning, Ability to Visualize Prioritization are considered for comparison.

Keywords ABC framework · Requirements prioritization · Multi-level framework · Software product development · Comparison

1 Introduction

Significant Research and empirical studies have taken place in the area of requirements prioritization [1]. Methods have evolved for prioritizing requirements based on different parameters—Value and Cost being prominent among them. Analytical Hierarchy Process—AHP [2] is based on pair wise comparison of requirements relative to each other on a scale at successive levels of hierarchy.

Cost-Value approach by Karlsson [3] takes the cost of implementation and value of requirements into consideration in pair wise comparison. Wiegers method [4] proposes risk weighted cost/value ratio for determining priority. Priority Groups method categorizes requirements based on ranking different parameters—mostly importance of requirements, and are put in groups.

S. Devulapalli (✉) · O.R.S. Rao · A. Khare
FMS, ICFAI University Jharkhand, Ranchi, India
e-mail: sitadpalli@yahoo.co.in

O.R.S. Rao
e-mail: orsrao.icfai@gmail.com

A. Khare
CSE, MVSR College of Engineering, Hyderabad, India
e-mail: khareakhil@gmail.com

N. Modi et al. (eds.), *Proceedings of International Conference on Communication and Networks*, Advances in Intelligent Systems and Computing 508,
DOI 10.1007/978-981-10-2750-5_61

Davis advises simplifying the process and advises Triage at successive levels, taking into account market realities [5]. Industry specific studies for software products meeting certain specific base parameters seem to have been very few [6]. This makes the conclusions and comparisons difficult to be applicable or reliable. Comparison of some of the methods for quality requirements is taken up by Karlsson [7].

ABC Framework [8] has been proposed by the author reflecting the practical aspects of the software development. The proposed framework takes into account different parameters considered during the course of "software in making" and links the prioritization to development process, release planning, change management, quality management. The paper looks at Priority grouping, Cost-Value method, Wiegers method and AHP and in comparison analyses the benefits of ABC Framework.

Brief description of the methods—AHP, Cost-Value, Priority Grouping, and Wiegers method is provided in Sect. 2. Section 3 describes ABC Framework. Comparison basis and merits are discussed in Sect. 4. Comparison is summarized in Sect. 5.

2 Requirements Prioritization Methods for Comparison

Laura Lehtola [9] puts prioritization approaches roughly into two categories—methods based on giving values to different factors of requirements and negotiation approaches. The methods based category is further subdivided into two subcategories, one with methods which process each requirement uniquely and the other with methods based on comparisons. Wiegers method, Priority grouping fit in first category and AHP, Karlsson's cost value pair wise comparison falls into second category. ABC falls in the second category.

2.1 AHP

Analytical Hierarchy Process (AHP) of Saaty [2] is a multi criteria decision making approach in which factors are arranged in a hierarchical structure that flows from overall goal to criteria to sub criteria and alternatives in successive levels. Hierarchy is expected to provide overview of the problem space and enable decision maker compare homogeneous elements in each level. Karlsson [3] illustrates using AHP for decision making that involves 4 steps for evaluating requirements using the criterion of value. A scale as defined by Saaty is used for pair wise comparison of the requirements—1, 3, 5, 7, 9 corresponding to equal value, slightly more value, strong value, very strong value and extreme value respectively. 2, 4, 6, 8 provide intermediate values when compromise is needed. In pair wise comparisons, reciprocal of assigned number of one requirement becomes the priority for the pair's other requirement.

For each criterion AHP's pair wise comparisons result in $n(n - 1)/2$ comparisons for n requirements. Assuming 4 requirements, Step 1 involves forming 4×4 matrix for pair wise comparison. Step 2 involves comparing each requirement with other one using the scale values. Step 3 involves deriving the priority matrix, which are Eigen values of the matrix arrived at by using averaging over normalized columns. Relative value is assigned to requirements based on the priority.

For 4 requirements and one criteria, there will be $4 * 3/2 = 6$ comparisons needed. If the number of criteria is 2, the number of comparisons will be $2 * 6 = 12$. For n requirements and c criteria the comparisons will be $c * n(n - 1)/2$. Then a step to correlate or combine the priorities across the criteria for a combined priority for each requirement to be arrived at. Estimating relative importance for each requirement in comparison with another one in the set is done using the scale. With different criteria at different levels, relative estimation on these criteria is required.

2.2 Cost Value

Karlsson and Ryan [3] proposed using implementation Cost and Value as the high level factors for requirements' pair-wise comparison as in AHP. Both Cost and Value based relative priorities for the requirements are arrived at as illustrated above and are plotted in a cost-value diagram, which can be used as a conceptual map for identifying requirements to be taken up for implementation. This information can also be utilized for strategizing release plan, according to Karlsson and Ryan. Here c is 2, hence the comparisons required for the 4 requirements will be $2 * 6 = 12$. For n requirements the comparisons will be $2 * n(n - 1)/2 = n(n - 1)$.

2.3 Priority Grouping

In this method requirements are not compared to each other based on a criteria, but are grouped into either three—low, medium, high priority groups/essential/conditional/optional groups or four—most needed, good to have, ok to have and not to have—priority groups based on importance of requirements. Each group can further be grouped within, to arrive at finer clusters of requirements. And this sub-classification can extend and form a hierarchy of levels. Whether the criteria at each level will be importance, which can be a combination of different criteria pre-determined or the criteria can be different for sub-grouping is not explicitly discussed in literature.

Taking the same 4 requirements, the number of decisions to be made will be 4—to decide which group the requirement will go to, for a single level grouping. For n requirements the decisions will be n. If successive grouping is done, the decisions would be n * c for c number of successive groupings. The decision making in classifying into groups is subjective in this method.

2.4 Wiegers Method

Wiegers semi quantitative, analytical approach distributes a set of estimated priorities across a continuum rather than grouping them into a few priority levels. Risk adjusted value/cost ratio is used to determine priority in this method. A features attractiveness is directly proportional to the value it provides and inversely proportional to its cost and technical risk of implementation. Wieger suggests applying this method to only negotiable features and not to core business functions or requirements that require compliance with Government regulations. Priority is calculated as value%/(cost% * cost weight + risk% * risk weight), where value is a weighted combination of value to customer and penalty of not implementing the requirement.

Since there are 4 criteria—value, penalty, cost, risk, to be estimated on a scale of 1 to 9, for 4 requirements, we will need 4 * 4 = 16 decisions to be made at the initial level. For n requirements, the decisions needed are n * 4. The requirements can be analyzed at subsequent levels for increased granularity. For c levels, the decisions required would be n * c * 4. Wieger indicates the method is not mathematically rigorous and is limited by the ability to estimate the 4 parameters for each requirement and suggests it should be used as a guideline to make trade-off decisions But this is the same limitation for all the methods using a scale to estimate on different criteria. Wieger points that the method can become unwieldy beyond several dozens of requirements and suggests initial and sub-lists analysis for ease of prioritization.

In this method Value includes the −ve value or penalty for not implementing. Cost is expected to take into account existing modules benefit, risk includes impacts.

2.5 ABC Framework

The Framework [8] is defined as 5 sets based on most used parameters in the sequence of priority determination. Each set is defined by three classes/bins defined by % value of the respective set parameters. Requirements are grouped into the classes in the sets in the process of prioritization. Prioritization sets—S1 to S5 and classes/bins—A, B, C within are described in Table 1.

Table 1 Framework—sets, classes

Sets	Classes/Bins—A, B, C
S1. Business Value(BV) in conjunction with Customer Base (CB)	A: 20 % of CB with 70 % BV
	B: 30 % of CB with 25 % BV
	C: 50 % of CB with 5 % BV
S2. Requirements Applicability with respect to product, where UW: User Interface, BI: Business Logic, CP: Core	A: 70 % UW, 30 % BI, 0 % CP
	B: 50 % UW, 40 % BI, 10 % CP
	C: 30 % UW, 50 % BI, 20 % CP
S3. Implementation Cost, where MI: Marginal Implementation, NI: New Implementation, IR: Impact Recovery.	A: 70 % MI, 25 % NI, 5 % IR
	B: 50 % MI, 40 % NI, 10 % IR
	C: 30 % MI, 50 % NI, 20 % IR
S4. Time Requirement, where L: 8–16 person weeks, M: 4–8 person weeks, S: 2–4 person weeks	A: 10 % L, 20 % M, 70 % S
	B: 15 % L, 25 %M, 60 %S
	C: 20 % L, 30 % M, 50 % S
S5. Resource Requirement, where RC: Core aware, RI: Industry aware, RT: Technology aware	A: 10 % RC, 20 % RI, 70 % RT
	B: 15 % RC, 25 % RI, 60 % RT
	C:20 % RC, 30 % RI, 50 % RT

When all sets are used for classification, 243 bins of requirements are formed. Based on the constraints and release theme, the bins can be selected in the order of preference for the releases. Requirements can be associated with their class membership at each level and a macro priority can be associated as well by associating weights to classes at each level and/or weights to each of the sets as illustrated by the author in [9]. With the unique numbering scheme, priority sequences can be generated for the requirements, based on class association in each set which help in visualizing basis of prioritization through the development process and visualizing requirements change implications. Tables 2, 3 illustrate the macro priorities and number sequences based on ABC framework for 3 requirements.

Table 2 Priority values with class (A-3/3, B2/3, C-1/3) and set weights

Requirements	S1-5/5	S2-4/5	S3-3/5	S4-2/5	S5-1/5	Pm
R1	A	A	B	A	A	$(5 * 4 * 3 * 2 * 1/5^5) * 3^4 * 2/3^5$ (or 0.0256)
R2	B	B	C	C	B	$(5 * 4 * 3 * 2 * 1/5^5) * 2^3/3^5$ (or 0.001264)
R3	C	C	C	B	C	$(5 * 4 * 3 * 2 * 1/5^5) * 1^4 * 2/3^5$ (or 0.000316)

Table 3 Unique priority sequences

Requirements	S1	S2	S3	S4	S5	Priority sequence
R1	0	0	0	0	0	00000
R2	1	0	0	1	2	10012
R3	2	1	0	2	1	21021

3 ABC Framework Comparison with Other Methods

For the 4 requirements ABC framework would require 4 * 5 = 20 decisions to be made, with all 5 sets utilized. For each set the number of decisions is same as in priority grouping that is 4. For n requirements the number of decisions will be n * 5.

ABC Framework adapts the idea of hierarchical structure of layers of AHP relevant to the problem space of software product development for requirements prioritization. The framework takes into account different aspects—business value, nature of implementation, and cost of implementation, including impacts, time needs and resource needs—encountered in the product development flow in a structured way and in a sequence of layers. The class boundaries are defined for intuitive decision making, and are adaptable to specific projects. The criteria encompass short term and long term benefit, cost aspects. Requirements prioritization is invariably linked to cost of development and benefit to be achieved in most of the methods proposed for prioritization. In general the cost factor is considered to the extent of time taken to develop or resources cost. Business value is normally understood to the extent of immediate revenue. Wieger included penalty of not implementing in value. Karlsson's cost-value are to be estimated a priori. Considering the "other than software world" projects and cost and benefit analysis done for taking up projects—Business value encompasses present value of future returns, indirect benefits, return on investment periods. The costs involve not just development costs, but also opportunity costs and impact costs. Wieger included impact costs in risk parameter.

ABC Framework does not pick up the AHP's scale or method of priority calculation. Typically Software requirements prioritization does not start or stop at one time or in one step. The prioritization of what will finally get into the product release goes through levels of decision making considering different aspects. Trying to club all the aspects into one or two parameters or trying to prioritize at one time considering all aspects generally results in suboptimal or not so well understood prioritization. The uncertainties in the input decision making related to determination of values of criteria or related to relative comparison, the author feels mathematical rigor is not warranted for determination of priorities. The classification is more akin to priority grouping at each level. ABC framework can be mapped to priority grouping with different criteria adopted at each level of hierarchy, which are not necessarily sub groups.

In Priority grouping, the grouping of high, medium, low is a subjective judgment. Same is the case with AHP scale, where scale values for comparison are subjective; ABC Framework attempts to define boundaries of subjective decision making, based on problem space of software development. The boundaries are adjustable as per the specific needs of a project. The criteria at each level in the ABC framework are intuitively defined based on practical aspects of software development. The criteria are not mutually exclusive strictly; they reflect the parameters considered as software development progresses.

In cost value method of Karlsson or in value-penalty-cost-risk method of Wieger, various aspects of software development are expected to be resulting in cost of development, value of requirement, so that decisions can be made on prioritization in terms pair-wise comparison or weighted grouping. ABC framework enables grouping into 3 classes at successive levels based on different criteria faced by the decision makers, without imposing a pair-wise comparison or estimation on a scale, yet resulting in the final outcome of relative priorities.

The framework enables visualization of relative prioritization of requirements at every level and in the final prioritization, instead of criteria getting lost in a mere prioritization number as in other methods. There is implicit cost and implicit value in each of the criteria and there are short term costs and values and long term costs and values with respect to each criteria and determining these is not a formalized science for requirements prioritization so far. Unlike in non-software industry, where project costs and project revenues are determined over projects life periods taking into account present and future revenue flows and costs to be incurred and opportunity costs, Software industry is still seen to be not amenable to this rigorous analysis.

ABC framework has criteria, at successive levels, which spawn out the development process and capture cost and value aspects implicitly. The decisions are to be taken based on the boundary values for the classes, which allows flexibility, adoption, approximation. It enables visualization of short term costs and value and also long term costs and value by virtue of the criteria and classes at successive levels, albeit implicitly, through the process and in final prioritization.

Prioritization is somewhat misconstrued concept in software development. It simply means what requirements can be picked up for now for a certain set of customers to provide a solution within a certain time period with the available resources and existing inventory (components/modules). And this scenario is subject to change. Under the changing scenario, it will be imperative to change the development course and it is needed to have as less impact as possible. How do we reconcile the changes to the current decisions on priorities of requirements? What were the parameters considered in the past and how do they change now? Visualization, ease of re-prioritization, impacts visibility on schedules, costs, value are needed. ABC framework provides ease of reprioritization [9], visibility to impacts of change, flexibility for re-planning, which is difficult with other methods.

Requirements are requirements and they need to be implemented at sometime or the other, they need to be spaced out and this spacing out needs to be visible all the time for dynamic decision making, or dynamic choices. The decision map and the criteria of decisions, nature of decisions needs to be visualized throughout the life cycle of the product/project. This is feasible with ABC framework through its unique representation of priorities and unique classification at successive levels with relevant criteria into A, B, C classes whose boundaries are predetermined.

Coming to scalability of the methods, methods based on pair-wise comparison—AHP and Cost-Value tend to be increasingly cumbersome. Wieger indicates to the unwieldiness of the method for large number of requirements due to estimation needs. Priority grouping is still the simplest and easiest, though approximate. ABC framework can be easily used for large number of requirements and number of decision grow only linearly with the number of requirements.

4 Summary of Comparison

Summarizing the comparative analysis in Sect. 3, ABC Framework offers the ease of Priority grouping method, adapts the hierarchical decision making concept of AHP and takes into account different aspects of practical relevance in software development space, which, in effect, are common with cost-value-penalty-risk. Any dynamic changes in priorities of requirements can be easily integrated, visualized and interpreted in ABC framework. The impacts on release plans and coming up with new release plans is similarly simple with ABC framework. Comparison of various aspects of the prioritization methods discussed in Sect. 3 is presented in Table 4.

Table 4 Comparison of various aspects

Method	AHP	Cost-value	Wieger	Priority grouping	ABC framework
Methodology	Pair-wise comparison	Pair-wise comparison	Independent assessment by estimation	Independent assessment	Independent assessment
Criteria	Importance. Can have multiple criteria	Cost, value	Value, penalty, cost, risk	Importance. Can have multiple criteria	Business value, nature of requirement, Implementation costs, development time, resources
Scale	1, 3, 5, 7, 9 2, 4, 6, 8 reciprocals of above	Same as AHP	1 (low) to 9 (high)	Grouping into 3 or 4 groups	Classifying into three classes in each set
Levels	As needed for other criteria	As needed for granularity	As needed for granularity	As needed for granularity	5
Number of decisions for n requirements	$n(n-1)/2$ for each criteria/level	$n(n-1) = 2 * n(n-1)/2$	4n for single level	n for single level	5n
Priority representation	Eigen values of comparison matrix	Eigen values	Value%/(cost % * weight + risk % * weight)	Group membership/ranking	Class membership in each set
Visualization of influencing factors in final priority	Relative priority	Cost-value diagram	Relative priority	Ranking in group	Class/set association sequence
Changes incorporation	Rework the process	Rework the process	Rework the process	Can be added/removed as needed	Can be added/removed as needed
Visualization of change impacts	–	–	–	–	Relative class sequence, macro priority
Release plan determination, changes in release plan visualization	based on relative priority	Based on cost-value diagram/correlation	Based on relative priority	Based on ranking	Based on release theme relevant class/set sequences

5 Conclusion

ABC framework can be seen as a hierarchy of levels with different criteria representing the software product development space that can be used to classify requirements similar to simple priority grouping method and taking into account cost and value and risk aspects as in cost-value and Wiegner's methods. In addition, it provides a unique representation for prioritization of the requirements. The framework enables understanding and interpreting prioritization in a visual and instant way. The Framework and priority representation enables simple and effective methodology for Requirements Prioritization for successive releases under dynamic changes and lead to better understanding and planning of releases.

It helps in prioritization of requirements and planning releases, streamlining the project deliveries to client's satisfaction without overworking the teams or missing time to market deadlines, providing dynamic prioritization throughout the process of software development.

References

1. Laura Lehtola, and Marjo Kauppinen: Suitability of Requirements Prioritization Methods for Market-driven Software Product Development. Software Process Improvement Practice 2006; 11: 7–19.
2. Thomas L. Saaty: How to make a Decision: The Analytical hierarchy Process. European Journal of Operations Research. 48 (1990) 9–26 North-Holland.
3. Karlsson J, Ryan K.: A cost-value approach for prioritizing requirements. IEEE Software 1997; 14(5): 67–74.
4. Wiegers KE.: Software Requirements, MicrosoftPress (1999): Redmont, DC.
5. Alan M. Davis, Ed Yourdon, Ann S. Zweig: Requirements Management Made Easy. 39–947. www.omni-vista.com.
6. Regnell, B., Höst, M., Natt och Dag, J., Beremark, P., Hjelm, T.: An Industrial Case Study on Distributed Prioritization in Market-Driven Requirements Engineering for Packaged Software. Requirements Engineering 2001, vol 6, no 1, pp 51–62.
7. Karlsson J, Wohlin C, Regnell B: An evaluation of methods for prioritizing software requirements, Inform. Software Technol. 1998, 39(14–15): 939-947.
8. Sita Devulapalli, Akhil Khare: A Framework for Requirement Prioritization for Software Products, IUJ Journal of Management, Vol 2, No. 1, May 2014.
9. Sita Devulapalli, Akhil Khare, ORS Rao: Mathematical treatment of ABC Framework for Requirement Prioritization, ICTIS Conference, Nov2015 [to be published in SIST].

Scalability Analysis of Medium Access Control Protocols for Internet of Things

Nurzaman Ahmed, Hafizur Rahman and Md. Iftekhar Hussain

Abstract The concept of Internet of Things (IoT) opens up a new vision for the future Internet where not only the users or computing systems but also the everyday objects are capable of processing, communicating, sensing, and actuating. Various IoT applications help in quality of living through the deployment of massive number of devices equipped with wireless communication capability. In supporting the requirements of such IoT applications with massive number of heterogeneous devices, Medium Access Control (MAC) protocol holds the key responsibility of optimal utilization of network bandwidth. This paper compares the performance of contention-based, reservation-based and hybrid MAC protocols in the context of large scale networks for IoT. Further, it provides a survey of the key requirements, technical challenges, and existing works on scalable MAC protocols for supporting efficient communications in IoT. We highlight the problems and prospects of existing MAC protocols and identify the factors for improvement and future direction.

Keywords Internet of things · Machine-to-machine communication · Hybrid MAC · Scalable MAC

1 Introduction

The Internet of Things (IoT) is a very popular expression these days, although still a fuzzy one mostly due to the large amount of concepts it encloses. In the vision of IoT, "things" are not only computers, people or mobile phones but also sensors,

N. Ahmed (✉) · H. Rahman · Md. Iftekhar Hussain
North-Eastern Hill University, Shillong, India
e-mail: nurzaman713@gmail.com

H. Rahman
e-mail: hafizjec@gmail.com

Md. Iftekhar Hussain
e-mail: ihussain@nehu.ac.in

N. Modi et al. (eds.), *Proceedings of International Conference on Communication and Networks*, Advances in Intelligent Systems and Computing 508,
DOI 10.1007/978-981-10-2750-5_62

actuators, refrigerators, TVs, vehicles, clothes, food, medicines, books, etc. These "things" can broadly be classified into three categories: (i) people, (ii) machine (e.g., sensor and actuator) and (iii) information (e.g., clothes, food, medicine, and books) [1]. "Things" are uniquely identifiable and capable of communicating with other networking devices.

Wireless networking technologies enable objects or things to interact and cooperate with each other using wireless links to ensure ubiquitous communications [2]. Communicating nodes might be sensors, Radio-Frequency IDentification (RFID) tags, actuators, or mobile phones, and many others. IoT is different from traditional homogeneous networks in many ways. Firstly, it encompasses a variety of wireless communication technologies such as WiFi, ZigBee, Bluetooth, etc., a variety of protocols stacks available in wired, wireless, and hybrid, and dynamic networking environments like tree, mesh, etc. Secondly, it consists of massive number of devices in a service coverage having different requirements with specific traffic characteristics. For example, a smart lighting networks can be very large with luminaires around 500 to 2000 [3]. In a smart parking application, packet generation rate depends on the vehicle's arrival and departure [4]. Thirdly, the processing and storage capability of such devices (things) are very low and are power-constraint in nature. Fourth, these networks have smart and cognitive gateway. Fifth, chances of coexistence of multiple types of networks is high. Finally, Machine-to-Machine (M2M) communications are highly involved in such networks.

Compared to traditional networks, IoT has characteristics such as low power requirement, low cost, low data rate, adaptive, and cognitive characteristics, and large scale deployment. The traditional MAC protocols like Carrier Sense Multiple Access with Collision Avoidance (CSMA/CA) and Time Division Multiple Access (TDMA) can't directly apply to IoT [5, 6]. The critical MAC layer challenges for IoT communications lies in facilitating channel access to extremely large number of low powered devices while supporting the diverse service requirements and unique traffic characteristics of devices in IoT networks [7]. In addition, MAC protocols for IoT should be efficient, scalable, consume low power, have low latency, and be implementable using low cost hardware. The MAC protocols adopting approaches like hybrid MAC (e.g. combination of CSMA/CA and TDMA), dynamic duty cycle, cognitive, etc., are better choices for enhancing scalability. There is a need for analyzing the performance of various types of MAC protocols in IoT. In this paper, we provide a simulation analysis and survey on the reported MAC protocols in the context of IoT compatibility. While doing so, the state-of-art proposals available in the literature have been considered based on their relevance in addressing various issues.

Rest of the paper is organized into five sections. Section 2 discusses the background study of this paper. Scalability analysis of major existing types of MAC protocols are presented in Sect. 3. Section 4 presents a detail of the existing scalable MAC protocols available for IoT indicating their advantages and disadvantages. Finally, Sect. 5 gives the conclusion to the paper.

2 Background

In IoT, huge number of tiny objects around us (electronic, electrical, and non electrical) connect and communicate within themselves to provide seamless connectivity and contextual services [1, 2, 8]. The development of objects like-RFID tags, sensors, actuators, mobile phones etc., make it possible to materialize the IoT which interact and co-operate each other to provide the service better and accessible anytime, from anywhere. One of the most popular IoT application is smart home (connected home), switching the home appliances ON and OFF remotely from Internet in order to avoid accidents and save energy. In a health-care application, keeping Bio-NanoThings [9] inside the human body which would collect health-related information, transmit it to an external healthcare service provider through the Internet, and execute commands from the same provider such as synthesis and release of drugs.

All of the IoT applications require smarter network which is designed by integrating various technologies and standards. Scalability is the most important property of IoT for forecasting and analyzing the network's capability of expansion. The traffic generation nature, technologies used, protocol standards applied etc., are the factors affect in scalability of network. It is useful to understand the traffic patterns of different applications for better scalability of network. Latest communication standard like 802.11ah [10, 11], which utilizes sub-1 GHz license-exempt bands to provide extended range WiFi networks, compared to traditional 802.11 standards based networks operating in the 2.4 and 5 GHz bands. The main features of this standard is lower energy consumption, allowing the creation of large groups of stations or sensors that cooperate to share the signal, and hence support the concept of IoT. RFID, ANT etc., are the most promising technologies for IoT. MAC protocols designed for these type of technologies need to be low power consuming and can enable large number of devices to communicate.

List of MAC protocols have been designed so far to provide reliable medium access services in wireless network. The MAC protocols designed for wireless networks can be divide into three categories—(i) *Contention:* The contention based CSMA/CA MAC protocols shows high collision rate at high traffic load. It is not scalable because of the possibility of transmission collisions is high when huge number of IoT devices are trying to grab the same medium all at once [5, 4], (ii) *Reservation:* In general, most of the TDMA based MAC protocols assign time slots in a centralized manner. Due to this, it not suitable for IoT network in terms of the scalability [5]. Similarly, fixed slot size in TDMA MAC protocol does not provide flexibility in the presence of dynamic network conditions (burst traffic which is unpredictable in nature). Further, it is inefficient at low loads due to the low transmission slot usage if only a small portion of devices have data to transmit. Scheduled slots can be fixed or on-demand, and (iii) *Hybrid:* The Hybrid MAC protocols use the usefulness of contention and reservation based MAC protocols have been brought into attention. This features proved to be solve the high collisions of packets and scalability problems in IoT network. The techniques commonly work in two phases-contention period and contention-free period. Again, if the node density is in order of magnitude or more, the hybrid MAC protocols may become bottleneck to prevents the network from achieving high utilization.

3 Scalability Analysis of MAC Protocols

Scalability gives the measure of ability of a protocol to process graceful increasing process loads as the size of the network increases or when the volume of traffic increases. The current MAC protocols performance does not scale well as the network size increases [12].

In the context of IoT communications, a key consideration for MAC protocols is scalability. IoT communications are expected to have a large number of nodes with nodes entering and leaving the network dynamically. Three major issues in provisioning scalability through MAC protocol are:

User population: The population in a network is the most important consideration for an IoT network to provide scalability. With the increasing active things in IoT, the MAC protocols should be able to control contention and collisions over the shared wireless medium to deliver stable performances
Physical-layer capacity: The advance physical layer techniques have greatly improved the link capacity in wireless networks. The maximum data rates provided by 802.15.4 is 255 Kbps. The initial 1 11 Mbps data rates specified in 802.11b standard have been improved to 54 Mbps in 802.11a/g, to 100 Mbps in 802.11n and up to Gbps in 802.11ac. The 802.11ah standard operating in 900 MHz range can provide at least 100 Kbps of data rate. Therefore, MAC layer throughput should be scale up accordingly for better performance.
Protocol overhead: Another important aspect for designing scalable wireless MAC is to minimize the protocol overhead as the network size and the physical layer capacity increases. The high overhead per packet and higher time consumptions due to backoff, beaconing, inter frame spacings, and hand-shakes should remain relatively small.

To analyze the scalability performance of existing types of MAC protocols (discussed in Sect. 2), we have consider contention, reservation and hybrid MAC protocol. Traditional CSMA/CA and TDMA MAC is used for contention and reservation type of MAC protocol respectively. For hybrid MAC protocol, Z-MAC [13] protocol is used. The main feature of Z-MAC is the adaptability in the network such that under low contention, it behaves like CSMA, and under high contention, like TDMA. A simulation sensor network topology with 100 nodes as shown in Fig. 1 is used. Throughput, delay, and packet loss is measured at the sink nodes with increasing number of connections from different stations. Constant Bit Rate (CBR) traffic with 100 byte of packet size is used as data transmitting from stations to the sink node.

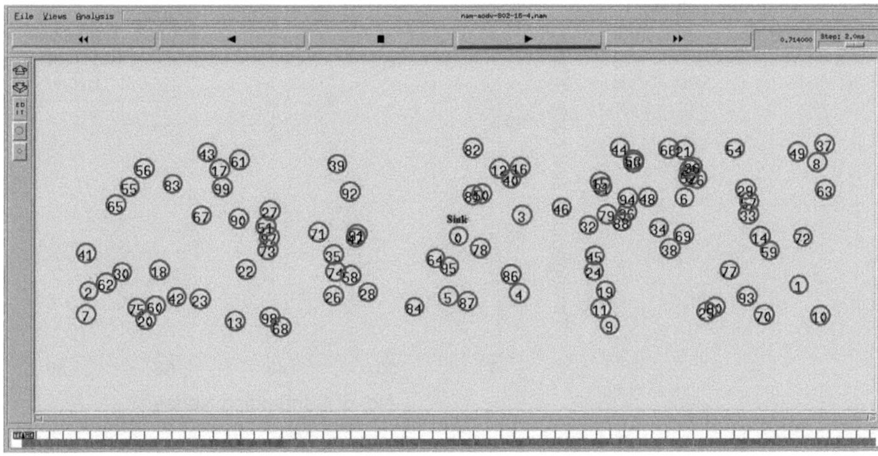

Fig. 1 Simulation topology

It is apparent from Fig. 2 that with the increase in number of sensor nodes, throughput decreases. The increase in number of sensor nodes leads to an increase in the probability of transmitting data at the same time which easily causes collision of sensor nodes transmission. That is the reason why throughput decreases as the number of sensor nodes increases. From Fig. 2, it can observed that some sensor nodes provide good performance whereas the others are not. The performance of hybrid MAC protocol is seen to be better than others as in varying channel conditions, failure of slot assignment and topology changes is common, as a result of which the node falls back from TDMA to CSMA mode. On the other hand, reservation based MAC protocols perform relatively well but a deviation is caused by the failure of time synchronization. Similarly Figs. 3 and 4 show delay and throughput performances of the MAC protocols. Similar type of trends can also be seen with respect to delay and packet loss. Sensor MAC (S-MAC) is one of an example of sampling MAC protocol which uses wake-sleep schedule to provide energy saving and scalability in sensor network. The network topology and traffic pattern is same with above experiments.

The throughput with different numbers of sensor nodes for the S-MAC protocol are given in Fig. 5. It can be observed that some sensor nodes provide good performance. In S-MAC protocol, it is possible that some of the nodes are in sleep mode and hence number of contention is lesser. In contrast, sensor node always chooses a smaller contention window in conventional CSMA/CA, so it transmits data without waiting long time, and hence, it has a low throughput. Figure 6 shows the delay performance. S-MAC achieves lesser delay than CSMA/CA. The access and queue delay is lesser as contenting devices in S-MAC is less (Fig. 7).

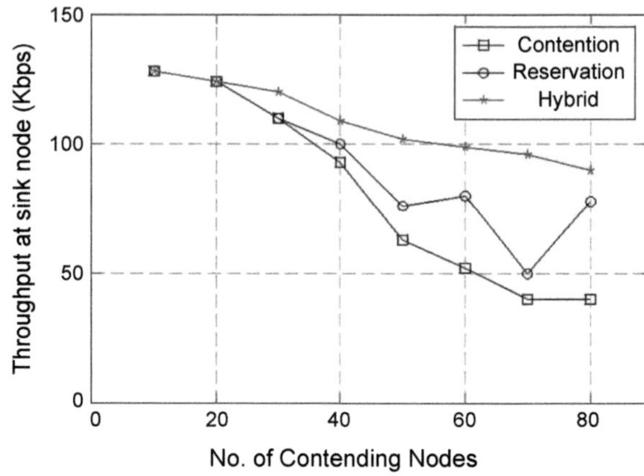

Fig. 2 Throughput achieved by contention, reservation and hybrid MAC protocol

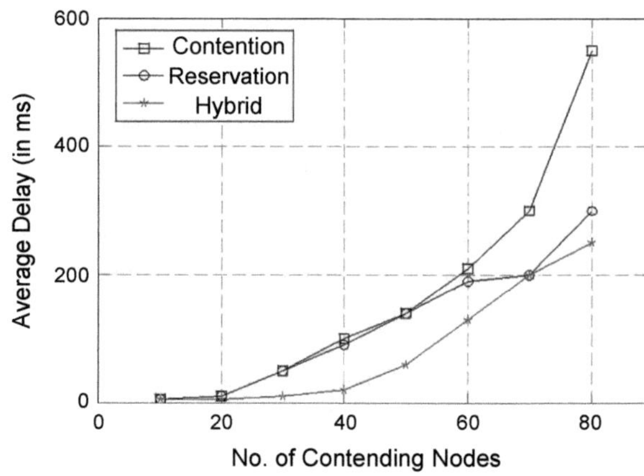

Fig. 3 Delay incurred by contention-based, reservation-based and hybrid MAC protocols

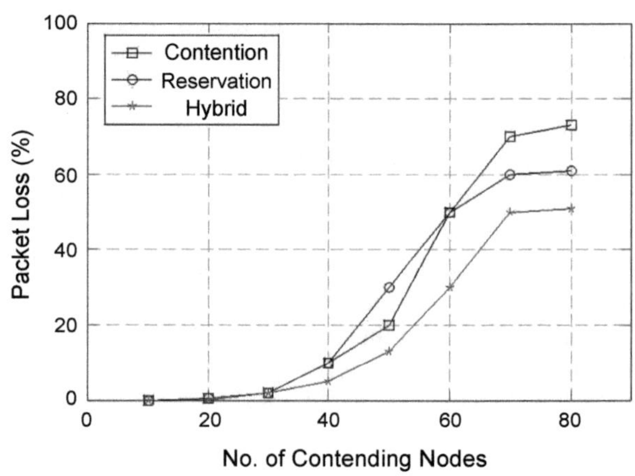

Fig. 4 Packet loss in contention-based, reservation-based and hybrid MAC protocols

Fig. 5 Throughput achieved by conventional CSMA/CA and sensor MAC protocol

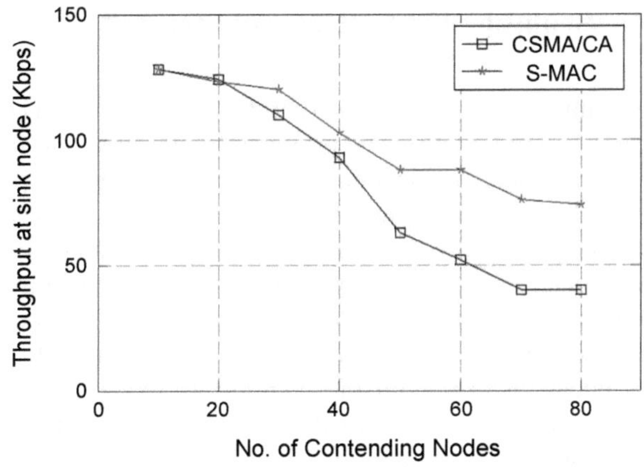

Fig. 6 Delay incur by conventional CSMA/CA and S-MAC protocol

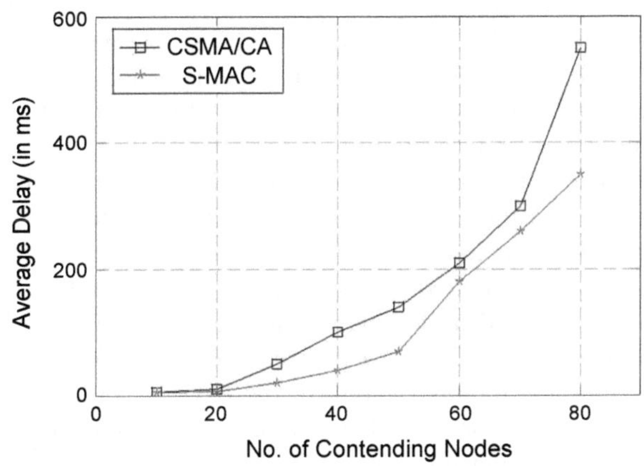

Fig. 7 Packet losses in conventional CSMA/CA and S-MAC protocol

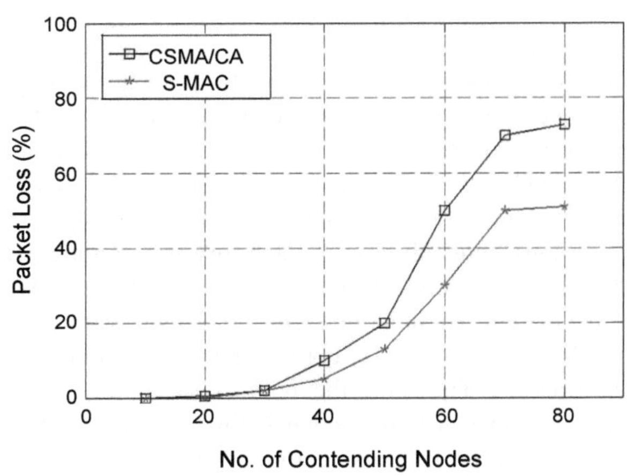

4 Scalable MAC Protocols for IoT

To improve scalability, a hybrid of CSMA and TDMA based MAC protocol is proposed in [14, 15]. The protocol divides time into frames and each frame consists of four fields: notification period (NP), contention period as COP, announcement period as AP, and transmission period as TOP. Base station (BS) announces for contention time to all node at the start of a frame i.e., NP. During the COP, nodes with data use p-persistent CSMA to send transmission requests to the BS. The nodes who contented successfully are allocated slots to transmit data in the TOP and the nodes are informed of their slots during the AP. The protocol in [14] is extended in [16] with the addition of QoS provisioning and fairness. The new protocol allows nodes to decide their contention probabilities according to their QoS priority. However, in [14, 16], due to the time required for COP and the need for contention, there is a tradeoff between the performance of CSMA and TDMA as well as incurs additional delay and energy consumption. Drx-MAC [17] protocol introduces MAC protocol for the latest low power chipsets. Drx-MAC is based on hybrid MAC communication scheme with automated slot allocation based on device identities. This protocol use ALOHA for channel sensing and TDMA slots are assigned accordingly.

The IEEE 802.11ah [18] based MAC protocol is known to be capable of IoT communication [19]. To facilitate the transmission of IoT traffic, IEEE 802.11ah uses beacons to divide time into frames and each frame is further divided into two sections: Restricted Access Window (RAW) and load traffic. Each RAW is divided into slots and a slot may either be allocated to a device by the AP or may be randomly selected by a device. The authors of [20] address the problem of estimating the required length of the RAW in order to facilitate the efficient channel access by the devices. This estimation is obtained by the AP using the probability of successful transmissions in the last frame.

Many IoT networks rely on preamble-sampling MAC protocols [21]. This kind of protocols require no prior knowledge, no negotiation and are thus scalable [22]. In such protocols, each node periodically wakes its radio up to sample the channel for incoming packets. An IoT compatible MAC protocol proposed in [22] with the aim of saving energy for nodes use heterogeneous MAC duty-cycle configurations among nodes in the network. The MAC protocol is based on Destination Oriented Directed Acyclic Graph (DODAG) [23], centered at the sink. To obtain an estimation of the expected traffic, each leaf must send a message up to the sink, containing the number of its sons (thus initially 0). Each forwarding node will then maintain a value corresponding to the number of nodes above in the DODAG construction and forward this value to their parent in the DODAG. This way, through cooperation between routing and MAC layers, a node come to know the number of nodes it will have to forward information. The more the number of nodes in the DODAG topology, shorter the low-power-listening value. In this way, each node selects a configuration that suits its needs. Table 1 shows a comparative analysis of the discussed MAC protocols with respect to provisioning scalability in IoT networks.

Table 1 Comparison of scalable MAC protocols for IoT

Protocol name	MAC protocols	Architecture	Scalability	Traffic requirements	Types of network	Energy saving
Hybrid [14, 15]	CSMA/CA, TDMA	Centralized	Moderate	No	Sensor/Backhaul	Moderate
Hybrid-QoS [16]	CSMA/CA, TDMA	Centralized	Moderate	Yes	Sensor/Backhaul	Moderate
Enhanced 802.11ah [20]	CSMA/CA, TDMA	Distributed	High	Yes	Sensor/Backhaul	Moderate
Drx-MAC [17]	ALOHA, TDMA	Centralized	Moderate	No	Sensor	High
Hetero-MAC [22]	CSMA/CA	Centralized	Moderate	Yes	Sensor	High

Although, a large amount of current research is focusing on hybrid MAC protocol design, but in many cases the nodes densities is very high and there are high collision at the time of slot assignment. Minimizing delay and energy consumption while switching from one MAC to another and finding an optimal solution of tradeoff between two MAC protocols is a challenging problem. In 802.11ah-based hierarchical network, there is a potential of poor quality of data reception due to long communication range, higher delay due to multi-hop transmission.

5 Conclusion

Considering the growing popularity and its speedy expansion in recent times, it can be expected that IoT will emerge more and more in the near future. Different IoT applications have their own requirements for smooth running. A communication network with efficient MAC protocol holds the major responsibility to support these requirements. In this paper, we have analyzed the major types of MAC protocols in supporting scalability. The discussed hybrid MAC protocols are found to be more scalable than conventional ones. Preamble-sampling MAC protocols are scalable for IoT network because of its simplicity and low energy consumption. Although, the current MAC schemes make the IoT concept feasible but do not fit well with the energy efficiency, scalability, and requirements of various envisaged applications.

Acknowledgment This work has been supported by the project titled “QoS Provisioning in Internet of Things (IoT)” (Ref No. 13 (7)/2015-CC&BT dated: 28/09/2015) funded by DeitY (CC & BT), Govt. of India.

References

1. The Internet of Things - Concept and Problem Statement(draft). https://tools.ietf.org/html/draft-lee-iot-problem-statement-00.
2. Want, R., Schilit, B.N., Jenson, S.: Enabling the internet of things. Computer (1), 28–35 (2015).
3. Dandelski, C., Wenning, B.L., Perez, D., Pesch, D., Linnartz, J.P.: Scalability of dense Wireless Lighting control Networks. Communications Magazine, IEEE 53(1), 157–165 (Jan 2015).
4. Lin, T., Rivano, H., Mou•el, F.L.: How to choose the relevant MAC protocol for wireless smart parking urban networks? In: Proceedings of the 11th ACM symposium on Performance evaluation of wireless ad hoc sensor, & ubiquitous networks (PE-WASUN). ACM (2014).
5. On, J., Jeon, H., Lee, J.: A Scalable MAC Protocol Supporting Simple Multimedia Traffic QoS in WSNs. International Journal of Distributed Sensor Networks 2011, 1–11 (2011).
6. Qian, Z., Wang, Y., Wang, X., Zhu, S.: M/I Adaptation Layer Network Protocol for IoT Based on 6LoWPAN. In: Internet of Things, pp. 208–215. Springer Science+Business Media (2012).
7. Rajandekar, A., Sikdar, B.: A Survey of MAC Layer Issues and Protocols for Machine-to-Machine Communications. IEEE Internet of Things Journal 2(2), 175– 186 (2015).
8. Miorandi, D., Sicari, S., De Pellegrini, F., Chlamtac, I.: Internet of things: Vision, applications and research challenges. Ad Hoc Networks 10(7), 1497–1516 (2012).
9. Akyildiz, I., Pierobon, M., Balasubramaniam, S., Koucheryavy, Y.: The Internet of Bio-Nano Things. Communications Magazine, IEEE 53(3), 32–40 (March 2015).
10. Aust, S., Prasad, R.V., Niemegeers, I.G.M.M.: IEEE 802.11ah: Advantages in standards and further challenges for sub 1 GHz Wi-Fi. In: 2012 IEEE International Conference on Communications (ICC). IEEE (jun 2012).
11. Zhou, Y., Wang, H., Zheng, S., Lei, Z.Z.: Advances in IEEE 802.11ah standardization for machine-type communications in sub-1 GHz WLAN. In: 2013 IEEE International Conference on Communications Workshops (ICC). IEEE (jun 2013).
12. Yuan, Y., Arbaugh, W.A., Lu, S.: Towards Scalable MAC Design for High-Speed Wireless LANs. EURASIP J Wirel Commun Netw 2007(1), 012597 (2007).
13. Rhee, I., Warrier, A., Aia, M., Min, J., Sichitiu, M.: Z-MAC: A Hybrid MAC for Wireless Sensor Networks. IEEE/ACM Transactions on Networking 16(3), 511–524 (jun 2008).
14. Liu, Y., Yuen, C., Chen, J., Cao, X.: A scalable Hybrid MAC protocol for massive M2M networks. In: 2013 IEEE Wireless Communications and Networking Conference (WCNC). IEEE (apr 2013).
15. Verma, P.K., Tripathi, R., Naik, K.: A Robust Hybrid-MAC Protocol for M2 M Communications. In: Computer and Communication Technology (ICCCT), International Conference on. pp. 267–271. IEEE (2014).
16. Liu, Y., Yuen, C., Cao, X., Hassan, N.U., Chen, J.: Design of a Scalable Hybrid MAC Protocol for Heterogeneous M2M Networks. IEEE Internet of Things Journal 1(1), 99–111 (feb 2014).
17. Bergamini, L., Corbellini, G., Mangold, S.: Resource-constrained Medium Access Control protocol for Wearable Devices. In: Wireless and Mobile Computing, Net-working and Communications (WiMob), IEEE 10th International Conference on. pp. 634–641 (Oct 2014).
18. Sun, W., Choi, M., Choi, S.: IEEE 802.11 ah: A long range 802.11 WLAN at sub 1 GHz. Journal of ICT Standardization 1(1), 83–108 (2013).
19. Adame, T., Bel, A., Bellalta, B., Barcelo, J., Oliver, M.: IEEE 802.11AH: the WiFi approach for M2 M Communications. Wireless Communications, IEEE 21(6), 144–152 (December 2014).
20. Park, C.W., Hwang, D., Lee, T.J.: Enhancement of IEEE 802.11ah MAC for M2M Communications. IEEE Communications Letters 18(7), 1151–1154 (jul 2014).

21. Cano, C., Bellalta, B., Sfairopoulou, A., Oliver, M.: Low energy operation in WSNs: A survey of preamble sampling MAC protocols. Computer Networks 55(15), 3351–3363 (oct 2011).
22. Beaudaux, J., Gallais, A., Noel, T.: Heterogeneous MAC duty-cycling for energy-efficient Internet of Things deployments. Networking Science 3(1–4), 54–62 (apr 2013).
23. Sha er, S., Vasseur, J.P., Shetty, S.J.: Dynamic reroute scheduling in a directed acyclic graph (DAG) (Jan 20 2015), uS Patent 8,937,886.

A Review on Comparison of Workflow Scheduling Algorithms with Scientific Workflows

Aditi Jain and Raj Kumari

Abstract Cloud computing is a technology that uses web and the clients can access the information by means of web programs. Rather than storing data on the desktop, it is stored on the cloud. Cloud as the name alludes is an abstraction of some complex infrastructure. Scheduling of tasks with minimum usage of resources and achieving maximum profit is an important concern in cloud computing. Load balancing is an important mechanism taken into account to handle the load on various dependent nodes in distributed environment. Due to large number of tasks in distributed environment, workflows are used for scheduling the tasks. In this paper, Scientific workflows are used to carry out the simulation on task scheduling algorithms with cloud resources. This review paper compares various scheduling algorithms on the basis of parameters like execution time and total cost that includes communication cost for input and output the data and computation cost. From simulation results, it is concluded that all algorithms shows different results depending upon the specific workflow due to varying size of their tasks.

Keywords Cloud computing · Workflows · Scientific workflows · Scheduling algorithms

1 Introduction

Cloud computing is an architecture that stores extensive collections of data and applications in one place i.e. on cloud and that can be accessed by any authorized clients from their private framework. Cloud computing is paradigm that allows users on demand network access to a large shared pool of resources on charge-per-utilization premise. Cloud computing makes use of computing resources

A. Jain (✉) · R. Kumari
Department of IT, UIET, Panjab University, Chandigarh, India
e-mail: aditijain6@yahoo.in

R. Kumari
e-mail: rajkumari@pu.ac.in

N. Modi et al. (eds.), *Proceedings of International Conference on Communication and Networks*, Advances in Intelligent Systems and Computing 508,
DOI 10.1007/978-981-10-2750-5_63

i.e. software and hardware, that are delivered on the network as a service to the client. Three sorts of services are provided by cloud: IaaS (Infrastructure as a Service), PaaS (Platform as a Service), and SaaS (Software as a Service).

Cloud computing is evolving as a model of a large distributed environment. To utilize the power of computing resources effectively, there are numerous task scheduling algorithms. Workflow is represented by a Directed Acyclic Graph (DAG) in which each computational task is represented by a node, and control dependency between those tasks is represented by a coordinated edge [1]. Scientific Workflows like Montage, Cyber Shake, Sipht etc. may composed of a large number of tasks and their execution may involve complex software. So because of complex software configuration, heavy processing of data; it requires larger execution time. Simulation based studies are widely accepted to evaluate their efficiency in workflow systems. Efficiency of Workflow based task scheduling is one of the issues in workflow execution. Scheduling is the mechanism of allocating specific resources to the tasks to carry out effective execution. The objective of this review paper is to compare the execution time and cost of some workflow scheduling algorithms in the WorkflowSim.

1.1 Workflows

A workflow application is represented as Directed acyclic graph (DAG) where each node represent a task and edges between tasks represents interdependencies between tasks. Workflow scheduling is an important issue in distributed systems.

Figure 1 demonstrates that the child nodes 1, 2, 3 are executed after execution of parent node 0. Task 0 is called entry node and task 7 is called as exit node. Entry node is the node having no parent node and exit node is the node having no child node i.e. the predecessor nodes will execute before execution of their successor nodes. In distributed systems, there are large number of tasks, all tasks work together to minimize the execution time. To find execution time of large number of tasks is a difficult job. So in distributed environment, workflow based task scheduling algorithms are used.

1.2 Scientific Workflows

Workflows are responsible for scheduling computational tasks on distributed environment and for managing dependencies among tasks [2]. A scientific workflow system is a specific type of a workflow management system that outlines the execution of series of computational tasks, in a scientific application [3]. These are extremely important for comparison of workflow based distributed systems.

Some of the scientific workflows are [4]:

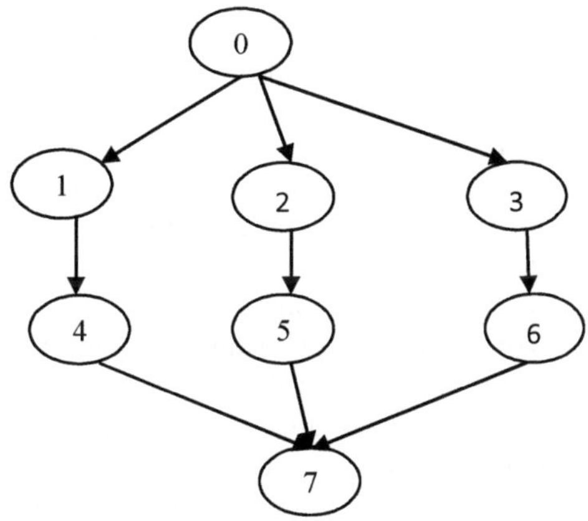

Fig. 1 DAG representation in workflow

1. **Montage**: The Montage application, created by NASA/IPAC, consolidates together numerous input images to make custom mosaics of the sky.
2. **Cybershake**: The CyberShake workflow is used to describe seismic hazards by the Southern California Earthquake Center.
3. **SIPHT**: The SIPHT workflow, from the bioinformatics project at Harvard, is used to computerize the search for untranslated RNAs (sRNAs) for bacterial replicons in the NCBI database.

2 Workflow Scheduling

Scheduling is the concept of allocating resource to a task so that overall completion time is minimized. To carry out the effective scheduling, various workflow scheduling algorithms are used. Load balancing is an important issue in workflows. As in distributed systems, some nodes are heavily loaded and some are underutilized. The target of load balancing is to disseminate the load among given nodes which is carried out by workflow scheduling. Two sorts of scheduling in workflow: Static scheduling and Dynamic scheduling. Static task scheduling pre-fetches the required data i.e. during compile time and pipelines tasks in the levels of their execution. It does not consider state of the current node. Hence, there is less overhead for computing the execution time and cost. Once a schedule is determined, tasks can be executed following that order [5]. Whereas Dynamic scheduling does not consider the past state of the system and no prior knowledge is needed, i.e., it relies only upon the current state of the system [6]. The allocation of tasks occurs during run time and thus, large overhead time is imposed.

There are many workflow scheduling algorithms. In this paper four of them are discussed:

1. **FCSS**: The simplest algorithm where resources are assigned to tasks on the basis of their submission time. The task having prior submission time is chosen first for its execution.
2. **Round-Robin**: This algorithm takes into account the concept of time quantum. Where time is partitioned into quanta's or slices and each task is assigned with a time slice and that task is allotted resources for that time slice and task performs its operations. After the first time slice gets expired, resources will be allotted to next task and so on.
3. **Min-Min**: This algorithm sorts the tasks in the increasing order of their execution times. Then the task with overall minimum completion time is scheduled to the corresponding resource. The scheduled task is removed from the set of unmapped tasks and the process is repeated until the unmapped task set gets void. Here, assigning smallest task first is the drawback. As small task will execute first then the larger ones and they have to wait until smaller ones get finished. So this algorithm is more efficient where the tasks are of smaller size.
4. **Max-Min**: This algorithm is same as that of min-min with the exception that here tasks are sorted in the increasing of their execution times but the task with overall maximum execution time (larger task) is scheduled first to the resource with minimum completion time (slowest resource) from the list of unmapped tasks. And after complete execution, it is removed from set of unmapped tasks. The algorithm is repeated until the list of unmapped tasks gets void. So, this algorithm does better for tasks with large sizes

All these scheduling algorithms comes under static scheduling.

3 Related Work

Rodrigo et al. [7] introduce cloudsim, a simulation toolkit, that does the simulation of cloud environment. With this, required number of virtual machines, datacenters, and cloudlets are created. But it deals with execution of only single flow of workloads.

Chen and Ewa [8] introduce workflowsim, a simulation toolkit, that does the simulation of cloud environment with multiple workloads in a distributed environment.

Magnan et al. [9] carried out the concept of mapping and scheduling of linear workflows in the cloud environment. In this, two model are used: one-port mode (without overlap) and multi-port model (with overlap). But the author arise the need of scheduling workflows with preemption.

Simsy Xavier, S.P. jeno Lovesum [10] compared various workflow scheduling algorithms on various factors. The author demands to implement the new scheduling algorithm that can further minimize the execution time.

Swachil patel, Upendra Boi [11] explained various priority based job scheduling algorithms in cloud computing on various parameters. With this there is a need to improve parameters such as arriving rate of jobs etc.

Chen et al. [11] introduces particle swarm optimization(PSO) based approach that is applied on the resource limited projects. In this, permutation based and priority based representations were described.

4 Methodology

Different scientific workflows are used to carry out the simulation: Montage with 25 tasks, Cybershake with 30 tasks, and Sipht with 30 tasks. Then all the resources like datacenter, host, virtual machines are allotted. After this, algorithms (FCSS, Max-min, Min-min, round robin) are applied to these workflows and they are compared with respect to execution time and cost to find the results.

Where execution time is the total time to execute tasks depth-wise in a workflow. First, task at the first level(depth) will be executed then the tasks at the next level and so on. Tasks that come under same depth are dependent on each other and they are scheduled according to the selected algorithm criteria.

Cost parameter includes communication cost and computation cost. i.e. the time to input and output task data on resource on virtual machine and time to perform computational task.

Cost = Communication cost + Computation cost = (data (both input and output) * unit cost of data + runtime * cpu cost) for each task.

5 Analysis

Simulation is performed on static workflow scheduling algorithms like FCSS, Max-Min, Min-Min, Round-Robin with the help of scientific workflows like Montage, Cybershake, Sipht and their execution time and cost are compared (Table 1).

After simulation, larger number of tasks are executed at depth (level) 1, 2 and 5 while at all other depths only one task is executed. So, it takes larger execution times to execute tasks at depth 1, 2 and 5 (Figs. 2 and 3, Tables 2 and 3).

In Sipht workflow for 30 tasks, maximum number of tasks are executed at depth 1 with comparatively larger execution time than that of Montage workflow. So, it takes larger execution time at depth 1 that at other depths. In this, FCSS shows very large execution time. Max. Min and round robin algorithms have approximately same execution times (Figs. 4 and 5, Tables 4 and 5).

In cybershake workflow, maximum number of tasks are executed at depth 2 so it takes larger execution times to execute. Here, min-min algorithm shows larger execution times but Round robin shows better results (Figs. 6 and 7, Table 6).

Table 1 Execution time calculation for 25 tasks of Montage workflow

Depth	FCSS	Max-Min	Min-Min	Round Robin
0	0.19	0.19	0.19	0.19
1	159.61	116.57	73.07	26.14
2	56.34	123.54	96.27	31.61
3	1.24	1.23	1.23	1.24
4	2.45	2.43	2.43	2.44
5	124.34	91	57.95	20.36
6	3.48	3.63	3.24	2.84
7	5.23	5.18	5.19	5.21
8	6.66	6.6	6.61	6.63
9	0.78	0.77	0.77	0.77

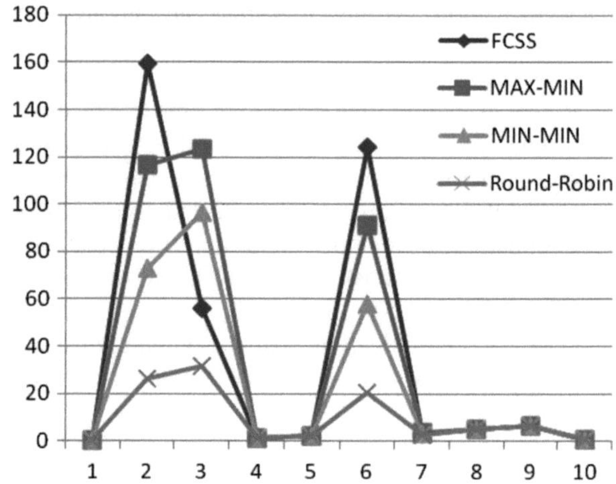

Fig. 2 Graphical representation of computation time for 25 tasks of Montage workflow

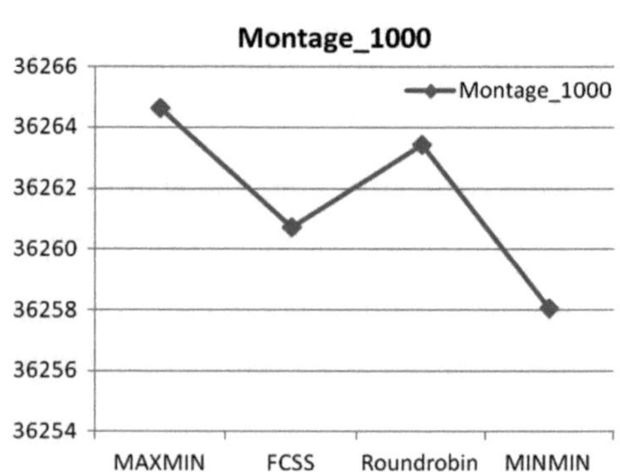

Fig. 3 Graphical representation of total cost for 1000 tasks of Montage

Table 2 Total cost for Montage_1000

Algorithm	Total cost
Max-Min	36264.65
FCSS	36260.73
Round Robin	36263.45
Min-Min	36258.04

Table 3 Computation time calculation for 30 tasks of Sipht workflow

Depth	FCSS	Max-Min	Min-Min	Round Robin
0	0.13	0.13	0.13	0.13
1	6724.23	2848.12	3649.9	2925.83
2	576.11	572	573.34	587
3	2751	1705.37	8.91	1725.77
4	38.64	42.61	38.42	39.39
5	2.61	2.56	2.64	2.61

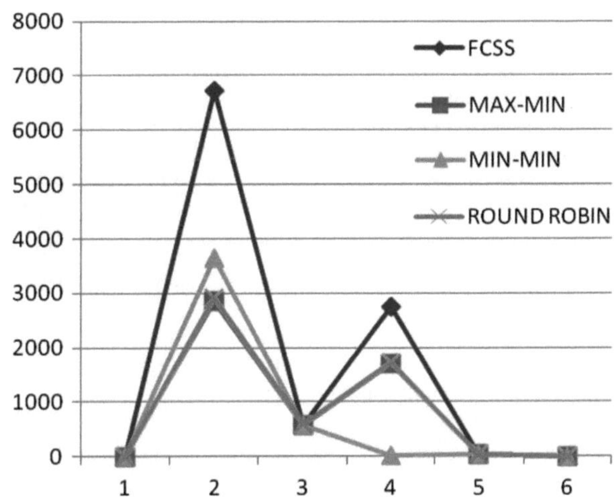

Fig. 4 Graphical representation of computation time for 30 tasks of Sipht workflow

The execution time for some of static workflow scheduling algorithms are compared for time and cost parameter in WorkflowSim. Max-Min algorithm in cybershake workflow yields better results over other workflows however not so good as compared to round robin. FCSS shows large execution times and Min-Min shows average results in Montage and Sipht workflows. But in case of Cybershake workflow, it shows negative result i.e. very large execution time. Cost factor adds up communication cost and computational cost. Cost parameter for max-min in Sipht is better than other workflows whereas for the min-min algorithm shows very high cost. FCSS and round robin shows average cost in all workflows. The reason behind variation in results is due to different sizes of tasks among these workflows.

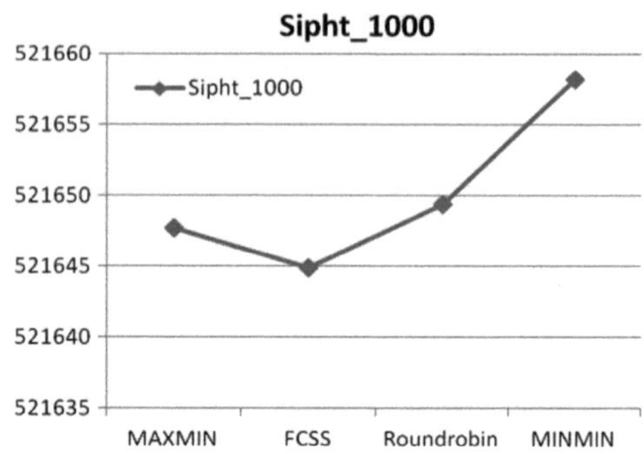

Fig. 5 Graphical representation of total cost for 1000 tasks of Sipht

Table 4 Total cost for Sipht_1000

Algorithm	Total cost
Max-Min	521647.7
FCSS	521644.9
Round Robin	521649.4
Min-Min	521658.18

Table 5 Computation time calculation for 30 tasks of Cybershake workflow

Depth	FCSS	Max-Min	Min-Min	Round Robin
0	0.13	0.13	0.13	0.13
1	230.71	226.5	262.06	262.42
2	415.44	287.17	1523.8	187.82
3	13.91	19.53	19.88	19.53
4	0.08	0.08	0.08	0.08

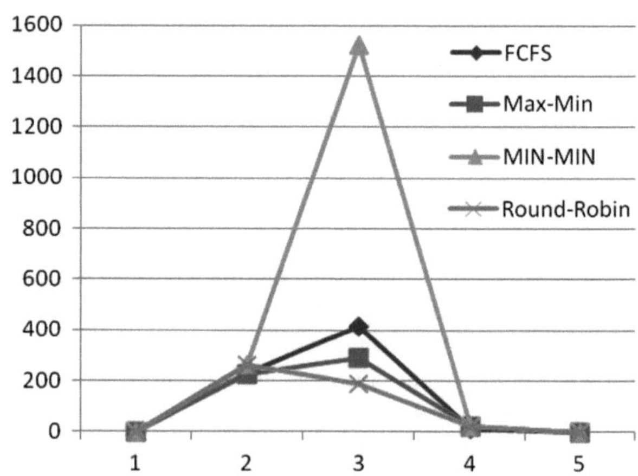

Fig. 6 Graphical representation of computation time for 25 nodes of Cybershake workflow

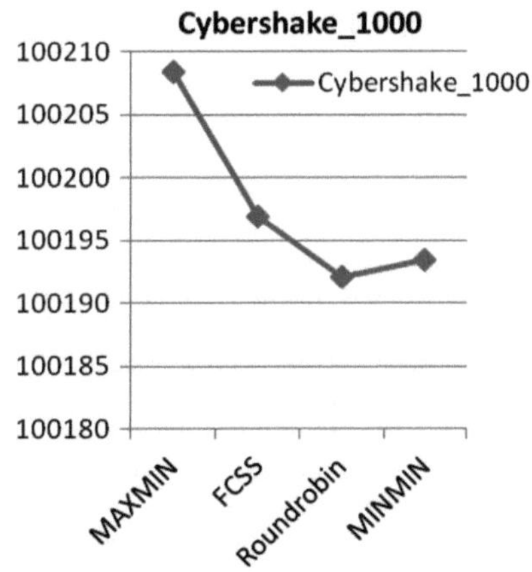

Fig. 7 Graphical representation of cost for 1000 tasks of Cybershake

Table 6 Total cost for Cybershake_1000

Algorithm	Total cost
Max-Min	100208.4
FCSS	100196.4
Round Robin	100192.1
Min-Min	100193.1

6 Conclusion

Scheduling is the most important task in cloud computing area. Scheduling focuses on utilizing the available resources efficiently and minimizes the execution time and total cost of scheduling. Workflow scheduling schedules large number of tasks in distributed environment. This review paper compared the different workflow scheduling algorithms based on the metrics of time and cost parameters with scientific workflows. Results shows that algorithms shows different time and cost values for different workflows depending upon the task size. As from the experimental results it is shown that FCSS algorithm shows negative results i.e. very large execution results in Montage and Sipht workflow while in cybershake workflow, FCSS has comparable less execution time but Round robin shows better results than others. Max-Min and Min-Min shows average results in both Montage and Sipht workflows. Whereas in cybershake workflow, it shows very large execution time. For cost, Min-Min algorithm shows least cost for Montage and cybershake workflows.

References

1. Bala, Anju, and Inderveer Chana. "A survey of various workflow scheduling algorithms in cloud environment." In 2nd National Conference on Information and Communication Technology (NCICT), pp. 26–30. 2011.
2. Xavier, Simsy, and SP Jeno Lovesum. "A Survey of Various Workflow Scheduling Algorithms in Cloud Environment".
3. Juve, Gideon, Ann Chervenak, Ewa Deelman, Shishir Bharathi, Gaurang Mehta, and Karan Vahi. "Characterizing and profiling scientific workflows."Future Generation Computer Systems 29, no. 3 (2013): 682–692.
4. https://confluence.pegasus.isi.edu/display/pegasus/WorkflowGenerator.
5. Mei, Jing, Kenli Li, and Keqin Li. "A resource-aware scheduling algorithm with reduced task duplication on heterogeneous computing systems." The Journal of Supercomputing 68, no. 3 (2014): 1347–1377.
6. Shoja, Hamid, Hossein Nahid, and Reza Azizi. "A comparative survey on load balancing algorithms in cloud computing." In Computing, Communication and Networking Technologies (ICCCNT), 2014 International Conference on, pp. 1–5. IEEE, 2014.
7. Calheiros, Rodrigo N., Rajiv Ranjan, Anton Beloglazov, César AF De Rose, and Rajkumar Buyya. "CloudSim: a toolkit for modeling and simulation of cloud computing environments and evaluation of resource provisioning algorithms."Softw. Pract. Exper 41 (2011): 23–50.
8. Chen, Weiwei, and Ewa Deelman. "Workflowsim: A toolkit for simulating scientific workflows in distributed environments." In E-Science (e-Science), 2012 IEEE 8th International Conference on, pp. 1–8. IEEE, 2012.
9. Agrawal, Kunal, Anne Benoit, Loic Magnan, and Yves Robert. "Scheduling algorithms for linear workflow optimization." In Parallel & Distributed Processing (IPDPS), 2010 IEEE International Symposium on, pp. 1–12. IEEE, 2010.
10. Bala, Anju, and Inderveer Chana. "A survey of various workflow scheduling algorithms in cloud environment." In 2nd National Conference on Information and Communication Technology (NCICT), pp. 26–30. 2011.
11. Chen, Ruey-Maw, Chung-Lun Wu, Chuin-Mu Wang, and Shih-Tang Lo. "Using novel particle swarm optimization scheme to solve resource-constrained scheduling problem in PSPLIB." Expert systems with applications 37, no. 3 (2010): 1899–1910.

Predictive Approach of CBR in Artificial Intelligence: A Case of Astrological Predictions About the Status of Person

Neelam Chaplot, Praveen Dhyani and O.P. Rishi

Abstract Astrological Prediction possesses the capability to create a great deal of interest in persons to know about their future. Most of the persons around the world believe in astrology hence it is necessary to establish validity of astrology. In astrology, prediction about a person's life is done based on the position of the stars in the horoscope at the time of the birth of the individual but these rules are different in different astrological prediction systems and hence lack a central design. This paper talks about various experiments performed for the purpose of identifying the profession of the person. The methodology used for this purpose is case base reasoning technique along with various classification methods such as Simple Cart, Decision Stump and Decision Table. Information of persons was collected to perform the research. Data for persons of profession Singer, Player and Doctors were collected. The information collected for this purpose were date of birth, place of birth and time of birth along with major events took place in person's life. The results generated by the research were satisfying and motivating.

Keywords Case base reasoning · Case base reasoning for astrological prediction · Classification · Astrological prediction system · Prediction of profession

N. Chaplot (✉) · P. Dhyani
Banasthali University, Banasthali, Rajasthan, India
e-mail: neelam.chaplot@gmail.com

P. Dhyani
e-mail: dhyani_p@yahoo.com

O.P. Rishi
University of Kota, Kota, Rajasthan, India
e-mail: omprakashrishi@yahoo.com

N. Modi et al. (eds.), *Proceedings of International Conference on Communication and Networks*, Advances in Intelligent Systems and Computing 508,
DOI 10.1007/978-981-10-2750-5_64

1 Introduction

Astrology means predicting the consequences of planets and heavenly bodies on the person. Astrology can be described as the study of planets and stars and their roles and impact on the life of a person. In astrology planets are accountable for prosperity and agony of any individual. Astrology has rules described to estimate effect of different planets on the life of a person. Every person has his/her personal astrological horoscope that depends on the position of planets and stars at the birth time. An astrological horoscope for a particular person is the chart that correlates the planets, zodiac and their positions at a particular time and place in 12 astrological houses. Using these astrological charts, personal forecasts can be done with precision. The prediction about upcoming events in someone's life is done in accordance with the relative positions of the planets. Different rules are defined by different astrologers for prediction of different aspects of person's life and events, hence there is no uniform method for forecast of person's life and its events is available in astrology.

Astrological forecasts can't be relied upon in some of the circumstances. However these forecasts make you familiar with future circumstances thus helping you to handle negative circumstances in a better way.

Astrological analysis can be done with a perspective to find academic prospective and career. Different planetary motions have a huge influence in deciding the profession of the person. Different distances between planets and their positions in astrological chart are significant in identifying different professions relevant to a person in particular. These planets have mutual relationship with each other and also have their own pattern. One can find time tested combinations which help us identify professional propensities of an individual through evaluation of astrological chart. The astrological combination makes an effort of identification, unification and incorporation of the potential of a person.

In existing paper case based reasoning method [1] is used along with various classification methods such as Simple Cart [2], Decision Table [3] and Decision Stump [4] for astrological prediction and analysis of career of a person. The case based reasoning strategy is based on the idea that the identical problems can be solved using the identical solution [5] hence the system understands from the mapping of cases stored previously in the case base storage of various persons to the new case to be predicted. The case base storage contains the birth data, birth horoscope and the positions of planets and zodiacs in the houses. For the purpose of prediction with regards to profession, one could possibly recycle the past cases according to new circumstances by simply reusing earlier cases in light of new conditions based on the similarity index. Some initial work for providing base and validity of astrology is done in paper [6] in the paper various Artificial Intelligent techniques were applied to perform prediction of astrological data.

2 Architecture of Astrological Prediction System

General architecture of Astrological Prediction system is provided as shown in Fig. 1. The person for whom prediction is to be done is known as the new case. The details of the planetary positions of the various planets at time of birth based on Indian astrology for new case are generated and inputted to the system through Input Interface. The data through Input Interface is converted into a form compatible to the data stored in the case based storage through Data Conversion module. The case based storage has collection of huge amount of previously already classified records. To provide prediction for the new case, data from case based storage is retrieved using Retrieval Module. The learning or the rules generated by Learning Module are used by Inference Engine on retrieved data to perform prediction for the new case. Learning Module keeps propagating learning, based on the cases stored and updated in the Case Based Storage. Various statistical, rule based and other methods are used to perform learning. Few of the methods used in this research are simple Cart, Decision stump, Decision Table and so on. The Prediction done by Inference Engine is provided to the user in the user understandable form through Output Interface. The cases can be collected and inputted into Case Based Storage through Case Base Interface which in turn uses Data Conversion Module to store data. The Retaining of the data in the case based storage is done based on the testing of the data through Test Module. Based on the prediction done by system the correctness of the solution is

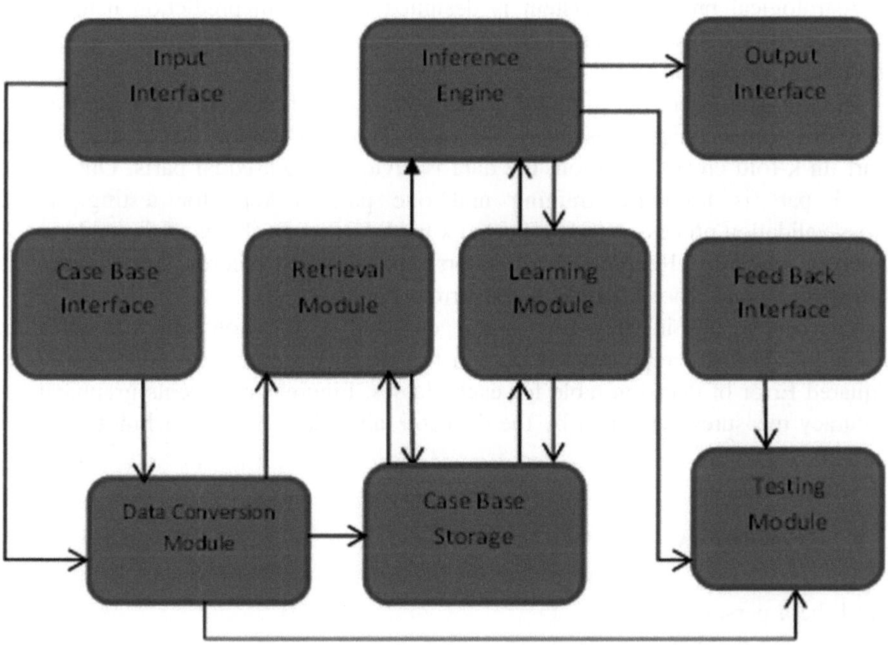

Fig. 1 Architecture of astrological prediction system using case base reasoning

identified. The information of correct/incorrect prediction is submitted by the user through feedback module. If testing is correct then information is stored in Case Base Storage after proper conversion.

3 Experiment and Results

Data of persons with professions such as Player, Singer and Doctor was collected. The professionals in these fields generally don't have overlapping career and belong to one profession only. The records are collected in two ways, firstly few records were collected personally from reliable sources through emails and by noting down the information on paper and pen. Secondly few records were collected from Astro Data bank. It is a non-profit community project and is a subdivision of Astrodienst AG, with no commercial interest. The data that were collected are date of birth, place of birth and time of birth along with details such as profession of the person and brief biography of the person. This information was then used to create the birth chart of the person. The birth chart was prepared based on Indian Astrology. Total 23 attributes Aries, Taurus, Gemini, Cancer, Leo, Virgo, Libra, Scorpio, Sagittarius, Capricorn, Aquarius, Pisces, Sun, Moon, Mars, Mercury, Venus, Jupiter, Saturn, Rahu, Ketu, Gender, and Class were selected and all of them were nominal data. The data was converted into .arrf format and then WEKA (Waikato Environment for Knowledge Analysis) [7] tool was used for performing the prediction.

Astrological prediction system is designed to perform prediction using Case Base Reasoning System. Various Supervised learning methods are used to perform the task of prediction of the data.

Results generated by different experiments are shown in Table 1. Various Classification techniques used are Decision Stump, Decision Table and Simple Cart. In k-fold cross validation, the data is divided into k equal parts. Out of that $k - 1$ part is used for training and one part is kept for testing. These cross-validation process is then repeated k times. Other attributes of the table are % correctly classified Records, % of incorrectly classified records, Mean Absolute Error(MAE) and Root Mean Squared Error (RMSE).

The best results in each class were generated by decision table classification technique. Figure 2 represents the graph for Mean Absolute error and Root Mean Squared Error of decision table for each classes. Figure 3 represents graphical the accuracy measures generated by the decision table classification technique.

3.1 Hypothesis

Let P be a person
Let n be the number of persons
So P1 to Pn are the persons whose actual status (output) is known

Table 1 Results of various classification techniques

Classification techniques	Class	K-Fold	MAE	RMSE	%Correctly classified	%Incorrectly classified
Decision Stump	Player	12	0.54	0.56	41.67	58.33
Decision Stump	Player	14	0.55	0.57	39.58	60.42
Decision Stump	Player	16	0.58	0.60	33.33	66.67
Decision Table	Player	12	0.48	0.5	62.50	37.50
Decision Table	Player	14	0.48	0.51	58.33	41.67
Decision Table	Player	16	0.46	0.48	66.67	33.33
Simple Cart	Player	12	0.54	0.63	43.75	56.25
Simple Cart	Player	14	0.51	0.60	43.75	56.25
Simple Cart	Player	16	0.55	0.65	37.50	62.50
Decision Stump	Singer	12	0.57	0.59	37.50	62.50
Decision Stump	Singer	14	0.53	0.54	45.83	54.17
Decision Stump	Singer	16	0.56	0.58	37.50	62.50
Decision Table	Singer	12	0.49	0.51	56.25	43.75
Decision Table	Singer	14	0.5	0.53	56.25	43.75
Decision Table	Singer	16	0.54	0.56	45.83	54.17
Simple Cart	Singer	12	0.57	0.65	39.58	60.42
Simple Cart	Singer	14	0.6	0.68	33.33	66.67
Simple Cart	Singer	16	0.57	0.68	37.50	62.50
Decision Stump	Doctor	12	0.54	0.57	43.75	56.25
Decision Stump	Doctor	14	0.53	0.56	45.83	54.17
Decision Stump	Doctor	16	0.54	0.56	43.75	56.25
Decision Table	Doctor	12	0.49	0.54	52.08	47.92
Decision Table	Doctor	14	0.51	0.54	47.92	52.08
Decision Table	Doctor	16	0.53	0.56	43.75	56.25
Simple Cart	Doctor	12	0.53	0.59	45.83	54.17
Simple Cart	Doctor	14	0.51	0.59	45.83	54.17
Simple Cart	Doctor	16	0.48	0.57	41.67	58.33
Decision Stump	Three Class	12	0.47	0.51	26.39	73.61
Decision Stump	Three Class	14	0.46	0.50	25	75
Decision Stump	Three Class	16	0.46	0.50	27.78	72.22
Decision Table	Three Class	12	0.44	0.49	30.56	69.44
Decision Table	Three Class	14	0.44	0.49	29.17	70.83
Decision Table	Three Class	16	0.46	0.50	26.39	73.61

(continued)

Table 1 (continued)

Classification techniques	Class	K-Fold	MAE	RMSE	%Correctly classified	%Incorrectly classified
Simple Cart	Three Class	12	0.47	0.56	29.17	70.83
Simple Cart	Three Class	14	0.48	0.53	16.67	83.33
Simple Cart	Three Class	16	0.47	0.55	20.83	79.17

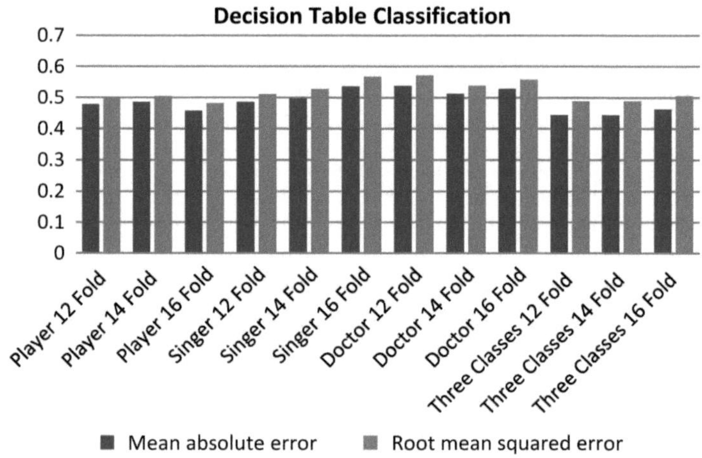

Fig. 2 Error measures of decision table classification technique

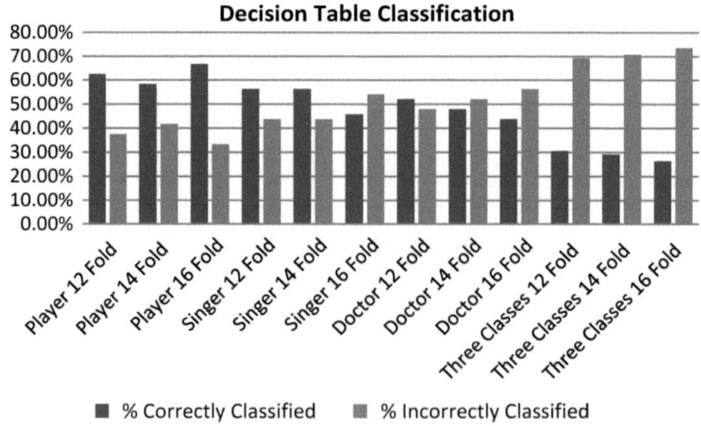

Fig. 3 Accuracy measure of decision table classification technique

Let O1 to On be the output values of persons P1 to Pn
Op1 to Opn be the predicted outputs of the persons P1 to Pn
L1, L2, L3 are the parameter related to date, time and place of birth of person.
A1 to A21 are the astrological attributes prepared based on parameters L1, L2, L3
Out of which A1 to A12 are zodiac attributes
A13 to A21 are planet attributes
A22 is gender of the person and
C1 to Cx be x number of classification techniques.

The hypothesis is described diagrammatical in Fig. 4. It states that data is collected for n persons that are P1 to Pn. Each record of person consist of 23 attributes, out of these attributes, A1 to A12 attributes hold the house number from 1 to 12. The house number is the position of a particular zodiac in the horoscope. A13 to A21 attribute holds the house number from 1 to 12. The house number is the position of a particular planet in the horoscope. A22 is the gender of the person it can hold value male or female. The actual profession is known and the value is stored in output O1 to On for each person. Op1 to Opn are the predicted output generated after applying classification techniques C1 to Cx on Attributes A1 to A22.

If the similarity of predicted output Op1 to Opn ≧ 90 % with O1 to On then it is validated that the life events of human are in full relevance with the Astrology and Horoscope.

If the similarity of predicted output Op1 to Opn ≦ 40 % with O1 to On then it is validated that the life events of human has no relevance with the Astrology and Horoscope.

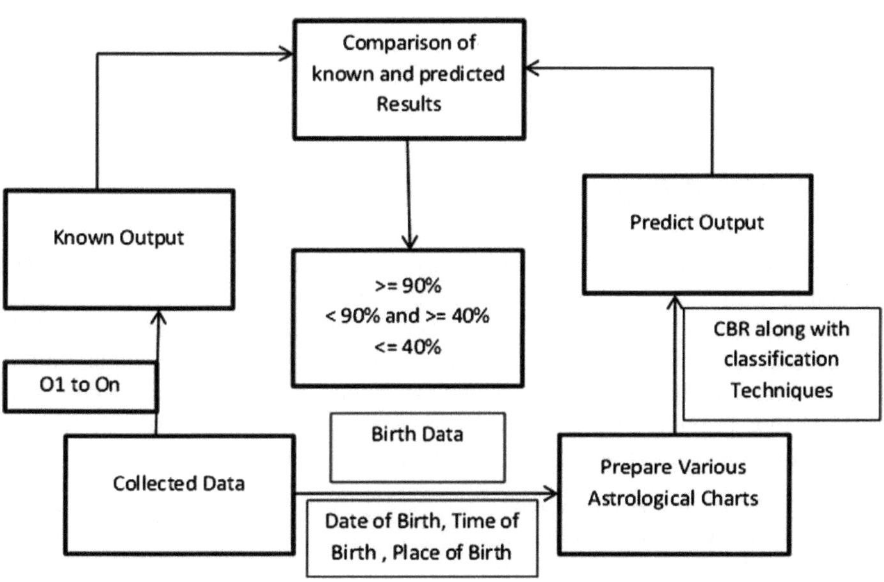

Fig. 4 Diagrammatical representation of hypothesis

If the similarity of predicted output Op1 to Opn < 90 % and ≧ 40 % with O1 to On then there is some relevance in Profession of the person with the Astrology and Horoscope. Hence improvement in the prediction and classification techniques is required.

3.2 Results of Profession Player

Total 48 records were used out of which 24 records were of profession player and remaining records were from other profession. Various methods were implemented the significant ones were Simple Cart, Decision Table, Decision Stump using 12, 14 and 16 fold classification. Best results were produced by Decision Table with 16 fold classification. This method provided results with an accuracy of 66.66 % hence we can predict that there is some relevance in life events taking place in person's life and the rules generated by this method were as below

Moon = 3 then Class is NonPlayer
Moon = 8 then Class is NonPlayer
Moon = 5 then Class is NonPlayer
Moon = 2 then Class is NonPlayer
Moon = 10 then Class is NonPlayer
Moon = 4 then Class is NonPlayer
Moon = 9 then Class is Player
Moon = 11 then Class is Player
Moon = 7 then Class is Player
Moon = 1 then Class is Player
Moon = 6 then Class is Player

3.3 Results of Profession Singer

Total 48 records were used out of which 24 records were of profession Singer and remaining records were from other profession. Various methods were implemented the significant ones were Simple Cart, Decision Table, Decision Stump using 12, 14 and 16 fold classification. Best results were produced by Decision Table with 12 and 14 fold classifications. This method provided results with an accuracy of 56.25 % thus providing some relevance in the life events taking place in person's life and the rules generated by this method were as below

Moon = 9 then Class is NonSinger
Moon = 1 then Class is NonSinger
Moon = 11 then Class is NonSinger
Moon = 6 then Class is NonSinger

Moon = 3 then Class is Singer
Moon = 2 then Class is Singer
Moon = 4 then Class is Singer
Moon = 4 then Class is Singer
Moon = 8 then Class is Singer
Moon = 5 then Class is Singer
Moon = 10 then Class is Singer

3.4 Result of Profession Doctor

Total 48 records were used out of which 24 records were of profession Doctor and remaining records were from other profession. Various methods were implemented the significant ones were Simple Cart, Decision Table, Decision Stump etc. using 12, 14 and 16 fold classification. Best results were produced by Decision Table with 12 fold classification. This method provided results with an accuracy of 52.0833 % thus providing some relevance in the life events taking place in person's life and the rules generated by this method were as below

Moon = 11 then Class is NonDoctor
Moon = 3 then Class is NonDoctor
Moon = 7 then Class is NonDoctor
Moon = 10 then Class is NonDoctor
Moon = 1 then Class is NonDoctor
Moon = 4 then Class is NonDoctor
Moon = 12 then Class is Doctor
Moon = 8 then Class is Doctor
Moon = 2 then Class is Doctor
Moon = 5 then Class is Doctor
Moon = 9 then Class is Doctor

3.5 Result of Three Classes Doctor, Singer, Player

Total 72 records were used out of which 24 records were of profession singer, 24 were of player and 24 were of doctor. Various methods were implemented the significant ones were Simple Cart, Decision Table, Decision Stump etc. using 12, 14 and 16 fold cross validation. Best results were produced by Decision Table with 12 fold cross validation. This method provided results with an accuracy of 30.5556 % thus providing no relevance in the life events taking place in person's life and the rules generated by this method were as below

Saturn = 6 then Class is Doctor
Saturn = 4 then Class is Doctor

Saturn = 8 then Class is Doctor
Saturn = 9 then Class is Player
Saturn = 3 then Class is Player
Saturn = 7 then Class is Player
Saturn = 1 then Class is Singer
Saturn = 10 then Class is Singer
Saturn = 2 then Class is Singer
Saturn = 5 then Class is Singer
Saturn = 11 then Class is Singer
Saturn = 12 then Class is Singer

4 Conclusion

In above experiments we used classification techniques that are Simple Cart, Decision Stump and Decision Table. The best results were produced by Decision Table algorithm for predicting the profession Player, Singer and Doctor. Hence for prediction of profession the decision Table algorithm can be used. Also as the methods tested were limited other classification techniques can also be checked for improvement of the results or hybrid methods can be generated to increase the accuracy of the results.

In above experiment the size of training data set is less so by increasing the number of records for training data set the accuracy of prediction can be increased.

In above experiments we performed prediction for various professions, similarly we can predict the basic nature of the person, the attitude of the person, money and other aspects of persons life.

Choice of attributes also affects the accuracy of the results generated by classification techniques. The attributes used in above experiments were only the parameters generated from basic horoscope of the person's birth chart hence we can try to perform experiments using other combinations of attributes.

References

1. Agnar Aamodt and Enric Plaza, AICom-Artificial Intelligence Communications, IOS Press, Vol 7(1), pp. 39–59 (1994).
2. Leo Breiman, Jerome H. Friedman, Richard A. Olshen, Charles J. Stone. 1984. Classification and Regression Trees. Wadsworth International Group, Belmont, California.
3. Kohavi R., The Power of Decision Tables, Proceeding European Conference on Machine Learning. (1995).
4. Iba, Wayne and Langley, Pat. Induction of One-Level Decision Trees. ML92: Proceedings of the Ninth International Conference on Machine Learning, Aberdeen, Scotland, San Francisco, CA (1992).
5. Kolodner, J.L. Case-Based Reasoning. California: Morgan Kaufmann, 1993.

6. Chaplot N., Dhyani P., Rishi O.P., Astrological prediction for profession using classification techniques of artificial intelligence. International conference on Computing, Communication and Automation, Noida, pp. 233–236 (2015).
7. G. Mark Hall, Eibe Frank, Geoffrey Holmes, Bernhard Pfahringer, Peter Reutemann, and Ian H. Witten, The WEKA data mining software. ACM SIGKDD Explorations Newsletter. vol. 11(1), pp. 10–18 (2009).

Machine to Machine Sensor Network Implementation for Securing Railway Transport

Chitra Suman, Lokesh Tharani and Saurabh Maheshwari

Abstract Majority of world population prefer Railways for transportation of goods and public. It works as a collection of several sub-systems of machine which together work in a synchronized manner to perform a safe running of trains. Most subsystems of the railway networks are manually operated which lacks reliability and real time nature. To avoid manual errors we are hereby proposing a machine to machine communication network which will connect all critical machinery and subsystems. The idea is to let the machines communicate their working status to a decision making unit. The collected data on the basis of threat's priority is used to decide whether to stop the train or let the train move. Whole critical machinery will be having their local error detection and correction mechanisms through sensors and transducers. A global sink placed off the train where the decision will be made by calculating a risk score. If the risk score is higher than the threshold then the train is signaled to stop immediately. This is first ever implementation of machine to machine network for railway security.

Keywords Railway security · Machine to machine network · Data aggregation · Security · Clustering · Critical systems

C. Suman (✉)
Department of Computer Science and Engineering,
University College of Engineering, RTU, Kota, India
e-mail: sumanchitra.suman3@gmail.com

L. Tharani
Department of Electronics & Communication Engineering,
University College of Engineering, RTU, Kota, India
e-mail: tharani123@gmail.com

S. Maheshwari
Department of Computer Science and Engineering,
Government Women Engineering College, Ajmer, India
e-mail: saurabh.maheshwari.in@ieee.org

N. Modi et al. (eds.), *Proceedings of International Conference on Communication and Networks*, Advances in Intelligent Systems and Computing 508,
DOI 10.1007/978-981-10-2750-5_65

1 Introduction

As known so far, rails were first utilized in England in form of wooden rails in sixteenth century as guiding and supporting structure [1]. In further development, around 1789 wooden tracks began to be replaced with L shaped cast iron plate rails to reduce friction loss. In nineteenth century, political, economic and social developments were indirectly affected by railways. At present, in countries like India, Russia and China the railway plays a crucial role where railways contains large infrastructure and have become a prime mode of transportation as it is cheapest as compare as to the other mode of transport [2]. Railways contribute a lot to both personal and commercial areas but safety and reliability is also a big issue till now. Accidents in railways increase due to the poor maintenance. Derailment is major cause of accidents; for human health and safety, this is potentially seriously hazardous [3]. This is basically caused due to the rail defects that include mechanical failure of tracks, such as broken rails, or the mechanical failure of wheels, weld problems, corrugations, worn out rails, some internal defects and rolling contact fatigue (RCF) [4] initiated problems such as head checks, squats, spalling and shelling. Root causes of accidents can be found and reduced by detailed research of advance technologies and better safety methods.

Various clustering techniques are used in WSN in which thousands of sensor nodes are used to sense the changes in external environment and transmit the collected data to the Base Station (BS). Limited battery power of sensor nodes affect the overall life time of network. Thus many clustering algorithms are introduced to enhance the network life time by reducing energy consumption.

Today Machine-to-machine (M2M) communications is emerging technology that offers omnipresent and direct connectivity between devices. Information is exchanged among M2M devices that are organized as a network and actions are performed without any human interference [5]. Being a part of Internet-of-Things (IoT), it contains wide range of applications. Wide coverage and low energy consumption are the important requirements of M2M networks because M2M devices are battery operated and could be located at wide variety of places in which some of places are not reachable by human. This paper is an attempt to design and simulate the feasibility of a machine to machine communication network.

2 Literature Survey

2.1 Identification of Breakage in Railway Track

The technique used in [6] is for obtaining quantifiable information about the depth and degree of declining of the ballast with minimum interference to the actual track bed and this is based on GPR. Automated video analysis technique used in [7] is to find missing clips and new or old clips based on their color in video sequence. By combining analysis methods and image processing, high performance is achieved in automated rail track inspection.

In [8], Long Range Ultrasonic Testing (LRUT) method is used to examine hard to access area on railway tracks and the foot of rail where associated fatigue cracking and corrosion is probably to occur, e.g. level crossings. In LRUT, the properties of guided waves are used in the three different parts of the rail section (head, web and foot) and capabilities of guided waves are also examined to detect defects in each part. A dynamic track gauge inspection method constructed using two laser sector lights and four CCD (Charge-Coupled Device) cameras based on computer vision is presented in [9], these devices are used for examining the inspection principle and corresponding calibration method of inspection system. Track gauge values are fast obtained with high precision and recurrence by using this method which enhances quality and safety of running by literally controlling the railway track gauge changes.

A crack detection algorithm is used in [10] that uses Light Emitting Diode (LED) which are attached to one side of rails and Light Dependent Register (LDR) to the opposite side, assembly in which faulty tracks exact location is identified. A GPS receiver receives the current latitude and longitude data to detect the current location and received information is communicated via GSM modem.

2.2 *Inspecting Railway Track Using Wireless Sensor Network*

Automatic Railroad Track Inspection [11] presents the use and availability of sensors which is limited to the track bed and the rails. This inspection method is used to test rails only for any fissures and flaws and not for the crossover error and misalignment. Using probabilistic selection method, the proposed system identifies high risk areas.

An idea to improve the current railway effective functions by using WSN is proposed in [12]. The system uses a fault tolerant multi-homing technique that does not require extra transmission energy to send multiple copies of the information to multiple homes. Sensors are used in [4] for detecting flaws in railway system, each sensor sends the information to nearby cluster-heads gathered from the environment. Using multi-level routing, the data packages are sent from lower to higher level, the final destination is the base station where decisions are made using fuzzy logic. The algorithm proposed in [13] is divided into movable and fixed algorithm. The fixed algorithm collects and investigates information about seismic data, crack, balance and pressure on the bridges. Movable algorithm displays the fixed sensor network collected information on the monitoring car using installed network. The system presented in [14] is based on IR Rays and Sensors for avoidance of collision in which sensors are fixed in train wheels and IR rays are transmitted the in the track. The trains coming from the opposite direction have the same option. If two trains are on the same track then the rays will get collided and get reflected back to the respective engines and Alarm blinks so the train can be stopped before collision occurs.

Multiple sensors are used in [15] for surveying track geometry, for this LVDT arrangement is used in the system and GSM is used to send the information immediately. Wireless module and sensors monitor the bridge damage status, when the sensor does not get the signal; the nearby wireless system immediately notifies and alerts the train coming on the track. Train collision avoidance system proposed in [16] uses Vibration sensor, a Proximity LASER Detector and the communication mechanism of PLC (Power Line Communication) to send any critical detail like breakage in the railway track or two trains are on same track in opposite direction, to the control room. Zigbee is used to communicate with the other sensor modules. In [17] vibration sensors are deployed on railway track to detect the exact location of track breakage and send it to the train engine. In this system, sensors, sink and train work as sender and receiver in multi-hop routing manner and communicate in full duplex mode. In [18], sensor network uses acoustic sensors to continuously monitor the railway track for detecting discontinuity in the track.

3 Proposed Method

In this section we will show the proposed architecture of the machine to machine communication network. We will describe the physical architecture of the sensor nodes, types of sensors that can be used for security, the forwarding and data aggregation network, decision making rules.

3.1 Entities

There are following entities in the complete network.

(1) Sensor Nodes
These are the local nodes equipped with problem detection sensors. The structure of the node is given in figure. It has a problem detection sensors connected with input pin of programmable microcontroller. The output of microcontroller is fed to an actuator and a wireless module. The role of actuator is to physically alarm the local station while the wireless module is used to transmit the sensed machine status from local node to local sink wirelessly.

(2) Local Sink
It is the collection node for all the sensing nodes. All the sensor nodes send their problem status whether OK or Problem to this local sink. It forwards the local sensor status to the collection poles. These sinks are placed either on the

train or in the wireless range of the machine sensors in the peripheral railway network. ZIGBEE 802.15.4 protocol has been chosen for local wireless communication as it requires very low power and the network life time is very large.

(3) Collection Poles
These are the fixed poles which collect the data from the local sinks. Local sinks send the status of all the sensing nodes to the nearest pole to which that local network is registered to. As the train moves this forwarding pole also keep changing. These poles are placed on both sides of the track. The train keeps on moving on the track and gets registered to nearest track. One pole is placed at every kilometer. The local sink and poles communicate using 3G communication with high data rate. A GSM module is attached for the purpose on both the poles and the local sinks.

4) Global Sink
It has the responsibility of complete network data collection, processing the data and decision making by calculating the risk score.

(5) Risk Score
It is the total score of all threats to the movement of the train x. It is calculated for each cycle and for all trains. Risk score must be always less than the predefined threshold else the train must be stopped.

(6) Unique IDs
Unique IDs are assigned to each train, track segment for which a breakage detections sensor is attached, signaling poles, platform clearance and for the railway crossings.

(7) Risk Priority Value
It is the priority value assigned to each risk type. Priority values of all the accumulated risks at the sink for a particular train at particular cycle are added to find the current risk score.

3.2 Local Machine to Machine Network

It is the network of the sensor nodes which are connected wirelessly to the local sink. These networks are installed at all critical locations where machinery is installed. This network consists of several sensor nodes which keep on monitoring the proper functioning of the particular system. It also checks the danger conditions if any. This is installed on and off the train viz. on the tracks, on the signaling poles, railway crossings, etc. All of these systems involve automatic, semiautomatic and manual machineries. Working of these machines properly and without damage is very critical for railway operations. The proper functioning and damage is monitored by available sensors. The final point of this network is the local sink which finally forwards the status packets received from all the sensing nodes after checking the status. Only those packets are forwarded which have the status as 'problem'. These packets have the sensor ID, machine ID and status bit.

3.3 Collection Pole Network

The status packets that have 'problem' status are forwarded to the nearest pole to the train/railway crossing/track location.

3.4 Data Aggregation

This is done at a global sink where complete network data is collected from all the machine networks. It means that proper functioning status of the machine on and off the train reaches to this sink. Here each coming packet is minutely scanned and checked for the type of machine facing the problem. Each type of machine is associated with a priority number, if the machine status is problem the priority number is added to train risk score. All the problems' score is added together for a particular train. If this priority risk score crosses a threshold then a command to stop the train is generated. Below n denotes the number of machines whose status is reported as 'problem'.

$$\text{Risk\_score} = \sum (\text{risk\_priority\_value}(\text{i})) \text{ where i} = 1 \text{ to n} \tag{1}$$

3.5 Decision Making

This is the final step taken at the global sink. All the data related to a particular train is collected at the sink. This is done for all trains with unique train ID for each cycle. Decision making is also done for each cycle. Final risk decision taken is not based on the single cycle but the stored for next four cycles.

3.6 Formal Algorithms

The modules discussed till now must work in a synchronized manner as per the algorithm discussed below.

Machine to Local Sink Network Processing

1. Sensors check the status of machinery working for each cycle.
2. The generated status either 'OK' or 'Problem' is forwarded to the microcontroller input pin.

3. Microcontroller processes the input and generates a programmed output at the output pin.
4. This output is fed to the actuator to convert the signals to physical output like alarm or blinking LED.
5. The other output line is fed to the wireless module where a status packet is created.
6. The status packet has the sensor ID, machine ID and status bit.
7. Status bit is 1 for problem and 0 if OK. This has been chosen to save power, as line will be ON only when problem.
8. The status packet is sent wirelessly to the local sink (Fig. 1).

Local Sink to Pole Forwarding

1. All the local sinks at all on and off train machine networks forward the collected problem status packets and prepare for forwarding to the poles.
2. All the local sinks are registered to unique pole at a particular time instant.
3. Stationary machine networks are registered to single pole statically whereas moving train keeps on registering with nearest pole while moving.

Pole to Global Sink Collection Network Processing

1. All the poles collect the instantaneous data from all registered sinks local machine networks.
2. Pole forward these status packets to the global sink for aggregation and decision making (Fig. 2).

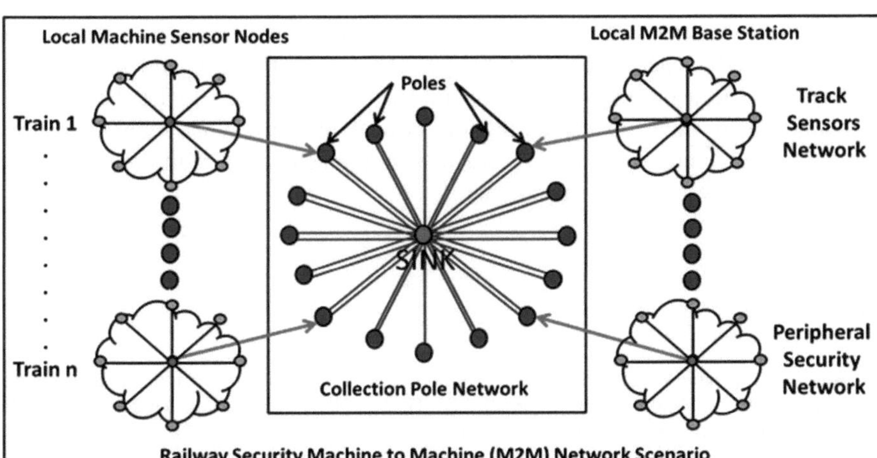

Fig. 1 Railway security M2M network scenario

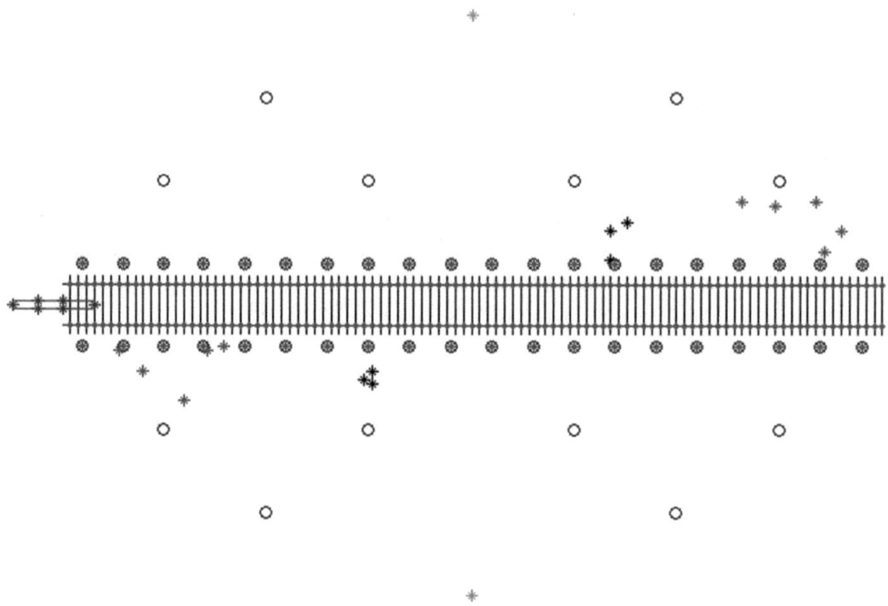

Fig. 2 Machine to machine network simulation diagram

4 Conclusion and Future Work

This paper is first ever mapping of machine to machine networks for railway security application. In this paper due to space constraints we could only describe the idea. The simulation of the presented algorithm will be detailed in subsequent presentation. This network can be of any scale from few sensors to thousands of sensors attached to each machine. In the age of internet of things growing very fast, this can become a burning application for all the researchers to let the machine communicate to each other securing the critical infrastructures such as railway transport. The future work lies in practical implementation of this network and evaluating against real time constraints.

References

1. Ahren, T., Waara, P., Larsson, "Technical and economic evaluation of maintenance for rail and wheels on Malmbanan" International Heavy Haul Association (IHHA) Specialist Technical Session, Session 5c Wheel/Rail Asset Life Extension, May 4–May 8 2003.
2. Cannon, D.F., Edel, K.O., Grassie, S.L. and Sawley, K. "Rail defects: an overview" Fatigue & Fracture of Engineering Materials & Structures, Volume 26, pp. 865–886. October 2003.
3. MuneedraRao Ch. and BalaJaswanth B.R., "Crack sensing scheme in rail tracking system", Vol. 4, Issue 1, pp. 13–18., January 2014.

4. Daliri Z.S., Shahaboddin Shamshirband, Mohsen AmiriBesheli, "Railway Security Through The Use of Wireless Sensor Networks Based on Fuzzy Logic" International Journal of the Physical Sciences Vol. 6(3), pp. 448–458, 4 February, 2011.
5. Xiong X., Zheng, K., Rongtao Xu, Xiang W, Chatzimisios, P., "Low power wide area machine-to-machine networks: key techniques and prototype", Communications Magazine, IEEE, Volume-53, Issue-9, pp. 64–71, 2015.
6. Nuaimy, W.A., Eriksen A. and Gasgoyne J., "Train-mounted GPR for high-speed rail trackbed inspection", IEEE Tenth International Conference on Ground Penetrating Radar, pp. 631–634, June 2004.
7. Singh M., Singh S., Jaiswal J., Hempshall J., "Autonomous Rail Track Inspection using Vision Based System". IEEE International Conference on Computational Intelligence for Homeland Security and Personal Safety. pp 56–59. October 2006.
8. Castellanos C.C., Gharaibeh Y., Mudge P., Kappatos V., "The Application of Long Range Ultrasonic Testing (LRUT) For Examination of Hard To Access Areas On Railway Tracks". IEEE Railway Condition Monitoring and Non-Destructive Testing (RCM 2011) Nov 2011.
9. Zheng S., Chai X., Xiaoxue A., Li L., "Railway Track Gauge Inspection Method Based On Computer Vision." IEEE International Conference on Mechatronics and Automation, pp (1292–1296). 2012.
10. Vanimiredd A., Kumari D. A., "Automatic Broken Track Detection Using LED-LDR Assembly" International Journal of Engineering Trends and Technology (IJETT), Volume 4 Issue 7, July 2013.
11. Hayre H.S., "Automatic Railroad Track Inspection" IEEE Transactions On Industry Applications, Vol. Ia-10, No. 3, May/June 1974.
12. Aboelela E., Edberg W., Papakonstantinou C, Vokkarane V. "Wireless Sensor Network Based Model for Secure Railway Operations", IEEE on International Performance, Computing, and Communications Conference, pp. 623–628, 2006.
13. Zinvandlorestani A., Mousavi S.A., "Monitoring Rail Traffic Using Wireless Sensor Network (WSN)", IJCSET, Vol 2, Issue 6, 1280–1282, June 2012.
14. Ramesh S., Gobinathan S., "Railway Faults Tolerance Techniques using Wireless Sensor Networks", IJECT Vol. 3, Issue 1, Jan., March 2012.
15. Kalaimathi M., Ilakya P., Sathiavathy E., "Innovative Railway Track Surveying with Sensors and Controlled by Wireless Communication", International Journal of Advanced Electrical and Electronics Engineering, (IJAEEE) pp. 2278–8948, Volume-2, Issue-3, 2013.
16. Bissa S.A., Jayasudha S., Narmatha R., and Rajmohan B., "Train Collision Avoidance System Using Vibration Sensors and Zigbee Technology", IJREAT International Journal of Research in Engineering & Advanced Technology, Volume 1, Issue 1, March, 2013.
17. Sharma, K., Maheshwari, S., Khanna, V., "Railway Track Breakage Detection using Vibration Estimating Sensor Network", 2014 International Conference on Advances in Computing, Communications and Informatics (ICACCI), IEEE, pp. 2355–2362, Sept. 24–27, 2014.
18. M.F. Islam, S. Maheshwari, Y. Kumar, "Energy Efficient Railway Track Security Using Vibration Sensing Network", IEEE International Conference on Signal Processing and Integrated Networks (SPIN), pp. 973–978, 2015.

Real Time Street Light System Using Low Power Microcontroller

Amey S. Laddad and Gayatri M. Phade

Abstract Automation the word cannot be defined by the just the definition. It's a continuous pattern involving of various systems to control and operating machines or processes in Industries, with interdependent networks of telephony, several other applications with minimized or significantly lesser intervention of humans. The Energy consumed by the street lights can be effectively conserved if we can control the traffic lights on the highways by glowing them only when needed. If road is full of traffic, and it is almost impossible to detect the arrival of a vehicle manually without the sensing of the vehicles. Under these situations we should think about a system which is capable of sensing the arrival of vehicle and Switch ON the lights and turn OFF as soon as the vehicle leaves the area. Also the intensity of the street light is controlled according to the surrounding light.

Keywords Auto intensity control · LDR · LED · MSP430 · PIR sensor

1 Introduction

The significance of this great increase is automation has greater advantages like consistency, conservation of energy, precision. The consumption of energy, the source of source of its origin and its conservation is very important in the present world scenario of ever demanding and increasing energy requirements. Such an approach of conservation has been proposed in the sector of Street Light Energy Conservation. Street lighting is an important and essential public service provided by public authorities at the municipal level, the main function of which is to illuminate the city's streets during the dark hours of the day, to reduce traffic

A.S. Laddad (✉) · G.M. Phade
Sandip Institute of Technology & Research Centre, Nashik, India
e-mail: laddadamey@gmail.com

G.M. Phade
e-mail: gayatri.phade@sitrc.org

N. Modi et al. (eds.), *Proceedings of International Conference on Communication and Networks*, Advances in Intelligent Systems and Computing 508,
DOI 10.1007/978-981-10-2750-5_66

accident and street crime [1]. So it is important that the on/off of the street light should be proper to avoid mishaps and also energy loss.

The smart controller used to switch street lighting on/off automatically is based on sunrise/sunset times or light intensity of controller surroundings. Although the smart controllers can automatically turn streetlight on and off, the cases of street lighting on at daytime and off at night occur because the light sensor is covered with dust. This increases the public dissatisfaction with improper streetlights and will lead to some potential danger for street vehicle and pedestrian. Main limitations of existing systems are uneasiness of handling and difficulty of maintenance [2]. Thus, we have suggested the system with PIR sensor and LDR. The ability to produce high power, high quality white light with LEDs is most recent and this technology is still evolving in a highly competitive market. Here large number of High voltage as well as low voltage devices is employed to fulfil different consumer as well as the employ needs. The controlling device of the whole system is a Microcontroller. This kind of automation has advantages like accuracy, energy conversation, and reliability and more over the automated systems do not require any human attention [3]. As the energy conversation is very important in the current scenario and should be done to a maximum extent where ever it is possible.

The system we have designed can be useful for the streets where the traffic of vehicle or human passing is very discrete. For a long time the current street lights remain ON for no use of it. The design we have proposed has a PIR sensor to detect motion around the street light in order to turn ON the output LED's when there is a movement in the region of Street light. We designed the system such that the intensity of output LED's will change automatically. The circuit includes LDR sensor to sense the status of daylight. The intensity of output LED's will change in accordance with output of LDR. When it's completely bright daylight the street light will be OFF. Vice versa when it's completely dark around the street light will be ON with highest intensity. We have controlled the output light intensity by generating PWM signal which controls the intensity of output LED light [4–6]. We can save the valuable electrical energy.

2 Proposed System

The block diagram of the system we designed is as shown in Fig. 1 it comprises of:

1. Regulated Power Supply.
2. PIR Sensor.
3. LDR sensor.
4. MSP 430 controller.
5. Output LED's.

Fig. 1 Block diagram of system

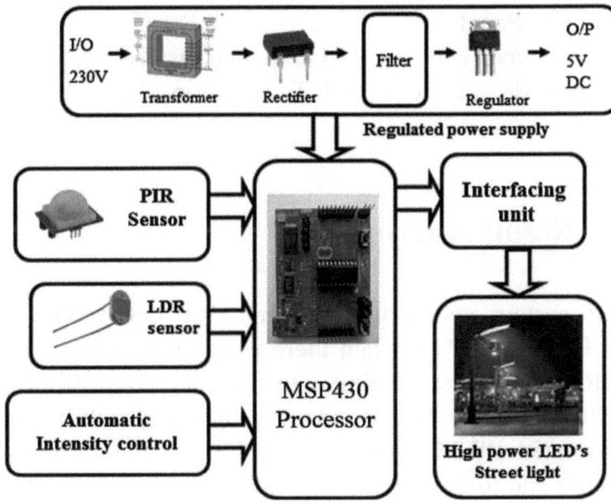

1. MSP 430 Controller:
   It is an ultra-low power mixed signal microcontroller. We have used the MSP430G2553 chip because of its easy availability and comparatively cheap prise. It has a 16 bit RISC CPU and has 24 I/O pins and has on chip ADC.
2. PIR Sensor:
   It is a motion sensor. It operates by sensing temperature difference between the moving object and surroundings. We have used the PIR sensor to sense the movement around the Street light to turn it ON only when it is needed [7].
3. LDR:
   Light Dependant Resistor is used in the system to sense the surrounding light intensity. The output of LED's is controlled through this data. During day time the street light will be OFF. During night time the street light will be ON with highest intensity. It operation is explained by the use of equation below.
   The equation to show the relation between resistance and illumination can be written as

$$R = A\,E^{\wedge}a \tag{1}$$

where:

E	is Illumination (lux)
R	is Resistance (Ohms)
A and a	are constants

The value of 'a' depends on the CdS used and on the manufacturing process. (Range 0.7–0.9) [8].

4. LED Panel:
   We have used two LED panels as the prototype of Street light. The LED's provide sufficient light. We have used two 1 W LED panels.

3 Result Analysis

The system operation is explained in detail. The circuit will only allow output LED's to turn ON when there is any movement around the street light. When the PIR sensor passes digital 1 to the microcontroller the system will take input from LDR and sense the surrounding light and will decide intensity of the output. The generated PWM signal passes through the microcontroller pin to the MOSFET circuit and to the LED's. The intensity level of output LED's are changed according to the light incident on the LDR. We can check the length of PWM signal changing from the circuit with the use of DSO. The DSO we have used is of rating of 25 MHz, 250 MS/s.

The Fig. 2a image shows the circuit at dark surroundings. The Fig. 2b, c images shows the characteristics of the PWM signal generated while there is dark surrounding. We find that the width of PWM signal is maximum during dark condition and intensity of the output LED's is highest.

The Fig. 3a image shows the circuit at Mid-range light surroundings. The Fig. 3b, c images shows the characteristics of the PWM signal generated while there is Mid-range light surrounding. We find that the width of PWM signal is medium during Mid-range light condition and intensity of the output LED's is medium.

The Fig. 4a image shows the circuit at bright surroundings. The Fig. 4b, c images shows the characteristics of the PWM signal generated while there is bright surrounding. We find that the width of PWM signal is minimum during bright condition and intensity of the output LED's is minimum.

In order to understand the readings of the system we have prepared tabular and graphical representation of light intensities of ambient light, LED output light, PWM voltage and LED voltage.

Table 1 shows the reading taken while transition from bight to dark ambient conditions. By observing the reading we can easily derive that the light falling on LDR is inversely proportional to the LED output intensity. From Eq. (1) we can find out that the LDR output is dependent upon the light falling on it (Graphs 1 and 2).

Table 2 shows the reading taken while transition from dark to bright ambient conditions. By observing the reading we can easily derive that the light falling on LDR is inversely proportional to the LED output intensity. From Eq. (1) we can find out that the LDR output is dependent upon the light falling on it.

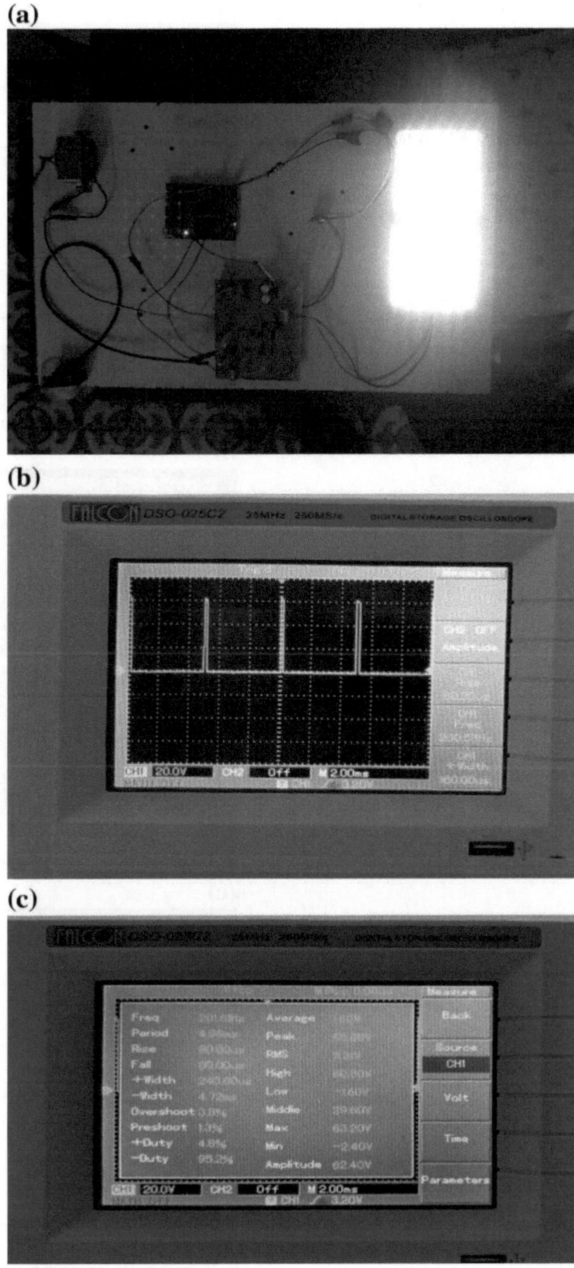

Fig. 2 **a** Highest LED intensity at *dark*. **b** Waveform of highest LED intensity at *dark*. **c** Parameters of highest LED intensity at *dark*

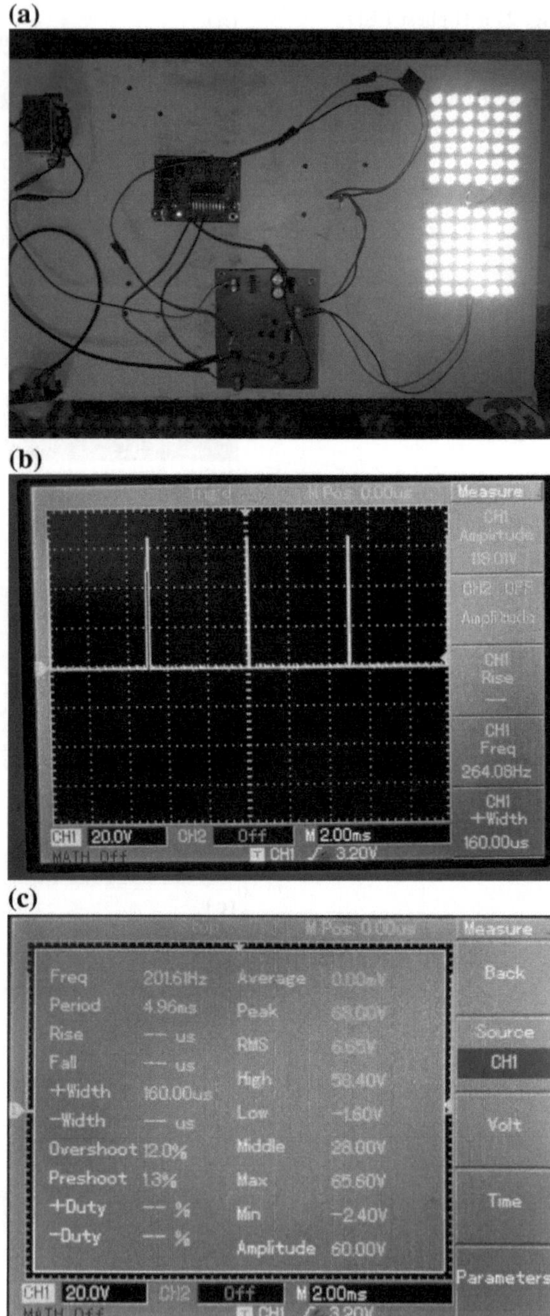

Fig. 3 **a** Mid-range LED intensity at *dull surrounding*. **b** Waveform of mid-range LED intensity at *dull surrounding*. **c** Parameters of mid-range LED intensity at *dull surrounding*

Fig. 4 **a** Low LED intensity at *bright surrounding*. **b** Waveform low LED intensity at *bright surrounding*. **c** Parameters of low LED intensity at *bright surrounding*

(a)

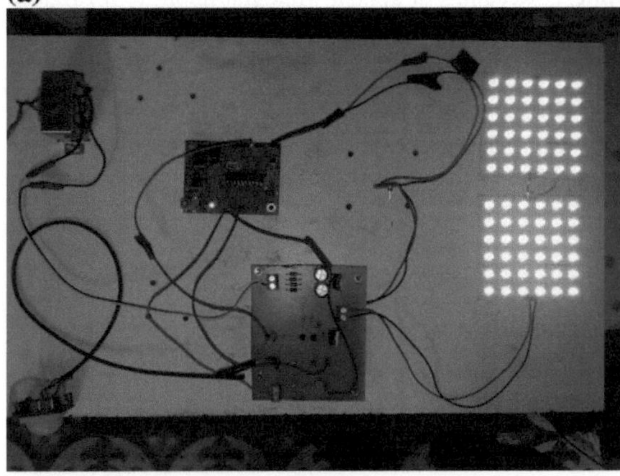

(b)

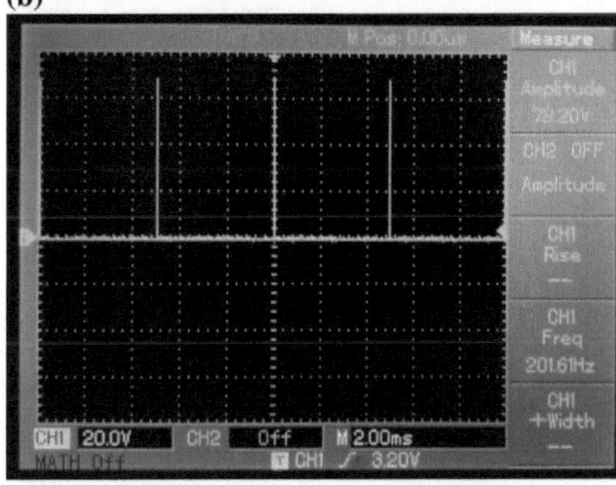

(c)

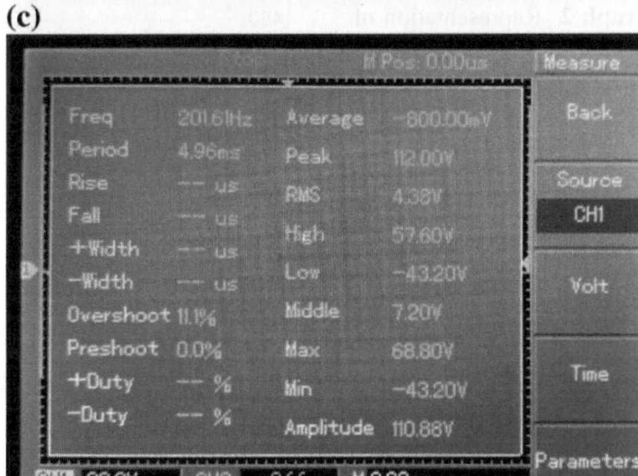

Table 1 Readings of system while transition from Bright to Dark

	Lighting condition	Luminance of ambient (lx)	Luminance of LED (lx)	PWM voltage (V)	Voltage at LED (V)
1	Bright	420	48	0.064	4.42
2	Mid-range	40	150	0.125	5.12
3	Dark	12	290	0.530	5.72

Graph 1 Representation of comparative readings from *bright* to *dark*

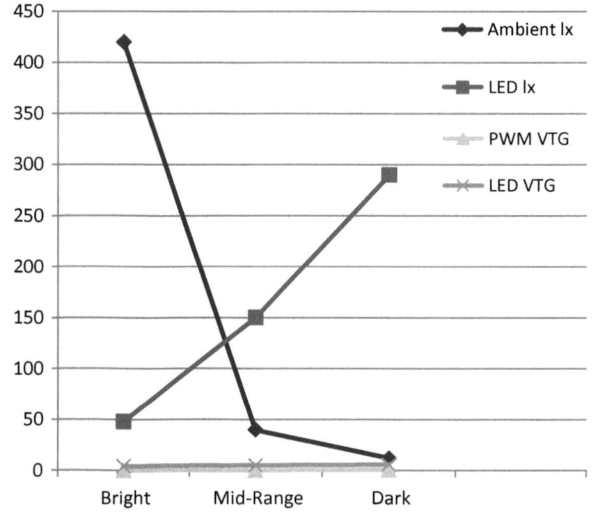

Graph 2 Representation of comparative readings from *dark* to *bright*

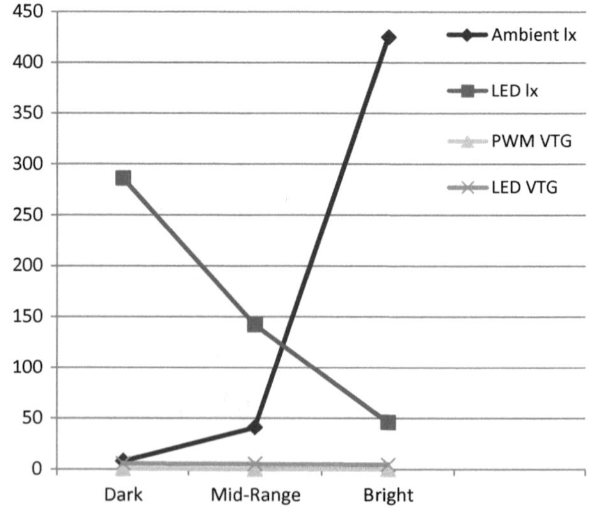

Table 2 Readings of system while transition from dark to bright light

	Lighting condition	Luminance of ambient (lx)	Luminance of LED (lx)	PWM voltage (V)	Voltage at LED (V)
1	Dark	8	286	0.531	5.69
2	Mid-range	41	142	0.122	5.25
3	Bright	425	46	0.063	4.45

4 Conclusion

We have implemented the system and its results are obtained as per the established theory. The results from above charts and tables show us that when minimum light is falling on the system, we get maximum intensity of output LED. The street lights only turns on when there is movement around it, and the intensity of output light is varied in accordance with the ambient light. We conclude that the system we have implemented can be useful in the future to reduce the power consumption of the street light system.

References

1. Gezer C, Niccolini M, and Buratti C.: An IEEE 802.15.4/ZigBee based wireless sensor network for energy efficient buildings.: IEEE 6th International Conference on Wireless and Mobile Computing Networking and Communications, art. No. 5645021, pp. 486–491 (2010).
2. WU Yue, SHI Changhong, ZHANG Xianghong, YANG Wei.: Design of New Intelligent Street Light Control System.: 8th IEEE International Conference on Control and Automation Xiamen, China, June 9–11 (2010).
3. Wu Yue, Shi Changhong, Zhang Xianghong, Yang Wei.: DESIGN of new intelligent street light control system.: 8th Conference on Control and Automation, ICCA, art. No. 5524348, pp. 1423–1427 (2010).
4. "White LED Driver Circuits for Off-Line Application using Standard PWM Controllers" intersil® Application Note AN1387.0 (2008).
5. Marco A.D. Costa, Guilherme H. Costa, Anderson S. dos Santos, Luciano Schuch, José R. Pinheiro.: A high efficiency autonomous street lighting system based on solar energy and leds.: 978-1-4244-3370-4/09, IEEE (2009).
6. Po-Yen Chen, Yi-Hua Liu, Yeu-Torng Yau, Hung-Chun Lee.: Development of Energy Efficient Street Light Driving System.: ICSET (2008).
7. "Home Alarm PIR Tech". Venice Locksmith—Home Security Technician's Notebook. (2012).
8. Lin Jianyi Jin Xiulong Mao Qianjie.: Wireless Monitoring System of Street Lamps Based on ZigBee.: 978-1-4244-3693-4/09, IEEE (2009).

Dealing with Indian Jurisprudence by Analyzing the Web Mining Results of a Case of Cybercrimes

M. Gouri Shankar, P. Usha Gayatri, S. Niraja and K. Chandra Sekharaiah

Abstract Earlier, in [1–5], we presented a case study of a cybercrime w.r.t. tripartite case of cybercrimes of (1) Violation of State Emblem of India Prevention of Improper Use Act 2005 (2) Cheating crime and (3) Sedition crime [6]. The cybercrimes were perpetrated by the culprit organization, JNTUHJAC, which was operational in the JNTUH University overtly and covertly. The case was based on web mining [7] results drawn from the web crawler tool, Wayback machine [8]. W.r.t. the case, the charge sheet was detailed in [5]. In this paper, we present the judgment in the aftermath of the chargesheet [4, 9]. We present that the Cyberabad police failed to make use of the cyberforensic evidence in handling the case. We present that the significance of Cyberethics in the case. Policing and judiciary are not just sufficient in handling cybercrimes. Cybercounseling and guidance is crucial in generating public opinion, national integration and academic development to remedy the flaws in handling the cybercrimes by organizational means such as police dept. and judiciary.

Keywords User interface design (UID) · Govt. of telangana (GoT) · National emblem of India-abusive GoT (NEIAGoT) · Fake · Cheating and seditious GoT (FCSGoT) · Threats to national integration in bharat (TsNI) · Culprit GoT which is involved in the twin crimes of online cheating and online sedition (TsCSGoT) · State emblem of india (Prohibition of improper use) act 2005 (SEIPIUA) · Jawaharlal nehru technological university hyderabad (JNTUH) · Joint action committee

M. Gouri Shankar (✉) · P. Usha Gayatri
Maturi Venkata Subba Rao Engineering College, Hyderabad, India
e-mail: m.gouris@gmail.com

P. Usha Gayatri
e-mail: ushagayatri@gmail.com

S. Niraja
MCE, Hyderabad, India
e-mail: neeraja.svce@gmail.com

K. Chandra Sekharaiah
Jawaharlal Nehru Technological University, Hyderabad, India
e-mail: chandrasekharaiahk@gmail.com

N. Modi et al. (eds.), *Proceedings of International Conference on Communication and Networks*, Advances in Intelligent Systems and Computing 508,
DOI 10.1007/978-981-10-2750-5_67

(JAC) · GoT whose appointed day is 2 jun 2014 after the parliament passed the A.P. reorganization bill (PGoT) · Underground GoT (UGoT) · Internet of things (IoT)

1 IoT Introduction

Internet of Things (IoT) has come to be in vogue side by side with Big Data. Also called as Internet of Objects, it means a network of physical "devices/objects/gadgets/things". It refers to the usage of intelligently connected micro-electronic devices and systems such as MEMS and ICT devices that control data gathered by embedded sensors and actuators in m/cs and other physical objects that are made "Smart". Nowadays, consequent upon diminishing size and price consciousness and declining energy consumption, processors, communications modules and other electronic components are integrated to make everyday objects "Smart". Smart cities use Smart objects which are IoT-based. Smart objects can be controlled remotely and can serve as physical access points to Internet services. IoT involves ubiquitous computing, as conceived by Mark Weiser in the early 1990s.

Umpteen IoT objects are interconnected through public as well as private IP networks. These can sense as well as communicate and share information. These interconnected objects have data regularly collected, analysed and used to initiate action, paving the way for web intelligence and collective intelligence for planning, management and decision making. IOT objects can be read, recognized, located, addressed, and/or controlled via the Internet, irrespective of the communication means (whether via RFID, wireless LAN, WANs etc.). Things in IoT include People, Location, Time Information as well as Condition of objects. The objects seamlessly integrate into the virtual world and facilitate anytime, anywhere connectivity. The term IoT was coined by Kevin Ashton in 1999. Typically, IoT is expected to offer advanced connectivity of devices, systems, and services that goes beyond M2M communications as well as cover a variety of protocols, domains, and applications. The interconnected embedded devices (including smart objects) are expected to usher in automation in all fields including Smart Grid and Smart Cities. Examples of IoT devices are smart thermostat systems as well as washer/dryers that use Wi-Fi for remote monitoring.

The scope and extent of IoT includes Architectures, Testbeds and Deployments; Devices Platforms, Sensors, Control and Actuators; Resource Constrained Nodes; Wireless Sensor Networks; Cognitive IoT; Coexistence and Interference; Network Clustering and Cooperative Schemes; Protocols and Algorithms; Energy Harvesting; Modeling and simulation; Software Defined Networking; Ubiquitous Computing; Big Data for IoT; Standards and Societal Impact; Fault Tolerance and Survivability; Safety and Reliability; Security and Privacy; Wellness and In body Sensors; Emotion Sensing; Industrial and Equipment Monitoring; Smart City Applications; Crowd Sensing and Management; Disaster Recovery and Management; Smart Grids and Energy Management; Intelligent Transport Systems and Applications; Agriculture and Green Houses; IoT based e-Commerce; Innovative Applications, Metrics, Measurement and Management; Scalability, Density, and heterogeneity Challenges; IoT-Cloud Integration; Context and Location-Aware Applications; Body Area Networks; Social IoT.

Some sub areas of IoT are as follows. Mobile IoT is an offshoot of state-of-the-art technologies applications of Smartphone technology. It generates and consumes big data. On-board sensors, making use of crowd sourced data, promoting e-commerce or enabling B2B, B2C and C2C connectedness are some examples of new applications of IoT.

IoT device or circuit design deals with the state-of-the hardware that enables field intelligence and data generation, innovative concepts that enable pervasive sensing. IoT applications and services deal with embedded field intelligence applications that challenge the state of the art in such fields as agriculture, automotive, civil and industrial Infrastructure monitoring, health care and retailing. Architecture and systems design deals with reliability, sustainability, scalability, profitability and scope for interoperability of an IoT ecosystem; novel business models and services enabled by IoT. Apart from other things, cloud computing and semantic web technologies deal with data security, storage and access, communication protocols, linked data and web semantics for interoperability among users and devices. Interfaces deal with adaptive UID informed actuation, and the interplay between the two. IoT analytics deals with innovative algorithms as well as data analysis techniques for extracting meaning and value from the IoT Internet of Things. IoT gives rise to scope for internet addiction disorder [10] and increased cybercrime incidence.

IoT is considered useful especially for accessing/storing data by perceptible sensor-driven IoT devices. It pioneers the development of Smart homes and onward to Smart cities. IoT is not altogether the solution to the current day problems of Cyber security. In this paper, digital crimes in the background presentation and digital exploitation in the foreground presentation are together made prevalent.

IT systems are under attack for political/commercial/reasons. In some cases, they are under attack to humiliate their targets. The Internet technology is exploited to perpetrate crimes. Criminals find an avenue for launching attacks with relative anonymity. The increased complexity of the communication and networking infrastructure is making investigation of cybercrimes difficult. Smart data is left untapped even as illegal activities are often buried in large volumes of data in order to detect crimes and collect clues of evidence.

The field of digital forensics and cybercrime investigation has become very important for law enforcement, national security, and information assurance. This is a multidisciplinary area that encompasses law, computer science, finance, telecommunications, data analytics, and policing. This IoT field brings together practitioners and researchers from diverse fields providing opportunities for business and intellectual engagement among attendees. The Cybercrimes of relevance to the case under consideration are highlighted in the paper notwithstanding the fact that for this kind of Cybercrimes, we do not have much to do with the IoT for the resolution of the Cybercrimes. These Cybercrimes are outside the possible domain of IoT. As such, IoT is of less potential to deal with the whole galaxy of Cybercrimes of the kind considered.

In this paper, we present the organization of our work as follows. In the forth coming section, we present about the prevalence of cybercrimes in higher educational institutions. In Sect. 3, we present the photocopy of the judgment related to the cybercrime. Section 4 concludes the paper in terms of the repercussions of the judgment and the cybercritical solutions suggested.

2 Cybercrimes in Higher Educational Institutions

Cybercrimes tend to increase by leaps and bounds in the present day ICT-based society. The higher educational institutions impart ICT in academics. On the other hand, rampant technology abuse is also taking place in the academia simultaneously. We present that mere technological knowledge could be baneful to the academics and society. Cybercritical solutions are not enough always. Social cognition should be modeled such that the technology usage only promotes the national development. Cybercrimes in academia is an index to it that the quality of national development is on the wane. This paper focuses on cybercrime in higher

Fig. 1 The overtly inviting flexi banner of JNTUHJAC (which is involved covertly in cybercrimes including Sedition, Cheating by Impersonation and violation of SEIPIUA during 2015 academic admissions time) for registrations by students in its cybercriminal website

educational institutions. In Fig. 1, the flexi banner fixed to a university main entrance gate is that of a cybercriminal organization, JNTUHJAC which is under NEIAGoT. It invites the academic stakeholders offline to get involved in its

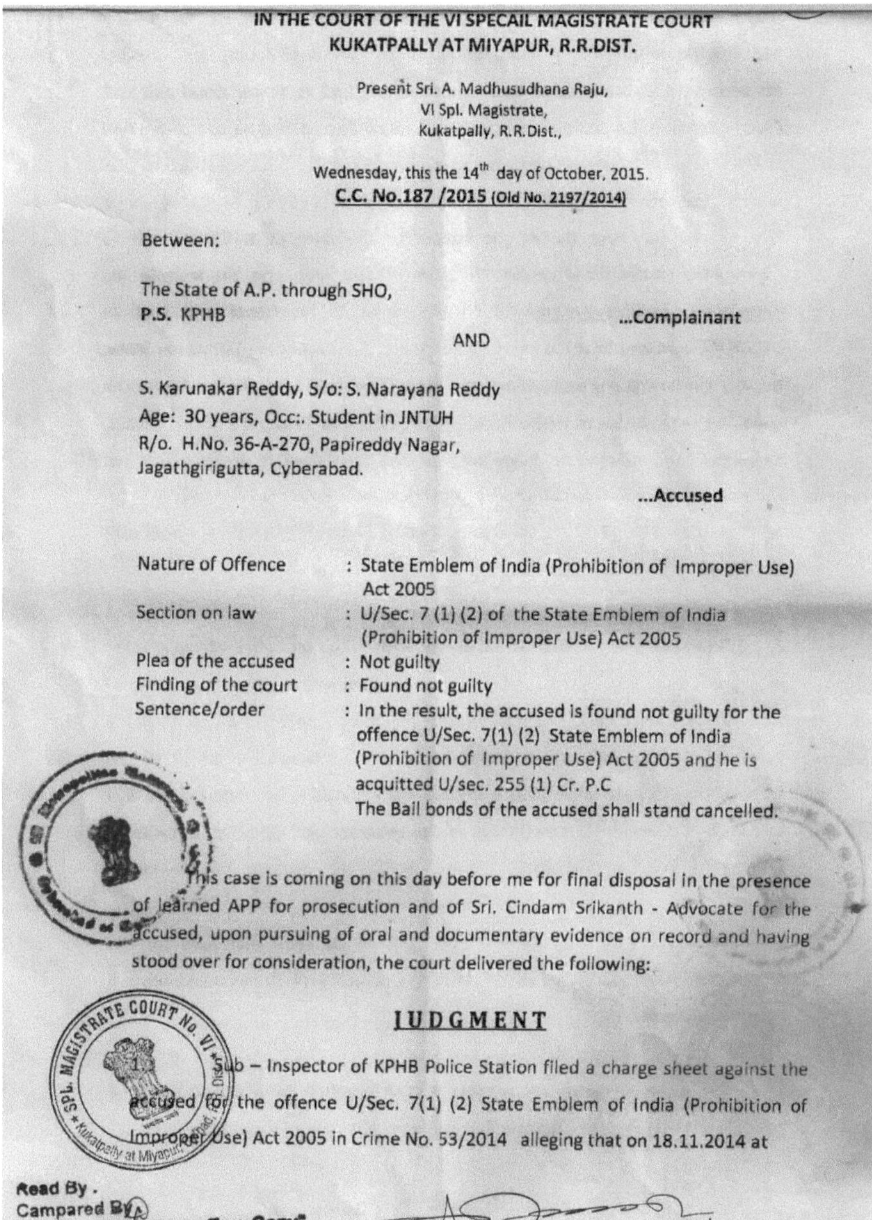

IN THE COURT OF THE VI SPECAIL MAGISTRATE COURT
KUKATPALLY AT MIYAPUR, R.R.DIST.

Present Sri. A. Madhusudhana Raju,
VI Spl. Magistrate,
Kukatpally, R.R. Dist.,

Wednesday, this the 14th day of October, 2015.
C.C. No.187 /2015 (Old No. 2197/2014)

Between:

The State of A.P. through SHO,
P.S. KPHB ...Complainant

AND

S. Karunakar Reddy, S/o: S. Narayana Reddy
Age: 30 years, Occ:. Student in JNTUH
R/o. H.No. 36-A-270, Papireddy Nagar,
Jagathgirigutta, Cyberabad.
...Accused

Nature of Offence	: State Emblem of India (Prohibition of Improper Use) Act 2005
Section on law	: U/Sec. 7 (1) (2) of the State Emblem of India (Prohibition of Improper Use) Act 2005
Plea of the accused	: Not guilty
Finding of the court	: Found not guilty
Sentence/order	: In the result, the accused is found not guilty for the offence U/Sec. 7(1) (2) State Emblem of India (Prohibition of Improper Use) Act 2005 and he is acquitted U/sec. 255 (1) Cr. P.C The Bail bonds of the accused shall stand cancelled.

This case is coming on this day before me for final disposal in the presence of learned APP for prosecution and of Sri. Cindam Srikanth - Advocate for the accused, upon pursuing of oral and documentary evidence on record and having stood over for consideration, the court delivered the following:

JUDGMENT

1. Sub – Inspector of KPHB Police Station filed a charge sheet against the accused for the offence U/Sec. 7(1) (2) State Emblem of India (Prohibition of Improper Use) Act 2005 in Crime No. 53/2014 alleging that on 18.11.2014 at

Read By.
Campared By
"Certified to be Xerox True Copy"

Fig. 2 Judgment page 1

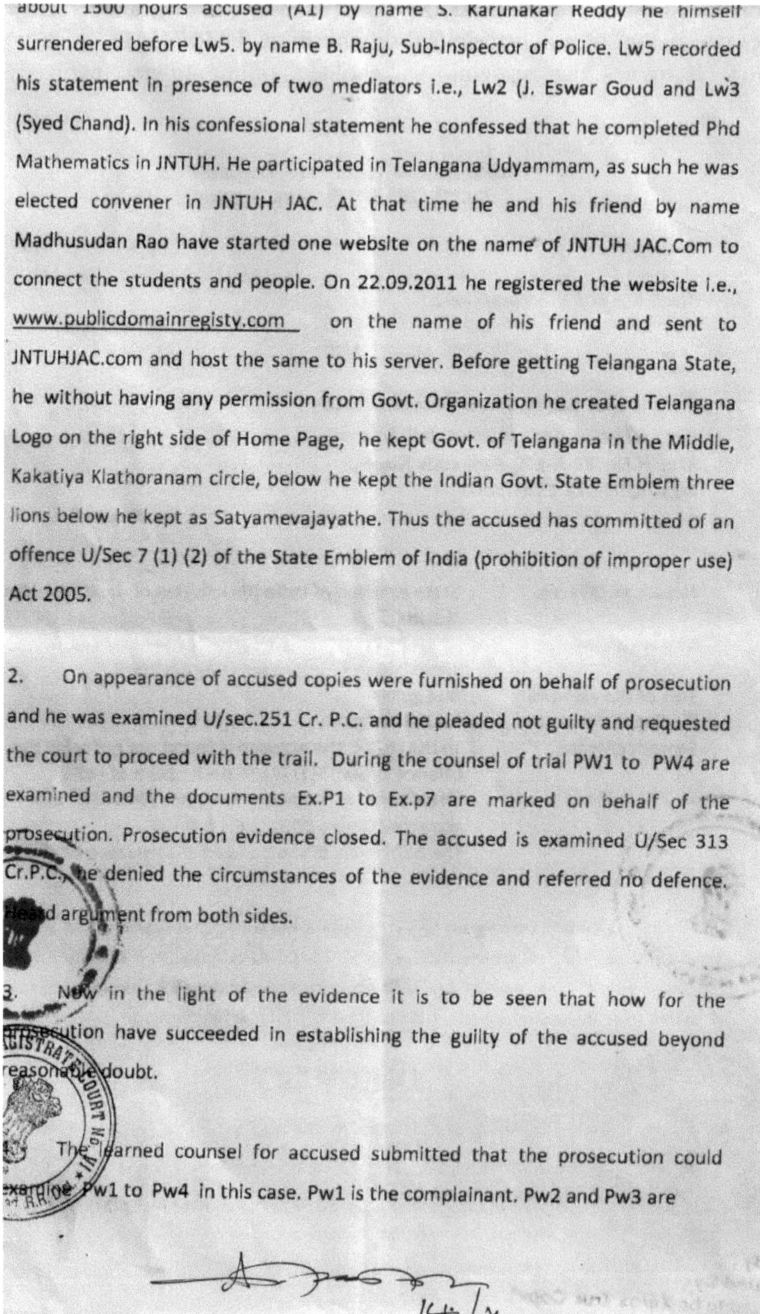

about 1500 hours accused (A1) by name S. Karunakar Reddy he himself surrendered before Lw5. by name B. Raju, Sub-Inspector of Police. Lw5 recorded his statement in presence of two mediators i.e., Lw2 (J. Eswar Goud and Lw3 (Syed Chand). In his confessional statement he confessed that he completed Phd Mathematics in JNTUH. He participated in Telangana Udyammam, as such he was elected convener in JNTUH JAC. At that time he and his friend by name Madhusudan Rao have started one website on the name of JNTUH JAC.Com to connect the students and people. On 22.09.2011 he registered the website i.e., www.publicdomainregisty.com on the name of his friend and sent to JNTUHJAC.com and host the same to his server. Before getting Telangana State, he without having any permission from Govt. Organization he created Telangana Logo on the right side of Home Page, he kept Govt. of Telangana in the Middle, Kakatiya Klathoranam circle, below he kept the Indian Govt. State Emblem three lions below he kept as Satyamevajayathe. Thus the accused has committed of an offence U/Sec 7 (1) (2) of the State Emblem of India (prohibition of improper use) Act 2005.

2. On appearance of accused copies were furnished on behalf of prosecution and he was examined U/sec.251 Cr. P.C. and he pleaded not guilty and requested the court to proceed with the trail. During the counsel of trial PW1 to PW4 are examined and the documents Ex.P1 to Ex.p7 are marked on behalf of the prosecution. Prosecution evidence closed. The accused is examined U/Sec 313 Cr.P.C, he denied the circumstances of the evidence and referred no defence. Heard argument from both sides.

3. Now in the light of the evidence it is to be seen that how for the prosecution have succeeded in establishing the guilty of the accused beyond reasonable doubt.

4. The learned counsel for accused submitted that the prosecution could examine Pw1 to Pw4 in this case. Pw1 is the complainant. Pw2 and Pw3 are

14/x

Fig. 3 Judgment page 2

panch witness of confessional statement. Pw4 is the I.O. Among these witness Pw2 and Pw3 who are said to be present as mediators at the time of reçording the confessional statement of the accused by the police have turned hostile and not supported to the prosecution case. As per their testimony both of them stated that the police called them to the police station and obtained their signatures on some white papers without disclosing any reason and no panchanama was conducted in their presence by the police. Both of them did not known anything about this case. Thus nothing has been elicited with regard to the factum of recording the confessional statement of the accused by the police in their presence. Therefore there is no weight in their evidence. Coming to the evidence of Pw1, he deposed that he is working as professor in JNTUH University, Hyderabad since from the year 2006. He know the accused as a scholar of the Ph.d student in the same JNTUH Universality. In the last quarter of the year 2012 he saw the website of **JNTUHJAC.COM** and he found one organizing group existing nearly two thousand people by violating of National Emblem of India (provision of improper act). Pw1 further stated that the said website bearing the above emblem is being under usage for the last four years which comes under the violation of the group organization. Pw1 found the photograph of the accused in the said website on base of which he lodged a complaint before the police. The contention of the counsel is that no doubt, Pw1 identified the accused photo in the said website to establish that fact. The I.O. did not collect any proper documentary proof to that affect and filed along with the charge sheet. Ex.P4 and Ex.P5 does not disclose as authenticated to believe the version of the prosecution case and simply appearance of the photograph of the accused does not mean to establish about his participation in the said mob, as stated by the Pw1. Apart of that the document Ex.P5 clearly disclose about the name of one Madhusudhana Rao who is also said to be participated and associated in the said mob. But the charge sheet does not disclose about his role to that affect. Now the prosecution relied upon the evidence of Pw4 who investigated the matter. Pw4 stated that he recorded the confessional statement of accused in presence of the mediators (Pw2 and Pw3) But in support of his version there is no such evidence through PW2 and Pw3 since they have turned hostile. Apart of that Pw4 accepted that

Fig. 4 Judgment page 3

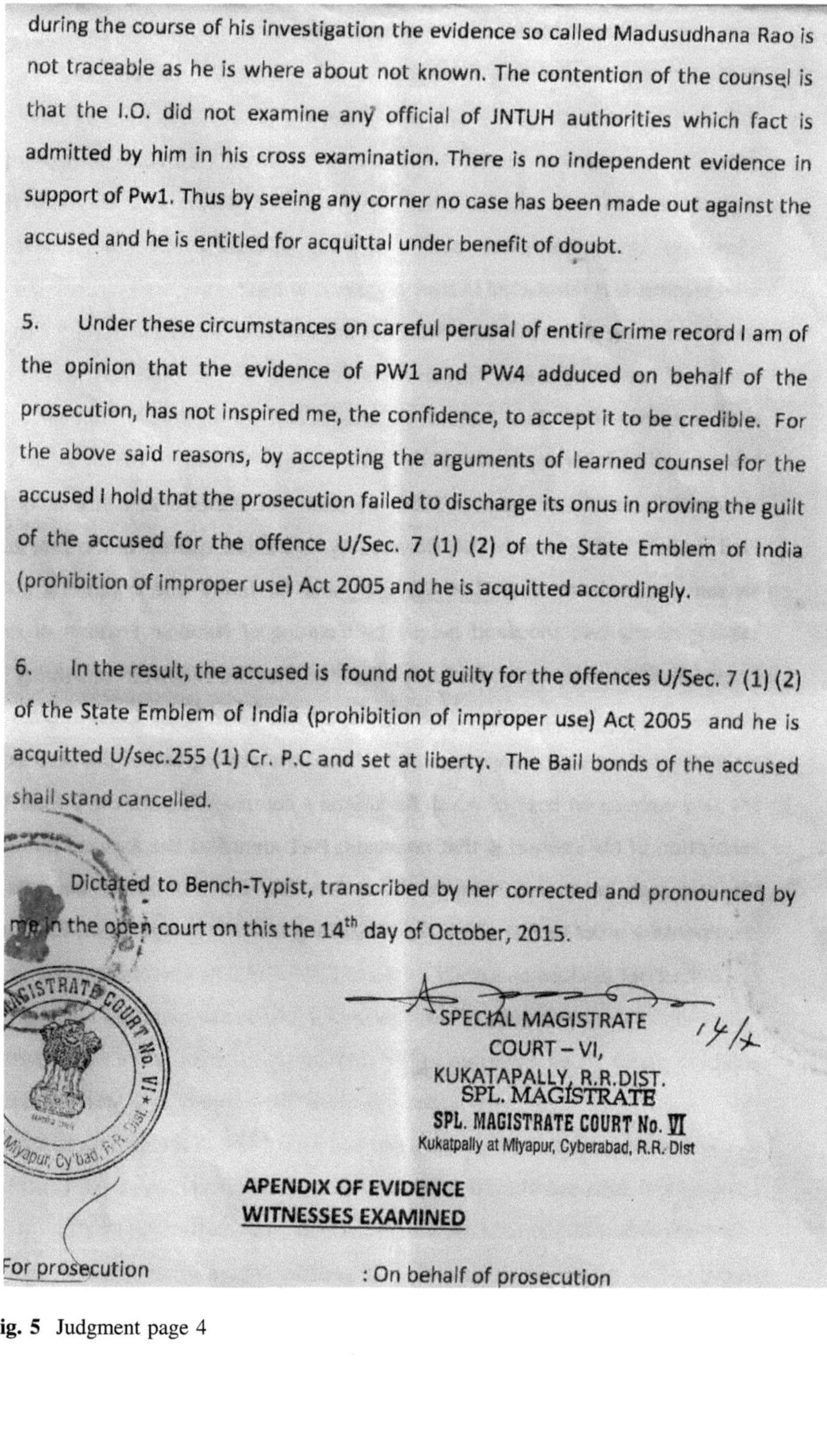

during the course of his investigation the evidence so called Madusudhana Rao is not traceable as he is where about not known. The contention of the counsel is that the I.O. did not examine any official of JNTUH authorities which fact is admitted by him in his cross examination. There is no independent evidence in support of Pw1. Thus by seeing any corner no case has been made out against the accused and he is entitled for acquittal under benefit of doubt.

5. Under these circumstances on careful perusal of entire Crime record I am of the opinion that the evidence of PW1 and PW4 adduced on behalf of the prosecution, has not inspired me, the confidence, to accept it to be credible. For the above said reasons, by accepting the arguments of learned counsel for the accused I hold that the prosecution failed to discharge its onus in proving the guilt of the accused for the offence U/Sec. 7 (1) (2) of the State Emblem of India (prohibition of improper use) Act 2005 and he is acquitted accordingly.

6. In the result, the accused is found not guilty for the offences U/Sec. 7 (1) (2) of the State Emblem of India (prohibition of improper use) Act 2005 and he is acquitted U/sec.255 (1) Cr. P.C and set at liberty. The Bail bonds of the accused shall stand cancelled.

Dictated to Bench-Typist, transcribed by her corrected and pronounced by me in the open court on this the 14th day of October, 2015.

SPECIAL MAGISTRATE
COURT – VI,
KUKATAPALLY, R.R.DIST.
SPL. MAGISTRATE
SPL. MAGISTRATE COURT No. VI
Kukatpally at Miyapur, Cyberabad, R.R. Dist

MAGISTRATE COURT No. VI ★ R.R. Dist. Miyapur, Cy'bad

APENDIX OF EVIDENCE
WITNESSES EXAMINED

For prosecution : On behalf of prosecution

Fig. 5 Judgment page 4

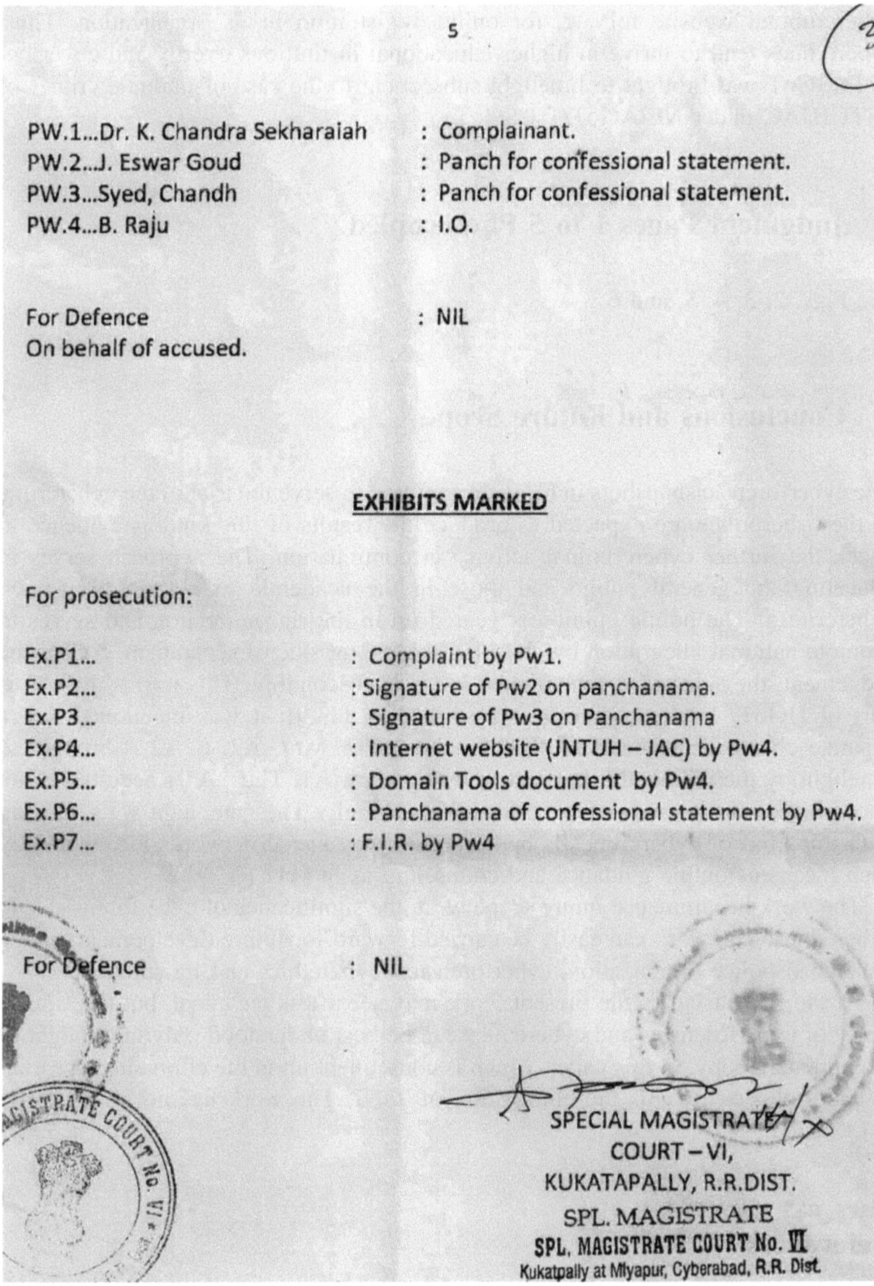

5

PW.1...Dr. K. Chandra Sekharaiah	:	Complainant.
PW.2...J. Eswar Goud	:	Panch for confessional statement.
PW.3...Syed, Chandh	:	Panch for confessional statement.
PW.4...B. Raju	:	I.O.

For Defence : NIL
On behalf of accused.

EXHIBITS MARKED

For prosecution:

Ex.P1...	:	Complaint by Pw1.
Ex.P2...	:	Signature of Pw2 on panchanama.
Ex.P3...	:	Signature of Pw3 on Panchanama
Ex.P4...	:	Internet website (JNTUH – JAC) by Pw4.
Ex.P5...	:	Domain Tools document by Pw4.
Ex.P6...	:	Panchanama of confessional statement by Pw4.
Ex.P7...	:	F.I.R. by Pw4

For Defence : NIL

SPECIAL MAGISTRATE
COURT – VI,
KUKATAPALLY, R.R.DIST.
SPL. MAGISTRATE
SPL. MAGISTRATE COURT No. VI
Kukatpally at Miyapur, Cyberabad, R.R. Dist.

Fig. 6 Judgment page 5

cybercriminal website for e.g. for online registration in its organization. Thus, cybercrimes tend to thrive in higher educational institutions overtly and covertly.

FCSGoT was brought to limelight subsequent to the case of multiple crimes of JNTUHJAC under NEIAGoT/TsCSGoT.

3 Judgment Pages 1 to 5 Photocopied

See Figs. 2, 3, 4, 5, and 6.

4 Conclusions and Future Scope

The cyberforensic snapshots in [4] were captured to serve the trial of the cybercrime as the cyberpolice are expected to produce the results of this kind as evidence to check the further cybercriminal activity in continuation. The approach serves in educating the general public and those in the academia to abstain from such cybercrimes. The public opinion is geared up in the right direction and serves to promote national integration by checking the wrong side of separatism. As per the judgement, the accused A2 mentioned in [4] is absconding. This means that A2 is part of UGoT. In the chargesheet described in [4, 9], it was mentioned that a separate chargesheet would be filed by the police w.r.t. A2. If A2 is brought to limelight by the police, the case may get dug w.r.t. A1. Thus, A1's acquittal is not altogether complete and not once for all, seemingly. The campaign w.r.t. dealing with this kind of cybercrimes is better possible by social networks means and by such means as online guidance and counseling as in [11].

The work has immense future scope w.r.t. the significance of cyberforensics and cyberethics. The work can easily be carried forward for future developments in the cyberabad police organization, cyberforensics, cyberethics and jurisprudence.

At the first instance, the present work may seem less technical, but the significance of cyberforensics and cyberethics can be best understood only in the light of the judgment copy, in this paper, which is consequent upon the efforts in our earlier research papers towards the elimination of TsNI. The work has not much to do w.r.t. PGoT.

References

1. UshaGayatri P., Neeraja S., Leela Poornima Ch., Chandra Sekharaiah K. and Yuvaraj M., "Exploring Cyber Intelligence Alternatives for Countering Cyber Crime", Proceedings of the 8th INDIACom; INDIACom-2014, International Conference on "*Computing for Sustainable Global Development*", 5–7March2014, BharatiyaVidyapeeth's Institute of Computer Applications and Management (BVICAM), New Delhi (INDIA).

2. B. Tirupathi Kumar, K Chandra Sekharaiah, P Mounitha, "*A Case Study Of Web Content Mining In Handling Cybercrime*", In Proceedings of 2nd International Conference on Science, Technology and Management, 27 Sep 2015, Delhi University, New Delhi, pp. 2290–2293.
3. B. Tirupathi Kumar, K Chandra Sekharaiah, P Mounitha, "*A Case Study Of Web Content Mining In Handling Cybercrime*", International Journal of Advance Research in Science and Engineering, Vol. No. 4, Special Issue (01), Sep 2015, ISSN 2319–8354, pp 665–668.
4. P. UshaGayatri, K. Chandra Sekharaiah, D. Radhika, G. Sruthi, K. Satish, S. Mounika, K. Shravani, Ambuja Kulshreshtha, "*Exploring Cyber Intelligence Alternatives for Countering Cyber Crime: A Continuing Case Study for the Nation*", in Proceedings of the CSI-2015 Intnl. Conf.@BharatiVidyapeeth's Institute of Computer Applications and Management (BVICAM), New Delhi, (INDIA), Dec. 2015 (*Presented*).
5. B. Tirupathi Kumar, K Chandra Sekharaiah, P Mounitha, "*Towards National Integration by Analyzing a Case Study of Cybercrimes*", In Second International Conference on Information and Communication Technology for Competitive Strategies (ICTCS-2016), March 4–5, 2016 (*Accepted*).
6. http://www.rmlnlu.ac.in/webj/sedition.pdf.
7. B. Tirupathi Kumar, K. Chandra Sekharaiah, B. Rajinikanth, "*Web Data Mining in E-commerce*", in Proceedings of National Conference on Technical Advancements in Computer Science and Engineering [NCTA-CSE-2014], pp. 5–8.
8. https://archives.org/web.
9. https://sites.google.com/site/chandraksekharaiah/miscellaneous333.
10. Kohli, V.; Saxena, S.; Patni, J., "A SURVEY—Academic demolition via internet addiction", 2nd International Conference on Computing for Sustainable Global Development (INDIACom), 2015, pp. 1769–1771.
11. https://sites.google.com/site/chandraksekharaiah/india-against-corruption-jntu.

New Approach for Performance and Security Enhancement in OCDMA Networks

Sumit Gupta and Aditya Goel

Abstract A new technique is proposed for the enhancement of security against the eavesdropper in Optical code division multiple access (OCDMA) Spectral amplitude coding (SAC) System. This technique is the combination of real and virtual user's spectral chip, according to the newly developed Zero cross correlation code. The define sequence assign to virtual users, is complement to each other in a group of two users. Probability of code detection of individual user decreases due to virtual user and complement of defined sequence of virtual user, removes the occurrence of single user's code in the channel. The simulation and analytical results show that the proposed scheme has a better performance than existing methods.

Keywords Optical code division multiple access (OCDMA) · Modified quadric code (MQC) · Multi diagonal code (MD) · Multiple access interference (MAI) · Spectral amplitude code (SAC) · Random diagonal code (RD) · Zero cross correlation code (ZCC)

1 Introduction

The main aim of CDMA system is to extract the information of a user in the presence of another user without affecting the security of system because other users interfere the desired user that is called the MAI (multiple access interference) [1]. That depends upon the cross-correlation between the users. It may be one or zero. In case of zero cross co-relation the MAI is reduced [2, 3]. Many methods exist with ideal in phase unity and zero cross correlation such as RD Code, MQC

S. Gupta (✉) · A. Goel
Department of Electronics and Communication Engineering,
Maulana Azad National Institute of Technology, Bhopal, India
e-mail: mtsumit.g@gmail.com

A. Goel
e-mail: Adityagoel2@rediffmail.com

N. Modi et al. (eds.), *Proceedings of International Conference on Communication and Networks*, Advances in Intelligent Systems and Computing 508,
DOI 10.1007/978-981-10-2750-5_68

and MD Code. At the decoder side in the MD method direct detection technique is applied for detecting the information [4–6]. In MD code every pulse represents the information [7]. If an eavesdropper detects the any pulse so it is easy for eavesdropper to retrieve the information and this can be avoided by reducing the amount of power of individual chip by increasing the number of weights of a user but this approach requires the wider spectral width [8]. So a technique is described to eliminate t above mention issues, as explain in following section. Section 2 constructs the proposed method, Sect. 3 Analyzes the performance of the system. Security and Result analysis is done in Sects. 4 and 5. Paper ends with the conclusion.

2 Proposed Method

Coding of the system is defined by the parameters (N_r, N_V, W_r, W_v, L, λ_c). Where N_r is the number of real users, W_r is their weight and N_V is virtual user, W_v is their weight, L is the length of the code and λ_c is the ideal in phase cross correlation between the users.

The basic structure of the technique is shown in Fig. 1. In this method the real users of N_r number, W_r weight sand virtual users of N_v number, W_v weights are transmitted by modulated spectral chip simultaneously as specific coding method.

A defined sequence is provided to each virtual user and the sequence is complements to each other in a group of two users. The length of code is given as

$$\text{Length of code } L = N_r W_r + N_v W_v \quad (1)$$

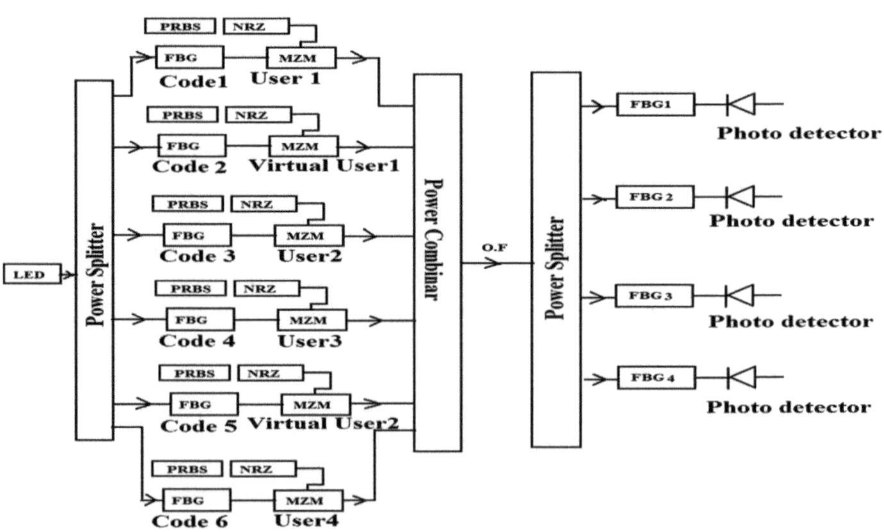

Fig. 1 Encoder and decoder design of the technique

Table 1 Code distribution of proposed method for $N_r = 6$, $W_r = 2$ and $N_v = 4$, $W_v = 2$

		λ_1	λ_2	λ_3	λ_4	λ_5	λ_6	λ_7	λ_8	λ_9	λ_{10}	λ_{11}	λ_{12}	λ_{13}	λ_{14}	λ_{15}	λ_{16}	λ_{17}	λ_{18}	λ_{19}	λ_{20}
Data U1	N_{r1}	0	0	0	0	1	0	0	0	0	0	0	0	0	0	1	0	0	0	0	0
10101010	N_{v1}	0	0	0	1	0	0	0	0	1	0	0	0	0	0	0	0	0	0	0	0
Data U2	N_{r2}	0	0	0	0	0	0	0	0	0	1	0	0	0	0	0	1	0	0	0	0
Data U3	N_{r3}	0	0	0	0	0	0	0	0	0	0	1	0	0	0	0	0	1	0	0	0
01010101	N_{v2}	0	0	1	0	0	0	0	1	0	0	0	0	0	0	0	0	0	0	0	0
Data U4	N_{r4}	0	0	0	0	0	0	0	0	0	0	0	1	0	0	0	0	0	1	0	0
Data U5	N_{r5}	0	0	0	0	0	0	0	0	0	0	0	0	1	0	0	0	0	0	1	0
10101010	N_{v3}	0	0	0	0	0	0	1	0	0	0	0	0	0	0	0	0	0	0	0	0
Data U6	N_{r6}	0	0	0	0	0	0	0	0	0	0	0	0	0	1	0	0	0	0	0	1
010101010	N_{v4}	1	0	0	0	0	1	0	0	0	0	0	0	0	0	0	0	0	0	0	0

The code formation is given in the following steps:

Step 1. The basic code is generated and assign to a first user. The basic code is formed by placing the zeros between the equal weights. The number of zeros is equal to the sum of virtual users and the real users'. The position of basic code (P_B) in code distribution is given by $P_B = \left(\text{Number of virtual user}(N_V) \times \left(\frac{W_v}{2}\right)\text{half weight of virtual users}\right) + 1$. Table 1 shows the code construction of the design.
Step 2. The basic code is arranged for first virtual user in position, that the half weight of code is placed in $2 \times \left(\text{Number of virtual user}(\mathbf{N_V})\right) + \left(\frac{W_v}{2}\right)\text{half weight of virtual users}\big)$ position and code for remaining virtual user shifted left with $\left(\frac{W_v}{2}\right)$ half weight of virtual users

3 Mathematical Analysis

For analysis of this system we use the Gaussian approximation in our calculation (3, 4). This system is based on the zero cross co-relation so, then we only consider the thermal noise (R_{th}) and shot noise (R_{sn}) in respect to PIIN. The SNR for electrical signal is the average signal power to noise power, SNR = [I/R]. Let $C_K(i)$ denotes the ith element of K user in this code ZCC than the following assumption is made

(a) Each light source spectrum is flat over the bandwidth [V_o – ΔV/2, V_o – ΔV/2] where V_o is central frequency and ΔV is the optical source bandwidth in Hertz.
(b) Each power spectral component has an identical spectral width.
(c) Each user has nearly equal power at the transmitter.
(d) Each user bit stream is synchronized.

The power spectral density (PSD) of the received signals can be given as

$$r(v) = \frac{P_{sr}}{\Delta v}\sum_{N=1}^{N} d_N \sum_{i=1}^{L} c_N(i)\text{rect}(i) \tag{2}$$

$$\text{rect}(i) = u\left[v - v_0 - \frac{\Delta v}{2L}(-L+2i-2)\right] - u\left[v - v_0 - \frac{\Delta v}{2L(-L+2i)}\right] = u\left[\frac{\Delta v}{L}\right] \tag{3}$$

where $u(v)$ is the unit step function expressed as:

$$u(v) = \begin{cases} 1, & v \geq 0 \\ 0, & v < 0 \end{cases} \tag{4}$$

$$\int_0^\infty G(v)\,dv = \int_0^\infty \left[\frac{P_{sr}}{\Delta v}\sum_{N=1}^{N} d_N \sum_{L=1}^{L} C_N(i)C_I(i)\mathrm{rect}(i)\right] dv \tag{5}$$

$$\int_0^\infty G(v)\,dv = \frac{P_{sr}}{\Delta v}\left[\sum_{N=1}^{N} d_N \cdot W_r \cdot \frac{\Delta v}{L} + \sum_{N\neq 1}^{N} d_N \cdot 0 \cdot \frac{\Delta v}{L}\right] \tag{6}$$

The value of $\sum_{N=1}^{N} d_N$ is equal to the 1 and for above than

$$\int_0^\infty G_{dd}(v)\,dv = \frac{P_{sr} W_r}{L} \tag{7}$$

The photo current I can be expressed as

$$I = I_{dd} = \Re \int_0^\infty G_{dd}(v)\,dv \tag{8}$$

The variation of photocurrent due to detection of an ideally un polarized thermal light can be expressed as

$$I = \Re\left[\frac{P_{sr}[W_r]}{L}\right] \tag{9}$$

$$I^2 = 2eB(I_{dd}) + \frac{4K_b T_n B}{R_L} \tag{10}$$

When all users transmitting 1 than probability of each user sending 1 is ½ than Eq. (11) becomes

$$I^2 = \frac{P_{sr} eB\Re}{L}[w_r] + \frac{4K_b T_n B}{R_L} \tag{11}$$

The signal to noise ratio of direct detection technique is given by following equation

$$\mathrm{SNR} = \frac{I_{dd}^2}{I^2} \tag{12}$$

When putting all equation than new formula for SNR will be

$$\mathrm{SNR} = \frac{\frac{\left(\Re^2 P_{sr}^2\right)}{N_r}\left(\frac{mn}{nm+1}\right)^2}{\left(\frac{P_{sr} eB\Re}{N_r}\right)\frac{mn}{(1+nm)} + \frac{4K_b T_n B}{R_L}} \tag{13}$$

This BER is

$$\mathrm{BER} = \frac{1}{2}\mathrm{erfc}\sqrt{\frac{\mathrm{SNR}}{8}} \tag{14}$$

Typical parameters used in the calculation as below:

Photo detector quantum efficiency (η)	0.6
Line-width broadband source (ΔV)	3.75 THz
Operating wavelength (k_O)	1552 nm
Electrical bandwidth (B)	311 MHz
Data bit rate (R_b)	622 Mbps
Receiver noise temperature (T_n)	300 K
Receiver load resistor (R_L)	1030 Ω

If n $\gg$1 then or m $\gg$1 then

$$\mathrm{SNR} = \frac{\frac{\left(\Re^2 P_{sr}^2\right)}{N_r^2}}{\left(\frac{P_{sr} eB\Re}{N_r}\right) + \frac{4K_b T_n B}{R_L}} \tag{15}$$

Where $w_r = mw_v$, $N_r = nN_v$.

4 Security Level Analysis

(i) Probability of presence of single real user in network =

$$1/2^{N_r} \tag{16}$$

At a time the probability of code detection of single user in group of virtual user =

$$\left(\left(\frac{N_V}{2}\right)^{w_V}\right)_{C_{(W_r)}} \tag{17}$$

Than the total probability of code detection of single user in group of virtual users is given as

$$P_S = \frac{1}{2^{N_r}}\left(\left(\binom{N_V}{2}W_V + W_r\right)_{C(W_r)}\right) \tag{18}$$

(ii) Probability of code estimation by whole code of a user

$$P_G = \left(\frac{W_r}{(L_r + L_v)}\right)\left(\frac{W_r - 1}{(L_r + L_v) - 1}\right) \times \cdots \times \left(\frac{1}{(L_r + L_v) - W_r}\right) \tag{19}$$

$$P_G = \left(\frac{W_r}{N_r\left(\frac{mn+1}{mn}\right)}\right)\left(\frac{W_r - 1}{N_r\left(\frac{mn+1}{mn}\right) - 1}\right) \times \cdots \times \left(\frac{1}{N_r\left(\frac{mn+1}{mn}\right) - W_r}\right) \tag{20}$$

5 Result and Discussion

The basic characteristics of OCDMA involve the security of the system against the eavesdropper. The security analysis is done in two modes, one is presence of single user and another is the presence of more than single users. As shown in Fig. 2 probability of code detection by unwanted user, in the presence of single user is reduced as compared to the presence techniques. The Code detection probability in presence of all users is calculated. Figure 3 shows the probability of W weight detection. The probability is lower compared to the existing coding method so confidentiality of the system is increased.

The block diagram of proposed scheme shown in Fig. 1. The simulation is done for 4 real user and 2 virtual with a weight of 2 as shown in Fig. 4. The width of each spectral chip kept 0.6 nm. The simulation is done in a practical environment in all, with all nonlinear effect is kept on. Simulation is performed at 1.25 Gbit/s for 20 km length of fiber with ITU standard single mode fiber (SMF). All the attenuation (α = 0.25 dB/km), Dispersion (18 ps/nm) is maintained. Decoder side after decoding the signal, the signal covert to electrical by passing to the photo detector and 0.75 GHz low pass Bessel filter (LPF). The dark current value was 5 nA, and the thermal noise coefficient was 1.8×10^{-23} W/Hz for each of the photo-detectors. The performance of the system was characterized by referring to the BER and eye pattern. A BER of the proposed Scheme shown that the better performance as the number of users are increased as shown in Fig. 5.

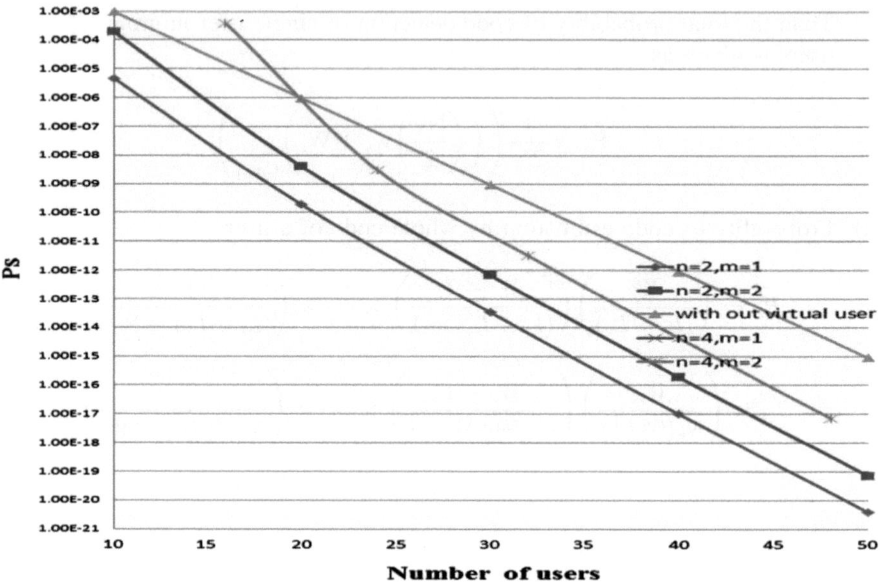

Fig. 2 Variation in probability of code detection of single user (P_S) with the number of users

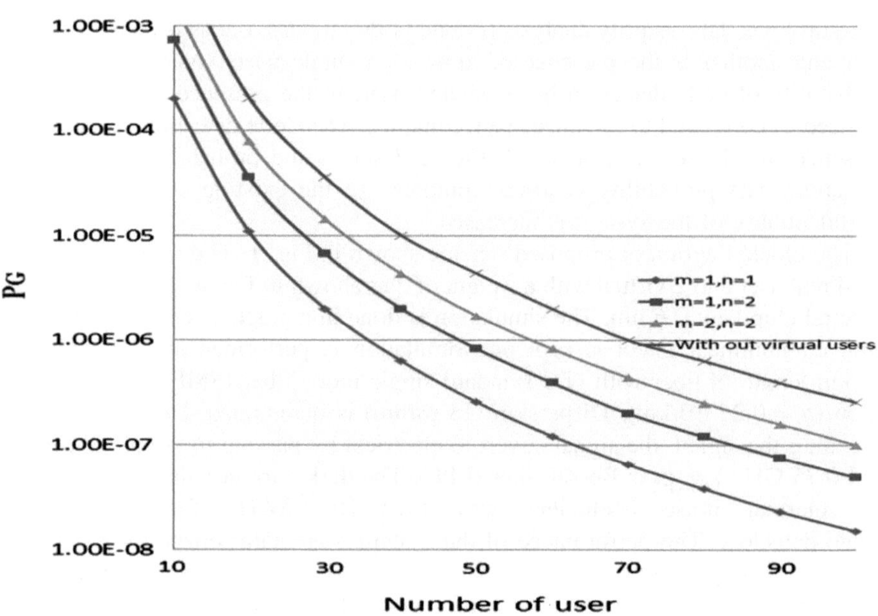

Fig. 3 Variation in probability of code detection with W weight (P_G), with number of users

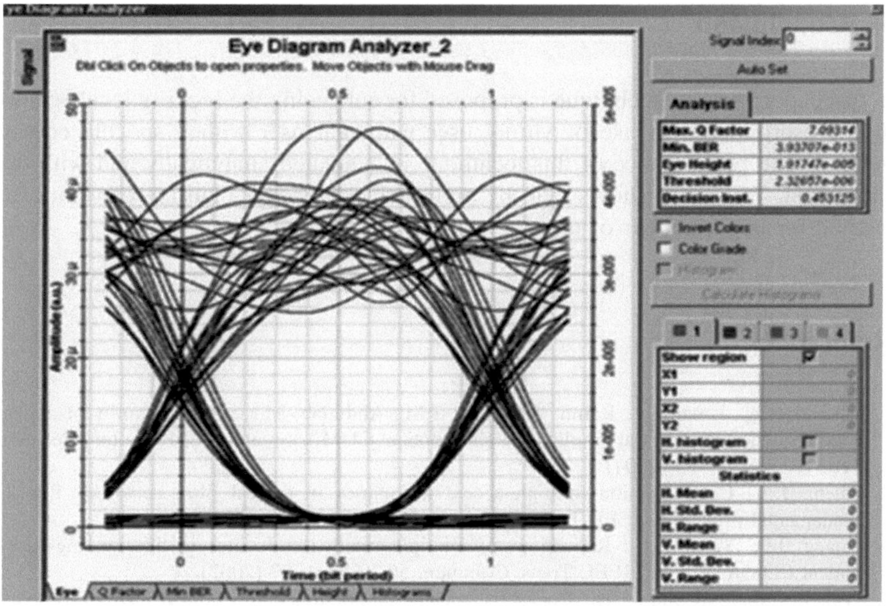

Fig. 4 Eye pattern for 4 real and 2 virtual users at 20 for 1.25 Gbits/s

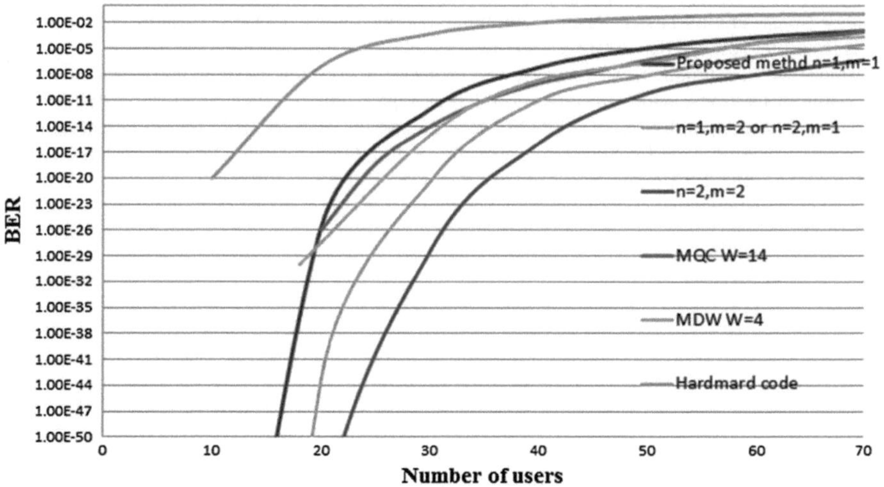

Fig. 5 BER variation with number of users at 622 Mbits/s

6 Conclusion

In this paper, the new technique is proposed for enhancing the security level against the eavesdropper with use of virtual user with real user with a specific coding method. The performance of this technique is evaluated analytically and with the simulation. The result shows that the performance is better than the conventional method for large number of users.

References

1. Aljunid, S.A., Ismail, M., Ramli, A.R., Ali, B.M., Abdullah, M.K.: A new family of optical code sequence for spectral amplitude coding optical CDMA systems. IEEE Photon. Technol. Lett. 16, 2383–2385 (2004).
2. Salehi, J.A.: Code division multiple-access techniques in optical fiber networks. Part 1: fundamental principles. IEEE Trans. Commun. 37 (1989).
3. Huang, J.F., Yang, C.C.: Reductions of multiple-access interference in fiber-grating-based optical CDMA network. IEEE Trans. Commun. 50, 1680–1687 (2002).
4. Wei, Z., Ghafouri-Shiraz, H.: Codes for spectral-amplitude-coding optical CDMA systems. J. Ligtwave Technol. 50, 1209–1212 (2002).
5. Fadhil, H.A., Aljunid, S.A, Badlisha, R.: Random diagonal code for spectral amplitude-coding optical CDMA system. Int. J. Comput. Sci. Network Secur. 7, 258–262 (2007).
6. Wei, X., shalaby, H.M.H., Ghafouri-Shiraz, H.: Modified quadratic congruence codes for fiber-Bragg grating-based spectral amplitude-coding optical CDMA systems. J. Lightwave Technol. 9, 1274–1281 (2001).
7. Anuar, M.S., Aljunid, S.A., Saad, N.M., Hamzah, S.M.: New design of spectral amplitude coding in OCDMA with zero cross correlation. Opt. Commun. 282, 2659–2664 (2009).
8. Anuar, M.S., Aljunid, S.A., Saad, N.M., Mohammed, A., Babekir, E.I.: Development of a zero cross-correlation code for Spectral-Amplitude Coding Optical Code Division Multiple Access (OCDMA). Int. J. Comput. Sci. Network Secur. 6, 180–184 (2006).

Decision-Based Spectral Embedding Approach for Identifying Facial Behaviour on RGB-D Images

Deepak Kumar Jain, Raj Kumar and Neha Jain

Abstract Automatic identification of various facial movements and expressions with high recognition value is important for human computer interaction as the facial behaviour of a human can be treated as an important factor for information representation as well as communication. A high deviation of human appearance and existence of noisy contextual background makes the human pose analysis is hard to achieve. A number of basic factors such as cluttered background, occlusion, and camera movement and illumination variations degrade the image quality resulting in poor performance for identifying different facial expressions. Moreover the identification of the automatic feature detection in facial behaviour requires high degree of correlation between the training and test images. Our proposed work tries to address the mentioned problems and resolve to some extent. In this methodology, a Decision-based Spectral Embedding approach combining appearance and geometry based features for head pose estimation and facial expression recognition by minimizing the objective function which leads to selection of optimal set of fiducial points. The method preserves the local information from different facial views for mapping neighbouring input to its corresponding output, resulting in low dimensional representation for encoding the relationships of the data. The proposed methodology is validated with the RGB-D data set and real depth images and compared with the state-of-art methods for analyzing the performance of recognition of facial behavior.

Keywords Geometry based features · Pose estimation · Facial expression · RGB-D data

---

D.K. Jain (✉)
University of Chinese Academy of Sciences, Beijing, China
e-mail: deepak.jain.juet@gmail.com

R. Kumar
SGTBIMIT College, Delhi, India
e-mail: raj_kashyap12@rediffmail.com

N. Jain
Jaypee University of Engineering and Technology, Raghogarh-Vijaypur, India
e-mail: neha.juet@gmail.com

N. Modi et al. (eds.), *Proceedings of International Conference on Communication and Networks*, Advances in Intelligent Systems and Computing 508,
DOI 10.1007/978-981-10-2750-5_69

1 Introduction

Facial expression and behaviour identification is the challenging task in computer vision. Facial expressions convey non-verbal cues, which play an important role in interpersonal relations. Automatic recognition of facial expressions can be an important component of natural human-machines interfaces; it may also be used in behavioural science and in clinical practice. Although humans recognize facial expressions virtually without effort of delay, reliable expression recognition by machine is still a challenge. The human pose estimation/recognition component is a key step in an overall human action recognition system. Our detection and recognition scheme must also be capable of tolerating variations in the faces themselves. The human face is not a unique rigid object. There are billions of different faces and each of them can assume a variety of deformations. Inter-personal variations can be due to race, identity, or genetics while intra-personal variations can be due to deformations, expression, aging, facial hair, cosmetics and facial paraphernalia.

Exceptions to this trend include a small number of works, In 2007 Zhao et al. [1] they are taking the problem of various facial expression prediction, the algorithm they used for their method is volume local binary pattern [2], they faced some problems, the work not well suitable for complex action prediction. In the actions with a complex articulated structure, the motions of the individual parts may be correlated. The relationship among these parts (or high-order features) is often more discriminative than the individual ones.

In 2010, Tong et al. [3], they are taking the problem of Spontaneous Facial Action Recognition, the method they used in their work was Shape based features-2D Local Facial Components Shapes, the main problem they were faced is Time Complexity.

In 2011 Diago et al. [4] they are taking the problem of various facial expression prediction using Fuzzy Quantized HNN Method, the problem with this paper is that they only used some limited databases.

In 2012 Huang et al. [5], they find the 3D facial Expression Prediction using Extended Local Binary Patterns, they faced some computational Complexity in their work.

In 2013, Rahulamathavan et al. [6], this paper presents a system that addresses the challenge of performing facial expression recognition when the test image is in the encrypted domain, they used Local Fisher Discriminant Analysis Method, Encryption also applied for security purpose. It may increase computational complexity.

2 Drawback of Existing Methods

In the existing approaches for the identification of facial behaviour, the automatic feature detection is a difficult task. These techniques are highly expensive and require a high degree of correlation between the training and the test images.

Moreover the dimensionality becomes another major factor for the detection process as the feature set in high dimensional data needs to be reduced for the selection optimizing the detection. Some methods perform the classification by majority schemes based on training examples, however the annotator labelling is dependent on the facial expression that transitively depends on correlation among expression that reduces the performance due to same labelling in some cases. The method of SIFT with K-NN Classifier when applied for occluded images and partial variations results in high dimensionality and improper results with local variations. The existing methods suffer from the extraction of proper facial feature points for identification of facial behaviour. Some feature extraction methods produces distinctive and repeatable scale invariance features. The change in image causes high sensitivity in the parameters resulting in low contrast between structural information and background pixels. The proposed method combines appearance and geometry based features for facial expression recognition that addresses the problem by achieving higher robustness and uniformly distributed relevant features in training and test set of images.

3 Proposed Methodology

The Flow Chart of Proposed Methodology is as: The Flow Chart of Proposed Methodology is given in Fig. 1.

In the proposed work, Decision-based Spectral Embedding for head pose estimation method combine the appearance and geometry based features for facial expression recognition that solves the decision of feature set obtained by minimizing the objective function which leads to selection of optimal set of fiducially points. The method preserves the local information from different views where

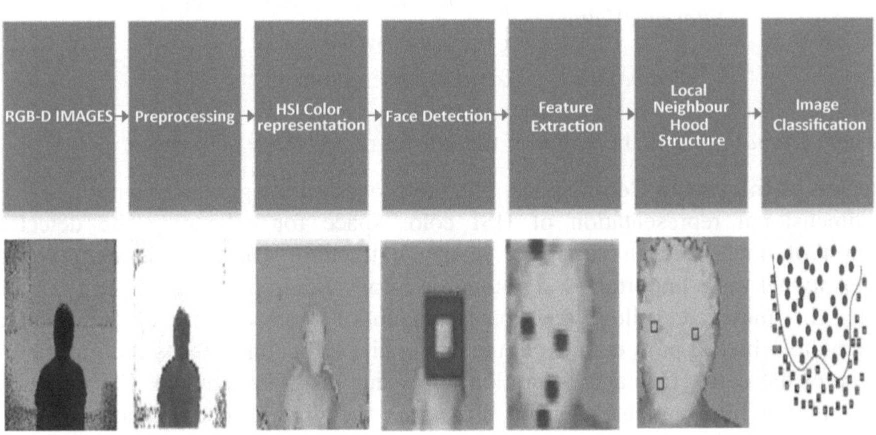

Fig. 1 Flow-chart of proposed methodology

vertices represent data and the edges represent neighbourhood relations. The basic idea is to preserve the local structures of the graph at each vertex for mapping neighbouring input to its corresponding output. The obtained low dimensional representation is directly related to the neighbourhood graphs for encoding the relationships of the data, then we perform the Classification using Single Value Decomposition (SVM).

4 Detailed Description of Proposed Methodology

4.1 Pre-processing Step

Noise-Removal A set of depth images with various facial expressions are considered as training set T_m for the pre-processing phase. The initial phase begins with the removal of noise from sequence of frames in Tm. The noise removal is done with the application of the median filter as it preserves the edges with local information. This is a linear filter that masks over each pixel where the average of the pixels under the mask is done to form single pixel. The output pixel contains the median value in a n-by-n neighbourhood around the corresponding pixel in the input image frame as shown in (1).

$$Medianfilter(x_1, \ldots, x_N) = Median\left(\|x_1\|^2, \ldots, \|x_N\|^2\right) \tag{1}$$

Background Subtraction The moving object detection can be done from processed frames in T_m with Background Subtraction method which obtains absolute difference between two consecutive frames $frameValue_k$ and $frameValue_{k-1}$ to detect moving region of interest based on the defined threshold value α as given in (2).

$$\begin{aligned} DifferenceValue_k &= \begin{cases} |frameValue_k - frameValue_{k-1}| \\ 0, \quad otherwise \end{cases} \\ if \quad &(frameValue_k - frameValue_{k-1}) > \alpha \end{aligned} \tag{2}$$

where $frameValue_k$ is the value of the kth frame in image sequences.

Region of Interest in Color Space The pre-processed images are transformed into 3-dimensional representation of HSI color space for enhancing the detected region of interest results in better feature extraction. The Hue, Saturation and Intensity are three important descriptors for colors. Hue represents the purity of the color (i.e. pure red, yellow, and green). Saturation represents the measure of the degree to which a pure color is diluted by white light. Intensity is the gray level value of the color. Hue and Saturation represents the color carrying Chrominance (Chromatic) information. Intensity represents the gray-level Luminance (achromatic) information.

4.2 Feature Extraction

Decision-based Scale Invariant Feature Transform obtains the set of feature points to identify the various facial expressions by locating fiducial points. The basic approach detects a set of interest points from obtained transformed sequence of frames T_m. The method is capable to capture the pixel-level distribution of the region of interest by means of local patterns extracted from a scale-space decomposition of an image frame. The scale space of an image is formed by performing convolution operation to it with a variable scale Gaussian factor $GaussianValue(x, y, \sigma)$ where σ is the scale parameter. The convolution result $ScaleSpace(x, y, \sigma)$ can be defined as:

$$ScaleSpace(x, y, \sigma) = GaussianValue(x, y, \sigma) * pixelValue(x, y) \tag{3}$$

where * denotes the convolution operation at coordinates (x, y) of image frame and

$$GaussianValue(x, y, \sigma) = \frac{1}{2\Pi\sigma^2}\left(e^{\frac{-(x^2+y^2)}{2\sigma^2}}\right) \tag{4}$$

In order to extract fiducial keypoints from obtained scale space, difference of Gaussian function of two nearby scale spaces convolved with the image frame as shown in Eq. (5).

$$ScaleSpace(x, y, \sigma) = (GaussianValue(x, y, K\sigma) - GaussianValue(x, y, \sigma))_{*pixelValue(x,y)}. \tag{5}$$

Where K is a constant scale factor.

The gradient modulus and orientation of each extracted key point can be computed. The methodology maps the obtained gradient magnitude into fuzzy region resulting in construction of the feature vector.

$$Features(x, y) = \frac{1}{1 + e^{-1(mag(x,y))}}$$

$$Orientation(x, y) = \tan^{-1}\left(\frac{scalespace(x, y-1 - scalespace(x, y-1}{scalespace(x+1, y) - scalespace(x-1, y}\right) \tag{6}$$

Laplacian matrices formed from the obtained set of feature vector[16–18] where the vertices represent various facial expressions and the edges represent their neighborhood relations preserving best local neighborhood structures exploiting class label information in the training set.

The local neighborhood structure is obtained on each view separately in such a way that the dissimilarities between the feature vectors and its corresponding[19, 20] neighbors become minimum and global structure is computed by summing up all the obtained local structures.

Given a dataset of N images with m different views represented as

$$X = \left\{X^{(i)} = \left[X_1^{(i)}, \ldots, X_1^{(N)}\right] \in R^{m_i \times N}\right\}$$

The lower dimensional representation is

$$Y = [y_i, \ldots, y_N] \in R^{d \times N}$$
$$where \quad d < \sum_{i=1}^{m} m_i \; where \; m_i \text{ is the ith view}$$

The local neighborhood structure is obtained by minimizing the objective function as described below:

$$obj = \min \sum_{j=1}^{k} \sum_{i=1}^{N} d_{ij}^2, \tag{7}$$

where

N is a vector of facial expressions under different head poses
k is the different class levels with facial expressions
d is the dissimilarities of different head poses.

The global structure is obtained as follows:

$$\arg\min_{Y,\alpha} \sum_{i=1}^{m} \alpha_i^r tr\left(YL^{(i)}Y^T\right),$$
$$\text{Such that} \tag{8}$$
$$YY^T = 1, \quad \alpha_i \geq 0, \sum_{i=1}^{m} \alpha_i = 1.$$

4.3 Classification

The classification is done in two steps:

The first classification scheme specifies that the smallest average dissimilarity decides the expression classification where the dissimilarity coefficients associated with its training expression are averaged over the whole training subjects. The major advantage of such classifier is that it reduces the time computation by obtaining the subset of the training data that reduces the number of comparisons. The second scheme is based on Support Vector Machines classifier [10, 11] which applied on the resultant subset obtained for finding the recognition rate. As this technique is sensitive to the presence of irrelevant or redundant features, the accuracy of classifier gets maximized[12–15] for action recognition as it extracts the relevant training feature set based on the similarity of the test data resulting in better performance of the classifier.

5 Results and Comparison

In this section we show our results using soft Kinetic camera to capture the depth Images and perform our algorithm.

See Figs. 2, 3, 4, 5 and 6.

- **Comparative Results**

See Table 1.

Fig. 2 Preprocessed depth images

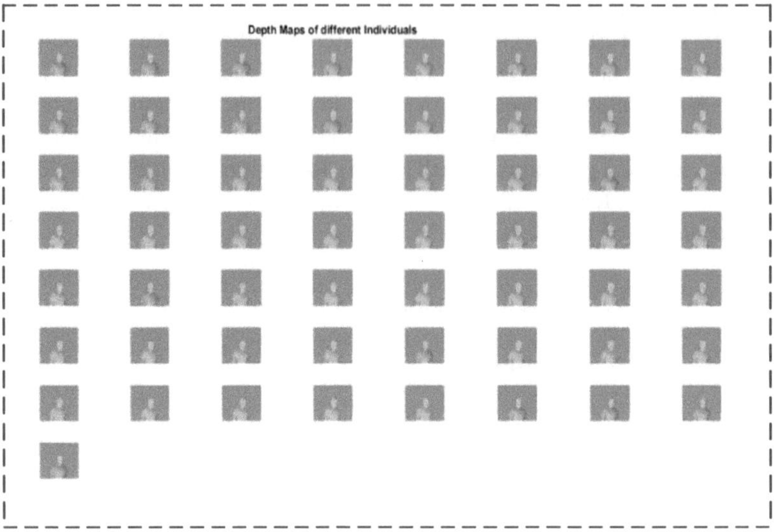

Fig. 3 Depth map of different individuals

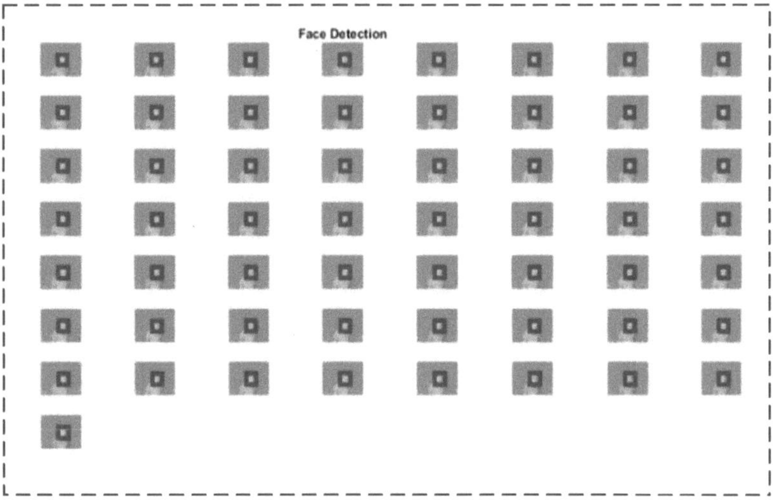

Fig. 4 Face detection of depth images

Fig. 5 Depth face having features

Fig. 6 Local neighbor hood structure

Table 1 Comparative results with previous reports

Recent work	Recognition rate (RR) %
Berretti [7]	77.5
Venkatesh [8]	81.7
Wang [9]	83.6
Proposed	90

6 Conclusion

In this paper we proposed a method to find or to recognize the facial expression in depth Images. We proposed a Decision-based Spectral Embedding approach combining appearance and geometry based features for head pose estimation and facial expression recognition by minimizing the objective function which leads to selection of optimal set of fiducial points. The method preserves the local information from different facial views for mapping neighboring input to its corresponding output, resulting in low dimensional representation for encoding the relationships of the data. According to these we get a good recognition results better than the previous ones. We also can extend this paper to reduce the time complexity and to predict the behavior in an accurate manner.

References

1. Guoying Zhao and Matti Pietikäinen "Dynamic Texture Recognition Using Local Binary Patterns with an Application to Facial Expressions" IEEE Transactions on pattern analysis and machine intelligence, vol. 29, no. 6, June 2007.
2. Caifeng Shan, Shaogang Gong, Peter W. McOwan. Facial Expression recognition based on local binary patterns: a Comprehensive Study, Image and Vision Computing, 27(2009) 803–816.
3. Yan Tong et.al. "A Unified Probabilistic Framework for Spontaneous Facial Action Modeling and Understanding" IEEE transactions on pattern analysis and machine intelligence, vol. 32, no. 2, February 2010.
4. Luis Diago, Tetsuko Kitaoka, Ichiro Hagiwara, and Toshiki Kambayashi "Neuro-Fuzzy Quantification of Personal Perceptions of Facial Images Based on a Limited Data Set" IEEE transactions on neural networks, vol. 22, no. 12, December 2011.
5. Di Huang, Mohsen Ardabilian, Yunhong Wang, Liming Chen, "3-D Face Recognition Using eLBP-Based Facial Description and Local Feature Hybrid Matching" IEEE transactions on information forensics and security, vol. 7, no. 5, October 2012.
6. Yogachandran Rahulamathavan et al. "Facial Expression Recognition in the Encrypted Domain Based on Local Fisher Discriminant Analysis" IEEE transactions on affective computing, vol. 4, no. 1, January–March 2013.
7. Steafano Berretti, Alberto Del Bimbo, Pietro Pala, A set of selected SIFT features for 3D facial expression recognition, 20th International Conference on Pattern Recognition, 2010, pp. 4125–4128.
8. Y. V. Venkatesh, Ashraf A. Kassim, O. V. Ramana Murthy, A novel approach to classification of facial expressions from 3D-mesh datasets using modified PCA, Pattern Recognition Letters, 30 (2009) 1128–1137.
9. Jun Wang, Lijun Yin, Xiaozhou Wei, Yi Sun, 3D facial expression recognition based on primitive surface feature distribution, 2006 IEEE Computer Society Conference on Computer Vision Pattern Recognition, 2 (2006) 1399–1406.
10. T. Senechal, V. Rapp, H. Salam, R. Seguier, K. Bailly, and L. Prevost, "Facial action recognition combining heterogeneous features via multikernel learning," IEEE Trans. Systems, Man, Cybern. B, Cybern., vol. 42, no. 4, pp. 993–1005, Aug. 2012.
11. C. Shan, S. Gong, and P. McOwan, "A comprehensive empirical study on linear subspace methods for facial expression analysis," in Proc. IEEE Conf. Comput. Vis. Pattern Recognit. Workshops, 2006, p. 153.

12. M. Turk and A. Pentland, "Eigenfaces for recognition," J. Cognitive Neurosci., vol. 3, no. 1, pp. 71–86, 1991.
13. S. Zafeiriou and M. Petrou, "Sparse representations for facial expressions recognition via l1 optimization," in Proc. IEEE Conf. Comput. Vis. Pattern Recognit. Workshops, 2010, pp. 32–39.
14. Z. Zeng, M. Pantic, G. I. Roisman, and T. S. Huang, "A survey of affect recognition methods: Audio, visual, and spontaneous expressions," IEEE Trans. Pattern Anal. Mach. Intell., vol. 31, no. 1, pp. 39–58, Jan. 2009.
15. S. Zhang, Y. Zhan, M. Dewan, J. Huang, D. N. Metaxas, and X. S. Zhou, "Towards robust and effective shape modeling: Sparse shape composition," Med. Image Anal., vol. 16, no. 1, pp. 265–277,2012.
16. Q. Zhao, D. Zhang, and H. Lu, "Supervised LLE in ICA space for facial expression recognition," in Proc. Int. Conf. Neural Netw. Brain, vol. 3, 2005, pp. 1970–1975.
17. Y. Zhu, F. De la Torre, J. Cohn, and Y.-J. Zhang, "Dynamic cascades with bidirectional bootstrapping for action unit detection in spontaneous facial behavior," IEEE Trans. Affective Comput., vol. 2, no. 2, pp. 79–91, Apr.–Jun. 2011.
18. T. Wu, N. Butko, P. Ruvolo, J. Whitehill, M. Bartlett, and J. R. Movellan, "Action unit recognition transfer across datasets," in Proc. IEEE Int. Conf. Autom. Face Gesture Recognit., 2011, pp. 889–896.
19. P. Yang, Q. Liu, and D. N. Metaxas, "Boosting encoded dynamic features for facial expression recognition," Pattern Recognit. Lett., vol. 30, no. 2, pp. 132–139, 2009.
20. P. Yang, Q. Liu, and D. N. Metaxas, "Dynamic soft encoded patterns for facial event analysis," Comput. Vis. Image Understanding, vol. 115, no. 3, pp. 456–465, 2011.

LTTC: A Load Testing Tool for Cloud

M.S. Geetha Devasena, V. Krishna Kumar and R. Kingsy Grace

Abstract Software testing is the process of software engineering to free the software from bugs. Load testing is one of the techniques in software testing and is used to find the maximum load that software can handle without affecting its performance. Load testing is used to test the cloud services that are running in a cloud. All the resources in a cloud are used by the cloud users based on their demand. Using cloud, it is easy to gather the required load for a particular application by forming clusters. If the required load is coming from different clusters and it is not known quantitatively then the problem of load balancing is raised. The proposed load testing tool avoids the problem of getting unequal loads coming from different clusters by distributing the same amount of load to all the clusters. Also the proposed load testing tool for cloud is used to find the maximum number of simultaneous users for a particular cloud system is to handle.

Keywords Load testing · Cloud testing · Load balancing

1 Introduction

The cost of testing web applications using traditional approaches is high for simulating user activity from different geographic locations [1]. Firewalls and load balancers testing cost includes hardware cost, software cost and maintenance cost. The web based load testing [2] is more effective in applications where rate of increase in numbers of users is not known and the differences are deployment

M.S. Geetha Devasena (✉) · V. Krishna Kumar · R. Kingsy Grace
Department of Computer Science and Engineering, Sri Ramakrishna Engineering College, Coimbatore 641022, India
e-mail: msgeetha@srec.ac.in

V. Krishna Kumar
e-mail: krishnakumar.v@srec.ac.in

R. Kingsy Grace
e-mail: kingsygrace.r@srec.ac.in

N. Modi et al. (eds.), *Proceedings of International Conference on Communication and Networks*, Advances in Intelligent Systems and Computing 508,
DOI 10.1007/978-981-10-2750-5_70

environment based on client requirements. Usually web application's behavior under normal and peak load conditions are tested using the basic type of load testing. While beginning the load testing, it is recommended to start with small numbers of virtual users and increase the load from normal to peak incrementally. The performance of the application during the gradual increase from normal to peak load is observed. Cloud based load testing is different from traditional load testing environments. Using cloud, the web applications are tested like real world usage of the web applications with geographically distributed users, and also variety of user scenarios. Cloud testing also ensures minimal start-up time and quality of assurance. The load testing for web application using cloud reduces the capital expenditure and improves scalability [3].

1.1 Load Testing

Giving demand to a system or device and measuring its response is called as load testing [4]. It is used to determine a system's behavior on both normal and peak load situations. The load testing is applicable in multi-user systems such as web servers. Usually the load testing simulates the actual use of software system and it is opposite to the testing in theoretical or analytical methods. The load testing is also used to measure the QoS performance of website's based on the user behavior. To test a website's load, the components such as script recorder, load generator are used to know the customer interaction in that website by recorded scripts and replaying those recorded scripts. The hardware and software statistics are monitored and gathered by the conductor to test the load of the website. The collected statistics includes the CPU, memory disk I/O of the server, response time and throughput of the System Under Test (SUT). All these statistics are analyzed and the corresponding load testing report is generated [5–8].

1.2 Cloud Testing

Cloud system is used for performing single small but complicated tasks with the help of coordinated resources in a network. Cloud testing is an emerging field in software testing. The various challenges in cloud testing are, limited test budget, meeting deadlines, high cost per test, large number of test cases, little or no reuse of test and geographical distribution of users. Usually traditional testing requires expensive dedicated infrastructure and resources but they are used sporadically. The business applications are increasing day by day; it is difficult to mimic real-time environments. In comparing in-house test facilities, cloud based infrastructure provides lower costs, and flexible collaboration. The service providers for cloud based testing provide a standardized infrastructure and predefined software images to reduce those errors in in-house testing environments.

1.3 *Operational Challenges for Testing in the Cloud*

The cloud based testing have different challenges than in-house testing that are listed as follows:

(i) There are no universal or standard solutions to combine public cloud resources with user's data centers. All the cloud provides have their cloud specific architecture, operations and pricing. The interoperability is very less also change to vendor is difficult.
(ii) The data and code related with testing may be stored in a remote location. Hence there is no security in the organization data because the legal and regulatory jurisdiction of the organization.
(iii) Also, it is difficult to create real-time test environments because of cloud configurations, technology, servers, storage, networking and bandwidth of a particular cloud provider.
(iv) Pricing of the cloud provider is a major issue in cloud based testing.

2 Literature Survey

Nowadays web applications are everywhere and used by large number of users. As web applications are used by more number of users, it has to be fast, reliable and up-to-date. Also load testing is necessary in web applications to ensure the best service to all the users. Even though there are open source and commercial tools are available for load testing web applications there is a chance for the development of new and efficient cloud based load testing tools. This section deals with some of the works done in the literature for load testing.

Meria et al. have proposed a peer-to-peer load testing in [1]. The main approaches used in [1] are based on a point-to-point connection between the test driver and the SUT. The test driver submits the load to the SUT interface. The proposed system in [1] checks the performance of the SUT for the given workload. The tools which are used to provide the test driver are: Hammerora4, Oracle Application Test Suit5 and AppPreface6. The two hypothesis used in the proposed system are (i) For creating realistic load conditions, the test driver must scale up by distributing the test tasks and sending concurrent requests to SUT. (ii) The SUT does not use all of the allocated resources on peak loads.

Draheim et al. [2] have proposed a new approach for performing load testing of web applications. The user behaviors are simulated with stochastic form-oriented analysis models. The form-oriented analysis is used for the specification of ultra-thin client based systems. Also form-oriented models are visualized using form charts.

Jerry GAO et al. have provided a comprehensive tutorial on cloud testing and cloud-based application testing [4]. This paper provides the solution to engineers and managers by explaining the concepts, discusses the special objectives, features, requirements and the need for cloud testing. Also this paper tells the difference between the web-based software testing and cloud-based application testing. The four objectives of the cloud testing are (i) functional services (ii) business processes (iii) performance (iv) scalability.

3 Problem Definition

Load testing is used to find out the change in behavior of the software upon changing the load to some higher order. Load testing can also be used to test the services that run in cloud. The resources in the cloud which are required by an application can be gathered from various clusters. The load balancing problem occurs when the load coming from various clusters are not known. The proposed system avoids the problem of unequal loads coming from the clusters by distributing the same amount of load to all the clusters.

4 Proposed System

As more number of users are migrating towards using the cloud services, it is very essential to make sure the maximum number of users a particular cloud service can respond. This gives the necessity of load testing the web applications. The proposed Load Testing Tool for Cloud (LTTC) helps to avoid the bottlenecks generated by the simultaneous usage of web applications by more number of users.

The overview of the approach followed to implement the initial cloud set up is given. The Fig. 1 shows the various phases involved in setting up a private cloud along with a distributed environment, cloud controller and a web Application Programming Interface.

The server system is setup using the CentOS. After the installation of CentOS server version, the network controller is configured. The current server version does

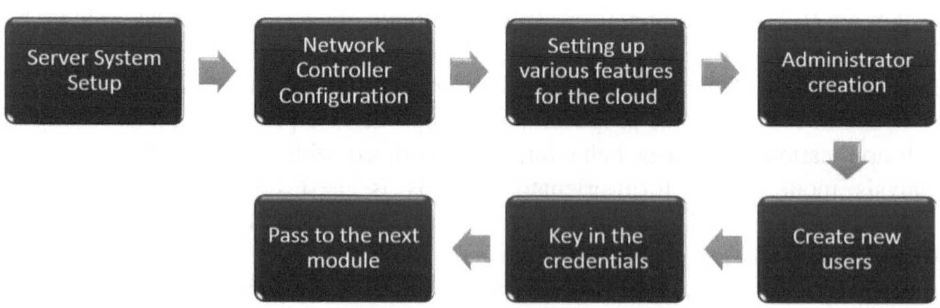

Fig. 1 Infrastructure setup

not contain features such PHP and Java. So various features required to run applications over the cloud are set up. An account for the administrator is created and the clients are also created. For each created client, the credentials such as username and password are keyed in.

5 Simulation

Normal load testing poses the problem of load simulation. Using cloud, it is easy to gather the required load easily in order to hit a particular application or a dataset by forming clusters. The proposed tool, LTTC, avoids the problem of unequal loads coming from the clusters and encourages load distribution. The number of simulated users are varied periodically and those simulated users are used simultaneously, to hit a particular application or a dataset that can be run in the cloud. This forms different load conditions under which the behavior of the cloud system is analyzed.

5.1 Dataset

The proposed load testing tool uses New York Stock Exchange (NYSE) dataset for load testing [9]. The dataset is stored in Apache Hive and which is accessed using HiveQL. The proposed tool analyzes the performance of Apache Hive by applying load over the NYSE dataset through the Cloudera distribution. Apache Hive comes preconfigured with CDH. CDH was configured in CentOS server which in turn installed as a virtual machine using VMWare workstation in the host OS. User simulation is done by creating independent hive processes as child threads. The users are simulated as per the given user increment factor, periodically. The queries submitted by the users are translated into relative HiveQL queries and are transferred to Apache Hive. Apache Hive then executes the queries over the stored dataset and returns the result set back to Hadoop distribution framework. Hadoop analyzes various parameters such CPU time taken, MapReduce time taken, and Cumulative CPU time and so on. The parameters can be fetched by the front end program and can be used to analyze the behavior of the cloud under various load conditions. The proposed tool also helps in visualizing the results obtained by generating the graph containing the number of users at X-axis and the time taken for execution at Y-axis.

5.2 Infrastructure Setup

The cloud technology used is Cloudera. Cloudera is based on open source Apache–Hadoop distribution. Cloudera comes in various variants. To test the load testing tool, Cloudera with a single node cluster is setup. In order to meet the testing essentials, Apache Hive is configured. Basically, Apache Hive is data warehouse software, and it facilitates querying and managing large datasets residing in distributed storage. Hive provides a mechanism to project structure onto this data and query the data using a SQL-like language called Hive Query Language (HiveQL). Apache Hive is used as an interface to work with Hadoop. SQL like queries of HiveQL is used in Hadoop to store, retrieve and modify the datasets that are compatible with Hadoop Distributed File System (HDFS). HiveQL queries are also used to fetch the information from the Hadoop framework such as the time taken for executing the query, MapReduce time taken and so on. After setting up Hive, a dataset has to be uploaded in the cloud for the testing process. The dataset chosen has to be intensive so that the execution time taken by the cloud to process submitted queries over the dataset is significant. The dataset chosen is the New York Stock Exchange details of the year 2001–2002. It contains 80,000 records and the size of the dataset is 42 megabytes. The dataset includes the details such as field names and the corresponding data types.

HQL commands are run from the Linux terminal. All sort of supported operations are performed through the terminal itself. Hive has also an open source user interface. It is called Beeswax. Beeswax is a user interface provided by Apache Hue for Hive. The Beeswax application enables the user to perform queries on Apache Hive, the data warehousing system designed to work with Hadoop. Hive tables are created and data is loaded to those tables, Hive queries are managed and executed over the tables and the results of the Hive queries are imported as a Microsoft Office Excel file or even as a comma separated file format. Since the tool is based on java, it makes advantageous that Hive queries are executed through Java language by creating a Hive process and assigning the Hive query as the task for the Hive process.

5.3 Result Observation

Profiling is a form of dynamic program analysis that measures, for example, the space (memory) or time complexity of a program, the usage of particular instructions, or frequency and duration of function calls. The most common use of profiling information is to aid program optimization. Profiling is achieved by incrementing either the program source code or its binary executable form using a tool called a profiler. A number of different techniques may be used by profilers, such as event-based, statistical, instrumented, and simulation methods. The details of the CPU Call Tree and the VM Telemetry for Threads/Loaded Classes, achieved by running the profiling operation over the project were given as screenshots in Figs. 2 and 3.

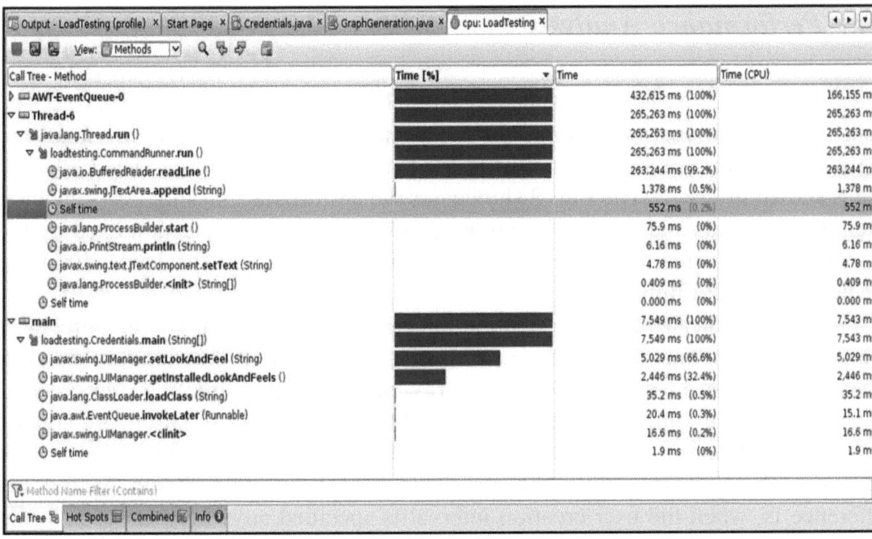

Fig. 2 CPU call tree

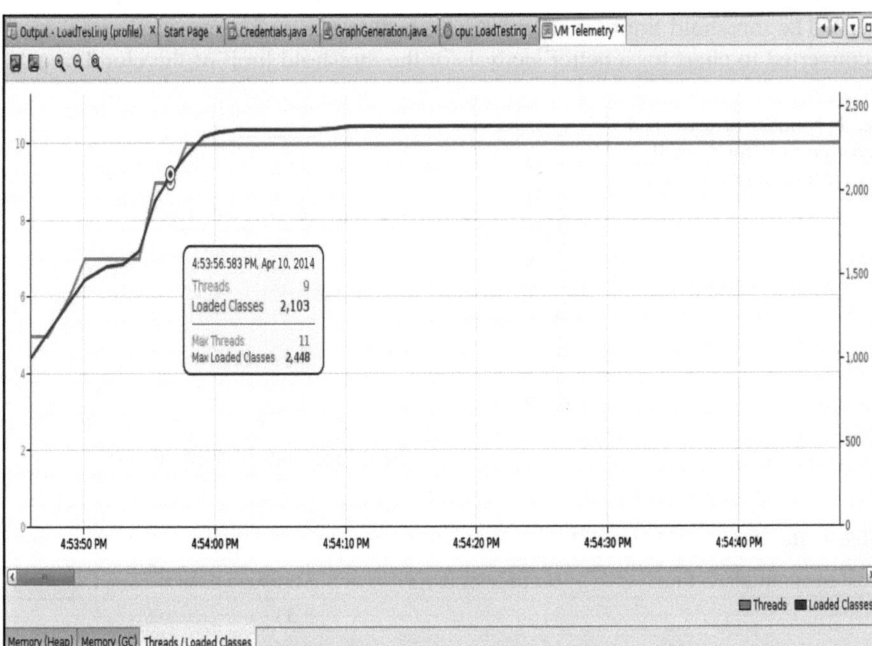

Fig. 3 VM telemetry for threads/loaded classes

5.4 Performance Analysis

The performance of the proposed tool is analyzed using the NYSE dataset. The result sets for various load conditions has been taken and the recorded values are aggregated to form charts for analyzing the performance of the tool. The CPU call tree is shown in Fig. 2. The Fig. 3 showcases the performance of the user's creation interval in seconds and the maximum number of users that can be simulated by the proposed load testing tool with the successful completion of the jobs within the specified threshold limit of 4 min.

The Fig. 4 indicates the user creation interval from 1 through 9 s, the number of users created, remains the same up to 11 s. This shows that the cloud system can efficiently process 11 users within 4 min (given threshold), while the user creation interval can be anything between 1 and 9 s. Meanwhile when the users are created with the time interval of 10 s, it will be enough for the cloud system to accomplish the tasks of 15 users, within the stipulated time limit. The reason behind the difference is, when the user creation interval is specified anywhere between 1 and 9 s, the users are incremented in the specified time interval, which increases the load over the system, as the time interval for incrementing the users is very low. Consider another result set containing the details in the same fashion as that of the previous case, except that, the user creation interval spans a long range (5, 10, 15...). The threshold limit for the result set is also 4 min. The result set in Table 1 is converted to chart for a better analysis of the threshold limit of the cloud system.

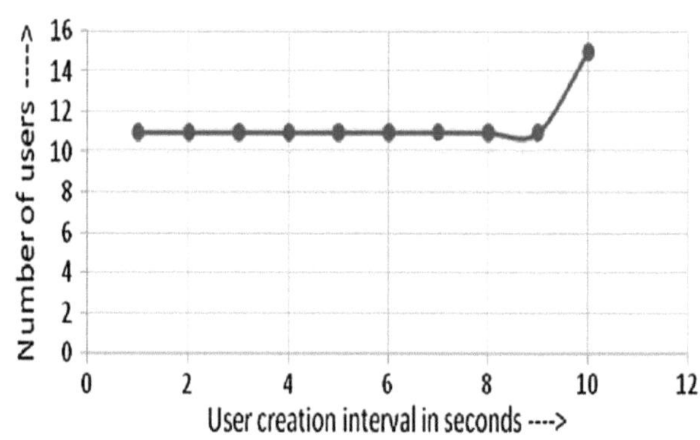

Fig. 4 Number of users and execution time for a small user creation interval range

Table 1 Result observation with large user creation interval range

User creation interval (s)	Number of users	Minutes taken to complete
5	11	3.53
10	15	3.45
15	15	3.9
20	11	3.83
25	7	2.93
30	7	3.41

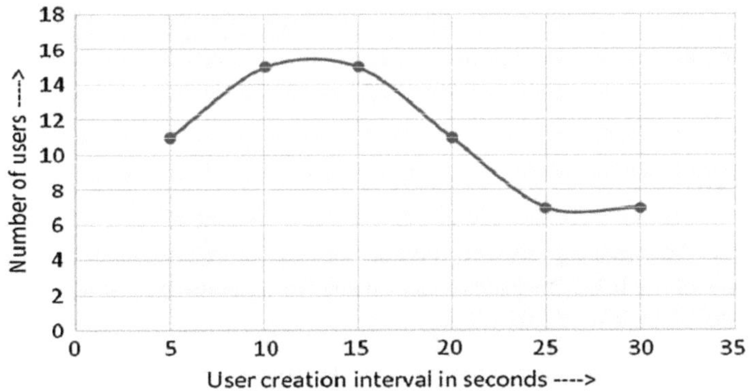

Fig. 5 Number of users and execution time for a large user creation interval range

Anyhow it will be difficult to analyze the behavior of the cloud without visualizing the observations.

The chart given as Fig. 5 clearly indicates that, when the user creation interval is 10 and 15 s, the maximum number of users created is 15. That is, when the user creation interval is between 10 and 15 s, some of jobs might get executed before the next cycle of user simulation. This reduces the load over the cloud, thereby increasing the total number users simulated. But when the user creation interval goes beyond 15 s (say 25 and 30 s), there will be a large amount of time between two cycles of user simulation, thereby reducing the total number of users created, within the specified time. So for the specified time limit of 4 min, the maximum number of users that can be accomplished by the cloud system is 15.

6 Conclusion and Future Enhancements

The proposed tool, LTTC, helps in analyzing the behavior of the cloud under different load conditions by forming virtual users and making those users to do the same job in the cloud distribution, simultaneously. The observations of the results are recorded as log files for graph generation and future processing. The parameters that are taken for analyzing the behavior of the cloud are number of users and the time taken for task completion. Any task, which takes beyond the specified time limit (threshold value) for completion, is said to be incomplete. Ultimately, the proposed tool determines the maximum number of users that the cloud can handle for a specified amount of time. The proposed load testing tool can be further extended with various web applications. The simulated users in turn initiate the web applications; thereby the capacity of the cloud as well as the scalability of the selected application is also determined.

Acknowledgment The authors would like to thank the Management, Principal and Head of the Department of Computer Science and Engineering of Sri Ramakrishna Engineering College, Coimbatore for their support towards the completion of this work.

References

1. Meira, J.A., de Almeida, E.C., Le Traon, Y., Sunye, G.: Peer-to-Peer Load Testing, proceedings of the IEEE 5th International Conference on Software Testing, Verification and Validation (ICST), 642–647 (2012).
2. Dirk Draheim, John Grundy, John Hosking, Christ of Lutteroth, Gerald Weber: Realistic Load Testing of Web Applications, proceedings of the 10th European Conference on Software Maintenance and Reengineering (CSMR), 11–70 (2006).
3. Pooja Ahlawat and Sanjay Tyagi.: A Comparative Analysis of Load Testing Tools Using Optimal Response Rate, International Journal of Advanced Research in Computer Science and Software Engineering (IJSEA), 3, 855–860 (2013).
4. Jerry Gao., Xiaoying Bai and Wei-Tek Tsai: Cloud Testing-Issues, Challenges, Needs and Practice, Software Engineering: An International Journal (SEIJ), 1, 9–23 (2011).
5. Arora, A., Sinha M., Web Application Testing: A Review on Techniques, Tools and State of Art, International Journal of Scientific & Engineering Research, 3, 1–9 (2012).
6. Jagadeesh, G., AnirbanBasu, Sandeep Aluri.: Cloud based testing: Need of testing in cloud infrastructures and cloud platforms, International Journal of Computer Science and Information Technology & Security (IJCSITS), 2, 398–401 (2012).
7. AtifFarid Mohammad, Hamid Mcheick.: Cloud Services Testing: An Understanding, proceedings of the 2nd International Conference on Ambient Systems, Networks and Technologies (ANT), 5, 513–520 (2011).
8. Shivangi Kaushal, Jagpuneet Kaur Bajwa.: Plug-in for analyzing web-based application load time testing, International Journal of Science and Engineering Applications (IJSEA), 1, 56–59 (2012).
9. www.stockhistoricaldata.com/nyse.

A Hybrid Approach to Enhance the Security of Automated Teller Machine

Sabarna Choudhury, Shreyasi Bandyopadhyay, Satyaki Chatterjee, Rahul Dutta and Sourjya Dutta

Abstract With the rise of crimes in Automated Teller Machines, the security of the ATM is at stake. The Traditional Security Methods such as passwords or pins had always been a cause of worry to the users because of it getting lost, stolen or forgotten. The biometric authentication uses one's unique features as password. The paper deals with one of the oldest biometric traits, Fingerprint Recognition. The paper moreover adds a sophisticated voice input system and a Password Breaking scheme. The end results of a highly secure and sophisticated ATM system enhances the efficacy of the system, providing customer satisfaction in banking sectors.

Keywords ATM · Crimes · Security · Biometric authentication · Fingerprint recognition · Voice input · Password breaking scheme

1 Introduction

The biometric authentication process has grown tremendously in recent years due to its ability to protect information from being hacked in many security systems with its unique identification of individuals. Biometrics is a measure of physical characteristics of an individual that can be captured and analyzed later on with another

S. Choudhury (✉) · S. Bandyopadhyay · S. Chatterjee · R. Dutta · S. Dutta
St. Thomas' College of Engineering and Technology, Kolkata, India
e-mail: sabarna.choudhury@gmail.com

S. Bandyopadhyay
e-mail: s.bando.93@gmail.com

S. Chatterjee
e-mail: satyakichaterjee14@gmail.com

R. Dutta
e-mail: rahulduttastcet@gmail.com

S. Dutta
e-mail: sourjyadutta1994@gmail.com

N. Modi et al. (eds.), *Proceedings of International Conference on Communication and Networks*, Advances in Intelligent Systems and Computing 508,
DOI 10.1007/978-981-10-2750-5_71

instance at the time of security checking. ATM provided a good platform to people for electronic banking and also it released the banking pressures. Customers don't have to wait in a long queue for withdrawal [1]; a swipe of card provides them the money. But ATM comes with its flaws. Crime at ATM has not only affected customers but also the banking operators. ATMs are provided with Traditional Security Methods which are based on Passwords and PINs but passwords and PINs can be forgotten or stolen. Thus this fear and threat have always left a section of the society feel unsafe and also deprived from the usefulness of the ATM. Use of biometrics has helped in solving these problems. The advantage of biometrics is that biometric identity is always carried by a person and it proves to be accurate as a password. So there is no chance of losing or forgetting it. Also, it is difficult to steal or decode any biometric identity. One of the most popular biometric trait to recognize a person is Fingerprint Recognition [2].

The proposed system is inspired by the concept of fingerprint recognition, voice input and password breaking system and thus has used a hybrid approach to enhance the security level of current Automated Teller Machine (ATM). The Sect. 2 deals with the fingerprint recognition algorithm, while Sect. 3 deals with the voice input algorithm. Section 4 describes about the Password Breaking System. The hybrid approach to the Automated Teller Machine is discussed in Sects. 5 and 6 deals with the performance measures and the results. The conclusions and the future scope are listed too in this section.

2 Fingerprint Recognition

The features obtained for each input fingerprint are to be matched with the database which contains preprocessed fingerprints. Fingerprint matching can be done in various ways such as Correlation method, Minutiae based technique and image based technique. Recently, apart from minutiae other techniques such as finger placement direction, ridge compatibility, ridge count, ridge length, ridge curvature direction and ridge type are used to match fingerprint some of which have motivated us in our proposed system [3]. Nonlinear distortion in fingerprints to be eliminated to improve system performance [3]. In our proposed system minutiae extraction from both gray scale and binarized image was carried on.

2.1 Preprocessing

The fingerprint image is enhanced by histogram equalization or Fourier transform. Figure 1 shows the database consisting of the fingerprint images. The original image along with the change in the image after histogram equalization has been shown in Fig. 2 whereas the Fig. 3 deals with enhancement by Fourier Transform.

Fig. 1 Database of fingerprints

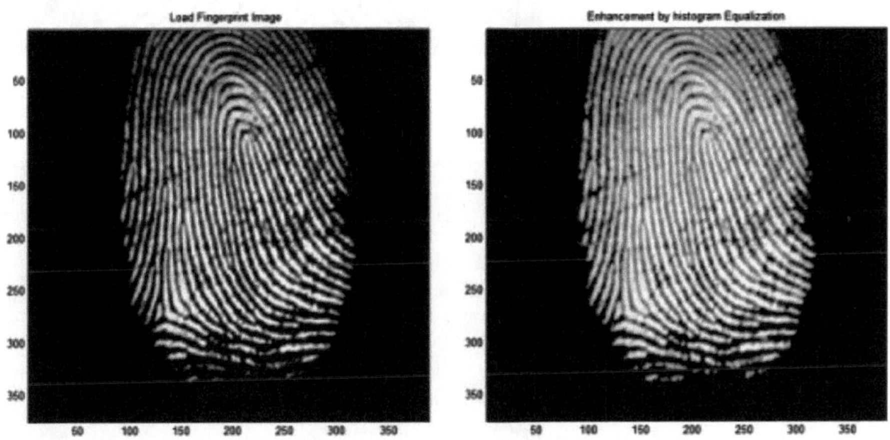

Fig. 2 Input enhanced by histogram equalization

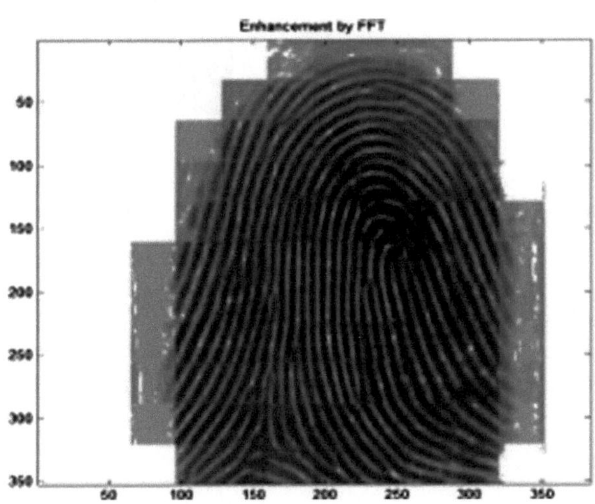

Fig. 3 Enhancement by Fourier transforms

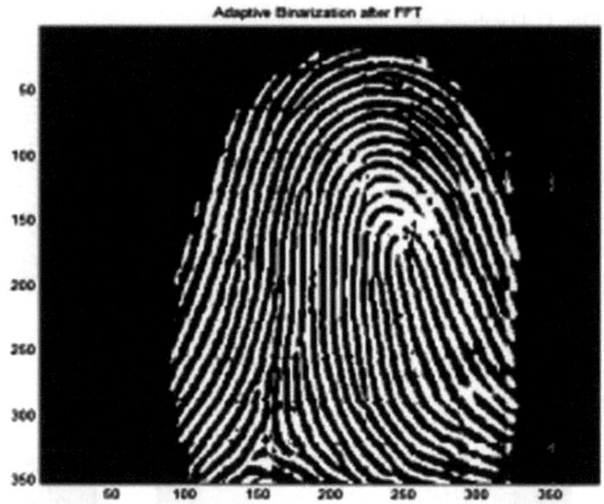

Fig. 4 Adaptive binarization

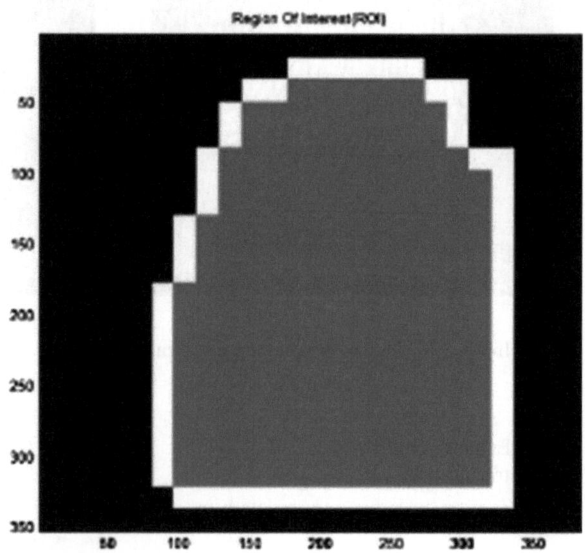

Fig. 5 ROI detected

2.2 ROI (Region of Interest) Extraction

The effective area of adaptive binarized image, Fig. 4 is obtained by discarding the background information of the ridges. (a) Block Direction Estimation and (b) Direction Variety check are implemented to obtain ROI, Fig. 5. The H-break of the original image is removed for further operations, Fig. 6.

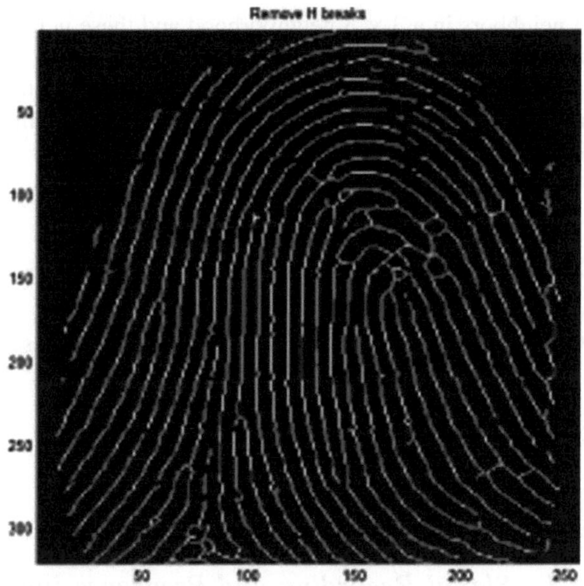

Fig. 6 H breaks removed

2.3 Extraction of the Minutiae and Minutiae Marking

Minutiae marking, based on mathematical morphology helps in determining all the different versions of minutiae present in a fingerprint image which further helps in the extraction of the true minutiae and also eliminate the false minutiae, Fig. 7. Ridge endings refer to the region in which the pixels have a single neighbor in a 3×3 neighborhood. Ridge bifurcations are detected by the pixels which have only

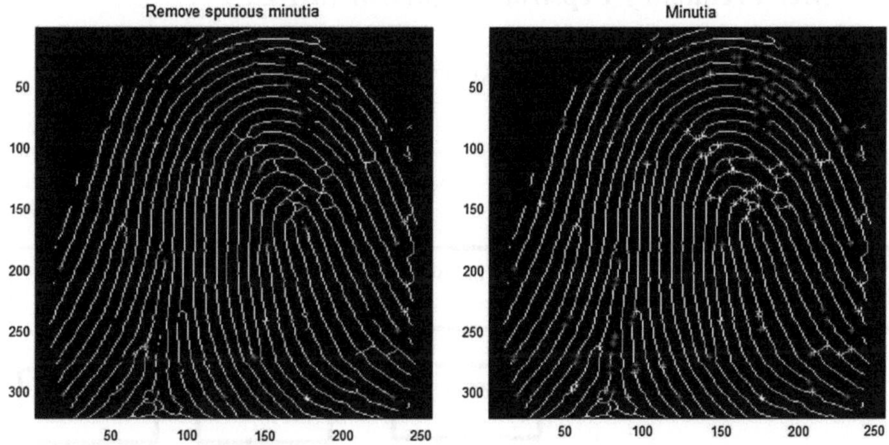

Fig. 7 True minutiae marked

3 neighbors in a 3 × 3 neighborhood and these neighbors should not be adjacent to each other [4–6].

If the fingerprint of the user is successfully matched, the system automatically moves into the withdrawal part. The input in this section is given by voice. Provision for key panel is also there.

3 Voice Input Technology

Our proposed technique comprises of 3 steps. These are:

i. Speech recording and preprocessing
ii. Feature extraction
iii. Feature identification/recognition

3.1 *Signal Recording and Preprocessing*

After the databases of the digits from 0 to 9 are made irrespective of gender with a time interval of 3 s, the signals are pre-processed.

The preprocessing stage consists of two steps:

1. Compute MFCC
2. Compute Weighted MFCC

3.2 *Mel Frequency Cepstral Coefficient (MFCC)*

The whole processing scheme is depicted in the Fig. 8.

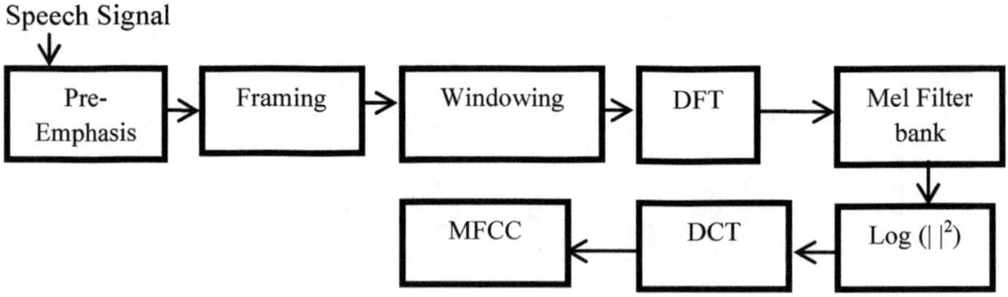

Fig. 8 Generation of MFCC

3.3 Feature Extraction

The sampled speech is framed into smaller segments of time length 20–40 ms [7]. Individuals of these frames are then windowed. This reduces the signal discontinuities occurring at the beginning and the end of each frame [7, 9]. Hamming window is the mostly used window in Automatic Speech Recognition (ASR) whose impulse response is defined as

$$\begin{aligned} \mathrm{w(n)} &= 0.54 - 0.46\cos\left(2\pi n/N - 1\right), \text{ where } 0 \le \mathrm{n} \le \mathrm{N} - 1 \\ &= 0 \text{ otherwise.} \end{aligned} \quad (1)$$

The side lobes in hamming window are also much lower than rectangular window which in turns reduces leakage [8, 9]. DFT of the output of the windowing stage leads to multiplication of x(n) and w(n) in the frequency domain, i.e., $Y(\omega) = X(\omega)\ W(\omega)$.

The output of the DFT is passed through Mel filter banks which are constructed by a set (usually 20–30) of triangular band pass filters since it simulates human auditory system [10]. There are more number of filters for the spectrum which is within 1 kHz, since it contains relevant information such as the first formant [10]. Figure 9 shows the Mel Filter bank. Speech input for digit 7 before and after pre-emphasis is shown in Fig. 10. For a given frequency f in Hz, the following approximated formula is used which computes mels for the frequency f. Mel (f) = 2595 $\log_{10}(1 + f/700)$ [10].

3.4 WMFCC (Weighted MFCC) Computation

To increase the efficiency and decrease dimensionality of the MFCC, the cepstral mean normalization is performed by the calculation of weighted MFCC features.

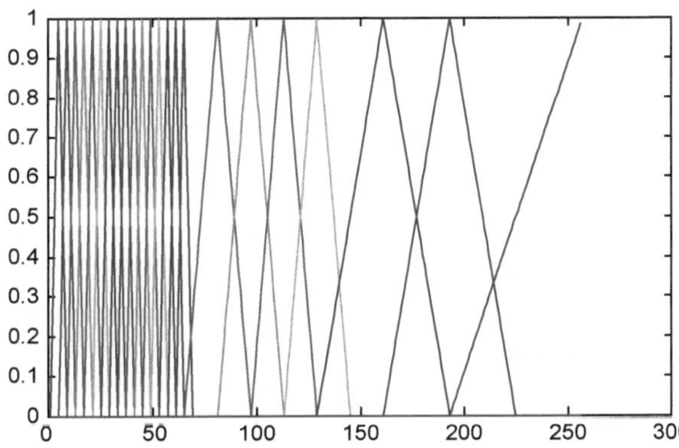

Fig. 9 Mel filter bank

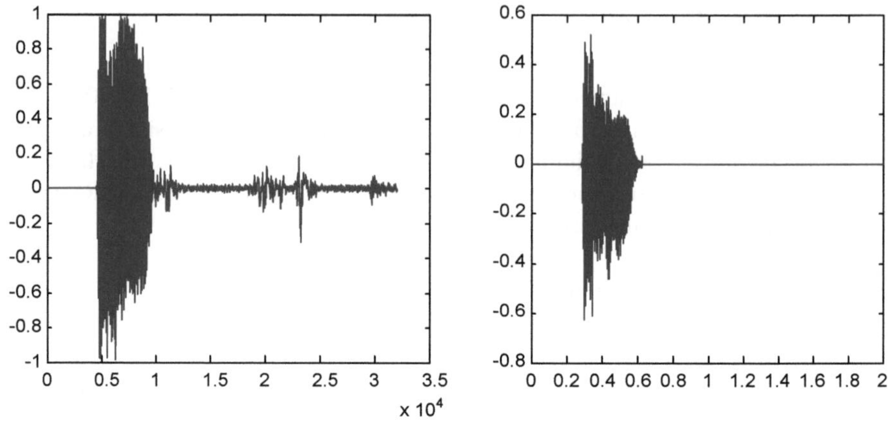

Fig. 10 Speech input for digit 7(*left*) and after pre-emphasis (*right*)

By analyzing frame to frame the tendency of speech in time can be lost. In order to recover them, delta, double-delta and triple-delta features are obtained by taking time derivative. Hence,

$$\Delta h(n) = \left(1/\sum\nolimits_{i=1}^{D} \cdot i^2\right) \sum\nolimits_{i=1}^{D} i\{h(n+i) - h(n-i)\} \tag{2}$$

where h(n) is the MFCC for each frame, D is the frame delay, set to 2. The derived features are then concatenated to the original cepstral features to obtain WMFCC

$$\mathrm{Wc(n) = h(n) + p \cdot \Delta h(n) + q \cdot \Delta\Delta h(n) + r \cdot \Delta\Delta\Delta h(n)} \tag{3}$$

where delta functions are weighted according to p, q and r, such as $1 > p > q > r$ [9].

3.5 *Dynamic Time Warping*

We assume that a few command words or digits would be recognized here. For an utterance of a word w which is TX vectors long, we get a sequence of vectors X = {x0, x1, …, xTX − 1} from the acoustic preprocessing stage. It is necessary to compute the "distance" between this unknown sequence of vectors X and known sequences of vectors W = {w0, w1, …, wTW}. They are the prototypes for the words we want to recognize.

Optimal Path: The time warping to suitable boundaries: The first vectors as well as the last vectors of X and W should be assigned to each other. For the time indices in between, any giant backward or forward leap in time is needed to be avoided and

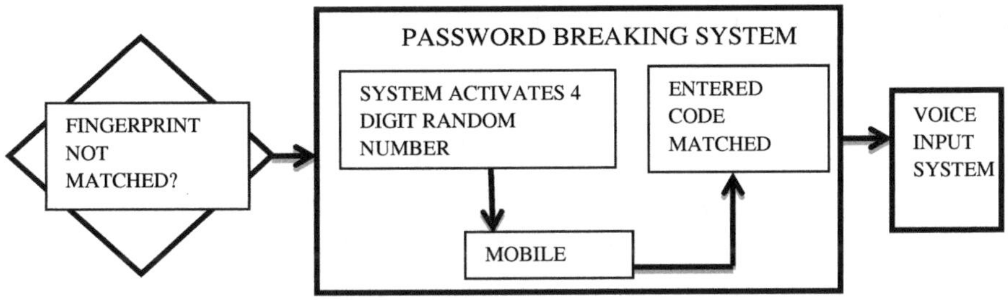

Fig. 11 Block diagram of password breaking system

the time warping is restricted in order to "reuse" the preceding vector(s) to warp the duration of a segment (short) of the speech signal locally.

4 Password Breaking System (PBS)

Password breaking system helps to retrieve any forgotten or unmatched passwords easily. In this paper the account of ATM card holder is secured by own finger print. But if that person had an accident or if ill, then it is possible for his family members to withdraw the money. Without the original fingerprint the account can still be unlocked. The chosen family member will go and use password breaking system in the ATM. The system in return will generate a 4 digit alpha numeric password which will be received by original user. By using that alphanumeric code the person can open that account and withdraw money and the amount of withdrawal will be immediately forwarded to the original user. To avoid the duplicity of having passwords at the same time the alphanumeric code will be valid for only few minutes and for only once it is usable. After that it will be stored into the used

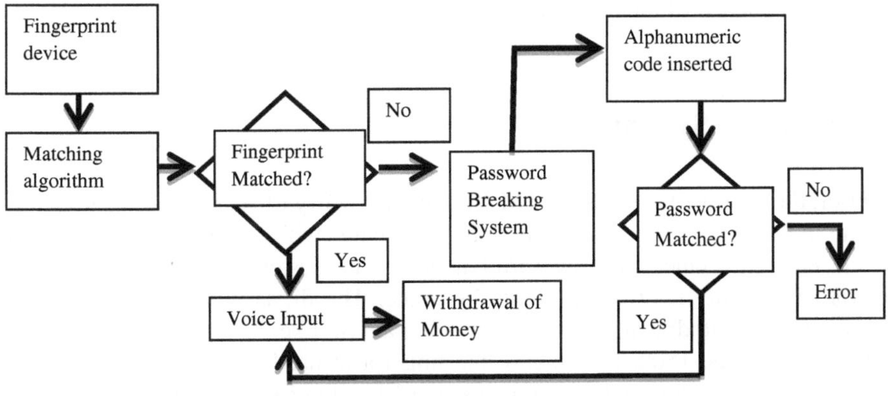

Fig. 12 Block diagram of the proposed ATM system

Table 1 Efficiency plot of spoken digits (0–9)

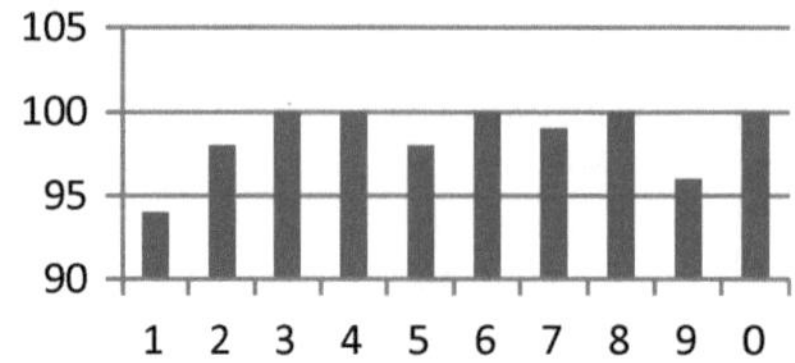

Table 2 Performance analysis of the ATM

Users	Fingerprint recognition	Password breaking system	Voice input	Transactions	Performance efficiency (%)
User 1	Yes (success)	No	Yes (success)	Success	100
User 2	Yes (success)	No	Yes (success)	Success	100
User 3 (fingerprint not in database but wants to use PBS with the permission of actual user)	No (failed)	Yes (success)	Yes (success)	Success	100
User 4 (Unknown person trying to access)	No (failed)	Yes (failed)	X	X	100

codes' database so that it can never be used by any other person in future. After few minutes the account will again revert back to original fingerprint password so that if there is no chance of any fraud case.

5 Proposed System

In brief, the user walks into the ATM without any worry of ATM card and scans his thumb in the fingerprint device which then compares with all the fingerprints stored in the database. Once the fingerprint matches the system will be automatically forwarded for the input of the digits by voice. The digits will be the amount of money to be withdrawn in isolated word pattern like "1–5–0–0" will lead to withdrawal of Rs. 1500. But in case the fingerprint used as the password doesn't match then the system will opt for password breaking system for withdrawal as shown in Fig. 11.

6 Results and Conclusion

The fingerprint module produces a reliable recognition technique with an efficiency as high as 100 % for people of age 10–50 and as high as 88 % for rest of the people. The voice input technology's efficiency plot of spoken digits is as shown below: Hence, this hybrid approach enhances the security of an ATM thereby making it more user friendly to common people without the worry of having any kind of tokens or swipe cards. The future scope of our proposed system may include the use of continuous words like "Two thousand and five hundred" can be recognized instead of saying "two five zero zero" (Tables 1, 2 and Fig. 12).

References

1. Ibidapo, O. Akinyemi, Zaccheous O. Omogbadegun, and Olufemi M. Oyelami.: Towards Designing a Biometric Measure for Enhancing ATM Security in Nigeria E-Banking System. In: International Journal of Electrical & Computer Sciences IJECS-IJENS Vol: 10 No: 06.
2. Prof. Selina Oko and Jane Oruh.: Enhanced ATM Security System Using Biometrics. In: IJCSI International Journal of Computer Science Issues, Vol. 9, Issue 5, No 3, September 2012.
3. K. Cao, X. Tao, X. Yang, P. Li, Y. Zang and J. Tian: Combining features for distorted fingerprint matching. In: Journal of Network and Computer Applications. Volume 33(2010) 258–267.
4. D. Maio and D. Maltoni: Direct gray-scale minutiae detection in fingerprints: IEEE Transactions on Pattern Analysis and Machine Intelligence, 19(1):27–40.
5. D. Maltoni and D. Maio: Neural network based minutiae filtering in fingerprints. In: Fourteenth International Conference Pattern Recognition, vol. 2, 1998, pp. 1654–1658.
6. X. Jiang, W.-Y. Yau, and W. Ser, Detecting the fingerprint minutiae by adaptive tracing the gray-level ridge, Pattern Recognition, vol. 34(5), 2001, 999–1013.
7. Wei HAN, Cheong-Fat CHAN, Chiu-Sing CHOY, Kong-Pang PUN: An Efficient MFCC Extraction Method in Speech Recognition.
8. Shivanker Dev Dhingra, Geeta Nijhawan, Poonam Pandit.: Isolated Speech Recognition Using Mfcc and Dtw. In: International Journal of Advanced Research in Electrical, Electronics and Instrumentation Engineering.
9. B. A. Dautrich, L. R. Rabiner, T. B. Martin.: On the effects of varying filter bank parameters on isolated word recognition. In: IEEE Trans. Acoust., Speech, Signal Processing, ASSP-31 (4):793–807,1983.
10. Davis,Paul Mermelstein, Steven B.: Comparison of Parametric Representations for Monosyllabic Word Recognition in Continuously Spoken Sentences. In: IEEE Transactions on Speech, Acoustics, and Signal Processing. Volume: ASSP-28, No. 4, August 1980.

A Novel Approach for Copy Move Forgery Detection Using Template Matching

Jyoti Yaduwanshi and Pratosh Bansal

Abstract Digital Photographs are most powerful and trustworthy media for conveying thoughts, emotions or message. Even only a single image is sufficient to reflect every situation or scenario. During past few years, various digital image manipulation tools came into picture and number is increasing. Editing software is available either at nominal rate or free of cost. Edit or alter any digital image for fun and other purposes is now a common practice. Sometimes need arises to check authenticity and originality of image. Digital image forensics plays an important role in this situation. Out of several image forgery methods Copy Move Forgery is one of the easy and effective method. Copy Move Forgery can be used with the intention of either to hide something in the image or to duplicate one region in an image. A study has been carried out to identify suitable scheme for detection of Copy Move Forgery especially in coloured digital images. The proposed scheme uses the concept of template matching.

Keywords Digital forensics · Digital image · Image forgery · Copy-move forgery · Digital image · Cloning

1 Introduction

Gone are the days when few people only used to have camera with camera film. Today every house has a camera, thanks to smart phones. Not every house, but all individuals who use smart phone has a camera. Digital camera facilitates quick and easy image generation. On the same time bundled software supports image processing with equal ease. With the ever-increasing growth in generation of digital

J. Yaduwanshi (✉)
SISTec-R, Bhopal, India
e-mail: jyoti12march@gmail.com

P. Bansal
IET-DAVV, Indore, India
e-mail: pratosh@hotmail.com

N. Modi et al. (eds.), *Proceedings of International Conference on Communication and Networks*, Advances in Intelligent Systems and Computing 508,
DOI 10.1007/978-981-10-2750-5_72

images; originality, authenticity, and security of digital data has become an important issue. So the field of digital image forensics is gaining importance very rapidly since last few years for ensuring integrity and authenticity of the images. Copy Move Forgery is very common kind of practice in image tempering. Detection of tempering and recovery of original image is part of Digital Forensics. In this paper, a new mechanism for Copy Move Forgery detection has been proposed.

1.1 Overview of Digital Forensics

Internet has made our life easy, but at the same time has opened new threats. Criminals are always in search for more sophisticated and latest technologies for making crime in unique way. Technology has changed way of crime and face of criminal. This situation is also putting challenges to police and law making agencies. Alone hacking can be a very harmful and dangerous crime for an individual and/or organization. Information security has become one of the most concerned research topic of this digital era. "*The present era use the modern cryptography techniques to facilitate the necessary information security features required by an application*" [1]. Still none of the application or computing device is completely safe in this world. The main role of Digital Forensics starts when security or authenticity got compromised. The digital forensics concentrates on finding the digital evidences against the criminals [2]. "*The Forensics is a combination of art and science*" [3]. Digital Forensic is now very promising discipline in digital security. Digital forensics can be defined as "*The scientific examination and analysis of digital evidence in such a way that the information can be used as evidence in a court of law*" [4].

1.2 Branches of Digital Forensics

Researchers have divided Digital Forensics Science into various branches. Some branches of Digital Forensics are [2]:

(a) Computer Forensics
(b) Mobile Device Forensics
(c) Network Forensics
(d) Multimedia Forensics (Audio–Video and Image Forensics) etc.

1.3 Overview of Image Forensics

Originality and authenticity of digital image should be checked before taking any decision on that image. Common users of ICT do not always check the originality and assume all the images correct as and when images comes to them. However, for professional, legal, political and security purposes originality of image should be verified. In this situation role of Digital Image Forensics starts and Image Forensics can be defined as "*a brand new research field which aims at validating the authenticity and integrity of images by recovering information about their history*" [5]. Digital images are the most powerful and trustworthy media for expressing our views. So, the field of digital image forensics has been growing very rapidly as a new research field [6]. To edit or alter any digital image has now become a common practice. It may also be possible to manipulate the image till such a level (extent) that it represents a wrong situation or leading wrong information. This is called forgery.

There are various ways to forge digital image and some of them are [7]:

(a) *CLONING (Copy Move Forgery)*
(b) *RESAMPLING*
(c) *SPLICING* etc.

The topic of Copy Move Forgery Detection mainly comes under the category of multimedia forensics and if we further sub classified multimedia Forensics, then it could be considered a part of image forensics. There is a term "CMFD" in image forensics, which is actually an acronym of "Copy Move Forgery Detection". This paper focuses on CMFD.

1.4 Copy Move Forgery

Copy move forgery is a special and most common kind of attack in image forensics for alteration of any digital image. Attacker performs this kind of attack with the intention of either to hide any particular content of digital image or to duplicate some portion of the image. These can be understood by Fig. 1.

For duplicating or hiding steps will remain same. To achieve copy move forgery, attacker has to perform four successive steps and these are mentioned as below:

(a) First step of copy move attack is to select a particular portion of the image that is to be duplicated on the other part of the same image.
(b) Secondly, Copy that selected portion by performing "copy" operation of editing software.
(c) In third step, move that copied part to new location.
(d) Final step of this attack is "paste" operation of editing software, in which copied portion (performed in 2nd step above) is pasted to new desired location (which is already decided in 3rd step).

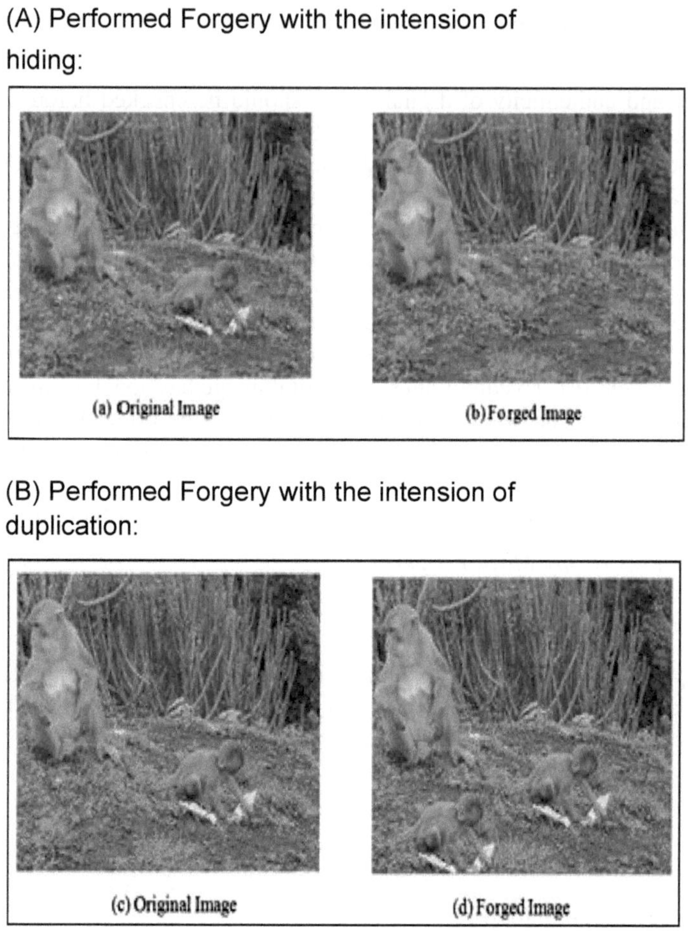

Fig. 1 Example of copy move forgery

This complete process is called as Cloning or Copy Move Attack or Copy Move Forgery. After performing this kind of attack no one can detect altered area of that image by normal observation. The reason behind is that the content of copied and pasted portion are taken from same image. The geometrical composition, colour and texture properties, pixel composition and also other properties are derived from the same image. Due to this, all such properties of both portions will be similar and will be very tough to detect those manipulated areas by normal human vision [8]. Attackers generally prefer those attacks most which does not left any clue about alteration and this attack provides all such things.

2 Related Work

Copy Move Forgery can be detected by various techniques. Methods suggested by various authors are discussed in this section. There are various methods to detect forgery namely DCT, DWT, SIFT, PCA, SURF, PCA-SIFT etc. Combination of these methods is also used sometimes to detect forgery. Various authors included some pre-processing and post-processing steps with above methods to improve already developed methods. The brief summary of methods of Copy Move Forgery detection is in Table 1.

There are some limitations associated with each of these methods. No method can work perfectly in all situations. SIFT is a method that is used frequently among

Table 1 Summary of related work on copy move attack

S. No.	Methods	Description
1	DCT (Discrete Cosine Transform) [9]	Copy move forgery can be detected by applying DCT. In this approach, input digital image is divided into appropriate overlapping blocks before applying DCT and then apply DCT on each block and sort the row lexicographically
2	DWT (Discrete Wavelet Transform) [10]	This paper is on detection of forged region by using DWT. Authors of this paper described a blind or passive forensic approach for CMFD. They first apply DWT to the input digital image and output of it produced a reduced dimension representation. Then phase correlation is computed to determine the spatial offset between copied and pasted region
3	SIFT (Scale Invariant Feature Transform) [11, 12]	Authors detected duplication of regions by the concept of SIFT features. They carried out their task into 4 steps, first they collected SIFT features, then matching and pruning is done. In third step estimate region transform was performed and finally they identified duplicated regions. This method is effective and robust even in presence of additive noise and different JPEG qualities [11]
4	SIFT and local features based integrated method [13]	Authors proposed an integrated method which is based on SIFT and calculations of local features. Local features include colour features and texture feature. Initially in their algorithm, similar feature points are paired by (or highlighted by) comparing SIFT features of image then they applied examination to eliminate false matched points, after that they first applied block colour feature inspection on image blocks and then applied block texture feature extraction

(continued)

Table 1 (continued)

S. No.	Methods	Description
5	Combination of SIFT and DCT [8]	Authors used hybrid method to detect copy move forgery. They used the combination of SIFT and DCT. First they converted RGB image into gray scale and then applied DCT and stored the intensity levels in a separate matrix. After that, they divided the image into a block size of 16 × 16. They applied SIFT on those blocks and stored the feature vector in rows of matrix and save all the coordinates value
6	Demonizing algorithm [14]	Authors presented an efficient non-intrusive method for CMFD. This method is based on segmentation of image. They applied demonizing algorithm on segmented image in which they used segmented image to estimate image noise and analysed noise pattern of each segment. At last they found the image is forged if the noise patterns of at least two image segments are similar

all described methods by various authors and also is used in combination with several different methods. But there are some limitations of SIFT as well, these are:

Unable to detect forgery from smooth surface. Variant to light color changes.

Variant to non-uniform illuminations etc.

All previously used methods have some of their own limitations. So an effort has been made here to detect forgery with a novel concept which doesn't found prior in literature of copy move forgery detection till now.

Apart from above literature study of copy move forgery detection, some study about template matching has been also made here to explain the concept:

Perveen et al. [15] provide the basic overview of template matching, its methodologies and its application areas. They described two types of template matching approaches, namely Features based approach and Area based approach. Technique of template matching can be used in image processing, computer vision, remote sensing, face detection and eye detection in facial image and in biological science as well and many more.

Kumar et al. [16] used the concept of template matching for object tracking. To successfully achieve this task they used logic of SAD (Sum of Absolute Difference) and SSD (Sum of Squared Difference).

3 Problem Statement

Literature shows that none of methods which are introduced and developed till now can work perfectly in every type of situation to detect copy move forgery. Also some of the methods are computationally complex. Therefore the performance improvement for available techniques is required by introducing a novel concept. So there is need to introduce and develop a new concept to detect copy move forgery which can be able to detect forgery efficiently in all type of situations like when forged part is rotated or resized and all.

4 Proposed Solution

Proposed Solution is based on the concept of template matching. "*A template matching in basic is matching the specific objects of the source image using a template image*" [16]. In order to provide an efficient solution, a template matching based novel approach is designed to implement for Copy Move Forgery Detection of coloured digital images. To detect this forgery we have to locate the altered (forged) area in forged image and also locate the source area of forged region in original image. The method of template matching is simple to implement and due to this reason it has less computational complexity. The architectural diagram of proposed solution is described in Fig. 2. Proposed solution for copy move forgery detection in coloured digital images involves following main modules and steps:

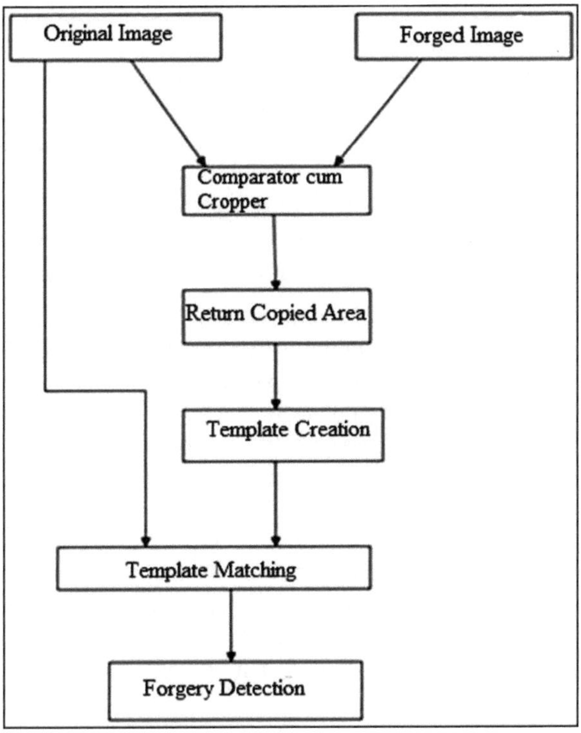

Fig. 2 Architecture of template matching scheme

i. **Input Images**: Two images (Original image and Forged Image) are taken as input.
ii. **Comparator cum Cropper**: It simply returns the different or dissimilar portion of both the input image (i.e. the forged image and original image). In order to find the object which is copied and then pasted in the same image is required to subtract the original image pixels from the forged image pixels. Thus if original image is denoted by I_o and the forged image is denoted by If then copied area can be calculated by the following formula:
   $C = I_o - I_f$; Where C represent copied area in forge image. Actually, the pixels which are same in both input images; we assigned black colour for those pixels and all the remaining pixels keep preserves which is basically represent dissimilar or different pixels (i.e. copied region). This module mainly extracts the copied area from forged image.
iii. **Template Creation**: The copied area (which we got by just previous step) is saved as Template. Template is also would be saved in the form of image which will play a role of template image.
iv. **Template matching**: Mainly this particular module is used to detect Copy Move forgery. In this step, original image and template image will be given as two inputs which perform the task of matching. This module will find out the area in original image which exactly match with template image (which contains copied area) by using SAD [17]. Template matching applications often use some simple similarity measures such as Sum of Absolute Difference (SAD) or Sum of Squared Differences (SSD) [17]. We used SAD to implement our solution.
v. **Forgery Detection**: In this step original part (Source of copied region) and copied part (Forged region) are displayed by drawing rectangle around those areas. To find the exact location of source of copied region (i.e. original part) is returned by template matching module in previous step (i.e. step iv) and to get the location where copied region is pasted (i.e. forged part) in forged image is captured by mouse motion listener and stored somewhere in memory by capturing coordinate of that location. In this way, we detected copy move forgery.

A screenshot of system is as in Fig. 3 for giving an understanding about the developed system. Initially two input images are taken and given to comparator. Comparator returned the copied area and saved in new image which plays a role of template image. Template image is shown on the right side of two input images which has black background. At bottom portion of Fig. 3, two more images are displayed. The source of copied area in original image (first image) and the destination area where copied content is pasted in forged image (second image) is displayed/highlighted with the help of two blue coloured rectangles by drawing rectangles around those regions. In this way proposed system detects forgery.

In proposed system, performance can be measured by accuracy, which shows how much exact detection results we are getting. The graph of accuracy of proposed system is shown in Fig. 4. This graph is between two parameters i.e. number of

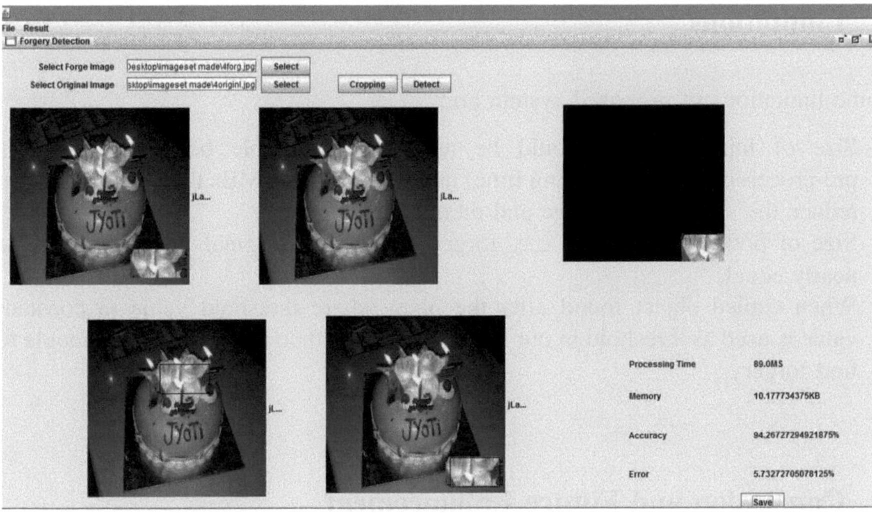

Fig. 3 Copy move forgery detection

Fig. 4 System performance

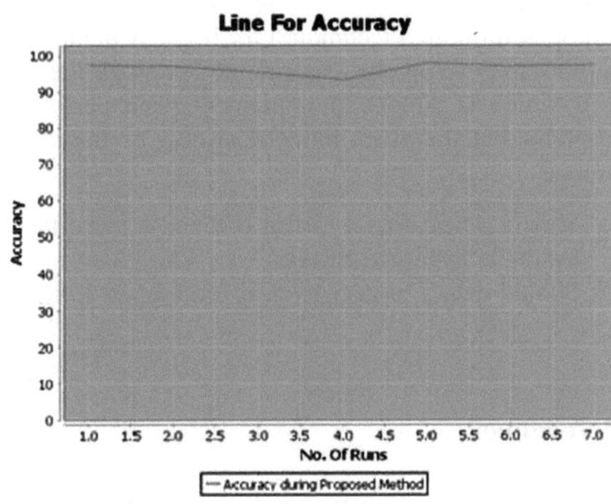

runs and accuracy of the system taken on X and Y axis respectively. Accuracy is calculated by the following formula:

Accuracy = [(Pixels covered in rectangle of forged image) ÷ (total number of pixels resides in actual source of copied part)] × 100.

5 Limitations

Some limitations of proposed system are:

- Size of input images should be as less as possible because it reduces pre-processing time (Cropping time) and if image is in MBs then need is to first reduce the size of that image and then use as input.
- Size of both input images (i.e. forged image and original image) should be nearly equal.
- When copied object found after the pixel where threshold value (a constant value is used as threshold in our system) reached, then this system will unable to find forgery.

6 Conclusion and Future Enhancement

Digital images acts like a natural and expressive communication medium for humans. In this work the coloured image based copy move forgery detection system is proposed (a blind technique) for implementation and performance improvement. In order to develop our research methodology the concept of template matching is used which is simple to implement. The performance of proposed systems is adaptable but the given model can also be improved by achieving the following issues:

(a) Should detect forgery even if there is more than one forged or copied region available in forged image.
(b) Should detect forgery even if forged part is resized or rotated at some angle.

References

1. H. Suryavanshi and P. Bansal, "Design and Implementation of an Improved Cryptographic Algorithm using UNICODE and Universal Colors," *Current Trends in Information and Technology*, vol. 3, no. 1, 2013.
2. G. Fenu and F. Solinas, "Computer forensics between the italian legislation and pragmatic questions," *International Journal of Cyber-Security and Digital Forensics*, vol. 2, no. 1, pp. 9–24, 2013.
3. V. Shah and P. Bansal, "CDCD-5 an Improved Mobile Forensics Model," *International Journal of Computer Science and Information Technology & Security,* vol. 2, no.4, pp. 739–741, Aug. 2012.
4. Digital Forensics, available at: http://www.cyberforensics.purdue.edu. Accessed on September, 2014.
5. Digital image forensics, available at: http://www.eurecom.fr/publication/3251. Accessed on November, 2014.

6. X. Pan and S. Lyu, "Region Duplication Detection Using Image-Feature Matching," *IEEE Transactions on Information Forensics and Security*, vol. 5, no. 4, pp. 857–867, Dec. 2010.
7. H. Farid "A survey of image forgery detection," *IEEE Signal Process. Mag.*, vol. 2, no. 26, pp. 16–25, 2009.
8. R. Singh, A. Oberoi, N. Goel, "Copy Move Forgery Detection on Digital Images," *International Journal of Computer Applications*, vol. 98, no. 9, pp-17–22, July 2014.
9. A. Gupta, N. Saxena, S.K. Vasistha, "Detecting Copy move Forgery using DCT," *International Journal of Scientific and Research Publications*, vol. 3, issue 5, May 2013.
10. J. Zhang and Z. Feng, Y. Su, "A New Approach for Detecting Copy-Move Forgery in Digital Images," *in Proc. IEEE International Conference on Computational Science,* 2008, pp-362–366, 2008.
11. X. Pan and S. Lyu, "Detecting image region duplication using sift features," *in Proc. IEEE International Conference on Acoustics Speech and Signal Processing*, pp. 1706–1709, 2010.
12. I. Amerini, L. Ballan, R. Caldelli, A. Del Bimbo and G. Serra, "A SIFT-Based Forensic Method for Copy–Move-Attack Detection and Transformation Recovery," *IEEE Transactions on Information Forensics and Security*, vol. 6, no. 3, pp. 1099–1110, Sep. 2011.
13. B. Liu and Chi-Man Pu, "A SIFT and Local Features Based Integrated Method for Copy-Move Attack Detection in Digital Image," *IEEE International conference on Information and Automation China*, pp. 865–868, August 2013.
14. N. Muhammad, M. Hussain, G. Muhamad, and G. Bebis, "A Non-intrusive Method for Copy-Move Forgery Detection," *Springer International Symposium on Visual Computing*, pp. 516–525, 2011.
15. N. Perveen, D. Kumar and I. Bhardwaj, "An Overview on Template Matching Methodologies and its Applications," *International Journal of Research in Computer and Communication Technology,* vol. 2, issue 10, pp. 988–995, October- 2013.
16. A. Kumar, A. Joshi, A. Kumar, A. Mittal and D R Gangodkar, "Template Matching Application In Geo-Referencing Of Remote Sensing Temporal Image," *International Journal of Signal Processing, Image Processing and Pattern Recognition,* vol. 7, no. 2, pp. 201–210, 2014.
17. A. Mahmood, S. Khan, "Correlation Coefficient Based Fast Template Matching Through Partial Elimination," *IEEE Transactions On Image Processing*, 11 August, 2010.

Analysis of Rule Based Expert Systems Developed and Implemented for Career Selection

Shaily Thakar and Viral Nagori

Abstract The use of Expert system in career selection offers better efficiency and correctness of decision. The paper presents the analysis of rule based expert systems developed and implemented for career selection. The analysis is done on the parameters: targeted users, development mode, tools used, system, knowledge base, inference engine, factors affecting career selection, and research gap. From the analysis, we are able to list down the important factors affecting the career selections. At the end, paper presents the guidelines for the design of rule based expert system for career selection.

Keywords Rule based expert system · Career selection · Influencing factors

1 Introduction

Expert system is used to simulate human intelligence. We can term an Expert system as a computer system made up of rules, knowledge base and user interface to solve problem of a particular domain, which requires expertise [1]. Expert systems have been developed, implemented and found to be used effectively in domains like agriculture, medical, industry, education and various other domains. For example, in Medical field it is used for diagnosis of disease, in education field it is used for effective teaching. In education area, selecting a right course/discipline is an important decision in student's life. Researchers have automated the decision making of selecting a right career through the use of expert systems. The paper here by reviews existing Expert systems developed and implemented for career selection. The main objectives of this review are:

S. Thakar (✉)
GLSICA, GLS University, Ahmedabad, India
e-mail: shaily.gls@gmail.com

V. Nagori
GLSICT (MCA), GLS University, Ahmedabad, India
e-mail: viral011@yahoo.com

N. Modi et al. (eds.), *Proceedings of International Conference on Communication and Networks*, Advances in Intelligent Systems and Computing 508,
DOI 10.1007/978-981-10-2750-5_73

1. To study and analyze existing expert systems for career selection
2. To list down factors that influences the career selection process from the literature.

The Expert Systems selected here for study are all are Rule based. The main reasons for selecting rule based system are because of inherent advantage of rule based provided as well as we are modeling our proposed expert system based on the rule base.

2 Summary of Literature Review of Expert Systems

We reviewed five expert systems. They are: (1) An Expert System for career guidance for students of Pondicherry. (2) IS-Advisor Expert system used at Ajman University, UAE. (3) iAdvice Expert system used at University of Moratuwa, Srilanka. (4) Rule based Expert system used at Sabanci University. (5) Automating Academic Advising and Course Schedule Planning with a CLIPS Expert System.

2.1 Design of an Online Expert System for Career Guidance—[2]

This is an online proposed Expert System that provides career guidance to higher secondary students for selection of undergraduate courses at colleges of Pondicherry. This system uses pattern matching and jSoup parsing technique to extract information from web pages of colleges in Pondicherry and knowledge base is constructed. The developed system will get details like student information, result, and college preference from the user as input and based on his requirements and eligibility criteria for the colleges, the user will be provided with best suitable college as a suggestion in the output screen. The system is made up of 5 modules:

1. Web information Extraction.
2. Structuring extracted information.
3. Developing Rule based knowledge Base
4. Details from User through a User Interface
5. Providing Required Information to the User.

The *knowledge base* has been constructed using ontology. It contains rules which can be classified into two categories: (1) University Admission Requirements Rules.

(2) Students' Preferences. The information collected is processed by *Inference engine.* It is based on rule based reasoning. So in each cycle, it attempts to pick an appropriate rule from its collection of rules. *Factors found that affect Career Selection are the* region student belongs to, Stream of student opted, preference,

financial support, 12th percentage, reserved category, hostel facility, current age. The *research gap* found in this System is that it can be extended to handle complex queries by making it accommodate all possible information apart from undergraduate courses alone. *The Outcome found in this system is that* the proposed Expert System generates assuring results by guiding higher secondary students to select undergraduate course and it reduces a great deal of human efforts in knowledge extraction. It provides students correct information which helps them in choosing the right path.

Some significant features we found in this system are the dynamic updating of college and student details, where the outdated contents of knowledge-base are automatically trimmed and new values are updated without any manual efforts hence increases accuracy and reliability of the system.

2.2 A Prototype Student Advising Expert System Supported with an Object-Oriented Database—IS-Advisor [3]

IS-Advisor a prototype rule based Expert System with object-oriented database used at Information Systems Department (IS), College of Information Technology Ajman University, UAE. The system provides guidance to High school students for selecting suitable courses for each semester towards academic degree. This system is a Standalone. The system uses Kappa PC Expert System shell. The system has graphical user interface and simple menus. The system performs three main steps (1) All courses that are offered can be registered by the student are stored in a list called Allowable Courses list. (2) Performs ranking process for the courses contained in Allowable courses list. (3) This a filtering step that generates the ordered list of recommended courses based on the contents of the list Ordered allowable courses. This step follows two models for advising (1) Prescriptive (2) Developmental. The rules of the *knowledge base* can be classified into two categories: Academic rules and student preference rules. Kappa-PC Expert System shell is used which supports both rule based reasoning as well as micro managing of the reasoning using classical programming techniques. IS-Advisor's *inference engine* uses both if-then rules processing and list processing techniques. *Factors found that affect Career Selection are* Advisor, preferences, AGPA, past academic performance. The *research gap* found in this system is that it can be connected with a University's student information system to automate the process of importing student's data. Advising students with exceptional cases and warned students can be added. The system can be improved to automate the process of lecture timings for a course offered which can be used by timetabling committee. *The Outcome found in this system is that* IS-Advisor is a prototype Expert System with an object oriented database for student academic advising. The system provides quick and easy way for course selection and evaluation of various alternatives. The system has a graphical user interface. IS-Advisor is unique than all other academic advising expert systems since it is based on the accumulated academic advising knowledge.

Some significant features we found in this system is that using of object oriented database that allows each student and each course to be modeled as a single object and the database is modeled as a collection of these objects. Kappa PC development tool supports Object oriented modeling. This Expert system not only gives output as list of desired courses to be selected but also gives explanation facility.

2.3 Artificial Intelligence Approach to Effective Career Guidance-iAdvice [4]

iAdvice is a Career advisory standalone Expert System used to guide the undergraduate students engaged in their higher education, to determine their career paths and to select the course subjects to be inline with their career goals. This system is used at Faculty of Information Technology at University of Moratuwa. This system is made using FLEX Expert System shell. iAdvice uses features such as reasoning ability, providing explanations, alternative solutions, uncertainty and probability measures, questioning ability. It is divided into two main subsystems (1) Career known subsystem and (2) Career unknown subsystem. iAdvice provides simple user interfaces, starting with login screen, to interact with the system. This system uses Flex Intelligent Server that supports rule based programming and data driven procedures fully integrated within a logic programming environment. The *knowledge base* is implemented using English like Knowledge Specification Language (KSL) using FLEX. *Inference Engine* of iAdvice was implemented using Flex inference engine which supports both forward and backward chaining, combined with business logic layer. *Factors found that affect Career Selection are* past examination performance, student preferences and skills, industry alignment with subjects. The *research gap* found in this system is to make this system as a web application, to incorporate Natural Language Processing. To develop a customizable solution to suit any educational institute requirement. *The Outcome found in this system is that* iAdvice will provide possible paths that one can take based on their past track records and their preferences also provide alternatives students. It recommends what is best career path for student. iAdvice also advices on the subject areas the student should improve if the performance is not up to the required standard. iAdvice has capability on working with incomplete information.

Some significant features we found in this system are it uses both chaining if in some case forward chaining does not work then uses backward chaining. iAdvice has been designed in modularized manner in order to maintain greater efficiency and become easy to maintain. It uses features such as reasoning ability, providing explanations, providing alternative solutions, providing uncertainty and probability measures, questioning ability are found in iAdvice.

2.4 *Rule-Based Expert System for Supporting University Students—[5]*

This rule based Expert System guides and recommends courses to undergraduate students of Manufacturing Systems Engineering Program students at Sabanci University using OPA (Oracle Policy Automation). This system is standalone system. The System is developed using OPA (Oracle Policy Automation). Course related data are collected from the University database using BannerWeb. Oracle Policy Automation saves all historical data related to user. This data is represented using yEd graph visualization software. OPA is used to collect and form database. When preparing the rule base *knowledge base*, rules related to prerequisite courses, student's GPA, information about which courses are available in a semester, area the student is specializing in, and courses details in which the student has readily registered to, was checked. *Inference Engine* of this system uses OPA and is capable of reading rules conveniently from spread sheet and word-processor file formats. OPA automatically forms problem solving or decision making application based on acquiring rule base. *Factors found that affect Career Selection are* Prerequisite courses, student's GPA, area of specialization, semester wise course, courses students have readily registered to. The *research gap* found in this system is to deploy the developed expert systems on the university intranet and/or on the Internet. The *Outcome found* in this system is that this Course Advising Expert system recommends courses to undergraduate students. List of courses the students can take is given as output.

Some *significant features* we found unique in this system are that the software used in this system is Oracle Policy Automation (OPA) which is capable of reading rules from spreadsheet and word processor file formats from this OPA easily forms rules. OPA also has Multilanguage support. If rules need frequent change then OPA is suitable and convenient.

2.5 *Automating Academic Advising and Course Schedule Planning with a CLIPS Expert System—[6]*

This Expert System is used to improve and streamline the advising process in Dept. of computer science, University of West Georgia for undergraduate students. The Expert System is used to satisfy three functional requirements. The first requirement is to help in reviewing the student's progress towards degree completion. The Second requirement is to help in planning a student's future course schedule. The Third requirement is to help in planning what courses should be offered in future terms that projects student's need. This Expert system is web-based and is developed in PHP and uses CLIPS (C Language Integrated Production System) Expert System Tool. CLIPS program is expandable, flexible and low cost. The approach used in this system is to implement two-tier mandatory advising process; it is centralized between professional staff, advisor with whom the student must first meet. Expert system uses a forward chaining, and uses a *knowledge base* for students' academic history, program requirements and projected course offerings. *Inference Engine* of this Expert System uses CLIPS engine to infer new rules based on what course does the student take in future and when the courses are expected to be offered. *Factors found that affect Career Selection are* Students' academic history, Program requirements and projected course offerings. The research gap for this system is not mentioned. *The Outcome found for this system is that* since 2007, the system has been successfully used and support is provided for advisement sessions and planning for course selection to undergraduate students.

Some significant features we found in this system are that it has two tier advising process with easy updating facility to update information like degree requirements and course pre-requisites while maintaining older ones. It also has analysis engine to be used with different user interfaces and reporting needs.

3 Comparison Among the Reviewed Papers

The Table 1 shows the comparison of all the five expert systems based on the following parameters: The comparative study of all the five Expert System based on various parameters like Targeted Students, Development Mode, Tools used, Knowledge base, Inference engine, factors used and research gap is done below.

4 Analysis of Reviewed Papers

From the above reviewed papers it is noted that expert system is used for guiding higher secondary, undergraduate's students for their career option/stream, major or minor course selection. Different factors affecting career selection are found such as.

Paper1: Region from which students come, Stream of student opted, Preference, finance support, 12th percentage, reserved category, hostel facility, current age.

Paper2: Advisor, Preferences, AGPA, past academic performance

Paper3: Past examination performance, student preferences, skills, industry oriented subjects

Paper4: Prerequisite courses, student's GPA, area of specialization, semester wise course, courses students have readily registered to.

Paper5: Students' academic history, Program requirements, projected course offerings.

There are common factors found in all the Expert Systems, they are Preference, past academic performance, Students Academic History.

Table 1 Comparative analysis of literatures based on parameters

Parameters	Paper1	Paper2	Paper3	Paper4	Paper5
Target students	Higher secondary school students	Higher secondary school students	Under graduate students	Under graduate students	Undergraduate students
Development mode	Web based	Stand alone	Stand alone	Stand alone	Web based
Tools used	Pattern matching, jSoup parsing technique	Kappa-PC expert system shell, object oriented database	Flex expert system shell	OPA, BannerWeb yEd graph visualization software	PHP, CLIPS expert system tool
Knowledge base	Made of ontology based on heuristic	Two categories. Academic rules and student preference rules	Uses FLEX intelligent server. It uses Knowledge Specification Language (KSL)	OPA is used to collect and form database	students' academic history, program requirements and projected courses offerings
Inference engine	Rule based reasoning	Uses if-then rules processing and list processing technique	Uses Flex's inference engine combined with business logic layer. It supports both forward and backward chaining	OPA forms rules	CLIPS engine is used to infer new rules. It uses forward chaining
Research gap	To accommodate all possible information apart undergraduate courses alone	To: Advise exceptional and warned students. Automate the process of lecture timings. Web-based	To: Make web application. Incorporate NLP. Develop customizable solution to suit any educational institute	To deploy the developed expert systems on the university intranet and/or the Internet	The future enhancement is not available

5 Guidelines Derived for Designing Expert System for Career Guidance

Based on the review and analysis of above expert systems we can derive the following guidelines to develop Expert System in Career Selection domain:

- It is desired that the Expert System should have explanation facility for the solution it generates.
- A true Expert System should never end up with system generated error message but should give feasible solution with all constraints in worst situations.
- Expert System can be developed that supports multiple languages, which might be convenient to users.
- Expert System should support both forward and backward chaining.
- It is desired that Expert System should have facility to automatically update the information from required sources so as to keep knowledge base up to date and generate new rules.
- It is easy for Web-based Expert System to connect and collect data from different sources
- The design of Expert system should be generic, so as to suit any educational institute.
- It is suggested that if Natural Language Processing is incorporated it might increase the usability of the system.

6 Future Scope

In future, we will review more Expert Systems for career selection to find the complete list of factors that influence the career selection for the students and also to find out research gap in this domain.

7 Conclusion

From review and analysis of the developed and implemented expert systems, it is found that Expert System is used for career selection targeted for higher secondary, undergraduate students to select their courses, career path or select major subjects in particular university. The analysis helped us to come out with the guidelines while developing our proposed rule based expert system for career selection. Based on the analysis, we are able to identify partial list of factors that influence the career selection decision.

References

1. Jackson, Peter (1998), Introduction To Expert Systems (3 ed.), Addison Wesley, p. 2, ISBN 978-0-201-87686-4.
2. S. Saraswathi1, M. H. (2014). DESIGN OF AN ONLINE EXPERT SYSTEM FOR CAREER GUIDANCE. JRET: International Journal of Research in Engineering and Technology eISSN: 2319-1163 | pISSN: 2321-7308, pp. 314–319, 2014.
3. Ahmar, M. A. (2011),"A Prototype Rule-based Expert System with an Object-Oriented Knowledge-base for University Undergraduate Major Selection". International Journal of Applied Information Systems (IJAIS) – ISSN: 2249–0868 Foundation of Computer Science, Volume 4– No. 8, pp. 38–42, December-2012.
4. Chathra Hendahewa, M.D. (2006). Artificial Intelligence Approach to Effective Career Guidance. Sri Lanka Association for Artificial Intelligence (SLAAI) pp. 32–42.
5. Gökhan Engina, B. A. (2014). Rule-based expert systems for supporting university students. Procedia Computer Science 31 (2014), pp. 22–31.
6. Edwin Rudolph, A. A. (2015.). Automating Academic Advising and Course Schedule. Int'l Conf. Artificial Intelligence| ICAI'15.

A Pragmatic Analysis of Security and Integrity in Software Defined Networks

Drashti Dave and A. Nagaraju

Abstract Virtualization is one of the key components during the performance evaluation of network enabled environment including distributed computing, cloud computing, grid computing or pervasive computing. In the classical aspects, it is difficult to perform the implementation at physical devices all the time in the research and development process. The network administrators and forensic teams are working on software defined networking (SDN) by which the network components can be controlled and managed via virtual infrastructure and suites. With the help of SDN, effective control including routing, scheduling, security and related algorithms can be implemented on real networks. Alongside the evolution of networking is the evolution of network attacks. With the introduction of SDN, new strategies for securing the traffic are needed. Even though many SDN systems are relatively new and SDN is still in the realm of the early adopters, but as the technology matures and is more widely deployed, it will become a target for attackers. In this paper, a pragmatic analysis of the security aspects and related dimensions is done. The vulnerabilities and associated factors on the network environment are underlined in this work, with that help the higher level security can be implemented.

Keywords Software defined network · Virtualization · Open source SDN suite · Openflow · Security

D. Dave (✉) · A. Nagaraju
Department of Computer Science & Engineering,
Central University of Rajasthan, Ajmer, India
e-mail: drashti2110@gmail.com

A. Nagaraju
e-mail: nagaraju@curaj.ac.in

N. Modi et al. (eds.), *Proceedings of International Conference on Communication and Networks*, Advances in Intelligent Systems and Computing 508,
DOI 10.1007/978-981-10-2750-5_74

1 Introduction

Software-defined network (SDN) is the unique and effective technology for the development and integration of agile as well as flexible network infrastructure using virtualization in the contemporary data centers. It is the unique approach towards computer networking which allows the network and digital resources administrators in managing the network services using higher level abstraction and effective functionality. This process is implemented with the use of decoupling of system which makes the decisions where data traffic is to be sent with the use of the control plane.

In the current scenario and implementations, OpenFlow [11] is deployed as software defined networking (SDN) technology. In SDN technology there is the decoupling of control and data planes in network [9]. A software based controller is responsible for managing the forwarding information of one or more switches; the hardware only handles the forwarding of traffic according to the rules set by the controller [13]. OpenFlow is an SDN technology proposed to standardize the way that a controller communicates with network devices in SDN architecture. It was proposed to enable researchers to test new ideas in a production environment [7].

OpenFlow provides the migration of layer for control logic from a switch into the controller. It presents a protocol for the effective communication between controller and network switches [3].

2 Overview of SDN Architecture

The idea of Software Defined Network architecture is a new and novel way of network management. Forwarding plane and Control Plane of conventional network are decoupled in software defined network. Figure 1 shows all the components of SDN Architecture.

In SDN, switches (*data plane*) do not process incoming packets. They look for a match of the incoming packet in their forwarding tables and if there is none, it will

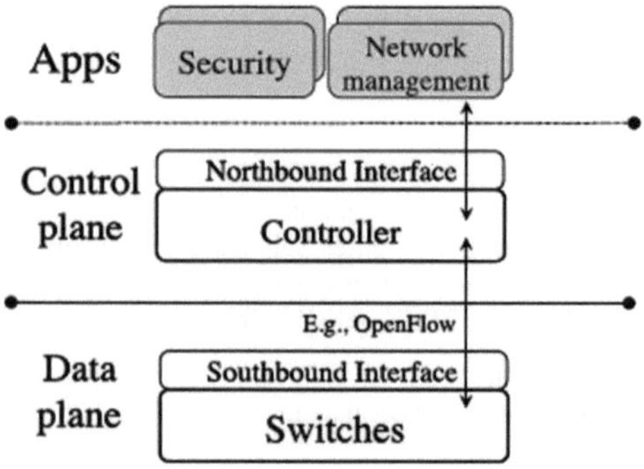

Fig. 1 Components of SDN [3]

be sent to the controller for processing. The controller (*control plane*) is the operating system of SDN. It processes the packets and decides whether the packet will be forwarded in the switch or will be dropped. By applying this procedure, SDN separates the forwarding and processing planes.

The API used to communicate between the SDN Controller and the services and applications running over the network are known as *North bound API* while the *South bound APIs* are used to communicate between the SDN controller and the switches and routers of the network (www.sdxcentral.com) [5]. The first south-bound protocol standard in SDN was considered the OpenFlow protocol [4]. It allows access to and manipulation of the data plane devices. Even though OpenFlow is not the only available protocol (e.g., Extensible Messaging and Presence Protocol XMPP [15]), it is considered as standard and supported by multiple companies in their SDN ready solutions [4].

3 Review of Existing Literature

SDN has great potential to change the way networks operate, and OpenFlow in particular has been touted as a "radical new idea in networking" [10]. The proposed benefits range from centralized control, simplified algorithms, commoditizing network hardware, eliminating middle boxes, to enabling the design and deployment of third-party 'apps' [13]. The idea of decoupling logic and programmable networks was introduced many years back. In 1995 the Open Signaling (OPENSIG) working group worked on "making ATM, Internet and mobile networks more open, extensible, and programmable" [2].

According to H. Farhady et al., Forwarding and Control Element Separation (ForCES) is an Internet Engineering Task Force (IETF) standard which provides an interface between control and data plane. Routing Control Platform (RCP), Path Computation Element, (PCE) Soft Router, and Intelligent Route Service Control Protocol (IRSCP), all foster a centralized approach to control the network. But the most widely used protocol for communication between the controller and data plane is OpenFlow. The immediate predecessor of OpenFlow is ETHANE/SANE. SDN and Open flow are the terms mostly used together and hence there is a misconception that both are equivalent. This is mainly due to the fact that the term SDN was coined after the introduction of OpenFlow and OpenFlow is one representative API to setup an SDN instance. The other few examples of APIs which can also be used for similar work are JunOS SDK and Cisco ONE.

According to Mehdi et al. [12], Shin et al. [17] SDN using the protocol of OpenFlow gives novel and interesting potentials which could be enhanced for solving the new challenges as well as to simplify the deployment of existing solutions.

It was stated by Kreutz et al. [8], Shin and Gu [16] that SDN is innovative and disruptive force in the industry of networking which influences every player encompassing equipment vendors, network operators, cloud and internet service providers. Further with SDN, configuration of low-level device and management could be handled by controller of centralized software which provides the upgrade of debugging and functionality. According to Kim and Femester [6] by distributing and managing the state in the network with a perspective of system, SDN frees the administrators to mine the difficult specifications of protocol with flexibility and agility for controlling the networks. At the same time, it was noted that SDN-enabled NFV (network functions virtualization) makes it probable for cloud service providers and internet for delivering their differentiation market advantage through enhancement of service in case of security and quality of service as illustrated by Mehdi et al. [12].

Alongside the evolution of networking is the evolution of network attacks. Even though there are many proposed detection methods and defense techniques, we cannot say that the security attacks are a solved problem or do not present an immense threat to the current Internet.

According to Ali et al. [1], categorize current SDN-based security research into two branches, research geared towards protecting the network, and providing security as a service. The first direction deals with security configuration and threat detection, remediation and verification using SDN.

As Rass et al. [14] says that one of the "gaps" in SDN is the capacity to ensure administration conveyance. While SDN devices can identify clogging, and at times can decrease blockage, they cannot avoid information misfortune.

According to Vizváry and Vykopal [19]. Denial of service (DoS) assaults is described by planned and deliberately considered endeavors to breaking point or keeps genuine clients from getting to network assets. Most handy DoS assaults include various target machines and numerous machines from which assaults are dispatched, suggesting disseminated foreswearing of administration assaults, marked DDoS assaults [18]. Given the omnipresence of PCs, DDoS assaults are rapidly turning into the most significant issue on the Internet.

As McKeown et al. [11] says if an assailant runs SDN scanner and gathers system data, he/she can explore whether an objective network is utilizing SDN or not through a straightforward measurable testing technique. In the event that the test outcomes demonstrate that an objective system is prone to utilize SDN, the assailant will further lead the asset utilization assault. Since, the assailant definitely knows the state of the own tenet for the objective system (with the assistance of SDN scanner), now he/she simply needs to send system bundles to devour SDN assets of the objective system.

4 Security and Integrity Challenges in Software Defined Networks

Security is the major concern and aspect that is mandatory for any network infrastructure. In SDN, as each and everything is dependent on the software and APIs part, it becomes necessary to enhance the layers of APIs and libraries so that any third party application or sniffer is not able to penetrate into the network.

In the SDN based architecture, following points should be considered and empowered so that the higher layers of security and privacy can be implemented.

1. Security of the Controller against sniffing attacks to escalate the quality of service.
2. Protecting the SDN Controller against DDoS Attacks for higher availability of the network.
3. Establishment of the Trust Architecture so that authenticated and genuine nodes can transmit the data as well as signals.
4. Creation of Robust and Strong Policy Guidelines as well Frameworks.
5. Incorporation of the Forensics as well as Remediation Measures to detect and push back any attack.

5 Weakness and Vulnerabilities in SDN

Even though many SDN systems are relatively new and SDN is still in the realm of the early adopters, but as the technology matures and is more widely deployed, it will become a target for attackers. Several attack vectors on SDN systems can be anticipated. The more common SDN security concerns include attacks at the various SDN architecture layers. Figure 2 shows the anticipated attacks that could occur at each of these layers. The SDN misuse/attack cases (including both surveyed and newly discovered) can be classified into three categories.

Control plane specific, which includes all misuse/attack cases against SDN control and application layer.

Control channel specific, which includes misuse/attack scenarios targeting to the interface (e.g., OpenFlow),

Data plane specific attacks, which comprises misuse/attack cases trying to torture network devices supporting SDN functions.

Model of SDN centralized control provides important advantages to the management of security and network, there are tradeoffs. Moreover logically physically distributed and centralized controllers of SDN are subject to unique set of threats and risks compared to traditional architectures of network. Centralized controller develops a capable single point of attack (SPoA) and failure which has to be safeguarded from threats. The south bound interface, between the controller and underlying devices of networking namely OpenFlow, which is vulnerable to issues which could degrade integrity, performance and availability of the network. At the same time, OpenFlow shows the adoption of transport layer security or user data protocol, either of which authenticates using encryption and certificates for securing the connection. In extra, security measures were required when these authentications fails; underlying infrastructure in the network has the potential to endure occasional periods where controller of SDN is not available; yet assure that any new flows would be synchronized when the devices rescue communications with SDN controller.

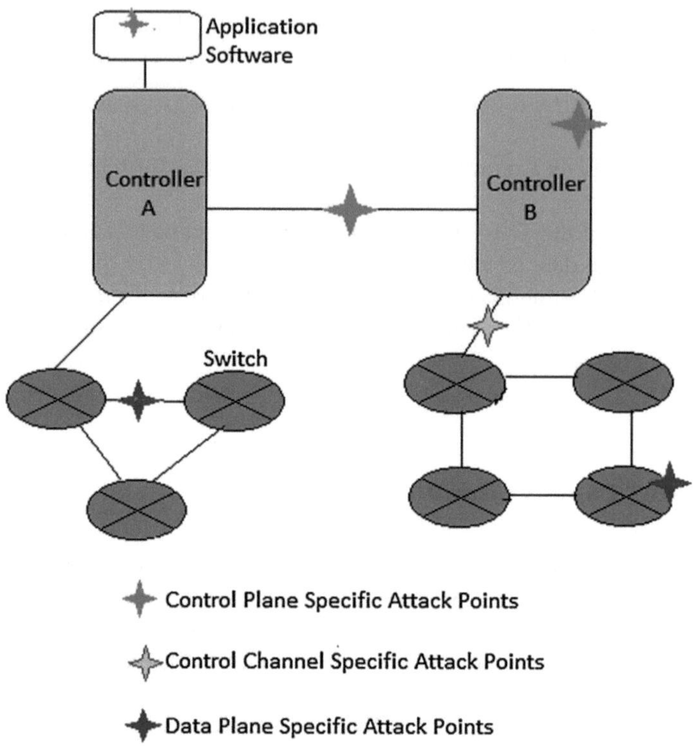

Fig. 2 Possible attack points in SDN environment

6 Proposed Model of Secured SDN Environment

Strengthening the SDN Environment is mandatory and thus a number of algorithmic approaches and techniques are implemented. Figure 3 shows the proposed SDN environment in secured mode. Following security measures can be integrated in the layers of SDN architecture to harden the security and penetration level.

1. Role Based Identification and Authentication of Nodes or Role Based Access Control.
2. State Table Management.
3. Conflict Resolution.
4. Use of TLS (Transport Layer Security) for authentication and encryption of the traffic between controller and device agent.
5. Configuration and Authentication of Tunnels.
6. Using OOB (Out of Bound) Network for controlling the traffic. It provides load balancing as well as security.
7. Flow Checking Approach to analyze the adoption of policies by different nodes.

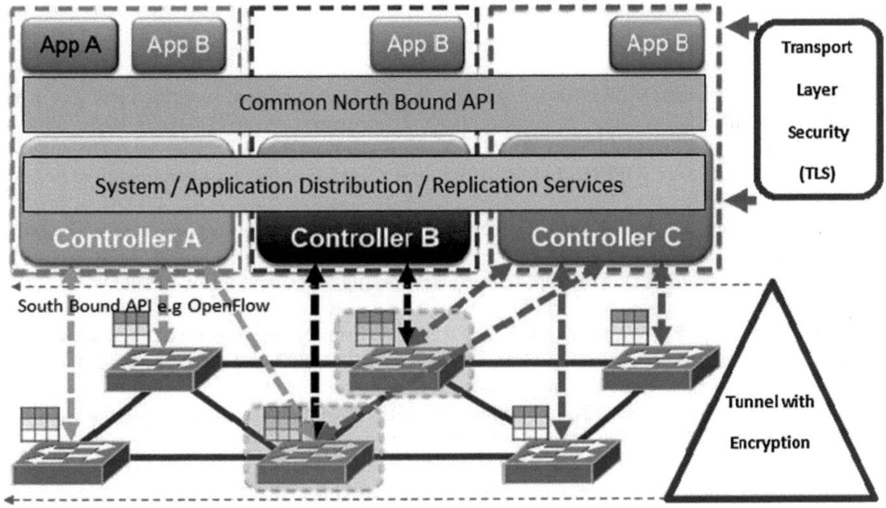

Fig. 3 Proposed SDN environment in secured mode

Alongside the evolution of networking is the evolution of network attacks. Even though there are many proposed detection methods and defense techniques, we cannot say that the security attacks are a solved problem or do not present an immense threat to the current Internet.

The research in the field of SDN and general security in SDN is still in its early phase. Moreover, every new technology and level of abstraction opens new attack vectors. However, we believe that the attributes of SDN can help in detection and remission of the attacks.

7 Conclusion

In this research manuscript, the analysis of assorted dimensions of SDN is done along with the security parameter in particular aspect. As security, privacy and integrity is utmost important for any network infrastructure, this paper underlines various attacks and vulnerabilities in the SDN environment. After detailed investigation of the attacks and security loopholes, an effective and secured model for the higher security can be developed and implemented so that the next generation SDN based environment can be made secured and cost effective.

References

1. Ali, S. T., Sivaraman, V., Radford, A., & Jha, S. (2015). A survey of securing networks using software defined networking. *Reliability, IEEE Transactions on*, *64*(3), 1086–1097.
2. Campbell, A. T., Katzela, I., Miki, K., & Vicente, J. (1999). Open signaling for ATM, internet and mobile networks (OPENSIG'98). *ACM SIGCOMM Computer Communication Review*, *29*(1), 97–108.
3. Farhady, H., Lee, H., & Nakao, A. (2015). Software-defined networking: A survey. *Computer Networks*, *81*, 79–95.
4. Fundation, O. N. (2012). Software-defined networking: The new norm for networks. *ONF White Paper*.
5. Gong, Y., Huang, W., Wang, W., & Lei, Y. (2015). A survey on software defined networking and its applications. *Frontiers of Computer Science*, *9*(6), 827–845. https://www.sdxcentral.com/resources/sdn/southbound-interface-api/.
6. Kim, H., & Feamster, N. (2013). Improving network management with software defined networking, *IEEE Commun Mag*, *51*(2), 114–119.
7. Kim, H., Kim, J., & Ko, Y. B. (2014, February). Developing a cost-effective OpenFlow testbed for small-scale Software Defined Networking. In *Advanced Communication Technology (ICACT), 2014 16th International Conference on* 758–761. IEEE.
8. Kreutz, D., Ramos, F., & Verissimo, P. (2013, August). Towards secure and dependable software-defined networks. In *Proceedings of the second ACM SIGCOMM workshop on Hot topics in software defined networking* 55–60. ACM.
9. Lantz, B., Heller, B., & McKeown, N. (2010, October). A network in a laptop: rapid prototyping for software-defined networks. In *Proceedings of the 9th ACM SIGCOMM Workshop on Hot Topics in Networks* 19. ACM.
10. Limoncelli, T. A. (2012). Openflow: a radical new idea in networking. *Queue*,*10*(6), 40.
11. McKeown, N., Anderson, T., Balakrishnan, H., Parulkar, G., Peterson, L., Rexford, J., & Turner, J. (2008). OpenFlow: enabling innovation in campus networks. *ACM SIGCOMM Computer Communication Review*, *38*(2), 69–74.
12. Mehdi, S. A., Khalid, J., & Khayam, S. A. (2011, September). Revisiting traffic anomaly detection using software defined networking. In *Recent Advances in Intrusion Detection*, 161–180 Springer Berlin Heidelberg.
13. Nunes, B. A., Mendonca, M., Nguyen, X. N., Obraczka, K., & Turletti, T. (2014). A survey of software-defined networking: Past, present, and future of programmable networks. *Communications Surveys & Tutorials, IEEE*, *16*(3), 1617–1634.
14. Rass, S., Rainer, B., Vavti, M., Göllner, J., Peer, A., & Schauer, S. (2014). Secure Communication over Software-Defined Networks. In *Internet of Things. IoT Infrastructures*, 211–221 Springer International Publishing.
15. Saint-Andre, P. (2011). Extensible messaging and presence protocol (XMPP): Core.
16. Shin, S., & Gu, G. (2012, October). Cloud Watcher: Network security monitoring using OpenFlow in dynamic cloud networks (or: How to provide security monitoring as a service in clouds?). In *Network Protocols (ICNP), 2012 20th IEEE International Conference on*, 1–6. IEEE.
17. Shin, S., Yegneswaran, V., Porras, P., & Gu, G. (2013, November). Avant-guard: Scalable and vigilant switch flow management in software-defined networks. In *Proceedings of the 2013 ACM SIGSAC conference on Computer & communications security*, 413–424. ACM.
18. Shu, Z., Wan, J., Li, D., Lin, J., Vasilakos, A. V., & Imran, M. (2016). Security in Software-Defined Networking: Threats and Countermeasures. *Mobile Networks and Applications*, 1–13. Springer Science + Business Media New York.
19. Vizváry, M., & Vykopal, J. (2014). Future of DDOS attacks mitigation in software defined networks. In *Monitoring and securing virtualized networks and services*, 123–127. Springer Berlin Heidelberg.

Modern Approach for Vehicle Traffic Monitoring and Signal Management System in ITS

Sagar Sukode and Shilpa Gite

Abstract Because of tremendous growth in number of vehicles, traffic monitoring and congestion control has become one of the crucial issues in road transportation. Internet of Things (IoT) and its applications like Intelligent Transportation System (ITS) can process current information status of traffic and useful for traffic control rooms to analyse and improve the traffic efficiency. Here, we have implemented a new and approach for analyzing, collecting, organizing, broadcasting and managing traffic as well as weather related information. Our proposed architecture: *Modern Approach for Vehicle Traffic Monitoring and Signal Management System in ITS* would perform well and facilitate the overall traffic management. Implementation of developed system turned out to be an easily adoptable solution for traffic congestion and pollution issues. This system provides a proficient yet accurate estimation of traffic density and weather condition along with signal management.

Keywords Intelligent transportation system (ITS) · Traffic congestion · Traffic monitoring system · Traffic signal management · Internet of things (IoT)

1 Introduction

Now a day's, IoT is an emergent technological trend and day by day growing very fast. In academia and computing industry, IoT has gained significant attention during the past decade. The term Internet of Things (IoT) was invented by Ashton

S. Sukode (✉) · S. Gite
Symbiosis Institute of Technology, Pune, India
e-mail: sagar.sukode@sitpune.edu.in

S. Gite
e-mail: shilpa.gite@sitpune.edu.in

N. Modi et al. (eds.), *Proceedings of International Conference on Communication and Networks*, Advances in Intelligent Systems and Computing 508,
DOI 10.1007/978-981-10-2750-5_75

in 1999. IoT has many more applications like: e-Healthcare, smart home, smart office, smart building, telecommunication and media, smart automotive, aerospace, smart disaster alerting, recycling, smart cities, smart environment, retail, smart water, smart meter, industrial control, logistics, smart agricultures, smart animal farming, transportation and logistics, etc. [1]. IoT has many applications, amongst them one of the most significant, important and advance application is nothing but ITS [2]. In the 21st century, there is rapid increase in the vehicles, conceivably traffic jamming on road also increases. In result, every day increases pollution and more time waste on road. Using the various embedded system components, we have developed a new system. We confine our work using some sensors, like: IR sensor, LDR sensor, sensor array, temperature sensor, gas sensor, etc.

The developed system is useful for broadcasting, monitoring and controlling the traffic as well as weather related information. This system includes a hardware kit, a server and two android applications. Other apps are not as useful as our new developed app. Other app is limited, e.g. Google maps. It only shows the direction, traffic density, routs, temperature, etc. and will not manage any traffic or jams on road. But our developed system will display traffic and weather related information and manages traffic signals depending on traffic density. For understanding the utility of developed application let us consider an example. Suppose, user want to go from one place to another and would like to know real-time traffic as well as weather condition. So, user simply first choose the location where user want to go and afterwards user can view the traffic density along with weather related information for selected location.

This paper is structured in six different sections. The survey of existing system describes in Sect. 2. Section 3 describes design and methodology. Tools and technologies are describes in Sect. 4. Implementation and results are described in Sect. 5. Finally in Sect. 6, conclude our developed work and given future direction.

2 Related Work

Li et al. [3] discussed vehicle detection, tracking and counting were very crucial issue for traffic monitoring, controlling and planning. Author developed a video-based solution applied with adaptive subtracted technology. This technology has a combination of virtual detector and blob tracking technologies. Also, they shown the results, implemented in Visual C++ code with OpenCV development kits.

Kanungo and Sharma [4] discussed and analyses present situation of traffic and its problem over road transport. Author proposed methods using live video feed from cameras at traffic junctions using video and image processing for real time traffic density calculation. It also works on algorithm for switching traffic lights according to vehicle density on road aiming to reduced traffic congestion and road accidents.

Waranusast et al. [5] discussed, tasks of automatic traffic monitoring was to count number of vehicles and vehicle classification and this tasks provided for planning transportation system. Author developed system in which, system extracts the moving objects and classify them as a car, motorcycle or other moving object based on KNN classifier.

3 Design and Methodology

Following Fig. 1 shows overall architecture of our new proposed system. Basically, this developed system consists of a hardware kit, a server and two android applications (one application is developed for hardware kit and another application is developed

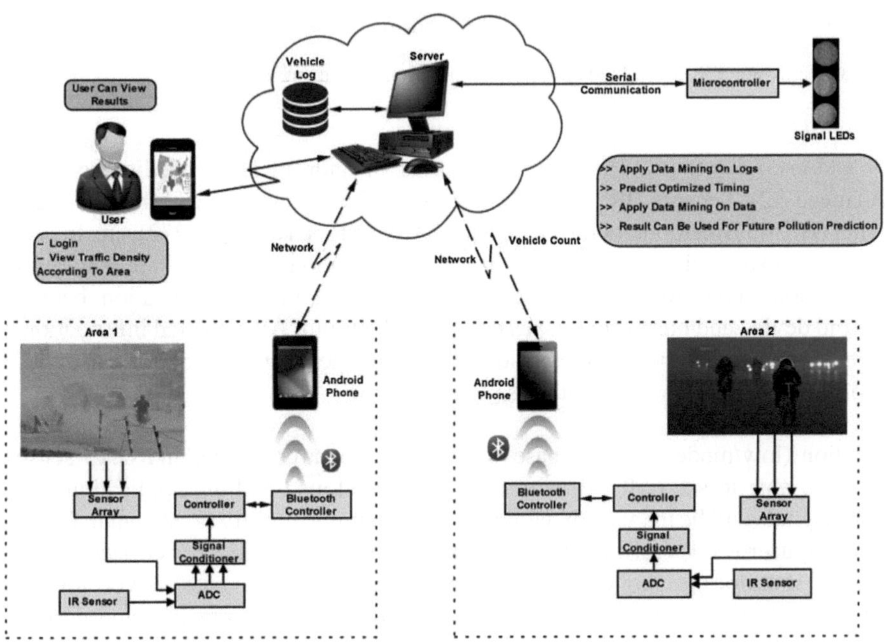

Fig. 1 System architecture

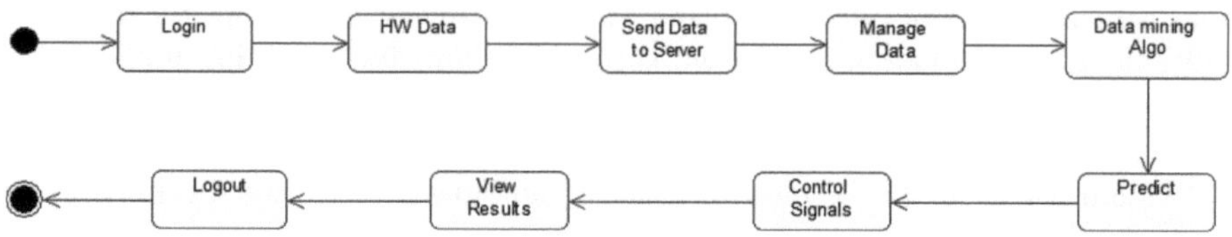

Fig. 2 System flow

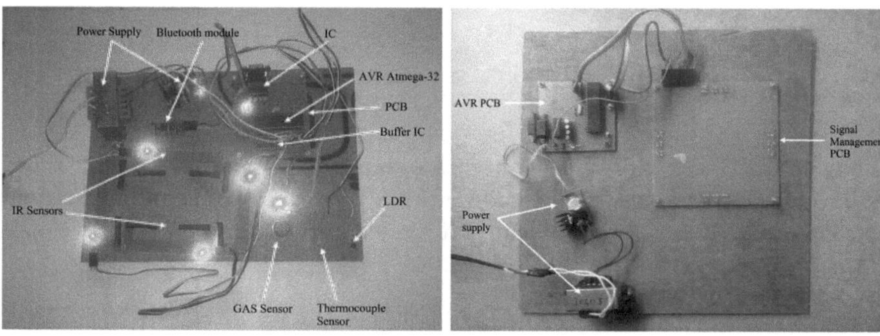

Fig. 3 Hardware Kit (Main hardware kit and signal management kit)

for end-user/client) [6]. Hardware kit consist of android Device, Microcontroller (ATmega-32), Sensors (IR, LDR, Temperature, Sensor array, etc.), Bluetooth controller (HC-05 Bluetooth module), Resistors, Capacitors, IC (MAX 232), etc. In given system, sensors will collect all sensor values/information and send that sensor information/values to server using android device and communication between android device and hardware kit was done via Bluetooth. All collected information is stored in sever machine. On server side, all mining operations will perform using ID3 algorithm. Here, ID3 algorithm is use to predict the

pollution (low/moderate/high) base values of temperature, LDR and GAS sensor. User can able to see traffic density using our developed android application. Also traffic signals will be manage dynamically base on traffic density (as shown in Fig. 2).

When user needs real-time traffic/weather related information, user need to open android monitoring application, that application request's web server for user requirement and server respond to end user with require information and user can see traffic density and traffic condition along with pollution/weather condition.

We have developed a hardware kit which consisting of various hardware components. This hardware components includes: Smart Device(smartphone),

Microcontroller (ATmega-32), Sensors (IR, LDR, Gas, Thermocouple sensor, etc.), Bluetooth (HC-05) module, Resistors, Capacitors, IC (MAX-232), etc. [7, 8] as shown in Fig. 3.

4 Tools and Technologies

We have used diverse software tools and techniques for designing and developing given system which includes: Embedded C, Eclipse, Android SDK, Web Server, Net-Beans, Express PCB, etc. Also, used Data mining and clustering technique to manage and mining receive big amount of data/information in a specific manner and finally prediction of pollution using ID3 algorithms. ID3 stands for Iterative Dichotomiser 3 and it is used to generate a decision tree.

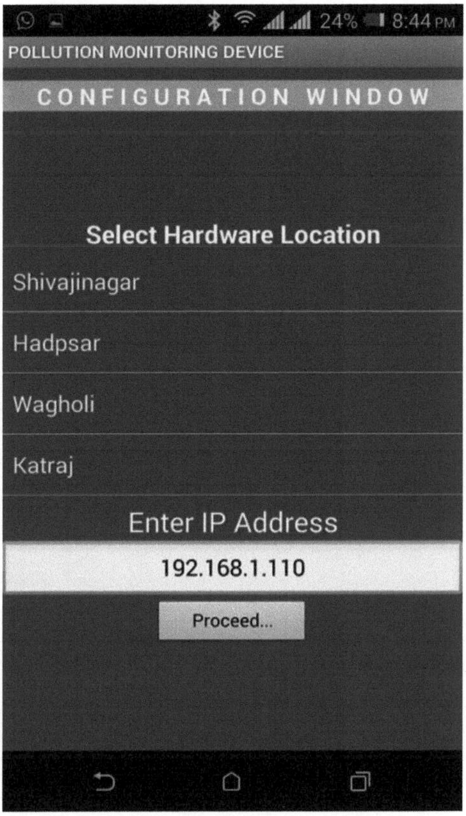

Fig. 4 Enter IP of server m/c

Android SDK is used to develop an android application. We have developed two applications. Web server which is a middleware between hardware module and end user. It is helpful for tracking, broadcasting, managing traffic and weather related information. Here, we have developed our own web server using NetBeans IDE. Embedded C is a extension of C programming language, useful for developing hardware module. Express PCB is use to design the circuit boards in a simple manner for beginner and professionals. We have design and developed circuit diagram and PCB layout using Express PCB software.

5 Implementation and Results

For implementation of proposed system, we have developed two android applications: one for hardware kit and another for end-user/client along with intelligent traffic signal management system.

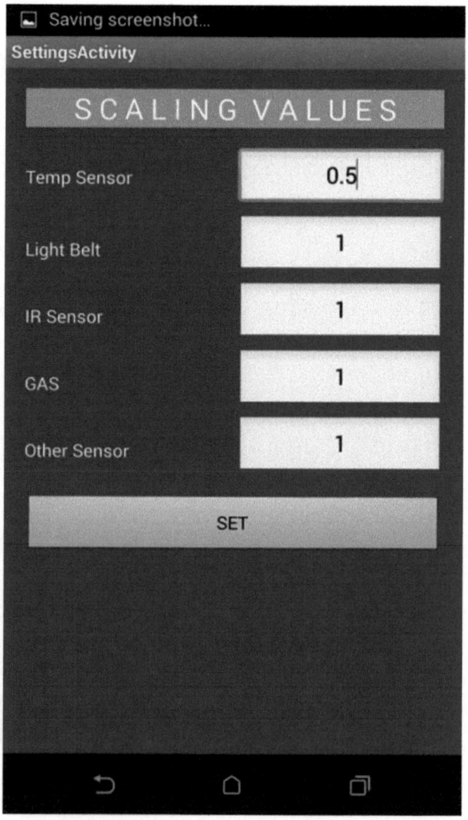

Fig. 5 Setting scaling factor

5.1 Android Application for Hardware Kit

First we have developed an android application for hardware kit to perform communication between hardware components and a server. Given android app communicates with hardware kit (in which all hardware components/sensors are present) via Bluetooth controller module (HC-05). The developed app collect all sensor values from hardware kit and sent to server (software module) via internet. Developed android app (which is developed for hardware kit) always active on any android device and it is also always connected to hardware kit or which is mounted with hardware kit.

The above Fig. 4 shows first screen (home screen) of our developed android application. On opening app, this screen appear and need to select required hardware location (Shivajinagar (area1), Hadpsar (area2), etc.) and need to enter the IP address of server machine to connect with web server and click on *proceed* button for further process. On clicking on *proceed* button a new screen will open called "setting activity screen" as shown in Fig. 5. Here, we need to give the scaling values to particular sensors. If we want temperature value in Degree Celsius or in

Fig. 6 Device lists

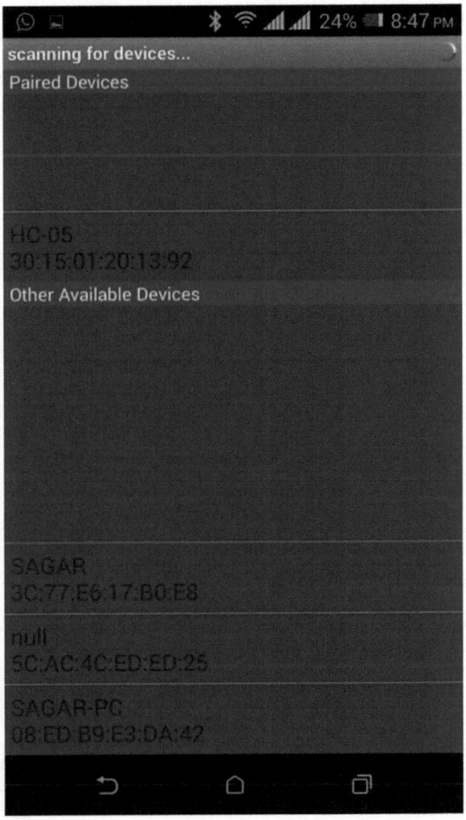

Fahrenheit then we need to put scaling value/scaling factor of that. Here, we put 0.5 to get temperature value in Degree Celsius.

After setting all scaling values of required sensor values, click on *SET* button for further process and a new screen will open i.e. "pollution monitoring device". On this screen two lists will be display. List shows all paired devices in upper half screen and in lower half screen, newly available devices, as shown in Fig. 6 and finally we need to select required one. In our case, we have HC-05 Bluetooth module, so we select HC-05 from given list and new screen will be open as shown in Fig. 7.

Figure 7 shows main "pollution monitoring device" screen consisting of three buttons i.e. *start button, stop button and logout button. Start button* is used to start monitoring activity which displays all sensors values (values read/received from hardware kit). *Stop button* is used to stop monitoring activity. And logout button is used to turn off the activity. Also, select reset timer is used to reset sensors values to update server machine. Reset time is provided to set reset time for updating the sensor values on server after a particular interval of time. Here, time is given in minutes, from 2 to 60 min as shown in Fig. 7. First we need to select reset timer and then click on *start button.*

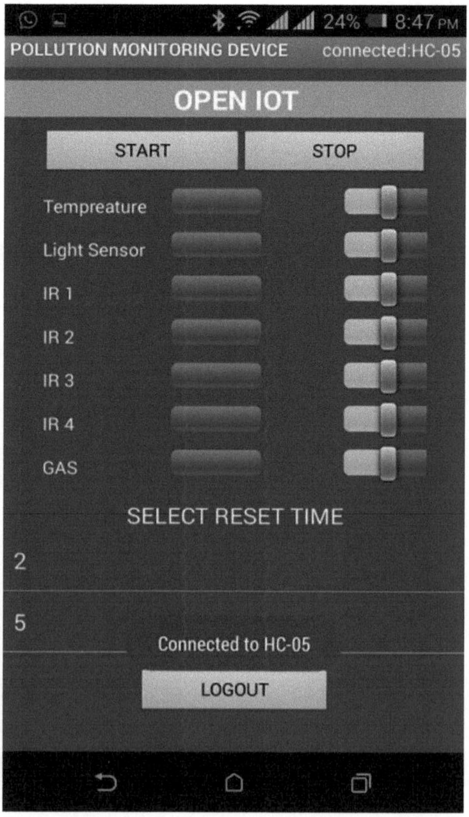

Fig. 7 Connected to HC-05

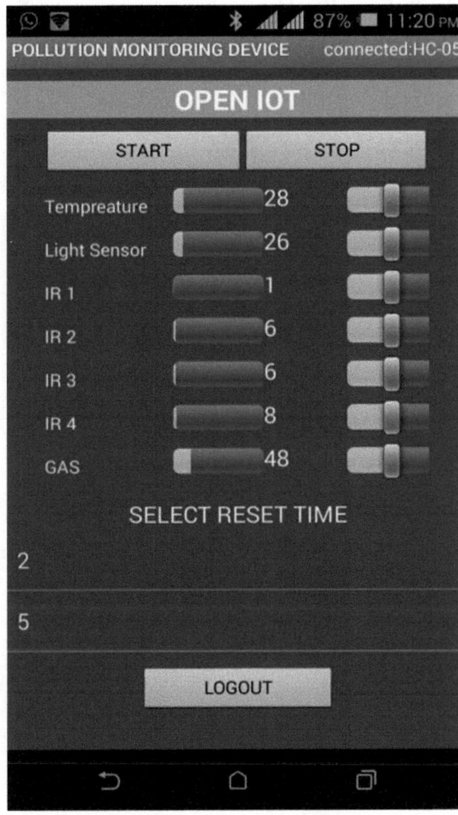

Fig. 8 Actual results

When we click on *start button*, all sensors values will be display i.e. light, temperature, IR (IR1, IR2, IR3, and IR4), Gas sensor, etc. as shown in Fig. 8. Also, Thresholding provided just beside to progress bar of sensor value to set threshold values like: normal, low, medium, high, etc. for respective sensor values.

5.2 Android Application for End-User/Client

We have developed another android app for the end-user/client. Use of this app is to monitor traffic density and pollution, based on the selected location. First we need to enter the IP address of server machine to connect server for fetching data/information from server. Figure 9 shows the welcome screen. After entering the IP address, need to click on *click here button* and new screen will open i.e.

Fig. 9 Enter IP address

"select hardware location" as shown in Fig. 10. Here, need to select require hardware location for which we want to monitor the traffic density and weather condition.

Here, need to choose require hardware location (Shivajinagar (area1), Hadpsar (area2), etc.). After selecting the require location, popup will displays the confirmation message which shows selected location, as shown in Fig. 10. Here we selected "*Shivajinagar*" (area1) as a hardware location and click on *proceed* button and new screen will be open i.e. "Monitoring Window" as shown in Fig. 11.

It include three buttons i.e. *start, stop* and *logout*. *Start button* is use to start the monitoring activity and then it will display all sensor values on the given screen (values read/received from the hardware kit). To stop monitoring activity *Stop button* is used. And, *logout button* is used to turn off the activity.

When we click on *start button*, all the sensors values will be display i.e. temperature, light, IR (IR1, IR2, IR3, and IR4), Gas sensor, etc. and pollution is predicted using the ID3 algorithm as shown in Fig. 11. Prediction will display in the form of: normal, low and high. In this manner all sensor values will display on developed android application.

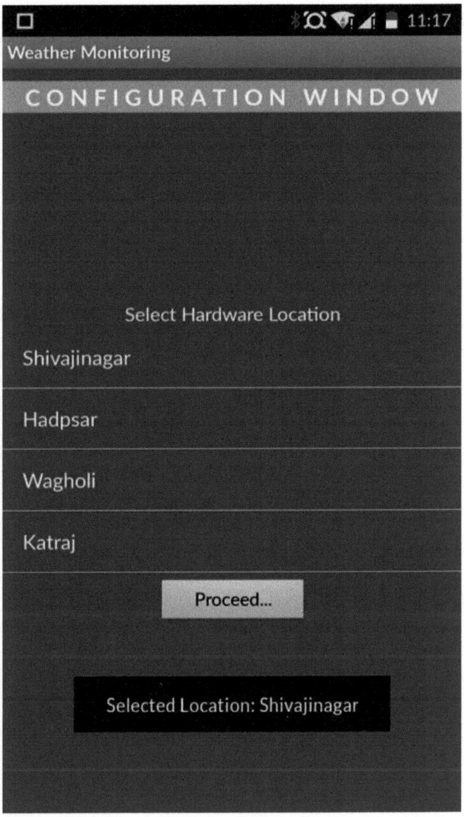

Fig. 10 Select hardware location

5.3 *Intelligent Traffic Signal Management System*

Intelligent traffic signal management system is system in which traffic signals will be manage dynamically depending on traffic density on road. As we discussed earlier about hardware kit, this kit can mounted at square/traffic signals. As, four IR sensors are use to count number of vehicles moving on road. When vehicle is moving on road, IR sensor cut's and count will store/save on server and displays on client app. By considering the number of vehicle counts, traffic signals can manages dynamically.

As the number of vehicle counts increases on particular road, according to that timing of green signal also increase. Figure 12 shows the dynamic management of traffic signals on road. We can see that, IR1 has count 4, IR2 count is 3, IR3 count is 7, and IR4 count is 1. So, as the number of vehicle count is maximum, timing for green signal will allocate more time than other 3 signals and if vehicle count is minimum then timing for green signal will allocate less time than other 3 signals as shown in Fig. 12 (GUI). In this manner, traffic signals can dynamically manages depending on vehicle counts and traffic density on road.

Fig. 11 Monitoring window

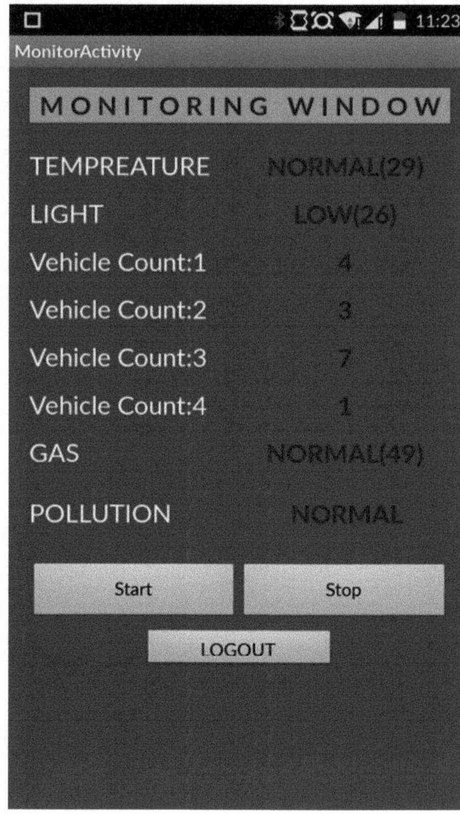

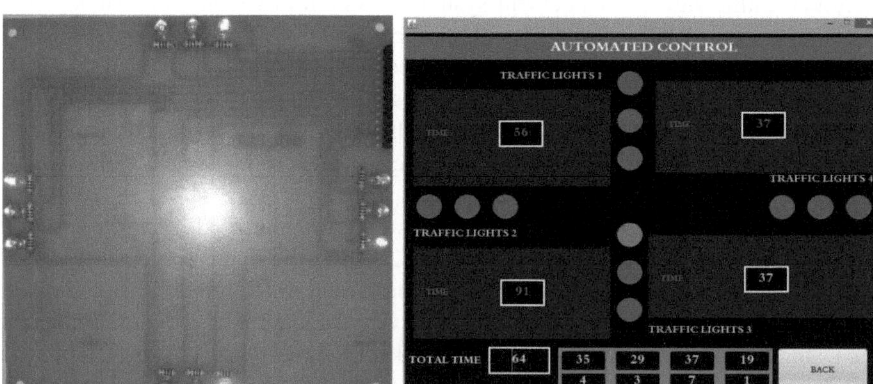

Fig. 12 Intelligent traffic signal management system (Hardware and GUI)

6 Conclusion and Future Scope

Modern approach for vehicle traffic monitoring and signal management system in ITS is a real-time system for analyzing, tracking, collecting, broadcasting and managing traffic along with weather related information. This system consists of a hardware kit, a server module and an Android application (client's app) for reporting and displaying traffic related information (traffic density, jams on road, traffic condition, pollution and weather condition, etc.). The interface of client application is intuitive and it is easy and safe to use while driving. Also, we have implemented the system where traffic signals can manages dynamically depending on traffic density.

For further research, we can make effective use of context aware systems. Context-aware computing allows us to store context data which is connected to sensor data, so classification can be done more simply, meaningfully. Also to perform machine-to-machine communication context is used easily as it is core element in the IoT environment. By using context-awareness, whole system can work very well without human interaction. Also, using ITS and IoT such new and emerging technologies, we can focus more on efficiency and reliability of traffic density.

References

1. Sukode, S., Gite, S., Agrawal, H.: Context Aware Framework in IoT: A Survey. IJATCSE-WARSE. 4, 1–9 (2015).
2. An, S., Lee, B., Shin, D.: A survey of intelligent transportation systems. In: Computational Intelligence, Communication Systems and Networks (CICSyN), 2011 Third International Conference on. IEEE, (2011).
3. Li, D., Liang, B., Zhang, W.: Real-time Moving Vehicle Detection, Tracking, and Counting System Implemented with OpenCV. IEEE. 631–634 (2014).
4. Kanungo, A., Sharma, A., Singla, C.: Smart Traffic Lights Switching and Traffic Density Calculation using Video Processing. In: Proceedings of 2014 RAECS UIET Panjab University Chandigarh. IEEE, (2014).
5. Waranusast, R., Timtong, V., Bundon, N., Tangnoi, C.: A Computer Vision Approach for Detection and Counting of Motorcycle Riders in University Campus. IEEE. (2014).
6. Sukode, S., Gite, S.: Vehicle Traffic Congestion Control & Monitoring System in IoT. IJAER. 10, 8, 19513–19523 (2015).
7. EngineersGarage, http://www.engineersgarage.com/electronic-componenets/atmega32-avr-microcontroller
8. RajGuru Electronics, http://www.rajguruelectronics.com/bluetooth-module.html

Author Index

N. Modi et al. (eds.), *Proceedings of International Conference on Communication and Networks*, Advances in Intelligent Systems and Computing 508,
DOI 10.1007/978-981-10-2750-5

MIX
Papier aus verantwortungsvollen Quellen
Paper from responsible sources
FSC® C105338

If you have any concerns about our products,
you can contact us on
ProductSafety@springernature.com

In case Publisher is established outside the EU,
the EU authorized representative is:
Springer Nature Customer Service Center GmbH
Europaplatz 3, 69115 Heidelberg, Germany

Printed by Libri Plureos GmbH
in Hamburg, Germany